Large Marine Ecosystems of the Pacific Rim: Assessment, Sustainability, and Management

On the cover

The satellite image of Pacific Rim productivity is provided by the SeaWiFS project, NASA/Goddard Space Flight Center and ORBIMAGE. The image is a composite of all SeaWiFS data collected between September 1997 and August 1998. The color-enhanced code depicts a shaded gradient of chlorophyll concentration from a high level of 30 mg/m^3 shown in reddish to orange color spectra to lesser concentrations in yellow and light blue down to an ecosystem low value of .03 mg/m^3 depicted in the darker blue. (Dr. Gene Feldman of NASA/Goddard Space Flight Center kindly provided the information on the chlorophyll gradients depicted in the image.)

Large Marine Ecosystems of the Pacific Rim: Assessment, Sustainability, and Management

Edited by

Kenneth Sherman

Director, Narragansett Laboratory, and Chief, Ecosystem Monitoring Branch, Northeast Fisheries Science Center, National Marine Fisheries Service, National Oceanic and Atmospheric Administration, Narragansett, Rhode Island; Adjunct Professor of Oceanography, Graduate School of Oceanography, University of Rhode Island, Narragansett, Rhode Island

Qisheng Tang

Director General, Yellow Sea Fisheries Research Institute, Chinese Academy of Fishery Sciences, Qingdao, People's Republic of China

Blackwell Science

Editorial Offices:
Commerce Place, 350 Main Street, Malden, Massachusetts 02148, USA
Osney Mead, Oxford OX2 0EL, England
25 John Street, London WC1N 2BL, England
23 Ainslie Place, Edinburgh EH3 6AJ, Scotland
54 University Street, Carlton, Victoria 3053, Australia

Other Editorial Offices:
Blackwell Wissenschafts-Verlag GmbH, Kurfürstendamm 57, 10707 Berlin, Germany
Blackwell Science KK, MG Kodenmacho Building, 7-10 Kodenmacho Nihombashi, Chuo-ku, Tokyo 104, Japan

Distributors:
USA
Blackwell Science, Inc.
Commerce Place
350 Main Street
Malden, Massachusetts 02148
(Telephone orders: 800-215-1000 or 781-388-8250; fax orders: 781-388-8270)

Canada
Login Brothers Book Company
324 Saulteaux Crescent
Winnipeg, Manitoba, R3J 3T2
(Telephone orders: 204-224-4068)

Australia
Blackwell Science Pty, Ltd.
54 University Street
Carlton, Victoria 3053
(Telephone orders: 03-9347-0300;
fax orders: 03-9349-3016)

Outside North America and Australia
Blackwell Science, Ltd.
c/o Marston Book Services, Ltd.
P.O. Box 269
Abingdon, Oxon OX14 4YN
England
(Telephone orders: 44-01235-465500;
fax orders: 44-01235-465555)

Acquisitions: Nancy Hill-Whilton
Production: Ellen Samia
Manufacturing: Lisa Flanagan
Typeset by Northeastern Graphic Services, Inc.
Printed and bound by Capital City Press
Printed in the United States of America
99 00 01 02 8 7 6 5 4 3 2 1

For further information on Blackwell Science, visit our website: www.blackwell-science.com

Library of Congress Cataloging-in-Publication Data

Large marine ecosystems of the Pacific Rim : assessment, sustainability, and management / edited by Kenneth Sherman and Qisheng Tang
p. cm.
Includes bibliographical references and index.
ISBN 0-632-04336-9 (alk. paper)
1. Marine ecology—Pacific Area. 2. Fishery resources—Pacific Area. 3. Ecological assessment (Biology)—Pacific Area. I. Sherman, Keneth, 1932– . II. Tang, Qisheng.
QH95.L37 1999
577.7′4—dc21 98-18549
CIP

Contents

Preface

In this volume, marine experts from Pacific Rim countries describe conditions within the large marine ecosystems (LMEs) of the Pacific Ocean. The countries of the region represent over a third of the world's population, many of whom are existing at or below the poverty level. Human interventions and climatic change have altered the productivity of the LMEs in the Pacific. Within the nearshore areas and extending seaward around the margins of the adjacent land masses, coastal ecosystems are being subjected to increased stress. Among the stressors are toxic effluents, habitat degradation, excessive nutrient loadings, harmful algal blooms, emergent diseases, fallout of aerosol contaminants, and episodic losses of living marine resources from the effects of habitat loss, pollution, environmental perturbations, and overexploitation. The LMEs of the North Pacific have also been subjected to both positive and negative effects from global climatic change. A growing awareness that multiple driving forces are affecting the quality of the Pacific LMEs has accelerated efforts to assess, monitor, and mitigate coastal stressors from an ecosystem perspective.

Within the Asian and South American LMEs, the potentials for economic growth through the development of coastal tourism, mariculture, mineral extraction, oil and gas production, and fisheries are being tempered by the need to ensure the long-term sustainability of marine resources. Case studies of ecosystem assessments presented in this volume by several authors illustrate the risks from continuing degradation of resources under the prevailing unmanaged conditions existing over most of the coastal areas of Asia and South America. Several authors explore the application of multidisciplinary ecosystem-based assessment and management strategies to the future development of the large marine ecosystems of the region. With the initiation of science-based management practices, the LMEs can provide a more sustainable source of economic growth, nutritional benefits, and improved food security than is presently being realized by the people living in coastal areas bordering the Pacific Rim.

International agencies are prepared to encourage private investment to assist developing countries in the region in collaborative, multinational management programs for marine resources within the LMEs. These agencies include the United Nations Development Program (UNDP), the UN Environmental Program (UNEP), the World Bank, and the Global Environment Facility (GEF). At present, the GEF is providing financial support to projects for the assessment and management of large marine ecosystems along the coast of West Africa. In the Pacific Rim region, planning efforts with the UNDP, UNEP, World Bank, and GEF are underway for the Yellow Sea, South China Sea, and Humboldt Current LMEs. Scientists, economists, and representatives from national resource ministries of China, Korea, the Philippines, Vietnam, Indonesia, Malaysia, Thailand, Chile, Peru, Australia, Canada, and the United States are participating in the process.

The willingness expressed by the marine interests of the Pacific Rim region for developing and participating in collaborative LME projects to assess and manage shared natural resources is an important step toward achieving greater global sustainability of marine natural resources. The present volume focuses on several of the more pressing transboundary issues in international waters to be addressed by the countries of the Pacific Rim.

K.S.
Q.T.

Background and Focus

In recognition of the need to improve prospects for the long-term sustainability of ocean resources, the United Nations Conference on the Environment and Development (UNCED) in 1992 issued a declaration on the oceans, recommending that nations of the globe (1) prevent, reduce, and control degradation of the marine environment so as to maintain and improve its life-support and productive capacities; (2) develop and increase the potential of marine living resources to meet human nutritional needs, as well as social economic, and development goals, and (3) promote the integrated management and sustainable development of coastal areas and the marine environment.

The UNCED oceans declaration represents an important milestone because it was endorsed by the leaders of 170 nations. The list of international conventions, declarations, treaties, and agreements for conserving natural resources is long (Wallace 1993). At present, however, no single international institution is empowered to reconcile the needs of individual nations with those of the global community of nations in taking management actions to ensure the long-term sustainability of marine resources and ecosystems (Myers 1990). In response to the need for improving global environmental conditions, the Global Environment Facility has emerged recently as both a facilitator and a funding mechanism for integrating global environmental concerns into a programmatic process for achieving the goals of several important international conventions and other focal areas identified by UNCED for global action, including the Conventions for addressing problems of biodiversity and ozone depletion and their associated linkage to global climatic change. To assist developing countries in overcoming what has commonly been a highly compartmentalized approach to marine resource assessment and management (e.g., pollution, fisheries, habitats, coastal zone management, shipping and transport, drainage basin effluents), the Global Environment Facility (GEF), established within the World Bank, is supporting programs aimed at improving assessments and management of shared marine resources in international waters that cross national boundaries and are largely contained within the geographic extent of 49 large marine ecosystems around the margins of the Atlantic, Pacific, and Indian Oceans.

The large marine ecosystems (LMEs) are relatively large regions of ocean space, on the order of 200,000 km^2 or greater, characterized by distinct bathymetry, hydrography, productivity, and trophically dependent populations. They extend from the nearshore areas, including river basins and estuaries, out to the seaward boundary of the continental shelf or the seaward margin of coastal current systems (AAAS 1986, 1989, 1990, 1991, 1993). To assist countries in overcoming highly compartmentalized approaches to reducing environmental and human-induced stress on coastal ecosystems, the Global Environment Facility has included LMEs as one of its focal areas for funding through its International Waters Projects.

The GEF operational strategy provides a planning framework for the design, implementation, and coordination of different sets of GEF International Waters Projects that can achieve particular global environmental benefits (GEF 1997). Through these programs the GEF is encouraging a paradigm shift from a sector by sector approach to a more comprehensive and integrated approach for the restoration and protection of international waters. The goal of the GEF and its international donor governments is to assist developing countries in making changes in the ways that human activities are conducted in different sectors so that a particular body of water and its multicountry drainage basin can sustainably support human activities.

Projects submitted to the GEF for funding consideration focus on seriously threatened bodies of water and the most imminent transboundary threats to their ecosystems, including pollution, overexploitation of living and nonliving resources, and habitat degradation, including that resulting from the introduction of nonindigenous species. The long-term objective of the program is to undertake a series of projects that involve helping groups of countries to work collaboratively with the support of implementing agencies in achieving changes in sectoral policies and activities so that transboundary environmental concerns degrading specific water bodies can be resolved. Among the short-term objectives of the International Waters program of the GEF is to initiate actions toward resolving transboundary environmental concerns with at least one freshwater basin project and one large marine ecosystem project in each of the world's five development regions, including sub-Saharan Africa, Asia, Latin America and the Caribbean, the Middle East/North Africa, and eastern Europe.

Various types of water bodies, with their varied ecological systems and varied economic values, will be the subject of GEF projects. Freshwater systems range from transboundary river and lake basins to transboundary groundwater systems. Marine waters are primarily addressed through large marine ecosystems. The 49 LMEs support about 95% of the world's annual fish catch. Water bodies selected for projects will be those wherein transboundary concerns create significant threats to the functioning of the ecosystem. In the case of LMEs, the marine, coastal zone, and relevant freshwater basin concerns are also addressed. The expected outcomes of the projects include a reduction of stress on the international water environments as a result of countries changing their individual sectoral policies, making critical investments, developing necessary programs, and collaborating jointly in managing transboundary water resources.

In recognition of the utility of the LME as an ecologically based unifying approach to the assessment, monitoring, and management of marine resources, a symposium was held in Qingdao, China, in October 1994 to (1) examine major driving forces causing large-scale changes in biomass yields and the health of LMEs of the Pacific Rim, and (2) consider the application of modular assessment strategies for addressing transboundary international waters issues related to the long-term sustainability of shared marine resources within LMEs of the region.

Coastal nations around the Pacific Rim have indicated a willingness to follow the lead of the GEF in bringing about a paradigm shift by moving away from sector by sector treatment of degraded coastal environments toward a more holistic, multisectoral approach to marine-based assessments of ecosystem health, productivity, and resource sustainability. Resource scientists and managers from these countries provided assessments to the symposium of the present state of the ecosystems off their coasts, the resources at greatest risk, and the sources or causes of the risks that need to be mitigated and managed from the perspective of ecosystem resources development and sustainability.

Several of these assessments, along with designations of high-risk issues and their causes, have been carefully selected from among the reports presented at the symposium. They have been peer reviewed and brought together in this volume with other pertinent studies of the Pacific Rim LMEs to provide a synthesis of their changing states in relation to the long-term sustainability of marine resources within their boundaries.

Regime Shifts

The first section of the volume is focused on the climate-induced oceanographic regime shifts in the Pacific. The opening chapter by A. Bakun documents the Pacific patterns of large-scale population increases and decreases of fish species in different LMEs around the margins of the Pacific

Ocean. He examines major fluctuations in sardine and anchovy populations of the Humboldt Current, California Current, Kuroshio Current, and Oyashio Current LMEs as being related to the El Niño–Southern Oscillation (ENSO) and broad-scale teleconnections of the ocean–atmosphere system of the Pacific. Comparisons are made to fluctuations in salmon populations of the Gulf of Alaska and East Bering Sea and in populations of bottom fish in the Gulf of Alaska with regard to ecosystem retention, concentration, and enrichment processes. Dr. Bakun's paper is a comprehensive treatment of El Niño effects on the fluctuations in the fisheries of the Pacific LMEs. The important driving forces appear to be related to interdecadal variability patterns of the ENSO system. In a study that follows closely on the El Niño focus of Bakun, the chapter by Beamish and Neville (Chapter 2) examines the effects of the oceanographic regime shifts on salmon populations of the northern North Pacific. The authors develop an explanation for the improved survival of pink, chum, and sockeye salmon in the North Pacific ocean in contrast to the reduced survival of chinook salmon in the coastal Straits of Georgia, off British Columbia. The authors describe the utility of using an index of the intensity of the Aleutian Low to reflect and possibly forecast salmon abundance levels in the open ocean and coastal ecosystems. The contribution by D. Lluch-Belda (Chapter 3) emphasizes the extent of regime shifts in the Pacific and the effect of the shifts on the changes in abundance levels of clupeoids. The author is explicit in his message that fisheries management should be more responsive to changing ecosystem states on the decadal scale. In consideration of the recurrence of large-scale regime shifts, he points to the need for improved climate forecasting systems. Shuntov et al. (Chapter 4) provide evidence from zooplankton, nekton, and marine bird time-series data that large-scale changes occurred in the northeastern North Pacific ocean regime in the Bering Sea and Sea of Okhotsk LMEs during the 1990s, the 1970s to 1980s, the 1940s to 1960s, and the 1920s to 1930s. This study serves to underscore the importance of considering multidecadal ocean regime variations in the development of studies of ecosystem structure and in the development of options for the management of fishing practices.

More recent papers, kindly made available for this volume by their authors and by the Coordinator and Editor of the California Cooperative Oceanic Fisheries Investigation, examine the changes in the oceanographic regime of the California Current ecosystem and the adjacent waters of the North Pacific. We are indebted to G. Hemingway and J. Olfe for allowing us to include these contributions.[1] The paper by Hayward acknowledges that atmospheric forcing has resulted in changes to the physical and biological structure of the upper layers of the North Pacific Ocean during the past two decades. The author emphasizes the need for improving mesoscale modeling of key processes affecting physical and biological structure so as to improve predictions of basin-scale changes in biological productivity within the California Current ecosystem. Nine specific recommendations are put forward toward reaching this objective. The paper by McGowan et al. advances the need to conduct large-scale spatial and temporal time-series studies that are relevant to observing changing states at scales comparable to the entire extent of the California Current or other large marine ecosystems. The biological feedback of zooplankton in the California Current LME to an 0.8flC warming of temperature in the upper 100 meters was very significant. Over a 42-year period, from the early 1950s to the early 1990s, the biomass of macrozooplankton decreased by 70% in response to increased thermocline stratification apparently related to basin-scale changes in wind forcing. The authors emphasize the need for a good definition of change that can be measured by ecosystem-wide time series of key ecological measurements. They provide the reader with their definition, which places the scientific evidence of change on the departure from some expectation or norm, its sign, amplitude, spatial extent, and temporal persistance. They offer the CalCOFI as a useful model for measuring the changing states of large marine ecosystems. The

[1] The papers by Hayward, McGowan et al., and Brodeur et al. appeared previously in CalCOFI Rep. Vol. 37, 1996.

last paper in the section by Brodeur et al., is a comparison of the large-scale changes ocurring in the North Pacific that is focused on changes in the abundance of zooplankton biomass in the Gulf of Alaska and the California Current LMEs. The authors present arguments for an inverse relationship in the biomass levels. They consider the implication of the differences in zooplankton biomass on the high trophic levels in both ecosystems.

Case Studies

The authors of the case studies have examined the large-scale changes in the biomass yields of the LMEs of the Pacific Rim for emergent trends indicative of decadal-scale biophysical or climatic forcing. In a study of the Eastern Bering Sea ecosystem, the focus is on the consumer groups in the ecosystem, including the offshore pelagic fish consumers, the inshore benthic fauna consumers, and the crab and fish consumers. In Chapter 5, Livingston et al. present convincing evidence of multidecadal fluctuations in species of marine birds, marine mammals, fish, and related benthic infauna consumers (starfish, crab species, and flounder species) and offshore pelagic fish consumers (fish-eating birds, seals, pollock, arrowtooth flounder, and turbot). However, they were unable to reach firm conclusions on the causes of the changes in abundance patterns in the absence of time-series measurements of key ecosystem health indicators. The authors stress the importance of long-term monitoring efforts for detecting mechanisms responsible for changing ecosystem states.

Among the other contributions from the North Pacific are several from the western Pacific. In a study of the Sea of Japan LME (Chapter 8), M. Terazaki presents the results of time-series studies on plankton, one of the key ecosystem indicators of change. The author documents shifts in the spawning and nursery grounds of Japanese sardine in relation to changes in the current patterns and the plankton of the Sea of Japan. Two case studies of the Yellow Sea LME are included. From the Korean perspective, the chapter (Chapter 6), by Zhang and Kim, reports on changes in the proportion of biomass yield by nations fishing in the Yellow Sea ecosystem. The authors focus attention on the need for an international effort to assess, monitor, and manage the resources of the ecosystem. Of particular concern is the declining trend of the catch per unit effort, indicating a large-scale decline in the demersal species and a significant change in the structure of the fish community of the Yellow Sea. In the next study (Chapter 7), by Tang and Jin, the change in the fish community structure of the ecosystem is documented in relation to an observed increase in the biomass of small pelagic species. The authors stress the importance of the predator–prey interactions of the fishery, particularly in relation to declines reported for the larger pelagic species, including Spanish mackerel, chub mackerel, and Pacific herring. The authors suggest that the ecological state of the Yellow Sea LME appears to have shifted to a system that favors the production of zooplankton and small pelagic fish, resulting in part from the decline in demersal and large pelagic fish species.

Further south, in the East China Sea LME, the study by Y. Chen and X. Shen (Chapter 9) describes a substantial shift in the fish community of the ecosystem during the past three decades. The shifting patterns of abundance in the East China Sea, from a target fishery of traditional demersal species in the 1960s and 1970s to so-called new species in the 1990s, are indicative of a major change in the state of the ecosystem. The East China and Yellow Sea case studies are suggestive of the limitations of the carrying capacity of LMEs in supporting traditional demersal fisheries. *Beddington (1995) reported that the levels of global primary production for supporting the "average levels" of global fish yields of the past several years may have been reached, and that any large-scale unmanaged increase in marine biomass yields would likely be in species closer to the lower end of the food chain. It would appear from the results provided by Chen and Shen of the East China Sea LME that any further increases in biomass yields would be based on fisheries that were targeted on "low food chain, fast-growing, and short-life-cycle species."

In the contribution by T. Piyakarnchana on the Gulf of Thailand ecosystem (Chapter 10), the author suggests that in addition to the effects of fishing, changes in land use in general, and the harvesting of mangrove forests in particular, are causes for declines in the biomass yields of the ecosystem. The author underscores the need for more rigorous enforcement of Thai conservation laws and management practices aimed at achieving long-term sustainability of the marine resources of the Gulf of Thailand. The concluding chapters are studies on the Humboldt Current LME and the South China Seas LME, reported in an earlier LME volume and reprinted here so as to include principal ecosystems of the Pacific Rim in which the fisheries biomass yields are important to the people of the region for food security and long-term socioeconomic benefit. The authors of the Humboldt Current case study (Chapter 11), J. Alheit and P. Bernal, provide evidence of the importance of El Niño as one of several environmental drivers of biomass yields in the ecosystem and suggest that the effects of fishing effort on changes in biomass levels is secondary to the natural environmental perturbations.

In the concluding chapter to the section, Bakun, Csirke, Lluch-Belda, and Steer-Ruiz provide an initial description of the Pacific Central American Coastal LME. An analysis is given of the physical setting, seasonal ocean current patterns, upwelling mechanisms, fisheries, and habitats. The authors discuss transboundary issues to be addressed by the nine nations sharing the resources of the ecosystem, including the increasing stress on the long-term productivity of the near-coastal waters of the ecosystem from the conversion of extensive mangrove areas into mariculture operations, principally for shrimp.

Methodological Studies

The implementation of assessments of the changing states of LMEs requires the application of cost-saving, cutting-edge methods for measuring changes in marine resource populations and marine resource environments over long-term temporal and spatial scales. Examples of pertinent methodology are included in this section. In the first contribution (Chapter 13), J. Rice describes the approach adopted by the International Council for the Exploration of the Sea for a more holistic ecosystem assessment approach toward full implementation of ecosystems management. The author focuses on the use of multispecies virtual population analyses (MSVPA) that include stomach contents data to improve understanding of predator–prey relationships at the ecosystem level. The importance of predation in structuring marine ecosystems is emphasized by N. Bax in his study on the prediction of the effects of predation in marine ecosystems (Chapter 14). The author stresses the utility of linear (direct) predation interactions and first-order interactions that have the potential to reverse the effects of direct interactions. The author provides an evaluation of the advantages and disadvantages of approaches for predicting predation impacts at the ecosystem level. He concludes that strategic management of complexity at the ecosystem level requires continued exploration of the properties of the managed system. New experimental management techniques are, in the opinion of the author, important to develop in cooperation with resource users, scientists, and managers.

In contrast to the use of information on predation in assessing the changing states of LMEs, the utility of hydroacoustics is described in the two following papers. Assessments of pelagic populations of fish and plankton are reported by Tang et al. (Chapter 15) based on surveys conducted in the Yellow Sea. Arenas and Robinson (Chapter 16) report on the results of studies of the shoaling strategy of small pelagic fish in the California Current LME. The methodological contributions that follow in the next three chapters represent cutting-edge applications of new sensing technology to obtain automated and continuous time-series measurements of changes in biogeochemical components of LMEs. A system that operates continuously using the seawater intake of a ferry boat to monitor nutrients, temperature, salinity, pH, CO_2, and phytoplankton and zooplankton particle size spectra, is described by Harashima et al. in Chapter 18. The system is in use collecting data from four transects per week between Pusan, Korea, and Kobe, Japan, across the southern boundary of the Sea of Japan LME.

The use of towed bodies outfitted with sensor payloads suitable for deployments from ocean-going commercial ships-of-opportunity is described in the report by Aiken et al. (Chapter 17). The system designated as the Undulating Oceanographic Recorder (UOR) can obtain time-series information concerning temperature, salinity, and chlorophyll from undulations made in the upper 70 m of the water column. When fitted with a newly developed pump and probe system, direct readings of primary productivity can be made. This towed body can carry an internal mechanism for sampling plankton onto rolls of silk mesh comparable to collections made for the past 50 years with the Hardy Continuous Plankton Recorder. In addition, the UOR can be fitted with an external optical plankton counter.

Indicators of Ecosystem Health

The stressors on coastal ecosystems from human population expansions of urban centers around the landward margins of LMEs is growing rather than diminishing, and is likely to become one of the most pressing issues for resolution during the next decade. Quantitative time-series measurements of this degradation will assist resource managers in effecting appropriate mitigation activities. The papers in this section represent a contribution toward this objective. In the paper by Y. Cui and Y. Song (Chapter 19), the results of assessing the nutrient loadings of three important embayments on the Bohai Sea are presented. The results depict an increase in nitrogen and a decline in phosphorus levels in the bays, attributed by the authors to increased use of nitrogen fertilizers in farming. The authors report that an increasing amount of industrial and domestic sewage is leading to the eutrophication of western Bohai Bay and to increases in the frequency and extent of harmful algal blooms. The importance of nutrient loading in affecting the carrying capacity of semi-enclosed LMEs is described in the report by Konovalov (Chapter 20). Among the LMEs in which ecosystem stress in localized areas is related to a reduction in numbers and biomass of autotrophic and heterotrophic species are the East China Sea, the Mediterranean Sea, and the Bay of Bengal. The LMEs most threatened by excessive nutrient loading and eutrophication are the Black Sea and the Sea of Azov, parts of the Yellow Sea adjacent to urban areas, parts of the North Sea, the Baltic, and the Adriatic Sea.

Other indicators, at the tissue level, are reported by Wendelaar Bonga et al. (Chapter 22) based on examinations of the gills and skin of fishes for evidence of disturbed ion regulation associated with reduced growth and reproductive potential. In the contribution by J. She (Chapter 23), a summary is given of the contaminant loadings in the Yellow Sea LME from the drainage basins and aerosol fallout. A listing is given of adverse impacts of pollution on the living resources of Dalian Bay, Jinzhou Bay, Bohai Bay, Liazhou Bay, Jiaozhou Bay, and Haizhou Bay, along with commentary on the increased frequency and extent of red tides in the coastal waters of China. The author concludes the report with the presentation of a strategy for quantitatively expanding the present monitoring efforts by China of coastal waters.

Management Linkages

Papers in the concluding section of the volume address strategies for improving assessment, monitoring, and management of LMEs in the region. The importance of addressing global climate change and the development of an ecological development plan in the coastal waters around Borneo are discussed in the paper by Arbain Hj. Kadri et al. (Chapter 25). The management system presently in place for the Great Barrier Reef LME is described in the contribution by J. Brodie (Chapter 24). A strategic approach for LME monitoring and assessment is given in the final chapter in the volume by K. Sherman (Chapter 26), followed by a list of recommendations endorsed by the Qingdao symposium participants.

Recommendations

The recommendations emphasize the importance of extending projects supported by the World Bank–Global Environment Facility to address transboundary concerns within the coastal and international waters of LMEs of the Pacific Rim. The participants were particularly concerned with the Yellow Sea and South China Sea LMEs. The transboundary marine resources within these LMEs are unmanaged and continue to be subjected to unsustainable human activities, including overexploitation of fisheries, coastal pollution, and the continuing loss of valuable coastal habitat. The imbalance between short-term economic gain and longer-term socioeconomic benefit is diminishing the value of the marine resources of the region. The International Waters focal area of the GEF was acknowledged by the symposium participants as an important opportunity for improving national infrastructure, promoting regional cooperation, and reversing the downward trend in resource sustainability.

K.S.
Q.T.

References

American Association for the Advancement of Science (AAAS). 1986. Variability and Management of Large Marine Ecosystems. AAAS Selected Symposium 99. Boulder, CO: Westview Press.

American Association for the Advancement of Science (AAAS). 1989. Biomass Yields and Geography of Large Marine Ecosystems. AAAS Selected Symposium 111. Boulder, CO: Westview Press.

American Association for the Advancement of Science (AAAS). 1990. Large Marine Ecosystems: Patterns, Processes and Yields. Washington, DC: AAAS Press.

American Association for the Advancement of Science (AAAS). 1991. Food Chains, Yields, Models, and Management of Large Marine Ecosystems. Boulder, CO: Westview Press.

American Association for the Advancement of Science (AAAS). 1993. Large Marine Ecosystems: Stress, Mitigation, and Sustainability. Washington, DC: AAAS Press.

Beddington, J. R. 1995. The primary requirements. Nature 374:213–214.

Global Environment Facility (GEF). 1997. GEF Operational Programs. Washington, DC: Global Environment Facility.

Myers, N. 1990. Working towards one world. [Book review]. Nature 344(6266):499–500.

Pauly, D., and Christensen, V. 1995. Primary production required to sustain global fisheries. Nature 374:255–257.

Wallace, R. L. 1993. (Compiled in collaboration with the Advisory Board for the Marine Mammal Commission Compendium.) The Marine Mammal Commission Compendium of Selected Treaties, International Agreements, and Other Relevant Documents on Marine Resources, Wildlife, and the Environment. 3 volumes. Washington, DC: U.S. Marine Mammal Commission.

Acknowledgments

The editors are indebted to the contributors for their willingness to take time out of busy schedules to provide the expert syntheses needed to move toward more sustainable marine ecosystems. We are pleased to acknowledge the interest and financial support of the International Union for the Conservation of Nature, known around the globe as the World Conservation Union. Without the contribution by the World Conservation Union, it would not have been possible to bring together in Qingdao marine resource experts from around the globe for the symposium on LMEs of the Pacific Rim. In the view of the editors, the robust exchanges of interpretations among the participants as to the causes and effects of the large-scale changes reported for the Pacific LMEs represent an important milestone in moving science-based ecosystem assessments toward the expectation of greater socioeconomic benefits from the marine resources of the Pacific. We gratefully acknowledge support of the Chinese government and the Yellow Sea Fisheries Research Institute in Qingdao for their excellent hospitality in hosting the symposium. The financial support of the U.S. National Oceanographic and Atmospheric Administration (NOAA) and the National Marine Fisheries Service (NMFS) is gratefully acknowledged. Other national and international institutions supporting the symposium whom the editors are pleased to acknowledge are the National Natural Science Foundation of China; the Chinese Fisheries Society; the Ministry of Agriculture, People's Republic of China; the North Pacific Marine Science Organization; the Food and Agricultural Organization of the United Nations; the Intergovernmental Oceanographic Commission, and the United Nations Environmental Program.

We are especially grateful to the Coordinator and Editor of the CalCOFI reports for their permission to include three key papers reporting on the regime shifts in the California Current and Gulf of Alaska ecosystems. Also, special thanks are extended to the American Association for the Advancement of Science for permission to include two important studies on the Humboldt Current and South China Sea LMEs, published earlier by the AAAS.

We are indebted to Noëlle Lewis of Narragansett, Rhode Island, for her extraordinary dedication, care, and expertise in technically editing the volume. We thank Jane Humphreys, Nancy Hill-Whilton, Ellen Samia, and Jill Connor of Blackwell Science for their care in the final production of the volume. We also thank Sally Adams of Saunderstown, Rhode Island, whose contribution in the later stages of production of the volume was invaluable.

K.S.
Q.T.

Contributors

Dr. J. Aiken
Plymouth Marine Laboratory
Plymouth, United Kingdom

Jürgen Alheit, Ph.D.
Senior Scientist
Biological Oceanography
Baltic Research Institute
Warnemünde, Germany

Dr. Virgilio Arenas
Fisheries Ecology Research Group
Instituto de Ciencias del Mar y Limnologia, UNAM
Mexico City, Mexico

Almah Bt. Awaluddin
Marine Science Department
University Kebangsaan Malaysia
Kampus Sabah
Kota Kinabalu Sabah, Malaysia

Dr. Andrew Bakun
Food and Agriculture Organization of the United Nations
Fisheries Department
Rome, Italy

Dr. Nicholas J. Bax
CSIRO, Division of Fisheries
Tasmania, Australia

Dr. Richard J. Beamish
Department of Fisheries and Oceans
Biological Sciences Branch
Pacific Biological Station
Nanaimo, Canada

Ian Bellan
Plymouth Marine Laboratory
Plymouth, United Kingdom

Dr. Patricio Bernal
Subsecretaria de Pesca
Ministry of Economics
Santiago, Chile

Dr. Chen Bijuan
Chinese Academy of Fisheries Sciences
Qingdao, People's Republic of China

Richard D. Brodeur, Ph.D.
Affiliate Assistant Professor
School of Fisheries
University of Washington
Seattle, Washington

Jon Brodie, M.Sc.
Director, Research and Monitoring Section
Great Barrier Reef Marine Park Authority
Townsville, Australia

Dr. Dudley B. Chelton
College of Oceanography
Oregon State University
Corvallis, Oregon

Professor Ya-Qu Chen
East China Sea Fisheries Research Institute
Shanghai, People's Republic of China

Dr. Alessandra Conversi
Marine Science Research Center
State University of New York
Stonybrook, New York

Dr. Jorge Csirke
FAO
Rome, Italy

Yi Cui
The China Society of Marine Environmental Science
Dalian, People's Republic of China

Dr. E. P. Dulepova
Laboratory of Applied Biocenology
Pacific Research Institute of Fisheries and Oceanography
Vladivostok, Russia

Dr. Robert C. Francis
School of Fisheries
University of Washington
Seattle, Washington

Dr. Bruce W. Frost
School of Oceanography
University of Washington
Seattle, Washington

G. Griffiths
Plymouth Marine Laboratory
Plymouth, United Kingdom

T. Hagiwara
Senior Research Scientist
Global Environmental Forum
Tsukoba, Ibaraki
Japan

Dr. A. Harashima
Chief of Ocean Research Team
National Institute for Environmental Studies
Japan Environment Agency
Tsukuba, Japan

Dr. Steven R. Hare
Quantitative Biologist
International Pacific Halibut Commission
Seattle, Washington

Dr. Thomas L. Hayward
Marine Life Research Group
Scripps Institution of Oceanography
University of California, San Diego
La Jolla, California

W. James Ingraham Jr.
Oceanographer
Resource Ecology and Fisheries Management Division
Alaska Fisheries Science Center
U.S. National Marine Fisheries Service, NOAA
Seattle, Washington

Dr. Li Jie
Postdoctoral Fellow
Department of Animal Physiology
University of Nÿmegen
Nÿmegen, The Netherlands

Xianshi Jin, Ph.D.
Deputy Chief, Living Marine Resources Division
Yellow Sea Fisheries Research Institute
Qingdao, China

Dr. Deng Jingyao
Chinese Academy of Fisheries Sciences
Qingdao, People's Republic of China

Dr. Arbain Hj. Kadri
Dean, Science and Natural Resources
University Kebangsaan Malaysia

Kampus Sabah
Kota Kinabalu Sabah, Malaysia

Dr. Suam Kim
Polar Research Center
Korea Ocean Research and Development Institute
Seoul, South Korea

T. Kimoto, B.S.
Research Chemist
Research Institute of Oceano-Chemistry
Osaka, Japan

Dr. S. M. Konovalov
Institute of Biology of the Southern Seas
Sevastopol, Ukraine

Dr. V. V. Lapko
Laboratory of Applied Biocenology
Pacific Research Institute of Fisheries and Oceanography
Vladivostok, Russia

P. A. Livingston, M.S., M.P.A
Task Leader, Resource Ecology and Ecosystem Modelling Task
Resource Ecology and Fisheries Management Division
Alaska Fisheries Science Center
U.S. National Marine Fisheries Service, NOAA
Seattle, Washington

Dr. Daniel Lluch-Belda
Centro de Investigaciones Biologicas del Noroeste SC
La Paz, Mexico

Dr. Rob Lock
Associate Professor of Animal Physiology
Department of Animal Physiology
University of Nÿmegen
Nÿmegen, The Netherlands

L.-L. Low, Ph.D.
Deputy Division Director
Resource Ecology and Fisheries Management Division
Alaska Fisheries Science Center
U.S. National Marine Fisheries Service, NOAA
Seattle, Washington

R. J. Marasco, Ph.D.
Division Director
Resource Ecology and Fisheries Management Division
Alaska Fisheries Science Center
U.S. National Marine Fisheries Service, NOAA
Seattle, Washington

Dr. John A. McGowan
Scripps Institution of Oceanography
University of California, San Diego
La Jolla, California

Dr. Chen Minshan
Chinese Academy of Fisheries Sciences
Qingdao, People's Republic of China

Mazlin B. Mokhtar
Chemistry Department
University Kebangsaan Malaysia
Kampus Sabah
Kota Kinabalu Sabah, Malaysia

Saleem Mustafa
Marine Science Department
University Kebangsaan Malaysia
Kampus Sabah
Kota Kinabalu Sabah, Malaysia

Chrys-Ellen M. Neville
Department of Fisheries and Oceans
Biological Sciences Branch
Pacific Biological Station
Nanaimo, Canada

T. Piakarnchana
Chulalongkorn University
Department of Marine Science
Chairman, Thailand National Committee on IGBP
National Research Council of Thailand
Bangkok, Thailand

R. Pollard
Plymouth Marine Laboratory
Plymouth, United Kingdom

Dr. Jake Rice
Department of Fisheries and Oceans
Pacific Biological Station
Nanaimo, Canada

Dr. Carlos Robinson
Fisheries Ecology Research Group
Instituto de Ciencias del Mar y Limnologia, UNAM
Mexico City, Mexico

Dr. Jun She
First Institute of Oceanography
State Oceanographic Administration
Qingdao, People's Republic of China

Dr. Xin-Qiang Shen
East China Sea Fisheries Research Institute
Shanghai, People's Republic of China

Kenneth Sherman, Ph.D.
Chief, LME Program
Narragansett Laboratory
USDOC, NOAA, NMFS, NEFSC
Narragansett, Rhode Island

Dr. V. P. Shuntov
Laboratory of Applied Biocenology
Pacific Research Institute of Fisheries and Oceanography
Vladivostok, Russia

Yunli Song
The China Society of Marine Environmental Science
Dalian, People's Republic of China

Dr. Rafael Steer-Ruiz
IOC-IOCARIBE
Cartagena de Indias, Colombia

Y. Tanaka, Ph.D.
Lecturer, Faculty of Agriculture
Kinki University
Nakamachi, Nara
Japan

Dr. Qisheng Tang
Director, Yellow Sea Fisheries Institute
Qingdao, People's Republic of China

Dr. Makoto Terazaki
Professor and Director
Center for International Cooperation
Ocean Research Institute
University of Tokyo
Tokyo, Japan

Dr. Sjoerd Wendelaar Bonga
Professor of Animal Physiology
Department of Animal Physiology
University of Nÿmegen
Nÿmegen, The Netherlands

Dr. R. Williams
Plymouth Marine Laboratory
Plymouth, United Kingdom

Dr. Jae Sam Yang
Department of Oceanography
Kunsan National University
Seoul, South Korea

Dr. Wonho Yih
Department of Oceanography
Kunsan National University
Seoul, South Korea

Dr. Chang Ik Zhang
Department of Marine Living Resources
Kunsan National University
Seoul, South Korea

Xianyong Zhao, M.Phil.
Scientist, Living Marine Resources Division
Yellow Sea Fisheries Research Institute
Qingdao, China

I

Regime Shifts

1

A Dynamic Scenario for Simultaneous Regime-Scale Marine Population Shifts in Widely Separated Large Marine Ecosystems of the Pacific

Andrew Bakun

Abstract

Impressive interdecadal regime shifts in marine fish populations have recently occurred simultaneously in widely separated large marine ecosystems (LMEs) of the Pacific. The period from the mid-1970s to mid-1980s saw phenomenal expansion of sardine populations in the northwestern and southeastern regions of the ocean as well as a modest recovery of the population in the northeastern region. The same period saw exceptional productivity in pollock and other extremely important groundfish stocks of the subarctic region. Alaskan salmon stocks also increased dramatically. Populations of lobsters, seabirds, and marine mammals in the central North Pacific appear also to have been extremely productive. Concurrently, the North Pacific albacore population has apparently experienced a sharp decline, as did many California Current salmon stocks. The need to understand the nature of such regime shifts, as well as their connotations for fishery management, is becoming recognized as one of the most crucial new problem areas in fisheries science.

An explanatory scenario is developed in which the dynamic systems in various regional LMEs of the Pacific Ocean experience intensification via climatic teleconnections in response to a relaxation of the coupled ocean–atmosphere system in the equatorial region during a decadal-scale period of enhanced El Niño characteristics. It is argued that such intensification would result in increased supply of nutrients through upwelling and mixing leading to increased primary organic production, as well as in enhancement of various mechanisms that may act to concentrate food organisms and thereby support efficient larval, juvenile, and adult feeding activity. In addition, intensified dynamic aspects (flow, turbulence, etc.) may favor sardines, which tend to be stronger swimmers, over the weaker-swimming, less migratory anchovies.

These considerations may serve to account for various aspects of the recent regime-shift experiences, including (1) the decline, during the decadal relaxation of the tropical Pacific system, of the albacore tuna population spawning near the equator, (2) the increase of groundfish and salmon stocks in the intensified subarctic Pacific system, and (3) the dramatic increases of sardine stocks and declines of anchovy stocks in the subtropical coastal LMEs in both the northern and southern hemispheres and on both the eastern and western sides of the Pacific Ocean.

The decade from the mid-1970s to the mid-1980s was a remarkable one for the large marine ecosystems (LMEs) of the Pacific Ocean. A pattern of population increases, followed by population declines after the mid-1980s, seems to have been remarkably widespread and consistent in a number of LMEs distributed around the Pacific and also in other parts of the world's oceans (Figure 1-1). For example, this mid-1970s to mid-1980s period was one of phenomenal productivity and growth of the major groundfish populations of the subarctic North Pacific, which sustained the massive expansions of the fisheries of that region through the period (FAO 1992). Conversely, since the mid-1980s these populations have been in decline in spite of continuing elaborate stock assessment activities and state-of-the-art fishery management efforts. Alaskan salmon stocks also increased dramatically during the period.

The period also appears to have been particularly productive in the tropical central North Pacific. Total chlorophyll in the water column appears to have increased north of Hawaii (Venrick et al. 1987). Lobsters, seabirds, seals, and coral reef fishes in the northwestern Hawaiian Islands all seem to have experienced increased production; conversely, since the period ended in the mid-1980s, lobster landings from this area have dropped by two-thirds (Fishing News 1993) and other biological populations are in a downward trend.

On the other hand, North Pacific albacore tuna appear to have suffered a steep population decline during the mid-1970s to mid-1980s period (FAO 1992). Many salmon stocks of the California Current LME also declined during this period (Pearcy 1992; Francis and Hare 1994).

The two largest sardine populations in the world have also followed the dome-shaped pattern shown in Figure 1-1. In the Peru-Humboldt LME off western South America, sardine populations grew during the mid-1970s to mid-1980s period to the point where nearly 6 million tons were taken in 1985, the highest yield of any sardine fishery to date (Lluch-Belda et al. 1992). After remaining extremely depressed for nearly three decades, the sardine population in the northwestern Pacific Ocean near Japan suddenly began to rebuild itself rapidly in the mid-1970s, leading to annual fishery landings in the 1980s of over 5 million tons. More recently, the population has experienced a second collapse. Now the population appears to be collapsing again. In fact, the very large populations of anchovies and sardines that dominate the fish biomass in the major eastern ocean upwelling regions of the world, as well as the northwestern Pacific off Japan, seem to have been rising and falling somewhat in phase (Kawasaki 1983). Both the Californian and Japanese fisheries grew during the 1920s and early 1930s and peaked in the mid- to late 1930s. (There were no corresponding landings off western South America because no significant fishing occurred.) Both populations remained at extremely low numbers for some three decades. The sardine fisheries in both regions then commenced sudden rapid growth near the mid-1970s, the same period in which enormous numbers of sardines appeared off South America, initiating a massive fishery in that region. Toward the latter part of the 1980s, the above-mentioned additional simultaneous reversals in trend occurred (Lluch-Belda et al. 1989, 1992).

Since the advent of substantial fisheries, anchovy populations have been generally out of phase with the sardine populations in the three regions. In the California Current LME, after a time lag of about a decade following the sardine collapse, the anchovy population increased to the point at which over 340,000 tons were taken in 1981. Off Japan, the anchovy catches grew during the period of low sardine abundance following the initial sardine collapse, attaining maximum levels of nearly half a million tons during the late 1950s and the 1960s. The anchovy catches then gradually declined as the sardine population proceeded in its rebuilding phase. Later, as the Japanese sardine population again declined, extremely large shoals of anchovy were reported (Lluch-Belda et al. 1992).

In the Peru-Humboldt Current LME, the fishery for anchovy peaked in 1970 at more than 13 million tons, constituting the largest fishery that has ever existed on earth. It then collapsed to less

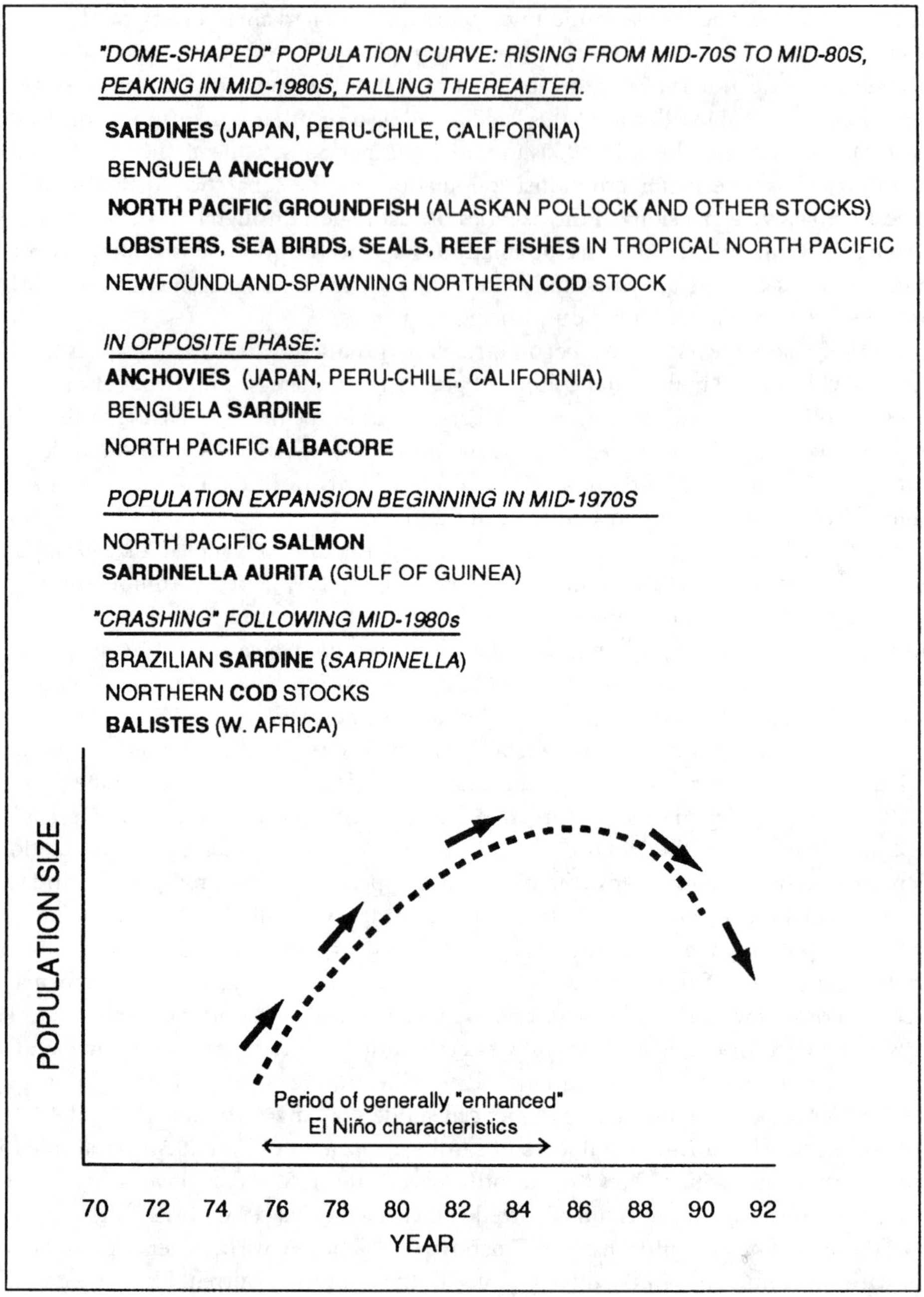

Figure 1-1. Diagram characterizing the pattern of variation, observed in many marine fish populations, showing a period of increased stock productivity and rapid population growth in the decade from the mid-1970s to the mid-1980s, followed by stock declines after the mid-1980s. Examples of various important fish populations whose stock fluctuations have been in phase with (or in opposite phase to) the pattern are listed. (Figure modified from Bakun 1996.)

than 1 million tons after the 1972 El Niño, rebounded briefly to about 2 million tons for several years, and then fell back following the 1976 El Niño to below 1 million tons and remained at this relatively low level to the mid-1980s. However, more recently, the anchoveta population has again explosively rebuilt to the point that by 1994, Peru had once again regained the status that it had lost in the 1970s of being the world's number one marine fishing nation (FAO, 1997).

Thus, we have a situation in which fish populations distributed in widely distant parts of the Pacific, and in other oceans of the world as well (Bakun 1996), appear to be fluctuating in synchrony. The populations are certainly far too widely separated to interact in any direct way. For example, the populations of sardines in the different corners of the Pacific were until very recently considered to be separate, distinct species (Parrish et al. 1989). But as one views the variability of more and more populations and notes the same apparent rhythms, the idea that the variabilities are somehow interlinked becomes rather compelling. However, this is very different from the way that many of us have been accustomed to think about fish stock variability, and so it is natural and understandable that the idea might strain the credulity of appropriately skeptical scientists. Of course, it would help the situation if there were available a plausible set of mechanisms to explain the long-range interconnections. [Note: Much of the material presented in the following sections of this chapter is abstracted from Bakun (1996); additional background material and expanded discussion may be found therein.]

Decadal-Scale El Niño

The waters of the Pacific Ocean constitute the greatest mass of heat storage capacity on earth. The Pacific is so large that its atmosphere as a whole is only minimally subject to continental effects and therefore is not nearly so strongly forced into seasonal climatic regularity as are the atmospheres of smaller, more continentally influenced oceans. The result is that the coupled ocean–atmosphere system of the Pacific wanders from year to year, in response to its own internal dynamics, to a much greater degree than do other ocean regions.

The most dramatic expression of this interyear variability of the Pacific system is the El Niño–Southern Oscillation (ENSO) phenomenon. El Niño episodes precipitate by far the strongest interannual fluctuations in conditions affecting marine populations in the equatorial and eastern boundary regions of the Pacific. These deviations from the norm tend to be so radical that they totally dominate any other mode of variability.

In normal, non–El Niño conditions, a vigorous tropical trade wind system acts to push equatorial surface waters from the eastern part of the ocean toward the western part. This results in a thickened, warmed surface layer in the west and a thinner, cooler surface layer in the east. Associated with this warming in the western Pacific are rising air, relatively low atmospheric pressure, release of latent heat, and copious precipitation (Figure 1-2). Conversely, the cooler waters of the eastern Pacific are associated with sinking air, high barometric pressure, and corresponding dry conditions. However, during El Niño events, the normal atmospheric pressure difference between the eastern and western sides of the equatorial Pacific collapses. Much of the equatorial band of the Pacific becomes anomalously warm. The zone of rising air and associated precipitation moves far eastward (such that floods may occur in normally arid Peru, for example). The Southern Oscillation Index (SOI) is a measure of the atmospheric pressure difference between the eastern and western parts of the ocean. Thus, a feature of El Niño events is a sharp drop in the Southern Oscillation Index (Figure 1-3). Within recent decades, important El Niño events have occurred in 1957–1958, 1966, 1972, 1976, and 1982–1983. Other recent El Niño–type episodes have occurred in 1986–1987, 1991–1992, and 1994–1995 and possibly, to a lesser degree,

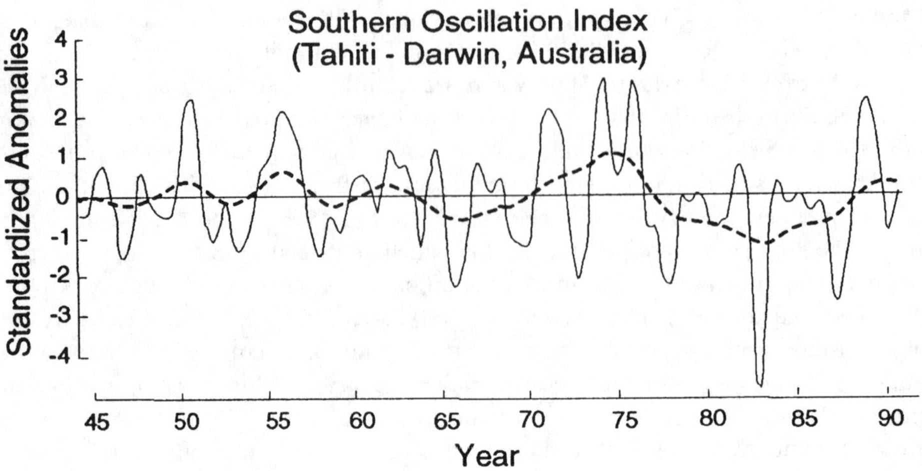

Figure 1-3. Southern Oscillation Index, standardized anomalies, 1945–1990 (redrawn from IOC 1992).

also in 1992–1993 (some scientists prefer to describe the entire 1991–1995 period as a single, essentially continuous, low to moderate intensity, several-peaked El Niño episode).

Because the Pacific is so large, the expansion and displacement of the area of rising air in the tropical Pacific have global effects on the earth's climate system. Accordingly, ENSO constitutes the dominant mode of short-term climatic variability throughout the world (Kiladis and Diaz

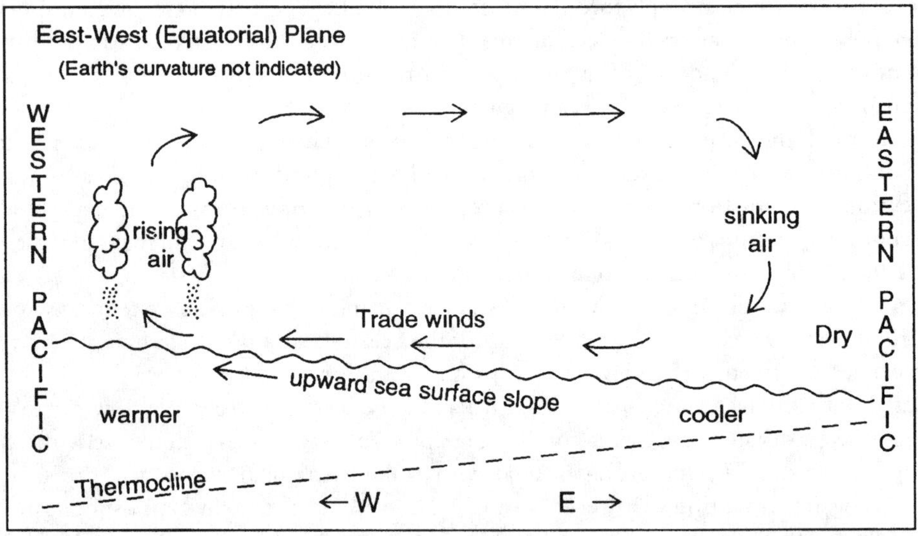

Figure 1-2. Diagram of the Walker circulation acting in the plane of the equator. Rising air in the western Pacific and sinking air in the eastern Pacific feed the surface trade wind circulation, which blows from east to west. This results in an upward slope of the sea surface from east to west along the equator and a much larger opposite downward slope of the thermocline. (Reprinted by permission from Bakun, A. 1996. Patterns in the Ocean: Ocean Processes and Marine Population Dynamics. University of California Sea Grant, San Diego, California, USA, in cooperation with Centro de Investigaciones Biolgicas de Noroeste, La Paz, Baja California Sur, Mexico.)

1989). El Niño events have been associated with drought and catastrophic wildfires in Indonesia (Leighton and Wirawan 1984; Quinn et al. 1978) and Australia (Martin 1983; Pittock 1984), with both drought and flooding in different parts of Brazil (Caviedes 1973; Rao et al. 1986), and with anomalies in monsoon rainfall and drought in India (Shukla and Paolino 1983; Kiladis and Sinha 1991). Apparent linkages with climatic anomalies in the central and eastern United States (Wilhite et al. 1987), western Europe (Palutikof 1987), Africa (Ogallo 1987, 1988; Glantz 1987), central Russia (Pitovranov 1987), China (Shao-wu and Mearns 1987), and Japan (Yoshino and Yasunari 1987) have also been indicated. The 1982–1983 El Niño has been held responsible for widespread incidents of drought, flooding, and severe storms on five continents, resulting in hundreds of deaths and damages estimated at $8.65 billion (Canby 1984; Lagos and Buizer 1992). If any climatic effect could be powerful enough to impart a degree of global synchrony to marine ecosystem processes, El Niño and the ENSO system would certainly seem to be the most likely candidate.

Being at the eastern end of the equatorial Pacific, the Humboldt Current large marine ecosystem off Peru is particularly affected by El Niño episodes. For example, during El Niño episodes this area tends to experience sharp increases in sea surface temperature, wind-driven offshore transport, and wind-generated turbulent mixing, as well as a local drop in barometric pressure (Figure 1-4). This pressure decrease reflects the collapse of the Walker circulation (diagrammed in Figure 1-2) and the resulting decrease in trans-Pacific atmospheric pressure difference that is represented in the Southern Oscillation Index (Figure 1-3). In Figure 1-4, the interannual El Niño signal is clearly dominant (e.g., exemplified by the individual El Niño episodes that occurred in 1957–1958, 1966, 1972, 1976, 1982–1983, etc.). However, one also sees an underlying interdecadal signal that appears as lower-frequency modulations of both the amplitude and frequency of interannual episodes. It is apparent that the mid-1970s to mid-1980s was a period in which the Pacific Ocean was characterized by a rather continuous state of elevated El Niño characteristics. Whether the period actually began with the 1972 El Niño or the 1976 El Niño is problematic. The 1972 event was intense and, from a visual inspection of the time series, seems to have established a new upper peak level that carried through the following decade and provided an elevated base out of which the monster 1982–1983 event arose. However, between the 1972 and 1976 events, the system fell back briefly into an opposite cool phase. After the 1976 event there were no more of what most would call "cool phases" until after the mid-1980s.

The 1972 event, while intense in the tropics, did not have much effect in the northern part of the Pacific. For example, temperature time series in the Gulf of Alaska show an abrupt shift from quite consistent negative anomalies in the first half of the 1970s to consistent positive anomalies in the second half of the 1970s and early 1980s (Royer 1983; Cole and McLain 1989). It is well known that the Aleutian low pressure system (the Aleutian Low), which is the main driving force for dynamic processes in the subarctic Pacific (see the section on Groundfish Populations of the Subarctic North Pacific), tends to be intensified during El Niño events (Namias and Cayan 1984). The anomalous rising air in the equatorial region apparently sets up a series of anomalous circulation cells in the global atmosphere (Figure 1-5), and the resulting intensification of the subarctic low pressure cell is probably the major linkage by which El Niño effects are transferred from the tropics to high latitudes of the Pacific. Bakun (1996) suggested that the main intensity of the 1972 event may have been phase-shifted from the northern hemisphere winter (Peruvian summer) toward spring and summer. Because the Aleutian Low rapidly loses intensity following the winter season, there may not have been enough energy remaining to transmit El Niño effects efficiently via the Aleutian Low teleconnection.

An extreme low-pass filtering (heavy dashed line in Figure 1-3) of the Southern Oscillation Index (SOI) yields a maximum in the mid-1970s followed by a minimum in the early 1980s (IOC

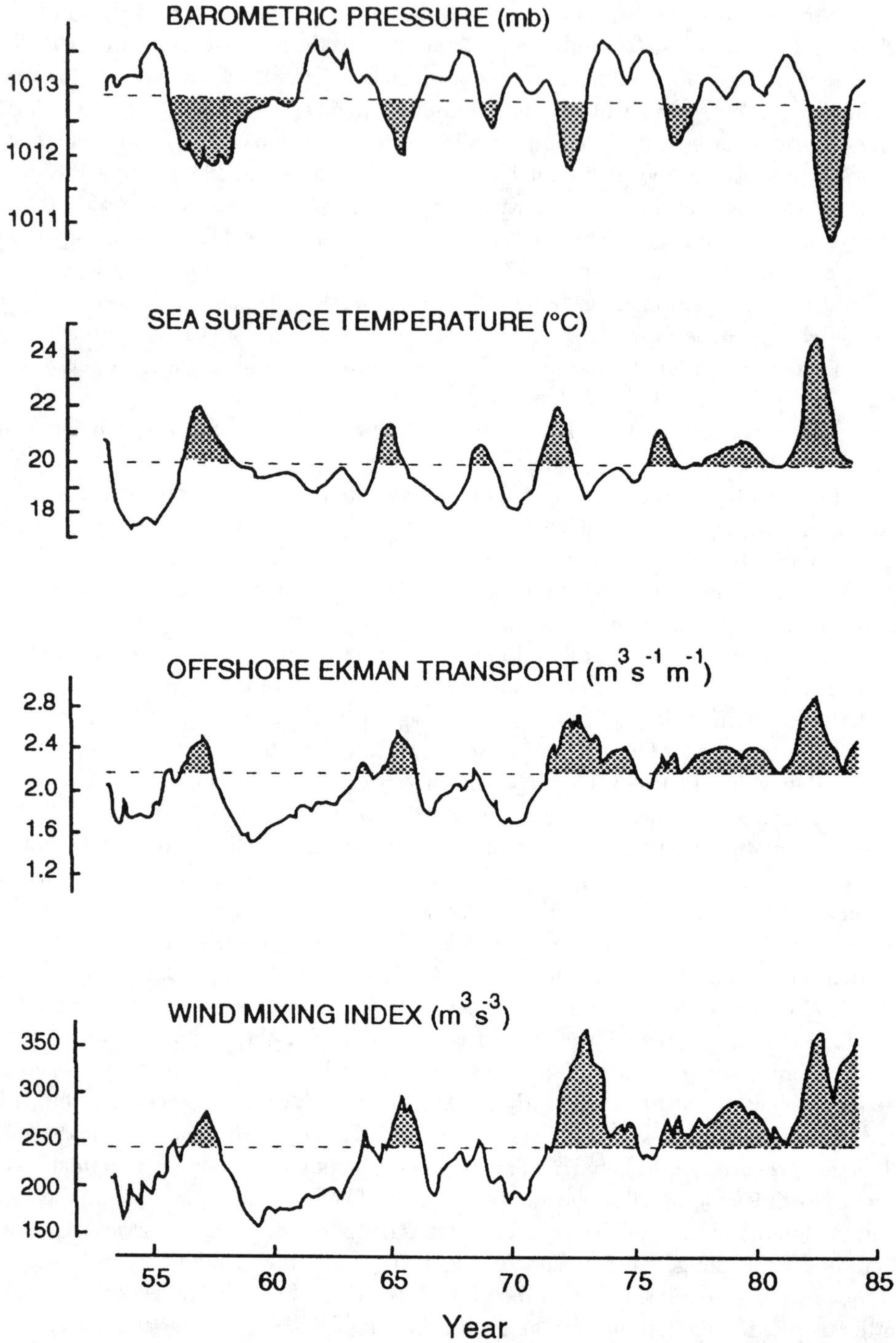

Figure 1-4. Low-frequency nonseasonal variations: 12-month running means of monthly averages produced from maritime reports from the ocean area off north-central Peru (redrawn from Bakun 1987).

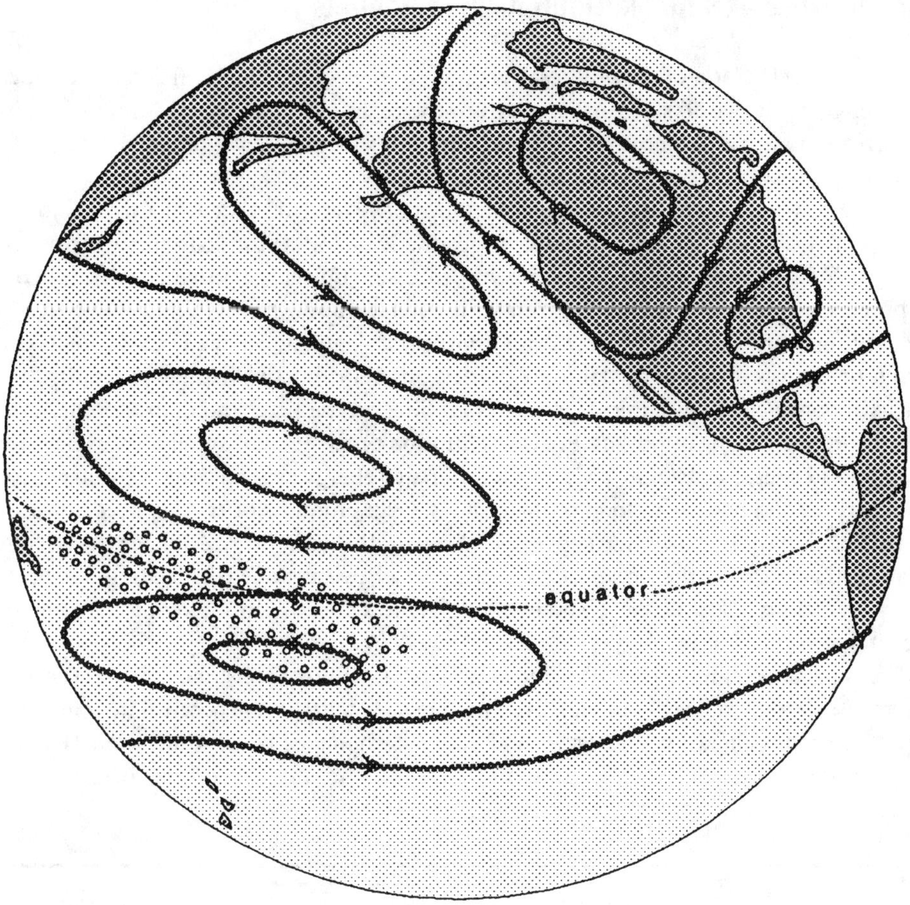

Figure 1-5. Circulation anomalies at the jet stream level during El Niño (northern hemisphere winter). The stippled area indicates the eastward extension of cloud cover and rainfall in the equatorial Pacific (redrawn from Horel and Wallace 1981).

1992). From this point of view, the mid-1970s to mid-1980s period resembles a decadal-scale analog to a typical annual-scale episode. Indeed, Inoue and O'Brien (1987) have reported a general decrease in strength of the trade winds in the tropical Pacific during the late 1970s.

However, whereas El Niño is characterized by a relaxed dynamic situation in the equatorial Pacific (diminished trade wind circulation, etc.), it tends to be countered by intensified situations in many other regions of the world. For example, the mid-1970s to mid-1980s period was characterized by general intensification of the Aleutian Low (Norton et al. 1985; Wooster and Hollowed 1995). As noted in Figure 1-4, wind-generated turbulence, offshore-directed Ekman transport, and other dynamics increased strongly off Peru during the period. There were also apparent decadal-scale intensifications of upwelling-favorable winds during the spring–summer upwelling seasons in the California Current, Canary Current, and Iberian LMEs during this period (Bakun 1990, 1992), and also off northern Chile (≈23°–30°S) (Yáñez 1991; Bakun 1992). Off central Chile (≈35°–38°S), although the mean upwelling rate apparently did not intensify, wind-induced turbulent mixing seems to have increased dramatically during this period (Yáñez et al. 1992).

Characteristics of Suitable Reproductive Habitats

So, how could relaxation of the physical dynamics in the equatorial Pacific and intensification in other regions of the world's oceans have driven the regime-change events summarized in Figure 1-1? For an answer one needs to consider the reproductive ecologies of the various species involved.

Comparative studies of fish habitat climatology have tended to identify three major classes of processes that combine to yield favorable reproductive habitats for coastal pelagic fishes as well as many other types of fishes: (1) enrichment processes (upwelling, mixing, etc.), (2) concentration processes (convergence, frontal formation, water-column stability), and (3) processes favoring retention within (or drift toward) an appropriate habitat. I have elsewhere (Bakun 1996) referred to these as a "fundamental triad" underlying reproductive habitat suitability.

The importance of enrichment processes is widely recognized and appreciated. Perhaps less widely appreciated is the importance of concentration processes. For small organisms, such as fish larvae, seawater represents quite a viscous fluid; major energy expenditures may be necessary just to move from food particle to food particle. Thus, large amounts of energy, needed for the rapid growth that is required for quick passage through the various size-related levels of intense predation that pervade the ocean environment, may be expended in feeding activity. Consequently, availability of processes whereby food particles are concentrated (e.g., Figure 1-6) tends to be vital to survival.

This is probably a major reason why various types of interfaces, or *ergoclines* (Legendre and Demers 1985), tend to be sites of enhanced biological activity in the ocean. These interfaces tend either to be maintained by, and/or to maintain, mechanisms of concentration (Bakun 1996). Ocean fronts (Figure 1-6) are obvious examples. The importance of processes occurring in or near

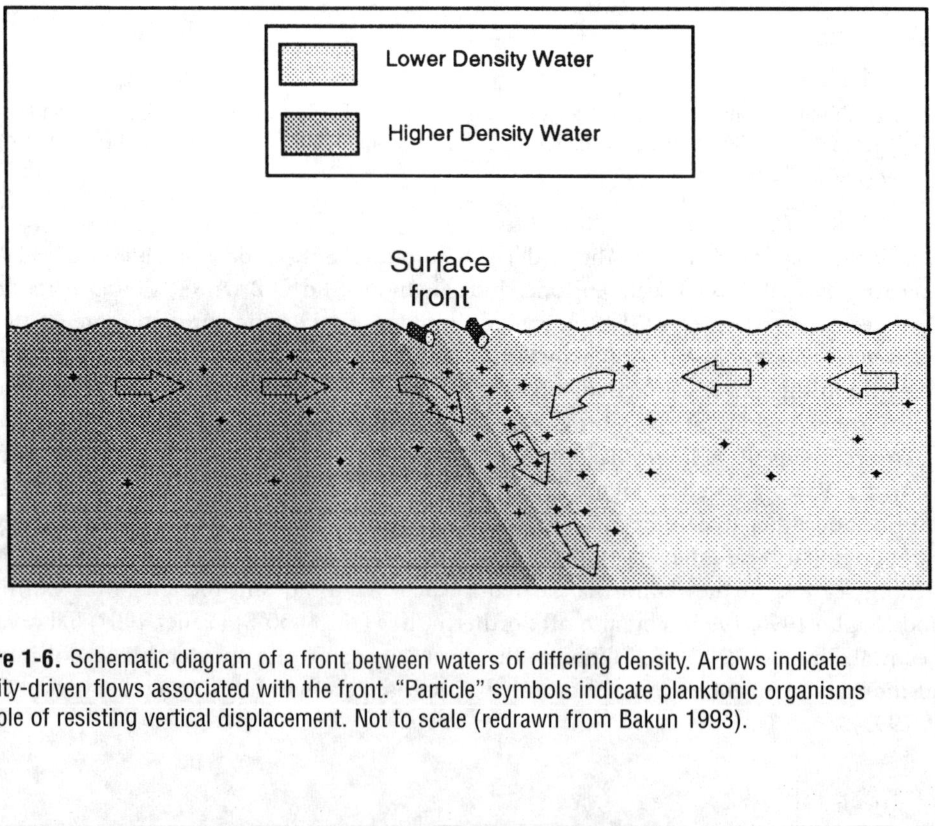

Figure 1-6. Schematic diagram of a front between waters of differing density. Arrows indicate density-driven flows associated with the front. "Particle" symbols indicate planktonic organisms capable of resisting vertical displacement. Not to scale (redrawn from Bakun 1993).

ocean fronts is suggested by the widespread attraction of fish and other marine animals to drifting objects. The actual convergent water motions associated with a front are probably too subtle to be directly sensed by pelagic organisms functioning in an environment devoid of fixed reference points. However, drifting objects tend to be carried into and to accumulate within frontal structures. An innate attraction to drifting objects serves to position the fish within the zones of enhanced biological activity and correspondingly improved feeding conditions (Bakun 1993, 1996). Conversely, turbulence is a dispersive process and so tends to act counter to concentration processes. Thus, intense turbulent mixing events have appeared to be detrimental to larval survival (see, for example, Lasker 1978, 1981a, 1981b). On the other hand, extremely small-scale turbulence may actually act like a concentration mechanism by increasing the encounter rate of small organisms with food particles (Rothschild and Osborn 1988; Sundby 1995).

The third factor in the triad is retention. Life cycles of marine organisms tend to include at least one stage of passive larval drift. Thus, in a dispersive fluid medium, loss of early life stages from the population habitat may represent serious wastage of resources. Consequently, fish populations tend to spawn in locations and seasons that minimize such losses (Parrish et al. 1981; Sinclair 1988).

The North Pacific Albacore

Pacific albacore tuna (*Thunnus alalunga*) are caught as juveniles in surface fisheries in the richly productive temperate-zone ecosystems both of the Kuroshio–Oyashi confluence of the western North Pacific and the California Current upwelling system of the eastern North Pacific. Adults are caught in longline fisheries over very large areas of the tropical and temperate North Pacific (Sund et al. 1981). These fish invest enormous energy expenditures in undertaking migrations of many thousands of kilometers to the tropical zone of the west-central North Pacific to place their voraciously predatory larvae in what is, in a large-scale spatial average sense at least, a region of quite low primary productivity (FAO 1972). It has been suggested (Bakun et al. 1990) that a major part of the explanation for this fairly puzzling behavior may lie in the fact that the region where the albacore spawns corresponds to the location where the core of maximum velocity of the northeast trade winds veers particularly close to the equator (Figure 1-7).

A feature of any persistent maximum in the sea surface wind-stress field is that it produces a closely linked "couple" of adjacent areas of divergence and convergence in the surface Ekman transport field (Figure 1-8). On the left side (when facing downwind in the northern hemisphere), the Ekman transport is directed toward the line of maximum stress and increases in magnitude as the distance from the line decreases. Thus, the Ekman transport field is diverging as it approaches the line of maximum stress. The result is upwelling, corresponding enrichment of the ocean surface layers, and associated enhanced photosynthetic production. As the water carried in the Ekman transport field crosses the line of maximum stress, Ekman transport begins to steadily lessen in magnitude with distance, and therefore the water is converging on that side of the wind-stress maximum. This leads to convergent frontal formation, which tends to concentrate food organisms (Figure 1-6) resulting from the planktonic succession occurring in the Ekman layer that is moving from the zone of enrichment on the other side of the stress maximum. Because of the inverse relationship of Ekman transport to Coriolis effect, these processes intensify rapidly as the equator is approached, without corresponding intensification of wind-related turbulent dispersion (Figure 1-9). The resulting concentrated distributions of planktonic food organisms may provide particularly favorable conditions for the tuna larvae, which need to feed voraciously and grow extremely rapidly to meet their life cycle needs.

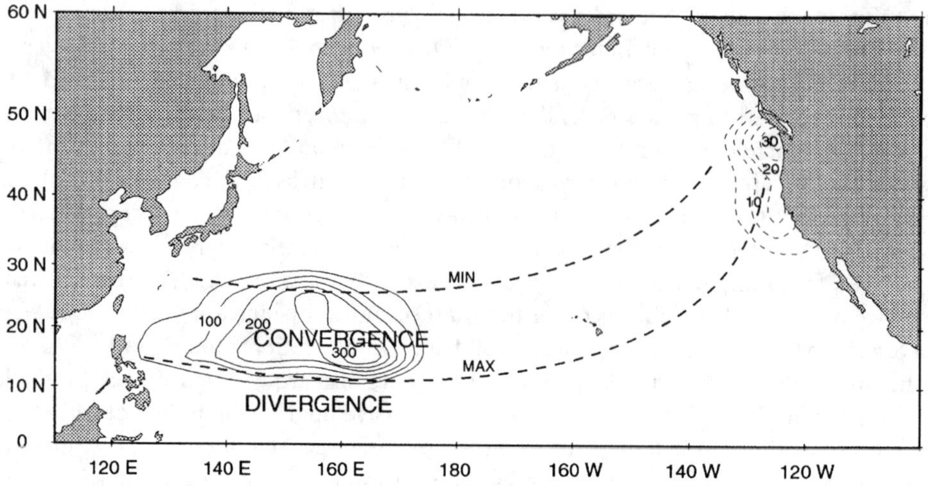

Figure 1-7. Distribution of albacore larvae (solid contours) and of 3-year-old fish (dashed contours). The heavy dashed lines indicate the average position of maximum intensity of westward-directed wind stress exerted by the northeast trade winds, and the corresponding "north side" minimum in westward-directed wind stress. There is Ekman divergence to the south of the wind-stress maximum, and convergence to the north. Larvae are distributed in the area of Ekman convergence in the zone where the stress maximum veers closest to the equator. Units: number of larvae per million cubic meters (data from Nishikawa et al. 1985); distribution of catch locations of 221,221 albacore (age 3, 55–67 cm) sampled in the North Pacific (R. H. Parrish, personal communication).

An additional enrichment and concentration process operates in such a situation. The respective zones of intense divergence and upwelling and of intense convergence and downwelling overlie corresponding zones of doming and of downwarping of the subsurface density structure (see the example transequatorial temperature section in Figure 1-10). The position of the maximum in wind stress thus marks the southern boundary of a zone of sharp dip in the density structure, which in this case corresponds to (and acts to reinforce) the high-velocity core of the North Equatorial Current. (Because of this response of the ocean density, surface current jets tend to underlie strong wind-stress maxima, and these become particularly intense at very low latitudes.) The southerly boundary or "wall" of the North Equatorial Current is a zone of cyclonic frontal shear that tends to produce cyclonic vortex action due to lateral friction. The dynamic processes involved in formation of such cyclonic vortices result in upwelling in the cores of the vortices (Figure 1-11). However, the incremental deceleration of the larger-scale flow by this frictional action has the effect of reducing the Coriolis force that acts to hold the surface water of the North Equatorial Current from encroaching southward. This effect allows encroachment of these waters into the frontal zone, forming a mixed water mass of greater density that flows under the North Equatorial Current surface water. This creates sinking motion in the frontal zone and the associated concentration mechanism displayed in Figure 1-6.

These two nested, mutually supportive sets of coupled processes of enrichment and concentration apparently act in concert with other conditions in the area to create sufficiently favorable circumstances for larval feeding and growth to warrant the massive migrational energy expenditures assumed by the adults (Bakun 1996).

However, retention, the third element proposed in the previous section as contributing to favorable reproductive habitat, is absent. Thus, this does not appear to be a suitable habitat for

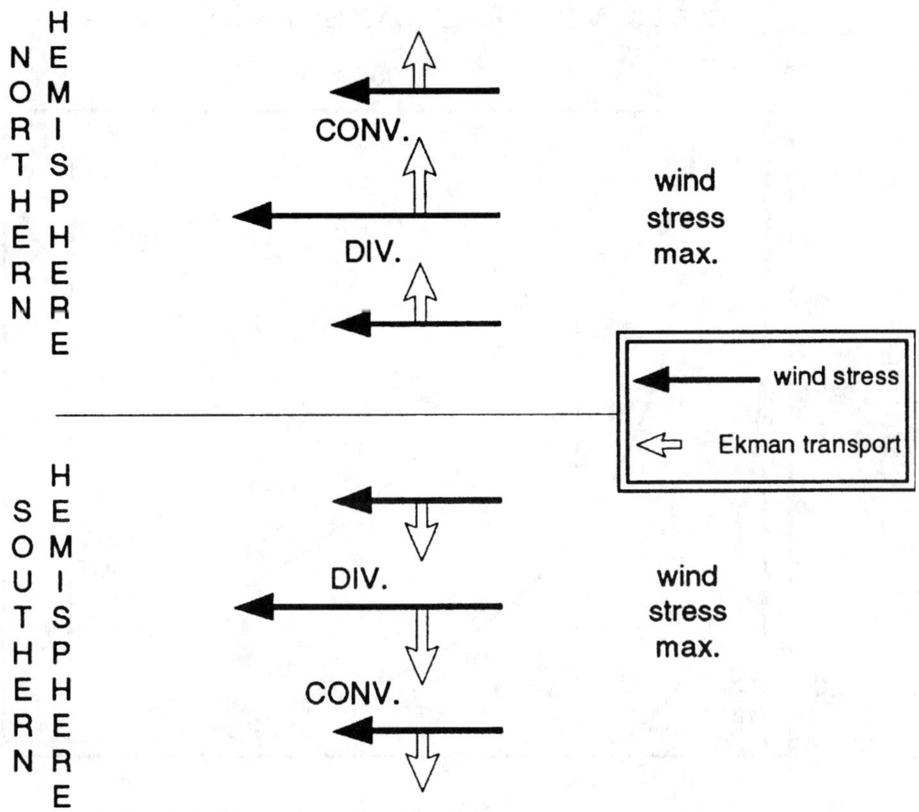

Figure 1-8. Diagram of zones of Ekman convergence and divergence on either side of a wind-stress maximum. Relative magnitudes are indicated by differences in lengths of the vector symbols. (Reprinted by permission from Bakun, A. 1996. Patterns in the Ocean: Ocean Processes and Marine Population Dynamics. University of California Sea Grant, San Diego, California, USA, in cooperation with Centro de Investigaciones Biolgicas de Noroeste, La Paz, Baja California Sur, Mexico.)

many types of fishes. The albacore solve the problem by, among other things, using the hydrodynamic advantages afforded by their growth to very large size (Bakun 1996). They thus are able to maintain the integrity of their population through sheer swimming power. In doing so, the albacore may manage to have their reproductive habitat largely to themselves, with many other types of fishes that would be of a size to compete with or to prey upon larval or early juvenile albacore having been precluded from the area by their inability to solve element three of the proposed triad.

The Groundfish Populations of the Subarctic North Pacific

An apparent burst of productivity of some of the most important groundfish stocks of the Gulf of Alaska and the Bering Sea LMEs occurred during the mid-1970s to mid-1980s period of enhanced El Niño characteristics, and a relative decline in productivity of these stocks appears to have occurred since that time. A burst of good recruitment years for Alaska pollock (*Theragra chalcogramma*) and other groundfish stocks in the late 1970s and early 1980s apparently acted to

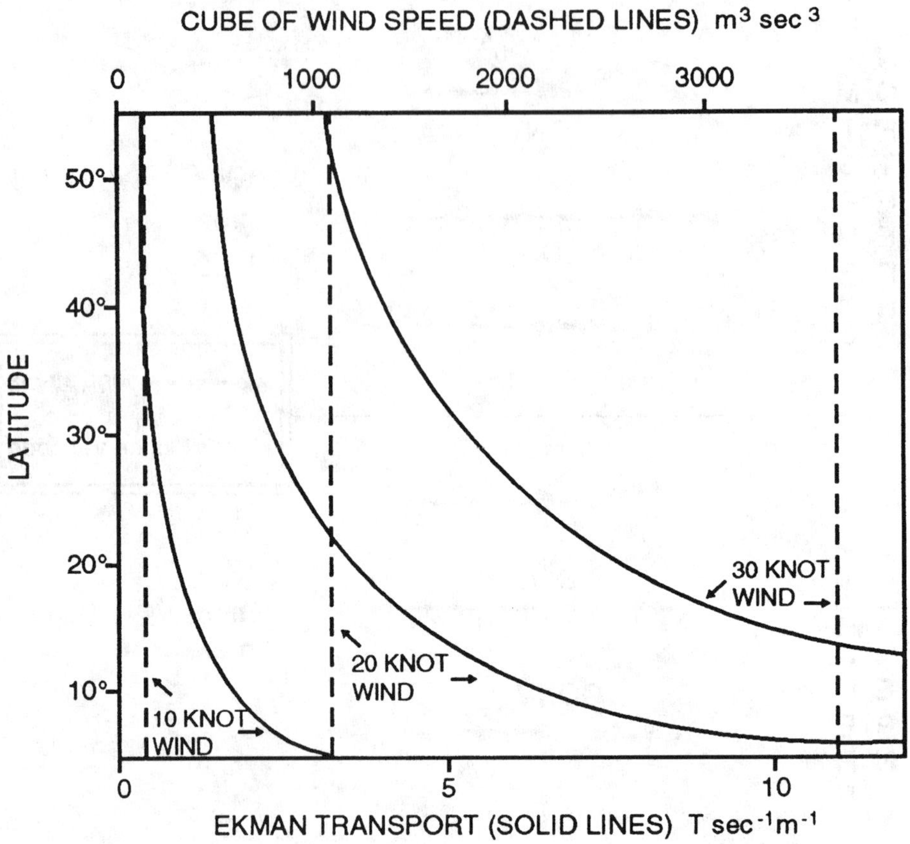

Figure 1-9. w^3 wind mixing index (third power, or cube of the wind speed), which is an indicator of turbulent mixing energy production, versus Ekman transport for several values of wind speed (after Bakun and Parrish 1982).

sustain the growth of the fisheries through much of the 1980s. However, since the mid-1980s the situation has changed drastically. Good recruitments have become much less frequent and these groundfish stocks have generally been in a state of gradual decline (NPFMC 1992a, 1992b). The king crab stock, which collapsed under extremely heavy fishing in the late 1970s and early 1980s, has remained at a depressed level in spite of stringent management efforts designed to rebuild it.

The Aleutian low pressure cell overlying these LMEs is one of the most intense quasi-permanent atmospheric systems on earth. It constitutes a center of violent atmospheric cyclogenesis during the winter half of the year, which exerts very powerful cyclonic wind-stress curl on the ocean surface. Accordingly, this wind-stress curl drives intense divergence in the ocean surface Ekman transport field, causing massive upwelling and abrupt upward ridging of the density structure (Favorite et al. 1976) in the interior of the gyre. This doming “ridge” structure supplies the baroclinic foundation in which the energetic ocean currents of the region are embedded. The Ekman transport, responding to the furious winds, drives the surface waters against the coastal boundary of the eastern and northern Gulf of Alaska (Figure 1-12), driving the density structure precipitously downward and further intensifying the flows along the boundary.

Thus, the ocean flow system in the Gulf of Alaska LME (Figure 1-13) is dominated by strong cyclonic (counterclockwise) circulation around its periphery. The northward flow on the eastern

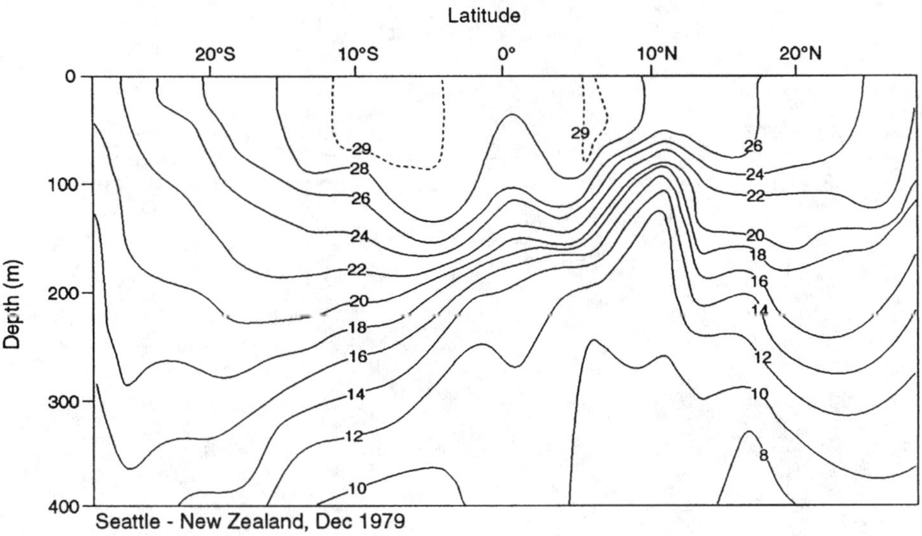

Figure 1-10. Transequatorial vertical temperature section, produced from expendable bathythermograph soundings taken from a ship of opportunity sailing from Seattle to New Zealand (i.e., crossing the wind-stress maximum somewhat to the east of its nearest approach to the equator). Strong doming of the thermocline, indicative of upwelling, appears near 10°N latitude. Abrupt deepening of the isotherms at the southern edge of the North Equatorial Current corresponds to the zone of anticyclonic wind-stress curl, and associated surface convergence and downwelling, existing to the north of the trade wind maximum (see Figure 1-7).

side of the Gulf is broad, diffuse, and rather slow. But as the flow veers around the northern boundary of the Gulf and turns southwestward, it narrows and intensifies to become the swift western boundary flow known as the Alaskan Stream (Favorite 1967; Thomson 1972). The Alaskan Stream flow then runs rapidly westward along the Aleutian Islands chain, with branches flowing into the Bering Sea through the deeper passes between islands. There are also intermittent exchanges with the Subarctic Current that flow eastward just to the south. As a result of the large inflows of surface waters into the Bering Sea through Aleutian Islands passes, the surface layer of the Bering Sea is composed primarily of waters from the Alaskan Stream and Subarctic Current systems (Favorite et al. 1976).

The other major driving force for the local system is the copious coastal runoff that occurs in the region (Royer 1979). The entire continental coastal region in many ways resembles a massive, continuous runoff plume spread along the coast. Geostrophy tends to confine the less dense, runoff-influenced surface waters against the coast, which adds to the upward sea surface slope from the ocean toward the coast produced by the wind-driven coastward Ekman transport and so further enhances the alongshore circulation. Much of the runoff in the region comes from summer melting of precipitation stored during the cold seasons as snow and ice. This continues to reinforce the flow system during the summer season when the wind nearly vanishes in comparison with its winter intensity level. The general counterclockwise advection around the Gulf brings large quantities of warmer water of southern origin into the northern part of the region, evident in the Alaskan Stream even during winter as a warm core at a depth of approximately 100 m to 300 m that extends out along the Aleutian chain well into the western part of the Pacific (Dodimead et al. 1963).

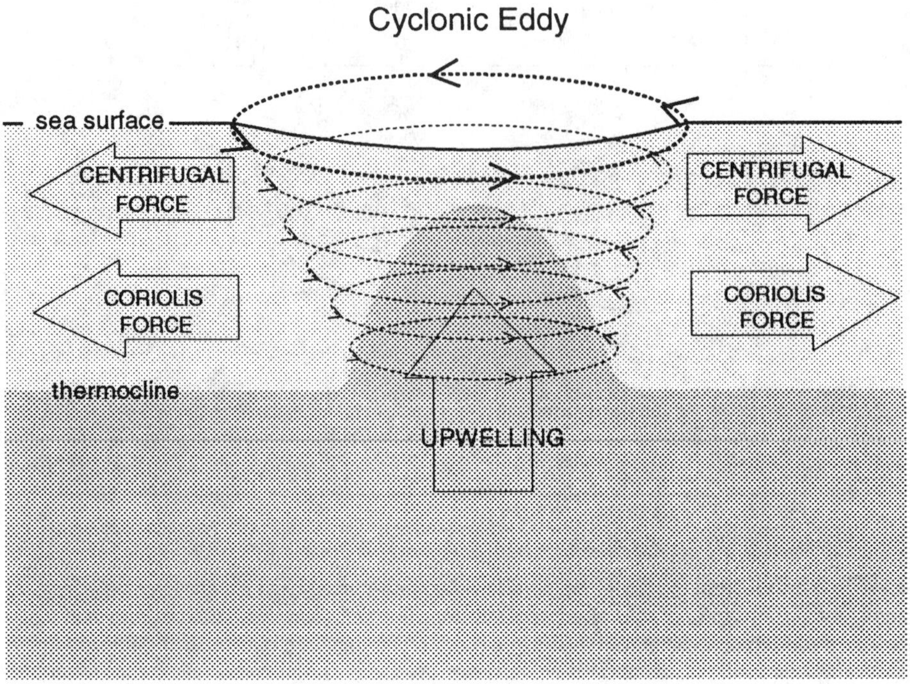

Figure 1-11. Upwelling occurring in a cyclonic eddy. Centrifugal and Coriolis forces due to the water motions are directed outward from the vortex center. In equilibrium (quasi-geostrophic balance), these are counteracted by inwardly directed pressure forces due to upwelling of deeper waters of greater density within the vortex core. (Reprinted by permission from Bakun, A. 1996. Patterns in the Ocean: Ocean Processes and Marine Population Dynamics. University of California Sea Grant, San Diego, California, USA, in cooperation with Centro de Investigaciones Biolgicas de Noroeste, La Paz, Baja California Sur, Mexico.)

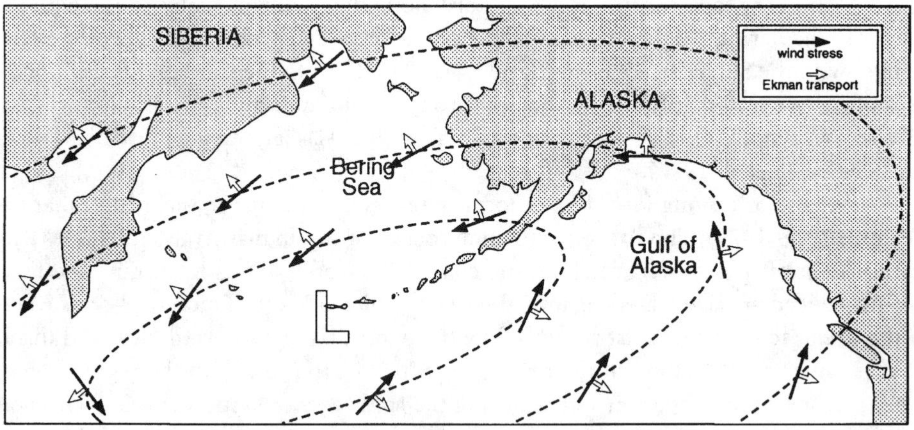

Figure 1-12. Diagram illustrating the effect of the Aleutian low pressure cell in driving coastward Ekman transport around the periphery of the Gulf of Alaska and other regions of the subarctic Pacific Ocean. Dashed lines indicate pressure isobars surrounding the low pressure center. (Note that whereas the higher-level geostrophic winds closely parallel isobars of pressure, the winds near the sea surface are deflected by friction to the left of the trend of the isobars.)

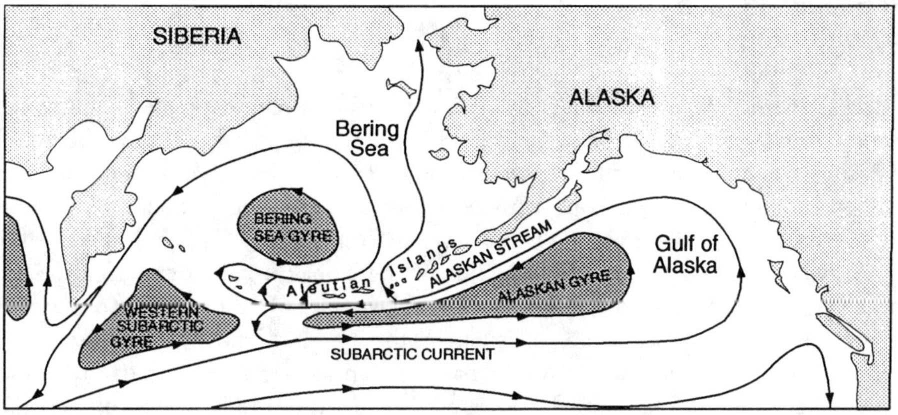

Figure 1-13. Schematic diagram of the surface circulation in the subarctic Pacific. Major cyclonic gyres, characterized by strong upward doming in the subsurface ocean structure, are indicated by shading (redrawn from Dodimead et al. 1963).

The El Niño–related intensification of the Aleutian Low would thus act to intensify the entire system, in that the increased cyclogenesis (storm production) that this implies would act to intensify both the wind forcing and the precipitation. One effect would be to increase the advection of warm waters into the northern part of the region, generally raising ocean temperatures all along the coast.

Enhanced enrichment of the system would occur through strengthened wind-stress curl–driven upwelling in the interior of the gyre, perhaps augmented by correspondingly increased winter wind mixing. The intensified Alaskan Stream flow would tend to increase the production of cyclonic frontal vortices, and associated vortex-driven upwelling, along its offshore edge. The tendency for eastward displacement of the anomalous intensified low pressure cell into the Gulf of Alaska proper during El Niño conditions would cause the Ekman transport pattern to be even more intensely advected toward the coast in the eastern and northern parts of the Gulf, and in fact extend the area of coastward Ekman transport much further westward toward the Aleutian Islands chain. This would tend to intensify the transport of enriched surface waters toward the coastal habitat.

Concentration mechanisms would also be enhanced by such an intensified system. Increased runoff would favor a stronger and more widespread frontal formation between runoff-influenced and oceanic surface waters. Intensified Alaskan Stream flow would generate increased frictional dissipation, which would act to break down geostrophy in the frontal zones, producing density-driven convergence both at the outer and inner edges of the current jet. Superimposed on this, the intensification and westward extension of the shoreward wind-driven Ekman transport pattern would act to increase the piling of oceanic surface waters against the coast, enhancing coastal downwelling and associated formation of convergent frontal structures.

The general pattern of coastward Ekman transport would carry larvae toward, and favor their retention within, the coastal habitat. The intensified geostrophic flow system would tend to rapidly spread larvae and juveniles along the coast and westward along the chain of Aleutian Islands, where those near the ocean surface would tend to be carried onto the shelf or into the proximity of island passes by the wind-driven surface Ekman transport. Once on the shelf, where the flows are somewhat decoupled from the flows off the continental shelf break, they might find a favorable juvenile habitat, or they might be injected through island passes and carried by the eastward coastal flow on the north side of the Aleutian chain onto the broad shelf habitat of the eastern Bering Sea.

North Pacific Salmon

Beamish and Bouillon (1993) note that trends in Pacific salmon catches from 1925 to 1988 have been similar to the smoothed trends in an index of the strength of the Aleutian Low. This correlation, in view of these linkages, indicates that salmon populations in the subarctic regions of the Pacific Ocean have generally done well during eras when the tropical Pacific has been under the influence of intensified El Niño activity. Conversely, salmon populations feeding in the subtropical coastal upwelling zone off the western United States have appeared to do poorly during El Niño episodes (Pearcy 1992). Thus, if the globally synchronous marine ecosystem regime changes are indeed related to interdecadal variability in the ENSO system, it is not surprising that salmon stocks would participate in the shared rhythms of variation.

Certainly, El Niño tends to warm the coastal ocean and adjacent coastlands of the north-eastern Pacific. Along the latitudinal gradient from Alaska south to California, northernmost stocks would tend to approach the cool limit of their range, whereas southerly stocks would be toward the warm limit. One would expect the optimum to be somewhat intermediate. In such a case, El Niño conditions would tend to push this temperature optimum northward.

While this author tends to believe that direct temperature effects are widely oversold as causes of fish stock variability (Bakun 1993, 1996), salmon may well be particularly sensitive in this respect. Salmon spend their earliest life stages in shallow freshwater habitats that are much more subject to drastic year-to-year temperature anomalies than are habitats within the ocean. For example, severely cold winter conditions in Alaska may even freeze salmon spawning streams completely down to the stream bed, thereby totally depleting the temperature-buffering capacity of the liquid phase of water and allowing the temperature to drop to lethal limits for the buried eggs. Extremely warm conditions in California may lead to early loss of the snowpack and reduced flow rates later in the season. Both reduced flow rates and higher temperatures would, among other things, tend to lower the dissolved oxygen levels available to egg and larval stages.

Beyond direct temperature effects, the decadal intensification of dynamic aspects of the subarctic system as described for groundfish in the previous section could have equally beneficial effects for growth and survival of salmon smolts, as well as of later life stages. Moreover, salmon enter the ocean after an extended period of early life in freshwater and have consequently grown to substantial size and associated swimming power by the time they enter the marine system. Thus they may be relatively well able to cope with the enhanced levels of turbulence and advection that typify an intensified dynamic system.

On the other hand, in the California Current LME, where salmon stocks did not fare well during the period, El Niño conditions tend to feature a thickened layer of warm, nutrient-depleted surface water such that upwelling is much less effective in enriching the trophic system. Thus, El Niño conditions are characterized by drastic reductions in primary production and zooplankton abundance (Bernal 1981; Miller et al. 1985). One might not be surprised that salmon smolts, being zooplankton feeders, might not survive well under these conditions.

Anchovy versus Sardine Dominance in Pacific Coastal LMEs

Perhaps the most striking regime change during the period was the sudden flip from general anchovy dominance to sardine dominance in the subtropical coastal LMEs of the Pacific.

In certain habitat configurations, such as described in the previous sections for the North Pacific albacore tuna and subarctic groundfish populations, the three elements of the reproductive habitat triad appear to be mutually supportive. (For additional examples of ways in which triad processes may interact to constitute suitable reproductive habitats for marine species, see Bakun 1996.) This is not unexpected. Natural selection must act quite immediately to reject, from various geographic and seasonal possibilities, spawning habitat configurations that would be unworkable for a given species. Thus, organisms that have access to mutually supportive combinations of processes would naturally make use of them in their reproductive behaviors. In such cases, intensification of the dynamic aspects of the physical system would appear to yield even more of what may generally be good things for reproductive success and resulting population growth. And indeed, the mid-1970s to mid-1980s period, during which the subarctic system seems to have experienced a general intensification (enhanced Aleutian Low, etc.), was very favorable for growth of the salmon and groundfish populations of the subarctic Pacific. Conversely, the relaxation of the near-equatorial system during this same period corresponded to a drastic decline in the abundance of North Pacific albacore tuna (Figure 1-1).

However, in the environmental configurations of coastal upwelling regions, the three elements of the triad tend not to be mutually supportive. Coastal upwelling occurs because wind-driven transport of surface water is directed offshore, causing a situation of flow divergence near the coast. Accordingly, wind-induced coastal upwelling is directly linked to loss of larvae from the coastal habitat. Likewise, convergence, which promotes concentration of food particle distributions, and upwelling, which is a result of flow divergence, cannot directly coincide. The strong winds that drive the upwelling also impart turbulent mixing energy to the water column. Such mixing may abet enrichment but destroy vital small-scale structure in food particle distributions (Lasker 1978).

Thus, upwelling systems appear to be difficult places for fish to reproduce successfully. However, the intensity of enrichment that occurs makes them so productive that the fishes that use them as feeding grounds represent some of the largest fish populations that exist in the oceans. Larger fishes tend to spawn elsewhere and migrate into coastal upwelling regions for feeding. For example, in the upwelling ecosystem of the California Current, major upper trophic level predators include the albacore, which spawn far away in the western part of the tropical Pacific (Figure 1-7), and the various species of salmon, which leave the ocean entirely to spawn in freshwater. Smaller fishes (sardines, anchovies, etc.) make use of various habitat features to reduce the impact of offshore transport (Bakun 1996). Nearly all important fish species appear to avoid attempting to reproduce in core zones of the strongest upwelling.

Over the past several decades, linear empirical analyses of the relationships between local wind effects and recruitment variability in different upwelling-zone pelagic fish populations often yielded differing and, therefore, rather unsatisfying results. Recently, effective nonlinear methods of empirical analysis have been introduced to marine ecology and fisheries science (Mendelssohn and Cury 1987; Mendelssohn and Mendo 1987). These methods have been applied to the Peruvian anchoveta, the California sardine, the California anchovy, the Moroccan sardine, and the Senegalese and Ivoirian sardinellas (Cury and Roy 1989; Cury et al. 1995). The result is a consistently domed-shaped relationship, or "optimal environmental window" (Figure 1-14), where reproductive success is highest at an intermediate wind intensity and decreases at both higher and lower wind intensities. In view of these results, it is not difficult to see why the previous results may have been inconsistent. In cases in which most of the data may have been on the "left flank" (low wind-speed side) of the optimal environmental window, a positive relationship with wind speed would be found. Conversely, if most of the data were on the "right flank" (high wind-speed side), an inverse relationship with wind speed would be found. If the data were rather evenly distributed across both flanks of the window, linear methods could pick up no relationship at all.

Constructing an interpretation of the optimal environmental window result is quite straightforward. The control on the right flank, that is, the high-wind side, could be exerted either by (1)

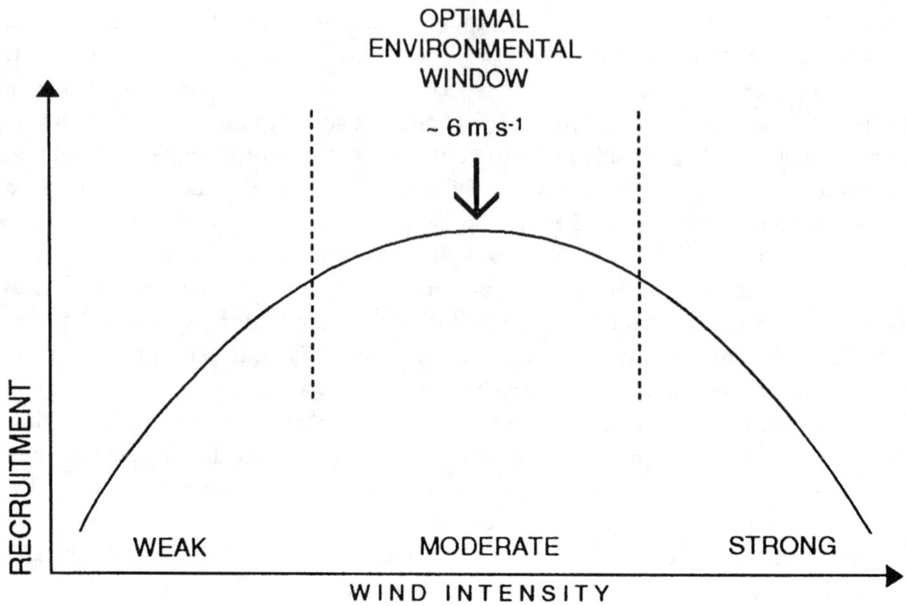

Figure 1-14. Optimal environmental window relationship between wind intensity and reproductive success in upwelling regions (redrawn from Cury and Roy 1989).

excessive offshore transport leading to offshore loss of pelagic larvae from the favorable coastal habitat (e.g., Parrish et al. 1981; Sinclair 1988) or by (2) overly intense turbulent mixing that could (a) disperse fine-scale concentrations of appropriately sized, nutritious food particles needed for successful larval feeding (Lasker 1978, 1981a, 1981b), (b) inhibit basic photosynthetic production by mixing phytoplankton cells beyond their critical depth (Sverdrup 1953; Steele 1974), or (c) impair a larva's ability to physically capture prey (MacKenzie et al. 1994). An obvious explanation for the left flank, or low-wind side, is a lack of nutrient enrichment by wind-induced upwelling or mixing, leading to inadequate production of appropriate larval food (Cushing 1969). Also, it is possible that under conditions in which the interaction of feeding behavior with stable fine-scale food particle structure may be less important than the energy savings produced by turbulent diffusion of food particles toward feeding larvae, the mechanism of Rothschild and Osborn (1988) may exert some control on the left flank by increasing larval survival toward the slightly higher wind speeds within the window. All in all, the interpretation of the optimal environmental window would seem to conform well to the triad framework, balancing the three elements, which are in basic opposition in coastal upwelling systems, in a mutual mid-intensity level trade-off.

Thus, both anchovies and sardines appear to be bound within the optimal environmental window, where either intensification or relaxation of the dynamic system could be a problem. If so, how might population variations and shifts in dominance between the two groups of fish species be interconnected and synchronized among different regional ecosystems by system intensifications related to decadal-scale variability in the tropical Pacific ENSO system?

No one has yet succeeded in convincingly demonstrating how sardines and anchovies might interact directly, one against the other, so that one might gain dominance while suppressing the other. Direct competition for resources does not seem to be the answer. And there has been little evidence found for other processes, such as selective predation of one species on the other's eggs or larvae, that could accomplish a similar result.

One significant difference is the fact that sardines have greater tendency and capabilities for migration than do anchovies (Parrish et al. 1981; Bakun 1993). Sardines in many cases migrate long distances between feeding and spawning habitats. The smaller, weaker-swimming anchovies tend to remain resident in their spawning habitats and appear in some ways to compensate by their greater adaptability for what the sardine may achieve by its migrational proficiency (Bakun 1996). Thus, it seems very reasonable to hypothesize that the stronger-swimming sardines could be more capable of successfully adapting to the increased flow speeds, enhanced levels of turbulence, and altered habitat geography resulting from system intensification. For example, in the very intense western boundary current situation near Japan, the anchovy population never really grows to more than a small fraction of the population sizes attained by the sardine during its high-abundance phases.

This leads us to the situation of the far eastern sardine stock that spawns in the extremely energetic situation near Japan. There is evidence that this stock may also respond to the subarctic low pressure system that characteristically intensifies during El Niño episodes and was in an anomalous state of general intensification in the decade following the mid-1970s. Parrish (1994) reported striking correspondences in anomalies in the wind and sea temperature fields in the region that appear to directly link an expanded, enhanced area of favorable sardine nursery habitat located off the west coast of Japan near the Kuroshio-Oyashio confluence during this period to the intensification of the Aleutian low pressure system.

Epilogue: After the Mid-1980s

In 1988–1989 the Southern Oscillation Index once again rose to levels not approached since the mid-1970s (Figure 1-3), indicating a return to non–decadal El Niño conditions. However, since 1990 it has again flipped back to being rather consistently well below the long-term seasonal mean levels. Thus, the period of decadal El Niño that began in the early to mid-1970s in some ways continues to this time.

This seems to be the case for the salmon of the subarctic Pacific, which, after the spectacular population expansions of the mid-1970s to mid-1980s period, began to decline after the mid-1980s. However, this decline appears as a mere hiccup in the general increasing trend. In the early 1990s these stocks have again resumed their dramatic aggregate rise toward unprecedented levels (Hare and Francis 1994).

On the other hand, the interruption of the decadal El Niño conditions appears to have shifted the balance in the small pelagic fish stocks of the subtropical coastal LMEs of the Pacific. Sardines now appear to be finished as the dominant fish populations of the subtropical coastal LMEs of the Pacific, with the enormous populations in both the northwestern and southeastern Pacific regions in continuing decline to mere fractions of their former sizes, while the Peruvian anchoveta experiences an enormous population explosion that may in the near future make it once again the largest exploited fish stock on earth.

So, while it is clear that the mid-1970s ushered in a dramatic period of synchronized change in the fish populations dominating many of the LMEs of the Pacific, it is not clear (at this point) whether the ecosystem regime changed in an equally dramatic way after the mid-1980s. For some stocks (sardines and anchovies) it evidently did. For others (salmon) it evidently did not. As a broad speculation, one might hypothesize that sardines have a replacement (i.e., the adaptable anchovy) available to displace them whenever they might falter due to a climatic shift from conditions to which their migrational and reproductive habits may have become adapted in a period of dominance (e.g., see discussion in Section 13.7 in Bakun 1996). On the other hand, there exists no readily apparent replacement for salmon in the trophic structures of the large marine ecosystems of the subarctic Pacific, and so the salmon may momentarily falter but manage to ride through a less favorable period (i.e., during a period of relative relaxation of the dynamic system) and then resume population growth when favorable (intensified) circumstances return.

References

Bakun, A. 1987. Monthly Variability in the Ocean Habitat off Peru as Deduced from Maritime Observations, 1953–84. In D. Pauly and I. Tsukayama (eds.), The Peruvian Anchoveta and Its Upwelling Ecosystem: Three Decades of Change. ICLARM Studies and Reviews 15, pp. 46–74. Instituto del Mar del Peru (IMARPE), Callao, Peru; Deutsche Gesellschaft für Technische Zusammenarbeit (GTZ), GmbH, Eschborn, Federal Republic of Germany; and International Center for Living Aquatic Resources Management (ICLARM), Manila, Philippines.

Bakun, A. 1990. Global climate change and intensification of coastal ocean upwelling. Science 247:198–201.

Bakun, A. 1992. Global greenhouse effects, multi-decadal wind trends, and potential impacts on coastal pelagic fish populations. ICES Mar. Sci. Symp. 195:316–325.

Bakun, A. 1993. The California Current, Benguela Current, and Southwestern Atlantic Shelf Ecosystems: A Comparative Approach to Identifying Factors Regulating Biomass Yields. In K. Sherman, L. M. Alexander, and B. Gold (eds.), Large Marine Ecosystems —Stress, Mitigation, and Sustainability. Washington, DC: American Association for the Advancement of Science, pp. 199–224.

Bakun, A. 1996. Patterns in the Ocean: Ocean Processes and Marine Population Dynamics. University of California Sea Grant, San Diego, California, USA, in cooperation with Centro de Investigaciones Biolgicas de Noroeste, La Paz, Baja California Sur, Mexico.

Bakun, A., and Parrish, R. H. 1982. Turbulence, transport, and pelagic fish in the California and Peru Current systems. CALCOFI Rep. 23:99–112.

Bakun, A., MacCall, A. D., and Parrish, R. H. 1990. Impacts of climate change on fisheries (Abstract A52A-07). EOS-Trans. Amer. Geophys. Un. 71:1263.

Beamish, R. J., and Bouillon, D. R. 1993. Pacific salmon production trends in relation to climate. Can. J. Fish. Aquat. Sci. 50:1002–1016.

Bernal, P. A. 1981. A review of the low-frequency response of the pelagic system in the California Current. CALCOFI Rep. 12:49–62.

Canby, T. Y. 1984. El Niño's ill wind. Natl. Geogr. 165(2):144–183.

Caviedes, C.N. 1973. Secas and El Niño: Two simultaneous climatological hazards in South America. Proceed. Amer. Geography 5:44.

Cole, D. A., and McLain, D. R. 1989. Interannual variability of temperature in the upper layer of the North Pacific eastern boundary region. NOAA Tech. Memo. NMFS-SWFC-125.

Cury, P., and Roy, C. 1989. Optimal environmental window and pelagic fish recruitment success in upwelling areas. Can. J. Fish. Aquat. Sci. 46:670–680.

Cury, P., Roy, C., Mendelssohn, R., Bakun, A., Husby, D. M., and Parrish, R. H. 1995. Moderate Is Better: Nonlinear Climatic Effect on California Anchovy. In R. J. Beamish (ed.), Climate Change and Northern Fish Populations. Canadian Special Publication of Fisheries and Aquatic Sciences 121.

Cushing, D. H. 1969. Upwelling and Fish Production. FAO Fish. Tech. Paper 84.

Dodimead, A. J., Favorite, F., and Hirano, T. 1963. Salmon of the North Pacific Ocean, Part 2. Review of oceanography of the subarctic Pacific region. Int. N. Pac. Fish. Comm. Bull. 13.

Food and Agriculture Organization (FAO).1997. Review of the State of World Fishery Resources: Marine Fisheries. FAO Fisheries Circular 920. Rome: FAO, 173 pp. (currently accessible from FAO Fisheries Dept. web site at HTTP://www.fao.org/fi).

Food and Agriculture Organization (FAO). 1972. Atlas of the Living Resources of the Sea. Rome: Food and Agriculture Organization.

Food and Agriculture Organization (FAO). 1992. Review of the State of World Fishery Resources. FAO Fish. Circ. 710, Rev. 8.

Favorite, F. 1967. The Alaskan Stream. Int. N. Pac. Fish. Comm. Bull. 21:1–20.

Favorite, F., Dodimead, A. J., and Nasu, K. 1976. Oceanography of the Subarctic Pacific region, 1960–71. Int. N. Pac. Fish. Comm. Bull. 33.

Fishing News International. 1993. Why lobster boom is over. March, p. 4.

Francis, R. C., and Hare, S. R. 1994. Decadal-scale regime shifts in the large marine ecosystems of the North-east Pacific: A case for historical science. Fish. Oceanogr. 3:279–291.

Glantz, M. H. 1987. Impacts of the 1982–83 Climate Anomalies in the West African Sahel. In M. H. Glantz, R. Katz, and M. Krenz (eds.), The Societal Impacts Associated with the 1982–83 Worldwide Climatic Anomalies. Boulder, CO: United Nations Environment Program, National Center for Atmospheric Research, pp. 62–64.

Hare, S. R., and Francis, R. C. 1994. Intervention Analysis: A Time Series Method for Detecting Interdecadal Climate Regime Shifts. Paper presented at the International Symposium on the Large Marine Ecosystems of the Pacific Ocean, Qingdao, China, 8–11 October 1994.

Horel, J. D., and Wallace, J. M. 1981. Planetary scale atmospheric phenomena associated with the Southern Oscillation. Mon. Weather Rev. 109:813–829.

Inoue, M., and O'Brien, J. J. 1987. Trends in sea level in the western and central equatorial Pacific during 1974–75 to 1981. J. Geophys. Res. 92:5045–5051.

Intergovernmental Oceanographic Commission (IOC). 1992. Oceanic Interdecadal Climate Variability. IOC Tech. Ser. 40. Paris: Intergovernmental Oceanographic Commission.

Kawasaki, T. 1983. Why Do Some Pelagic Fishes Have Wide Fluctuations in Their Numbers? Biological Basis of Fluctuation from the Viewpoint of Evolutionary Ecology. In G. D. Sharp and J. Csirke (eds.), Reports of the Expert Consultation to Examine Changes in Abundance and Species Composition of Neritic Fish Resources. FAO Fish. Rep. 291(1), pp. 1065–1080.

Kiladis, G. N., and Diaz, H. F. 1989. Global climatic anomalies associated with extremes in the Southern Oscillation. J. Climate 2:1069–1090.

Kiladis, G. N., and Sinha, S. K. 1991. ENSO, Monsoon, and Drought in India. In M. Glantz, R. W. Katz, and N. Nicholls (eds.), Teleconnections Linking Worldwide Climate Anomalies. Cambridge: Cambridge University Press, pp. 431–458.

Lagos, P., and Buizer, J. 1992. El Niño and Peru: A Nation's Response to Interannual Climate Variability. In S. K. Majumdaar, E. W. Miller, G. S. Forbes, and E. Schmalz (eds.), Natural and Technological Disasters: Causes, Effects and Preventative Measures. Easten, PA: The Pennsylvania Academy of Science.

Lasker, R. 1978. The relation between oceanographic conditions and larval anchovy food in the California Current: Identification of the factors leading to recruitment failure. Rapp. P.-v. Réun. Cons. Int. Explor. Mer. 173:212–230.

Lasker, R. 1981a. Factors contributing to variable recruitment of the northern anchovy (*Engraulis mordax*) in the California Current: Contrasting years 1975 through 1978. Rapp. P.-v. Réun. Cons. Int. Explor. Mer. 178:375–388.

Lasker, R. 1981b. The Role of a Stable Ocean in Larval Fish Survival and Subsequent Recruitment. In R. Lasker (ed.), Marine Fish Larvae: Morphology, Ecology and Relation to Fisheries. Seattle: University of Washington Press, pp. 80–87.

Legendre, L., and Demers, S. 1985. Auxiliary energy, ergoclines and aquatic biological production. Naturaliste Can. (Rev. Écol. Syst.) 112:5–14.

Leighton, M., and Wirawan, N. 1984. Catastrophic Drought and Fire in Bornean Tropical Rain Forest Associated with the 1982–83 ENSO Event. In G. T. Prance (ed.), Tropical Rain Forest and World Atmosphere. New York: AAAS.

Lluch-Belda, D., Crawford, R. J. M., Kawasaki, T., MacCall, A. D., Parrish, R. H., Schwartzlose, R. A., and Smith, P. E. 1989. World-wide fluctuations of sardine and anchovy stocks: The regime problem. S. Afr. J. Mar. Sci. 8:195–205.

Lluch-Belda, D., Schwartzlose, R. A., Serra, R., Parrish, R. H., Kawasaki, T., Hedgecock, D., and Crawford, R. J. M. 1992. Sardine and anchovy regime fluctuations of abundance in four regions of the world oceans: A workshop report. Fish. Oceanogr. 1:339–347.

MacKenzie, B. R., Miller, T. J., Cyr, S., and Leggett, W. C. 1994. Evidence for a dome-shaped relationship between turbulence and larval fish ingestion rates. Limnol. Oceanogr. 39:1790–1799.

Martin, B. R. 1983. The 1982 drought in Australia. Desertification Control Bull. 9:23–30 (UN).

Mendelssohn, R., and Cury, P. 1987. Forecasting a fortnightly abundance index of the Ivoirian coastal pelagic species and associated environmental conditions. Can. J. Fish. Aquat. Sci. 44:408–421.

Mendelssohn, R., and Mendo, J. 1987. Exploratory Analysis of Anchoveta Recruitment off Peru and Related Environmental Series. In D. Pauly and I. Tsukayama (eds.), The Peruvian Anchoveta and Its Upwelling Ecosystem: Three Decades of Change. ICLARM Studies and Reviews 15, pp. 294–306. Instituto del Mar del Peru (IMARPE), Callao, Peru; Deutsche Gesellschaft für Technische Zusammenarbeit (GTZ), GmbH, Eschborn, Federal Republic of Germany; and International Center for Living Aquatic Resources Management (ICLARM), Manila, Philippines.

Miller, C. B., Batchelder, H. P., Brodeur, R. D., and Pearcy, W. G. 1985. Response of Zooplankton and Ichthyoplankton off Oregon to the El Niño Event of 1985. In W. S. Wooster and D. L. Fluharty (eds.), El Niño North—Niño Effects on the Eastern Subarctic Pacific Ocean. Seattle: Washington Sea Grant Program, University of Washington, pp. 185–187.

Namias, J. R., and Cayan, D. R. 1984. El Niño: The implications for forecasting. Oceanus 27:41–47.

Nishikawa, Y., Honma, M., Ueyanagi, S., and Kikawa, S. 1985. Average distribution of larvae of oceanic species of Scombrid fishes, 1956–1981. Far Seas Fish. Res. Lab., S Series (12):1–99.

North Pacific Fishery Management Council (NPFMC). 1992a. Stock Assessment and Fishery Evaluation Report for the 1993 Gulf of Alaska Groundfish Fishery. Anchorage, AK: North Pacific Fishery Management Council.

North Pacific Fishery Management Council (NPFMC). 1992b. Stock Assessment and Fishery Evaluation Report for the Groundfish Resources of the Bering Sea/Aleutian Islands Regions as Projected for 1993. Anchorage, AK: North Pacific Fishery Management Council.

Norton, J., McLain, D. R., Brainard, R., and Husby, D. M. 1985. The 1982–83 El Niño off Baja and Alta California and Its Ocean Climate Context. In W. S. Wooster and D. L. Fluharty (eds.), El Niño North—Niño Effects on the Eastern Subarctic Pacific Ocean. Seattle: Washington Sea Grant Program, University of Washington, pp. 44–72.

Ogallo, L. J. 1987. Impacts of the 1982–83 ENSO Event on Eastern and Southern Africa. In M. H. Glantz, R. Katz, and M. Krenz (eds.), The Societal Impacts Associated with the 1982–83 Worldwide Climatic Anomalies. Boulder, CO: United Nations Environment Program, National Center for Atmospheric Research, pp. 55–61.

Ogallo, L. J. 1988. Relationships between seasonal rainfall in East Africa and the Southern Oscillation. J. Climatol. 8:31–43.

Palutikof, J. P. 1987. 1982–83 Climatic Anomalies over Western Europe and Their Impacts. In M. H. Glantz, R. Katz, and M. Krenz (eds.), The Societal Impacts Associated with the 1982–83 Worldwide Climatic Anomalies. Boulder, CO: United Nations Environment Program, National Center for Atmospheric Research, pp. 70–74.

Parrish, R. H. 1994. The World's Sardine and Anchovy Fisheries: Changing Yields in a Changing World. Paper presented at the First International CEOS Meeting, Monterey, California, 6–8 September 1994.

Parrish, R. H., Nelson, C. S., and Bakun, A. 1981. Transport mechanisms and reproductive success of fishes in the California Current. Biol. Oceanogr. 1:175–203.

Parrish, R. H., Serra, R., and Grant, W. S. 1989. The monotypic sardines, *Sardina* and *Sardinops*: Their taxonomy, distribution, stock structure, and zoogeography. Can. J. Fish. Aquat. Sci. 46:2019–2036.

Pearcy, W. G. 1992. Ocean Ecology of North Pacific Salmonoids. Seattle: Washington Sea Grant Program, University of Washington Press.

Pitovranov, S. E. 1987. The Impact of 1982–83 Weather Anomalies on Some Branches of the Economy of the USSR. In M. H. Glantz, R. Katz, and M. Krenz (eds.), The Societal Impacts Associated with the 1982–83 Worldwide Climatic Anomalies. Boulder, CO: United Nations Environment Program, National Center for Atmospheric Research, pp. 65–69.

Pittock, A. B. 1984. On the reality, stability and usefulness of southern hemisphere teleconnections. Australian Metorol. Mag. 32:75–82.

Quinn, W. H., Zopf, D. D., Short, K. S., and Kuo Yang, T. R. W. 1978. Historical trends and statistics of the Southern Oscillation—El Niño and Indonesian droughts. Fish. Bull. U.S. 76:663–678.

Rao, V. B., Satyamurty, P., and de Brito, I. V. 1986. On the 1983 drought in Northeast Brazil. J. Climatol. 6:43–51.

Rothschild, B. J., and Osborn, T. R. 1988. The effects of turbulence on planktonic contact rates. J. Plankton Res. 10:465–474.

Royer, T. C. 1979. On the effect of precipitation and runoff on coastal circulation in the Gulf of Alaska. J. Phys. Oceanogr. 9:555–563.

Royer, T. C. 1983. Commentary: Interannual Variability in the Abiotic Environment of the Bering Sea and Gulf of Alaska. In W. S. Wooster (ed.), From Year to Year: Interannual Variability of the Environment and Fisheries of the Gulf of Alaska and the Bering Sea. Seattle: Washington Sea Grant Program, University of Washington, pp. 134–138.

Shao-wu, W., and Mearns, L. O. 1987. The Impact of the 1982–83 El Niño Event on Crop Yields in China. In M. H. Glantz, R. Katz, and M. Krenz (eds.), The Societal Impacts Associated with the 1982–83 Worldwide Climatic Anomalies. Boulder, CO: United Nations Environment Program, National Center for Atmospheric Research, pp. 43–49.

Shukla, J., and Paolino, D. A. 1983. The southern hemisphere and long-range forecasting of the summer monsoon rainfall over India. Mon. Weather Rev. 111:1830–1837.

Sinclair, M. 1988. Marine Populations. An Essay on Population Regulation and Speciation. Seattle: Washington Sea Grant Program, University of Washington Press.

Steele, J. H. 1974. The Structure of Marine Ecosystems. Cambridge, MA: Harvard University Press.

Sund, P. N., Blackburn, M., and Williams, F. 1981. Tunas and their environment in the Pacific Ocean: A review. Oceanogr. Mar. Biol. Ann. Rev. 19:443–512.

Sundby, S. 1995. Wind Climate and Foraging of Larval and Juvenile Arcto-Norwegian Cod (*Gadus morhua*). In R. J. Beamish (ed.), Climate Change and Northern Fish Populations. Canadian Special Publication of Fisheries and Aquatic Sciences 121.

Sverdrup, H. U. 1953. On conditions for vernal blooming of phytoplankton. J. Cons. Int. Explor. Mer. 18:287–295.

Thomson, R. E. 1972. On the Alaskan Stream. J. Phys. Oceanogr. 2:363–371.

Venrick, E. L., McGowan, J. A., Cayan, D. R., and Hayward, T. L. 1987. Climate and chlorophyll a: Long-term trends in the Central North Pacific ocean. Science 238:70–72.

Wilhite, D. A., Wood, D. A., and Meyer, S. J. 1987. Climate-Related Impacts in the United States during the 1982–83 El Niño. In M. H. Glantz, R. Katz, and M. Krenz (eds.), The Societal Impacts Associated with the 1982–83 Worldwide Climatic Anomalies. Boulder, CO: United Nations Environment Program, National Center for Atmospheric Research, pp. 75–78.

Wooster, W. S., and Hollowed, A. B. 1995. Decadal Scale Variations in the Subarctic Pacific: A. Winter Ocean Conditions. In R. J. Beamish (ed.), Climate Change and Northern Fish Populations. Canadian Special Publication of Fisheries and Aquatic Sciences 121.

Yáñez, E. 1991. Relationships between Environmental Changes and Fluctuating Major Pelagic Resources Exploited in Chile (1950–1988). In T. Kawasaki, S. Tanaka, Y. Toba, and A. Taniguchi (eds.), Long-term Variability of Pelagic Fish Populations and Their Environment. Oxford: Pergamon Press, pp. 301–309.

Yáñez, E., Barbieri, M. A., and Santillán, L. 1992. Long-term Environmental Variability and Pelagic Fisheries in Talcahuano, Chile. In A. I. L. Payne, K. H. Brink, K. H. Mann, and R. Hilborn (eds.), Benguela Trophic Functioning. S. Afr. J. Mar. Sci. 12, pp. 175–188.

Yoshino, M. M., and Yasunari, T. 1987. Climatic Anomalies of El Niño and Anti-El Niño Years and their Socioeconomic Impacts in Japan. In M. H. Glantz, R. Katz, and M. Krenz (eds.), The Societal Impacts Associated with the 1982–83 Worldwide Climatic Anomalies. Boulder, CO: United Nations Environment Program, National Center for Atmospheric Research, pp. 50–54.

2

Large-Scale Climate-Related Changes in the Carrying Capacity of Salmon in the Strait of Georgia and Northern North Pacific Ecosystems

Richard J. Beamish and Chrys-Ellen M. Neville

Abstract

Changes in the ecosystem in the Strait of Georgia are linked to changes in the ecosystem in the northern North Pacific Ocean. The intensification of the Aleutian low pressure system that began in the late 1970s was associated with a decrease in snowpack, a decrease in total annual discharge from the Fraser River, and an increase in the water temperatures in the Strait. The changes in the northern North Pacific were associated with an intensification of the Aleutian Low, a cooling of the mid-Pacific, and increased plankton production, possibly as a consequence of mid-ocean upwelling. Increases in salmon production occurred in the mid-Pacific, whereas the changes in the Strait of Georgia appear to be associated with decreases in production of chinook salmon and decreases in the survival of hatchery-reared chinook and coho salmon. The apparent opposite effects of a large-scale climate shift on salmon survival in a coastal and mid-ocean ecosystem highlight the importance of understanding the linkages between marine ecosystems and salmon production when attempting to manage fisheries for these species.

Many people in the North Pacific associate marine ecosystems with Pacific salmon. However, fluctuations in the abundance of Pacific salmon are usually associated with the changes in abundance of spawners in freshwater. The number of spawners and the number of young produced in freshwater are believed to be closely related to the numbers of salmon that are caught in the fisheries or return to spawn. Even though Pacific salmon, in general, spend more time in the ocean than they do in freshwater, the perception is that abundance trends are regulated in freshwater. There is evidence to support this view (Ricker 1975), and this evidence has been used to propose rebuilding programs for Pacific salmon with the objective of restoring salmon abundance to historic levels of abundance (Canadian Department of Fisheries and Oceans 1991).

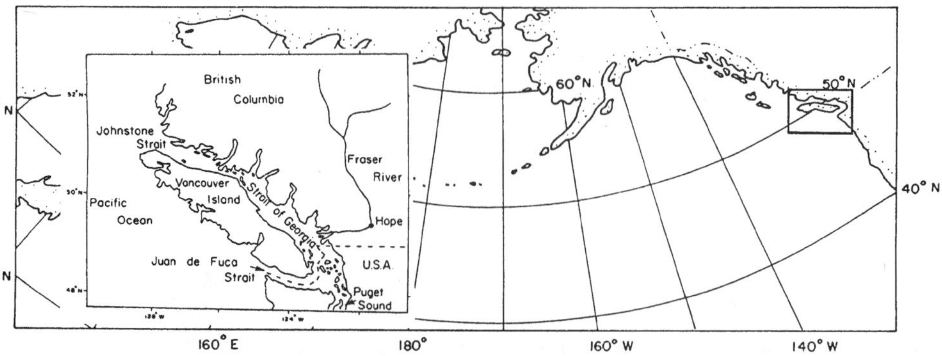

Figure 2-1. The Strait of Georgia and its location in the northern North Pacific Ocean.

If salmon abundance is largely controlled in freshwater, then plans to rebuild stocks to historic high levels, and presumably keep them there, might be theoretically possible (Henderson and Healey 1993). However, if the marine environment also regulates abundance in a nonrandom manner, then managing Pacific salmon requires an understanding of both the marine and freshwater capacities to support salmon. Plans to rebuild salmon populations to the highest levels ever recorded by increasing smolt abundance may be noble, but they may also create unfulfilled expectations if the marine ecosystem cannot support the increased releases.

In this study we show that in two important marine ecosystems there were shifts in the ocean regimes that were associated with changes in the abundance trends of Pacific salmon. We propose that the changes in abundance in one of these ecosystems may be opposite in phase to the other.

We are not the first to study this linkage. Ebbesmeyer et al. (1991) identified a decadal-scale linkage between the winter climate pattern over the subarctic Pacific and the circulation in a Pacific Northwest fjord (Puget Sound). We examine the same linkage, but using the Strait of Georgia (immediately north of Puget Sound) (Figure 2-1) and the Fraser River in relating the ecosystem changes to salmon abundance changes.

Coastal Ecosystem

The Strait of Georgia, located between Vancouver Island and the British Columbia mainland (Figure 2-1) is an important marine ecosystem on Canada's west coast and one of the most important salmon-producing areas in the Pacific Ocean. There are five species of Pacific salmon that are resident in or move through the Strait of Georgia: pink (*Oncorhynchus gorbuscha*), chum (*O. keta*), sockeye (*O. nerka*), coho (*O. kisutch*), and chinook (*O. tshawytscha*). Most of the production of these five species comes from the Fraser River, which produces more Pacific salmon than any other river in British Columbia. For the 5-year period 1986–1990, Fraser River salmon stocks produced approximately 32% of the average total Canadian catch of Pacific salmon, or approximately 41% of the total catch in a year of major pink salmon returns to the Fraser River.

The Strait of Georgia is 220 km long, 33 km wide, has a surface area of about 6900 km^2, an average depth of 155 m, and a maximum depth of 420 m (Thomson 1981). It is connected to the Pacific Ocean in the north by Johnstone Strait and in the south by Juan de Fuca Strait. Approximately 80% of the freshwater entering the Strait of Georgia comes from the Fraser River. The total volume of the freshwater discharge reduces the surface salinity, and because of the reduced salinity and the semienclosed nature of the Strait, it could be considered a large estuary.

Wind mixing is the dominant physical mechanism that entrains nitrates from the nitrate-rich deep water into the surface layer in the Strait of Georgia. A model of the impact of wind events on nitrate fluxes (St. John et al. 1993) in the southern Strait showed that nitrate entrainment in the surface layer was reduced when runoff from the Fraser River was high because greater energy was required to break down the more buoyant surface layer. Increases in primary production lagged behind the wind-mixing events and were lower and occurred earlier when runoff was high. An earlier spring bloom occurred because the increased stability of the more buoyant surface layer resulted in a shallower depth of mixing for plankton and effectively increased the light available for photosynthesis at a time that light limits plankton growth. When Fraser River flows were reduced, wind mixing was more effective and productivity increased. It is this period from April to September, when plankton production is periodically nutrient limited, that is most important for salmon survival. Low plankton production in April through June, when salmon migrate into the Strait, will reduce growth and increase the risk of mortality, which is already very high (approximately 90%).

Temperature

Vertical temperature profiles were taken 8 to 20 times each month in the deep water at the Nanoose Bay Naval Underwater Weapons Test Range approximately midway between the north and south ends of the Strait of Georgia. The temperature profiles were collected using several different continuous temperature and depth instruments. Data were checked for spikes and unrealistic parameter values, which were then deleted (Fissel et al. 1991). In addition, the temperature data were compared with data collected once each month using reversing thermometers; the two data sets agreed well (Fissel et al. 1991).

Average monthly surface temperatures have been reported to vary from 5° to 19°C throughout the year (Thomson 1981). In our study the range of surface temperatures (0.5–1.0 m) was 5.5° to 20.4°C. The average temperature from 1970 to 1977 was 10.5°C and from 1978 to 1991 was 11.2°C. An intervention analysis (Noakes 1986) indicated that the step shift in temperature in 1978 was significant (step intervention = 0.337, S.E. = 0.165, $p < 0.05$).

Bottom temperatures were taken at approximately 325 m and were summarized as mean monthly values (Figure 2-2). In the Strait of Georgia, bottom temperatures typically vary throughout the year from approximately 8° to 10°C (Thomson 1981), with the winter temperatures higher than the summer temperatures. In our study the range of bottom temperatures was 7.2° to 10.0°C. The low temperatures in the annual cycle occur around June, when flows of bottom water entering the Strait from offshore as a consequence of the estuarine circulation are greatest.

From 1970 to 1977 the average annual temperature was 8.7°C. In 1978 and 1979 the annual pattern of temperature change was disrupted by extreme fluctuations. The annual pattern was reestablished in 1980 at a warmer average annual monthly temperature of 9.2°C. An intervention analysis identified 1980 as the year that a significant increase in temperature occurred (step intervention = 0.207, S.E. = 0.086, $p < 0.05$).

Fraser River Flow

Trends in flows of the Fraser River were examined using a cumulative sum analysis (Murdoc 1979). This analysis is useful for detecting changes in the trends of flows. From the early 1920s to the late 1940s the flows were below average (Figure 2-3). The pattern of flows changed in 1945, and from 1946 to 1956 flows were average. From 1957 to 1976 flows increased and were above

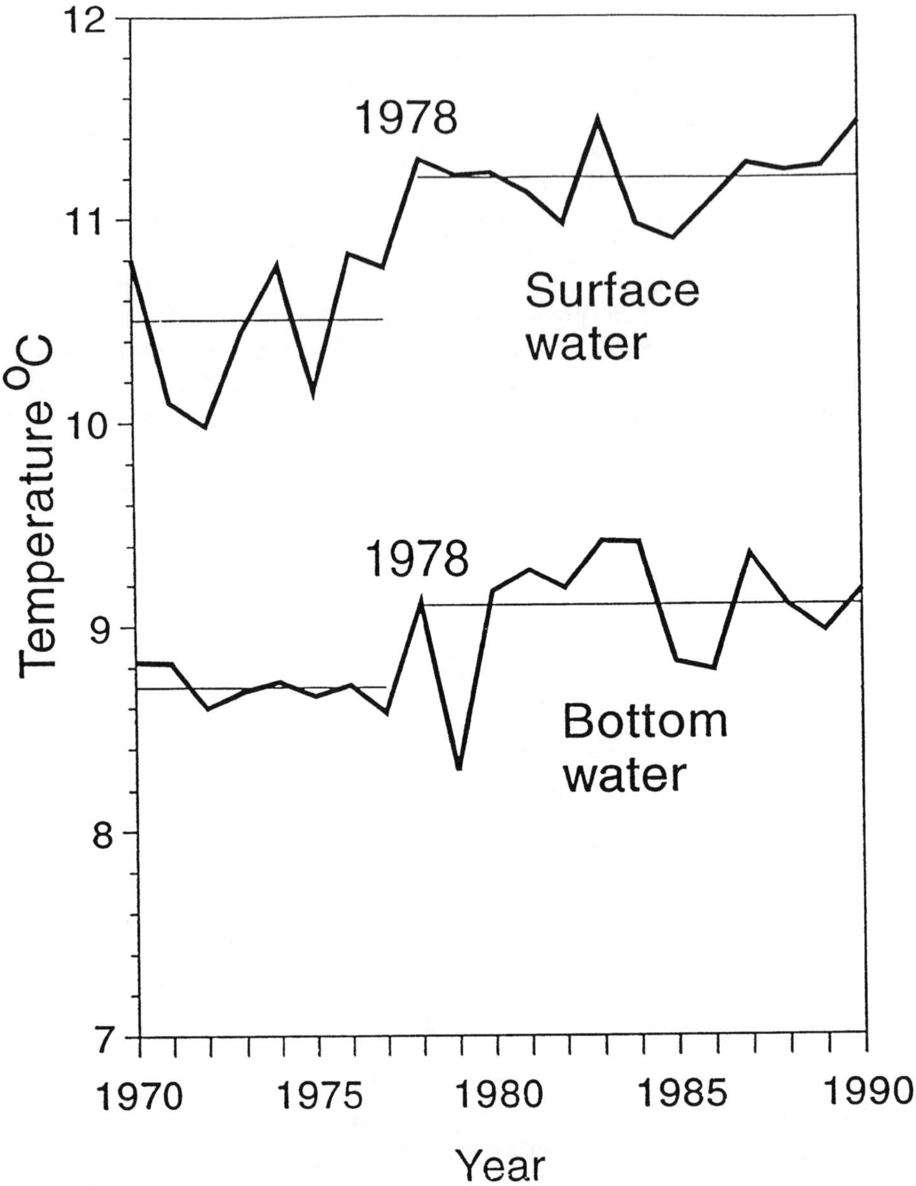

Figure 2-2. Changes in the average annual surface and bottom temperatures (°C) in the Strait of Georgia. Data are average daily values summarized by year. Note the abrupt increase that occurred in 1978.

the long-term average. This period ended abruptly in 1977 when a period of reduced flows began. This period of reduced flows has continued since 1977.

Moore (1991) compared the mean surface air temperatures for the northern hemisphere (Jones et al. 1986) with Fraser River flow patterns. During the extreme cold period (1963–1972), both maximum and minimum flows were higher than during an earlier warm period (1938–1947). The timing of the minimum flows was about the same for both periods, but the maximum flow occurred earlier in the spring during the warm decade.

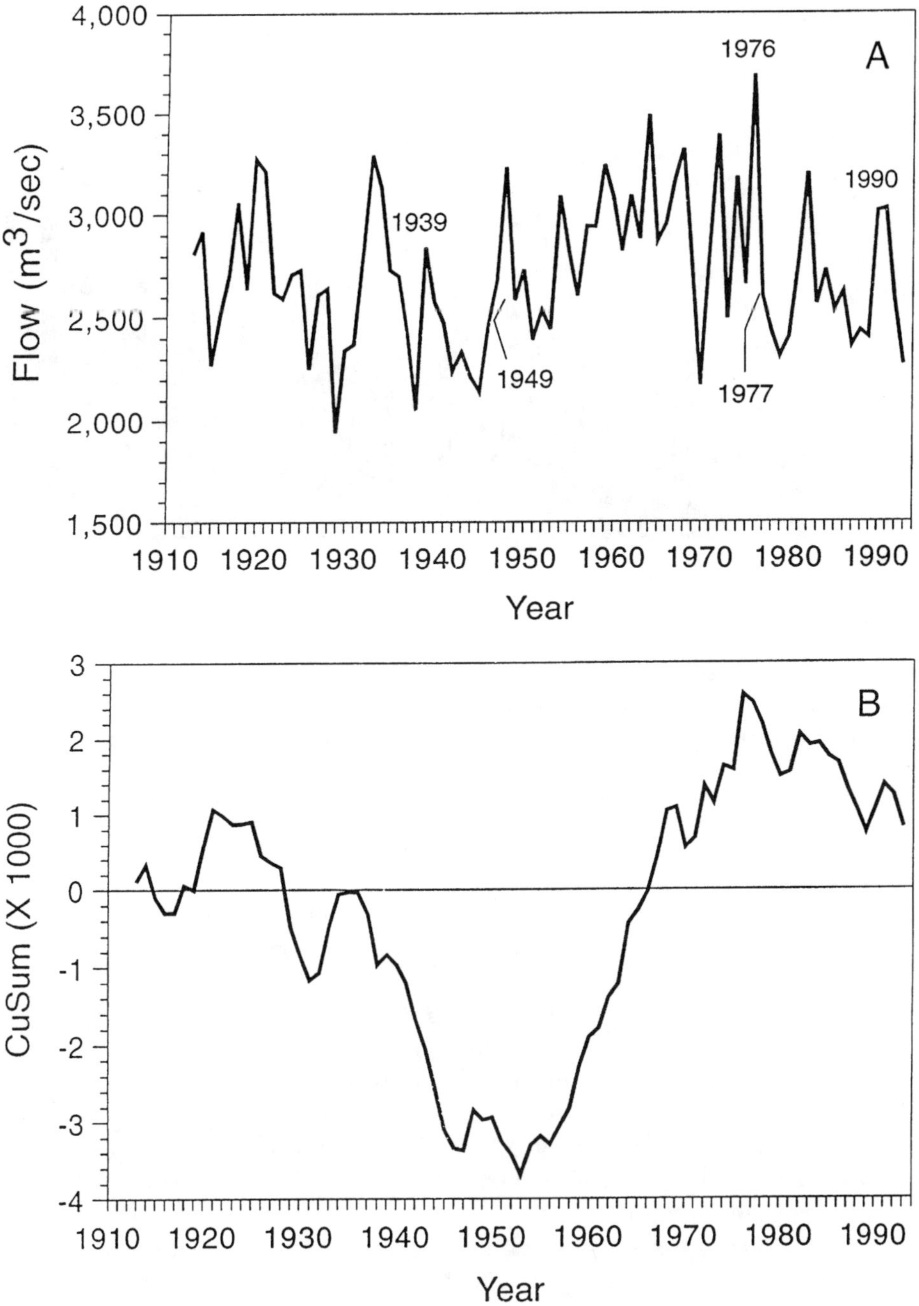

Figure 2-3. A. Average annual daily discharge rate of the Fraser River at Hope from 1912 to 1993. The highest discharge in the time series occurred in 1976. B. Cumulative sum (CuSum) analysis of the average annual daily discharge of the Fraser River at Hope showing alternating trends of below-average (decreasing slope) and above-average (increasing slope) flows.

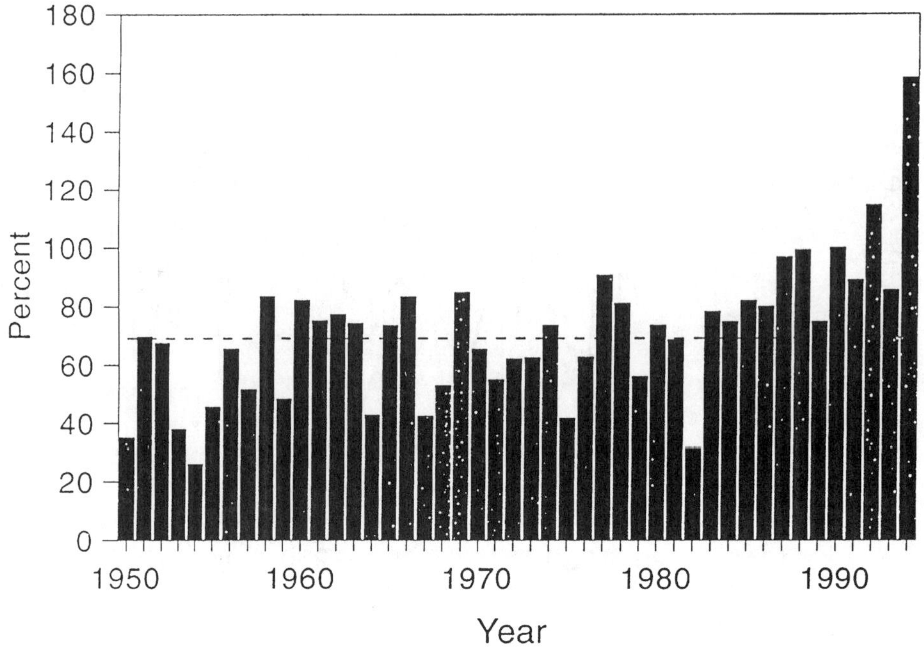

Figure 2-4. Average daily Fraser River discharge in April presented as a percentage of the average daily discharge for the year. For 1994, average daily discharge in April is presented as a percentage of the average daily discharge for 1993 and is preliminary. The dotted line represents the average of the entire time series. Since 1983, the trend in the April discharge has been above the long-term mean and increasing.

Moore (1991) identified a relationship between mean annual Fraser River flow and maximum annual snow accumulation, indicating that the depth of the snowpack is related to Fraser River flow. Because the decline in the average snow depth in the Fraser River drainage from 1977 to 1988 relative to the 1956 to 1976 period (22%) was similar to the declines in flows (28%) for the same period, Moore (1991) proposed that the climate change beginning in 1977 caused the decrease in snow levels and the declining trend in flows.

We looked at the pattern of flow in April because earlier increased flows can result in an earlier spring bloom, as discussed previously (St. John et al. 1993). When flows in April were expressed as an average daily rate in April divided by the average daily rate for the year, there was a clear trend of increasing flows in April beginning about 1989, with April 1994 being the highest in the time series since 1950 (Figure 2-4). It is clear that the recent decline in total annual flow has been associated with higher flows in April, particularly since 1989.

Estuarine Circulation

In the winter of 1976–1977, there was a shift in the climate over the Pacific (Ebbesmeyer et al. 1991; Beamish 1993; Miller et al. 1994; Polovina et al. 1994). As described previously, synchronous with this shift was the beginning of the declining trend in annual snowpack levels, a declining trend in Fraser River discharge, and an abrupt increase in the temperature of the Strait of Georgia. The estuarine circulation is related to the amount and pattern of Fraser River discharge. The circulation dominates the supply of nitrates to the Strait of Georgia. Variations in river flow, therefore, will strongly affect the nutrient supply to the Strait of Georgia (Harrison et al. 1994). Because the maximum river flow occurs in the summer, the maximum flows of deep water into the Strait of Georgia occur at the same time. The bottom water that enters the Strait of Georgia at this time is California undercurrent water that upwells along the outer coast at this time of year (Freeland and Denman 1982). In this report we have not reviewed the long-term trends in the oceanography of this source bottom water, but it is possible that the major climate events of 1976–1977 also affected the chemical and physical composition of this water. For example, the temperature increases in the Strait of Georgia probably are closely related to temperature increases in this source water. Increases in temperature would also be associated with decreases in salinity and thus in nitrate concentration (Harrison et al. 1994).

We also have not determined how the biological processes were altered in the Strait of Georgia at this time. Because of the sensitivity of marine organisms to temperature (Regier et al. 1990), it is highly probable that changes in the survival and distribution of some marine organisms occurred in the Strait when the regime shift occurred. The relative increase in flows in April could also affect the timing of biological events, such as an earlier spring bloom (St. John et al. 1993). An earlier spring bloom could improve the survival of young fishes that reproduce early in the year and have an adverse effect on the survival of salmon smolts that enter the Strait in May and June. This effect could have increased chinook salmon (*Oncorhynchus tshawytscha*) mortality, resulting in the decline in chinook salmon abundance that started at the same time as the shift in the ecosystem.

Impact of the Regime Shift on Chinook Salmon in the Strait of Georgia

Catches of chinook salmon in the Strait of Georgia began to decline in 1979 (Figure 2-5A). Catches and escapements decreased until the mid-1980s, when regulations were established to stop the decline. At almost the same time that catches started to decline, there was an abrupt change in the survival of chinook salmon released from hatcheries into the Strait of Georgia (Figure 2-5B). The average marine survival of these hatchery fish from 1974 to 1977 was 4.8%, compared with 0.65% from 1978 to 1990.

The decline in chinook catch and escapement in the Strait of Georgia could have resulted from overfishing or from decreases in the capacity of the Strait to support chinook, or both. If overfishing was the principal reason for the decline, there would be a gradual reduction in smolt production beginning at the time the declines occurred.

The largest catches and spawning escapements occurred in the late 1970s. We used the estimated escapements and a cohort analysis to estimate wild smolt production (Beamish et al., unpublished). Hatchery production was obtained from the Canadian Mark Recover Program database (Kuhn et al. 1988). An average of 29 million wild and 2 million hatchery smolts were produced each year from 1975 to 1979.

In recent years, there has been a large increase in the number of chinook produced in hatcheries. We were able to identify these hatchery-reared chinook using the daily microstructure in the otolith (Zhang et al. 1995). Because the number of hatchery-produced chinook was known, we were able to estimate wild smolt abundance by determining the percentages of hatchery and wild fish in samples of chinook smolts collected throughout the Strait. We estimated that approximately 50% of the smolts in 1992 were from hatcheries. The total releases were 31 million; thus, there were approximately 31 million wild smolts in the Strait, for a total of 62 million. These estimates must be considered preliminary because the percentages of hatchery and wild smolts in our catches may need to be weighted for sampling biases relating to the time of entry into salt

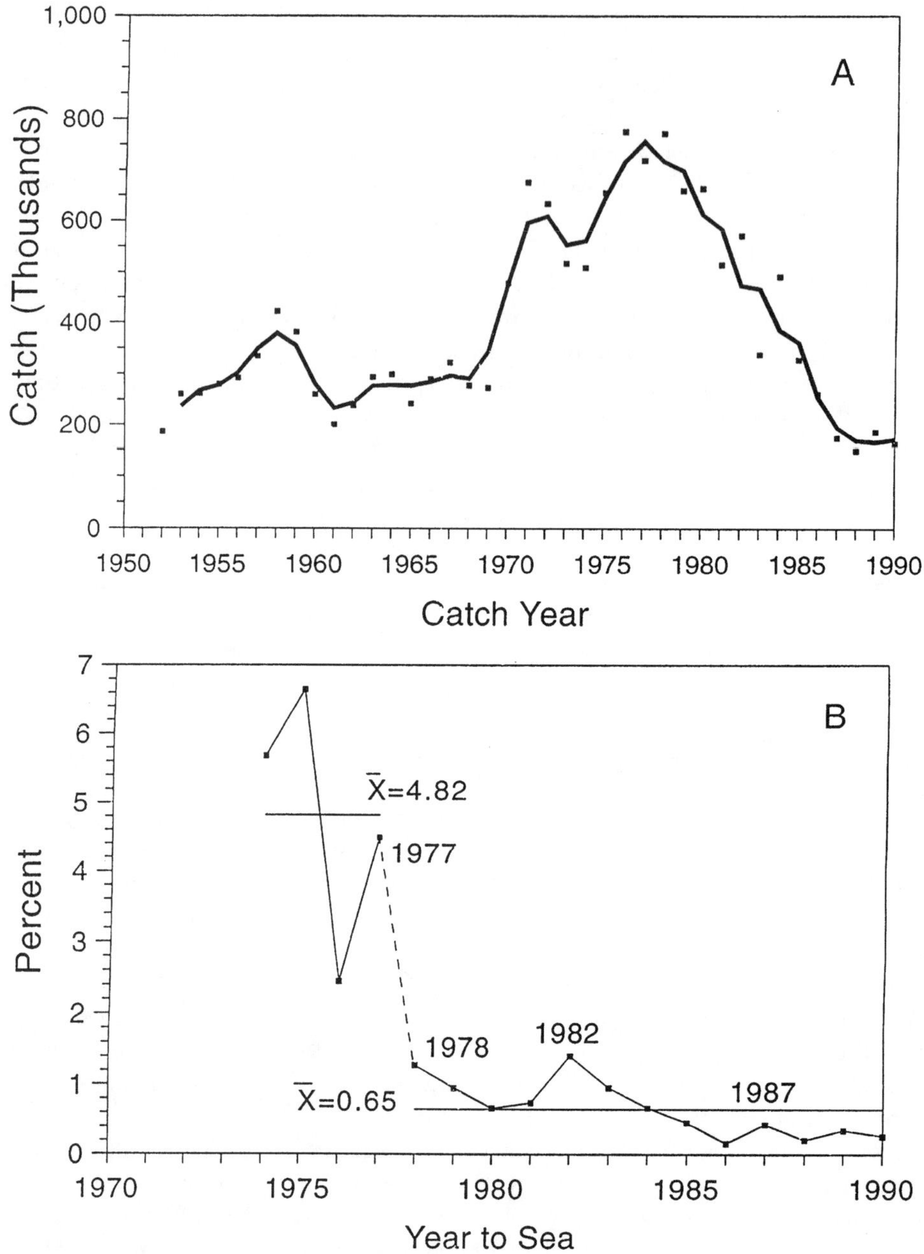

Figure 2-5. A. Total catch (number of fish) of chinook salmon in the Strait of Georgia commercial and sports fisheries. Points represent catches and the solid line is the 3-year running average through the time series. B. Marine survival of combined releases into the Strait of Georgia and Fraser River of hatchery-reared chinook salmon, showing the abrupt drop in survival in 1978. Data from the Canadian Mark Recovery database (Kuhn et al. 1988).

water; however, we consider that the estimates approximate the number of smolts entering the Strait. The total number of smolts in 1992 was approximately double the total number in the late 1970s, yet the abundance of adults in 1992 was about 25% of the abundance in the late 1970s. This could only occur if there were an eightfold decrease in marine survival, which is very close to the decline observed in hatchery survival. Thus, we believe that the synchrony of the beginning of the decline in catch and the regime shift indicates that the two events were related. We cannot, however, prove this at this time. We are not proposing that fishing effects were unrelated to the decline, because such effects could exacerbate the rate of decline.

Regime Shifts in the Northern North Pacific Ocean

The climate changes in the winter of 1976–1977 had a major impact on the climate over the subarctic Pacific. Beamish (1993) summarized some of the changes that occurred. During December 1976 through May 1977, an area of intense low pressure dominated most of the northern North Pacific Ocean. It was the most intense low since 1940–1941 and initiated a change in the pattern of winter (November–March) low pressures (Trenberth 1990). Trenberth showed that from 1965 to 1976, the mean sea-level pressure over the North Pacific Ocean was approximately 1012.8 mbar and that after 1976 (1977–1988) it was approximately 1010.8 mbar. In addition to this drop of 2 mbar over the North Pacific, the center of the Aleutian Low moved farther east and became deeper by an average of 4.3 mbar for the winter months and deeper by 7 to 9 mbar in January.

Changes in sea surface temperatures were another indication of a climate change over the North Pacific in the late 1970s. There are two general patterns of sea surface temperatures in the Northeast Pacific Ocean. The first has below-normal sea surface temperatures along the coast of North America and above-normal temperatures farther offshore. The second pattern is the opposite (McLain 1984). The patterns are most obvious in the winter. The pattern oscillated over the years 1957–1976, but in the summer of 1976 it changed from having below-normal sea surface temperatures adjacent to shore to having above-normal sea surface temperatures adjacent to shore (McLain 1984). This pattern of increased sea surface temperature persisted until at least the mid-1980s and was most dramatic in the Northeast Pacific Ocean (Chelton 1984).

Ebbesmeyer et al. (1991) identified the abrupt change in climate in 1976 by combining records of 40 environmental variables that reflected environmental conditions for 1968–1984 in the vicinity of Puget Sound, off the west coast of North America. The changes were also identified in the Gulf of Alaska (Royer 1989) and off California (Bakun 1990) and began a new climate regime over a vast area of the Pacific Ocean and North America (Kerr 1992).

The impact of this regime shift on Pacific salmon was studied by Beamish and Bouillon (1993) by comparing salmon abundance trends with trends in the Aleutian Low. The Aleutian Low is usually centered over the Aleutian Islands and typically begins forming during the autumn months, intensifies during the winter months, and then breaks down by early summer. Beamish and Bouillon (1993) produced mean seasonal sea-level pressure maps for the North Pacific Ocean between latitude 20° and 70°N and between longitude 120°E and 120°W. For the winters (December to February) and the springs (March to May) of 1899–1990, they calculated the area of the North Pacific Ocean that was covered by the Aleutian low pressure system of less than 100.5 kPa. They then used the sum of the winter and spring Aleutian Low values as an index of historical weather over the North Pacific Ocean to compare with salmon production trends (Figure 2-6). The Aleutian Low pressure indices were smoothed using a LOWESS smoother with a common band width, f, of 0.20.

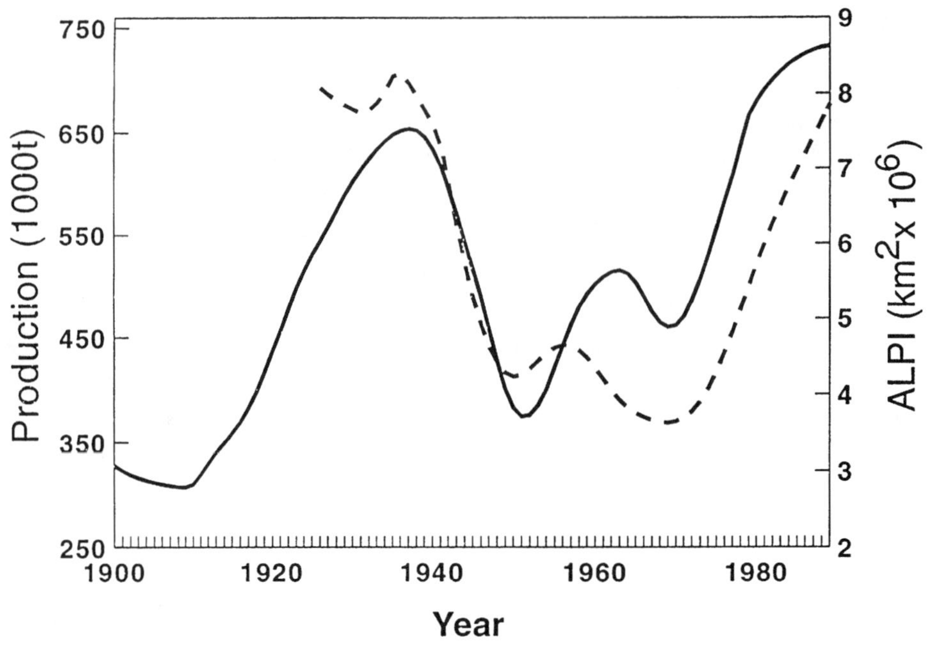

Figure 2-6. Comparison of the smoothed Aleutian Low Pressure Index (solid line, 1900 to 1989; LOWESS smoother with a band width, *f*, of 0.20) and the combined all-nation catches of pink, chum, and sockeye salmon (broken line, 1925 to 1989).

The Aleutian Low Pressure Index for the years 1900–1989 indicated that sea-level pressure was high (a low index represents high pressures, or less intense lows) over a vast area of the northern North Pacific Ocean at the turn of the century. The Aleutian Lows intensified steadily until the late 1930s (a high index represents low pressures, or intense lows). From about 1940 to 1950 the index dropped, indicating a change to higher pressures. The period from 1950 to 1970 was characterized by high pressure, or weak lows, except for a small period of intense lows in the late 1950s and early 1960s. Beginning in the early 1970s, the intensity of the lows again increased, reaching the most intense levels in the time series in the 1980s. In the late 1980s, there was a weakening of the Aleutian Low, but it is too early to determine if the trend has changed.

The intensification of the Aleutian Low increases mid-ocean upwelling under the center of the Low due to surface-layer Ekman divergence caused by the cyclonic winds in the Aleutian Low system. The subarctic Pacific planktonic food chain depends on this upward transport of nutrients into the euphotic zone. There is evidence of a large increase in plankton biomass in the 1980s in the subarctic Pacific (Brodeur and Ware 1992), confirming that the intensification of the Aleutian Low resulted in an increase in productivity. Beamish and Bouillon (1993) also observed a close relationship between the Aleutian Low Pressure Index and plankton production (Figure 2-7).

The combined all-nation catches of pink, chum, and sockeye salmon averaged 673,100 tons from the mid-1920s to the early 1940s (Figure 2-6). Maximum catches occurred in the late 1930s, with a high catch of 837,400 tons in 1939. A period of low catches occurred from the mid-1940s until the mid-1970s, except for a small increase around the mid-1950s (the average from 1944 to 1975 was 374,100 tons). The lowest total salmon catch of 275,600 tons occurred in 1974. In

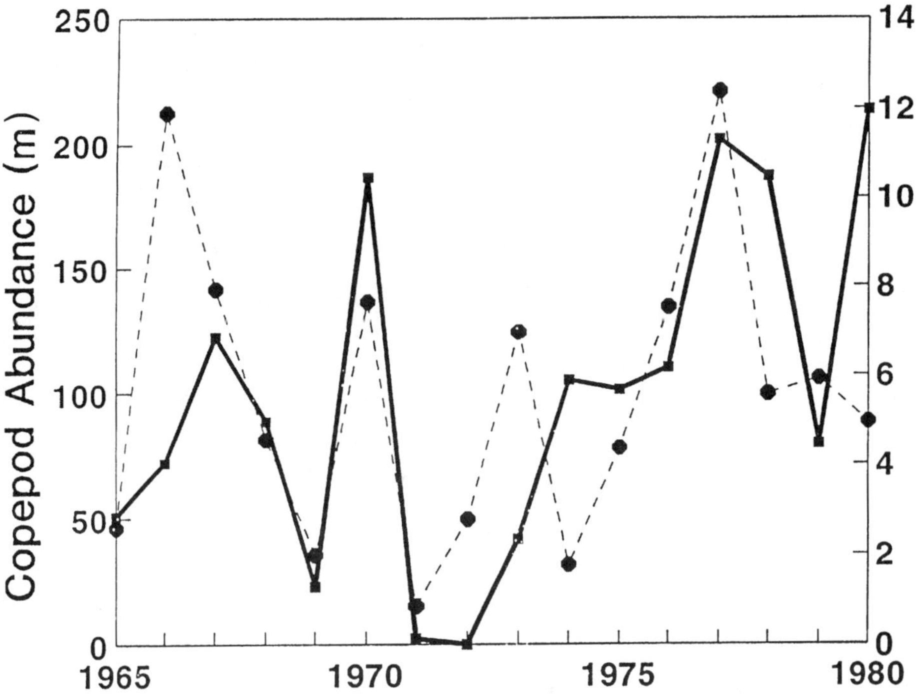

Figure 2-7. Comparison between copepod abundance at Ocean Station P (50°N, 145°W, broken line) and the Aleutian Low Pressure Index (solid line), from Beamish and Bouillon (1993).

the most recent years, production increased, with the historic high catch of 857,000 tons occurring in 1991. The patterns of the all-nation catches for each of the three species were similar (Figure 2-8).

The pattern of the smoothed Aleutian Low Pressure Index corresponded closely to the trend in combined all-nation salmon catches (Figure 2-6). The declines in catches from the late 1930s to the late 1940s correspond to a sharp decline in the Aleutian Low Pressure Index. Similarly, the increases in catches beginning in the mid-1970s followed an increase in the index.

It appears that the trends in salmon production from 1925 to 1989 were strongly linked to environmental factors. The beginning of this shift in salmon production occurred in the late 1970s, at the same time that the climate changed throughout the Pacific. The robustness of the general trend in the three species for all countries over such a vast geographical area, the continuous rather than precipitous change in abundance, and the close association of the abundance trends with trends in the Aleutian Low suggest that this vast low pressure system is associated with the changes in salmon abundance.

Conclusion

Changes in the ecosystem in the northern North Pacific Ocean were synchronous with changes in the ecosystem in the Strait of Georgia. The changes in the northern North Pacific Ocean improved the marine survival of pink, chum, and sockeye salmon, whereas the changes in the Strait of Georgia were closely associated with reduced marine survival of chinook salmon. The

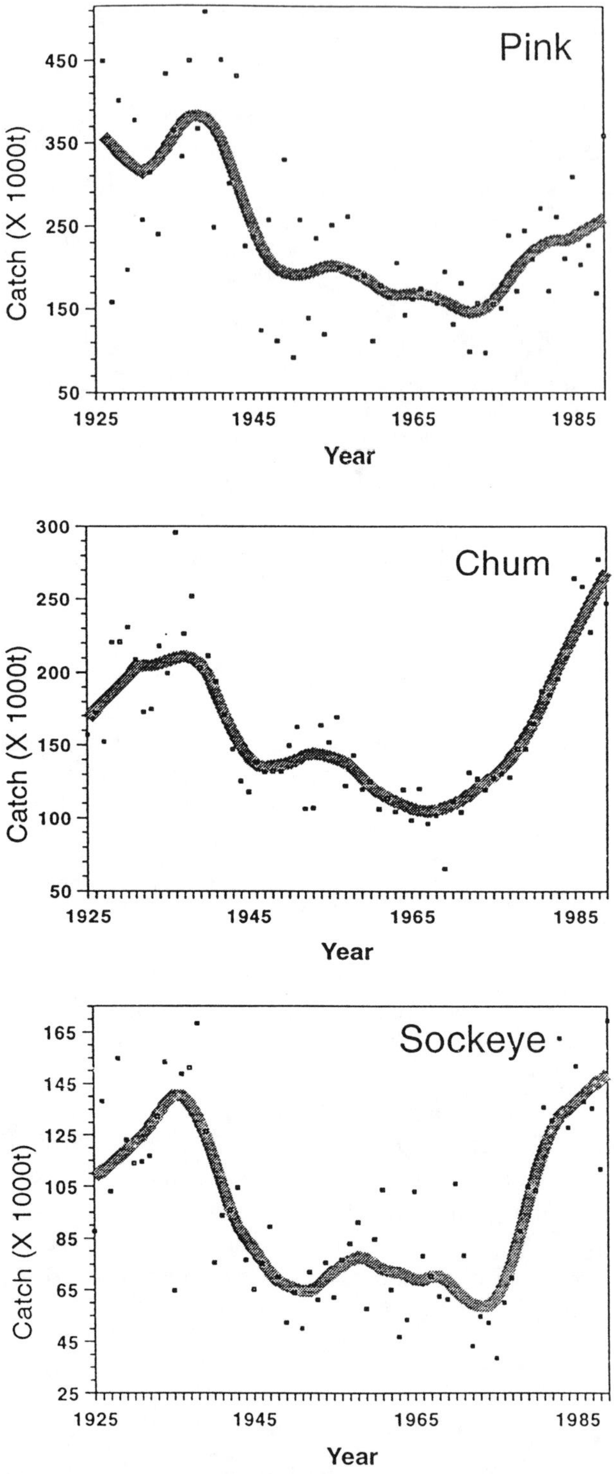

Figure 2-8. All-nation catches of pink, chum, and sockeye salmon from 1925 to 1989. Solid squares are the total annual catches and the solid line is the LOWESS smoother with a band width, f, of 0.20.

common factor affecting both of these ecosystems was the Aleutian Low. The intensification of the Low beginning in the late 1970s increased salmon production in the northern North Pacific, possibly through increases in mid-ocean upwelling. The linkages with production in the Strait of Georgia are not as clear, but involved decreases in coastal precipitation that resulted in decreases in snowpack and subsequently in Fraser River discharge. At the same time as the reduced flows, there was a warming of the Strait of Georgia. There is a relationship between chinook salmon catches and extremes in interannual Fraser River flow (Beamish et al. 1994), indicating a linkage between chinook salmon catch and the oceanography of the Strait. The dynamics of the linkage with chinook survival are unclear because the factors regulating the abundance of chinook in the Strait (the carrying capacity for chinook) are poorly understood. However, the decline in survival may result from an earlier spring bloom, changes in temperature, and changes in the oceanography of the bottom water entering the Strait.

Ebbesmeyer et al. (1989) identified decadal shifts that occurred in the climate over the North Pacific Ocean and showed that these changes affected the oceanography of Puget Sound. They identified a relationship between the position of the Aleutian Low and the regime shifts in Puget Sound. In a subsequent study, Ebbesmeyer et al. (1991) showed that one of the decadal shifts occurred in the winter of 1976–1977. In our study we used an index of Aleutian Low intensity rather than an index of the position of the Low because the index of intensity more closely reflected salmon catch trends. Both our study and the Ebbesmeyer et al. (1989) study show that an ocean and a coastal marine ecosystem are linked by large-scale climatic events. The linkage of the open ocean and coastal ecosystems is important for salmon because the environmental effects on the population dynamics of pink, chum, and sockeye that rear in the open ocean appear to be out of phase with the dynamics of chinook salmon that rear in the Strait of Georgia. The apparent opposite impacts on survival in the two ecosystems indicate the complexity of the linkages. The existence of these linkages indicates that we have a lot to learn about the interrelationships between salmon population dynamics and marine ecosystems if we are to adapt to the speculated changes resulting from a general warming of our climate.

References

Bakun, A. 1990. Global climate change and intensification of coastal ocean upwelling. Science 247:198–201.

Beamish, R. J. 1993. Climate and exceptional fish production off the west coast of North America. Can. J. Fish. Aquat. Sci. 50(4):2270–2291.

Beamish, R. J., and Bouillon, D. R. 1993. Pacific salmon production trends in relation to climate. Can. J. Fish. Aquat Sci. 50(2):1002–1016.

Beamish, R. J., Neville, C. M., Thomson, B. L., Harrison, P. J., and St. John, M. 1994. The effect of Fraser River discharge on interannual production of Pacific salmon and herring in the Strait of Georgia. Can. J. Fish. Aquat. Sci. 51(2):2843–2855.

Brodeur, R. D., and Ware, D. M. 1992. Long-term variability in zooplankton biomass in the subarctic Pacific Ocean. Fish. Oceanogr. 1:32–38.

Canadian Department of Fisheries and Oceans. 1991. The Fraser River Salmon Rebuilding Program. Vancouver, BC: Canadian Department of Fisheries and Oceans.

Chelton, D. G. 1984. Commentary: Short-Term Variability in the Northeast Pacific Ocean. In W. G. Pearcy (ed.), The Influence of Ocean Conditions on the Production of Salmonids of the North Pacific. Oregon State University Sea Grant College Program ORESU-W-83-001, pp. 87–99.

Ebbesmeyer, C. C., Cayan, D. R., McLain, D. R., Nichols, F. H., Peterson, D. H., and Redmond, K. T. 1991. 1976 Step in the Pacific Climate: Forty Environmental Changes between 1968–1975 and 1977–1984. In J. L. Betancourt and V. L. Tharp (eds.), Proceedings of the Seventh Annual Pacific Climate (PACLIM) Workshop, April 1990. Calif. Dept. Water Resour. Interagency Ecol. Stud. Program Tech. Rep. 26, pp. 129–141.

Ebbesmeyer, C. C., Coomes, C. A., Cannon, G. A., and Bretschneider, D. E. 1989. Linkage of ocean fjord dynamics at decadal period. Geophys. Monogr. 55:399–417.

Fissel, D. B., Birch, J. R., and Chave, R. A. J. 1991. Measurements of Temperature, Salinity and Sound Velocity at the Nanoose Bay Naval Underwater Weapons Test Range, 1967–1984 [updated March 1991 to include 1985–1989, Appendix 4]. Sidney, BC: Arctic Sciences Ltd.

Freeland, H. J., and Denman, K. L. 1982. A topographically controlled upwelling center off Southern Vancouver Island. J. Mar. Res. 40(2):1069–1093.

Harrison, P.J., Mackas, D. L., Frost, B. W., Macdonald, R. W., and Crecerus, E. A. 1994. An Assessment of Nutrients, Plankton and Some Pollutants in the Water Column of Juan de Fuca Strait, Strait of Georgia and Puget Sound and Their Transboundary Transport. In R. C. H. Wilson, R. J. Beamish, F. Aitkens, and J. Bell (eds.), Review of the Marine Environment and Biota of the Strait of Georgia, Puget Sound and Juan de Fuca Strait. Proceedings of the BC/Washington Symposium on the Marine Environment, January 13–14, 1994. Can. Tech. Rep. Fish. Aquat. Sci. 1948, pp. 138–174.

Henderson, M. A., and Healey, M. C. 1993. Doubling sockeye salmon production in the Fraser River—Is this sustainable development? Env. Manag. 17(6):719–728.

Jones, P. D., Raper, S. C. B., and Wigley, T. M. L. 1986. Southern hemisphere surface air temperature variations: 1851–1984. J. Clim. Appl. Meteorol. 25(2):1213–1230.

Kerr, R. A. 1992. 1991: Warmth, chill may follow. Science 255:281.

Kuhn, B. R., Lapi, L., and Hamer, J. M. 1988. An Introduction to the Canadian Database on Marked Pacific Salmonids. Can. Tech. Rep. Fish. Aquat. Sci. 1649.

McLain, D. R. 1984. Coastal Ocean Warming in the Northeast Pacific, 1976–83. In W. G. Pearcy (ed.), The Influence of Ocean Conditions on the Production of Salmonids of the North Pacific. Oregon State University Sea Grant College Program ORESU-W-83-001, pp. 61–86.

Miller, A. J., Cayan, D. R., Barnett, T. P., Graham, N. E., and Oberhuber, J. M. 1994. The 1976–77 climate shift of the Pacific Ocean. Oceanogr. 7(1):21–26.

Moore, D. R. 1991. Hydrology and Water Supply in the Fraser River Basin. In H. J. Dorcey (ed.), Water in Sustainable Development: Exploring Our Common Future in the Fraser River Basin. Vancouver, BC: Westwater Research Centre, University of British Columbia, pp. 21–40.

Murdoc, J. 1979. Control Charts. London: MacMillan Press Ltd.

Noakes, D. J. 1986. Quantifying changes in British Columbia dungeness crab (*Cancer magister*) landings using intervention analysis. Can. J. Fish. Aquat. Sci. 43(1):634–639.

Polovina, J. J., Mitchum, G. T., Graham, N. E., Craig, M. P., Demartini, E. E., and Flint, E. N. 1994. Physical and biological consequences of a climate event in the central North Pacific. Fish. Oceanogr. 3(1):15–21.

Regier, H. A., Holmes, J. A., and Pauly, D. 1990. Influence of temperature change on aquatic ecosystems: An interpretation of empirical data. Trans. Am. Fish. Soc. 119(2):374–389.

Ricker, W. E. 1975. Computation and Interpretation of Biological Statistics of Fish Populations. Bull. Fish. Res. Board Can. 191.

Royer, T. C. 1989. Upper ocean temperature variability in the Northeast Pacific Ocean: Is it an indicator of global warming? J. Geophys. Res. 94:18175–18183.

St. John, M. A., Marinone, S. G., Stronach, J., Harrison, P. J., Fyfe, J., and Beamish, R. J. 1993. A horizontally resolving physical-biological model of nitrate concentration and primary productivity in the Strait of Georgia. Can. J. Fish. Aquat. Sci. 50(3):1456–1466.

Thomson, R. E. 1981. Oceanography of the British Columbia coast. Can. Spec. Publ. Fish. Aquat. Sci. 56.

Trenberth, K. E. 1990. Recent observed interdecadal climate changes in the Northern Hemisphere. Bull. Am. Meteorol. Soc. 71:988–993.

Zhang, Z., Beamish, R. J., and Riddell, B. E. 1995. Differences in otolith microstructure between hatchery-reared and wild chinook salmon, *Oncorhynchus tshawytscha.* Can. J. Fish. Aquat. Sci. 52:344–352.

3

The Interdecadal Climatic Change Signal in the Temperate Large Marine Ecosystems of the Pacific

Daniel Lluch-Belda

Variability is a fundamental characteristic of living systems, from cells to ecosystems. Often, when trying to understand the dynamics of their aggregations, we tend to simplify this variability—among many other things—to reduce the enormous complexity of their functioning to manageable dimensions. Sometimes this simplification leads to incorrect conclusions.

At the natural population level, fisheries have been managed based on the assumption that in the absence of harvesting they would follow a sigmoid growth function until reaching their carrying capacity and then remain there with minor variations around a mean level. Any changes in their abundance and in the associated parabola-shaped yield curve were attributed solely to the effect of the fishery. Indeed, there are many natural populations for which this supposition must be close to being true. Random interannual variations in this context may be interpreted as noise, and these signals are recurrent in many fisheries. Some authors have recognized that a part of such noise is environmentally induced.

Many authors have oversimplified the complexity of ecosystem functioning and suggest that we should aim at keeping these variable superorganisms in a stationary state. No matter what we do, however, they have their own variable paths and the efforts to prevent them from following these tracks are mostly useless.

It has become clear that nonrandom naturally induced environmental variability has had a major effect on the abundance of some resources, among which are some of the most abundant species in the world's catches: the small pelagic fishes, mackerels, and so forth. Most of the catch of these species originates in a few coastal marine ecosystems, mainly those ones in which varying degrees of mixing of a cold and a warm current occurs. Figure 3-1 shows the variation of sardine and anchovy catches from the main systems of the Pacific Ocean.

Naturally induced variability has been recognized by Clark and Marr (1955) in examining the California sardine fishery. Soutar and Isaacs (1969) analyzed the prefishery variability of fish populations as revealed by scale deposition rates, which have been consistently validated in other areas of the world by many other authors. Detailed study of changes in the marine ecosystem by

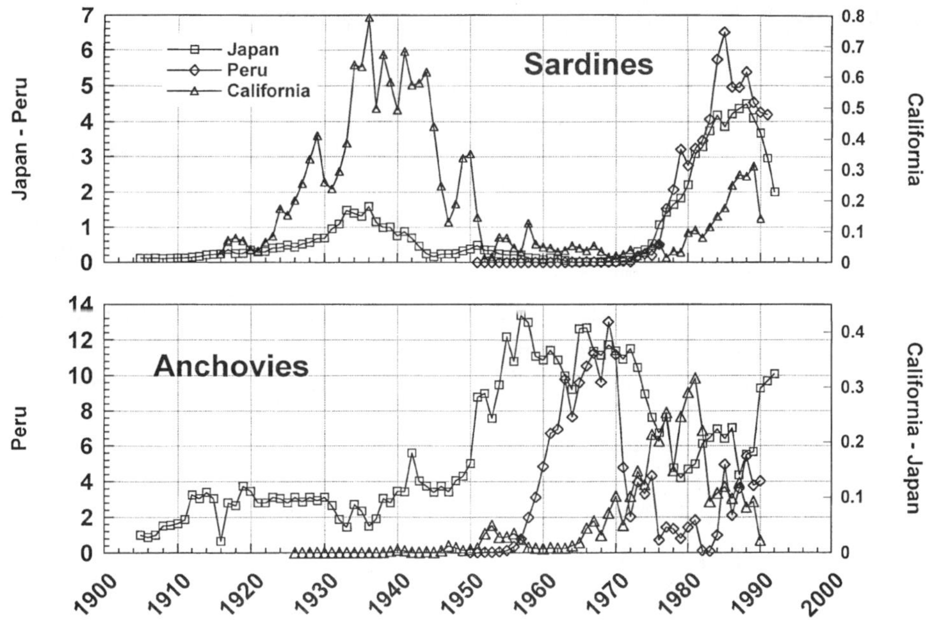

Figure 3-1. Variation of sardine and anchovy catches from the main systems of the Pacific Ocean.

Russell and Demir (1971) and the description of the wide variability in the Japanese sardine fishery as accounted in ancient historical records (Kondo 1986) also point to natural variability. These fluctuations operate on a decadal time scale, resulting in prolonged periods of high or low abundance of certain populations, and are nonrandom in nature.

The Regime Problem

The synchronous, very significant variations in the abundance of the major clupeid populations have led to the conceptualization of the "regime problem." This problem has been defined as follows:

1. Coherent fluctuation over a time scale of decades exists in the productivity of fish and the structure of their ecosystems; transitions between stages are typically abrupt.
2. Worldwide coincidences of such regime changes imply links to global climate.
3. Regime transition is presently occurring in several major oceanic ecosystems.
4. Regime change poses severe problems for sustainability of economic development.
5. Regime changes, of far greater magnitude than interannual variation, present fundamentally different problems than are usually considered by fisheries science; existing approaches are inadequate for the management of sardine and anchovy fisheries and associated economic development.

The effect of environment on population abundance (or availability) seems to operate directly through several locally different mechanisms, although it is probable that a limited number of closely interrelated factors may be the driving force. Indeed, many local events must be related to large-scale climatic and oceanic changes.

The Variability of the Pacific Large Marine Ecosystems

The Intergovernmental Oceanographic Commission (IOC) has defined the main large marine ecosystems (LMEs) of the Pacific Ocean (IOC 1993). These include mainly coastal cold, temperate, and tropical areas. The variability of these areas may be partially analyzed by using sea surface temperature (SST) records as an index. The COADS (Comprehensive Ocean–Atmosphere Data Set) database has a large number of observations grouped in 2° × 2° areas. We may thus analyze the spatial and temporal variability of the parameter in groups of areas within the delimited LMEs of the Pacific.

The total spatial and temporal variabilities are shown in Figure 3-2 as a function of average SST. The variance is considerably larger in the temperate ecosystems than in either the cold or the tropical ones, the peak being at about 15°C. The seasonal variation of each area (Figure 3-3) is

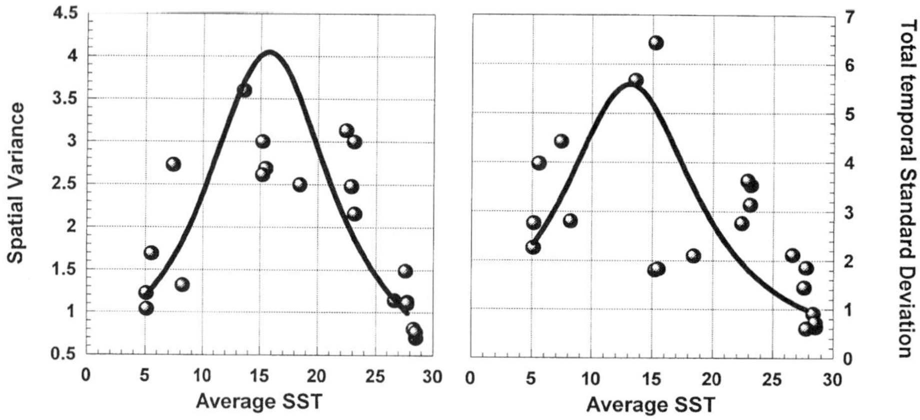

Figure 3-2. Total spatial and temporal variabilities as a function of average sea surface temperature (SST).

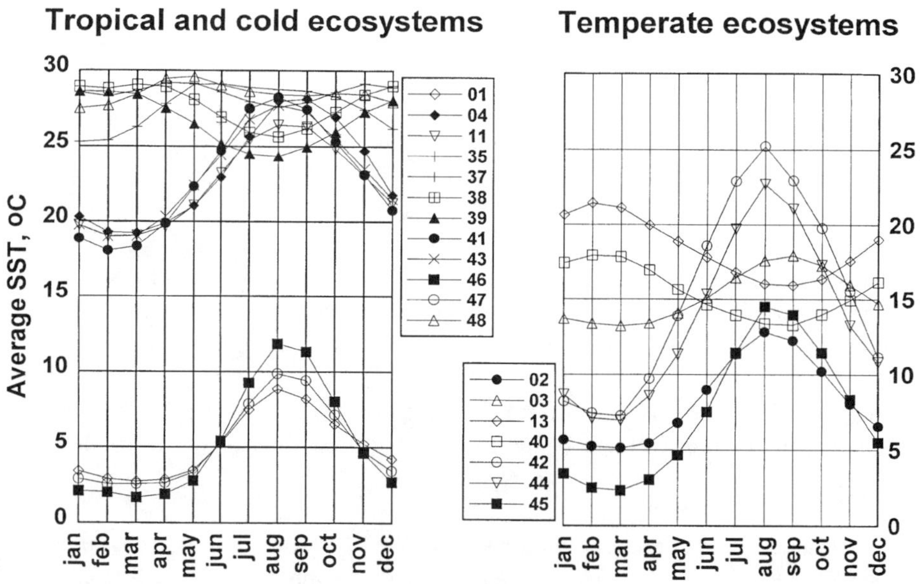

Figure 3-3. The seasonal variation of the tropical, cold, and temperate ecosystems.

shown, separating the tropical, cold, and temperate ecosystems. Total and seasonal variations are closely correlated.

What remains after removing the seasonal variation is the interannual variation plus the noise. This variability is quite small in the tropical LMEs and increases considerably toward the cold areas, particularly when the high-frequency noise is eliminated by smoothing (Figure 3-4). It is clear that interannual variability is much more important in the cold ecosystems and that its weight diminishes as SST increases.

There are many indications of a particular regime shift, such as the one that occurred during the mid-1970s. This event has been documented, and its signal is clear in several environmental indices, such as the Global Surface Air Temperature series, the Global Energy Balance, and the CO_2 residuals, among others (Gunn 1991) (Figure 3-5). The smoothed average anomalies for tropical, warm-temperate, temperate, and cold LMEs of the Pacific also reflect the change, particularly the temperate and cold ones (Figure 3-5). This event has not been unique. Although data series are not available, similar regime shifts have occurred before. During this century, there was a regime of high intensity peaking around 1940; there have been regimes of low intensity during 1900–1930 and 1950 to the mid-1970s.

If sardine abundance is a reliable proxy for regime shifts (as it seems to be), there have been cycles of about 60 years in regime changes during the last 2,000 years (Baumgartner et al. 1992).

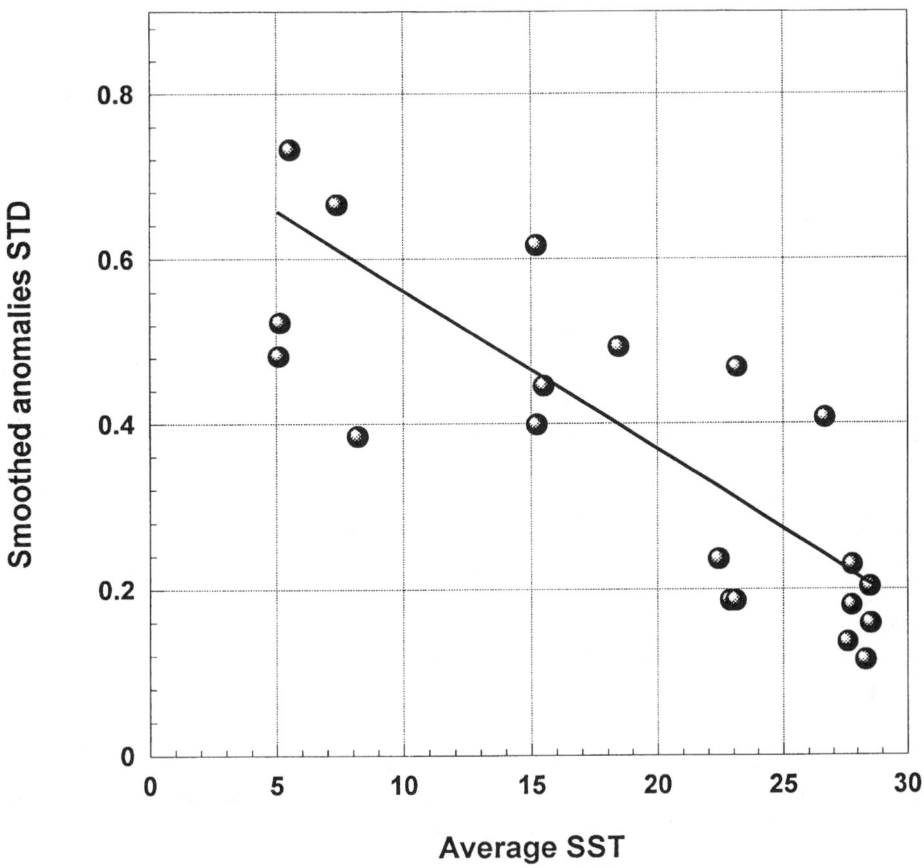

Figure 3-4. Average sea surface temperature (SST) versus standard deviations of smoothed anomalies.

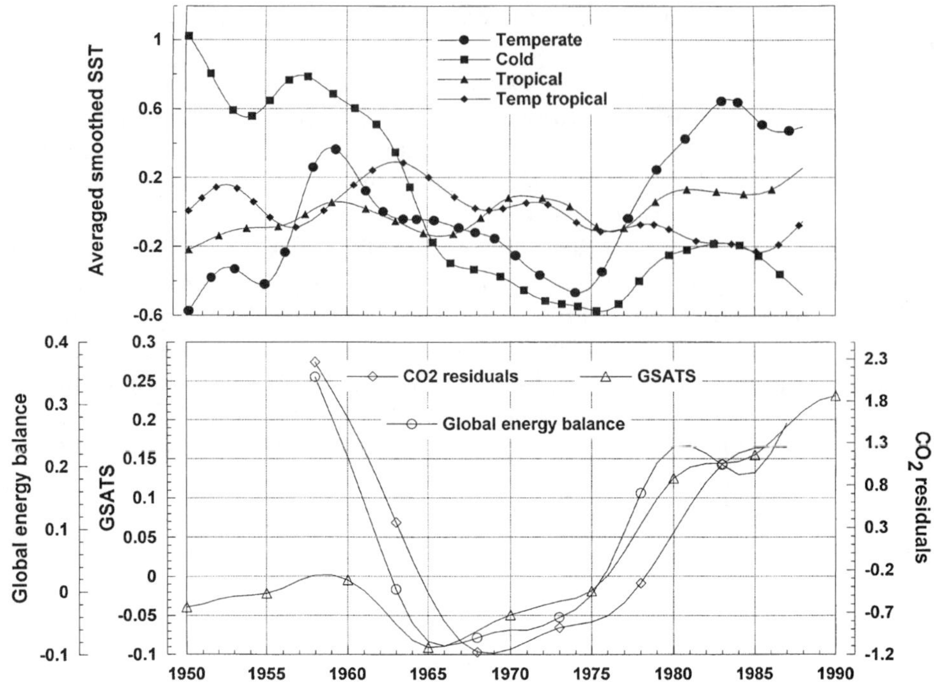

Figure 3-5. Top. The smoothed average anomalies for tropical, warm-temperate, temperate, and cold LMEs of the Pacific. Bottom. Global Surface Air Temperature Series (GSATS), the Global Energy Balance, and the CO_2 residuals from 1950–1990.

Not only that, but sardine abundance indicates that since roughly 1985 we have been in the process of a new shift from a high to low regime, even though the normal delay in environmental data publishing does not show it as well as the former regime shift.

The Consequences of Regime Changes

As the regime problem becomes evident, management of fisheries and whole ecosystems will become more complicated because a natural variability component will have to be taken into account. Recognizing that some populations will grow or collapse even with the best management is indeed an added difficulty, one that will require a considerable amount of research effort to cope with.

The problem of managing fisheries (or ecosystems) is increasingly a matter of efficiency; if we are not able to forecast the changes, we will not be able to prepare for the utilization of a temporally abundant resource and to shift from it to something else when it collapses. There is an urgent need to change from day-to-day managing schemes, based on the assumption that all the variations in the utilized resource derive from the amount of stress we place on it, to a long-range, decadal to secular management strategy, based not only on artificial stress but also in the recognition of a surmounting natural variability that we cannot control.

We should learn that nature is variable and that we cannot keep populations, communities, or ecosystems in a stationary state. We badly need a reappraisal of the dynamic equilibrium concept, not as a mean with minor random variations but as an array of possible levels with probabilities in time attached to them. Our capacity has to extend from the managing of steady-state aggregations to the managing of changing ones, not only during the high or low regimes but also—and perhaps even more important—during the shifts.

ACKNOWLEDGMENTS

Participation from Sergio Hernandez V., Daniel Lluch Cota, and Salvador Lluch Cota is acknowledged.

References

Baumgartner, T. R., Soutar, A., and Ferreira-Bartrina, V. 1992. Reconstruction of the history of Pacific sardine and northern anchovy populations over the past two millennia from sediments of the Santa Barbara Basin, California. Calif. Coop. Oceanic Fish. Invest. Rep. 33:24–40.

Clark, F. N., and Marr, J. C. 1955. Population dynamics of the Pacific sardine. Calif. Coop. Oceanic Fish. Invest. Progress Rep. 4:11–48.

Gunn, J. 1991. Influences of various forcing variables on global energy balance during the period of intensive instrumental observation (1958–1987) and their implications for paleoclimate. Climatic Change 19:393–420.

Intergovernmental Oceanographic Commission. 1993. Assessment and Monitoring of Large Marine Ecosystems. Intergovernmental Oceanographic Commission INF-942.

Kondo, K. 1986. Relationships between Long Term Fluctuations in the Japanese Sardine *Sardinops melanostictus* (Temmnick and Schlegel) and Oceanographic Conditions. In T. Wyatt and M. G. Larraneta (eds.), Long Term Changes in Marine Fish Populations. Vigo, Spain: Institute de Investigaciones Marinas de Vigo, pp. 365–392.

Russell, F. S., and Demir, N. S. 1971. On the seasonal abundance of young fish. XII. The years 1967, 1968, 1969 and 1970. J. Mar. Biol. Ass. 44:343–364.

Soutar, A., and Issacs, J. D. 1969. History of fish populations inferred from fish scale in anaerobic sediments off California. Calif. Coop. Oceanic Fish. Invest. Rep. 13:63–70.

4

Long-Term Variations in Pelagic Communities in the Large Marine Ecosystems of the Northwest Pacific Ocean

V. P. Shuntov, E. P. Dulepova, and V. V. Lapko

Since the 1990s there have been detectable changes in pelagic communities in the large marine ecosystem of the Northwest Pacific Ocean that may be linked to climatic and oceanographic changes (Shuntov 1993a, 1993b; Shuntov et al. 1993a). In this chapter, we present updated data from our research surveys performed in 1993.

Materials and Methods

Our research surveys were carried out in the western Bering Sea, Sea of Okhotsk, and Pacific waters off Kamchatka and the Kuril Islands. These surveys were conducted in 1993 by three research vessels that have similar characteristics and used similar collection and processing methods.

Plankton samples were collected by vertical hauls in the 0–200 m water layer with Gedy nets. The nets had a mesh size of 0.168 mm and a square mouth opening of 0.1 sq m. Zooplankton were processed into three size groups: small (<1.5 mm), medium (1.5–3.5 mm), and large (>3.5 mm). Plankton abundance was estimated using the following catchability coefficients: 1.5 for the small-sized group, 2.0 for the medium-sized group, and 5.0 for the large-sized group (euphausiids and chaetognaths, amphipods, and copepods). Catchability coefficients for euphausiids and chaetognaths were 2.0 (for sizes <10 mm), 5.0 (for sizes 10–20 mm), and 10.0 (for sizes >20 mm). Catchability coefficients for amphipods were 1.5 (for sizes <5 mm), 3.0 (for sizes 5–10 mm), and 5.0 (for sizes >10 mm). For copepods the coefficients were 2.0 and 5.0 for sizes less than and larger than 5 mm, respectively.

Nekton were caught by trawl nets with a vertical dimension of 50 m and horizontal dimension of 70 m. The mesh size of the cod-end (last 10 m of net) was 10 mm. Trawl hauls were made mainly in the top 50 m water layer. Tows were for 1 hour at 4.5–5.0 knots (8.33–9.26 km/hr). In 1991 and 1993, 214 tows and 393 tows were done, respectively. Nekton biomass and quantity were estimated using the following catchability coefficients: 0.1 (for juveniles, mesopelagic fish, squid, and jellyfish), 0.3 (for salmon and sharks), 0.4 (for walleye pollock and herring), and 0.5 (for smooth lumpsucker, large mesopelagic fish, and demersal fish).

Results

Zooplankton

The abundance of the small-sized group (juvenile stages of crustacea, mainly copepods) and the large-sized group of zooplankton decreased simultaneously in both the Sea of Okhotsk and Pacific waters off the Kuril Islands (Table 4-1). In the 1990s, copepod abundance decreased, euphausiid biomass increased, and chaetognath biomass remained stable (Table 4-2). In 1993, the abundance of the large-sized zooplankton (major macroplankton) decreased in Pacific waters.

We do not believe that the reasons for these marked changes were strongly linked. There were multiple processes causing the changes. The main process was the change in abundance of common planktivorous fishes, which may have had feeding impacts on the plankton composition and abundance. We do not have data to comment on changes to the plankton community from the 1980s to the 1990s for Pacific waters off Kamchatka. It appears, however, that there was not a significant difference in this area between 1991 and 1993 (Tables 4-1 and 4-2).

The situation in the Bering Sea differed considerably (Tables 4-3 and 4-4). The biomass of the small-sized group of zooplankton decreased appreciably from the 1980s to the 1990s. Biomass for the other two size groups (medium and large) also decreased, but to a lesser degree. This was reflected by the total zooplankton biomass, which decreased from 113 million tons to 78 million tons from 1989 to 1993.

Table 4-1. Biomass (in thousands of metric tons) of zooplankton by size groups in the Sakhalin-Kurils region in summer

Region	Date	Size groups			Total
		Small	Medium	Large	
Sea of Okhotsk	1986	8,190	10,070	104,700	122,960
	1991	9,898	5,340	77,251	92,489
	1993	4,078	10,445	78,543	93,066
Pacific waters	1991	15,705	6,566	146,597	168,868
Kuril Islands	1993	9,935	4,943	99,760	114,638

Table 4-2. Interannual changes of macroplankton biomass (in thousands of metric tons) in the Sakhalin-Kurils region in summer

Region	Year	Euphausiids	Amphipods	Copepods	Chaetognaths	Others
Sea of Okhotsk	1986	24,050	4,190	60,080	14,530	1,850
	1991	39,698	2,654	17,354	16,866	679
	1993	40,471	4,855	13,820	16,665	2,733
Pacific waters	1991	32,685	5,317	31,175	71,611	5,809
Kuril Islands	1993	12,415	4,519	22,131	56,039	4,654

Table 4-3. Biomass of zooplankton (in thousands of metric tons) by size groups in the western Bering Sea and Pacific waters off Kamchatka in summer

		Size groups			
Region	Date	Small	Medium	Large	Total
Bering Sea	1989	31,074	5,481	76,973	113,528
	1991	7,301	1,909	33,485	42,695
	1993	8,098	3,072	67,127	78,297
Pacific waters off Kamchatka	1991	7,519	3,313	36,493	47,325
	1993	7,413	3,955	30,751	42,119

Table 4-4. Interannual changes in biomass of macroplankton (in thousands of metric tons) in the western Bering Sea and Pacific waters off Kamchatka In summer

Region	Year	Euphausiids	Amphipods	Copepods	Chaetognaths	Other
Bering Sea	1989	4,665	645	56,735	14,114	802
	1991	2,891	457	6,985	20,777	2,374
	1993	6,131	1,000	18,281	39,201	2,514
Pacific waters off Kamchatka	1991	4,781	516	6,148	23,054	1,995
	1993	3,353	886	7,951	16,110	2,452

In terms of species components, copepod biomass decreased considerably, from 56.7 million tons in 1989 to 18.2 million tons in 1993 (Table 4-4). However, chaetognath biomass increased proportionally. Considering that copepods are smaller than chaetognaths and are more acceptable as prey to juvenile fish, the inverse biomass trend for copepods and chaetognaths appears to be related to feeding pressure. However, such changes in the plankton community may not be so simple because they are related to complicated changes in oceanographic conditions and the nekton community.

Nekton

In most areas, the total biomass of fish decreased. However, changes to the Sea of Okhotsk side and the Bering Sea side of Kamchatka varied considerably. In the southern Sea of Okhotsk and Pacific waters off the Kuril Islands (Table 4-5), the total fish biomass in 1993 (546,000 mt) was appreciably less than in 1991 (823,000 mt). The decrease was lower in the southern Sea of Okhotsk than in Pacific waters. The biomass of pink salmon (*Oncorhynchus gorbuscha*), walleye pollock (*Theragra chalcogramma*), sardine (*Sardinops sagax melanosticta*), and anchovy (*Engraulis japonicus*) decreased sharply; whereas that of northern smooth tongue (*Aptocyclus ventricosus*) increased sharply.

Fish abundance changes for different reasons. The decrease in salmon abundance in 1993 corresponded to their natural population dynamics and ocean environment. Our 1993 data support the idea that a period of weak survival began in the early 1990s after a favorable period in the 1970s to 1980s (Chigirinsky 1993; Beamish and Bouillon 1993). In addition, density-dependent factors could have caused a decrease in pink salmon abundance. In the Sakhalin-Kuril

Table 4-5. Composition and biomass of fishes (in thousands of metric tons) in the upper epipelagic water layer (0–50 m) in the Sakhalin-Kuril region in summer

	Southern Sea of Okhotsk		Pacific waters off Sakhalin-Kurils	
Species and groups	1991	1993	1991	1993
Pink salmon	163.3	59.6	133.9	90.7
Chum salmon	14.3	12.9	5.8	12.6
Coho salmon	0.5	0.7	3.3	3.4
Chinook salmon	3.3	1.3	2.8	1.6
Sockeye salmon	0.3	0.2	2.2	1.3
Walleye pollock	145.6	21.4	3.8	0.16
Herring	0.1	15.1	—	—
Sardine	307.7	—	13.6	2.6
Northern smooth tongue	29.4	359.6	3.1	16.5
Myctophids	0.1	0.1	233.4	108.3
Other mesopelagic fishes	0.1	0.1	48.4	26.2
Deal-fish	—	—	36.9	25.9
Sharks	14.7	2.1	1.3	0.4
Pacific pomfret	—	—	21.3	7.7
Japanese anchovy	6.8	0.1	47.1	11.1
Pacific saury	—	—	22.6	1.8
Other fishes	9.5	30.6	29.9	1.2
Total	695.7	503.8	609.4	311.5

region, pink salmon were abundant in 1991 but scarce in 1993. In 1994, pink salmon were very abundant near western Kamchatka, due probably to better survival in 1992. We believe that dominant populations of salmon will vary over time according to ocean environment and density-dependent dynamics.

The 1993 surveys also indicated that recruitment of walleye pollock was low in the southern Sea of Okhotsk; thus, abundance would decrease in the future. The abundance of pollock had already been on a general decrease since the early 1990s (Shuntov et al. 1993b). Sardine biomass decreased very sharply from 1991 to 1993. In 1993 there was almost no sardine fishing in the southern Sea of Okhotsk and its adjacent Pacific waters.

The distribution of Pacific herring (*Clupea harengus*) in the Sakhalin-Hokkaido region has moved about 150 miles off the eastern Sakhalin coast. We believe that this expansion in spatial distribution indicates growth of the herring population, which has been depressed for the past 10 years. The Shelikhov Bay herring population appeared stable.

The composition and biomass of fishes in the western Bering Sea and off eastern Kamchatka are presented in Table 4-6. The abundance of most fishes was lower in 1993 than in 1991. The total biomass of fish dropped from 369,000 mt in 1991 to 177,000 mt in 1993 in the western Bering Sea; the biomass east of Kamchatka dropped from 176,000 mt in 1991 to 97,000 mt in 1993. Decreases in salmon and walleye pollock biomass seemed supportive of recent reports that suggested their population dynamics were greatly affected by global climatic–oceanographic events such as those noted by Beamish and Bouillon (1993).

The average biomass of nekton and groups of epipelagic species, and their estimated predation of food organisms for the Bering Sea, are presented in Table 4-7. The data indicate that walleye pollock and salmon constituted the largest components of the nekton biomass and accounted for most of the predation of epipelagics.

Table 4-6. Composition and biomass of fishes (in thousands of metric tons) in the upper epipelagic water layer (0–50 m) in the western Bering Sea and Pacific waters off Kamchatka in summer

	Western Bering Sea		Pacific waters off Kamchatka	
Species and groups	1991	1993	1991	1993
Pink salmon	95.40	37.61	45.99	13.82
Chum salmon	54.41	40.05	75.02	13.24
Sockeye salmon	2.54	11.65	42.48	16.21
Chinook salmon	4.42	2.59	0.89	1.62
Coho salmon	—	0.26	2.66	0.68
Dolly varden	4.45	11.34	0.37	—
Walleye pollock	155.53	30.89	0.14	0.32
Herring	17.89	2.78	0.01	—
Salmon shark	2.61	—	—	2.03
Capelin	12.89	1.33	0.04	0.01
Atka mackerel	0.53	20.32	—	35.61
Northern smooth tongue	0.30	3.93	2.00	5.65
Other mesopelagic fishes	4.29	9.32	6.52	3.50
Other fishes	3.92	4.63	0.19	4.53
Total	369.25	176.70	175.76	97.22

Table 4-7. Average biomass of main nekton and groups of Bering Sea epipelagic fishes and their annual predation of food organisms (in million metric tons) in the second half of the 1980s to the early 1990s

	1980s		1990s	
Nekton species and groups	Biomass	Predation	Biomass	Predation
Walleye pollock	22.5	220.0	6.0	66.0
Mesopelagic fishes[a]	9.3	46.0	10.5	51.9
Squids	0.6	5.9	0.65	6.4
Herring	0.7	10.6	0.9	12.9
Capelin	0.36	2.0	0.5	2.8
Smooth lumpsucker	0.25	4.0	0.25	4.0
Pink salmon[b]	0.09	0.1	0.1	0.11
Chum salmon	0.1	0.56	0.126	0.71
Sockeye salmon	0.09	0.51	0.1	0.57
Chinook salmon	0.03	0.14	0.03	0.14
Coho salmon	0.005	0.02	0.003	0.01
Dolly varden	—	—	0.008	0.01

[a]Average assessment of group share that ascends into epipelagic zone at night.
[b]Salmon biomass in summer.

Marine Birds

Some changes have been noted in the seabird community. The composition, concentration, and abundance of seabirds differed appreciably from 1991 to 1993 in the northern Kuril area. In 1993 the abundance of seabirds was estimated to be less than 1.8 million. Shearwaters (*Puffinus tenuirostris, P. griseus*) numbered less than 1.6 million. They were distributed more southward and closer to the ocean, near the northern Kuril Islands. The mottled petrel (*Pterodroma inexpectata*), which was common within the survey area in 1991, was practically absent in 1993. Black-footed albatross (*Diomeda nigripes*) was scarce, and McCormick scua (*Catharacta maccormicki*) was practically absent.

Some species were distributed in the outerlying areas of the western Pacific, in the Sea of Okhotsk and the Bering Sea. For example, the abundance of northern fulmar (*Fulmarus glacialis*), fork-tailed storm petrel (*Oceanodroma furcata*), laysan albatross (*Diomedea immutabilis*), red-legged kittiwake (*Rissa brevirostris*), glaucous-winged gull (*Larus glaucescens*), murres (*Uria* spp.), and other species of seabirds increased in these areas. Westward distribution of fork-tailed storm petrel and northern fulmar was especially evident, and they were common in most parts of the Sea of Japan. Distribution of fork-tailed storm petrel was similar to that in the 1950s and 1960s. Further observations are needed to show which examples reflect long-term trends and which simply interannual variability in seabird distributions.

Discussion

The data presented indicate that changes in the biota of the far-eastern seas were considerable from 1991 to 1993 and differed by region. In the Bering Sea, long-term oscillations of walleye pollock and herring occurred at opposite times (Naumenko et al. 1990). In the Sea of Okhotsk, both walleye pollock and herring populations were still significant. We believe that these large biomasses were due to density-dependent factors in the Bering Sea and by more abundant plankton resources in the Sea of Okhotsk (Shuntov and Dulepova 1991).

The total fish abundance decreased in both the southern Sea of Okhotsk and its adjacent Pacific waters. Simultaneously, squid and jellyfish biomass increased (Table 4-8). In the Bering Sea, despite increases in some fish species, total fish biomass and its density decreased, mainly due to walleye pollock and Pacific salmon declines. The total fish biomass decreased by about half for both the Bering Sea and Pacific waters (Table 4-9).

Since fish life cycles, in general, are longer than those of planktonic organisms, the succession of different species takes more time. The abundance of walleye pollock and herring in the Bering Sea seemed to oscillate alternately. The 1991–1994 period is a transitional period because walleye pollock abundance has decreased considerably, while herring abundance is beginning to increase. We think that in the near future, the Bering Sea fish biomass will grow due to increases in the abundance of herring, capelin (*Mallotus villosus*), and atka mackerel (*Pleurogrammus monopterygius*); however, these increases will not compensate for the loss of walleye pollock. The decrease in planktivorous fish abundance is one of the most important factors affecting plankton communities in this region. We think that this decrease has initiated a growth of predatory

Table 4-8. Nekton and jellyfish density (in metric tons per sq km) in the upper epipelagic zone of the Sakhalin-Kurils region

	Sea of Okhotsk		Pacific waters off Sakhalin-Kurils	
Year	1991	1993	1991	1993
Fishes	1.70	1.00	1.16	0.46
Squids	0.10	0.16	0.33	0.44
Total nekton	1.80	1.16	1.49	0.90
Jellyfish	0.08	0.21	0.19	0.45

Table 4-9. Nekton and jellyfish density (in metric tons per sq km) in the upper epipelagic zone of the western Bering Sea and Pacific waters off Kamchatka and the Commander Islands in summer

	Western Bering Sea		Pacific waters	
Year	1991	1993	1991	1993
Fishes	0.98	0.47	0.39	0.22
Squids	0.21	0.25	0.56	0.35
Total nekton	1.19	0.72	0.95	0.57
Jellyfish	0.02	0.56	0.02	1.43

plankton abundance, that in its turn, has caused the decrease in the biomass of small-sized plankton. Our interpretations are supported by the observations that predatory plankton were less abundant in areas with large fish populations and that the ratio of size and trophic groups in plankton was more stable in areas with few fish. We emphasize that future ecosystem investigations are very important, especially taking into consideration the cyclic nature of ecosystem structure.

Conclusion

We believe that in the beginning of the 1990s, a climatic–oceanographic regime occurred similar to the one in the 1940s to 1960s. It replaced the regime in the 1970s and 1980s that was similar to one in the 1920s and 1930s. This regime shift affected the composition and structure of the plankton and nekton communities. The situation in the far-eastern seas is similar, but not identical, to that in the 1940s to 1960s because the initial states of the ecosystems are different. The early to middle 1990s is considered a turning point of both the 22- and 60-year cycles of climatic–oceanographic conditions in the North Pacific (Davydov 1986; Shuntov 1993a, 1993b). However, 1993 differed considerably from previous years. In 1993, the Oyashio Counter Current was not strong, and ocean water that entered the Bering Sea was mainly from the Alaska Current. Ocean waters did not enter into the Sea of Okhotsk through the northern Kuril Straits, and the western Kamchatka and eastern Sakhalin currents were very weak. We believe that these and other changes in oceanographic conditions should be considered as large environmental perturbations that affected the distribution of plankton and nekton communities.

References

Beamish, R. J., and Bouillon, D. R. 1993. Pacific salmon production trends in relation to climate. Can. Jour. Fish. Aquatic Sci. 50(5):1002–1016.

Chigirinsky, A. I. 1993. The global natural factors, fishing and quantity of Pacific salmon (in Russian). Rybnoe Khozyaistvo 2:19–22.

Davydov, I. V. 1986. On the Nature of Prolonged Changes of the Number of Fishes and Possibility of Their Foresight. Dynamics of the Number of Fishery Objects in the Far-Eastern Seas (in Russian). Vladivostok: TINRO, pp. 3–7.

Naumenko, N. I., Balykin, P. A., Naumenko, E. A., and Shaginjan, E. R. 1990. Long-term fluctuations in abundance and structure of pelagic fishes in the western Bering Sea (in Russian). Izvestia, TINRO 111:49–57.

Shuntov, V. P. 1993a. Once more about the problem of global warming and its influence on biota of the far-eastern seas (in Russian). Rybnoe Khozyaistvo 6:39–41.

Shuntov, V. P. 1993b. Combined expedition to the far-eastern seas in summer 1992 (in Russian). Vestnik DVO RAN 2:68–74.

Shuntov, V. P., and Dulepova, E. P. 1991. Ecosystems of the Bering Sea and Okhotsk Seas (in Russian). Rybnoe Khozyaistvo 6:25–27.

Shuntov, V. P., Dulepova, E. P., Radchenko, V. I., and Temnykh, O. S. 1993. On the beginning of the large reformation in communities of plankton and nekton of the far-eastern seas (Abstract). North Pacific Mar. Sc. Organ. (PICES) Second Annual Meeting, 25–30 October 1993, Seattle, Washington, p. 35.

Shuntov, V. P., Volkov, A. F., Temnykh, O. S., and Dulepova, E. P. 1993a. Walleye Pollock in Ecosystems of the Far-Eastern Seas (in Russian). Vladivostok: TINRO.

Long-Term Change in the North Pacific Ocean: A Consideration of Some Important Issues[1]

Thomas L. Hayward

Introduction

This report summarizes the results of a meeting held in La Jolla on October 12–13, 1995, which was organized by Tom Hayward at the request of the CalCOFI Committee. The purpose of the meeting was to consider recent observations of long-term changes taking place in the North Pacific Ocean, and to make recommendations about whether these observations merit a response by CalCOFI.

Meeting participants: Tim Barnett, Tim Baumgartner, Ed Brinton, Dan Cayan, Francisco Chavez, Karen Forney, Doyle Hanan, Tom Hayward, John Hunter, Ron Lynn, John McGowan, Rick Methot, Art Miller, Dean Roemmich, Frank Schwing, Paul Smith, Mia Tegner, and Amy Weinheimer.

The following people provided helpful written input: Alec MacCall, Dave Mackas, Michael Mullin, and Jeff Polovina.

Observations and Model Results

There is no doubt that important changes have been taking place in atmospheric forcing and in the physical and biological structure of the upper layers of the North Pacific Ocean in the past few decades. Zooplankton biomass and the abundance of pelagic fish and squid in the subarctic Pacific were doubled in 1980–89 in comparison with 1956–62 (Brodeur and Ware 1992). There was also a positive correlation between the intensity of winter winds and zooplankton biomass measured during the subsequent summer in the subarctic Pacific. In the subtropical North Pacific, a doubling of the vertically integrated chlorophyll starting in the mid-1970s is associated with a cooling of SST and an increase in winter winds (Venrick et al. 1987). Decade-scale trends in atmospheric pressure patterns, in mixed layer depth, and in the productivity of several trophic levels has been observed in the subtropical North Pacific (Polovina et al. 1994, in press). Macrozooplankton biomass off Peru declined sharply in the mid-1970s (Pauly et al. 1989), and this was associated with an increase

[1]Previously published in CalCOFI Report, Vol. 37, 1996. Reprinted with permission of the CalCOFI Coordinator and the author.

in the strength of upwelling-favorable winds (Bakun 1990). In the California Current, a secular rise in coastal sea level in the last few decades is partly due to warming (and hence expansion) in the top 300 m (Roemmich 1992). A decrease in macrozooplankton biomass (Roemmich and McGowan 1995a, b) and in the abundance of pelagic seabirds (Veit et al., in press) in the California Current is associated with a warming of the surface waters and an increase in stratification.

Much progress is being made in describing and modeling the physical structure of the system, at least in terms of understanding what has taken place. For example, isotherms in the thermocline off the western coast of North America were observed (in XBT data) to deepen in the mid-1970s while the mixed layer concurrently shallowed. This shift in basic-state structure was studied by Miller et al. (1994) in a basin-scale model driven by observed heat flux and wind-stress anomalies. New results from the analysis of that model show that the thermocline deepening off the California coast is part of a basin-scale shift in thermocline structure which is driven by a basin-scale change in wind stress curl. A concomitant weakening of the subtropical gyre (by about 10 percent) occurs in the model as well but with little long-term change in the California Current transport. Definition of scale is an important issue, since shorter-term interannual changes in thermocline structure associated with El Niño provide a large local signal but a finite lifetime for biological response.

Prediction is still some way away, especially on the mesoscale. Models and mechanisms that may link physical and biological structure on the variety of time and spatial scales involved here need to be developed. Such models must include variations in upwelling and nutrient inputs due to interannual changes in the upwelling-favorable wind field (Bakun 1990), variations in the nutrient input due to a secular trend of increasing stratification (Roemmich and McGowan 1995a, b), and variations in primary production associated with interannual changes in mixed layer depth (Polovina et al., in press). Physical models are improving more rapidly than biological models, and it is highly probable that predictive ability in physical structure will precede that in biological structure.

Some Important Questions

Basic Science versus Policy and Management Issues

Our meeting emphasized basic science over policy and management issues, probably due to the mixture of those in attendance. In spite of this, it was recognized that there is a real need to make the basic scientific findings of long-term change in the California Current accessible to policy makers and managers in a form that is compelling and useful to them. There was discussion that policy and management issues in the coastal region of California generally seem to be focused upon point-source effects and regional anthropogenic change, while the largest changes we have observed in ecosystem structure are those associated with long-term change which are correlated with changes in the large-scale physical structure. There was general consensus that improving our understanding of the basic science issues will lead directly to increased value for management and policy issues. The group also felt that making the scientific and management communities more aware of these issues would help in gaining research support and seeing more practical application of the findings.

How Is Physical Structure Linked to Biological Structure, Both at the Level of Biomass and Species?

An understanding of how physical structure is linked to biological structure is at the heart of understanding the nature of the changes that have taken place in the California Current, and central to predicting future change. Recent studies have suggested that changes in several different aspects of environmental structure, including mixed layer depth, stratification, and the strength of upwelling-favorable winds, are the cause of changes in primary production. These mechanisms assume that changes in, at least, the biomass of higher trophic levels are forced by bottom-up forcing. Although these physical changes are related, they are also conceptually different mechanisms which may have different regional patterns or different time series. Since time series of several aspects of physical structure are correlated with each other, it is difficult to separate causal mechanisms via time series analysis. Furthermore, the effects of in situ, bottom-up forcing may be masked by apparent changes in population size due to advection, changes in range or preferred habitat of mobile species, and top-down forcing (where changes in predator or grazer populations regulate the abundance of the lower trophic levels). Separating these different mechanisms will be a difficult challenge, especially at the level of individual species. Physical-biological model studies could help to sort out these causalities.

Did Species Structure Change during the Period of Declining Macrozooplankton Biomass?

There are really two questions here. What changes took place in species structure of trophic levels (phytoplankton, macrozooplankton)? How did the population size of selected individual species change? When trophic levels or broad categories such as phytoplankton or macrozooplankton are considered, it is unknown whether species structure changed in addition to biomass. For a few individual species, large changes in population size have been documented. However, in these cases it is difficult to know what caused the changes and whether they are related to long-term environmental change. Most of the discussion focused upon the need to know whether species structure of the macrozooplankton changed, because this is an important response of the ecosystem and it may affect the population dynamics and species structure of higher (and lower) trophic levels.

Are the Changes We See Due to Natural Cycles or Global Change?

One of the most important questions is whether the changes in macrozooplankton biomass are part of a natural cycle or due to anthropogenic global warming. The time rate of change is so small that it is unlikely that this question can be answered by direct measurement of changes in rate processes. However, long time series (10^2–10^3 year) preserved in the sedimentary record can be used to extend the temporal coverage of CalCOFI and estimate scales of natural variability. The paleooceanographic record shows that changes of similar magnitude have occurred in the past. For example, deposition rates of sardine scales over the past 2,000 years show interdecadal–century variability of greater magnitude than has occurred since the initiation of CalCOFI (Baumgartner et al. 1992). Improvement in our understanding of how physical structure changed and how it is coupled to biological structure will help to answer this question.

Can the Present Sampling Scheme Be Improved within the Scope of the Currently Available Resources?

Discussion of the CalCOFI sample scheme focused upon the quarterly time-series cruises. The present scheme does a poor job of characterizing the annual pattern of events. It was observed that attempting to fit the cruises into a specific time window leads to a phase lock, which biases construction of a mean annual cycle. Sampling from south to north on the cruises leads to two-week gaps between sampling the southern and northern lines, and the present scheme leads to gaps in ADCP coverage of the perimeter of the pattern. It was suggested (by Roemmich) that the pattern could be modified to sample the perimeter at the end of the cruise; this would have the advantages of complete ADCP coverage and completing the sampling in a few days. This change would greatly improve the ability to construct a control volume (e.g., Roemmich 1989) and constrain fluxes into and out of the sample grid.

It was also observed that changes within the CalCOFI pattern are forced by physical processes on larger scales, and that populations in different regions may respond differently to changes in environmental structure. Collecting time series of similar data at sites off northern California and Baja California would help to provide a larger-scale view of forcing and indicate whether populations respond to change in the same way.

Recommendations

Nine specific recommendations were made at the meeting. In preparing this list we recognized that recommendations had the greatest chance of being implemented if the costs were modest and if some reasonable source of support were feasible. We did not set priorities on the recommendations because we recognized that priorities will vary depending on the enthusiasm of individuals to pursue the recommendations and on funding priorities of the federal agencies. However, we can state that recommendations with the highest priority were 1 and 2, the general areas of improving our understanding of changes at the species level and increasing modeling efforts in physical models and their coupling to biological structure.

1. Improve Resolution of Zooplankton Species Changes

There was general agreement on the need to understand how species structure of at least some of the major components of the macrozooplankton has changed. It was recognized that several approaches to this—such as counting pooled samples, counting selected years or regions, counting only selected taxonomic groups, or scanning for key species—could reduce the effort required to a more manageable level.

2. Start Putting Biology in Physical Models and Test the Relation to See If the Decline in Macrozooplankton Biomass Can Be Reproduced

One way to gain understanding of the processes linking physical and biological structure will be to include biological processes in physical models of the California Current which are now being developed. The macrozooplankton biomass time series will be ideal for testing such models, and the ongoing data collection program can be used to evaluate predictive ability as it evolves. As a first step in the model-testing process, a very simple model of primary productivity was proposed to be studied by using the mixed-layer depth changes hindcast by Miller et al. (1994) as the sole input, with the output compared with historical data on ecosystem productivity.

3. Gain Access to the Output of "Operational" Physical Models Being Developed with Navy Resources

A set of "operational" models is being developed and will be used in the future by the navy for an up-to-date estimate of the upper ocean circulation. The navy will use the model output for its own purposes, but does not have plans to distribute or archive the model output. If the navy is willing to release these data, it would be valuable to distribute the output in a timely manner and to archive the data.

4. Establish a Sample Program off Northern California and Baja California

It would be useful to establish CalCOFI-type field programs off northern California and Baja California. Establishing these would require local resources. It is clear that they would not need to be of the same scope as the present CalCOFI program to be valuable, but it is not clear what the minimum number of stations would need to be (especially to understand changes in macrozooplankton species structure). The CalCOFI data could be used to answer this question.

5. Improve Resolution of Annual Pattern

The annual pattern is poorly resolved by the quarterly sampling on the CalCOFI survey cruises. Additional information about the annual cycle comes from the historical data when cruises were more frequent, ship-of-opportunity cruises made by the cooperating agencies and other regional research programs, coastal shore station data, and coastal mooring data. The annual pattern could be better resolved by including other data sets in the analysis and by establishing moorings at a few selected sites.

6. Have a Meeting on CalCOFI Sampling

People at this meeting felt that there was neither time nor the right group of participants to consider the CalCOFI sample scheme in great detail and to make any recommendations for change. It was recommended that a future meeting deal with this issue. Some focused data analysis may be needed prior to such a meeting. The agenda could include (a) consideration of the present CalCOFI cruise sample scheme; (b) location and sample scheme for additional field programs (e.g., northern California, Baja California); (c) coordination of data distribution and analysis with other sample programs; and (d) location and value of moorings.

7. Better Link the Paleo-oceanographic Record to Water-Column and Current Oceanographic Conditions

There was not sufficient time at the meeting to develop specific recommendations on how to do this, but there was a strong sentiment that it is important to calibrate the paleo-oceanographic record with current water-column conditions. This may require the use of sediment traps and water-column sampling coupled with benthic sampling. It was noted that the Marine Life Research Group has recently increased its level of effort in these areas.

8. Retrospective Data Analysis

The group recognized, as have several other meetings, that much more can be learned by analyzing the existing data from the California Current and at larger scales influencing the California Current (e.g., North Pacific). One specific area which was highlighted was the need to determine if the relation between the nutrient concentration (or dissolved oxygen as a proxy) and temperature/salinity structure has varied in the past few decades. This will be important in understanding how changes in upper ocean physical structure might have affected the nutrient input and primary production.

9. Prepare a Group Synthesis Paper

Most at the meeting were surprised to learn about some aspects of the changes which are taking place in the California Current and felt that the oceanographic community is not generally aware of these changes. It was recommended that a synthesis paper be prepared for a major journal. The paper should focus upon describing change in physical, chemical, and biological (biomass and species) structure in the California Current on a range of space and time scales. This paper would be a useful background citation in grant proposals and a general introduction to the issues. There is no reason why this could not be started now.

Literature Cited

Bakun, A. 1990. Global climate change and intensification of coastal ocean upwelling, Science 247:198–201.

Baumgartner, T. R., A. Soutar, and V. Ferreira-Bartrina, 1992. Reconstruction of the history of Pacific sardine and northern anchovy populations over the past two millennia from sediments of the Santa Barbara Basin, California. Calif. Coop. Oceanic Fish. Invest. Rep. 33:24–41.

Brodeur, R. D., and D. M. Ware. 1992. Long-term variability in zooplankton in the subarctic Pacific Ocean. Fish. Oceanogr. 1:32–38.

Miller, A. J., D. R. Cayan, T. P. Barnett, N. E. Graham, and J. M. Oberhuber. 1994. Interdecadal variability of the Pacific Ocean: model response to observed heat flux and wind stress anomalies. Clim. Dynamics 9:287–302.

Pauly, D., P. Muck, J. Mendo, and L. Tsukayama. 1989. The Peruvian Upwelling System: dynamics and interactions. ICLARM Conference Proceedings 18, IMPRPE, Callao, Peru, 438 pp.

Polovina, J. J., G. T. Mitchum, N. E. Graham, M. P. Craig, E. E. Demartini, and E. N. Flint. 1994. Physical and biological consequences of a climate event in the central North Pacific. Fish. Oceanogr. 3:15–21.

Polovina, J. J., G. T. Mitchum, and G. T. Evans. In press. Decadal and basin-scale variation in mixed layer depth and the impact on biological production in the central and north Pacific 1960–88. Deep-Sea Res.

Roemmich, D. 1989. Mean transport of mass, heat, salt and nutrients in southern California coastal waters: implications for primary production and nutrient cycling. Deep-Sea Res. 36:1359–1378.

———. 1992. Ocean warming and sea level rise along the south-west United States coast. Science 257:373–375.

Roemmich, D., and J. McGowan. 1995a. Climatic warming and the decline of zooplankton in the California Current. Science 267:1324–1326.

———. 1995b. Sampling zooplankton: correction. Science 268:352–353.

Veit, R. R., P. Pyle, and J. A. McGowan. In press. Ocean warming and long-term change in pelagic bird abundance within the California Current System. Mar. Ecol. Prog. Ser.

Venrick, E. L., J. A. McGowan, D. R. Cayan, and T. L. Hayward. 1987. Climate and chlorophyll-a: long-term trends in the central North Pacific Ocean. Science 238:70–72.

Plankton Patterns, Climate, and Change in the California Current[1]

John A. McGowan, Dudley B. Chelton, and Alessandra Conversi

Introduction

The Problem

A serious problem facing oceanographers is evaluating the climate's role in the regulation of populations and communities. There can be no doubt that entire ecosystems respond to climate perturbations and that these responses can be large (Chelton et al. 1982; CPR survey team 1992; Frances and Hare 1994; Barry et al. 1995; and this symposium). While evidence for these effects has accumulated, no real theory has developed about the role of climate in population or community dynamics.

Recognizing that climate is important, most marine biologists have emphasized the role of seasonal changes in the regulation of populations. But species tend to adapt well to predictable seasonality. It is the unpredictable anomalies from seasonal norms that we do not understand and that seem to affect populations so strongly. Nor do we understand the driving forces behind such anomalies. It is in the study of climatic and biological anomalies and their interrelationships that a rich field awaits us. After all, we have pretty much known since Copernicus (in 1540) the cause of seasonality.

Our ability to define what is meant by the word *change* is at the root of the problem of understanding climate's role. Populations, communities, temperature, and water movement do vary on a seasonal basis, but they also vary on many other frequencies as well, both higher and lower. These other frequencies of change are of special interest now. Because humans have altered the mix of gases in the atmosphere, there are good reasons to believe that the heat budget of the earth may also be altered and that the consequences will be more variable atmospheric and oceanographic climates. It is expected that these changes will be evident and be felt by ecosystems sometime in the next 10 to 100 years (Martinson et al. 1995).

Defining such temporal changes and understanding their consequences to biological systems are important tasks. But such large environmental temporal changes involve large spatial dimensions as well, so large-scale spatial/temporal ecosystem studies will be necessary because effects due to local changes simply cannot be extrapolated outward to include large areas without evidence for doing so. We can

[1]Previously published in CalCOFI Report, Vol. 37, 1996. Reprinted with permission of the California Cooperative Oceanic Fisheries Investigation, University of California, San Diego, Marine Life Research Group, Scripps Institution of Oceanography, 9500 Gilman Drive, La Jolla, CA 92093-0227.

also expect that some changes in time may well include range shifts, and this is a spatial problem as well.

The issue, then, is how will climate change affect marine biological systems? How and in what direction will they respond? There are two basic approaches: (1) local, short-term, process-oriented research, and (2) long-term time-series monitoring. These need not be mutually exclusive, but usually are.

Approaches to the Climate Problem

PROCESS STUDIES. All plants and animals constantly interact with other plants and animals of the same species and other species. Many of those others are competitors, predators, prey, symbionts, or degraders. Because of close spatial/temporal species associations, many biologists have come to believe that at least some of these groupings of species are recurrent and that over time a skein of interactions between species has evolved through natural selection so that there are persistent feedback loops which determine community structure, and regulate energy flow and population dynamics. The introduction of foreign species and the subsequent observed disruption of food webs and established species structure are taken as evidence for this, as are numerous case history studies of coevolved species-species interactions.

If this concept of community is even partially true, then to fully understand the population dynamics of a single species—say the anchovy (*Engraulis mordax*)—some reference to the entire system in which it lives is necessary. But the word *community* and its companion concept *stability* are used in ambiguous ways, and there are still no clear, generally accepted, meanings (Schrader-Frachette and McCoy 1993).

Because of this vagueness of concept and because of the great complexity of even simple communities (that is, many linkages and cross-linkages between co-occurring species populations) many ecologists have chosen to make simplifying assumptions in order to study them. The most common of these assumptions is that various species populations can be aggregated into functional groups. But even here the definition of a functional group is frequently elusive. Many functional designations are synonymous with trophic level. Here the definitions are clear even if the measurements in nature—of, say, the energy flux through an entire trophic level—are at best dubious. This is especially true if the research objectives are to measure the flux through several or all the levels to some top predator, like the anchovy, in order to determine its potential yield (figure 1).

There are also practical considerations in measuring flux or even just the biomass of entire trophic levels. In order to apply the trophic-dynamic theory of Lindemann or Odum, one generally measures the flux of energy through "representative" species and assumes that the rest of that level is pretty much the same. Alternatively, the respiration of the biomass of what can be captured with a certain mesh-size net is used. But not all herbivores metabolize at the same rates even under identical conditions, and a zooplankton biomass sample may contain as many as three trophic levels.

Trophodynamic studies have not falsified many important hypotheses nor been useful in predictions nor helped us understand the role of climate. But we have learned much from the efforts to understand flux between components, the most important "fact" being the almost universal finding that these fluxes are nonlinear and density dependent. The systems are usually not known to be in steady state when measured. This, of course, makes modeling very difficult.

Ecosystem models are popular, but present models may or may not have much to do with nature. At their worst they are full of untested and untestable assumptions and preconceived ideas. At their best, they are simple quantitative approximations of what we believe we have observed about complex interactions. As such, they stimulate us to think about the potential consequences of such interactions and what might perturb them. Such models should have predictive value and lead to potentially disprovable hypotheses, although there is little evidence for this.

Many contemporary studies in biological oceanography seem to have the mostly unstated objective of providing data for filling in one or another functional role in flux models; i.e., the arrows between the boxes in ecosystem box model depictions (such as figure 1). But this rationale is often unclear. These studies are often referred to as "process oriented" or "case histories," and some of the most decipherable models are based on such observations (e.g., Ross et al. 1993). Many such process studies are reductionist in the extreme, and there are few attempts at true replication, or randomization or controls, and almost never any manipulation of the variables. The more successful of the ecosystem models, based on such studies, have managed to duplicate a seasonal biological signal, but not the sort of responses to anomalies in seasonality that are observed. Such case history or process studies have no doubt

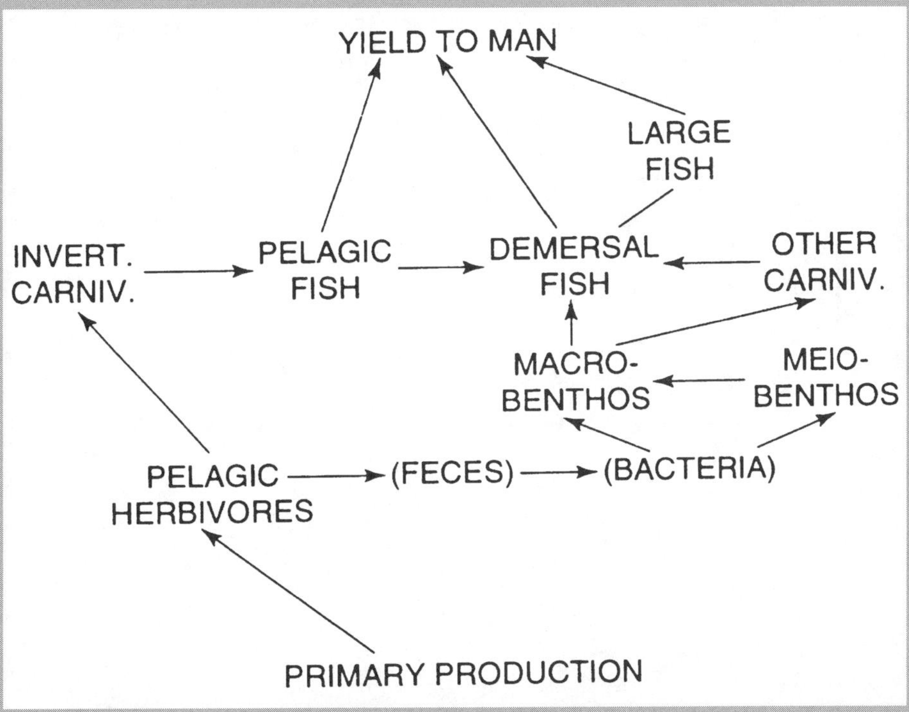

Figure 1. A typical box model of a pelagic food web in which the *arrows* indicate the transfer or "flux" of energy (usually as calories) or biomass or nitrogen or some other index. The emphasis in almost all studies based on this sort of concept is to measure one or more of the rates of transfer either directly or indirectly (as respiration, for example). Some box models are more complex, others are simpler, but very few measure all fluxes at once, so the assumption must be made that the system is in steady state (after Steele 1974).

enhanced our understanding of some of the details of how the ocean functions and have clear advantages, some of which are:

1. They describe aspects of the dynamics of parts of larger systems and are therefore mechanistic.
2. They can test some specific hypotheses such as: The clearing rate of copepods is not a linear function of food concentrations.
3. They can sometimes provide advice to managers of single-species fish populations.
4. They can sometimes circumscribe or set limits on some rates of transfer of energy or material.
5. They can sometimes provide rough generalizations such as: The rate of growth of individuals is a positive function of temperature; or Phytoplankton growth is limited by nutrients.
6. They are compatible with short funding cycles.
7. They are simple conceptually and therefore popular.

Process studies also have shortcomings, which are:

1. In the context of ecosystem response to climate change, process studies cannot define the word *change* in any meaningful way because of the short duration of the data set.
2. There is no quantitative algorithm for scaling them up in space or forward in time.
3. They are often so highly focused that the relation to the rest of the system is problematical.
4. They are often not really experimental, in the sense that there are usually no replicates, randomization, controls, or manipulation.
5. The ecosystem from which the sample is drawn is seldom in steady state, and the fluxes between boxes are almost always nonlinear.
6. They are often tautological; as such, they are derivative, repetitious questions and add little to our understanding.
7. Conclusions drawn from a process study at one location are not always applicable to other locations; for example, *Calanus* life histories in the North Atlantic versus the North Pacific.
8. There are many untidy system effects. For example, predator switching is a common phenomenon. It has been observed that exogenous perturbations can make prominent players out of system components which were not response variables in the first place.

TIME SERIES. Time series are merely the repetitive measurement of some variable over time. The objective is to determine how it varies in magnitude with respect to time in order to define the temporal scales of variability—that is, the frequency spectrum.

Apart from selecting the variable to be measured and the location, time series are designed with few prior assumptions about the behavior of the variable or what may or may not influence its frequency spectrum of variability. Time series are, or should be, empirical. Because they assume little prior knowledge of rates, directions, amplitudes, or frequencies of variation, the sampling should be as frequent as possible for as long as possible. High-frequency changes are just as likely to be the "important" ones as low-frequency under this empiricism.

The main questions asked by a time-series study have been given by Wiebe et al. (1987). What are the temporal scales of variability of basic community ecosystem properties? Are there some frequencies that show the largest and therefore the most ecologically "important" changes? If various components of the biogeochemical system interact to influence each other's magnitude, there should be detectable statistical relationships between them, even in the presence of a large amount of noise. What are the cross-correlations or coherences between system components? We need to know more about what types of atmospheric or hydrographic perturbations affect biogeochemical systems and what types do not. How and in what direction is the response to different kinds of physical events? It seems likely that the mean state of ecosystems is set by the cumulative effects of variability on many space and time scales. But if there are multiple steady states, how are shifts between them brought about? Is it necessary for forcing to change on all frequencies for a shift from one system state to another?

Can a long time series of measurements of the products of nonlinear interactions help us understand the limits of predictability and modeling? Predictions based on models of complex systems about which we have only limited knowledge have not enjoyed great success. But the product of these flexible webs of functional rates, namely changes of the main components of community structure with respect to time, can be easily measured. Changes in structural state can be quantitatively described. Testable null hypotheses can be proposed. We can then construct a sort of inverse model, not based on "first principles" (whatever they are) but on empiricism: what kind of a structure and what kinds of dynamic interactions can result in the system we have observed?

As with the process study approach, there are advantages and shortcomings with the time-series approach. The advantages are:

1. The research is empirical.
2. This research can define what is meant by the word *change*.
3. Dominant frequencies of change can be determined. Frequency spectra can be examined.
4. Time series are on the correct scale for studying ecosystem change on interannual and decadal time scales.
5. If a number of properties at a number of locations have been measured, spatial coherence information can be determined.
6. With a number of well-spaced stations, spatial averaging can be done, thus allowing a smoothing of noisy, high-frequency spatial patchiness.
7. Such series can generate and test higher-order null hypotheses such as: Zooplankton productivity does not respond to changes in coastal upwelling index.
8. Empirical results obtained from time series provide important guidance for designing and developing process studies and ecosystem models. The data can also be used to validate or invalidate models.

The disadvantages of time series are:

1. The results of time series studies provide little information on the intermediary mechanisms of change.

2. Insight into cause and effect depends on correlations.
3. If correlations "fail" there are usually no explanations.
4. Time series are tricky; they can easily be done improperly and are often complicated by practical considerations. If there are frequent gaps or changes in frequency of sampling or if the frequency of sampling is much less than the frequency of variability, the data can be badly aliased, which can lead to misleading results. Single stations do not necessarily tell us about spatial extent of variability or change.
5. Time series require sustained effort over a long period of time.
6. Collecting time series is routine and uninteresting until enough time has elapsed so that natural phenomena can be resolved.

In spite of these shortcomings, time series remain the only way to obtain information on ecosystem variability, and they almost always reveal important new and unexpected results.

The California Current

The California Cooperative Oceanic Fisheries Investigations (CalCOFI) program was designed in the late 1940s as a time series, chiefly to study the Pacific sardine and its physical and biotic habitat. There were preconceived notions in the selection of variables to be measured and the time and space scale on which to measure them, but these had little or nothing to do with an assumed architecture or dynamics of the system. The program was driven mostly by practical considerations—the decline of the sardine, and what time-space resolution could be afforded that was consistent with understanding the habitat, i.e., the problem posed by Stommel (1965). A grid plan was selected and adhered to; methods, while changing somewhat over time, were intercalibrated and found to be reasonably consistent.

Several of the many advantages of the CalCOFI plan are that it covers a large area that includes most of the spawning area of the sardine and some other fish; important biogeographic and physical features can be seen; and large-sector spatial averaging of the time series can be used to generalize about real temporal changes.

This grid plan is well known (figure 2), as are the long-term spatial means of temperature, salinity, density, and dynamic heights (figure 3). These show considerable spatial variability and strongly imply that time series and spectral analyses are called for in order to investigate the temporal nature of this variability.

Spatial Patterns

The main purpose of the net tows conducted since 1949 by CalCOFI was to obtain samples for estimating the patterns of abundance of fish eggs and larvae. Since sampling zooplankton at the same time could hardly be avoided, a very large time series of zooplankton has accumulated. Figure 4 shows the number of plankton tows per 1° square over the period 1950–88. Each of these is accompanied by a hydrographic cast to at least 500 m, so that relations between biological and physical variables can be examined on the same frequencies. Although it is clear that there have been wide variations in the intensity of measurement per 1° square, a very large

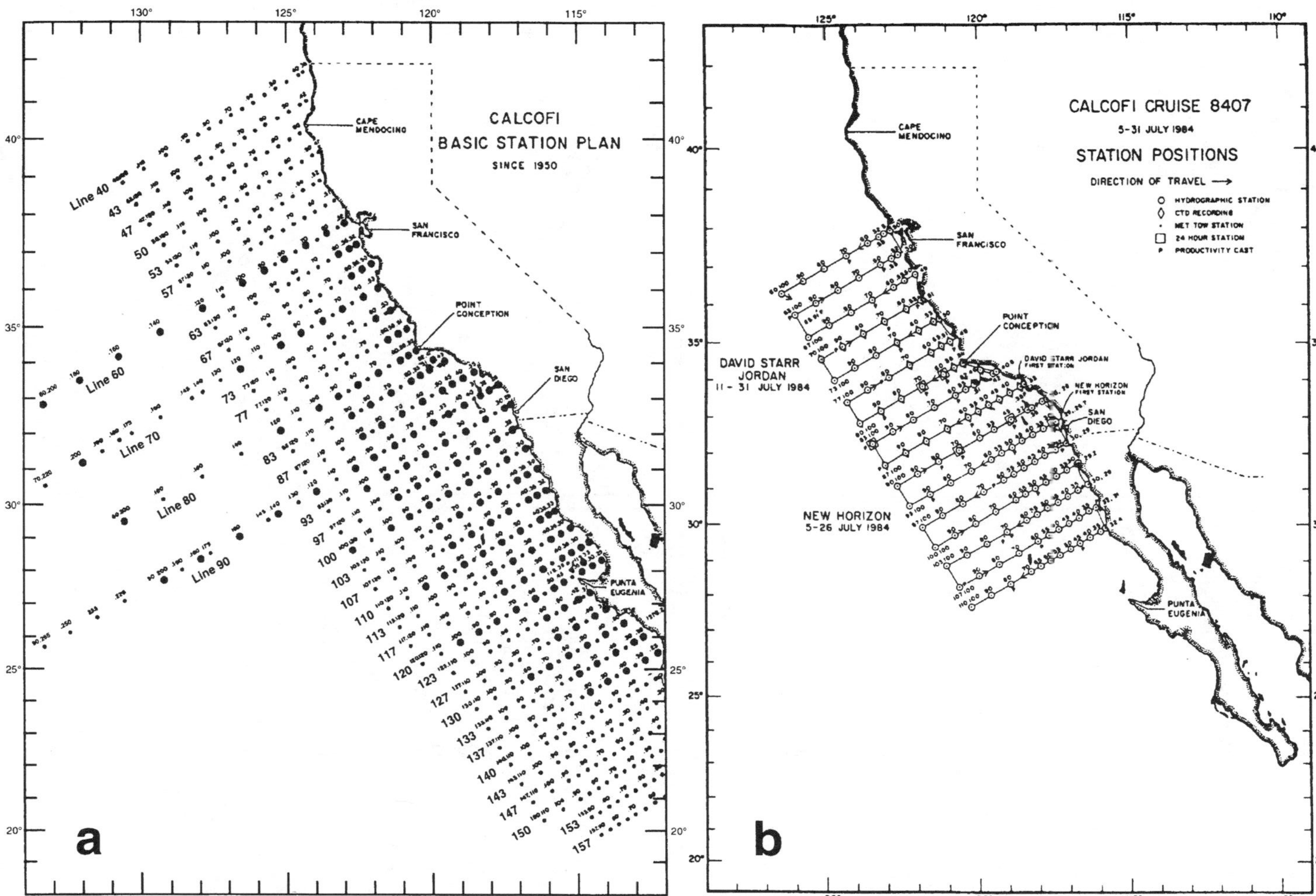

Figure 2. *a*, The basic CalCOFI plan. No one cruise covered this area (*small dots*). The most frequently sampled stations (1950–79) are shown also by *large dots* (after Lynn et al. 1982). *b*, A typical cruise pattern from the 1970s and early 1980s, showing the type of sampling done on each station (SIO Cruise Report).

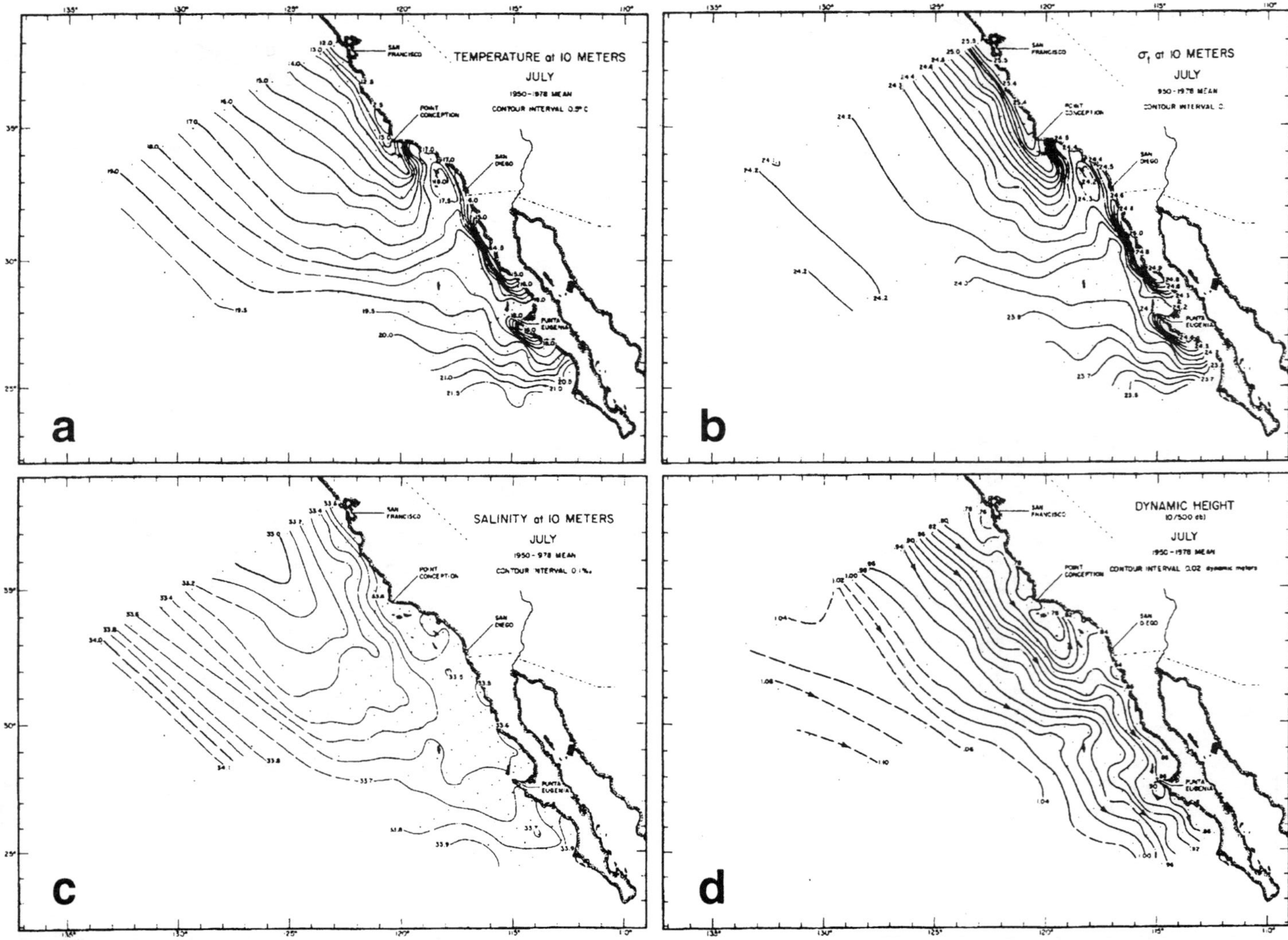

Figure 3. *a*, The 1950 to 1978 mean 10 m temperature for the summer season (July–Sept.). Longshore and onshore-offshore gradients are evident. After about 1977 the California Current warmed significantly. *b*, The 1950 to 1978 mean summer salinities at 10 m. *c*, The mean 10 m summer density patterns, 1950–78. *d*, The mean summer dynamic heights 10/500db for summer. Separation of isopleths indicates speed of current (close together, faster; widely separated, slower). There is a longshore jet in midstream (*a–d* from Lynn et al. 1982).

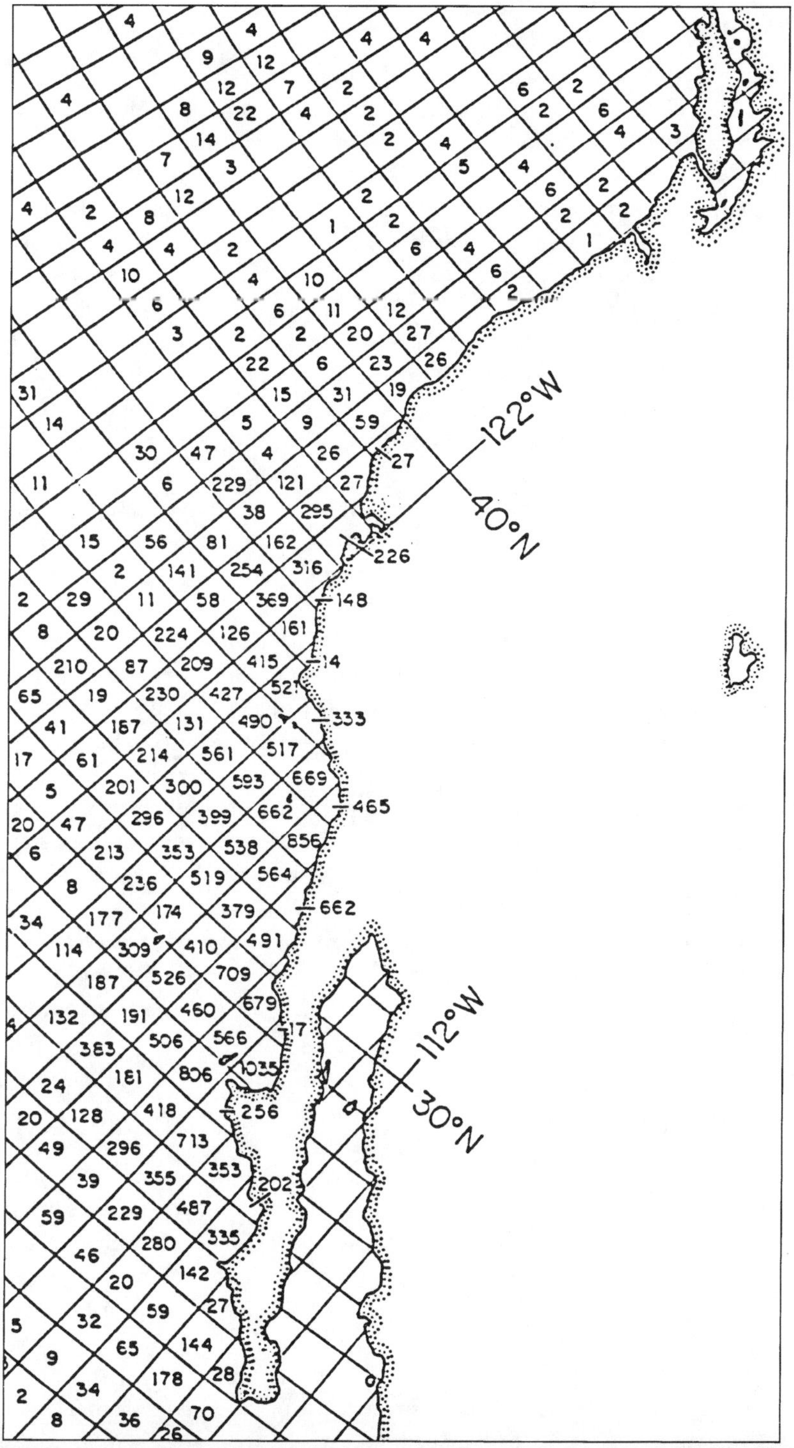

Figure 4. The number of large oblique zooplankton tows per 1° square 1950–88 (see Kramer et al. 1972 for methods).

number of measurements do occur in the central sector (the main area of fish spawning), especially on lines 80 and 90.

The long-term spatial mean of zooplankton biomass (figure 5) shows an extraordinarily patchy field despite the long-term temporal averaging. This seems especially true where the sampling intensity is high. Further, most of the plankton is concentrated in a relatively small fraction of the time-spatial domain. But it is clear that there is a north-south gradient off most of California and a well-defined offshore maximum south of Point Conception. The low values off Oregon and Washington may be due to sparse sampling in that region.

In an effort to smooth the data further we used the coefficient of variation (SD/*X*) since the magnitude of the variance is dependent on the mean in this case (figure 6). This did smooth the patchiness somewhat and also had the effect of shifting the main area of apparent high spatial variability farther south. The frequency distribution of zooplankton biomass (figure 7) shows many large outliers 4 to 10 times the overall mean. These must represent very dense patches which are episodic in space and time, thus the failure of our averaging and coefficient of variation to effectively smooth the long-term data. This seems clear evidence of the highly variable nature of zooplankton and strongly implies that large but episodic and spatially limited ecological events are features of the California Current System.

The area of high biomass variability is also one of highest species diversity (McGowan and Miller 1980). It is here that species whose populations are continuous with those of the Subarctic, the Transition Zone, and the Central Gyre, and with warm-water cosmopolites mix together (figures 8, 9, 10). This zone of high variability in biomass and diversity (figure 11) is almost certainly due to physical mixing of waters and their faunas from different source regions. Species dominance shifts in the plankton here occur very rapidly and over very short distances as compared to the Central Gyre (McGowan 1990). Such rapid and large shifts in rank order of abundance are unlikely if competitive displacements are the main cause. It seems much more likely that simple physical mixing of water parcels from different sources with different faunas is the cause behind the nonsignificant species-tank correlation coefficients between samples from this area (McGowan and Walker 1993).

Interannual Patterns

The mean spatial distribution of variables shows the California Current System to be very complex, especially the zooplankton biomass field. But there are reproducible patterns, especially on the large spatial scales. There are also patterns in temporal variations (Bernal 1979; Bernal and McGowan 1981; Brinton 1981; Chelton et al. 1982; Roesler and Chelton 1987; Roemmich 1992; Roemmich and McGowan 1995).

The seasonal temperature signal in the mixed layer was well resolved by the early period of the CalCOFI surveys, when cruises were at approximately monthly intervals (figure 12). It was, and still is, a strong and unambiguous signal. But as the time series progressed, it became evident that there were strong interannual variations as well. Standard deviations of 1.5°C in the monthly mean temperature were common (figure 13). A large El Niño occurred in 1958–59 (Sette and Isaacs 1960). As further low-frequency events occurred, these temperature standard deviations became larger. Eventually it became apparent that a long-term trend of increased

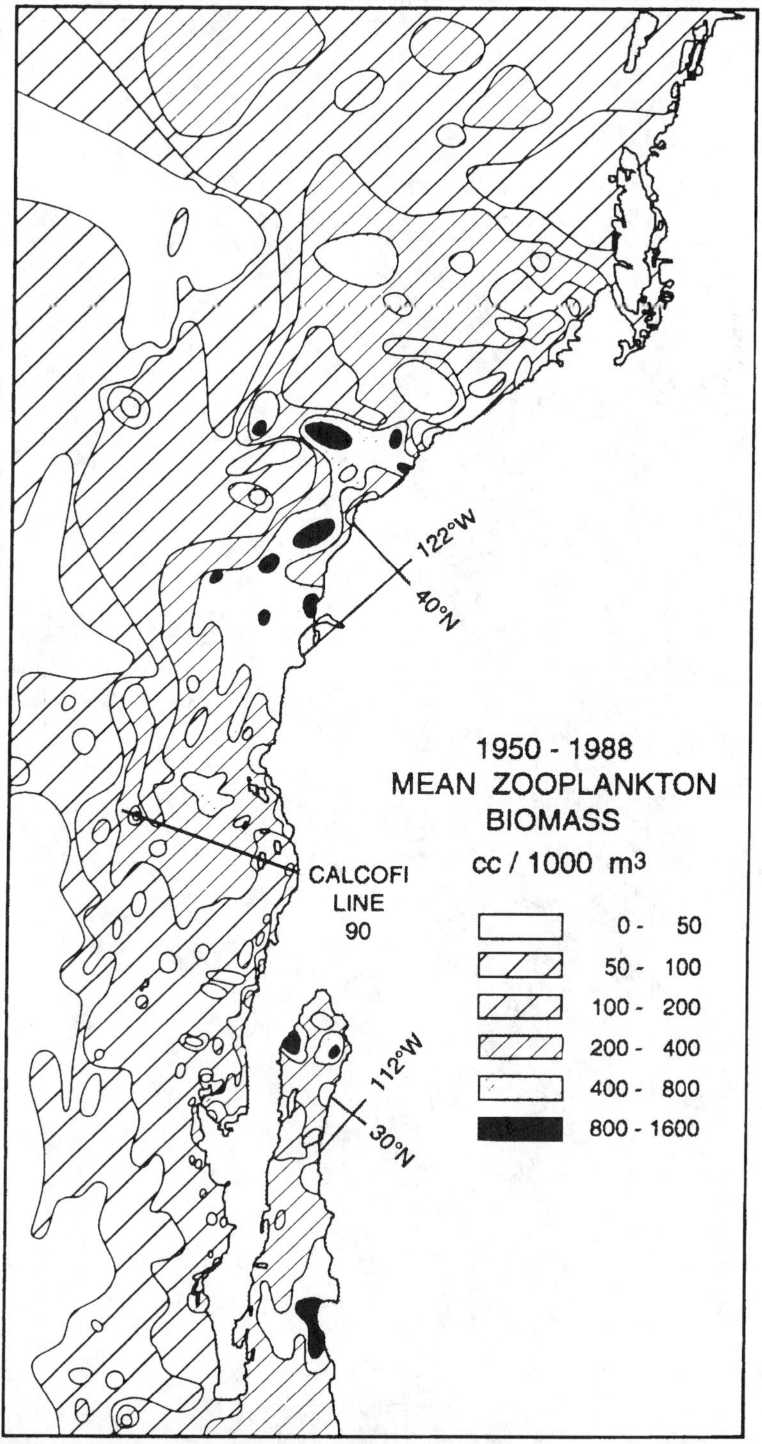

Figure 5. The long-term mean zooplankton biomass. There are onshore-offshore gradients and north-south gradients.

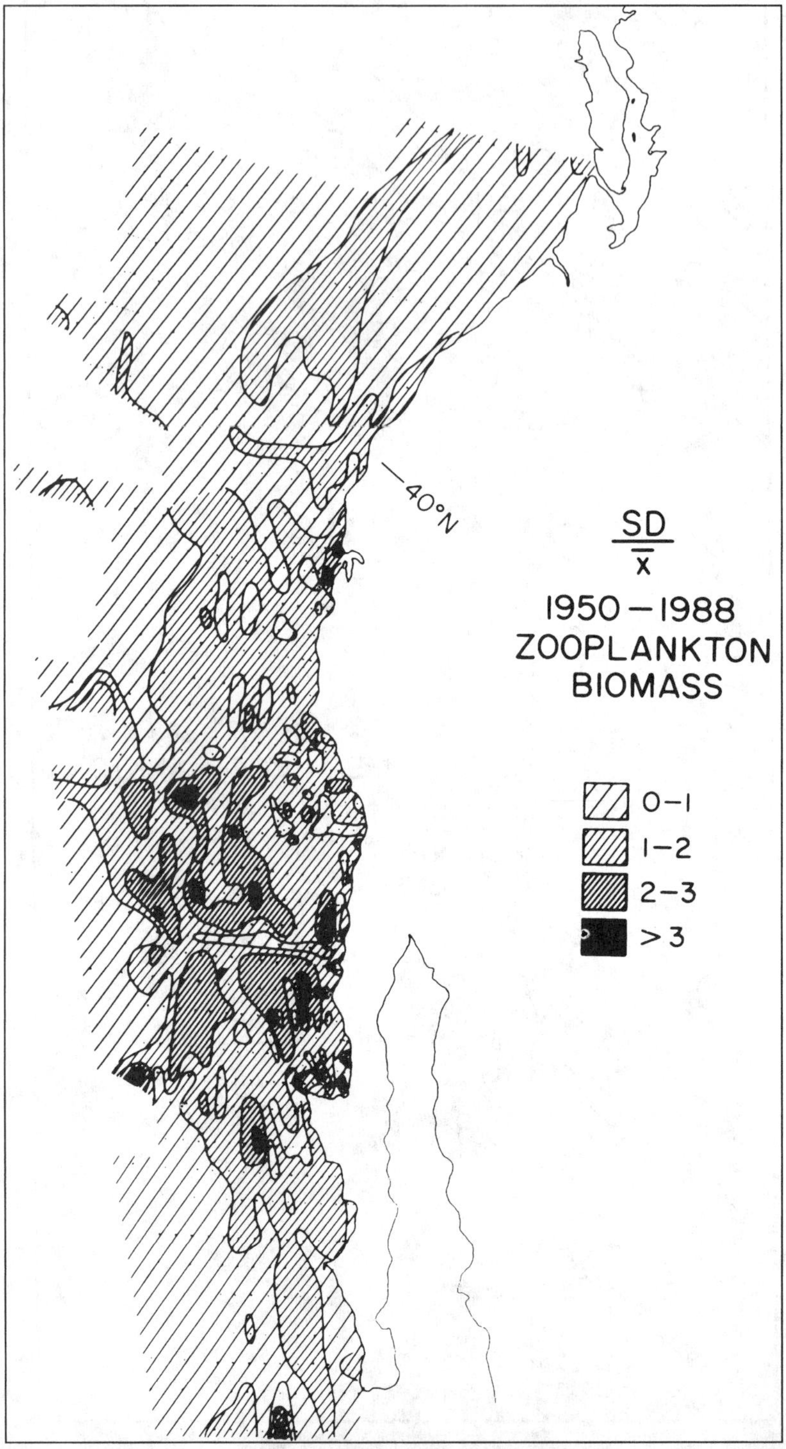

Figure 6. The index of variation (standard deviation divided by the mean) of the zooplankton biomass data in figure 5.

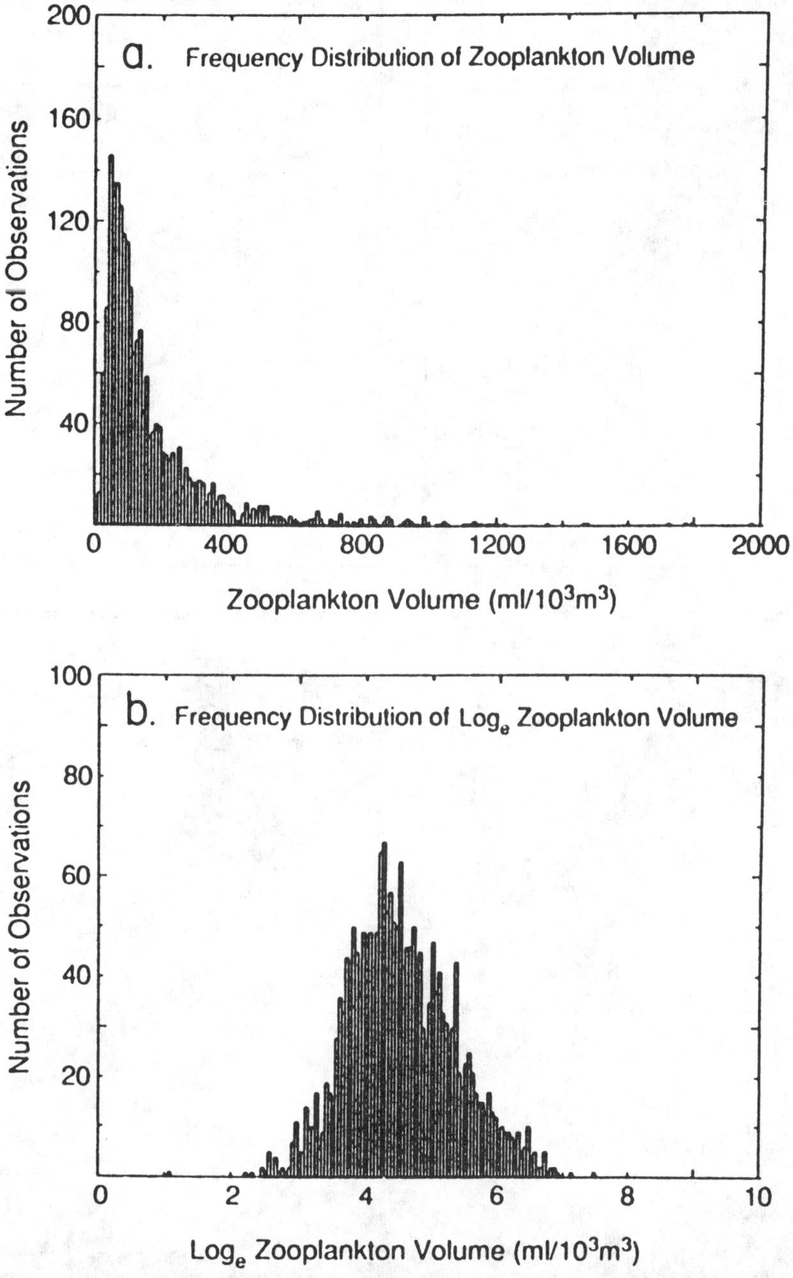

Figure 7. *a*, The frequency distribution of CalCOFI zooplankton biomass estimates; *b*, $\log_e$-transformed data.

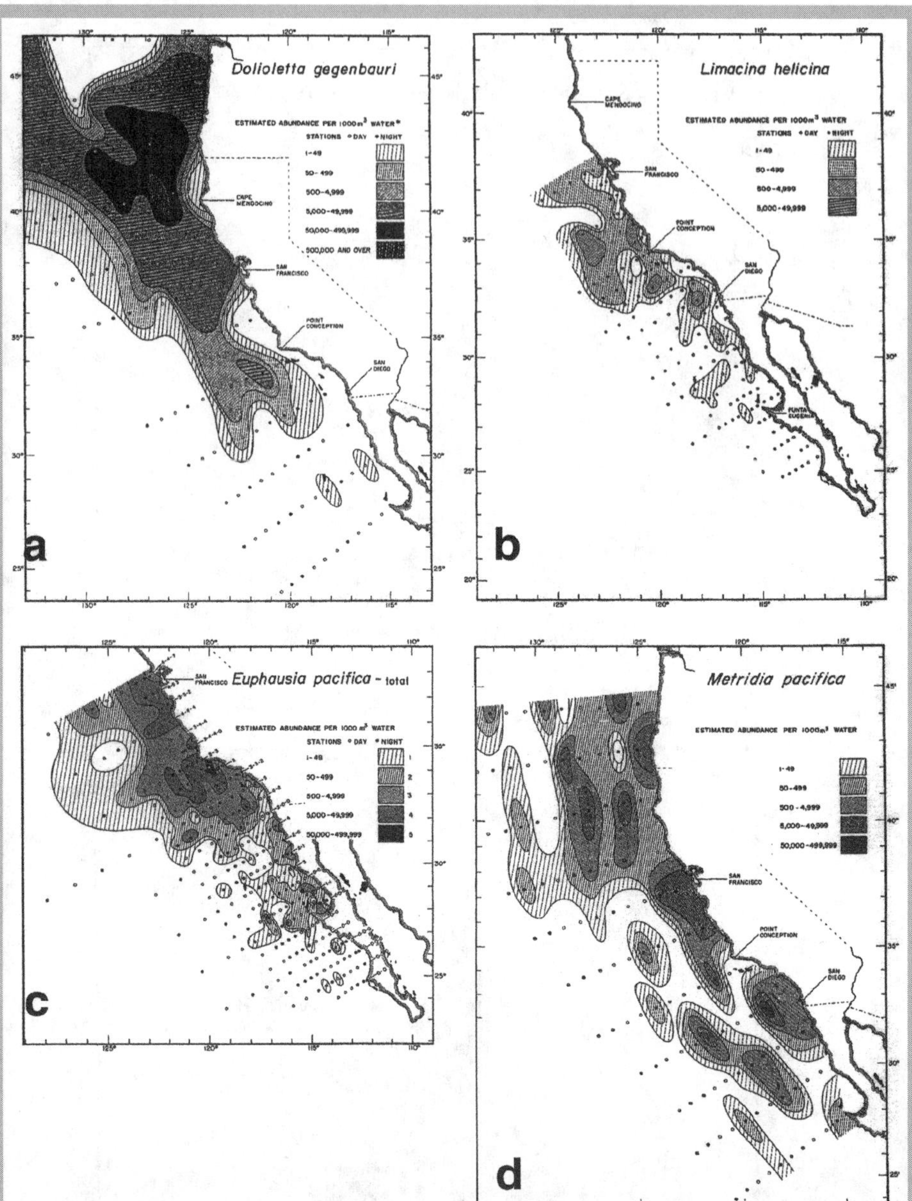

Figure 8. The distribution and abundance of four species of zooplankton that are common in the Gulf of Alaska and the Subarctic North Pacific. *a*, *Dolioletta gegenbauri* from cruise 4907 (from Berner 1967); *b*, *Limacina helicina*, a pteropod from cruise 5204 (from McGowan 1967); *c*, *Euphausia pacifica*, a euphausiid from cruise 4907 (from Brinton 1967); *d*, *Metridia pacifica* from cruise 4909 (from Bowman and Johnson 1973).

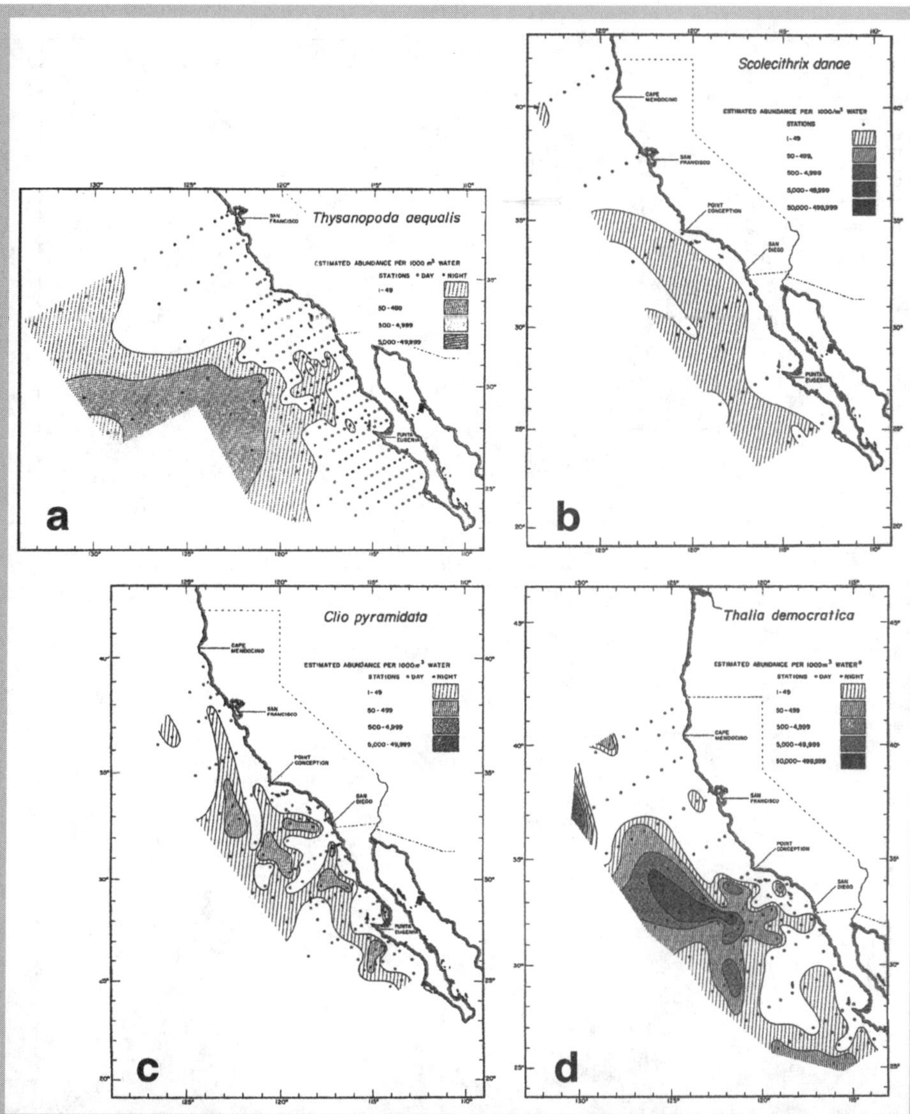

Figure 9. The distribution and abundance of four species that are common in the Central Gyre of the North Pacific and intrude into the California Current; from the west. *a*, *Thysanopoda aequalis*, a euphausiid from cruise 6304 (after Brinton 1967); *b*, *Scolecithrix danae*, a copepod from cruise 5807 (after Fleminger 1964); *c*, *Clio pyramidata*, a pteropod from cruise 5206 (after McGowan 1967); *d*, *Thalia democratica*, a thaliacean from cruise 5005 (after Berner 1967).

temperature of the mixed layer had occurred, but this required about 40 years of data in order to be detected unambiguously (Roemmich 1992). The entire upper layer of the area surveyed up to 1992 had warmed by at least 0.8°C between 1950 and 1990.

Early in the program Reid et al. (1958) showed a negative relationship between zooplankton biomass and 10 m temperature over time. Perhaps because his data

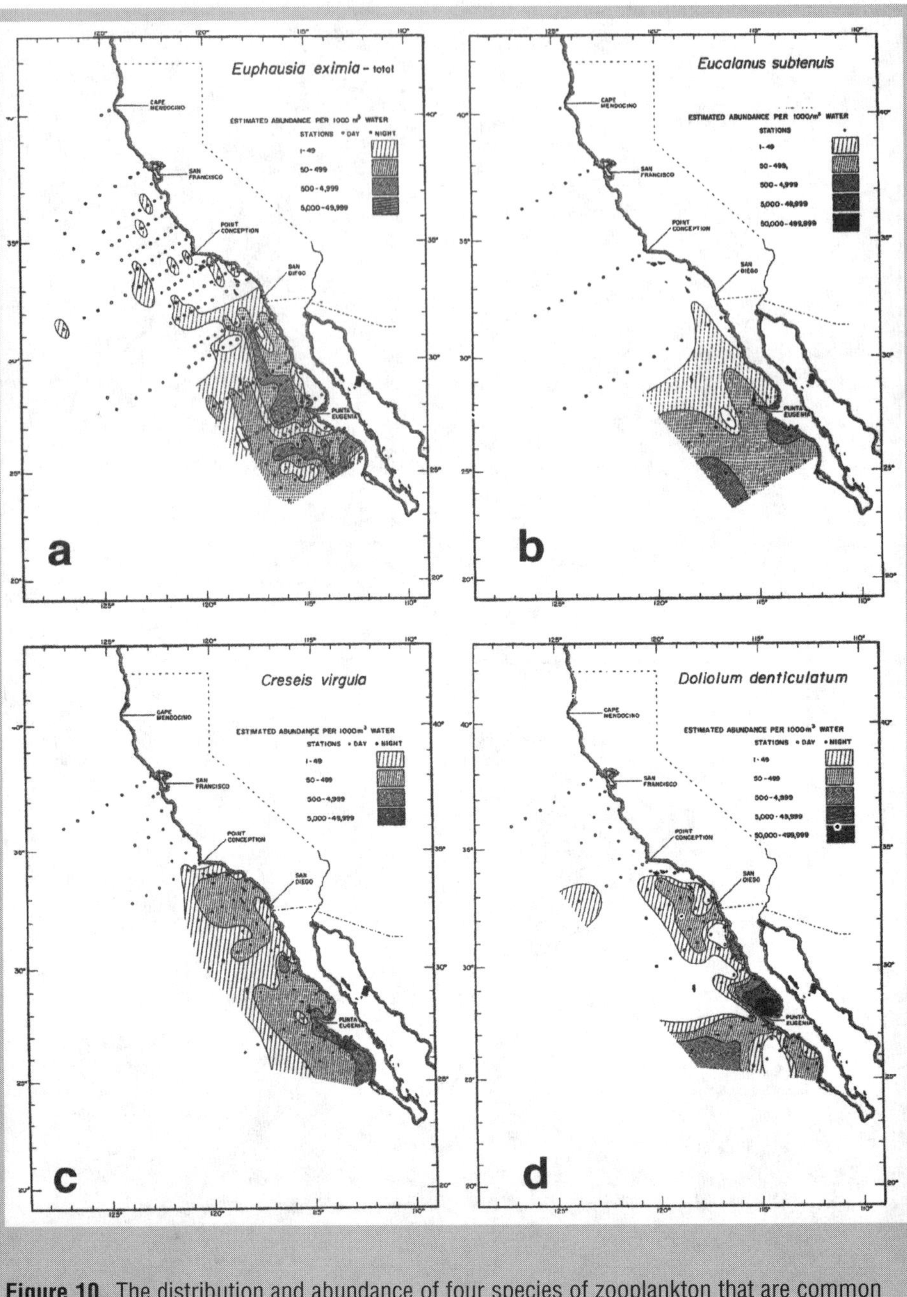

Figure 10. The distribution and abundance of four species of zooplankton that are common in the eastern tropical Pacific and intrude into the California Current from the south. *a*, *Euphausia eximia* from cruise 5804 (after Brinton 1967); *b*, *Eucalanus subtenuis* from cruise 5804 (after Fleminger 1964); *c*, *Creseis virgula* from cruise 5210 (after McGowan 1967); *d*, *Doliolum denticulatum* from cruise 5210 (after Berner 1967).

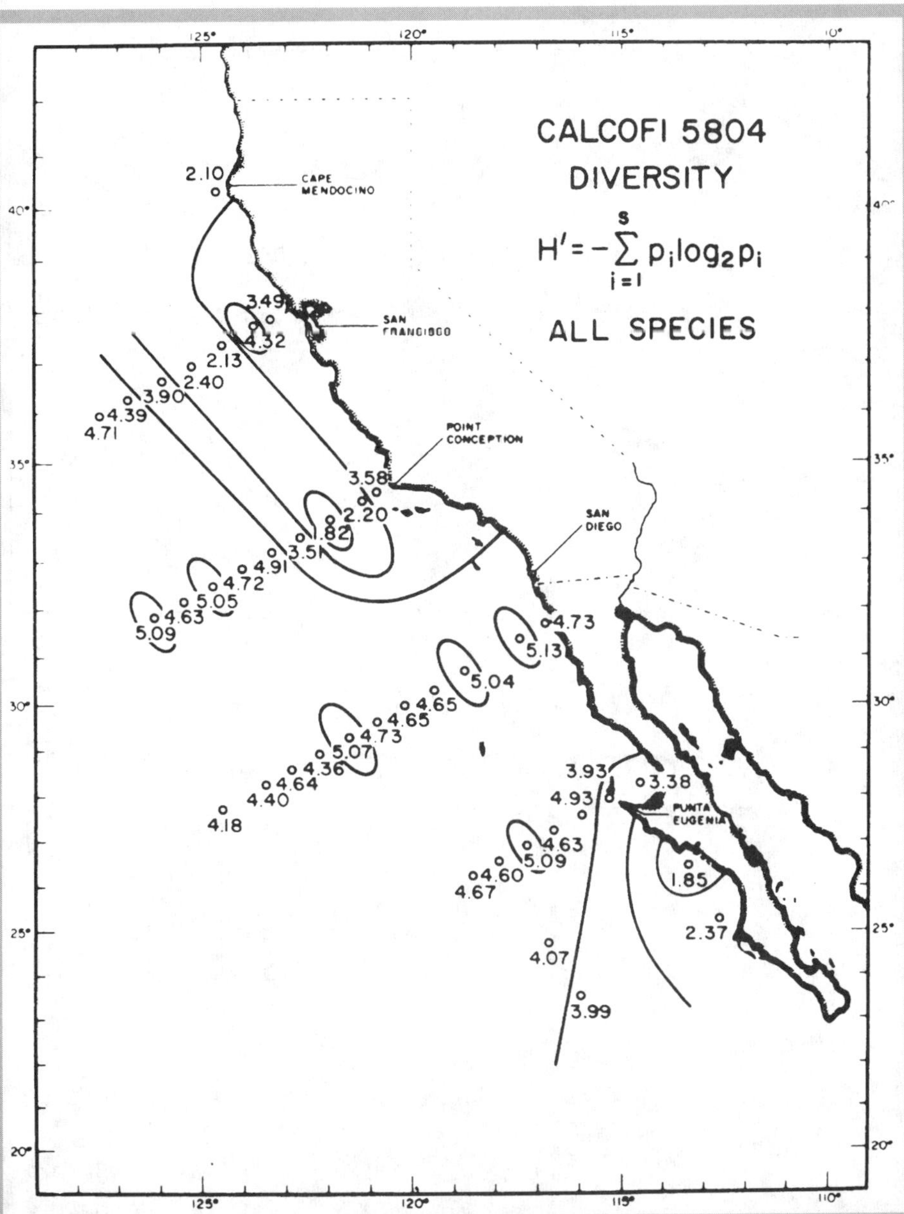

Figure 11. The spatial pattern of zooplankton species diversity in the California Current, from cruise 5804. Over 100 species were used in this study (from McGowan and Miller 1980).

set was limited, this observation did not receive much attention, and ecosystem modelers continued to use temperature increase to drive higher production. With the much larger data set now available, we can confirm Reid's point (figure 14). Experimentalists have shown that the growth rate of individuals increases with temperature. This observation has led many to believe that somehow population growth also increases with temperature because larvae and juveniles pass through their vulnerable early stages more rapidly and therefore survivorship is enhanced. The

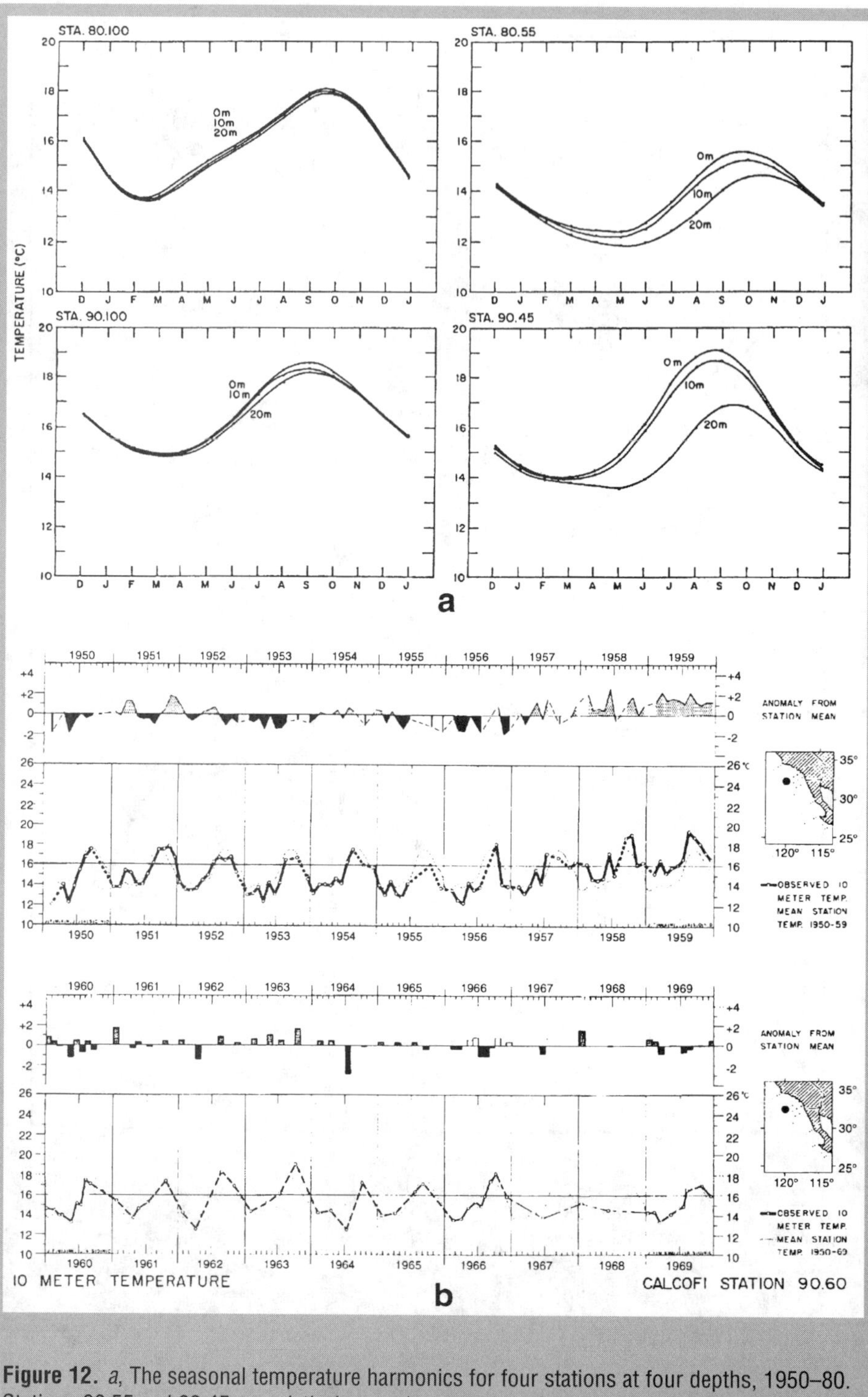

Figure 12. *a*, The seasonal temperature harmonics for four stations at four depths, 1950–80. Stations 80.55 and 90.45 are relatively nearshore; stations 80.100 and 90.100 are offshore. See figure 2 for locations. *b*, The 10 m temperature and temperature anomalies at station 90.60 in the Southern California Bight (see inset) for 1950–69 (after Anonymous 1963 and Wyllie and Lynn 1971).

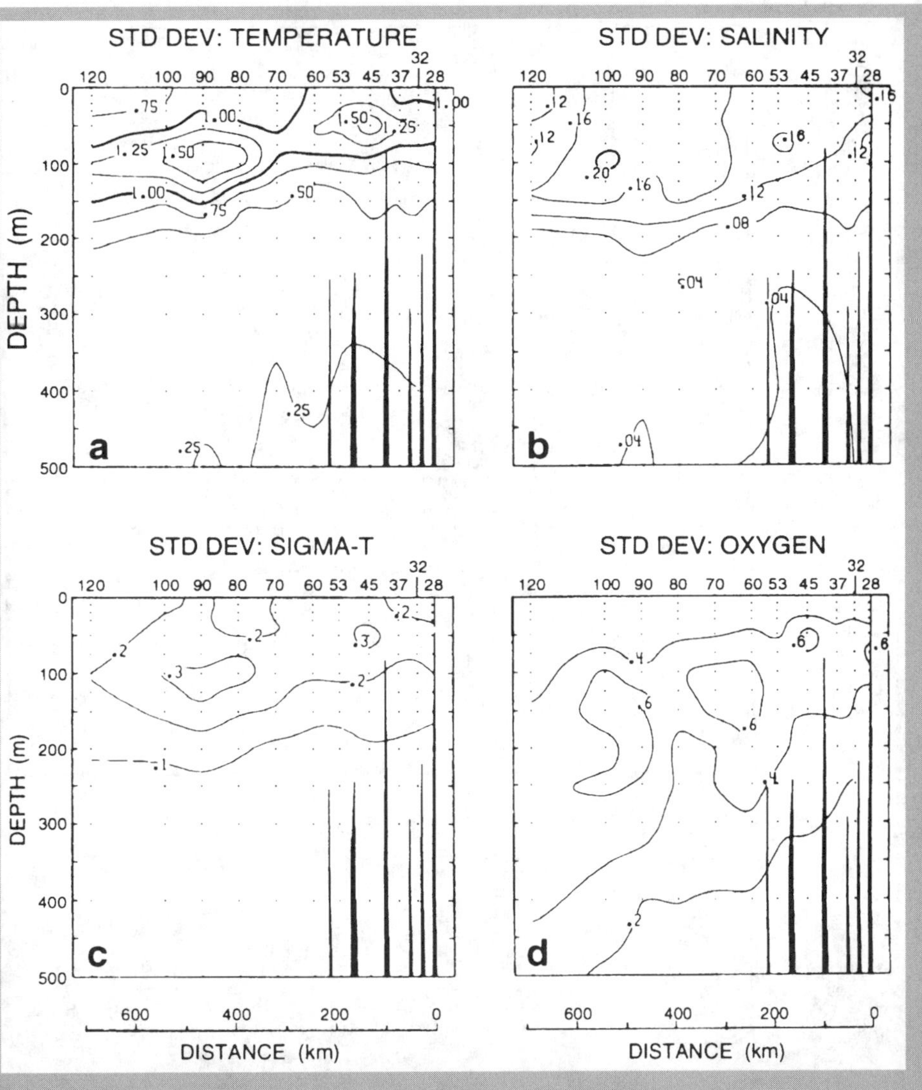

Figure 13. The standard deviation of temperature, salinity, density, and oxygen along line 90 (1950–76, after Eber 1977).

CalCOFI temperature biomass data indicate that this hypothesized relationship is not the dominant mechanism controlling zooplankton abundance in the California Current.

Whereas zooplankton biomass and larval fish have seasonal variations of abundance (figure 15), Chelton et al. (1982) have shown that there are large interannual variations as well and that the magnitude of interannual variability is comparable to or larger than that of the seasonal variability. It is quite clear now that zooplankton biomass and larval fish abundance in the California Current System are both dominated by interannual variability (figure 16). The nonseasonal variations (anomalies from the seasonal signal) are of particular interest since (unlike the causes of the seasonal cycle) the causes of interannual variations are not yet well understood. These

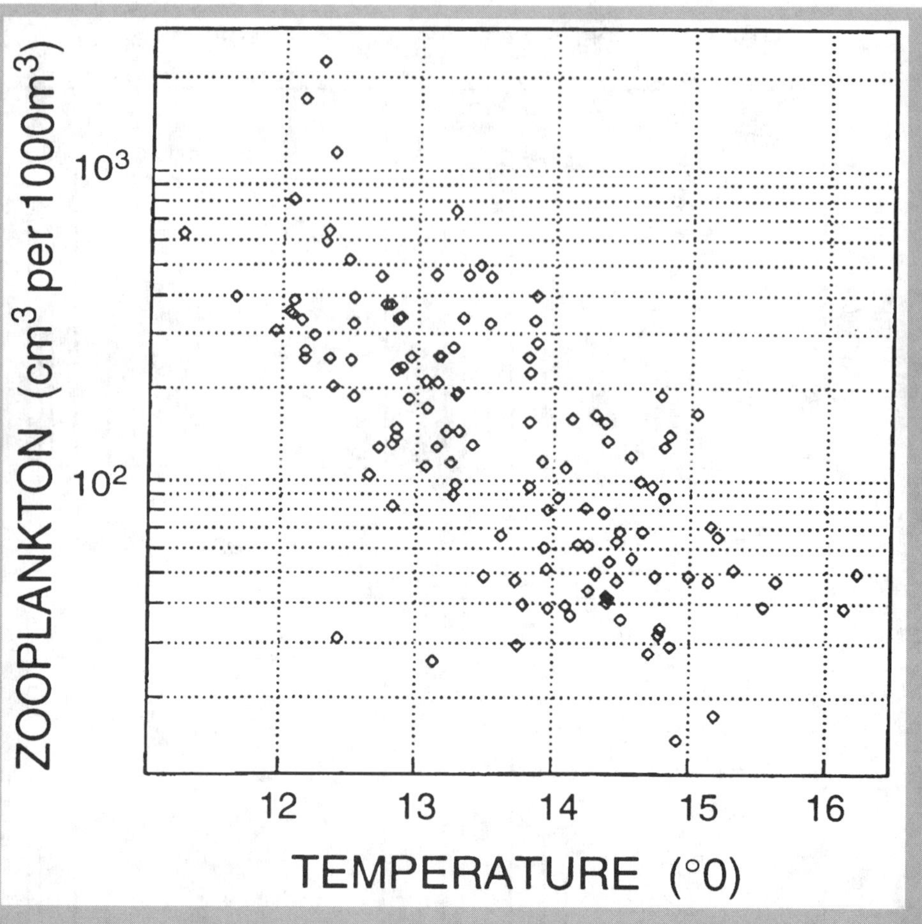

Figure 14. Average California Current zooplankton biomass per cruise, 1950–93, from stations 30 to 120 on line 90 (see figure 2) and the average temperature per cruise of the upper 100 m on that line. There were 10 or 11 stations on this line per cruise.

interannual variations are indicators of short-term climate changes in the California Current System.

Chelton et al. (1982) have examined lower-frequency changes in both hydrography and plankton. They found that interannual anomalies in temperature, salinity, zooplankton, and mass transport from the north rise and fall together throughout the entire system. The anomalously high zooplankton biomass episodes were correlated with colder, low-salinity episodes which occur during periods of anomalously high transport from the north (figure 17).

The high productivity of the California Current System is a well-known feature in reviews and textbooks, and this richness is almost routinely attributed to the fact that coastal upwelling of deeper water brings nutrients to the surface near shore, enhances primary production, and thus enriches the rest of the food web. But direct measures of nutrient upwelling have proven elusive, and the surrogate often used by biologists is the Bakun index of coastal upwelling (Bakun 1973). This index, based on Ekman volume transport dynamics, is reported in both absolute terms of

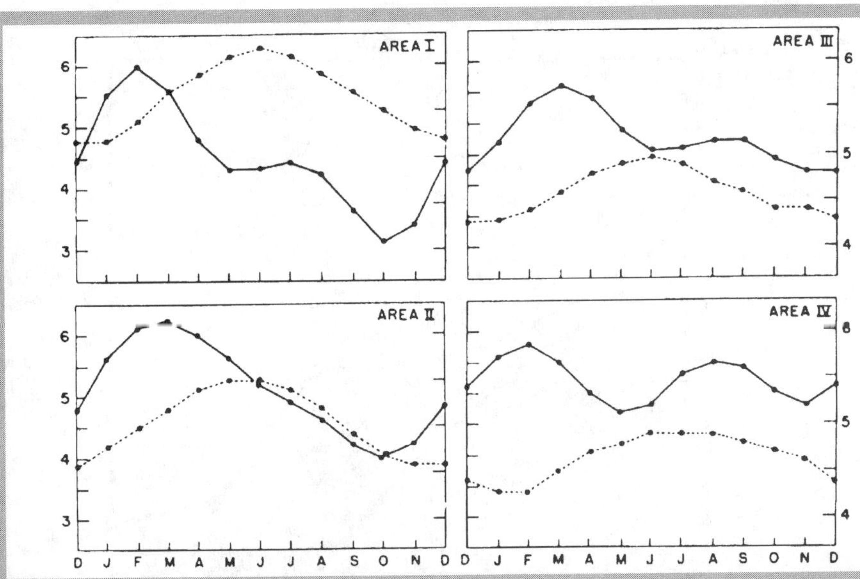

Figure 15. Seasonal cycles of zooplankton displacement volumes in 10^{-3} ml/m^3 (*dashed lines*) and larval fish in units of 10^{-3} individuals/m^3 (*solid lines*) 1950–70. The seasonal cycle is defined here to be the sum of an annual and a semiannual harmonic and a constant offset. The amplitude and phases of these three constituents have been determined by least squares regression.

"metric tons per second per 100 m of coastline" and in anomalies. Thus the null hypothesis can be stated: H_0: "Nonseasonal zooplankton biomass anomalies are driven by coastal upwelling." This hypothesis was disproved from time-lagged cross-correlations (figure 18; Bernal and McGowan 1981). This correlation failure is largely due to the fact that the time scales associated with the physical and biological fluctuations are much longer than those of the local wind stress used in calculating the upwelling index.

An alternative null hypothesis is: H_0: "Zooplankton biomass anomalies are driven by mass transport from the north." This hypothesis could not be disproven, because the cross-correlations between transport anomaly and zooplankton are positive and significant (figure 18). Whatever the climatic forcing responsible for interannual changes in the mass transport from the north (otherwise known as the strength of the California Current), they clearly are associated with interannual zooplankton biomass variations off central and southern California. Some of these variations are associated with episodic El Niño events, but some are not.

The nature of the underlying causal mechanisms of the changes in advection are probably quite complex and are not yet understood, but they seem clearly related to climatic changes. Roesler and Chelton (1987) have more closely examined this biological response to physical forcing. They suggest that the timing of the biomass variations relative to variations in southward advection indicates that they are controlled by two processes: (1) the response of local zooplankton to advection of biomass—the dominant process in the north—and (2) the response of local populations to changes in local conditions caused by changes in advection. This latter

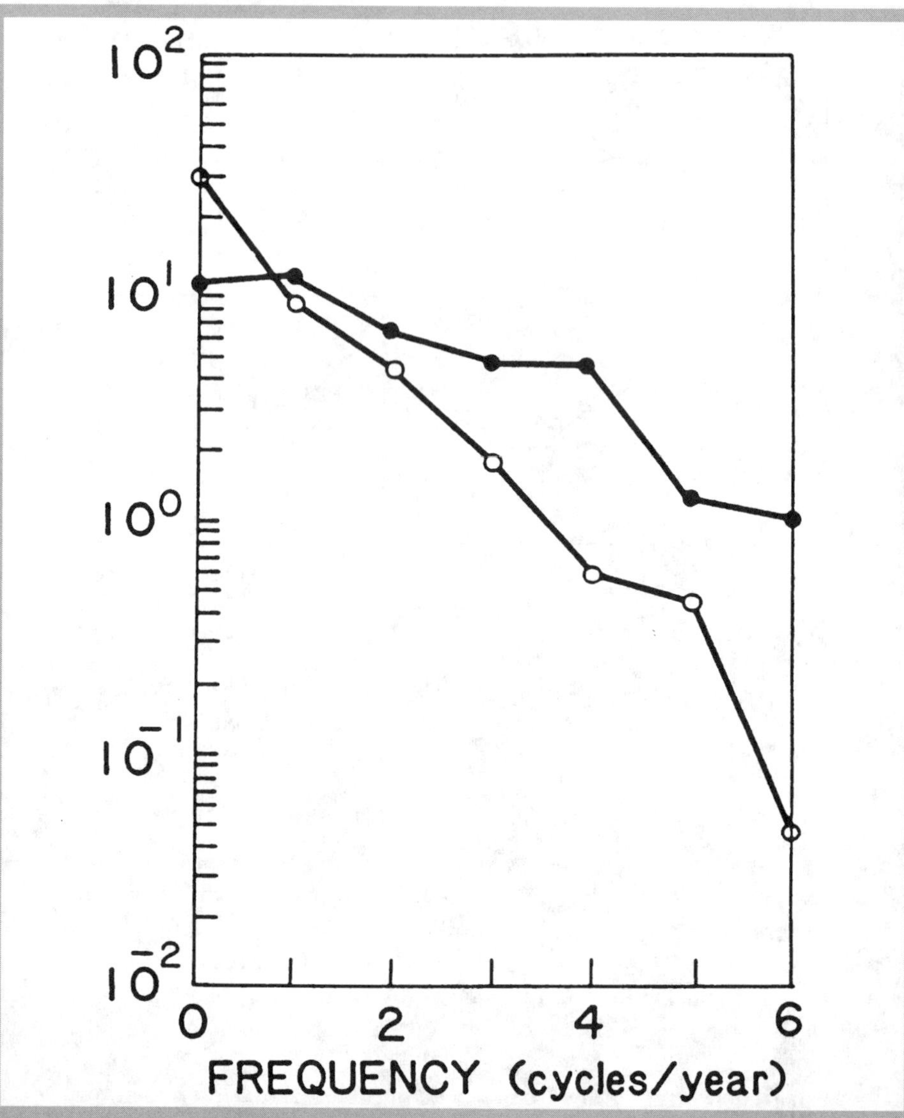

Figure 16. Normalized power spectrum, showing the average, nonseasonal $\log_e$-transformed zooplankton biomass (*open circles*) and the nonseasonal $\log_e$-transformed larval fish abundance (*solid circles*). The fish power spectrum is much more regular than the zooplankton spectrum, and there seems to be a residual broad-bandedness. The small peak of one cycle per year in seasonally corrected fish data may be due to fish having a sharply peaked spawning period but one that varies from year to year.

process becomes increasingly dominant from north to south. Biogeographic species boundary shifts also indicate that these mechanisms control the interannual variability of zooplankton biomass.

The CalCOFI zooplankton samples were taken for the main purpose of capturing fish eggs and larvae. There is thus a 45-year record of variations in larval fish abundance accompanied by a zooplankton biomass volume for each sample. It is possible

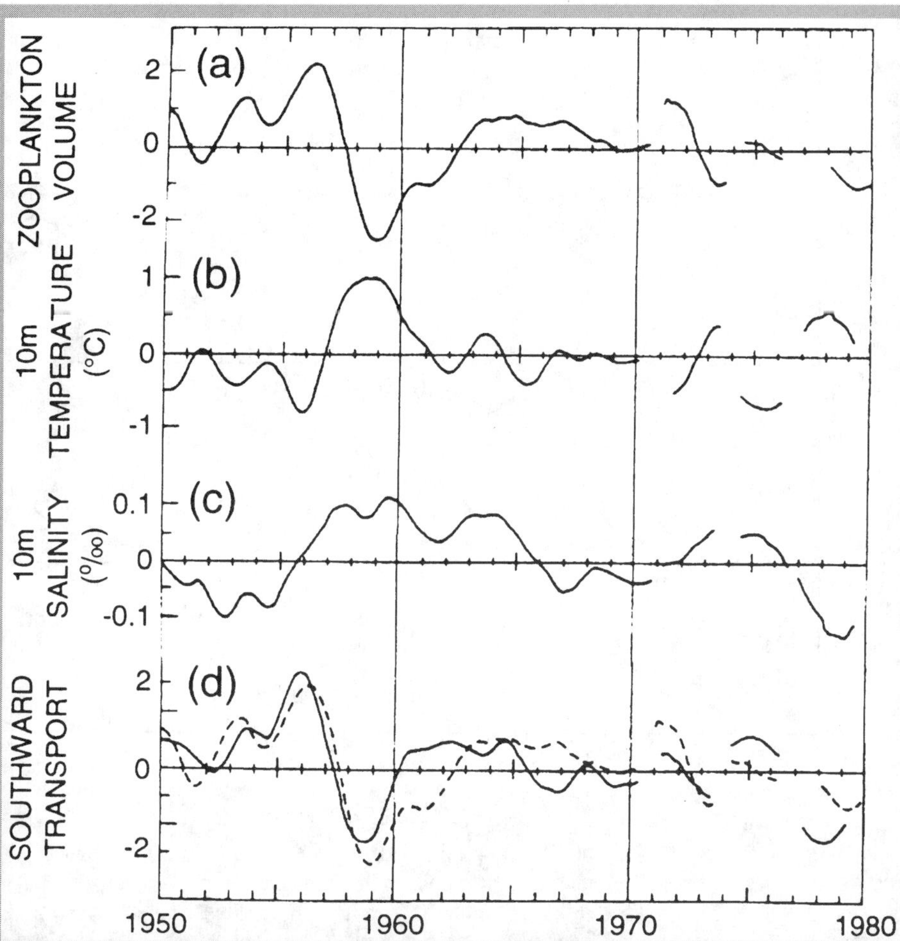

Figure 17. Time series of nonseasonal values of four variables averaged over the entire area: *a*, zooplankton in $\log_e$ were standardized; *b*, average 10 m temperature over the area; *c*, average 10 m salinity; *d*, the amplitude time series of the principal EOF of 0/500 steric height. The raw time series were filtered with a double 13-month running mean. *Dashed line* in *d* represents the zooplankton time series shown in *a* (from Chelton et al. 1982).

now to compare temporal variations in larval fish to temporal variations in zooplankton.

Theilacker (1987) and Theilacker and Shen (1993) have documented that sea-caught larval fish may often be badly undernourished, and Houde (1987), Lo et al. (1995), Kendall et al. (1996), and others have shown that the feeding larval and early juvenile stages of many—perhaps most—species of pelagic fish have larger rates of, and variations in, mortality than any other life-history stage. This variability is thought to be one of the important contributions to variations in recruitment. Since larval fish are about the same size as zooplankton, occur together with zooplankton, and frequently have similar diets, it is not unreasonable to assume that they compete with or are preyed upon by members of the macrozooplankton community. This is especially true in view of the fact that many individual species of macrozooplankton,

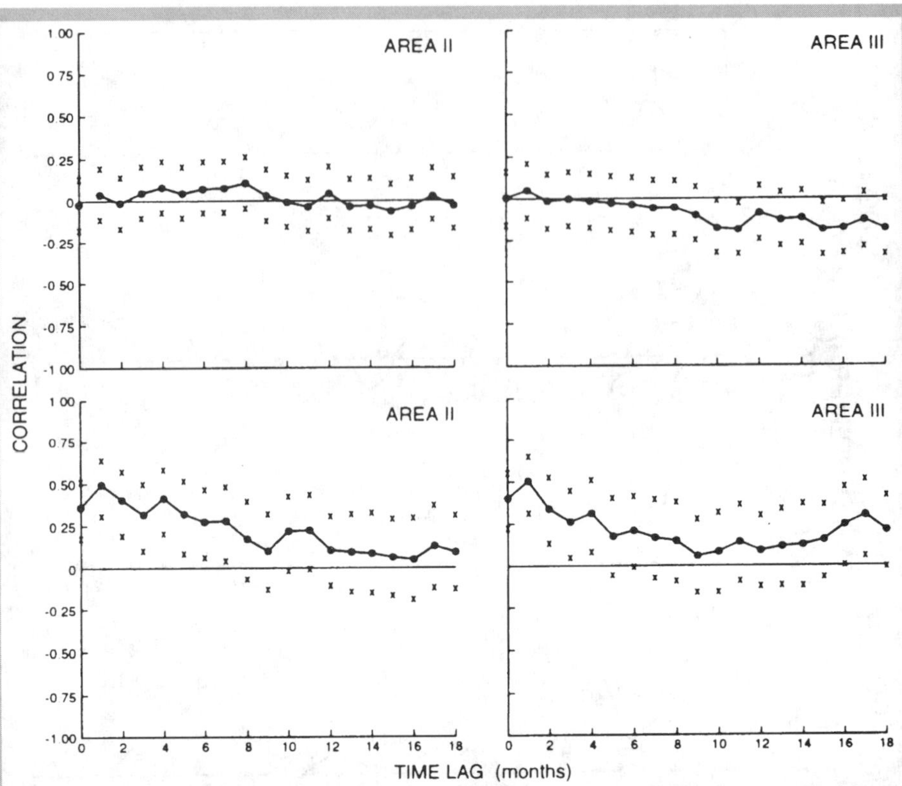

Figure 18. Time-lagged cross-correlations from 0 to 18 months. *a*, Non-seasonal Bakun's upwelling index in Area I leads nonseasonal zooplankton biomass in Area II. *b*. Nonseasonal upwelling index in Area II leads zooplankton biomass in Area III. *c*, Nonseasonal transport from the north (across line 80) leads zooplankton in Area II. *d*, Transport across line 80 leads zooplankton in Area III. *X*s are 95% confidence limits (after Bernal and McGowan 1981).

as caught by CalCOFI, are far more abundant than the co-occurring larval fish (McGowan and Miller 1980). But those same macrozooplankton are the diet of the adults that gave rise to the fish larvae. Further, the crustacean zooplankton (at least) give rise to naupliar larvae, which are the food for many fish larvae. The interactions between trophic levels are clearly very complex and nonlinear.

MacCall and Prager (1988) have estimated "Historical changes in abundance of six fish species off Southern California, based on CalCOFI egg and larva samples" by "developing annual indices of larval abundances of the northern anchovy, Pacific sardine, Pacific mackerel, jack mackerel, and Pacific whiting (hake)." These are the most abundant pelagic species in the California Current System. Three of these five (the sardine, anchovy, and Pacific mackerel) had larval abundance indices that agreed well with other estimates of variations in adult population size. The other two species have apparently shifted their spawning range to the north because of the warming trend that began in the mid-1970s. But Hollowed (1992), in an extensive study of hake abundance, has concluded that "production of strong year classes of Pacific hake was regulated by factors influencing survival during early and late larval periods."

Thus there are good reasons to suspect that variations in macrozooplankton abundance can affect larval fish populations in several ways: (1) Adult fish fecundity may vary because of changes in the food supply; (2) Macrozooplankton provide food for larval fish through their own larvae, especially crustacean nauplii; (3) Macrozooplankton may compete for food with fish larvae; and (4) some species of macrozooplankton are known to prey on larval fish. In cases 1 and 2 we expect macrozooplankton and larval fish abundance to be positively correlated, perhaps with some time lag. In cases 3 and 4 larval abundance changes should be negatively correlated with zooplankton. Large variations in larval abundance should result in variations in recruitment. Therefore large-scale interaction between trophic levels should be occurring. But it is well known that variations in the abundance of different species of harvested fish are not highly cross-correlated. That is, peak abundances occur at different times, and there appears to be a succession of species abundance over time (McCall and Prager 1988; Sund and Norton 1990; MacCall 1996). It has been suggested several times that interspecific competition for food is a possible explanation of these results. As a consequence, repeated shifts of species dominance occur (Skud 1982), although the standing crop of total "fish" might remain fairly constant.

An important question to be addressed is whether variations in the trophic level represented by larval fish are somehow associated with variations of the trophic level represented by macrozooplankton. It is quite clear that selecting an individual fish species for the test may not yield interpretable results. The same is true if we wish to test whether or not larval abundances are affected by climate. Although variations of one or several species may be correlated with zooplankton or SST or wind-speed variations, these relationships cannot be generalized. If most pelagic fish do replace one another in some sort of competitive succession, then "total larval fish" is likely to be a more appropriate variable to consider.

In view of the evidence that mortality rates are highest during larval stages, that variations in the abundance of larval fish agree with estimates of adult fish, that recruitment depends on survivorship during larval periods, and that fish larvae may often be found to be malnourished, the assumption is made here that "total larval fish" is an index of the status of the next higher trophic level above macrozooplankton. Since the abundance of total fish larvae has been measured over time (figure 19), we can now test two more hypotheses.

Of these two, perhaps the most interesting is the null hypothesis: H_0: "Fish larvae variations in abundance over time are not related to macrozooplankton variations over time in the California Current System." Here we are asking whether both fish and their larvae are food-limited . . . and whether variations in the amount of food available to adults affect the rate of production and/or survivorship of larvae and juveniles. We may test this by asking if temporal variations in larval fish abundance are correlated at any time lag with macrozooplankton over the period 1950 to 1970, when sampling was frequent and before the warming trend to be described. The time lagging is important here because, if larval fish abundance declines occur well after zooplankton, it may mean that poor nutrition of adults in the preceding months has affected their fecundity. If, on the other hand, larval declines are virtually synchronous with zooplankton declines, it is likely that the availability of nauplii is important. Note that this latter result does not really distinguish between an instantaneous response of the adult fish to poor nutrition, and food limitation of fish larvae. If the

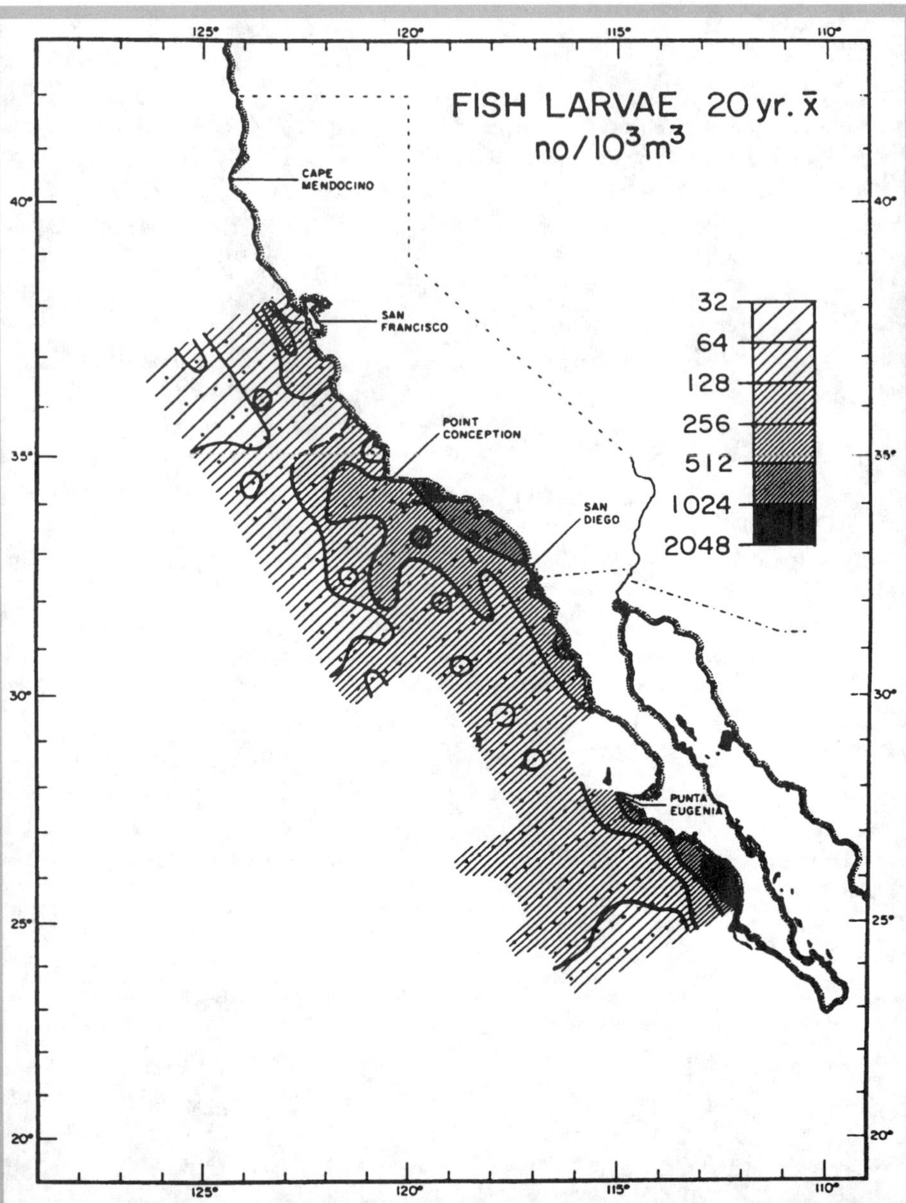

Figure 19. The spatial distribution of estimated number of larval fish of all specieis averaged per station 1950–70.

correlation tends to be negative, then perhaps competition or predation are responsible.

Since nonseasonal, total larval fish variations throughout all four sectors of the California Current System are intercorrelated at low frequencies, we can space-average the estimates throughout the entire system (figures 20–22 and table 1). Although fish and zooplankton abundance both exhibit energetic interannual variability (figure 23), there is little consistent relation between the two time series. This observation is

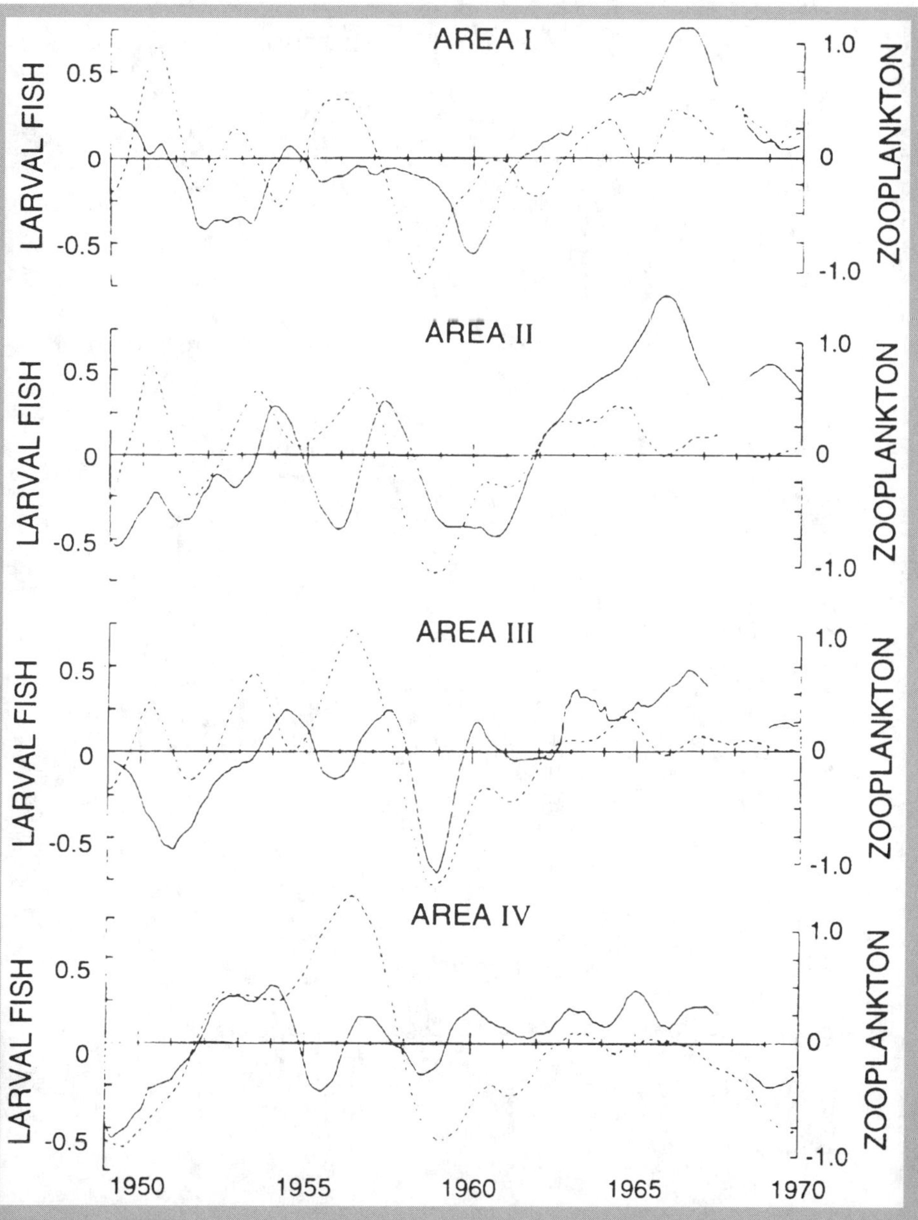

Figure 20. The stations used in the time-series analysis in figures 21, 23, 24, 25. The breaks between areas are not entirely arbitrary but are based on biogeographic and hydrographic patterns (after Bernal and McGowan 1981).

illustrated by time-lagged correlations of the two data sets (figure 24). The maximum correlation between fish and zooplankton occurs when zooplankton leads fish by 4 to 5 months in all sectors but IV. These correlations are small (about 0.3 to 0.4) but are statistically significant even when the statistical significance is adjusted to account for the effective degrees of freedom using the method described by Chelton (1983). Auto-correlations of fish and zooplankton show very different decorrelation time scales

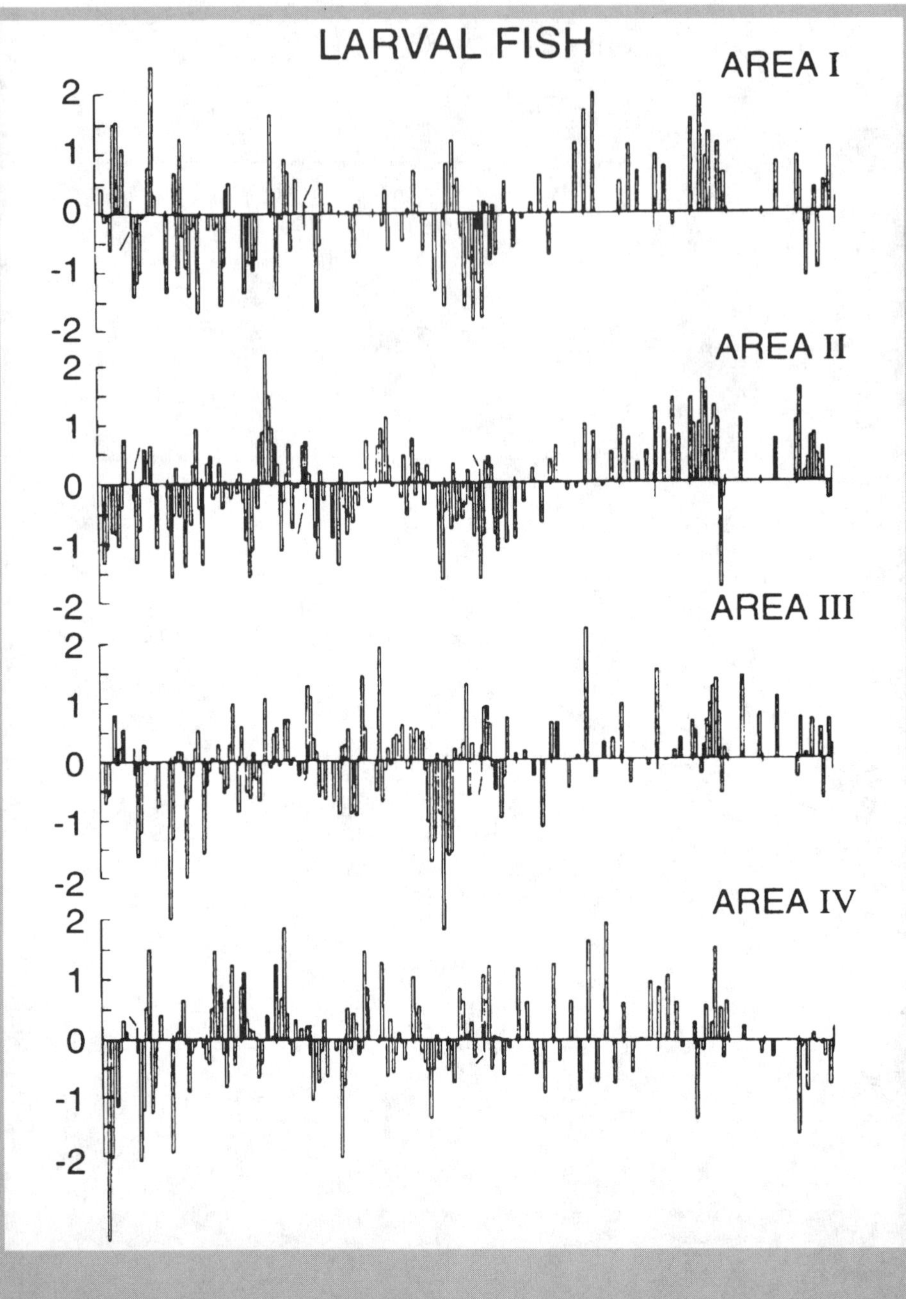

Figure 21. Time series of nonseasonal, $\log_e$-transformed, larval fish data abundance for each of the four areas. Seasonal cycles have been removed, and the time series for each area has been normalized to have a standard deviation of one. *Tic marks* are centered on January of each year.

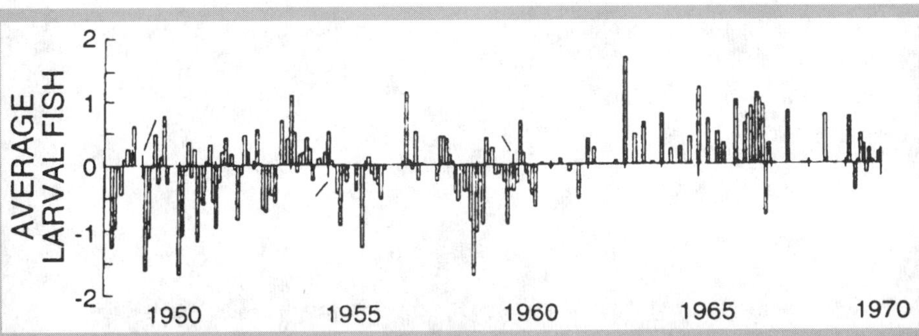

Figure 22. Time series of nonseasonal, transformed, larval fish as in figure 21, but with all four areas averaged.

(figure 25). That is, fish have a much shorter "memory" of past states than do zooplankton. Although the cross-correlations are weak and the autocorrelations are very different, we have been unable to formally disprove the null hypothesis of no relation between larval fish abundance and that of zooplankton. Further, it appears that the correlation does lag in the direction to be expected if poor nutrition of adult fish has affected their fecundity. But a visual examination of the overall, smoothed fish–zooplankton data shows just how weak this relationship is (figure 26).

Since zooplankton is well correlated, in general, with temperature, salinity, and transport, it seems unlikely that fish will be also, since they are poorly correlated with zooplankton. But we may ask if they are somehow influenced by some more general climate indicator. We may now propose the null hypothesis: H_0: "There is no difference between the time series larval fish and the time series longshore winds." Longshore wind anomalies are used here because they form the basis for the coastal upwelling model. The reasoning behind its test is that there is a general consensus that coastal upwelling brings nutrients to the euphotic zone, thus stimulating new production and enhancing production in the entire food web. If this mechanism is important, larval fish survival should benefit from upwelling events. On the other hand, Lasker (1975) has proposed that windiness and upwelling may disturb and disperse the layers of food that some larval fish depend on, in which case their survivorship should be reduced. In the first case larval fish abundance should, at some time lag, be positively correlated with longshore winds. In the second case they should, at some time lag, be negatively correlated. Figure 27 shows the time-lagged correlations between local longshore wind and fish larvae. The two are uncorrelated, and both of the hypotheses are therefore disproved. We thus conclude that the variations in the abundance of zooplankton do not strongly affect adult fish population fecundity or larval mortality. Nor does coastal upwelling affect larval fish survivorship in any simple manner.

The most abundant larvae in the CalCOFI samples are those of the harvested species. Harvesting of adults is an unusual intervention in the normal life histories of pelagic fish, where most natural population regulation seems to come from mortality of larvae. Intensive predation of adults (harvesting) at this point in the life history may so alter normal population dynamics that food limitation or climate-driven perturbations are simply not unambiguously resolvable in a time-series test unless

Table 1. A. Cross-correlations of the nonseasonal, $\log_e$-transformed time series of larval fish abundance in Areas I–IV (see figure 20)[a]

	Area I	Area II	Area III	Area IV
Area I		0.46 (0.25)	0.31 (0.27)	0.10 (0.30)
Area II	0.69 (0.63)		0.50 (0.27)	0.24 (0.26)
Area III	0.41 (0.47)	0.74 (0.40)		0.44 (0.25)
Area IV	−0.20 (0.73)	0.29 (0.50)	0.51 (0.21)	

B. Cross-correlations between the nonseasonal, $\log_e$-transformed zooplankton biomass data in Areas I–IV[a]

	Area I	Area II	Area III	Area IV
Area I		0.57 (0.43)	0.42 (0.43)	0.19 (0.41)
Area II	0.75 (0.54)		0.74 (0.31)	0.31 (0.36)
Area III	0.69 (0.47)	0.91 (0.25)		0.55 (0.40)
Area IV	−0.23 (0.57)	0.49 (0.47)	0.71 (0.50)	

C. Correlations between the large, spatial-scale average, nonseasonal $\log_e$-transformed larval fish time series with those of each of the four areas

	Unfiltered[b]	Interannual[b]
Area I	0.71 (0.27)	0.72 (0.54)
Area II	0.77 (0.23)	0.93 (0.35)
Area III	0.80 (0.28)	0.87 (0.36)
Area IV	0.65 (0.23)	0.43 (0.29)

D. Correlations between the large, spatial-scale average, nonseasonal $\log_e$-transformed zooplankton biomass time series with those of each of the four areas

	Unfiltered[b]	Interannual[b]
Area I	0.75 (0.45)	0.72 (0.46)
Area II	0.85 (0.31)	0.90 (0.36)
Area III	0.88 (0.27)	0.98 (0.35)
Area IV	0.68 (0.44)	0.75 (0.51)

[a]Values above the diagonal are for the unfiltered data; those below are correlations of data filtered with a low-pass double 13-month running mean to emphasize the interannual signal. Parentheses show values required for 95% significance levels with the method described by Chelton (1983).
[b]Values in parentheses are 95% significance levels.

they are large signals. But two further tests of hypotheses are suggested: Do the larvae of nonharvested fish respond to plankton changes or weather? A third and unlikely possibility is that larval fish abundance shifts are normally uncoupled from adult population variability.

Interdecadal Patterns

It has become clear that large, spatial-scale environmental changes occur on an interannual (year-to-year) basis and that these variations are often larger than seasonal, monthly, or weekly changes. We are now learning that longer-period variations are

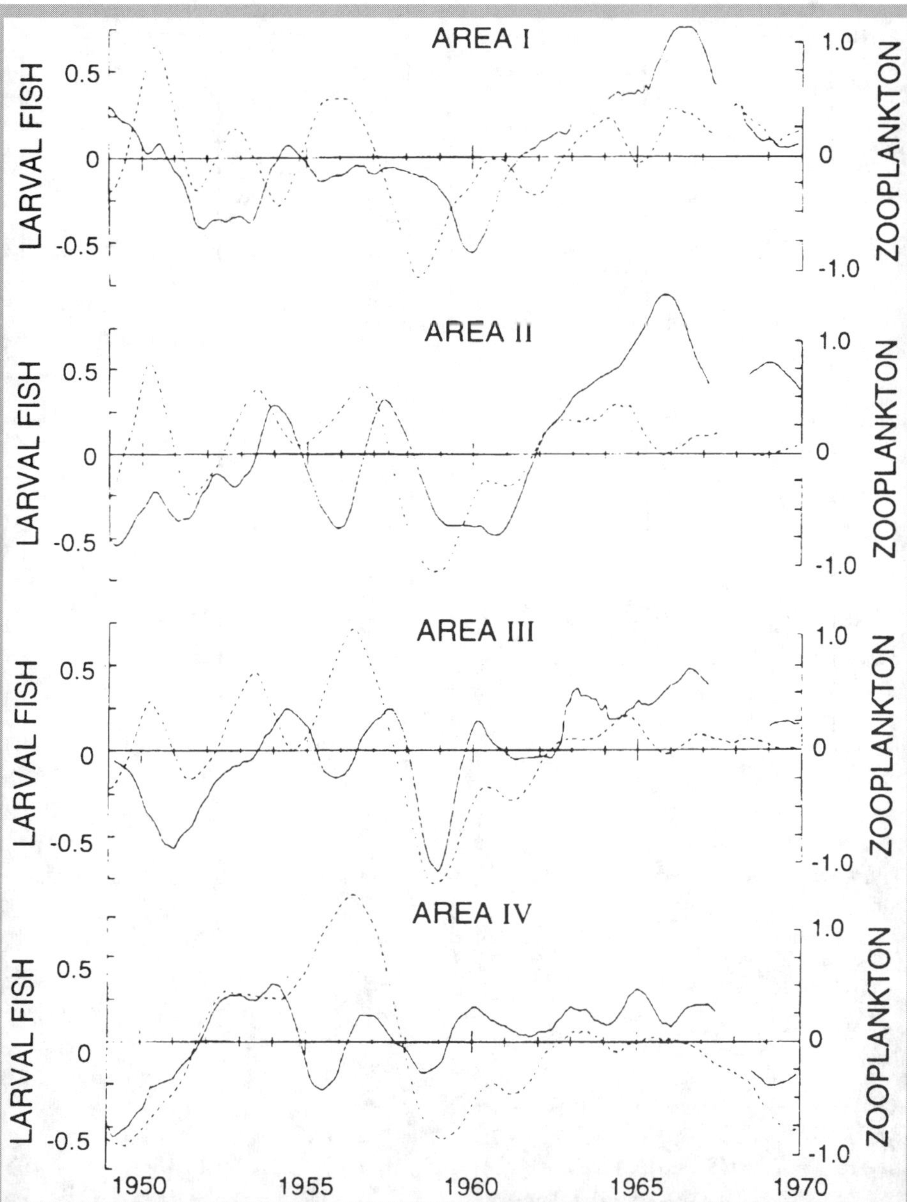

Figure 23. Low-passed filtered, time series of nonseasonal, $\log_e$-transformed larval fish (solid line) and zooplankton (dashed line) for each of the four areas. The raw, nonseasonal, time series were filtered with a double 13-month, objective, running mean, which effectively extracts the interannual variability.

also important—the interdecadal. Of course, paleontologists have long known that the thousand- or ten-thousand-year periods or Pleistocene epochs are important ecological and evolutionary milestones. But paleontologists cannot usually resolve any marine fossil frequency of change of much less than a thousand years except in the special circumstances of varved cores. The interdecadal-to-century frequencies of climate

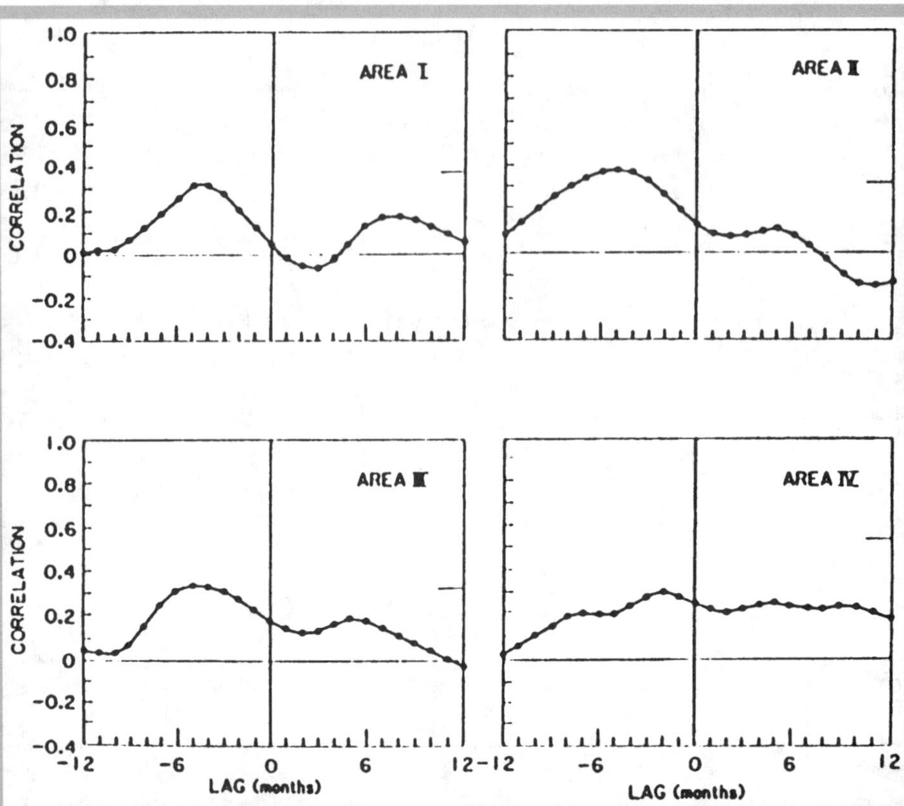

Figure 24. Time-lagged cross correlations between nonseasonal $\log_e$-transformed larval fish at time *t*, and nonseasonal $\log_e$-transformed zooplankton at time (t + lag). Negative lags indicate zooplankton leads larval fish. *Tic marks* on right axis represent 95% significance levels.

change are important in that the magnitudes can be large and there are demonstrable consequences to many environmental variables (Ebbesmeyer et al. 1991).

The 45-year data set of CalCOFI's physical-biological program can, however, be used to investigate interdecadal variability. Roemmich (1992) has shown that over the 42-year period 1950–91 temperature increased by 0.8°C in the upper 100 m of the California Current. An increase is clearly detectable even to depths of 300 m. Concurrently, steric sea level rose by 0.9 ± 0.2 mm per year. The fact that the entire upper layer has warmed is of particular significance to marine life, for here is where all the primary productivity takes place and where the greatest fraction of the biomass resides. Roemmich compared the period 1950–56 to the period 1985–91 for changes in steric height. The onshore–offshore steric heights were spatially uniform between the two time periods, but differed systematically from one another by about 3 cm. This spatial uniformity indicates that the large-scale warming and sea-level rise were not accompanied by significant changes in the strength of the surface geostrophic circulation.

Nearshore, daily surface temperature records from Scripps Pier have been taken since 1916. These show a similar warming trend between 1950 and 1991 (Roemmich

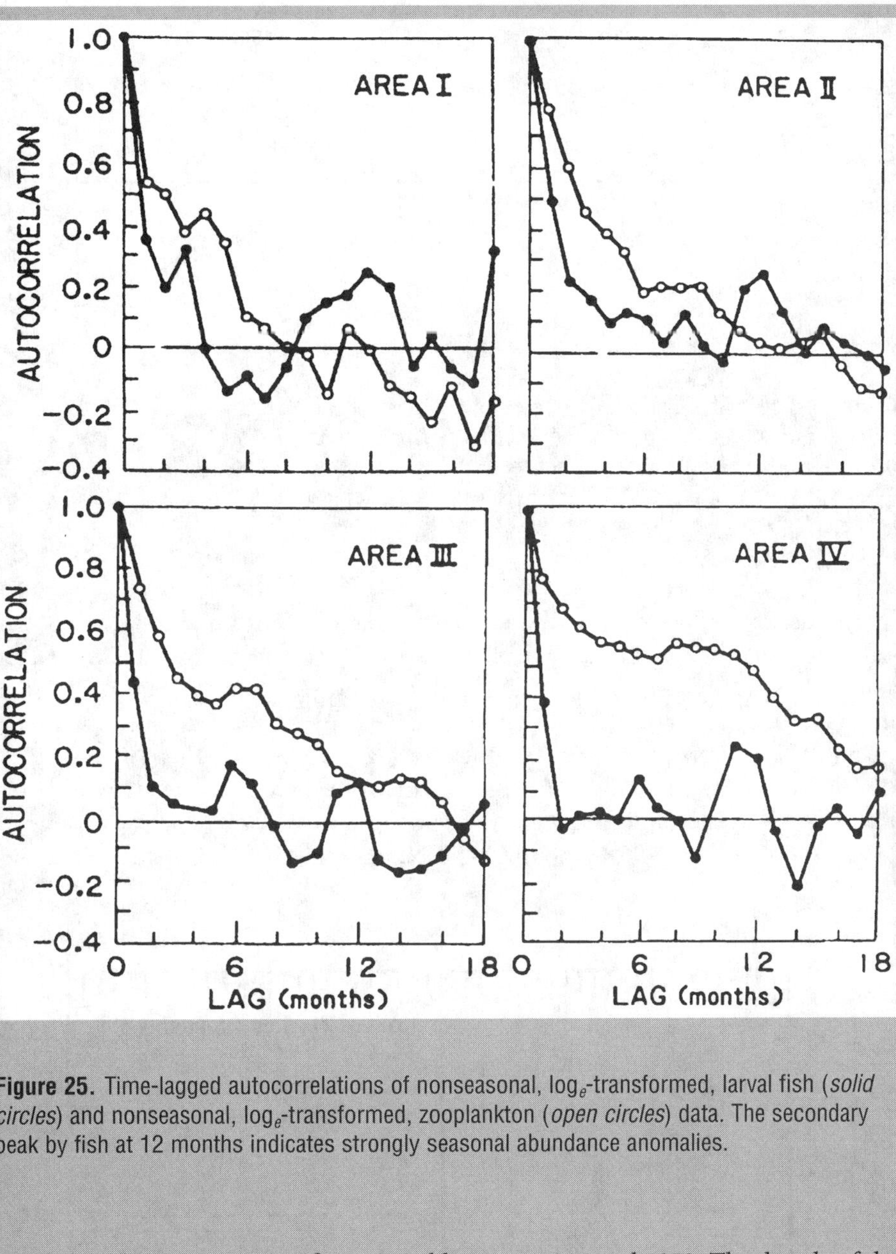

Figure 25. Time-lagged autocorrelations of nonseasonal, $\log_e$-transformed, larval fish (*solid circles*) and nonseasonal, $\log_e$-transformed, zooplankton (*open circles*) data. The secondary peak by fish at 12 months indicates strongly seasonal abundance anomalies.

1992), but there was no significant trend between 1916 and 1950. The decade of the 1980s was the warmest on record.

Roemmich and McGowan (1995) have reported that the warming of the California Current did affect the plankton populations (and no doubt others) of the upper 200 m. Since 1951, the biomass of macrozooplankton in southern California waters has decreased by over 70% during the same period that the surface layer warmed (figure 28). The temperature difference across the thermocline increased, and this increased stratification resulted in less lifting of the thermocline by wind-driven upwelling. A shallower source of upwelled waters would provide fewer inorganic nutrients for new production and hence support a smaller zooplankton population.

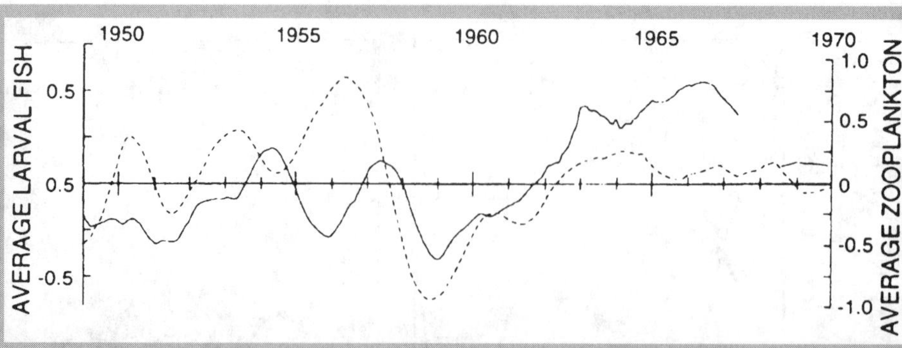

Figure 26. Low-passed filtered time series of nonseasonal $\log_e$-transformed larval fish numbers (*solid line*) and zooplankton displacement volume (*dashed line*) averaged over the four areas shown in figure 20. The nonseasonal time series were filtered with a double, objective running mean. The 1958–59 El Niño effects can be seen clearly. The decline of larval fish abundance tended to lag the decline of zooplankton.

Because of the high energetic interannual variability, it is uncertain whether the zooplankton decline occurred gradually over the entire time series or more rapidly since the 1970s. At both the northern (line 80) and southern (line 90) zones, the onshore–offshore plankton gradients were approximately constant (figure 2a), as was the north–south gradient (figure 29; Roemmich and McGowan 1995). Geostrophic velocity at line 90 showed an increase in the wind-driven recirculation—greater northward flow nearshore and greater southward flow offshore—but no change in net transport. Bakun (1990), analyzing a combination of directly estimated winds and

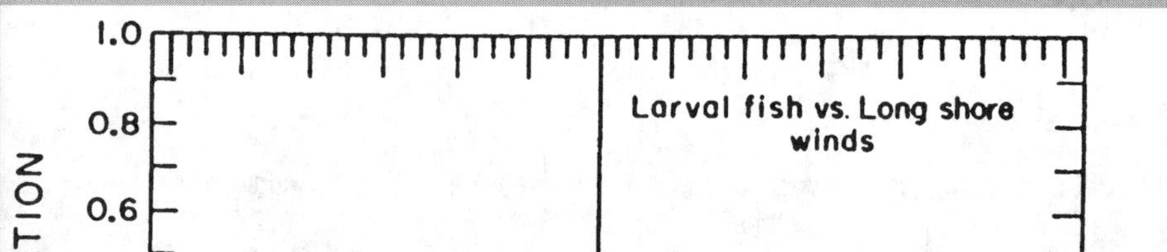

Figure 27. Time-lagged cross-correlation of mean larval fish abundance and longshore winds in Area II, 1950–69. Negative values indicate that winds lag fish. The upper 95% confidence limits are shown.

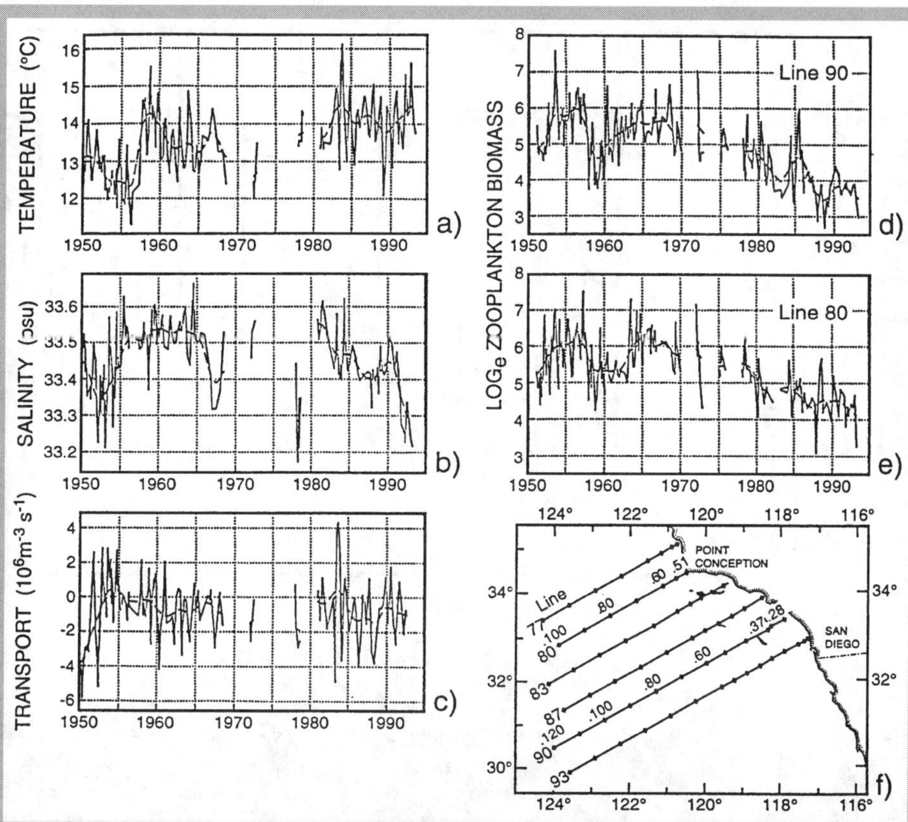

Figure 28. *Left,* Time series of the upper 100 m (*a*) temperature, (*b*) salinity, and (*c*) transport from the north. *Right,* time series of $\log_e$-transformed zooplankton volume per 1,000 m^3 of water filtered by the net on CalCOFI line 90 (*d*) and line 80 (*e*) (After Roemmich and McGowan 1995).

onshore–offshore barometric pressure differences, concluded that coastal winds off California favorable to upwelling increased substantially from 1946 to 1988. Although his reported increase in alongshore wind is consistent with the increase in recirculation noted above, his predicted consequences—SST cooling and increased production—are the opposite of the observed increased temperature and decreased zooplankton biomass.

The observed rate of decrease in zooplankton biomass is more than $5 \times 10^{-10}\ s^{-1}$ over 43 years. On long time scales, biomass is controlled by the net effect of decreases due to excess mortality over births versus increases through advection from the north. But advective input by currents is estimated to be $\sim 5 \times 10^{-8}\ s^{-1}$ (Roemmich and McGowan 1995), far larger than the observed trend. Thus the long-term trend is a small residual of much larger terms. It cannot be isolated by studies of advection or by process-oriented studies of reproductive rates or mortality. Nevertheless, it is a dramatically large signal when accumulated over 43 years.

These observed trends in the California Current may be related to basin-scale changes in wind forcing. A strengthening of North Pacific winter atmospheric circulation began in the late 1970s (Trenberth 1990; Graham 1994), near the time when

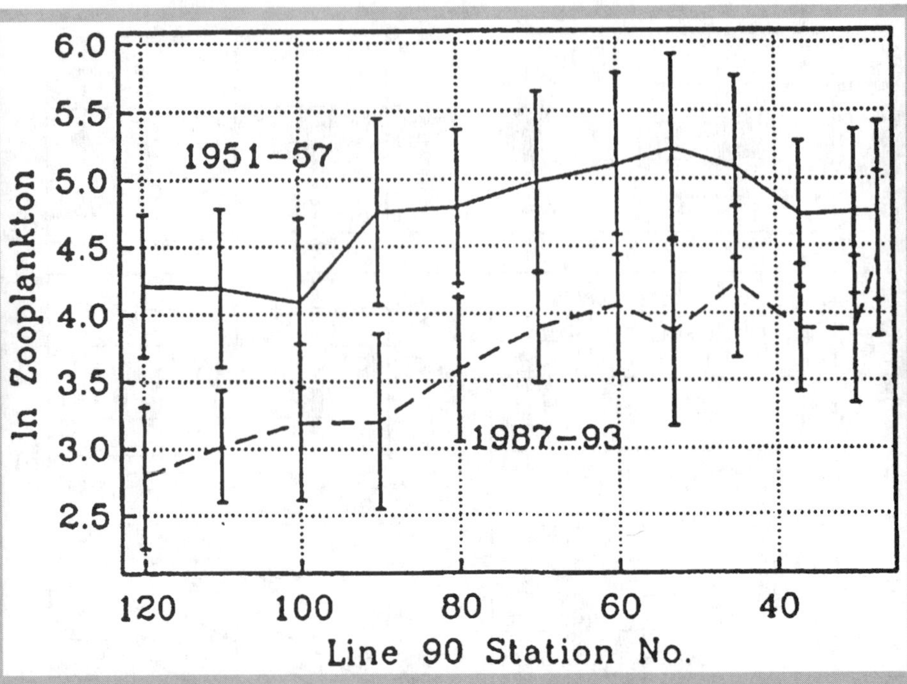

Figure 29. Temporal average and standard deviation (*vertical bars*) of $\log_e$-transformed zooplankton volumes for all cruises in 1951–57 and 1987–93 along line 90. Stations are numbered from nearshore (40) to offshore (120) (from Roemmich and McGowan 1995).

the temperature increase and zooplankton decline began to be clearly seen. A number of other effects have been observed in relation to this basin-scale climate shift (Venrick et al. 1987; Miller et al. 1994).

In addition to the plankton decline, there has been a 90% decline in an apex marine predator—the sooty shearwater (*Puffinus griseus*), once the most abundant pelagic seabird in the California Current. Other seabirds have declined as well (figure 30; Veit, Pyle, and McGowan, in press), although not so dramatically (Veit et al., in press).

Conclusions

Determining the ecological consequences of climate change, whether it is natural or anthropogenic, is one of the most serious problems facing oceanographers today. At the heart of this problem is our limited ability to quantitatively define the word *change*. Because of the paucity of long time series of ecosystem variables, what does this word mean in terms of ecosystem structure and function?

Two Approaches

The departure from some expectation or norm, its sign, amplitude, spatial extent, and temporal persistence, is surely an essential aspect of the concept of change. A

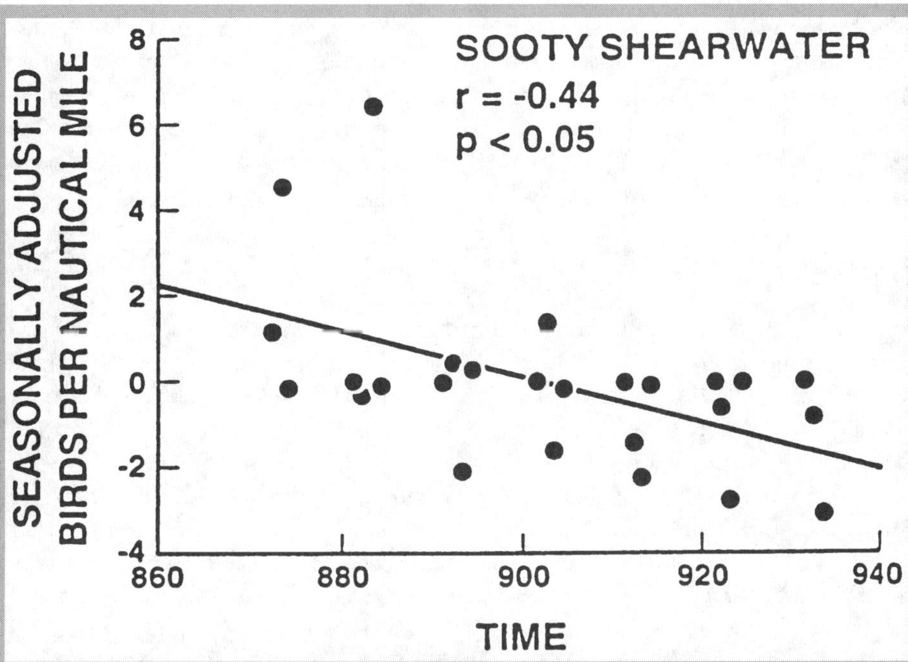

Figure 30. The seasonally adjusted counts of sooty shearwater abundance per cruise from 1987 to 1993. Counts were made during all daylight hours while the ships were under way. Each cruise lasted about two weeks; cruises were made approximately quarterly (from Veit et al., in press).

common practice of studying change is to extrapolate from relatively short-term, local, "process-oriented" studies with the expressed goal of providing values for "fluxes" of energy or materials between aggregated categories (sometimes called functional groups) of community ecosystems. These fluxes are assembled in the familiar ecosystem box model. Such inductive models have been used for many years in marine ecology but have not proven useful for understanding variability in system structure and function. In particular, they have provided little insight into the effects of climatic perturbations on system behavior or on the state of its populations. Pelagic community–ecosystem structure is complex, enigmatic, and vacillatory. There are many unknowns, and it has been necessary, therefore, to include untested and often untestable assumptions in such models. But such process-oriented, model-driven research has given us some valuable information. These studies have shown that most, if not all, of the rates of interaction between structural entities are nonlinear and density dependent. The studies have also provided many engaging details about the maze of intermediary mechanics in community function.

It is clear that time series, if they are long enough, can define the word *change*, its sign, amplitude, and persistence. They can also determine the spatial dimensions of change if the correlation length scales are adequately sampled. Anomalies from the baseline seasonal cycle can be calculated and compared with climatic signals by means of the classical methods of null hypothesis testing. However, correlations between climatic anomalies and biological anomalies cannot provide information on the innumerable intermediary mechanisms that link community response to climate change. Although time series, of themselves, do not provide predictive capability, they do offer a wealth of information for testing and validation of ecosystem models. Unless the models can reproduce the observed time series, they cannot be used for predictive purposes.

The California Current Example

The CalCOFI time series has been used to construct long time series over a large portion of the California Current. In spite of highly variable and patchy data, these maps show persistent onshore, offshore gradients and north–south gradients of virtually all physical, chemical, and biological properties. Subarctic and Transition Zone species intrude from the north, along with cool, low-salinity waters. Warm-water cosmopolites come in from the west with warm, high-salinity Central Gyre waters. Equatorial species intrude from the south. These species are mixed together in the central sector of the California Current. It is here that plankton species diversity is the highest of any sector of this current. It seems reasonable to infer that the spatial maximum in diversity is due to simple physical mixing rather than to ecological processes such as resource allocation or disturbance.

The 45-year monitoring study has also allowed time-series and spectral analyses, even though gaps exist in the data. All ecosystem variables show large, nonseasonal, low-frequency variations. Macrozooplankton biomass changes are clearly correlated with interannual, climate-driven changes in mass transport of water from the north. Larval fish are also highly variable in time, but these variations are seldom clearly related to changes in zooplankton (the diet of the adult fish) and, in our data, never to anomalies in longshore winds, the basis of coastal upwelling indices.

Zooplankton also responded to another lower frequency of climate change, the interdecadal. This is manifested as a long-term warming of the entire upper 300 m water column and a coincident decline in zooplankton production. The very large decline in zooplankton (over 70%) with such a "small" temperature signal (~0.8°C) indicates that the organisms were not responding directly to temperature with, say, a change in their thermally regulated metabolism, but rather that the warming is a surrogate index for some greater environmental change. A candidate mechanism is an increase in water column stratification due to the warming (Roemmich and McGowan 1995). This would lead to a decrease in vertical eddy diffusivity, mixing, and upwelling and therefore a lesser input of inorganic nutrients from the deep water. The overall effect would be less primary and secondary productivity, consistent with the observations.

There are several steps in this proposed mechanism as yet to be verified by analyses of the existing observations. But if it is substantially correct, it differs greatly from the mechanism proposed for interannual changes. This mechanism had to do with horizontal advective input to the system proposed for interannual changes (Bernal and McGowan 1981; Chelton et al. 1982). The interdecadal trend in upper-layer temperature may be related to basin-scale wind forcing. A strengthening of the North Pacific winter atmospheric circulation began in the late 1970s (Graham 1994), near the time when the CalCOFI trends began to be clearly seen (figure 28). A number of other effects have been observed in relation to this basin-scale change (Venrick et al. 1987).

If the interdecadal decline in zooplankton is part of a natural climatic cycle that reverses itself in coming years, then the impact will be similarly transient. But if the decline is anthropogenic or a natural trend of longer duration, then the magnitude of the decline is a cause for concern (Roemmich and McGowan 1995). For example, if large-amplitude physical perturbations work the way we suggest and if fish populations generally become food limited during such periods, then we may anticipate large-scale declines in the populations that eat macrozooplankton or whose larvae depend upon their larvae.

Thus there are different forcings, amplitudes, and consequences depending on time scales (daily, weekly, monthly, seasonal, interannual, and interdecadal). This demonstrates the dangers of induction. Climate studies dominated by short-term, process-oriented work cannot simply be extrapolated to interannual or decadal time scales where the balance of terms is different from monthly or seasonal balances.

The continuous plankton recorder studies in the North Atlantic (CPR Survey Team 1992), other studies in this volume, and the CalCOFI time series have all shown that the most energetic biotic changes in space and time are of low frequency; that is, the frequency spectra of biological variables are dominated by low-frequency variability. These low-frequency variations in macrozooplankton are all closely associated with large-scale climatic trends in both the Pacific and Atlantic.

If large changes in biomass over time and over large areas are considered to be important ecological and evolutionary events, then the spectra dominated by low-frequency variability have great underlying meaning; there must also be structural and functional changes occurring, and they cannot be understood or predicted by short-term, process-oriented studies. If the consequences of climate and climatic change are of interest, then biotic measurements must be maintained over periods of many decades in order to detect, let alone understand, these consequences.

All of this argues for a new outlook to guide research on marine communities and populations. Apparently little can be inferred about kinematics and dynamics on scales much larger than the scales of measurement. This may be because density-independent forces extrinsic to the internal workings of communities are very important and yet essentially unpredictable from a knowledge of internal dynamics. Further, short-term perturbations such as weather often do not have large or lasting effects on ecosystems. The concepts of stability and resilience are based on this fact.

Since there is legitimate concern that large climate changes are possible in the near future, the time is now to direct more of our effort to defining, detecting, and understanding "change."

Many ecologists are not yet accustomed to thinking about larger scales or doing large-scale research. So the new challenge will require a different outlook and some new approaches. One such approach is to use time series to test some hoary old hypotheses such as zooplankton vs. upwelling index, larval fish vs. zooplankton, or larval fish vs. longshore winds.

It is very clear that there are large interannual changes in total larval fish abundance. This is a fact of considerable interest, for it appears that the entire complex of species is responding over a large area to some forcing other than seasonal. While microscale, fine-scale, or mesoscale studies can suggest some plausible scenarios for these large changes (see Mullin 1993 for a review) the number of potentially testable and falsifiable hypotheses derived from these works, relevant to the climate scale, is small, and the actual number of formal tests of hypotheses is practically nonexistent. Species populations exist on very large spatial scales, and their largest and therefore most energetic variations seem to be interannual or interdecadal. There is no evidence to support the notion that the results of micro- to mesoscale studies can be extrapolated upward and outward to the climate-change scale. Time-series measurements can be made on the appropriate scales. They do, however, take time.

Some of the pragmatic details of this new research canon are not well formulated, and we are ill prepared to design a large-scale, long-term monitoring system, since most of our thinking and effort has been dominated by reductionism. But perceiving the problem of scale is necessarily the first step in formulating a new kind of marine ecology.

ACKNOWLEDGMENTS

We are grateful to the technicians and scientists for the existence and high quality of the CalCOFI data. The larval fish data were provided by H. Geoffrey Moser and Paul Smith of the Southwest Fisheries Center, National Marine Fisheries Service (NOAA), and we thank them for it. This study was supported by the Marine Life Research Group of Scripps Institution of Oceanography, University of California, San Diego.

Literature Cited

Anonymous, 1963. CalCOFI atlas of 10 meter temperatures and salinities 1949 through 1959. CalCOFI Atlas No. 1, State of Cal. Mar. Res. Comm. (unpaginated).

Bakun, A. 1973. Coastal upwelling indices, West Coast of North America, 1946–71. Washington, D.C.: NOAA Tech. Rep., Nat. Mar. Fish. Ser. SSRF-671.

Bakun, A. 1990. Global climate change and intensification of coastal ocean upwelling, Science 247:198–201.

Barry, J. P., C. H. Baxter, R. D. Saqarin, and S. E. Gilman. 1995. Climate-related, long-term faunal changes in a California rocky intertidal community. Science 267:672–675.

Bernal, P. A. 1979. Large-scale biological events in the California Current. Calif. Coop. Oceanic Fish. Invest. Rep. 20:89–101.

Bernal, P. A., and J. A. McGowan, 1981. Advection and upwelling in the California Current. *In* Coastal upwelling, F. A. Richards, ed. Amer. Geophys. Union, Washington, D.C.: pp. 381–399.

Berner, L. D. 1967. Distributional atlas of *Thaliacca* in the California Current region. CalCOFI Atlas No. 8, State of California, Marine Research Committee, pp. 1–322.

Bowman, T. E., and M. W. Johnson, 1973. Distributional atlas of calanoid copepods in the California Current region, 1949 and 1950. Calif. Coop. Oceanic Fish. Invest. Atlas No. 19.

Brinton, E. 1967. Distributional atlas of Euphauseacia (Crustacea) in the California Current, Part I. Calif. Coop. Oceanic Fish. Invest. Atlas No. 5.

———. 1981. Euphausiid distributions in the California Current during the warm winter–spring of 1977–78, in the context of a 1949–1966 time series. Calif. Coop. Oceanic Fish. Invest. Rep. 22:135–154.

Chelton, D. B. 1983. Effects of sampling errors in statistical estimation. Deep-Sea Res. 30:1083–1103.

Chelton, D. B., P. A. Bernal, and J. A. McGowan, 1982. Large-scale inter-annual physical and biological interaction in the California Current. J. Mar. Res. 40:1095–1123.

CPR Survey Team. 1992. Continuous plankton records: the North Sea in the 1980s. ICES Mar. Sci. Symp. 195:243–248.

Ebbesmeyer, C. C., D. R. Cayan, D. R. McLain, F. H. Nichols, D. H. Peterson, and K. T. Redmond. 1991. 1976 step in the Pacific climate: forty environmental changes between 1968–1975, and 1977–1984, Proc. Seventh Ann. Pac. Climate (PACLIM) Workshop. Calif. Dep. Water Resources. Interagency Ecol. Studies Prog. Tech. Rep. 26:115–126.

Eber, L. E. 1977. Contoured depth-time charts (0 to 200 m 1950 to 1966) of temperature, salinity, oxygen and sigma-τ at 23 CalCOFI stations in the California Current. Calif. Coop. Oceanic Fish. Invest. Atlas No. 25.

Fleminger, A. 1964. Distributional atlas of calanoid copepods in the California Current region, Part I. Calif. Coop. Oceanic Fish. Invest. Atlas No. 2.

Frances, R. C., S. R. Hare. 1994. Decadal scale shifts in the large marine ecosystems of the North-East Pacific: a case for historical science. Fish. Oceanogr. 3(4):279–291.

Graham, N. E. 1994. Decadal scale variability in the 1970s and 1980s: observations and model results. Clim. Dynamics 10:60–70.

Hollowed, A. B. 1992. Spatial and temporal distributions of Pacific hake, *Merluccius productus*, larvae and estimates of survival during early life stages. Calif. Coop. Oceanic Fish. Invest. Rep. 33:100–123.

Houde, E. D. 1987. Fish early life dynamics and recruitment variability. Am. Fish. Soc. Symp. 2:17–89.

Kendall, A. W., J. D. Schumacher, and S. Kim. 1996. Walleye pollock recruitment in Shelikof Strait: applied fisheries oceanography. Fisheries Oceanography Supp. 1, Fisheries oceanography of walleye pollock in Shelikof Strait, Alaska, pp. 4–18.

Kramer, D., M. J. Kalin, E. G. Stevens, J. R. Thrailkill, and J. R. Zweifel. 1972. Collecting and processing data on fish eggs and larvae in the California Current region. U.S. Dep. Commerce. N.O.A.A. Tech. Rep., NMFS Circ-370, pp. iii-iv, 1–38.

Lasker, R. 1975. Field criteria for survival of anchovy larvae: the relation between inshore chlorophyll maximum layers and successful first feeding. Fish. Bull. 73:453–462.

Lo, N. C., P. E. Smith, and J. L. Butler. 1995. Population growth of northern anchovy and Pacific sardine using stage-specific matrix models. Mar. Ecol. Prog. Ser. 127:15–26.

Lynn, R. J., K. A. Bliss, and L. E. Eber. 1982. Vertical and horizontal distributions of seasonal mean temperature, salinity, sigma-τ, stability, dynamic height, oxygen and oxygen saturation in the California Current, 1950–1978. Calif. Coop. Oceanic Fish. Invest. Atlas No. 30, State of California.

Martinson, D. G., K. Bryan, M. Ghil, M. M. Hall, T. R. Karl, E. S. Sarachik, S. Sorooshian, and L. D. Talley. 1995. Natural climate variability on decade-to-century time scales. Washington, D.C.: Nat. Res. Com., Nat. Acad. Press.

MacCall, A. D. 1996. Patterns of low-frequency variability in fish populations of the California Current. Calif. Coop. Oceanic Fish. Invest. Rep. 37 (this volume).

McCall, A. D., and M. H. Prager. 1988. Historical changes in abundance of six fish species off southern California, based on CalCOFI egg and larva samples. Calif. Coop. Oceanic Fish. Invest. Rep. 29:91–101.

McGowan, J. A. 1967. Distributional atlas of pelagic molluscs in the California Current. Calif. Coop. Oceanic Fish. Invest. Atlas No. 6.

———. 1990. Species dominance-diversity patterns in oceanic communities. *In* The earth in transition, G. W. Woodwell, ed. Cambridge Univ. Press, pp. 395–421.

McGowan, J. A., and C. B. Miller. 1980. Larval fish and zooplankton community structure. Calif. Coop. Oceanic Fish. Invest. Rep. 21:29–36.

McGowan, J. A., and P. W. Walker. 1993. Pelagic diversity patterns. *In* Species diversity in ecological communities, R. E. Rickleffs and D. Schulter, eds. Chicago: Univ. Chicago Press, pp. 203–214.

Miller, A. J., D. R. Cayan, T. P. Barnett, N. E. Graham, and J. M. Oberhuber. 1994. The 1976–77 climate shift of the Pacific Ocean. Oceanography 7:21–26.

Mullin, M. M. 1993. Webs and scales: physical and ecological processes in marine fish recruitment. Books in recruitment fishery oceanography, Wash. Sea Grant Prog., Wash. Univ. Press. Seattle, pp. 1–135.

Reid, J. L., G. I. Roden, and J. G. Wyllie, 1958. Studies of the California Current System. CalCOFI Progress Report 1 July 1956 to 1 Jan. 1958, 27–57.

Roemmich, D. 1992. Ocean warming and sea-level rise along the southwest U.S. Coast. Science 257:373–375.

Roemmich, D., and J. A. McGowan. 1995. Climatic warming and the decline of zooplankton in the California Current. Science 267:1324–1326.

Roesler, C. S., and D. B. Chelton. 1987. Zooplankton variability in the California Current, 1951–1982. Calif. Coop. Oceanic Fish. Invest. Rep. 28:59–96.

Ross, A. H., W. S. Gurney, M. R. Heath, S. J. Hay, and E. W. Henderson. 1993. A strategic simulation model of a fjord ecosystem. Limnol. Oceanogr. 38:128–153.

Sette, O. E., and J. D. Issacs, eds. 1960. Part II Symposium. The changing Pacific Ocean in 1957 and 1958. Calif. Coop. Oceanic Fish. Invest. Rep. 7:13–211.

Shrader-Frechette, K. S., and E. D. McCoy. 1993. Method in ecology: strategies for conservation. New York: Cambridge Univ. Press.

Skud, B. E. 1982. Dominance in fishes: the relation between environment and abundance. Science 216:144–149.

Steele, J. H. 1974. The structure of marine ecosystems. Cambridge: Harvard Univ. Press. pp. 1–128.

Stommel, H. 1965. Some thoughts about planning the Kuroshio survey. Proc. of Symp. on the Kuroshio. Tokyo, Oct. 29, 1963. Oceanographical Soc. Japan and UNESCO, pp. 22–33.

Sund, P. N., and J. G. Norton. 1990. Interpreting long-term fish landing records: environmental and/or exploitation? *In* Proc. Sixth Annual Pacific Climate (PACLIM) workshop, Asilomar, CA. J. L. Betancourt and A. M. Macky, eds. Tech. Rep. 23, Interagency Ecol. Stud. Prog. for the Sacramento-San Joaquin Estuary. Calif. Dep. Water Resources, Sacramento, CA 94236.

Theilacker, G. H. 1987. Starvation induced mortality of young sea caught jack-mackerel *Trachurus symmetricus*, determined with histological and morphological methods. U.S. Fish. Bull. 84:1–17.

Theilacker, G. H., and W. Shen. 1993. Calibrating starvation induced stress in larval fish using flow cytometry. Am. Fish. Soc. Symp. 14:85–94.

Trenberth, K. E. 1990. Recent observed interdecadal climate changes in the Northern Hemisphere. Bull. Am. Meteorol. Soc. 71:988–993.

Veit, R. R., P. Pyle, and J. A. McGowan. In press. Ocean warming and long-term change in pelagic bird abundance within the California Current System. Mar. Ecol. Prog. Ser.

Veit, R. R., J. A. McGowan, D. G. Ainley, T. R. Wahl, and P. Pyle. In press. Apex marine predator declines 90% in association with changing oceanic climate. Global Change Biology.

Venrick, E. L., J. A. McGowan, D. R. Cayan, and T. L. Hayward. 1987. Climate and chlorophyll-a: long term trends in the central North Pacific Ocean. Science 238:70–72.

Wiebe, P. H., C. B. Miller, J. A. McGowan, and R. A. Knox. 1987. Long time series study of oceanic ecosystems. EOS 68:1178–1190.

Wyllie, J. G., and R. J. Lynn. 1971. Distribution of temperature and salinity at 10 meters, 1960–1969, and mean temperature, salinity and oxygen at 150 meters, 1950–1968, in the California Current. CalCOFI Atlas No. 15, State of California, Marine Research Committee. pp. 1–188.

Interannual Variations in Zooplankton Biomass in the Gulf of Alaska, and Covariation with California Current Zooplankton Biomass[1]

Richard D. Brodeur, Bruce W. Frost, Steven R. Hare, Robert C. Francis, and W. James Ingraham Jr.

Abstract

Large-scale atmospheric and oceanographic conditions affect the productivity of oceanic ecosystems both locally and at some distance from the forcing mechanism. Recent studies have suggested that both the Subarctic Domain of the North Pacific Ocean and the California Current have undergone dramatic changes in zooplankton biomass that appear to be inversely related to each other. Using time series and correlation analyses, we characterized the historical nature of zooplankton biomass at Ocean Station P (50°N, 145°W) and from offshore stations in the CalCOFI region. We found a statistically significant but weak negative relationship between the domains. We investigated whether such a relationship arises from different forcing mechanisms or as an opposite response to the same mechanism. We found that the seasonal peak of both data sets occurred in the summer but that the CalCOFI data lagged the Ocean Station P data. A surface-drift simulation model showed that winter trajectories started at Ocean Station P and along 145°W drifted more into the California Current before the 1976–77 regime shift, and more into the Alaska Current after the 1976–77 shift. We examined physical and biological conditions which may lead to this inverse relationship between the two ecosystems, and we discuss the implications of these results for higher trophic levels.

Introduction

Substantial temporal and spatial heterogeneity occurs in the production of oceanic ecosystems. Much of this heterogeneity results from seasonal and geographic vari-

[1]Reprinted with permission from its authors and from the Coordinator of the CalCOFI Report. Original printed in the CalCOFI Report published by Scripps Institute, U. Cal. San Diego, 37:80–99.

ations in nutrient availability, mixed-layer depths, or solar radiation. Processes that enhance productivity (e.g., upwelling, wind and tidal mixing) tend to be localized and transient. Physical forcing in the form of the large-scale circulation pattern redistributes the elevated production to areas less favorable for in situ production. Thus production at any location can be affected by both local and remote processes, and it is often difficult to distinguish their relative contributions (Wickett 1967; Chelton et al. 1982; Roessler and Chelton 1987).

It has become increasingly apparent that atmospheric and oceanic conditions are likely to change due to a buildup of greenhouse gases in the atmosphere (Graham 1995). Although there has been much interest in predicting the effects of climate change, especially on fisheries resources (e.g., see papers in Beamish 1995), different scenarios exist for future trends in basic physical processes such as upwelling (Bakun 1990; Hsieh and Boer 1992). Biological processes are more laborious to monitor and difficult to predict because of their inherent complexity.

There are numerous examples showing that large-scale physical and biological changes have occurred throughout much of the Northeast Pacific Ocean over the last few decades (Francis and Hare 1994; Miller et al. 1994). Indices which showed these changes include atmospheric (Trenberth 1990; Trenberth and Hurrell 1994), oceanographic (Royer 1989; Hsieh and Boer 1992; Miller et al. 1994; Lagerloef 1995; Polovina et al. 1995), productivity (Venrick et al. 1987; Polovina et al. 1994), and biomass of various trophic levels (Brodeur and Ware 1992, 1995; McFarlane and Beamish 1992; Beamish 1994; Hare and Francis 1995; Roemmich and McGowan 1995a, b). A number of studies have suggested that biological changes occurred rather suddenly sometime around 1976–77, concurrent with a dramatic shift in physical regimes (Francis and Hare 1994; Miller et al. 1994).

Documenting the effects of climate change on marine ecosystems requires long time series of sampling to examine low-frequency periodicity. In the Northeast Pacific Ocean, two series are notable not only for their length, but also for the broad suite of biological and physical measurements made at each location. The first of these, Ocean Station P (50°N, 145°W; hereafter called Station P), nominally represents a subarctic oceanic ecosystem that was sampled almost continuously from 1956 to 1980 but only sporadically since then. The second, the CalCOFI grid, is an eastern boundary current ecosystem that has been sampled since the late 1940s, although not always with the same geographic and temporal intensity. Although Ekman pumping of deep, nutrient-rich water is a feature common to both these ecosystems, the mechanisms behind biological production in the two systems differ (Ware and McFarlane 1989).

Over the past few decades, there has been much speculation on the nature of the eastern bifurcation of the Subarctic Current into the Alaska Current and California Current and its possible effects on biological production in these two large ecosystems (figure 1). Wickett (1967), Chelton and Davis (1982), and Chelton (1984) have speculated that the intensities of the flows in the Alaska and California Currents fluctuate in opposition to one another. They hypothesized that north–south shifts in the bifurcation of the Subarctic Current (West Wind Drift) could be forced by physical factors occurring in the western or central Pacific Ocean.

Using environmental indices and fish-recruitment data, Hollowed and Wooster (1992) and Francis (1993) have characterized two alternating interdecadal states of atmospheric and oceanic circulation in the Northeast Pacific Ocean which result in

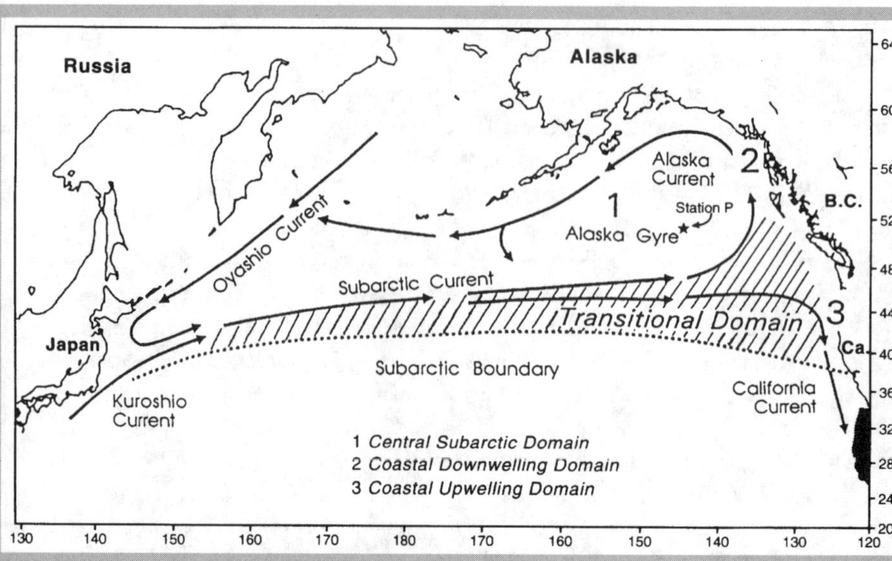

Figure 1. Large-scale near-surface circulation in the North Pacific Ocean, and Ocean Station P (*star*) and the CalCOFI sampling region (*shaded area*). Also shown are the different ocean production domains of Ware and McFarlane (1989) and Pearcy (1991).

very different components of fisheries production (e.g., groundfish, salmon) in the Alaska Current and California Current domains. Hollowed and Wooster (1992) have characterized these states as lasting 6 to 12 years each. A cold era (Type A) is associated with a weak Aleutian Low, relatively weak circulation in the Alaska Gyre, strong upwelling inshore of the California Current, negative sea-surface temperature (SST) anomalies throughout the coastal Northeast Pacific Ocean, and positive SST anomalies in the central North Pacific Ocean (centered at 40°N). A warm era (type B) is associated with a strong Aleutian Low, strong gyral circulation, and reduced upwelling and high temperatures to the south (figure 2).

Francis (1993), Francis and Hare (1994), and Hare and Francis (1995) find similar although longer (20–30 year) periods of oscillating "warm" and "cool" regimes, which they relate to the production dynamics of Alaska salmon. In addition, Francis (1993) speculated that the interdecadal variations in salmon production in these two oceanic domains are inversely correlated.

This paper examines factors related to long-term changes in production in the subarctic Pacific Ocean, using examples mainly from Station P. We examine, in particular, trends in zooplankton biomass as an indicator of changes in productivity in this oceanic ecosystem. Using time-series analysis, we then compare these trends to those evident in the offshore region of the California Current to examine the hypothesis of Wickett (1967) that production in both systems may be affected by an influx of water from the same source—the Subarctic Current. We examine physical data relevant to regime-shift changes in flow between the Alaska Current and California Current. Finally, we discuss implications of regime shifts to higher trophic levels and suggest new hypotheses and further studies that could be undertaken to address these hypotheses.

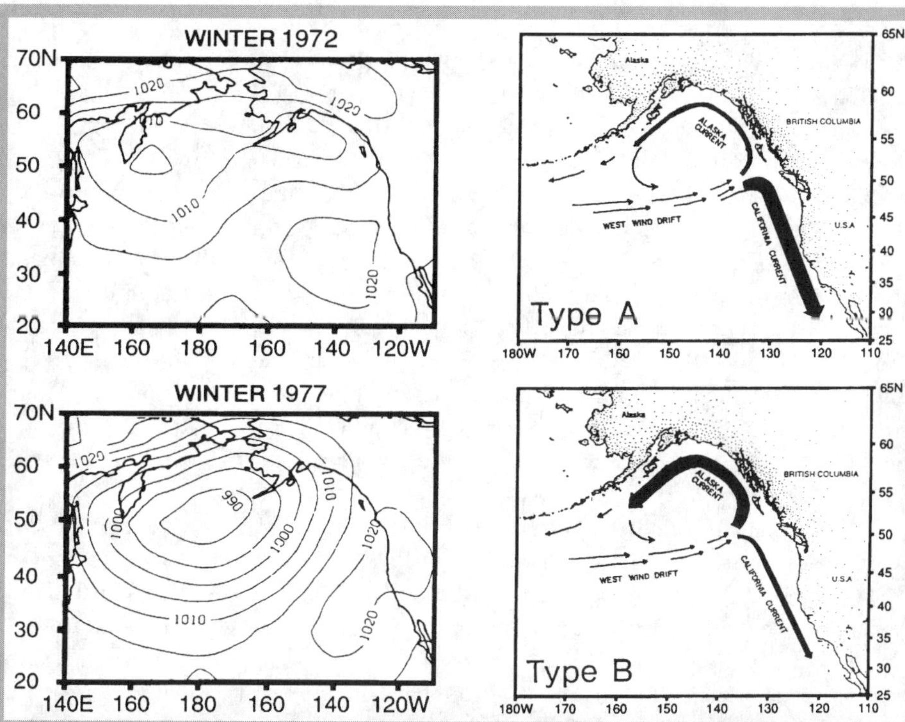

Figure 2. Winter mean sea-level pressure from Emery and Hamilton (1985) for two contrasting years, and the alternate states of atmospheric and oceanic circulation patterns in the eastern North Pacific Ocean proposed by Hollowed and Wooster (1992).

Data Sources and Methods

Zooplankton Data Sets

Zooplankton biomass data from 24 years (1956–80) of vertical net sampling at Station P were used in our analyses (figure 1). Before 1969, sampling was conducted over alternate 6-week periods, but after this time, sampling was continuous (Fulton 1983). Sampling frequency varied from 1 to 29 samples per month over 8 to 12 months of the year (Frost 1983). Hauls were mainly done during daytime, and all were from 150 m to the surface. Sampling gear was changed from a 0.42-m-diameter NORPAC net to a 0.57-m SCOR net in August 1966, although the mesh size remained the same (0.351 mm). Fulton (1983) estimated that the catching efficiency of the SCOR net was 1.5 times greater than the NORPAC net, based on a series of intercalibration tows. But a new estimation based on the original data presented by Fulton (1983) suggests that the correction factor should be higher, somewhere in the range of 1.6–2.1×, with 1.77× being the most likely value (Waddell and McKinnell 1995; Frost, Ware, and Brodeur, unpubl. data), which is what we used in this analysis.

Zooplankton displacement volumes from the central part of the CalCOFI grid (lines 77–93) over a longer time frame (1951–94) were provided by Paul Smith (NMFS, SWFC, La Jolla). The gear and maximum haul depths changed during this period from a bridled 1-m-diameter ring net fished obliquely to 140 m (1951–68) or to 212 m (1969–78) to an unbridled bongo net fished down to 212 m (1978–94). An analysis by Ohman and Smith (1995) has determined that the deeper tows with the 1-m net are low by a factor of 1.366 compared with the shallower 1-m nets and the deep bongo tows without bridles, and we adjusted the data accordingly.

Since we were interested in advective rather than local production processes (Chelton et al. 1982), we trimmed the data set to include only those stations farther than 60 km offshore. We initially aggregated the data in several ways, such as by the northern (north of CalCOFI line 80), middle (line 80 to line 90), and southern (south of line 90) parts of the region, as well as by working with only the most frequently sampled transects (lines 80 and 90). We found, however, that the time series utilizing all the data was highly correlated with almost all other combinations of data. Therefore, the CalCOFI data for a given year were combined spatially for all analyses. In addition, the number of missing data points was reduced, though large gaps still existed.

Additional zooplankton biomass data from oceanic areas of the Northeast Pacific Ocean besides Station P exist for two time periods (1956–62 and 1980–89, except for 1986), and are more fully described in Brodeur and Ware (1992). In this analysis, we extended the geographic range of values to 40°N to include the transition region south of the subarctic boundary (Pearcy 1991) for both time periods (see figure 3 for sampling locations). Contour maps of zooplankton biomass (g/1,000 m^3) were generated for both time periods with a raster-based GIS program (Compugrid, Geospatial Ltd.). Yearly interpolated means were computed for each year. Since the earlier analysis, an additional 5 years (up to 1994) of data collected by Hokkaido University have become available. Although there is not enough coverage for this later period to map the overall distribution of biomass, many of the same transects were sampled each year so that the interannual variability can be examined. As before, only biomass data collected from 15 June to the end of July and in the same geographic area described by Brodeur and Ware (1992) were included.

Time Series Analyses

We examined the temporal relationship between the Station P zooplankton data and the offshore CalCOFI data using time series cross-correlation analyses (Box and Jenkins 1976). We investigated monthly, seasonal, and annual lagged relationships. The time series of available data for each region shows incomplete temporal overlap, especially for the CalCOFI region, which was intermittently sampled during the 1970s and 1980s (figure 4). The distribution of both the Station P and offshore CalCOFI zooplankton data was highly skewed, and a log-transformation of the data was performed before the time-series analysis. In addition, a pronounced seasonal signal is evident in each region (figure 4). This signal was removed by calculating a yearly average biomass for each month and season (spring = March–May; summer = June–August; fall = September–November; winter = December–February). Both monthly time series exhibited substantial lag-1 autocorrelation, which can result in spurious cross-correlations (Box and Jenkins 1976; Myers et al. 1995). Therefore it was necessary to filter both series by a process known as prewhitening (*sensu* Box and Jenkins 1976) before computing the cross-correlation function (CCF) at lagged periods.

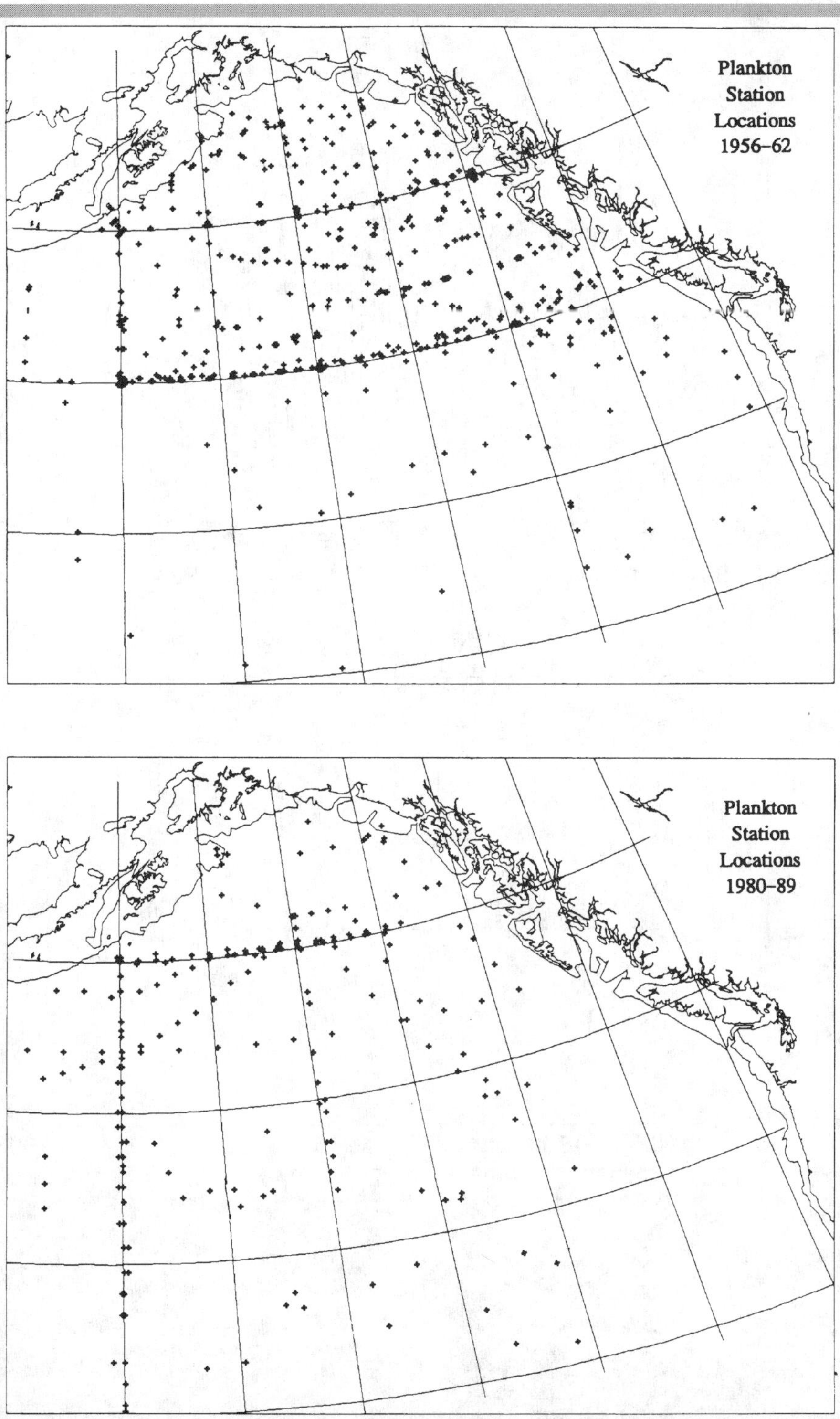

Figure 3. NORPAC zooplankton sampling stations (*plus signs*) during 1956–62 and 1980–89.

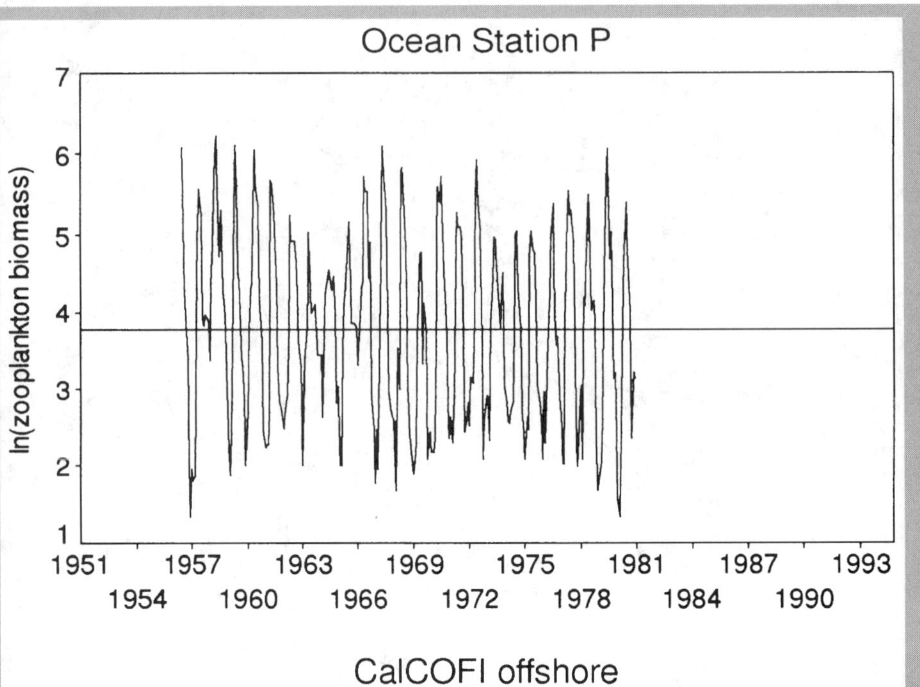

Figure 4. Log-transformed zooplankton biomass time series for Ocean Station P and offshore CalCOFI sampling areas.

Ocean Surface Current Simulations

Due to a lack of time series of open-ocean current data, we used a model developed for the North Pacific Ocean which provided a continuity of surface mixed-layer currents through space and time. The OSCURS (Ocean Surface CURrent Simulations) model uses grided daily sea-level pressure fields to compute daily winds, and from them to compute daily ocean surface currents (Ingraham and Miyahara 1988). The long-term mean geostrophic current vectors computed from existing temperature and salinity versus depth data are added vectorially to these daily currents, and the model is then tuned to existing satellite-tracked drifter data.

Production Trends in the Central Subarctic Domain

Phytoplankton and Primary Productivity

Nearly 30 years of observations at Station P indicate no major seasonal variation in phytoplankton biomass (figure 5; see also Wong et al. 1995). There are generally no phytoplankton blooms, and the mean phytoplankton biomass is always very low, averaging about 0.4 mg chlorophyll m^{-3} (figure 5; Banse 1994). There are no indications of long-term increases in chlorophyll over this time period (figure 6) similar to those documented for the central North Pacific Ocean by Venrick et al. (1987).

However, as model studies show (Frost 1993; Fasham 1995), phytoplankton biomass is not necessarily a sensitive indicator of either seasonal or interannual variations of phytoplankton productivity. Depth-integrated productivity at Station P shows a pronounced seasonal cycle, increasing in early spring with the seasonal increase in insolation and peaking in midsummer at levels 4–8 times as high as in winter (Sambrotto and Lorenzen 1987; Welschmeyer et al. 1993; Boyd et al. 1995; Wong et al. 1995). Estimates of phytoplankton production rate at Station P also

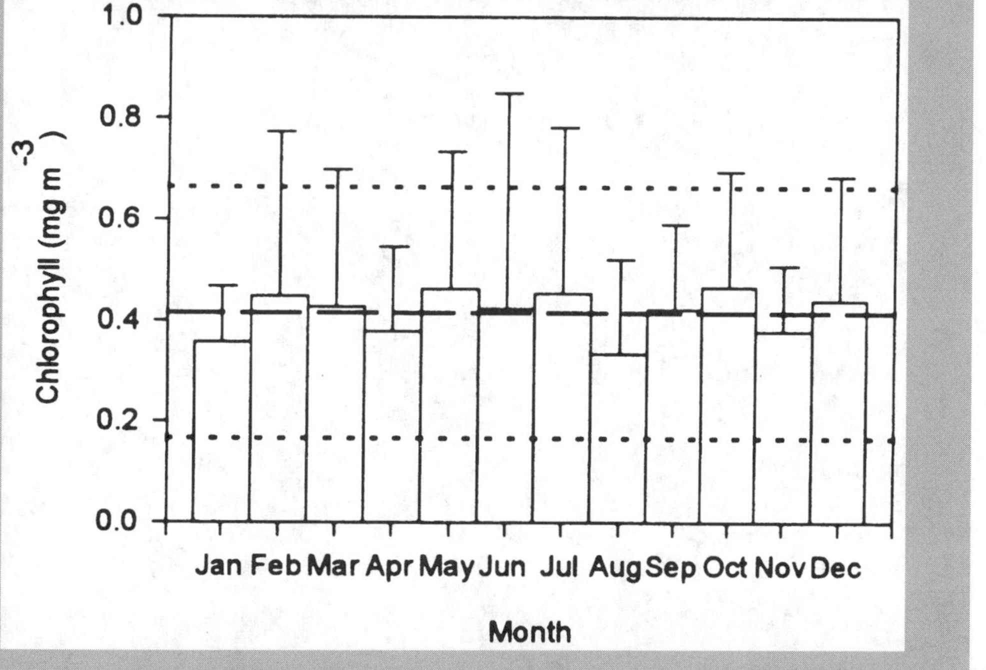

Figure 5. Monthly mean surface chlorophyll a concentration at Ocean Station P. *Error bars* are standard deviations (s.d.) of the mean. Also shown is the overall annual mean (*dashed line*) and ±1 s.d. of the annual mean (*dotted lines*).

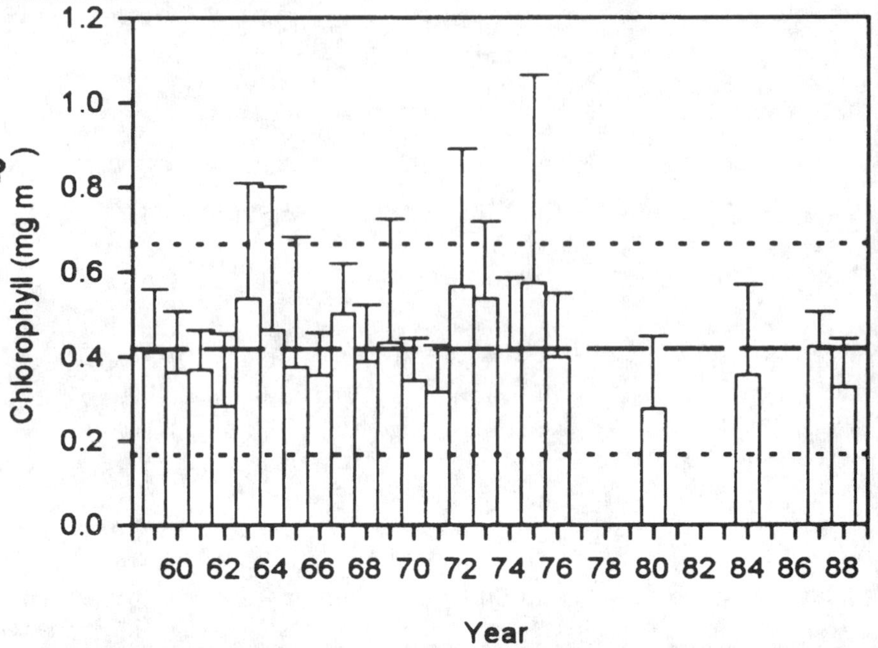

Figure 6. Annual mean and standard deviation of surface chlorophyll a concentration at Ocean Station P. Means for the period 1959–76 are based on data in McAllister et al. (1959). McAllister (1962), and Stephens (1964, 1966, 1968, 1977); two very high values in 1975 (21 June, 26 October) were excluded from the analysis because of lack of supporting evidence from either subsurface depths or preceding and subsequent dates. Sources of other data: 1980 (C. B. Miller, pers. comm.); 1984 (C. J. Lorenzen, unpubl. data rep.); 1987 and 1988 (N. A. Welschmeyer, pers. comm.). Also shown are the long-term mean value (*dashed line*) and ±1 s.d. of the mean (*dotted lines*) for the entire data set.

suggested higher rates in the 1980s than in the 1960s and 1970s, but this is considered inconclusive evidence for interdecadal variation in phytoplankton production rate because of methodological problems prior to the 1980s (Welschmeyer et al. 1993; Wong et al. 1995). Moreover, indirect evidence of changes in phytoplankton productivity based on nitrate removal during the spring–summer period provides no suggestion of interannual trends (C. S. Wong, IOS, Sydney, B.C., pers. comm.). Wong et al. (1995) noted that because availability of dissolved iron may affect phytoplankton production rate at Station P (Martin and Fitzwater 1988), interannual variations in atmospheric input of iron could produce parallel changes in phytoplankton production, but any significant interdecadal changes in phytoplankton production rate induced by this mechanism also should be evident in seasonal nitrate removal.

Zooplankton

SEASONAL VARIABILITY. Zooplankton biomass is highly seasonal at Station P, peaking in May and June and decreasing almost an order of magnitude from late fall to early spring (October to March; figure 7). This contrasts with the situation for the offshore CalCOFI grid, which shows some seasonality but only about a doubling between

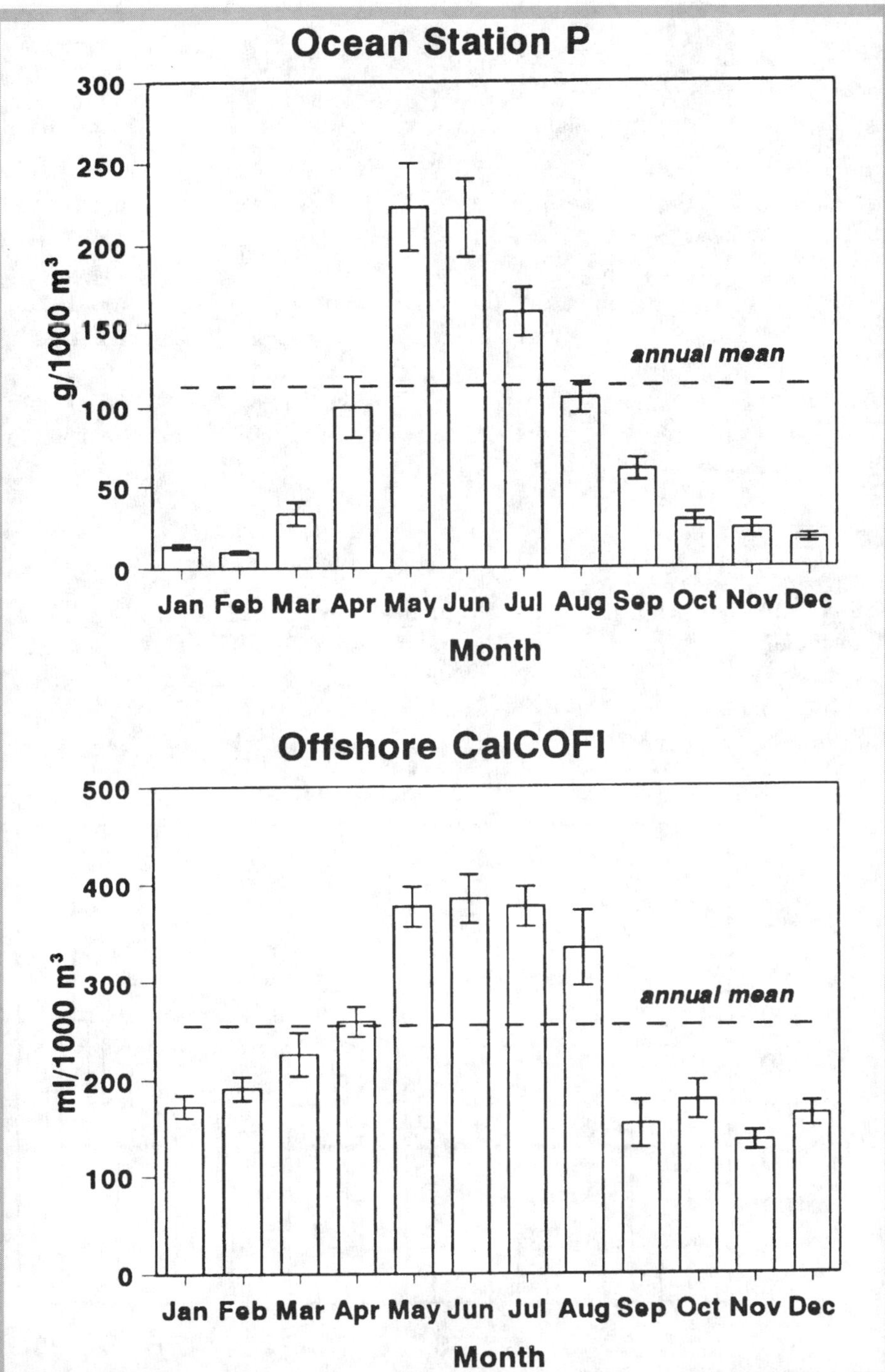

Figure 7. Monthly values of zooplankton biomass at Station P and in the offshore region of CalCOFI. The *bars* represent the means, and the *error bars* represent ±1 standard error. We calculated the Station P data as means of the monthly means for each year (17–23 values per month) rather than using all data within a month (range 104–339 observations per month). The annual means for each area are shown as a *dashed line.*

winter and summer (figure 7). With each month given equal weighting, the annual mean weight of zooplankton at Station P is 82 g/1,000 m^3. The annual mean displacement volume (260 ml/1,000 m^3) for the CalCOFI region converts to a higher wet weight (using wet weight = 0.8 displacement volume; Wiebe 1988) than that seen for Station P.

INTERANNUAL VARIABILITY. In the entire data set of zooplankton biomass from Station P (figure 4), there appears to be no discernible trend over the 26-year period. However, Brodeur and Ware (1992) found a significant long-term increase in biomass for a 6-week period in early summer, when biomass is near its annual peak. We reanalyzed these data using the new gear-correction factor (see Methods) and still found a significantly positive relationship ($p = 0.003$) despite the high annual variability and the fact that so few years are represented beyond the 1976–77 regime shift (figure 8). For the large-scale sampling (figure 9), a discernible shift in zooplankton biomass was observed between the earlier period (1956–62) and most of the 1980s (up to 1988). The mean zooplankton biomass for the years after 1988, although above the long-term mean, was generally intermediate relative to the other periods examined, and showed more interannual variability (Figure 9).

GEOGRAPHIC VARIABILITY. The large-scale zooplankton biomass distributions showed dramatic differences between the two periods not only in the overall biomass, as described previously, but also in the distribution patterns for each time period (figure 10). High biomass in the 1956–62 period occurred mainly in the central part of the Gulf of Alaska, with much lower concentrations nearer to the coast. Conversely,

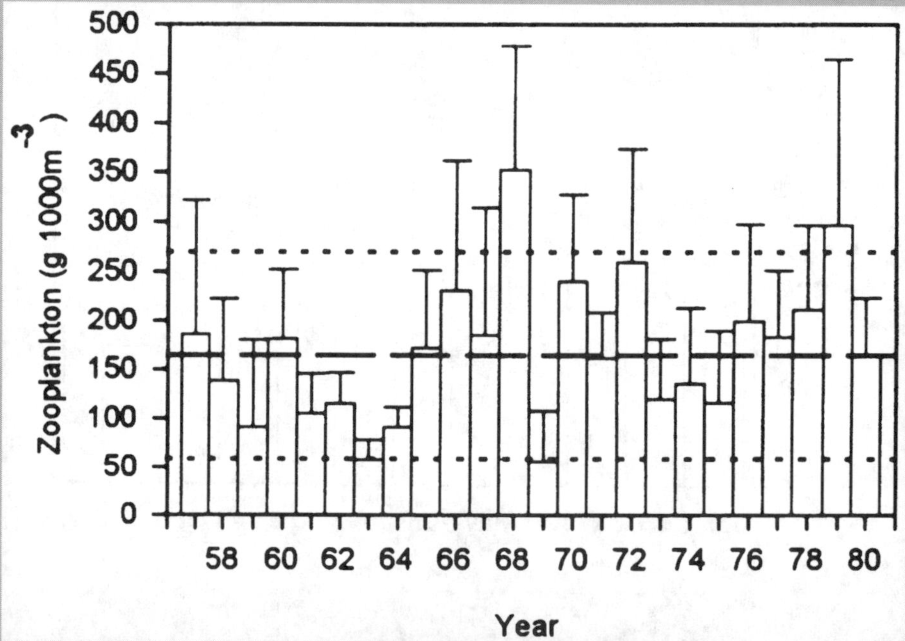

Figure 8. Annual values of zooplankton biomass (means +1 s.d.) at Station P from 1956 to 1980 for the 6-week period beginning June 1. Also shown are the long-term mean value (*dashed line*) and ±1 s.d. of the mean (*dotted lines*) for the entire data set.

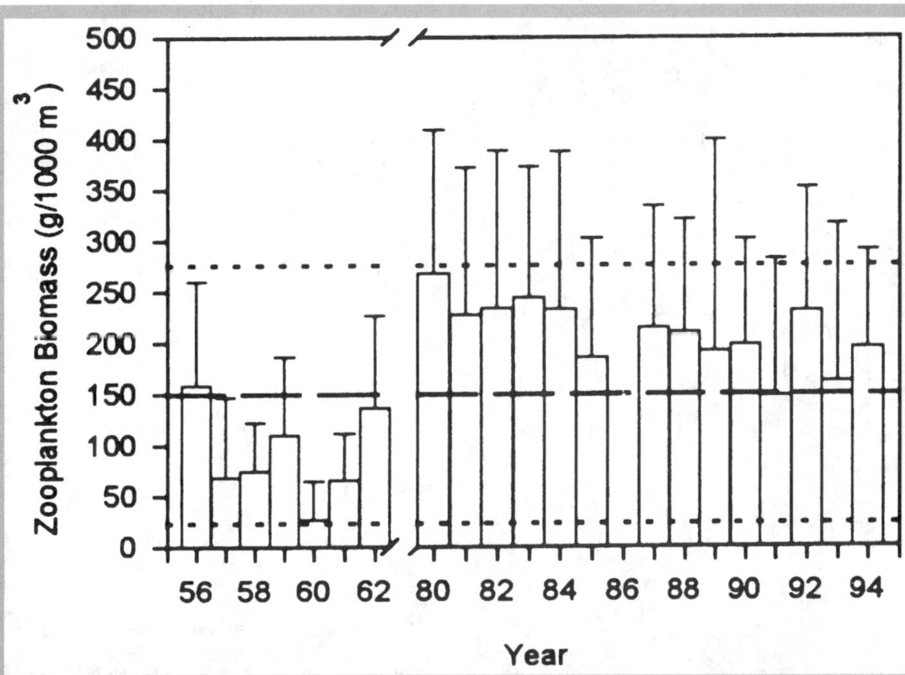

Figure 9. Annual values of zooplankton biomass for the large-scale sampling in the subarctic and transition regions of the North Pacific. Shown are the annual mean (*histograms*) and +1 s.d. (*error bars*). The *dashed line* is the mean for all years, and the *dotted lines* are +1 standard deviation of the long-term mean.

during 1980–89 elevated biomass occurred in a continual band from the Transition Zone up to the northern coast of Canada and Southeast Alaska, and then extended westward along the shelf edge (figure 10). The overall pixel distribution of biomass between the periods showed that almost all the biomass values in the later period were above the mean value for the earlier period (figure 10).

Analysis of Covariation in Station P and CalCOFI Data Sets

The magnitude of anomalous values was similar in both regions, but the CalCOFI data showed an apparent long-term decline in zooplankton biomass over the last two decades (figure 4). Autocorrelation plots were examined to determine whether serial autocorrelation exists within each region (figure 11). Both monthly time series show significant autocorrelation. The CalCOFI data are significantly autocorrelated up to 12 months, and all correlations up to 24 months are positive. Conversely, the Station P data show that anomaly events are much more short-lived (lasting 2–3 months), and little temporal pattern is evident after this time.

The cross-correlation between the autocorrelated data sets shows highly significant lagged correlations in both directions (figure 12). The presence of autocorrelation within both time series, however, renders this relationship highly suspect (Katz

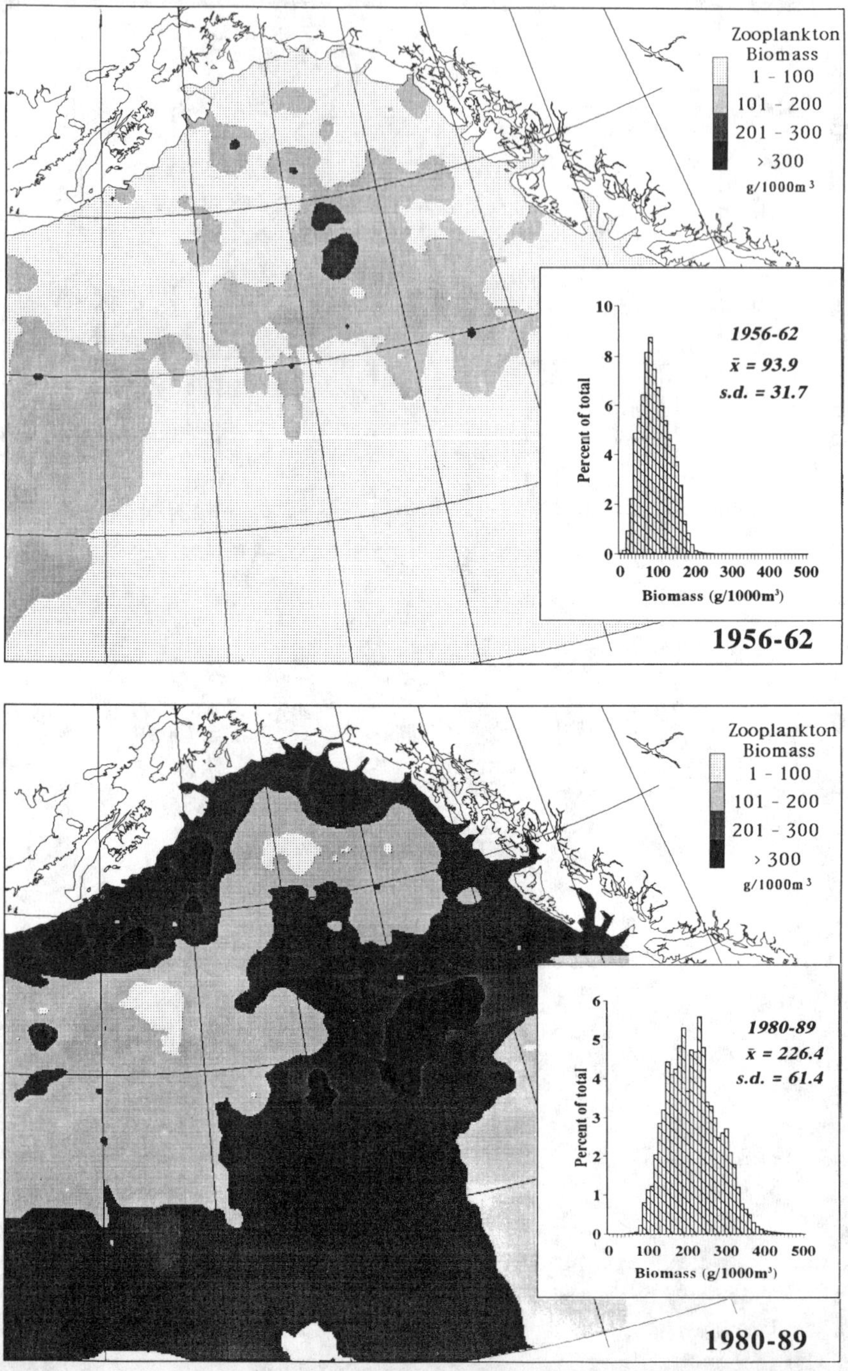

Figure 10. Large-scale distribution of zooplankton biomass from sampling during the 6-week period beginning June 1 for 1956–62 and 1980–89. See figure 1 for locations of sampling stations for each period. The insets show the zooplankton biomass pixel distributions as a percentage of the total number of pixels for each time period. The overall mean and standard deviation of biomass for the time period are given.

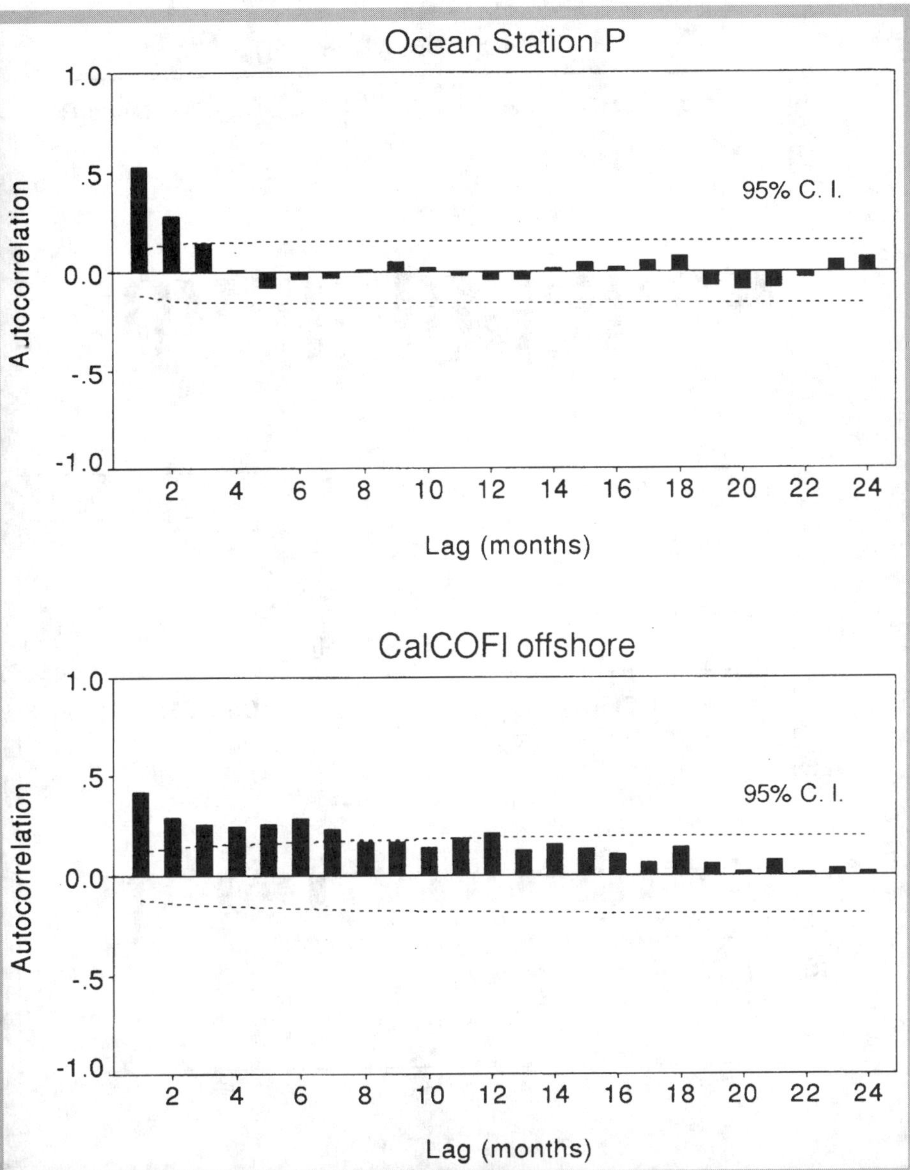

Figure 11. Serial autocorrelation of Station P and offshore CalCOFI zooplankton-biomass at various time lags. Upper and lower 95% confidence intervals are indicated as *dashed lines.*

1988; Newton 1988). To ascertain whether there is a real relationship, it is necessary to prewhiten both series—that is, remove the autocorrelation structure—and then plot the CCF of the residual series. Both series were adequately described by a lag-1 autoregressive (AR1) model. Separate filters were used for the two series ("double prewhitening"), and the coefficients were approximately equal to the lag-1 autocorrelation values—0.412 for CalCOFI and 0.524 for Station P. After filtering, almost all the significant lag correlations disappeared (figure 12). The only one that remained

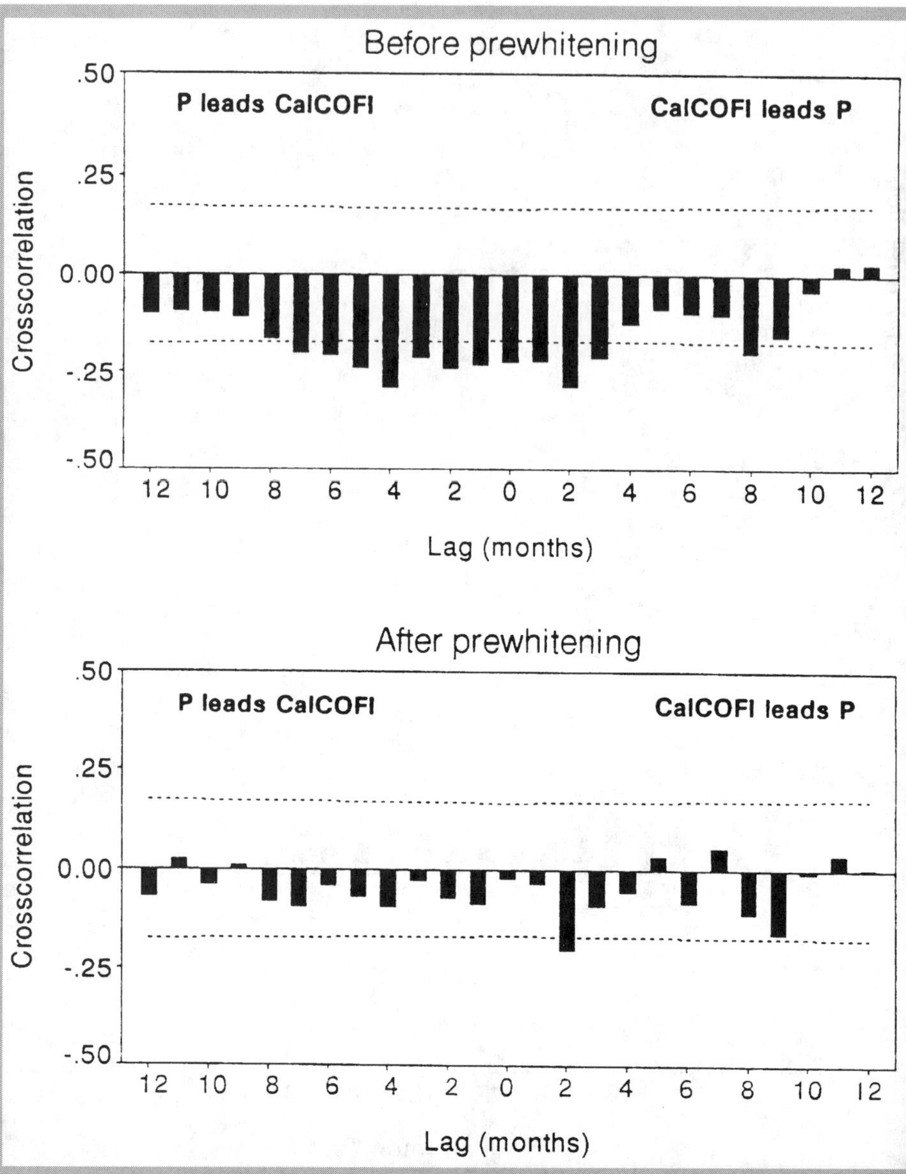

Figure 12. Cross-correlation between Station P and offshore CalCOFI zooplankton biomass at various time lags before and after prewhitening. The upper and lower 95% confidence intervals are indicated as *dashed lines.*

(CalCOFI leading Station P by 2 months) was quite small (−0.203) and possibly spurious.

The effect of prewhitening, while statistically justified, may also have the side effect of overcompensating for autocorrelation and removing evidence of an actual signal. As an alternative to prewhitening, an "effective degrees of freedom" is sometimes employed when calculating the confidence intervals around the cross-correlation estimates (Trenberth 1984). In the case of two AR1 time series, the true standard deviation at lag 0 is inflated over the no autocorrelation case by

$$f = [(1 + \phi_X{*}\phi_Y)/(1 - \phi_X{*}\phi_Y)]^{0.5}$$

where ϕ_X and ϕ_Y are the lag-1 autocorrelation coefficients (Katz 1988). Thus at lag 0, the standard deviation is approximately 1.25 times greater than that computed on an assumption of independent data points.

Assuming an approximately equal inflation at other lags, only lags −4 (Station P leads CalCOFI by 4 months) and +2 (CalCOFI leads Station P by 2 months) are significant at the 0.05 level. The −4-month lag is interesting in that it is interpretable in terms of ocean physics. A parcel of water originating upstream in the Subarctic Current would arrive at Station P approximately 4 months ahead of a particle diverted to the CalCOFI region (see next section). The +2-month lag may or may not be spurious, but it has no obvious physical explanation (though see below).

We next examined within- and between-area lag correlations for the yearly 3-month seasonal averages. The seasonal data exhibited almost no autocorrelation among years, a result that can be deduced from the monthly correlations, which die out by 12 months. Therefore, the cross-correlations between seasonal averages did not require a prewhitening step. Most within-region correlations were positive (10 of 12), with the CalCOFI data showing much stronger between-season coherence than the Station P data (table 1).

We examined cross-region correlations lagged up to a year in each direction (table 2). Out of a total of 36 separate correlations, 32 were negative. Of those 32, 5 were significant at the 0.05 level. There are a few details worth noting about the pattern of correlations. Those where Station P leads CalCOFI tend to be stronger than the converse set, particularly for the winter and spring seasons. If we focus only on the spring months, which is the time when these ecosystems are most likely to respond to variations in the north–south diversion from winter Aleutian Low intensity (figure 13), there is indication of a strong, lag-0 inverse relationship between Station P and offshore CalCOFI ($r = -0.62$; $n = 16$; $p = 0.011$). This relationship remains strong

Table 1. Within-region correlation matrices of Ocean Station P and offshore CalCOFI seasonal biomass time series (log-transformed and zero-centered)

	Winter	Spring	Summer	Fall
Ocean Station P				
Winter	1.000	0.221	0.005	0.387
Spring		1.000	<u>0.474</u>	−0.163
Summer			1.000	−0.032
Fall				1.000
Offshore CalCOFI				
Winter	1.000	<u><u>0.498</u></u>	0.242	0.351
Spring		1.000	<u><u>0.590</u></u>	<u>0.376</u>
Summer			1.000	<u>0.411</u>
Fall				1.000

Single underline indicates significance at the 0.05 level, double underline at the 0.01 level.

Table 2. Cross-region lag-correlation matrices of Ocean Station P and offshore CalCOFI seasonal biomass time series: Cross-region correlations are shown up to 4 seasons later

Ocean Station P leads CalCOFI	C winter	C spring	C summer	C fall	C winter + 1[a]	C spring + 1	C summer + 1	C fall + 1
P winter	−0.205	−0.315	−0.343	−0.039	−0.182			
P spring		−0.615	−0.397	−0.463	−0.218	−0.172		
P summer			−0.342	−0.506	−0.269	−0.055	0.256	
P fall				−0.009	−0.210	−0.376	−0.617	−0.062

CalCOFI leads Ocean Station P	P winter	P spring	P summer	P fall	P winter+ 1	P spring+ 1	P summer+ 1	P fall+ 1
C winter	−0.205	−0.263	0.121	−0.011	−0.018			
C spring		−0.615	−0.309	−0.026	−0.004	−0.217		
C summer			−0.342	−0.123	−0.322	−0.571	−0.412	
C fall				−0.009	−0.112	−0.610	−0.068	0.328

[a] "+ 1" indicates a season in the following calendar year.
Underlined values are significant at the 0.05 level. The cross-correlations with no lag are repeated in each section of the table for consistency (first diagonal row).

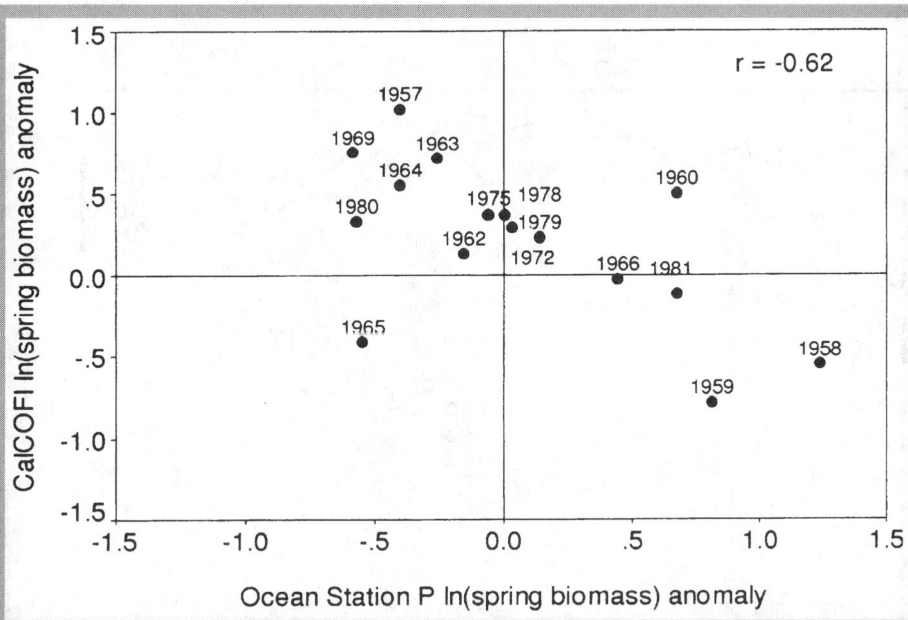

Figure 13. Relationship between Station P and offshore CalCOFI biomass anomalies for the spring quarter (March–May) by year.

even after the weak autocorrelation in the California Current data is removed ($r = -0.58$).

Comparison to Environmental Variables

In order to examine variability of surface flow into the Gulf of Alaska, we started OSCURS model runs at Station P on December 1 and ran them for 3 months for all years from 1946 to 1994. A substantial divergence of the tracks occurred in the eastern part of the gulf, with some tracks going north into the Alaska Current and others heading into shore or turning south. When the tracks are partitioned into pre-1975 and 1975 and later, some decadal changes become evident (figure 14). In the earlier period, flow into the Alaska Current was more relaxed, and a substantial number of the trajectories veered southward. The later period appeared to show stronger flow into the Gulf of Alaska, and relatively few trajectories went south. The occurrence of either northward or southward flow tends to run in series of various lengths (table 3). Simulations started 5° south of Station P (45°N, 145°W) showed more directed eastward and southward flow but again showed some differences between the two time periods (figure 15).

The model was then used to simulate the north–south divergence of the Subarctic Current along its eastern boundary for equivalent 5-year time periods before and after the regime shift (1971–75 vs. 1976–80) in order to assess changes in circulation between the two regimes. Long-term mean flow tracks begun in January along 145°W showed a more southern diversion before the regime shift than after, with the primary differences in the trajectories starting at or north of 48°N (track 5 in figure 16).

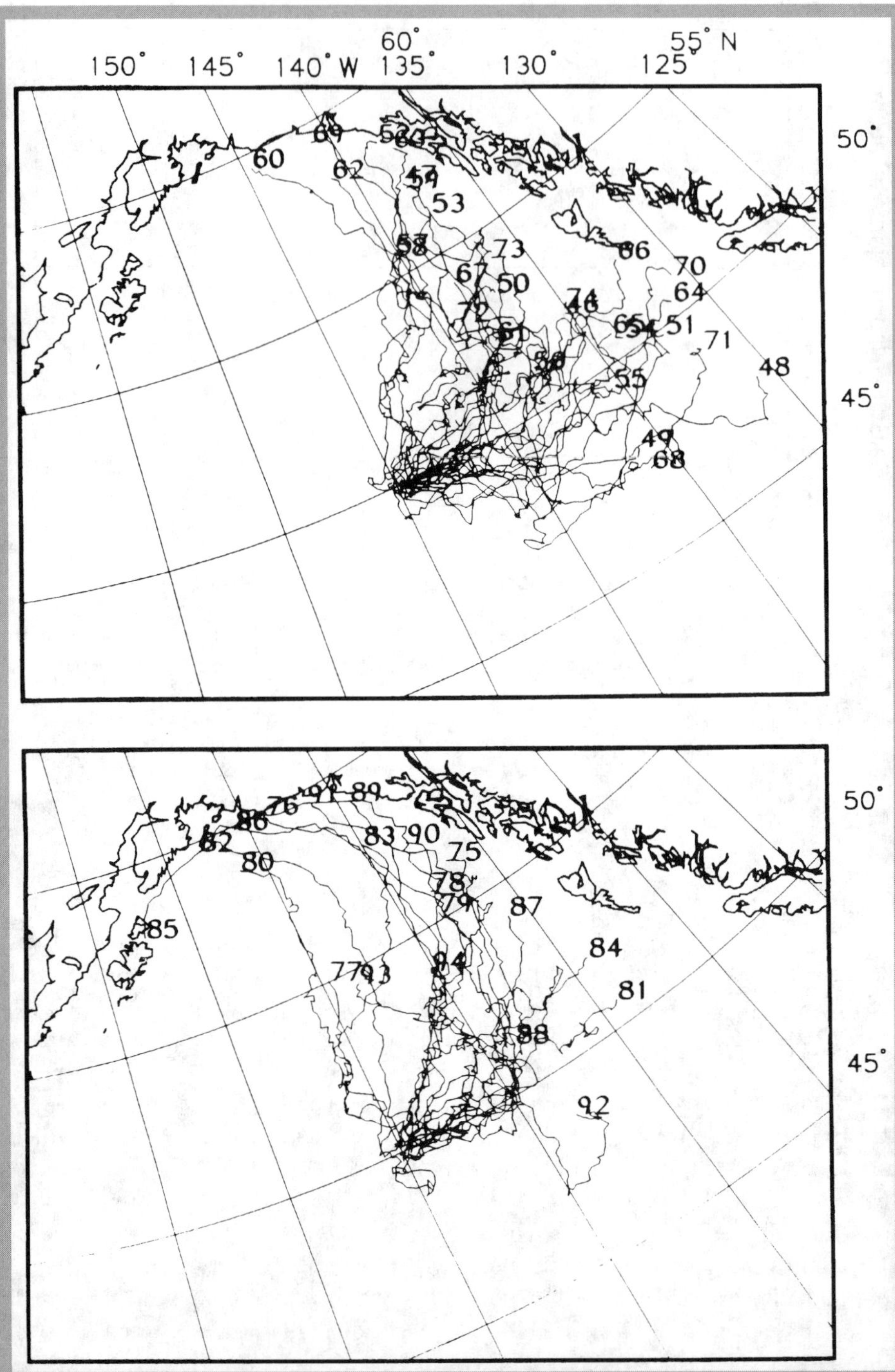

Figure 14. Simulated flow trajectories for the OSCURS model runs from 1946 to 1974 (*top*) and 1975 to 1994 (*bottom*). Each trajectory was started at Station P (50°N, 145°W) on December 1 and run for 3 months, incrementing the positions daily. Each year is marked at the endpoint of its trajectory.

Table 3. Yearly anomalies of simulated OSCURS trajectories from Ocean Station P north or south of the long-term mean trajectory, in three-month (Dec.–Feb.) trajectories for 1946 to 1994

North	South	North	South
	46		71
47			72
	48		73
	49		74
	50	75	
	51	76	
52		77	
53		78	
	54	79	
	55	80	
	56		81
57		82	
58		83	
59			84
60		85	
	61	86	
62			87
63			88
	64	89	
	65	90	
	66	91	
67			92
	68	93	
69		94	
	70		

Discussion

For oceanic waters of the eastern subarctic Pacific Ocean, direct evidence for interannual and decadal variations in biological production—that is, phytoplankton production—is weak. However, zooplankton standing stock, when viewed on a basinwide scale, has varied and seems to be higher since the 1976–77 regime shift. Recognizing that primary production might not have changed between the regimes, Brodeur and Ware (1992) hypothesized a causal link between increased wind stress and increased zooplankton stock and production, in that intensified wind mixing would result in deeper mixed-layer depths (MLD) in winter. This would slow the growth rate of the phytoplankton, retard the spring increase in primary production, and allow grazers to make more efficient use of phytoplankton production. Experiments with an ecosystem model (Frost 1993) do not provide support for such a mechanism. Indeed, just the opposite effect should occur. A deeper mixed layer in winter should result in decreased balance between phytoplankton growth and grazing in spring, when the surface layer restratifies. Phytoplankton production should be less efficiently utilized

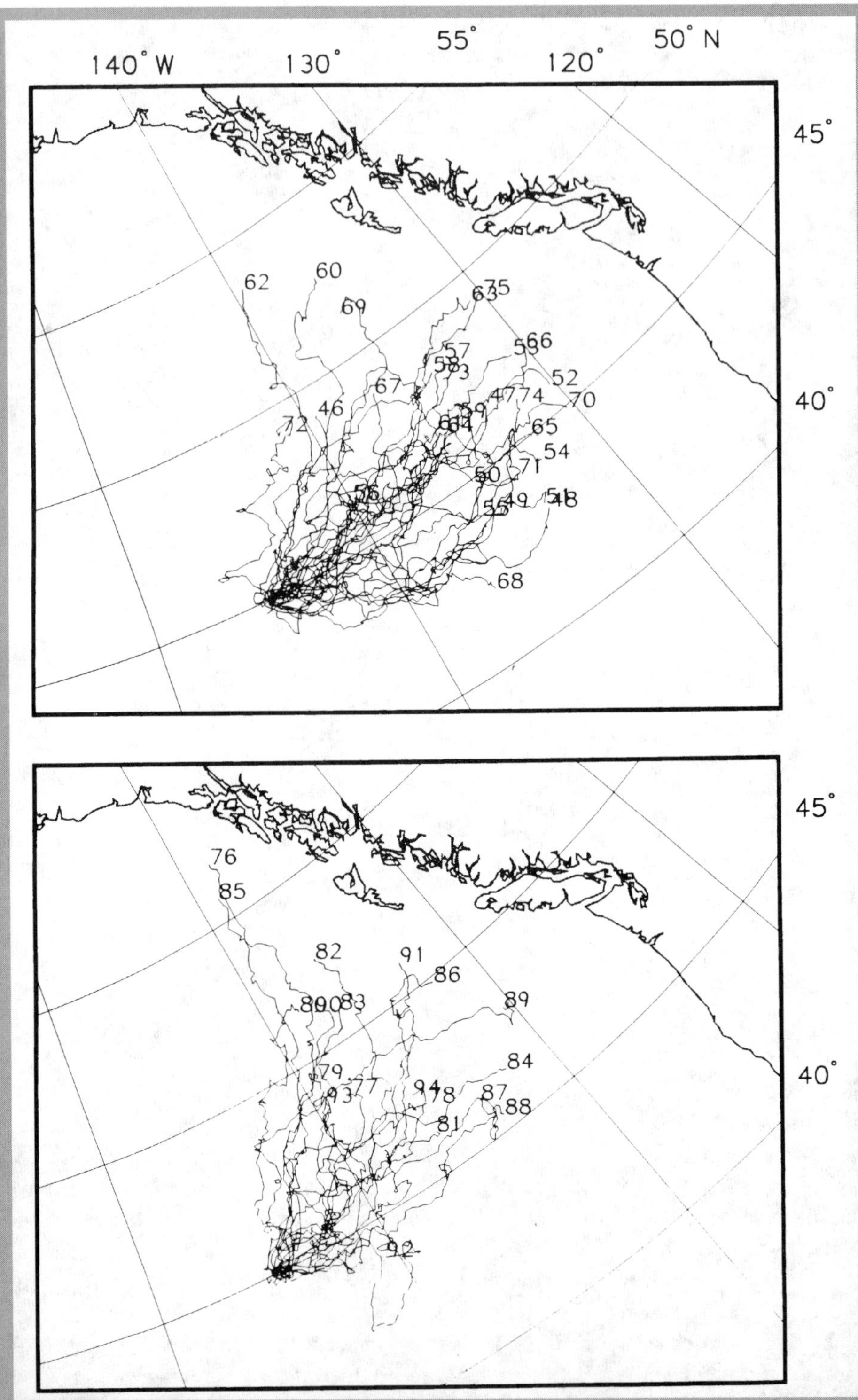

Figure 15. Same as Figure 14 but with the simulations started 5° south of Station P (45°N, 145°W).

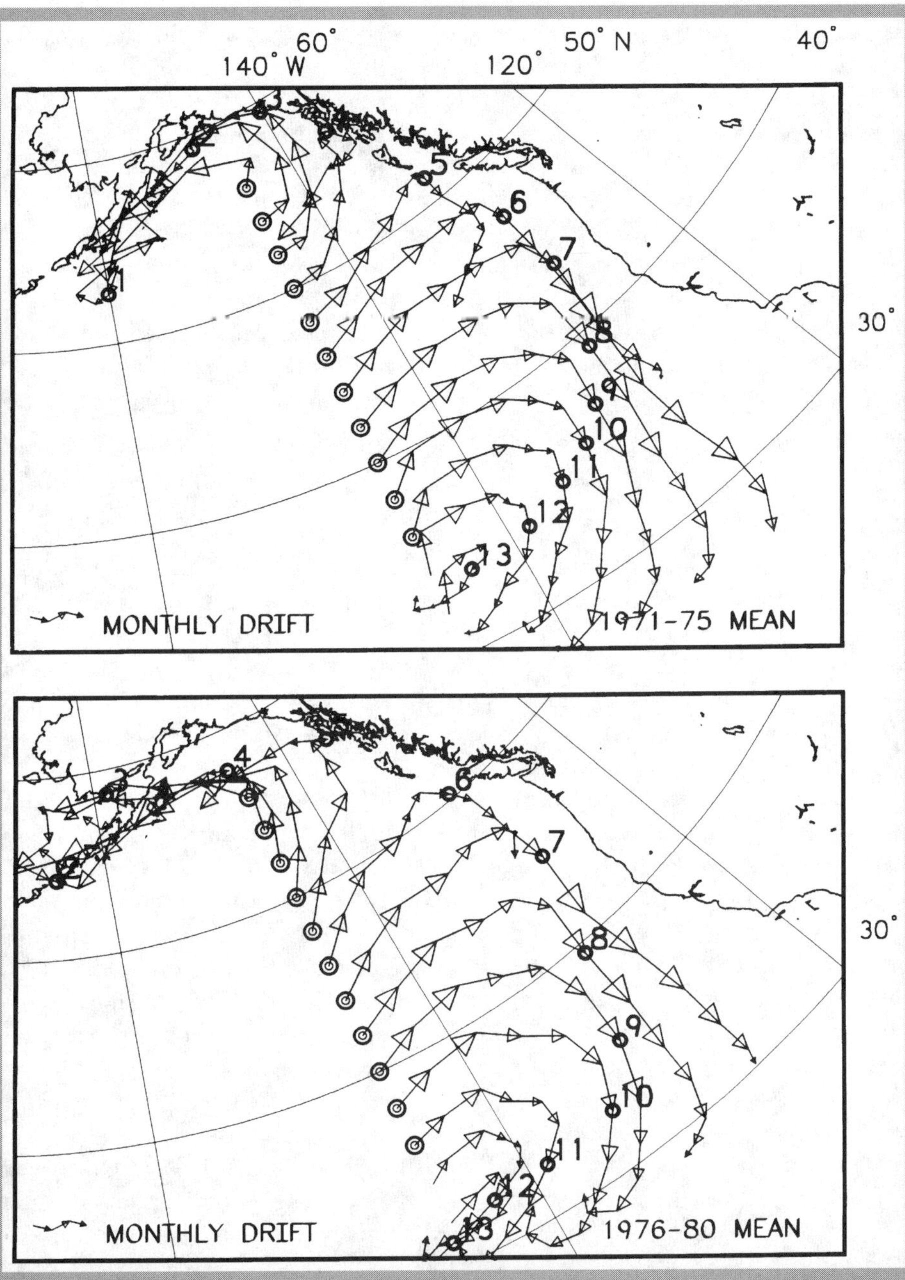

Figure 16. Mean simulated flow trajectories for OSCURS model runs for the 5-year period before (*upper*) and after (*lower*) the regime shift. Model runs were started at 145°W on January 1 and run for 12 months. The size of the arrowhead and the length of the shaft indicate the relative current speed. Each trajectory is marked with a circle after 6 months.

by grazers, and more production should be lost to mixing and sinking below the surface layer.

But, in fact, there is not much interannual variation in MLD at Station P because of the halocline, and over the observed interannual range of winter MLD (80–130 m), Frost's (1993) model suggests that such variations would have relatively little effect on biological production or zooplankton biomass at Station P (Frost, unpubl. data). Using a 1-D dynamic mixed-layer model (modified Garwood model) coupled to a nitrate-phytoplankton-zooplankton (NPZ) model, McClain et al. (1996) found surprisingly little interannual variation in phytoplankton production rate at Station P.

It is likely that events at Station P are not generally representative of the entire open Gulf of Alaska, due both to the singularity of the station and its location. Polovina et al. (1995) reported on interannual changes in winter–spring mixed-layer depth throughout the North Pacific Ocean. There was considerable spatial variation in change in MLD, with the region of Station P showing little long-term change but other areas showing rather large changes (e.g., the northeast Gulf of Alaska). Polovina et al. (1995) used an NPZ model to look at the effects of their modeled changes in MLD. The model predicts that large changes in MLD will have little effect on phytoplankton stock, but potentially large effects on phytoplankton production rate and zooplankton stock in the mixed layer. Thus, spatial variation in the processes controlling production rate may explain the apparent increase in zooplankton stock evident on a large spatial scale in the eastern subarctic Pacific Ocean during the 1980s. This hypothesis is difficult to test because there are no basinwide data on nutrient concentrations and biological observations.

To our knowledge, there is no extensive time series of phytoplankton standing stock for the California Current region comparable to the data set from Station P. However, the analysis by Fargion et al. (1993) of Coastal Zone Color Scanner data and in situ chlorophyll measurements indicates little seasonal variation in phytoplankton stock in the offshore area represented by the zooplankton data presented here (30° to 35°N). Moreover, phytoplankton pigment concentrations are similar to those at Station P (cf. Chavez 1995). Because the California Current region covered by the zooplankton data is at considerably lower latitude than Station P, it is probable that higher phytoplankton production accounts for the higher zooplankton standing stock observed (figure 7).

Two factors hinder definitive identification of a lead/lag or "out-of-phase" relationship between zooplankton productivity at Station P and the CalCOFI region. First, as noted earlier, there is a great deal of missing data in both regions. Secondly, the high observed autocorrelation, the robust estimation of which is also hindered by data gaps, needs to be removed. Despite these difficulties, there is evidence that the two regions are negatively related. It is less clear whether there is a consistent lag time between the two regions. The evidence points more toward anomalies at Station P leading anomalies at CalCOFI. It is possible that the inverse relationship is due to differential (i.e., inverse) flow from the Subarctic Current, but that the flow speed within each regime is highly variable, hence the lack of a consistent lag relationship.

El Niño–Southern Oscillation (ENSO) events are another factor that may play a role in the timing and intensity of anomalous zooplankton production. During ENSO events, positive SST anomalies are propagated poleward in the form of coastally trapped Kelvin waves. Roemmich and McGowan (1995a, b) attributed the decline in CalCOFI region zooplankton biomass to sea surface warming, part of which resulted from a large number of ENSO events since the mid-1970s. While sea surface temperature anomalies associated with ENSO events have occurred in the Gulf of Alaska (Wooster and Fluharty 1985), as often as not, there has not been any North Pacific Ocean response (Freeland 1990; Bailey et al. 1995). Thus, depending on the magnitude and northward extent of the ENSO event and the Station P zooplankton response to surface warming, there is the potential for a Station P response to lag a CalCOFI response. At the very least, the ENSO factor serves to cloud the relative effects on zooplankton productivity from variations in the Subarctic Current.

It may not be coincidental that the increase in zooplankton shown here in the Subarctic Pacific is opposite to the trend for California Current zooplankton reported by Roemmich and McGowan (1995a, b). Although there is some indirect biological evidence for the ocean-circulation model proposed by Hollowed and Wooster (1992), the hydrographic evidence is more limited. However, several recent papers shed some light on the issue and suggest considerable modification to the model. Tabata (1991), in reexamining the Chelton and Davis (1982) premise, found a correlation between the coastal component of the Alaska Current and California coastal sea levels, particularly during El Niño years. He attributed this correlation, however, to the coastal currents being in phase from Canada to California rather than to changes in the bifurcation of the Subarctic Current. Kelly et al. (1993) analyzed sea-surface height anomalies for the Northeast Pacific Ocean over a 2.5-year period. Their results tended to support those of Chelton and Davis (1982) that the California and Alaska Current systems fluctuate "out of phase," coinciding with variations in wind-stress curl in the North Pacific Ocean and subsequent diversion of flow from the Alaska Gyre into the California Current, as well as with some correlation with ENSO dynamics. Van Scoy and Druffel (1993), in an analysis of tritium (^3H) concentrations in seawater from Ocean Station P and a station in the southern California Current, suggest that there is increased advection of subpolar water into the California Current during non–El Niño years and that ventilation of the Alaska Gyre (intensification) occurs during El Niño years.

Lagerloef (1995), in his analysis of dynamic topography in the Alaska Gyre during 1968–90, suggested that after the well-documented climatic regime shift of the late 1970s, the Alaska Gyre was centered more to the east and that its circulation appeared weaker after the shift than before. The implication is that the intensification of the winter Aleutian Low associated with the regime shift did not result in a spin-up of the Alaska Gyre.

Finally, Miller (1996) reviews some recent advances in large-scale modeling of the California Current and its interaction with basin-scale circulation and forcing. He reports the significant deepening of the thermocline off California after the 1976–77 regime shift similar to that described by Roemmich and McGowan (1995a) and attributes this to basin-scale changes in wind-stress curl. This is achieved at two time scales—the first at the decadal and North Pacific Gyre scale, forced by significant deepening and weakening of the Aleutian Low, and the second at the interannual ENSO scale, forced by waves propagating through the ocean from the tropics. Miller (1996) also reported that after the 1976–77 regime shift there appeared to be a stronger than normal northward flow into the central Gulf of Alaska but little change in the flow into the California Current system.

If zooplankton biomass is advected preferentially to either region, as the current-simulation model suggests, then this allochthonous biomass should be higher than that produced locally for our results to be valid. There are few comparable measurements of zooplankton biomass in both the Transition Domain and Subarctic Domain. Our data for the large-scale sampling during the 1980s suggest that levels were high in the Transition Domain and are somewhat higher than in the central part of the Alaska Gyre. Data taken in summer for several years from north–south transects in the western subarctic (155°E, 170°E, 175.5°E, and 180°E) show elevated zooplankton wet weights in the transition zone compared with those in the subarctic (Shiga et al. 1995). Sampling along 180° and in the Gulf of Alaska during June and July of 1987 revealed higher zooplankton biomass in transition zone waters than in the central Subarctic Domain, especially in the 150–300-m depth strata (Kawamura 1988).

An alternative explanation for the inverse relationship in zooplankton might be that similar large-scale changes in thermal structure of the western North Pacific (Venrick et al. 1987; Royer 1989; Roemmich and McGowan 1995a; Miller 1996) could have radically different effects on biological production in the two regions. At Station P, the slightly warmer temperature of the mixed layer could directly affect increased zooplankton production rate and standing stock, as suggested by Conversi and Hameed (1996). The same warming and associated deepening of the upper mixed layer (Miller 1996) could cause decreased zooplankton production and standing stock in the California Current region by impeding the supply of nutrients to the surface layer (Roemmich and McGowan 1995a). Our data are not sufficient to allow examination of this alternate hypothesis.

Implications for Higher Trophic Levels

The dramatic increase and change in distribution of mesozooplankton biomass seen in the subarctic Pacific Ocean between the periods 1956–62 and 1980–89 would be expected to have important ramifications for higher trophic levels dependent on these food sources. Brodeur and Ware (1995) documented substantial increases in the catch rates of most pelagic nekton (fishes, squids, and elasmobranchs) caught in research gill nets over roughly the same time periods. The only species that showed a decline in catch rates (jack mackerel, *Trachurus symmetricus*) is primarily a California Current species which migrates into the Gulf of Alaska only during periods of peak abundance. Although these authors were not able to convert catch rates to abundance or biomass because of the paucity of collaborative time series of abundances for the noncommercial species, Brodeur and Ware (1995) estimated that total salmon abundance nearly doubled between these two periods.

For the present study, we combined catch data of the 14 species examined by Brodeur and Ware (1995) and plotted nekton catch-rate distributions for roughly the same two time periods over the same geographic range examined previously for zooplankton. Although there are differences between them, the nekton distribution plots (figure 17) showed some similarities to the zooplankton distribution in that most concentrations are offshore in the Alaska Gyre during the 1950s and occur in a band around the outside of the gyre in the 1980s. The magnitude of the increase in catch rate (figure 17 inset) is also similar to that of the plankton. Although this is not cogent evidence of a strong link between these trophic levels, since there is often an additional trophic level (macrozooplankton and micronekton) between the meso-

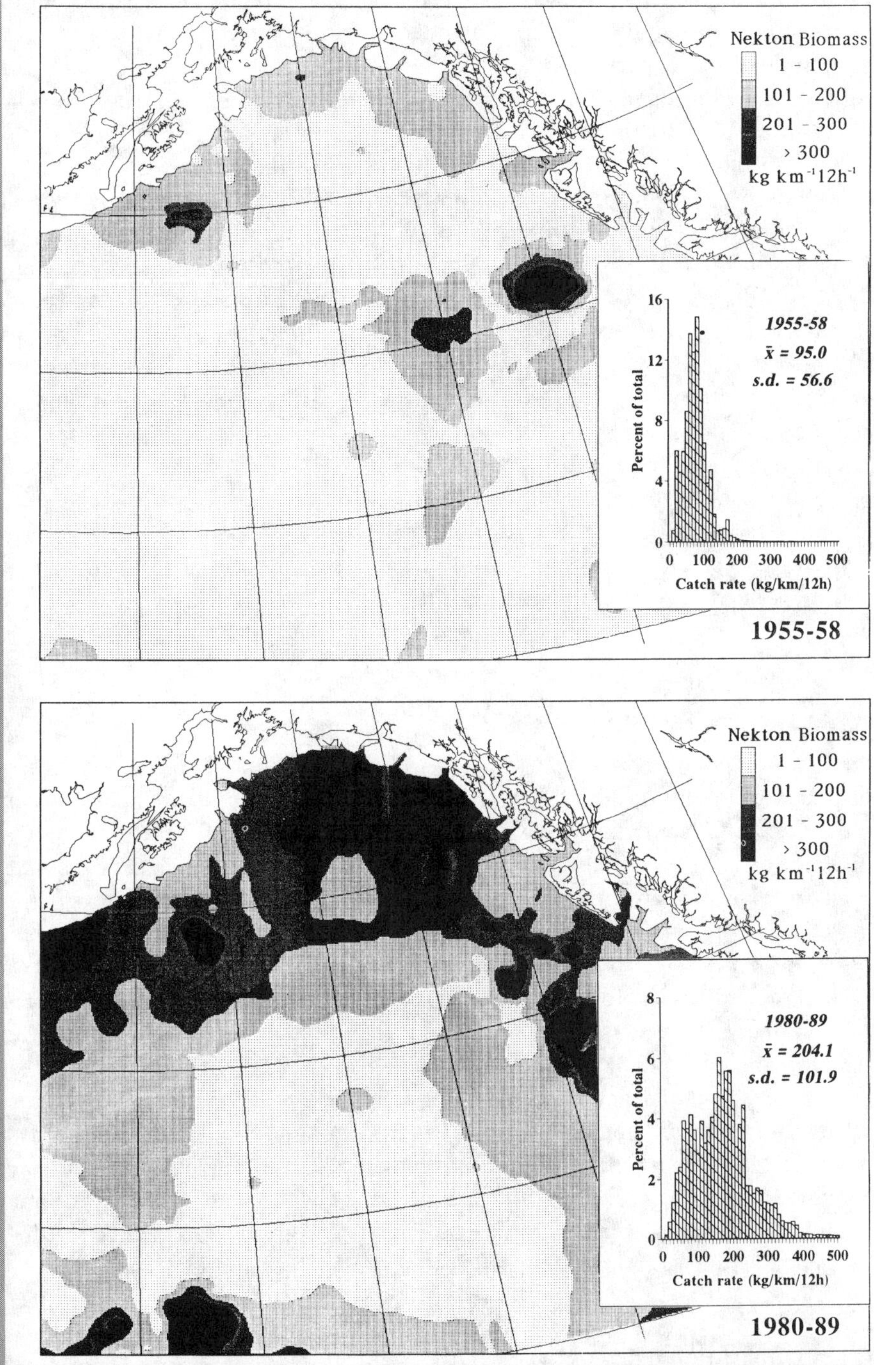

Figure 17. Large-scale catch-rate distribution for 14 species of nekton commonly caught in research gill nets during the periods indicated. See Brodeur and Ware (1995) for sampling methodology, locations of sampling stations, and species included. Insets show nekton biomass pixel distributions as a percentage of the total number of pixels for each time period. The overall mean and standard deviation of biomass for the time periods are given.

zooplankton and the larger nekton, there is enough commonality in the distribution patterns to suggest that the distribution and abundance of zooplankton is positively related to that of higher-level predators.

Coastal fishes in the Gulf of Alaska would be expected to benefit most from the increase in zooplankton biomass that we observed during the 1980s. High rates of upwelling in the center of the Alaska Gyre would push nutrients and subsequent phytoplankton and zooplankton production onto the shelf along the edge of the gulf, thereby stimulating coastal production. Cooney (1986) has suggested that large oceanic species of copepods (*Neocalanus* spp. and *Eucalanus bungii*) are transported onto the shelf in the northern Gulf of Alaska, providing rich food resources for the coastal community. A direct link between atmospheric circulation, oceanic copepod production, and sablefish (*Anoplopoma fimbria*) recruitment has been hypothesized by McFarlane and Beamish (1992), but such mechanisms have not been explored for other demersal fishes.

For an investigation of ocean effects on fish species, Pacific salmon are an attractive group to study, since they have a relatively short life span, show substantial interannual variability in marine survival, and can be reliably censused at least several times during their life history. As discussed previously, Pacific salmon stocks have substantially increased in abundance since the mid-1970s in Alaska waters, whereas southern stocks have shown opposite trends (Pearcy 1992; Beamish 1994; Hare and Francis 1995). In some cases, the inverse relation between stocks in the two domains is striking (Francis and Sibley 1991). Our analyses suggest that zooplankton biomass in the subarctic region is inversely related to that in the California Current region. A combination of increased transport into the Alaska Current and advection of nutrients and zooplankton onto the shelf would probably increase the carrying capacity for juvenile salmon entering Alaska coastal waters (Cooney 1984).

By studying time lags between atmosphere/ocean and salmon statistics, Francis and Hare (1994) indicated that this regime-scale effect on Alaska salmon production is most likely to be felt during the early ocean life history. If salmonid production and survival are limited by factors occurring early in their marine life history, then the relative flow into the California Current and Alaska Current may profoundly affect their dynamics by enhancing prey production for smolts in the coastal zone. However, the increasing number of salmon surviving to maturation in the open ocean after the regime shift may have imposed an excess burden upon the oceanic zooplankton, which did not appear to increase, as dramatically as those in the coastal zone (figure 10). It is likely that the amount of zooplankton available per individual salmon has decreased over this period, as suggested by Peterman (1987), which may be manifested in the long-term decreases in size at age and the older age of maturity witnessed in several salmon stocks (Ishida et al. 1993; Helle and Hoffman 1995).

Suggestions for Further Study

1. Examine taxonomic composition of zooplankton over time to see if shifts in species composition have occurred along with the decadal-scale biomass shifts. This objective has been facilitated by the entry of the entire Station P detailed zooplankton data set in digital format that may be amenable to analyses (Waddell and McKinnell 1995).

2. Construct more spatially-explicit coupled physical and NPZ models to account for geographic variability in ocean conditions, nutrient input, and phytoplankton and zooplankton species composition (e.g., as in McGillicuddy et al. 1995).
3. Use models to examine potential top-down control on phytoplankton and zooplankton populations, extending—if possible—some of the presently available models (e.g., Frost 1993) to include nekton.
4. Establish new oceanic sampling sites for comparison with Station P to see whether processes occurring at Station P are representative of the subarctic region as a whole.
5. Continue any present time-series sampling, and—if possible—revive discontinued sampling. It is imperative that the methodology does not change substantially during any time series. If it becomes necessary to make changes, then at least a sufficient number of intercalibration studies between old and new methodologies should be conducted to provide a seamless time series.
6. Examine factors that control the production of phytoplankton in the open subarctic Pacific. A major uncertainty concerns the rate of supply of iron, which may stimulate the growth rate of large phytoplankton species, enhance the growth rate of large zooplankton, and produce favorable feeding and growth conditions for pelagic fish.

ACKNOWLEDGMENTS

We thank Michael Mullin of the Scripps Institution of Oceanography for inviting us to participate in the symposium. Paul Smith of the La Jolla NMFS laboratory generously provided the CalCOFI data. Earlier drafts of the manuscript were reviewed by Art Kendall and Michael Mullin. This is NOAA's Fisheries Oceanography Coordinated Investigations Contribution FOCI-0273.

Literature Cited

Bailey, K. M., S. A. Macklin, R. K. Reed, R. D. Brodeur, W. J. Ingraham, J. F. Piatt, M. Shima, R. C. Francis, P. J. Anderson, T. C. Royer, A. B. Hollowed, D. A. Somerton, and W. W. Wooster, 1995. ENSO events in the northern Gulf of Alaska, and effects on selected marine fisheries, Calif. Coop. Oceanic Fish. Invest. Rep. 36:78–96.

Bakun, A. 1990. Global climate change and intensification of coastal ocean upwelling. Science 247:198–201.

Banse, K. 1994. Grazing and zooplankton production as key controls of phytoplankton production in the open ocean. Oceanography 7:13–20.

Beamish, R. J. 1994. Climate change and exceptional fish production off the west coast of North America. Can. J. Fish. Aquat. Sci. 50:2270–2291.

Beamish, R. J., ed. 1995. Climate change and northern fish populations. Can. Spec. Pub. Fish. Aquat. Sci. 121, 739 pp.

Box, G. E. P., and G. M. Jenkins, 1976. Time series analysis, forecasting and control. San Francisco: Holden-Day, Inc., 575 pp.

Boyd, P. W., F. A. Whitney, P. J. Harrison, and C. S. Wong, 1995. The NE subarctic Pacific in winter; biological rate processes. Mar. Ecol. Prog. Ser. 128:25–34.

Brodeur, R. D., and D. M. Ware, 1992. Long-term variability in zooplankton biomass in the subarctic Pacific Ocean. Fish. Oceanogr. 1:32–38.

———. 1995. Interdecadal variability in distribution and catch rates of epipelagic nekton in the Northeast Pacific Ocean. *In* Climate change and northern fish populations. R. Beamish, ed. Can. Spec. Pub. Fish. Aquat. Sci. 121:329–356.

Chavez, F. P. 1995. A comparison of ship and satellite chlorophyll from California and Peru. J. Geophys. Res. 100:24,855–24,862.

Chelton, D. B. 1984. Short-term climate variability in the Northeast Pacific Ocean. *In* The influence of ocean conditions on the production of salmonids in the North Pacific, W. G. Pearcy, ed. Corvallis: Oregon State Univ. Sea Grant Rep. W-83-001, pp. 87–99.

Chelton, D. B., and R. E. Davis. 1982. Monthly mean sea level variability along the western coast of North America. J. Phys. Oceanogr. 12:757–784.

Chelton, D. B., P. A. Bernal, and J. A. McGowan. 1982. Large-scale interannual physical and biological interaction in the California Current. J. Mar. Res. 40:1095–1125.

Conversi, A., and S. Hameed. 1996. Quasi-biennial component of zooplankton biomass variability in the subarctic Pacific. J. Geophys. Res.

Cooney, R. T. 1984. Some thoughts on the Alaska Coastal Current as a feeding habitat for juvenile salmon. *In* The influence of ocean conditions on the production of salmonids in the North Pacific. W. G. Pearcy, ed. Corvallis: Oregon State Univ. Sea Grant Rep. W-83-001, pp. 256–268.

———. 1986. The seasonal occurrence of *Neocalanus cristatus. Neocalanus plumchrus,* and *Eucalanus bungii* over the shelf of the northern Gulf of Alaska. Cont. Shelf Res. 5:541–553.

Emery, W. J., and K. Hamilton. 1985. Atmospheric forcing of interannual variability in the Northeast Pacific Ocean: connections with El Niño. J. Geophys. Res. 90:857–868.

Fargion, G. S., J. A. McGowan, and R. H. Stewart. 1993. Seasonality of chlorophyll concentrations in the California Current: a comparison of two methods. Calif. Coop. Oceanic Fish. Invest. Rep. 34:35–50.

Fasham, M. J. R. 1995. Variations in the seasonal cycle of biological production in subarctic oceans: a model sensitivity analysis. Deep-Sea Res. 42:1111–1149.

Francis, R. C. 1993. Climate change and salmonid production in the North Pacific Ocean. *In* Proceedings of the Ninth Annual Pacific Climate (PACLIM) Workshop, K. T. Redmond and V. L. Tharp, eds. Calif. Dep. Water Res. Tech. Rep. 34, pp. 33–43.

Francis, R. C. and S. R. Hare. 1994. Decadal-scale regime shifts in the large marine ecosystems of the North-east Pacific: a case for historical science. Fish. Oceanogr. 3:279–291.

Francis, R. C. and T. H. Sibley. 1991. Climate change and fisheries: what are the real issues? NW Environ. J. 7:295–307.

Freeland, H. J. 1990. Sea surface temperatures along the coast of British Columbia: regional evidence for a warming trend. Can. J. Fish. Aquat. Sci. 47:346–350.

Frost, B. W. 1983. Interannual variation of zooplankton standing stock in the open Gulf of Alaska. *In* From year to year: interannual variability of the environment and fisheries of the Gulf of Alaska and the Eastern Bering Sea, W. S. Wooster, ed. Seattle: Washington Sea Grant, pp. 146–157.

———. 1993. A modelling study of processes regulating plankton standing stock and production in the open subarctic Pacific Ocean. Prog. Oceanogr. 32:17–56.

Fulton, J. 1983. Seasonal and annual variations of net zooplankton at Ocean Station "P". 1956–1980. Can. Data Rep. Fish. Aquat. Sci. 374, 65 pp.

Graham, N. E. 1995. Simulation of recent global temperature trends. Science 267:666–671.

Hare, S. R., and R. C., Francis. 1995. Climate change and salmon production in the Northeast Pacific Ocean. *In* Climate change and northern fish populations, R. Beamish, ed. Can. Spec. Pub. Fish. Aquat. Sci. 121:357–372.

Helle, J. H., and M. S. Hoffman. 1995. Size decline and older age at maturity of two chum salmon (*Oncorhynchus keta*) stocks in western North America, 1972–92. *In* Climate change and northern fish populations, R. Beamish, ed. Can. Spec. Pub. Fish. Aquat. Sci. 121:245–260.

Hollowed, A. B., and W. S. Wooster. 1992. Variability of winter ocean conditions and strong year classes of Northeast Pacific groundfish. ICES Mar. Sci. Symp. 195:433–444.

Hsieh, W. W., and G. J. Boer. 1992. Global climate change and ocean upwelling. Fish. Oceanogr. 1:333–338.

Ingraham, W. J., Jr., and R. K. Miyahara. 1988. Ocean surface current simulations in the North Pacific Ocean and Bering Sea (OSCURS - Numerical Model). U.S. Dep. Commer., NOAA Tech Memo. NMFS F/NWC-130, 155 pp.

Ishida, Y., S. Ito, M. Kaeriyama, S. McKinnell, and K. Nagasawa. 1993. Recent changes in age and size of chum salmon (*Oncorhynchus keta*) in the North Pacific Ocean and possible causes. Can. J. Fish. Aquat. Sci. 50:290–295.

Katz, R. W. 1988. Use of cross correlations in the search for teleconnections. J. Climatology 8:241–253.

Kawamura, A. 1988. Characteristics of the zooplankton biomass distribution in the standard NORPAC net catches in the North Pacific region. Bull. Plankton Soc. Jpn. 35: 175–177.

Kelly, K. A., Caruso, M. J., and J. A. Austin. 1993. Wind-forced variation in sea surface height in the Northeast Pacific Ocean. J. Phys. Oceanogr. 23:2392–2411.

Lagerloef, G. S. E. 1995. Interdecadal variations in the Alaska Gyre. J. Phys. Oceanogr. 25:2242–2258.

Martin, J. H., and S. Fitzwater. 1988. Iron deficiency limits phytoplankton growth in the Northeast Pacific subarctic. Nature 331:341–343.

McAllister, C. D. 1962. Data record, photosynthesis and chlorophyll *a* measurements at Ocean Weather Station "P", July 1959 to November 1961. Fish. Res. Board Can., Manuscript Rep. Ser. (Oceanogr. Limnol.), No. 126, 14 pp.

McAllister, C. D., T. R. Parsons, and J. D. H. Strickland. 1959. Data record, oceanic fertility and productivity measurements at Ocean Weather Station "P", July and August 1959. Fish. Res. Board Can., Manuscript Rep. Ser. (Oceanogr. Limnol.). No. 55, 31 pp.

McClain, C. R., K. Arrigo, K. Tai, and D. Turk. 1996. Observations and simulations of physical and biological processes at Ocean Weather Station P. 1951–1980. J. Geophys. Res. 101:3697–3713.

McFarlane, G. A., and R. J. Beamish. 1992. Climatic influence linking copepod production with strong year-classes in sablefish. *Anoplopoma fimbria.* Can. J. Fish. Aquat. Sci. 49: 743–753.

McGillicuddy, D. J., Jr., A. R. Robinson, and J. J. McCarthy. 1995. Coupled physical and biological modelling of the spring bloom in the North Atlantic (II): three dimensional bloom and post-bloom processes. Deep-Sea Res. 42:1359–1398.

Miller, A. J. 1996. Recent advances in California Current modeling: decadal and interannual thermocline variations. Calif. Coop. Oceanic Fish. Invest. Rep. 37 (this volume).

Miller, A. J., D. R. Cayan, T. P. Barnett, N. E. Graham, and J. M. Oberhuber. 1994. The 1976–77 climate shift of the Pacific Ocean. Oceanography 7:21–26.

Myers, R. A., G. Mertz, and N. J. Barrowman. 1995. Spatial scales of variability in cod recruitment in the North Atlantic. Can. J. Fish. Aquat. Sci. 52:1849–1862.

Newton, H. J. 1988. TIMESLAB: a time series analysis laboratory. Pacific Grove, Calif.: Wadsworth & Brooks/Cole, 623 pp.

Ohman, M. D., and P. E. Smith. 1995. A comparison of zooplankton sampling methods in the CalCOFI time series. Calif. Coop. Oceanic Fish. Invest. Rep. 36:153–158.

Pearcy, W. G. 1991. Biology of the Transition Region. *In* Biology, oceanography, and fisheries of the North Pacific Transition Region and Subarctic Frontal Zone, J. A, Wetherall, ed. NOAA Tech. Rep. NMFS 105:39–55.

———. 1992. Ocean ecology of North Pacific salmonids. Seattle: Univ. Wash. Press, 179 pp.

Peterman, R. M. 1987. Review of the components of recruitment of Pacific salmon. Amer. Fish. Soc. Symp. 1:417–429.

Polovina, J. J., G. T. Mitchum, N. E. Graham, M. P. Craig, E. E. DeMartini, and E. N. Flint. 1994. Physical and biological consequences of a climatic event in the central North Pacific. Fish. Oceanogr. 3:15–21.

Polovina, J. J., G. T. Mitchum, and G. T. Evans. 1995. Decadal and basin-scale variation in mixed layer depth and the impact on biological production in the Central and North Pacific, 1960–88. Deep Sea Res. 42:1701–1716.

Roemmich, D., and J. McGowan, 1995a. Climatic warming and the decline of zooplankton in the California Current. Science 267:1324–1326.

———. 1995b. Sampling zooplankton: correction. Science 268:352–353.

Roessler, C. S., and D. B. Chelton. 1987. Zooplankton variability in the California Current, 1951–1982. Calif. Coop. Oceanic Fish. Invest. Rep. 28:59–96.

Royer, T. C. 1989. Upper ocean temperature variability in the Northeast Pacific Ocean: is it an indicator of global warming? J. Geophys. Res. 94:18175–18183.

Sambrotto, R. N., and C. J. Lorenzen. 1987. Phytoplankton and primary production. *In* The Gulf of Alaska: physical environment and biological resources, D. W. Hood and S. T. Zimmerman, eds. Ocean Assessments Division, Mineral Management Services, pp. 249–282.

Shiga, N., K. Numata, and K. Fujita. 1995. Long-term variations in zooplankton community and biomass in the Subarctic Pacific transition domain. Mar. Sci. 17:408–414 (in Japanese).

Stephens, K. 1964. Data record, productivity measurements in the Northeast Pacific with associated chemical and physical data, 1958–1964. Fish. Res. Board Can., Manuscript Rep. Ser. (Oceanogr. Limnol.) No. 179, 168 pp.

———. 1966. Data record. Primary production data from the Northeast Pacific, January 1964 to December 1965. Fish. Res. Board Can., Manuscript Rep. Ser. (Oceanogr. Limnol.) No. 209, 14 pp.

———. 1968. Data record. Primary production data from the Northeast Pacific, January 1966 to December 1967. Fish. Res. Board Can., Manuscript Rep. Ser. (Oceanogr. Limnol.) No. 957, 58 pp.

———. 1977. Primary production data from weatherships occupying Ocean Station "P" 1969 to 1975. Fish. Mar. Serv. Data Rep. 38, 88 pp.

Tabata, S. 1991. Annual and interannual variability of baroclinic transports across Line P in the Northeast Pacific Ocean. Deep Sea Res. 38 (Suppl. 1):S221–S245.

Trenberth, K. E. 1984. Some effects of finite sample size and persistence on meteorological statistics. Part I: autocorrelations. Mon. Weather Rev. 112:2359–2368.

———. 1990. Recent observed interdecadal climate changes in the Northern Hemisphere. Bull. Am. Meteorol. Soc. 71:988–993.

Trenberth, K. E. and J. W. Hurrell. 1994. Decadal atmosphere-ocean variations in the Pacific. Clim. Dynamics 9:303–319.

Van Scoy, K. A., and E. R. M. Druffel. 1993. Ventilation and transport of thermocline and intermediate waters in the Northeast Pacific during recent El Niños. J. Geophys. Res. 98:18,083–18,088.

Venrick, E. L., J. A. McGowan, D. R. Cayan, and T. L. Hayward. 1987. Climate and chlorophyll a: long-term trends in the central North Pacific Ocean. Science 238:70–72.

Waddell, B. J., and S. McKinnell. 1995. Ocean Station "Papa" detailed zooplankton data: 1956–1980. Can. Tech. Rep. Fish. Aquat. Sci. 2056, 21 pp.

Ware, D. M., and G. A. McFarlane. 1989. Fisheries production domains in the Northeast Pacific Ocean. *In* Effects of ocean variability on recruitment and an evaluation of parameters used in stock assessment models. R. J. Beamish and G. A. McFarlane, eds. Can. Spec. Publ. Fish. Aquat. Sci. 108:359–379.

Welschmeyer, N. A., Strom, R. Goericke, G. DiTullio, M. Belvin, and W. Petersen. 1993. Primary production in the subarctic Pacific Ocean: Project SUPER. Prog. Oceanogr. 32:101–135.

Wickett, W. P. 1967. Ekman transport and zooplankton concentration in the North Pacific Ocean. J. Fish. Res. Board Can. 24:581–594.

Wiebe, P. H. 1988. Functional regression equations for zooplankton displacement volume, wet weight, dry weight, and carbon: a correction. Fish. Bull. U.S. 86:833–835.

Wong, C. S., F. A. Whitney, K. Iseki, J. S. Page, and J. Zeng. 1995. Analysis of trends in primary productivity and chlorophyll-a over two decades at Ocean Station P (50°N, 145°W) in the Subarctic Northeast Pacific Ocean. *In* Climate change and northern fish populations, R. Beamish, ed. Can. Spec. Pub. Fish. Aquat. Sci. 121:107–117.

Wooster, W. S., and D. L. Fluharty. 1985. El Niño north. Seattle: Washington Sea Grant, Univ. Wash., 312 pp.

II

Case Studies

5

Eastern Bering Sea Ecosystem Trends

P. A. Livingston, L-L. Low, and R. J. Marasco

Abstract

The eastern Bering Sea is a biologically productive marine environment containing large populations of marine mammals, birds, crabs, and fish. Changes in species dominance and abundance in some ecological groups, particularly those containing flatfish and marine mammals, have raised questions concerning ecosystem health. Results of an examination of abundance trends and status indicators are described.

Monitoring the changing states of large marine ecosystems (LMEs) is important for understanding and predicting the effects of stress on those systems. Declines in marine mammal and bird populations in the eastern Bering Sea ecosystem prompted the formation of a National Research Council committee to investigate the causes of those declines. Other changes in species composition in the Bering Sea have also been noted, particularly the decrease in the red king crab (*Paralithodes camtschatica*) population size concomitant with the increase in abundance of shallow-water flatfish species that co-occur with red king crab (Otto 1993). Fishing removals or fluctuations in the physical environment, or both, are thought to be the main forces behind the observed changes. Although fishing removals are considered an external stress on ecosystems and normally occurring environmental fluctuations are not, it is necessary to examine the nature of the changes in an ecosystem over time to understand the possible role that fishing or environment might have in producing the observed change. Most analyses performed to date have looked at the nature of the changes from a single-species perspective, comparing changes in one species with changes in another (Bakkala 1993). A more holistic approach to monitoring change is required to infer what the changes suggest about ecosystem status or health.

At present, not enough is understood about the levels of natural variability in many of the species abundances in the eastern Bering Sea, primarily because of the limited time series of unbiased observations. Many components, such as phytoplankton, zooplankton, and benthic infauna, are not even being monitored in any systematic or regular fashion. This makes even more difficult the task of determining whether a perturbation has caused a significant change in a component. The suggested minimum time period for observing a population or community in order to make judgments about stability is at least one complete turnover of all individuals (Connell and Sousa 1983). For some flatfish species in the eastern Bering Sea, this period is at least 20 years, a period longer than the existing unbiased time series, which began circa 1979.

Detecting the effects of perturbation on the eastern Bering Sea ecosystem with any assurance will likely require a long-term monitoring of sensitive ecosystem variables (Schindler 1987; Suchanek 1994). However, examining existing time series could at least begin the process of defining levels of variability and identifying causative factors in ecosystem change.

The purpose of this analysis is to develop indicators of eastern Bering Sea ecosystem change using existing time series on pinnipeds, birds, groundfish, crab, and other benthic epifauna. The focus will be on indicators regarding the status of three fairly distinct species groups: (1) the offshore group of fish, mammals, and birds that consume small pelagic fish; (2) the inshore group of fish, crab, and other benthic epifauna that consume mainly benthic infauna; and (3) the more ubiquitous group that feeds on crab and fish. As mentioned previously, these groups have exhibited changes in species composition and abundance that could be due to stress induced by fishing or to environmental change, or both. Change in various indicators will be compared with fluctuations in fishing removals and the environment to determine the possible role each of those influences might have in causing the observed change.

Change Indicators

Because population declines have been noted for individual species in each of the groups mentioned, it is important to look at indicators that also include their competitors to see whether these show the same or opposite trend. Grouping species into trophic guilds could provide extra information about what factor(s) might be responsible for the change and assist in prediction of community change (Austen et al. 1994). For example, if we knew that fishing removals of one species in a trophic guild were very large we might expect to see increases in other members of that guild as competitive pressures were eased. Similarly, if fishing removals of an important prey to all members of the guild were large, we might see an overall decrease in abundance of species in the guild and perhaps even decreased mean size at age. Alternatively, if the factor inducing observed change was an environmental one, we might see trends in mean size at age or abundance that correlate positively or negatively with temperature trends depending on whether the species in question was near its northern or southern distribution limit. Thus, using the trophic guild concept is logical to help explain observed trends in change indicators.

Stress applied to an ecosystem will tend to reduce stability, diversity, and average size of dominant species, and may also cause a shift in species composition towards shorter-lived species (Rapport et al. 1985; Odum 1985). Thus, those are the variables of interest in assessing the health of ecosystems. Methods proposed to assess ecosystem health range from development of indicators at the species, guild, or community levels (Rapport et al. 1985) to more complex integrated indices that combine measures of system-level health components (Costanza 1992). There is a general recognition of the difficulty of deriving an overall measure of system health. The data requirements are extensive (and nonexistent for many systems), and much more modeling is required. Furthermore, because of the homeostatic nature of ecosystems, system-level measures may not be the most sensitive indicators for the early detection of ecosystem distress (Odum 1985; Rapport et al. 1985; Schindler 1987).

Reliable data for most system-level measures are not available for the eastern Bering Sea, so the emphasis here will be on indicators at the species and trophic guild levels. Variables important in assessing the complexity of an ecosystem include diversity indices that measure species evenness and richness (Sherman 1993; Costanza 1992; Rapport et al. 1985). Although some recent work on extracting trends in multispecies community change using minimum/maximum autocorrelation factor analysis (Solow 1994) looks promising, the method is fairly time-consuming and does not provide an index that could be compared across systems. There is general agreement that the Shannon-Wiener index and Simpson's index are reasonable tools for comparing diversity over time and space (Suchanek 1994; Hill 1973). For this study we were concerned mainly with detecting changes in the importance of the more abundant (i.e., commercially fished) species, so Simpson's indices of species richness and evenness were used, since those indices are more sensitive to changes in the most abundant species than is the Shannon-Wiener index (Peet 1974). Simpson diversity indices measuring species richness (SR) and species evenness (SE) for each trophic guild were derived for each year in the time series of biomasses of the members (B_i) of each guild using the following formulas:

$$\text{SR} = 1 / \Sigma_i p_i^2$$

$$\text{SE} = 1 - \Sigma_i p_i^2$$

where

$$p_i = B_i / \Sigma_i B_i$$

Examining temporal trends in the biomass of individual species and the diversity of guilds may provide extra information about the effect of fishing removal on members of each guild. Decreases in guild diversity corresponding with increases in exploitation rates of a guild member may be the first sign that fishing removals are affecting guild composition. Decreases in biomass of exploited species in the guild coinciding with biomass increases of nonexploited species might also be indicative of a fishing-induced change in guild composition.

Other key measures of health at the species level are growth and reproduction (Chapman 1992). A measure of importance in assessing the nature of change would include temporal trends in mean size at age for a particular species to determine whether those trends are more related to temperature changes in the area or to population increases in that species or its competitors that might be large enough to cause growth-rate reduction by limiting food resources. Decreases in mean size at age could increase the age at maturity and effectively decrease spawning stock size. Competition for limited benthic food sources has been named as a possible reason for lack of increase in the red king crab population in the face of rapidly increasing flatfish competitor populations. Recruitment indices for individual species are also important for determining whether changes in population abundance at age are due to environment, predation, or other factors.

Finally, measures that indicate the nature of the stress being inflicted on the system need to be developed. Fishing removals of species have been cited as the major stress on the eastern Bering Sea ecosystem. Therefore, we need to look at those removals in a more detailed fashion. Information on the total catch, the species composition of the catch, catch rates relative to exploitable biomass, trophic level of the catch, and amount and proportion of pelagic species in the catch over time might all be important indicators of how fishing patterns and amounts induce stress on marine ecosystems.

Species Groups

Offshore Pelagic Fish Consumers

This species group consists of fish, mammals, and birds that primarily reside and feed in the 80–250 m bottom depth zone of the eastern Bering Sea. Species were chosen as members of this group if small pelagic or bathypelagic fish (such as herring, capelin, juvenile walleye pollock, or myctophids) constitute over 25% of their diet by weight and if the main feeding grounds or distribution are in the offshore zone. Walters and McPhail (1982) defined fish and invertebrate species groups in the eastern Bering Sea based solely on species co-occurrence in bottom trawl catches. The present analysis takes those groups (their Site Group 3) and further restricts them based on similarity of diet. The resulting assemblage would be considered a trophic guild—animals that may not necessarily be taxonomically related but which utilize a common food resource. The species included in this guild are northern fur seal (*Callorhinus ursinus*), northern sea lion (*Eumetopias jubatus*), black-legged kittiwake (*Rissa tridactyla*), red-legged kittiwake (*Rissa brevirostris*), northern fulmar (*Fulmarus glacialis*), common murre (*Uria aalge*), thick-billed murre (*Uria lomvia*), adult walleye pollock (*Theragra chalcogramma*), arrowtooth flounder (*Atheresthes stomias*), and Greenland turbot (*Reinhardtius hippogloissoides*).

Inshore Benthic Infauna Consumers

This group consists of fish, crab, and other benthic invertebrates found at bottom depths of less than 100 m whose diet is predominantly bivalves, polychaetes, and echiuroid worms (infauna) and brittle stars (epifauna). As with the offshore group, members were chosen using Walters and McPhail's (1982) assemblage grouping (their Site Group 2) modified to include only species that are of the same trophic guild. The species included in this guild are yellowfin sole (*Pleuronectes asper*), longhead dab (*P. proboscidea*), rock sole (*P. bilineatus*), Alaska plaice (*P. quadrituberculatus*), flathead sole (*Hippoglossoides elassodon*), blue king crab (*Paralithodes platypus*), red king crab (*Paralithodes camtschatica*), tanner crab (*Chionoecetes bairdi*), snow crab (*C. opilio*), and starfish (primarily *Asterias amurensis*).

Crab and Fish Consumers

The members of this group are groundfish species from Walters and McPhail's (1982) Site Group 3 with a depth range of 80–250 m further restricted by food habits (Brodeur and Livingston 1988; Livingston et al. 1993) to include only species that consume crab and fish. Species included in this guild are Pacific cod (*Gadus macrocephalus*), Pacific halibut (*Hippoglossus stenolepis*), two genera of sculpins (*Myoxocephalus* sp., *Hemilepidotus* sp.), and the skate family (*Rajidae*).

Biomass, Catch, and Environment Trends

Offshore Pelagic Fish Consumers

The biomass of the offshore pelagic fish consumer guild is dominated by walleye pollock (Figure 5-1). It has undergone at least one cycle of biomass increase and decrease during the 1979–1993 period. During this period the shelf component of the Greenland turbot population declined, while the arrowtooth flounder population on the shelf increased (Figure 5-2). The amount of increase in arrowtooth flounder population biomass is much larger than the observed decline in Greenland turbot biomass. The most abundant piscivorous bird population in this guild is the thick-billed murre (Figure 5-3), with the other bird species at much lower relative abundances. All the piscivorous bird populations declined in abundance from 1976 to around 1985, but most appeared to have stabilized by the mid- to late 1980s to the present. Measures of hatching success on St. George Island for kittiwakes were low during the 1980s and higher in the 1970s and early

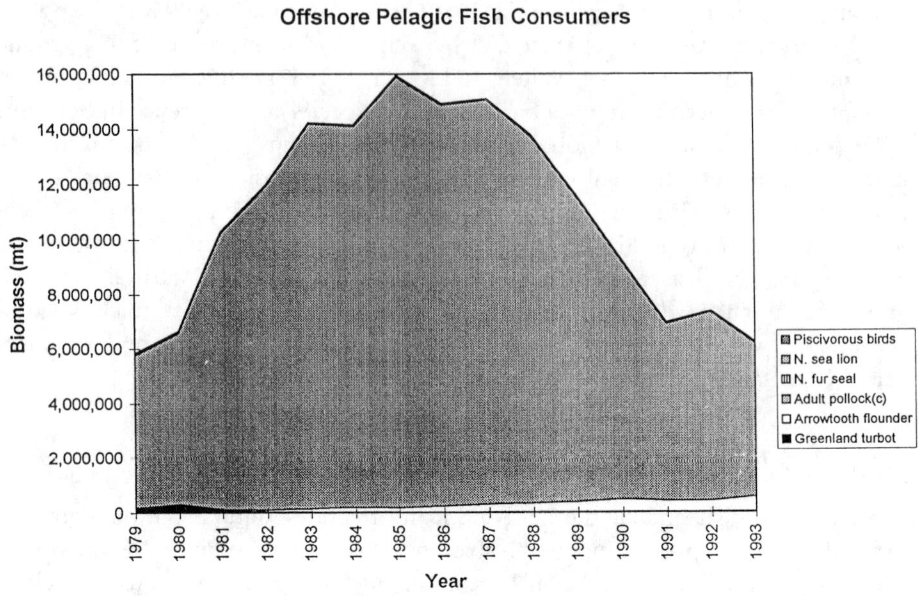

Figure 5-1. Biomass of the offshore pelagic fish consumer guild in the eastern Bering Sea, 1979 to 1993.

1990s, whereas murre productivity did not have a trend (Dragoo and Dragoo 1994). The pinniped populations also showed a decline in the beginning of the time period and, at least for northern fur seals, a stabilization by the mid-1980s (Figure 5-4). Species diversity and evenness measures over time in this guild have been fairly constant over time (Figure 5-5). Because of the dominance

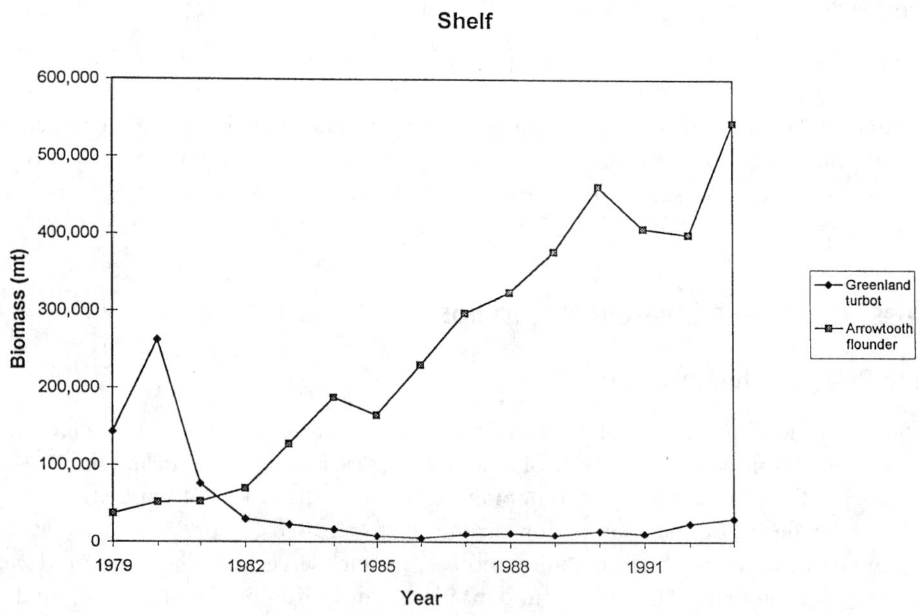

Figure 5-2. Biomass of arrowtooth flounder and Greenland turbot populations on the eastern Bering Sea shelf from 1979 to 1993.

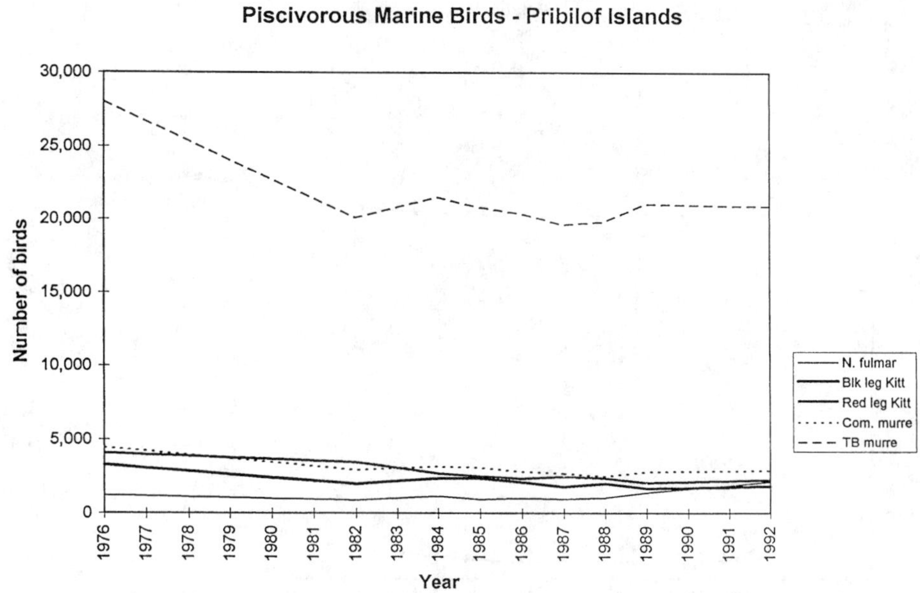

Figure 5-3. Abundance of piscivorous marine bird populations on the Pribilof Islands in the eastern Bering Sea from 1976 to 1992.

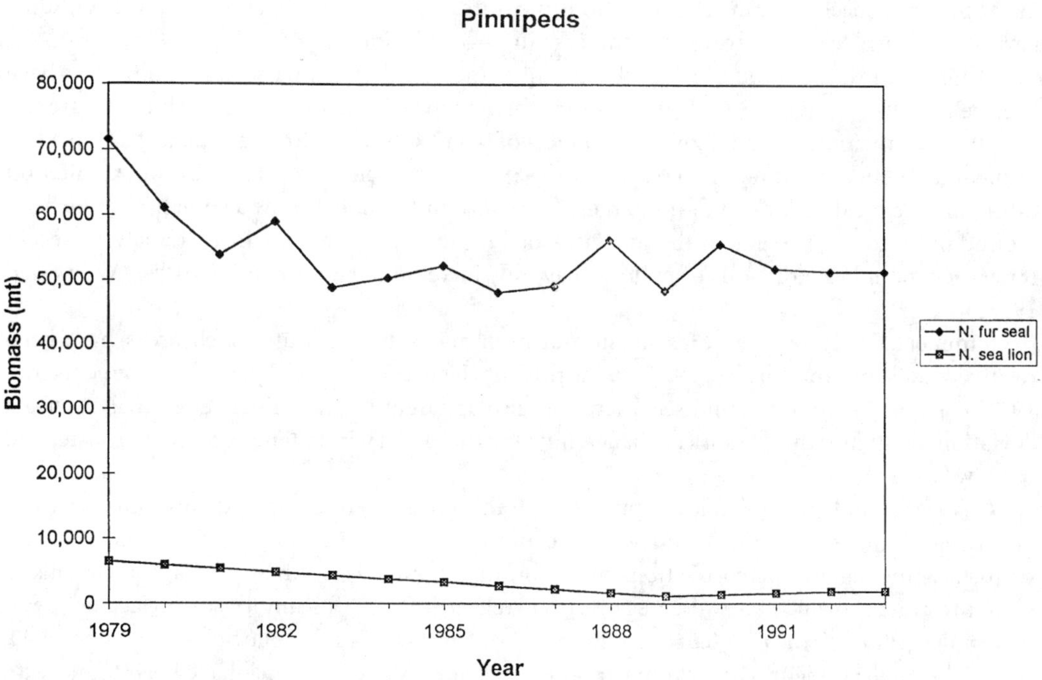

Figure 5-4. Northern fur seal and sea lion biomass in the eastern Bering Sea, 1979 to 1993.

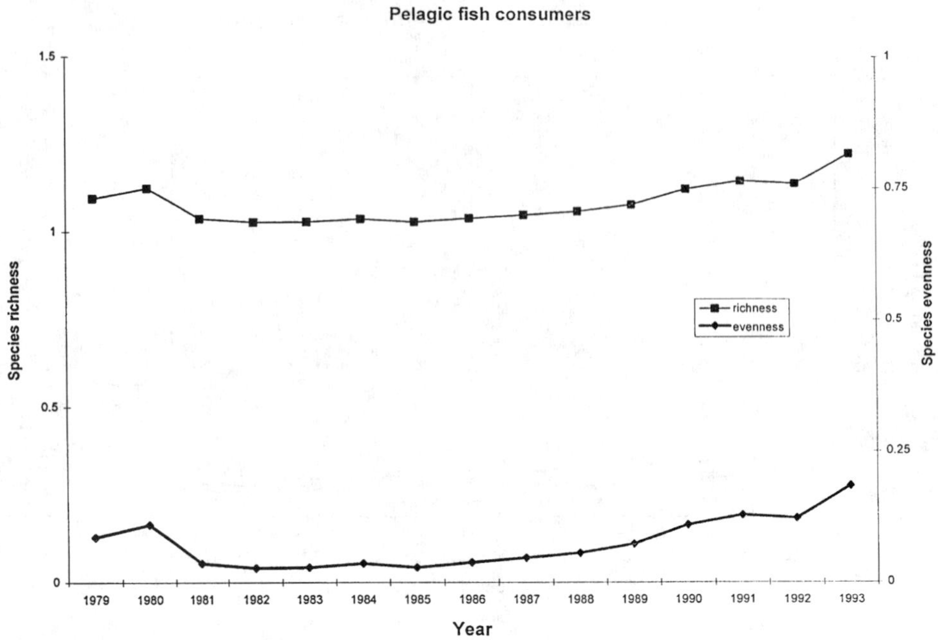

Figure 5-5. Species richness and evenness indices for the pelagic fish consumer guild in the eastern Bering Sea from 1979 to 1993.

of pollock in the total biomass of the guild, the effective number of species in the guild is close to one. Similarly, species evenness is low. Both measures show a slight decline during the middle part of the time period, mirroring the increase in pollock biomass.

Fishing removals of groundfish in this group as measured by biomass exploitation rates have been relatively low (Figure 5-6). Estimates of landed catch as a fraction of mature biomass are less than 0.2 over the time period. Exploitation rates of pollock declined from a peak in 1974 to a low in the mid-1980s, and show an increase from 1989 to the present. Greenland turbot exploitation rates (and the incidental catch of nontargeted arrowtooth flounder) show a straight decline from around 1984 to the present. Although discards are not accounted for in this analysis, recent observer data indicate discards of pollock may only increase catch estimates by 10% (Wespestad 1993).

Summer surface temperatures in the northern and southern outer shelf areas have been relatively constant from the late 1970s to the present (Figure 5-7). A cold period was evident from 1971 to 1976. The cooling trend seen here was also apparent in other time-series analyses using deviations from monthly means (Niebauer and Day 1989) and winter temperatures (Wooster and Hollowed 1995).

Changes in individual species populations in this group were compared with other changes occurring in the environment and with the most important member of the group, walleye pollock. Some significant correlations were found. Abundance estimates of age-1 Greenland turbot recruits (Ianelli 1993) were negatively correlated with both summer sea surface temperature in the year of birth ($r^2 = 0.36$, $n = 22$, $p < 0.05$) and with adult pollock biomass ($r^2 = 0.42$, $n = 23$, $p < 0.05$) (Figure 5-8). Stomach samples of adult walleye pollock have been found with very small (20–40 mm) Greenland turbot (Livingston et al. 1993), and this might explain the negative correlation between age-1 Greenland turbot recruits and pollock biomass. Young Green-

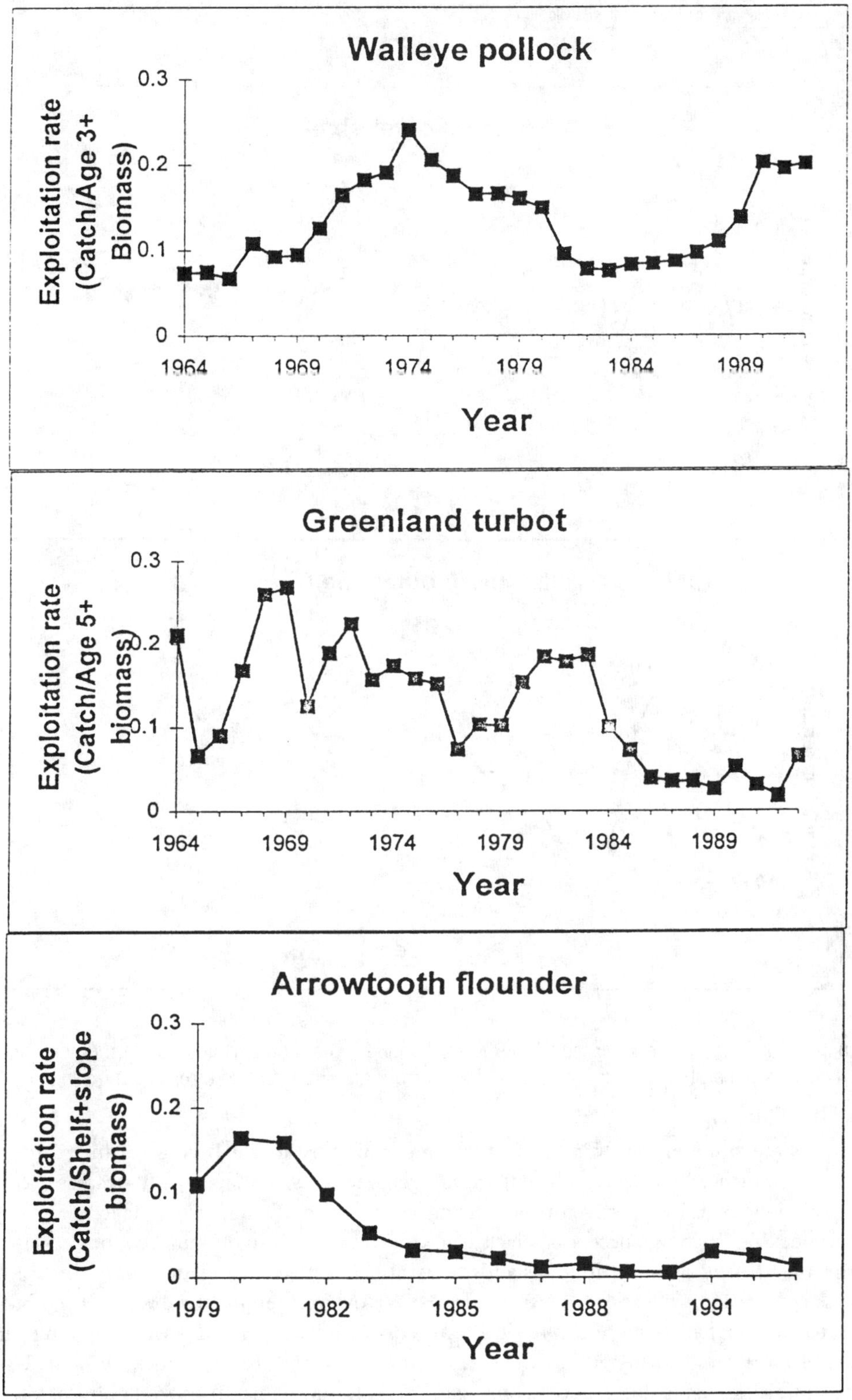

Figure 5-6. Fishery exploitation rates of walleye pollock, Greenland turbot, and arrowtooth flounder in the eastern Bering Sea. Note the different time scales on the *x*-axes.

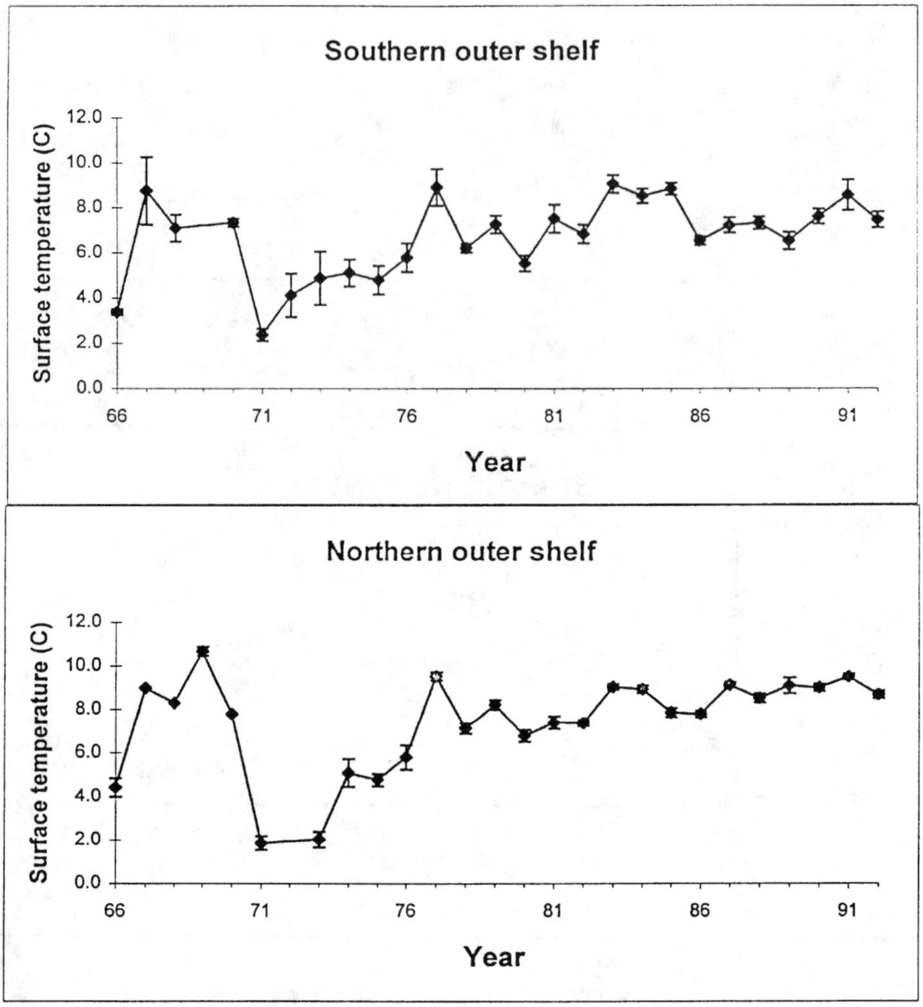

Figure 5-7. Summer surface water temperatures (°C) in the southern (upper) and northern (lower) outer shelf regions of the eastern Bering Sea, 1966–1992. (Error bars = 2 standard error of the mean.)

land turbot are found primarily in the northern outer shelf area of the Bering Sea and recruitment of this more northern species may be enhanced in colder years, perhaps due to decreased overlap with adult pollock or to other factors enhancing survival of early life history stages.

An index of hatching success of black-legged kittiwakes (Dragoo and Dragoo 1994) on St. George Island (one of the Pribilof Islands) was also negatively correlated with adult pollock biomass ($r^2 = 0.46$, $n = 17$, $p < 0.05$) (Figure 5-9). However, temperature was not significantly correlated with hatching success. No significant correlations were found with hatching success indices of murres or red-legged kittiwakes and either temperature or pollock biomass. Using a shorter time series of hatching success, Springer (1992) found significant correlations of kittiwake productivity with both pollock biomass and surface temperature. The negative correlation found in the present study between adult pollock biomass and kittiwake hatching success could be due to competition for juvenile pollock, a common food item of the two groups. Livingston (1993, 1994) estimated predation mortality rates due to cannibalism for young pollock and found

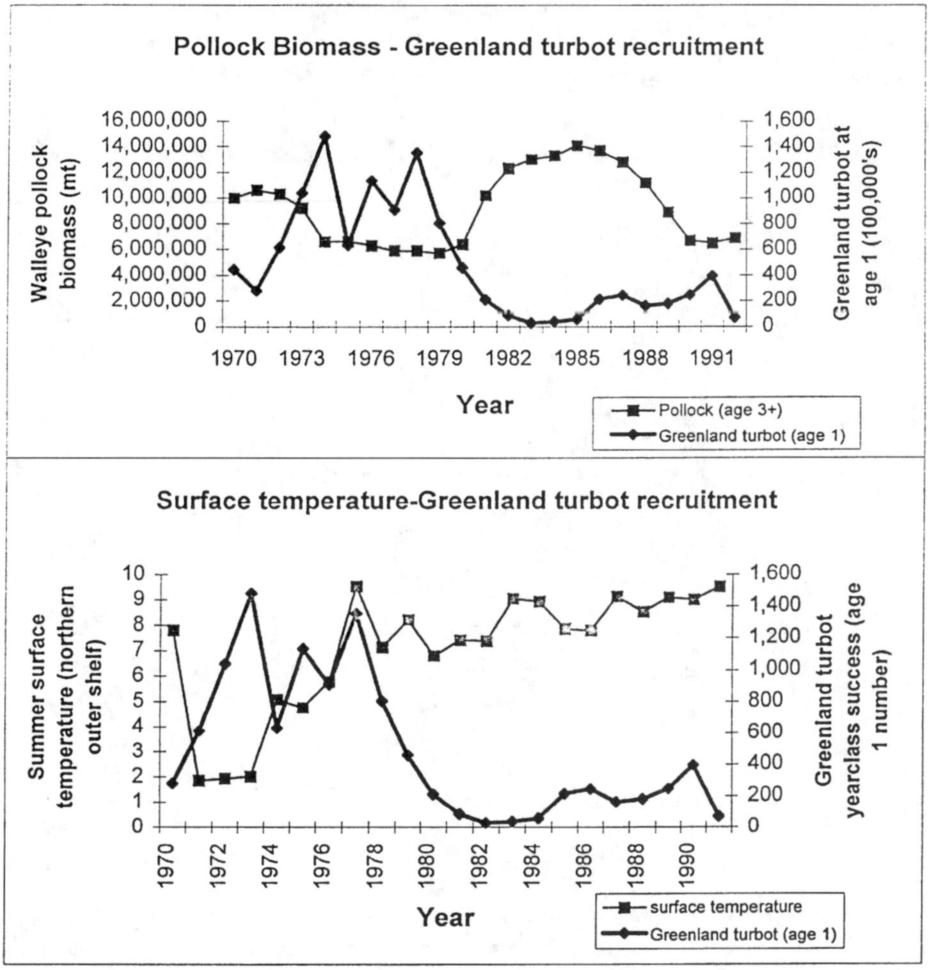

Figure 5-8. Greenland turbot recruitment (number of age-1 recruits) compared with walleye pollock biomass (upper) and summer surface temperatures (lower), 1970–1992.

mortality rates, particularly for age-1 pollock, to be very high during the period from 1983 to 1990 when the very abundant 1978 year-class of pollock attained a cannibalistic age (size). Pollock recruitment appears to be dampened at high spawning stock sizes due to cannibalism, and this may result in lower availability of juvenile pollock to other predators at high pollock population sizes. Kittiwakes are surface feeders and may be more affected by changes in prey abundance and distribution than diving species such as murres. Also, estimates of kittiwake hatching success are relatively variable across time even in adjacent years (see estimates for 1986–1990, Figure 5-9), and there are likely other factors influencing kittiwake population size.

No significant correlations were found with total fur seal or sea lion biomass and pollock biomass or sea surface temperature in the Bering Sea. Although Springer (1992) attempted to link pollock biomass and catch with sea lion declines in the Bering Sea, he does not cite a significant correlation between any of those factors. A more specific index of sea lion recruitment is probably needed to obtain any idea of what factors may be influencing survival. York (1991) found a significant difference in survival rates to age 2 for northern fur seal cohorts born between 1950

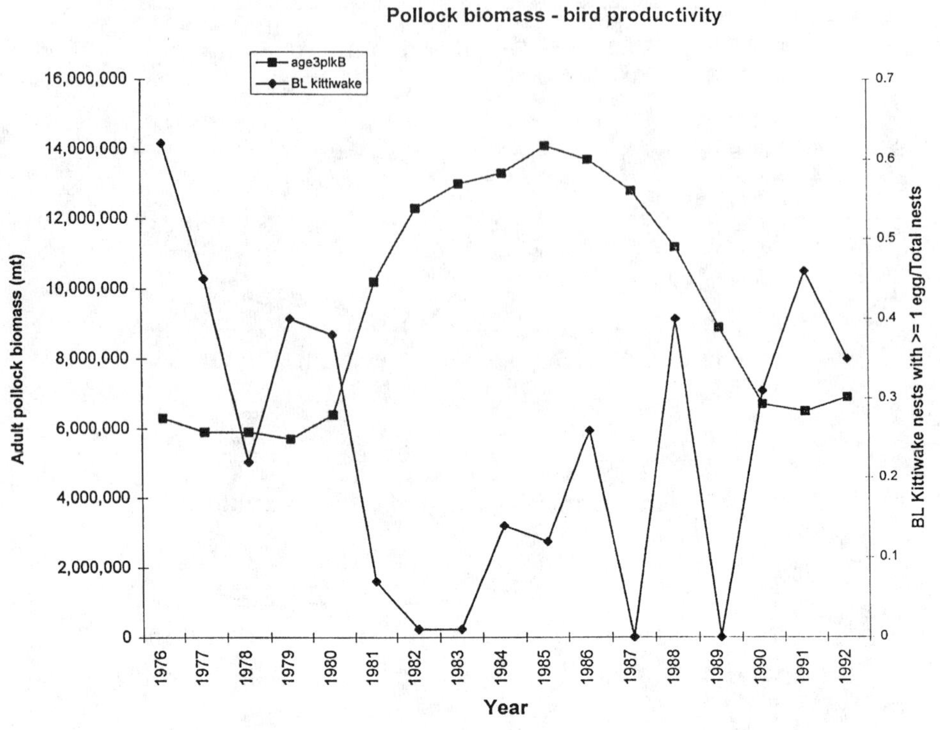

Figure 5-9. Black-legged kittiwake hatching success on St. George Island compared with adult walleye pollock biomass in the eastern Bering Sea, 1976–1992.

and 1979 and sea surface temperatures off British Columbia, the center of the known distribution of young animals. Thus, factors external to the eastern Bering Sea ecosystem may also influence population trends, especially in highly migratory animals such as fur seals. Lack of detailed information on the movements and habits of young sea lions may hinder our understanding of what factors are most influential in determining their survival. The data presented here at least appear to rule out the possibility that pollock fisheries in the eastern Bering Sea have reduced the overall amount of prey available to these groups. However, localized prey depletion by fisheries could still be a potentially important factor to animals such as sea lions or birds that have limited feeding ranges during particular times of year.

Inshore Benthic Infauna Consumers

The total biomass of the benthic infauna consumer group has increased from 1979 to 1993 (Figure 5-10). The biomass of yellowfin sole, the dominant species, has been fairly constant following a large increase after heavy exploitation during the 1960s (not shown), while the next most abundant species, rock sole, has been steadily increasing due to good recruitment throughout the 1980s (Wilderbuer and Walters 1993). Flathead sole biomass has also increased. Although crabs do not dominate the biomass in this guild, they are the most important in terms of their economic yield. Snow crab (*C. opilio*) are the most abundant of the crab species and have undergone apparent cyclic abundance changes from 1979 to 1993 (Figure 5-11). Tanner crab (*C. bairdi*) also show the same pattern of cylic abundance change as snow crab. Red and blue king

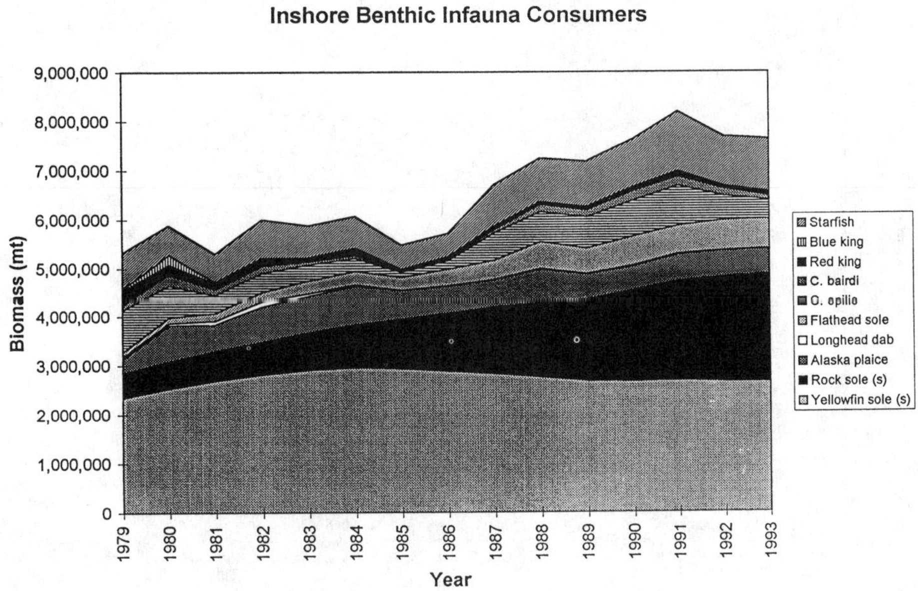

Figure 5-10. Biomass trends in the inshore benthic infauna consumer trophic guild from 1979 to 1993 in the eastern Bering Sea.

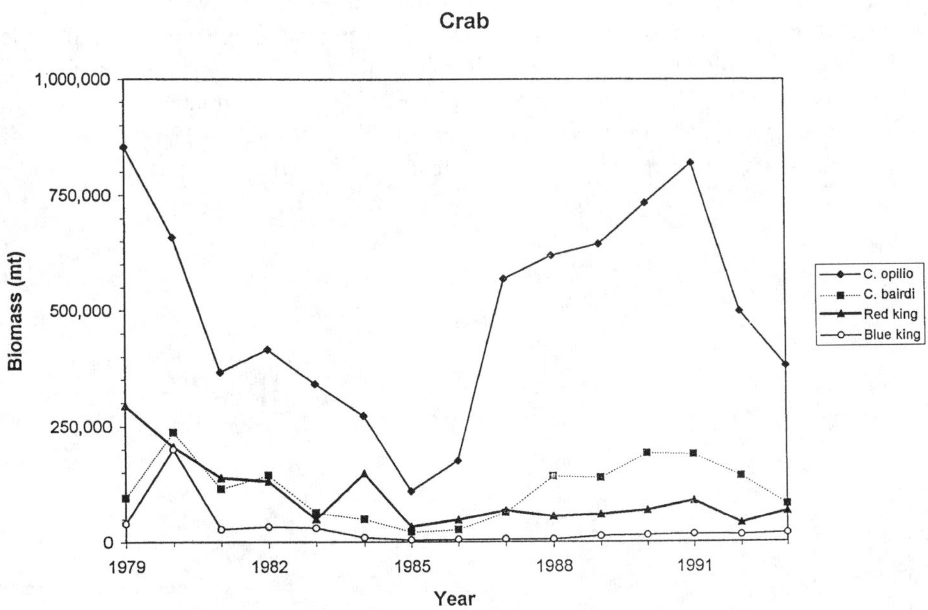

Figure 5-11. Biomass trends of crab species in the inshore benthic infauna consumer guild from 1979 to 1993 in the eastern Bering Sea.

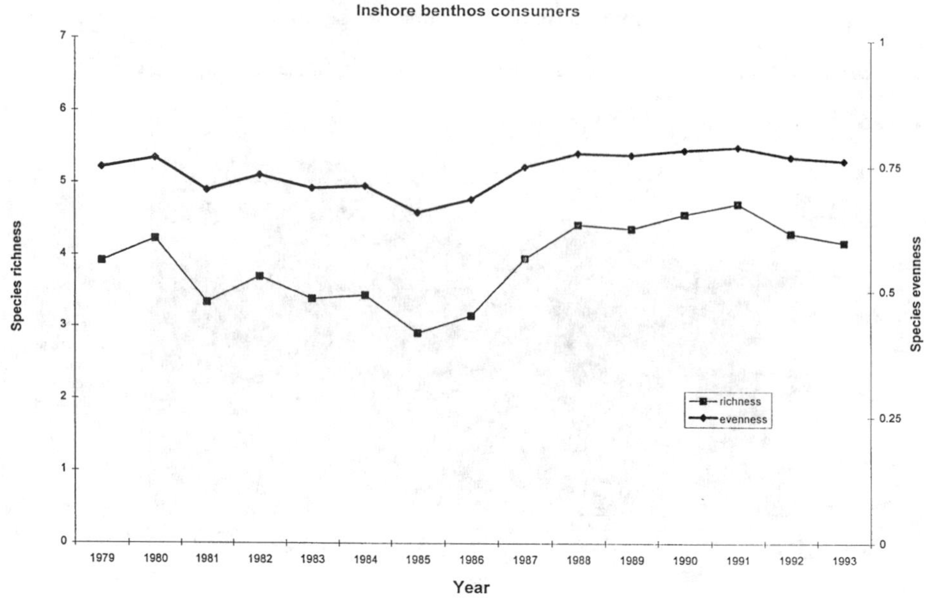

Figure 5-12. Species richness and evenness indices for the inshore benthic infauna consumer guild from 1979 to 1993 in the eastern Bering Sea.

crab, however, were more abundant in the beginning of the time period and appear to have stabilized at lower abundance levels from 1985 to 1993. Species richness and evenness indices over the time period for the guild as a whole were stable or slightly increasing (Figure 5-12).

Fishery exploitation rates were apparently low for most species when measured as catch relative to mature or exploitable stock for crab and the other flatfish group (which includes rock sole) (Figure 5-13). Yellowfin sole, however, had higher rates of fishing removals from 1964 to 1971 compared with the other species in this guild. Flatfish catch estimates include estimates of discards, whereas crab catch estimates are based on landed catch. Reeves (1993) estimates discards of red king crab from crab fisheries alone to be substantial, constituting up to 40% of the mature

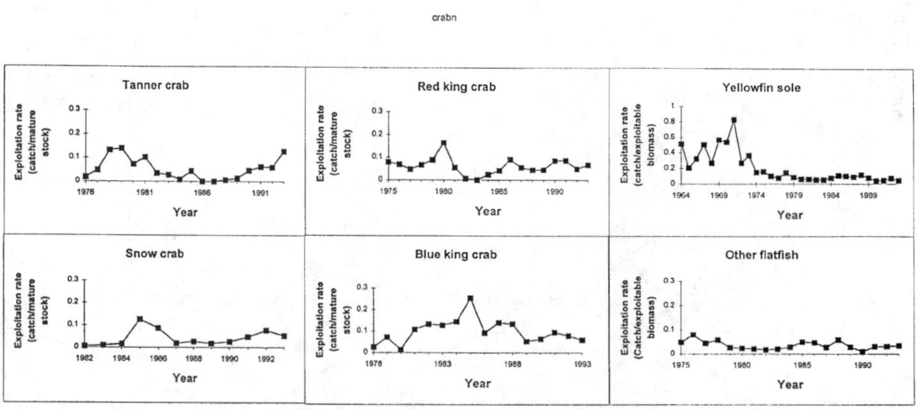

Figure 5-13. Fishery exploitation rates of crab and fish species in the inshore benthic infauna guild in the eastern Bering Sea (note different time scales on the *x*-axes).

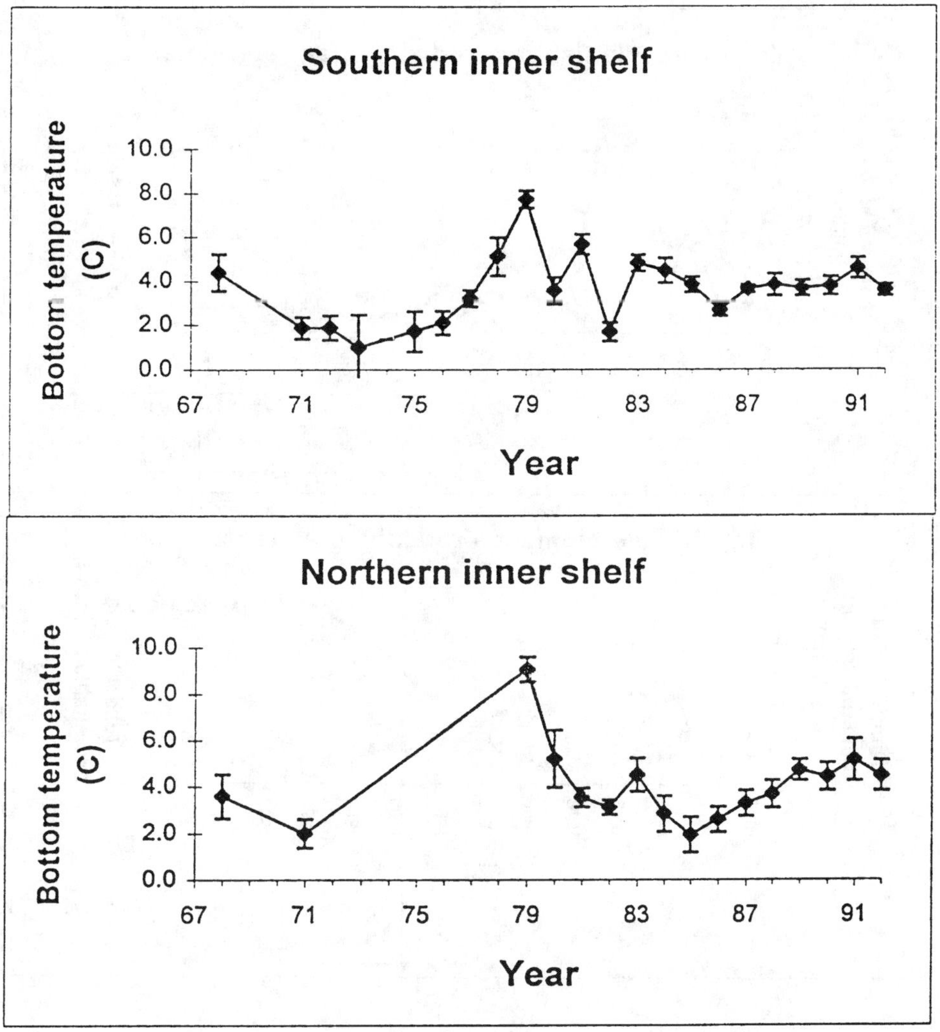

Figure 5-14. Summer bottom temperatures (°C) in the southern (upper) and northern (lower) inner shelf areas of the eastern Bering Sea from 1967 to 1992. (Error bars = 2 standard error of the mean.)

stock. If the handling mortality of these crabs is high, then actual exploitation rates of the red king crab stock may be much higher than estimates based on landed catch.

Summer bottom temperatures in the inner shelf area were lower in the period from 1971 to 1976, increasing and variable from 1977 to 1984, and moderate from around 1985 to 1992 (Figure 5-14). This again confirms trends in other temperature indices (i.e., Wooster and Hollowed 1995) that show 1971–1976 as a cooler period.

Individual species recruitment indices in this group were not significantly correlated with bottom temperatures in year of birth. A significant positive correlation, however, was found between a recruitment index of red king crab (number of age-6 males) and exploitation rate of yellowfin sole in red king crab year of birth ($r^2 = 0.59$, $n = 24$, $p < 0.05$), and a significant negative correlation between yellowfin sole biomass lagged 6 years behind the abundance of age-6 red king crab males ($r^2 = 0.38$, $n = 24$, $p < 0.05$) (Figure 5-15). There was no significant relationship between large female abundance and recruitment of age-6 males 6 years later. It has been

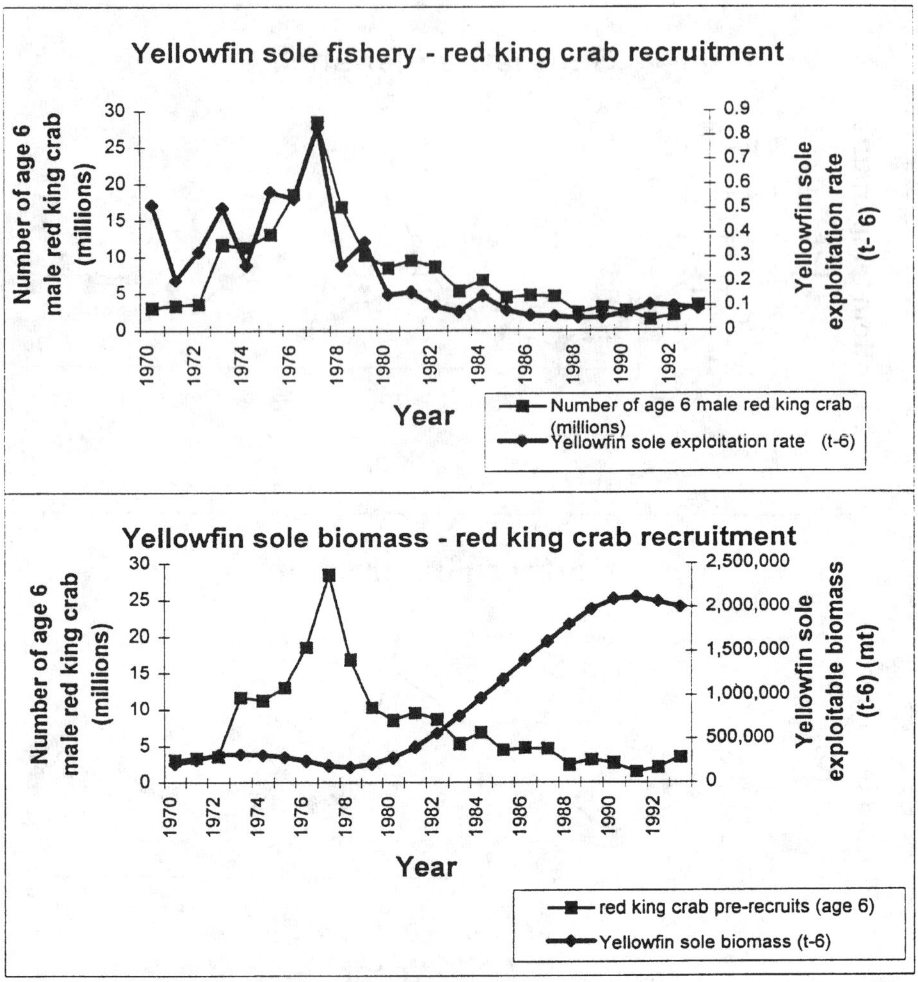

Figure 5-15. Red king crab recruitment compared with yellowfin sole fishery exploitation rates (upper) and yellowfin sole biomass (lower) in the eastern Bering Sea, 1970 to 1992.

documented that yellowfin sole consume crab megalops larvae (Haflinger and McRoy 1983; Livingston et al. 1993) and this could be the mechanism affecting red king crab recruitment. In this case, the high fishery exploitation rates of yellowfin sole would have had the effect of reducing the standing stock of megalops predators and thus increasing the average survival rate of king crab. However, as discussed above, actual exploitation rates of red king crab when discards are accounted for may be much higher than rates based on landed catch. This factor could also explain the current low level of red king crab abundance.

Trends in mean size at age across time were observed for yellowfin sole and rock sole (Figure 5-16). There appeared to be a decline in the mean size at age from 1979 to 1992, particularly for rock sole. An analysis of variance (Table 5-1) showed that the factor that explained most of the deviation after correcting for age was rock sole biomass. The next most important was guild biomass, a factor highly correlated with rock sole biomass. Summer bottom temperature in the southern inner shelf was also significant when entered after age. However, once the most

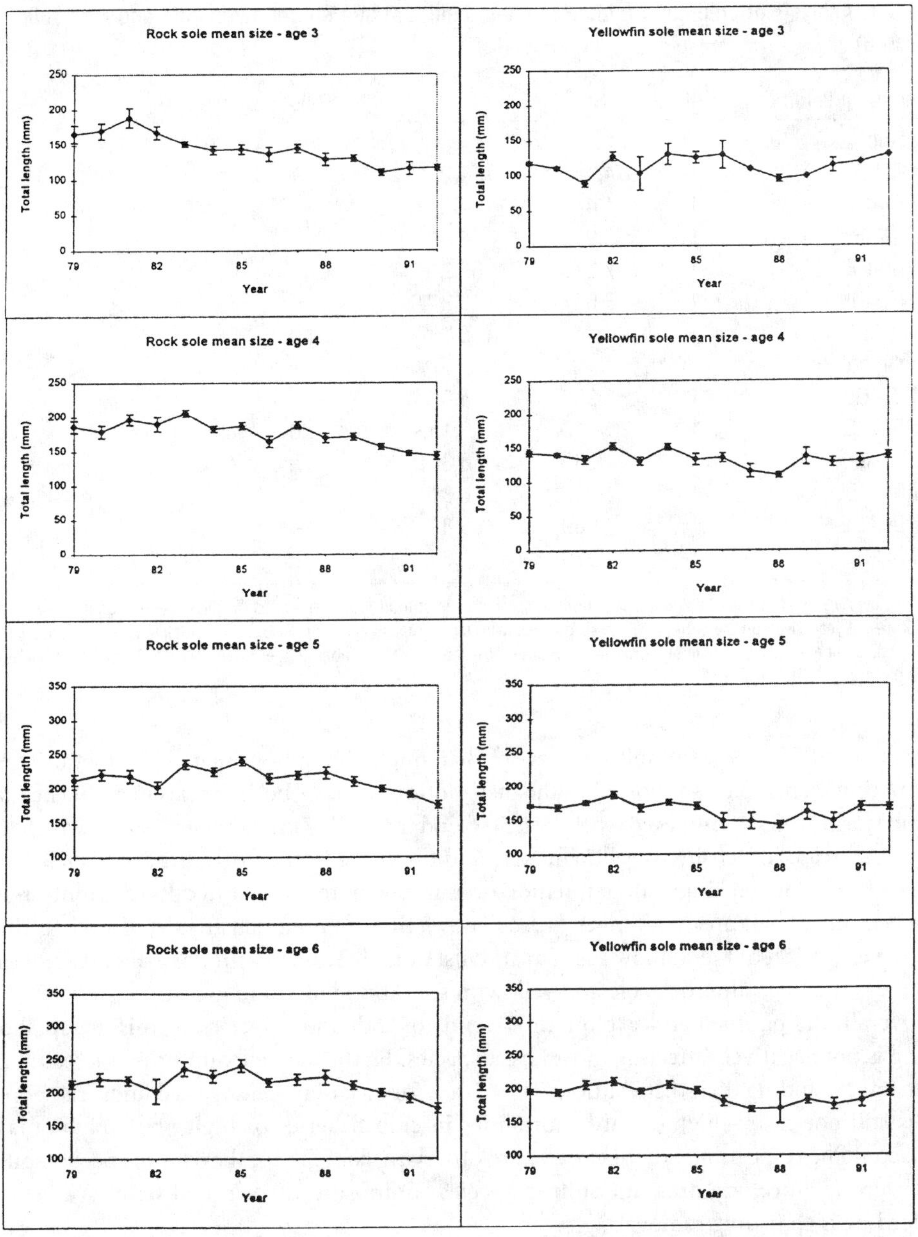

Figure 5-16. Trends in mean size at age for rock sole and yellowfin sole (ages 3–6) from 1970 to 1992 in the eastern Bering Sea. (Error bars = 2 standard error of the mean.)

significant factor (rock sole biomass) was entered after age, no other factor was significant. Thus, the observed declines in mean size at age could be a density-dependent growth response to increases in rock sole biomass, implying that benthic food resources available to both yellowfin sole and rock sole might be at a level that is unable to support the full growth potential of all the benthic infauna consumers in that area. An alternative explanation is that moderate bottom temperatures encountered in recent years have allowed a northward expansion of the distribution

Table 5-1. Analysis of predictors related to mean length at age for yellowfin sole and rock sole (ages 3–6)

Source of variation	df	SS	F	Relation
Rock sole				
AVGT/AGE	1	4,151	14.2**	+
YFSB/AGE	1	1,449	4.2*	+
RSB/AGE	1	8,968	44.5**	−
GUIB/AGE	1	7,979	36.2**	−
RSNA/AGE	1	3,937	13.2**	−
YFNA/AGE	1	44	0.1 ns	0
Yellowfin sole				
AVGT/AGE	1	953	5.6*	+
YFSB/AGE	1	18	0.1 ns	0
RSB/AGE	1	1,339	8.3**	−
GUIB/AGE	1	1,056	6.3*	−
RSNA/AGE	1	969	5.8*	−
YFNA/AGE	1	329	1.8 ns	0

Predictor Key: AVGT = Average of the summer bottom temperatures in stratum 10 encountered by each species age group during the total number of years lived on the bottom; YFSB = Total yellowfin sole biomass in a year; RSB = Total rock sole biomass in a year; GUIB =Total inshore benthos consumer biomass in a year; RSNA = Rock sole number at age; YFNA = Yellowfin sole number at age; $^*p < 0.05$; $^{**}p < 0.01$.

of rock sole and yellowfin sole into areas that might be suboptimal for growth. This latter explanation seems to be supported in the distribution maps of both species in 1979 (Bakkala and Wakabayashi 1985) compared with 1991 (Goddard and Zimmerman 1993), which show a definite increase in density of yellowfin sole and rock sole in the northern inner shelf area from 1979 to 1991. The red king crab population does not seem to have shifted its distribution into the northern inner shelf area. Haflinger (1981) shows that infauna densities are lower in that area, with the exception of the shallow areas in and just outside Kuskokwim Bay—an area not sampled in surveys that measure rock sole and yellowfin sole abundance and growth.

Overall, the biomass and exploitation trends of individual species in this group show that fishing is not negatively affecting most target species. Flatfish are encountering increased recruitment and expanding their distribution. The reason for lack of increase in red king crab population size is still not clear—high discard mortalities in crab fisheries or high predator biomasses, or both, could be responsible. It appears we are observing some natural cycles in species abundance in this group and not a dramatic shift in species composition that might be indicative of stress induced by high fishing removals.

Crab and Fish Consumers

Total biomass in this trophic guild was highest during the middle part of the 1979–1993 period, due to the increase in cod biomass (Figure 5-17). Skate biomass began increasing around 1986, peaked in 1990, and has stabilized or decreased slightly from 1991 to 1993. Cod biomass is undergoing another increase, due to above-average recruitment from the 1989 and 1990 year-classes (1991 and 1992 also may be strong; Thompson and Methot 1993). Cod and skates are the two most dominant members of this guild; the species richness and evenness indices reflect mainly changes in their biomass (Figure 5-18). Both indices increased over the time period mainly because of the decrease in cod biomass and increase in skate biomass.

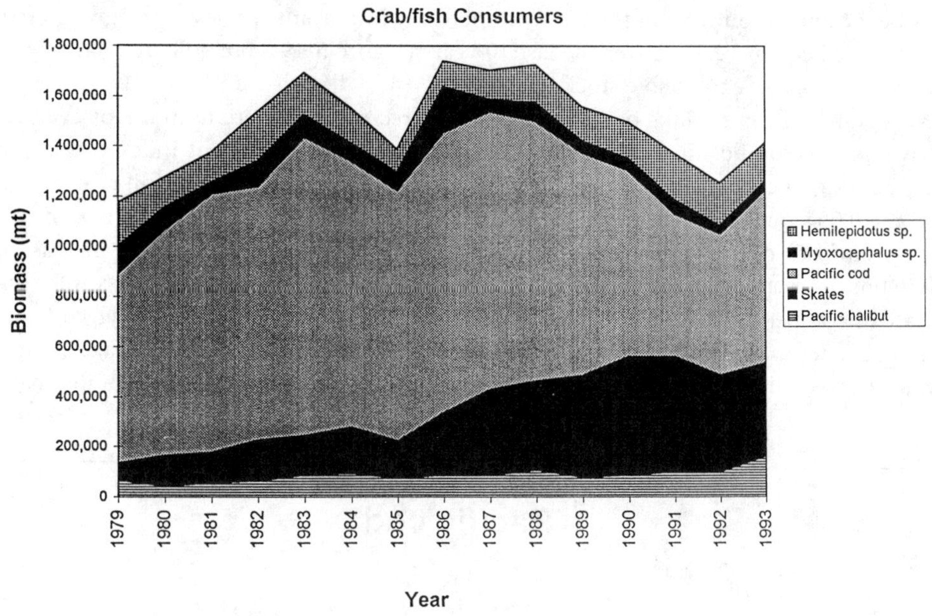

Figure 5-17. Biomass trends of the crab and fish consumer guild from 1979 to 1993 in the eastern Bering Sea.

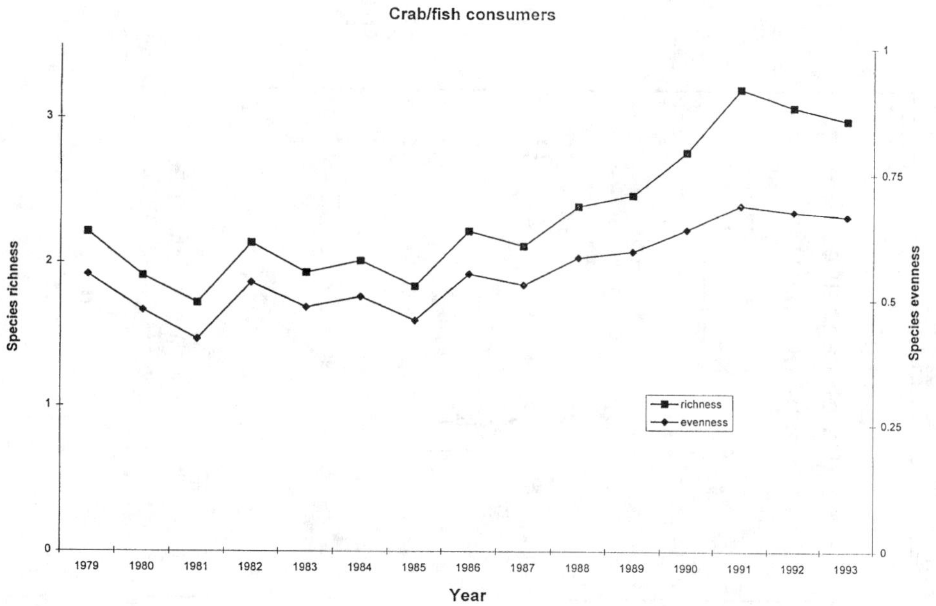

Figure 5-18. Species richness and evenness indices for the crab and fish consumer guild from 1979 to 1993 in the eastern Bering Sea.

Fishery exploitation rates of the two commercially important species in the groups (cod and halibut) have been increasing over the last 10 years (Figure 5-19), but still are relatively conservative proportions of exploitable stock.

No significant correlations of temperature with recruitment were found. However, because we have no size composition or age data on skates, a skate recruitment index is not available. There was a significant correlation between skate biomass and cod exploitation rate ($r^2 = 0.66$, $n = 14$, $p < 0.05$), but there was not a significant correlation between the increase in skate biomass and cod biomass decrease. Also, the increase in skate biomass preceded the increase in cod exploitation rates and decrease in cod biomass. If the increase in skate biomass were due to a decrease in competitor biomass, then the skate biomass increase should have lagged behind the cod biomass decrease. Until more information on skate life history and size or age composition is available, we will have little chance to improve our understanding of causes of fluctuations in

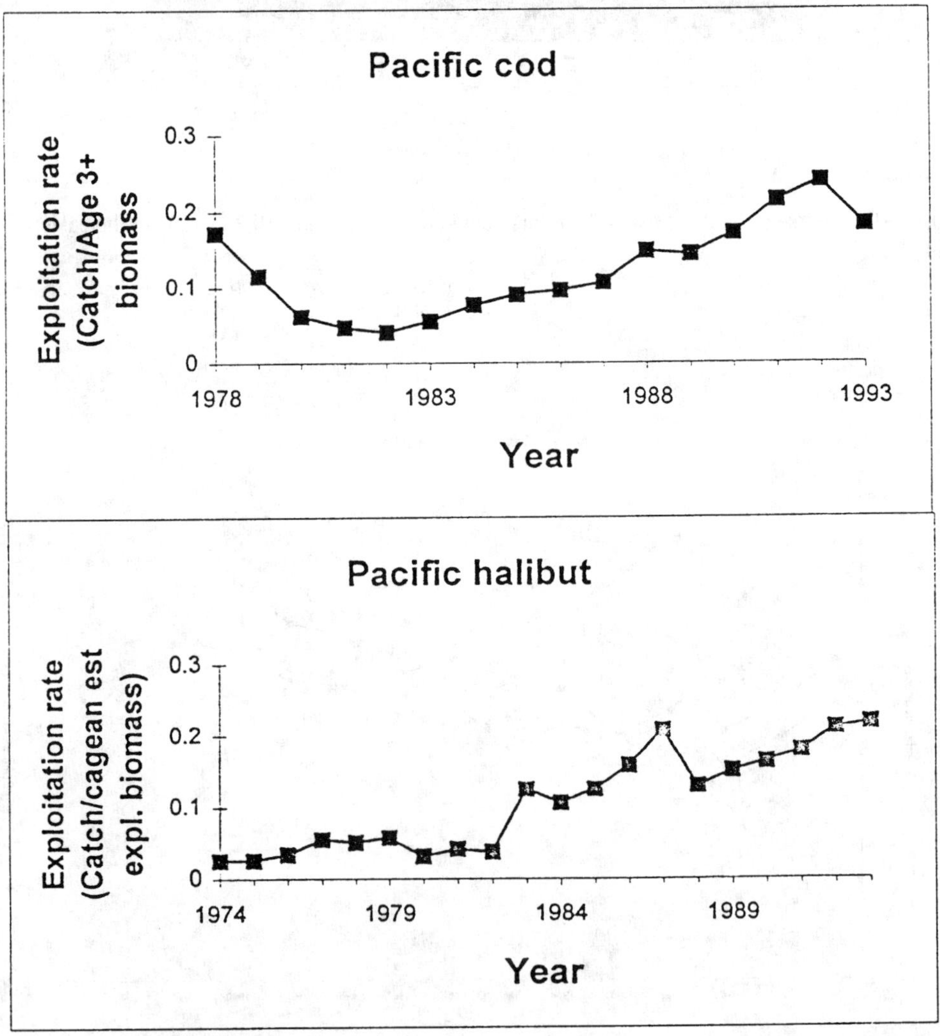

Figure 5-19. Fishery exploitation rates of Pacific cod and halibut in the eastern Bering Sea. Note the different time scales on the *x*-axes .

their population size. However, the abundance variations observed in this guild as a whole from 1979 to 1993 appear to be natural abundance cycles unrelated to fishing pressure.

Overall Trends in Fishing Removals

Total removals of groundfish, herring, Pacific halibut, and crab from the eastern Bering Sea were summarized with respect to weight, pelagic fraction, and trophic level over time (Figure 5-20). Trends in any of these measures may be another indicator of fishing-induced stress on the ecosystem. For instance, a decline in trophic level of the catch would indicate a shift of fishing pressure to prey of upper trophic level predators, potentially affecting the population status of those predators. Also, a shift in the distribution of catch between pelagic and benthic fish could similarly cause a shift in energy pathways in the system. However, the trophic level and pelagic fraction of the catch appear to have been constant from 1964 to the present. The total catch and

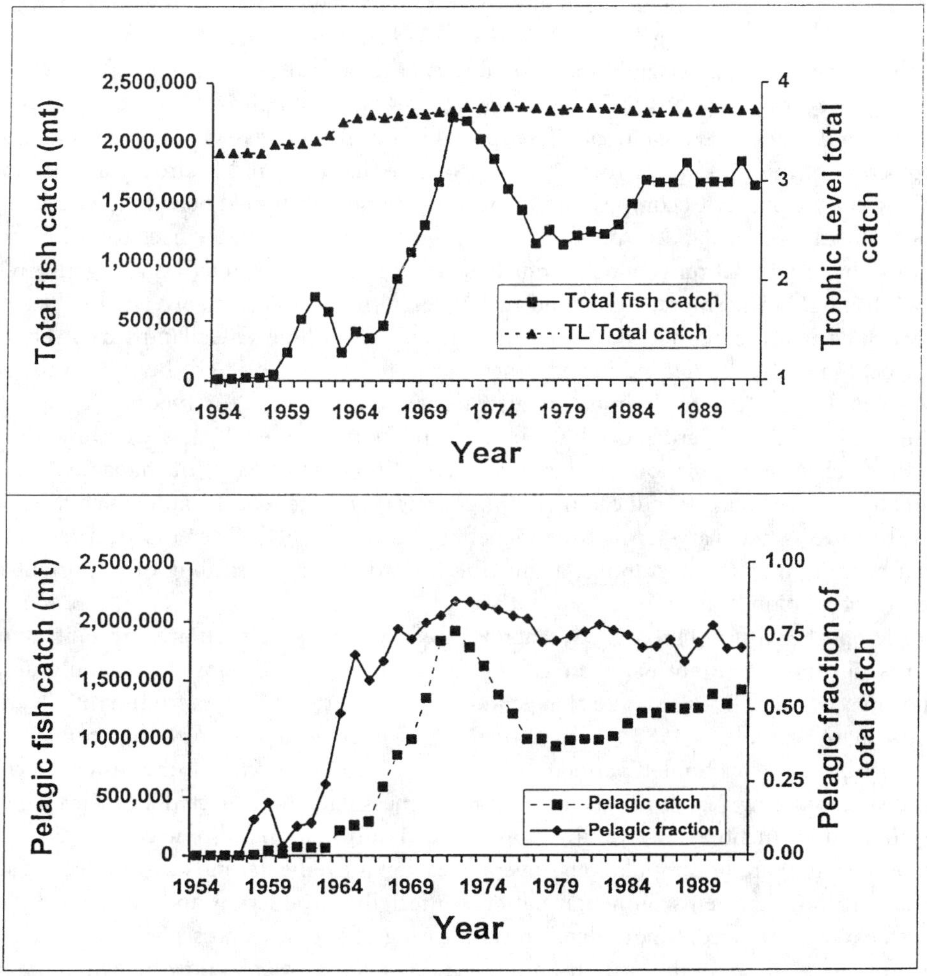

Figure 5-20. Total fishery catch and trophic level of the catch (upper) and pelagic fish catch and pelagic fraction of total catch (lower) in the eastern Bering Sea from 1954 to 1993.

pelagic catch amounts from 1964 to the present are dominated by walleye pollock. We have already shown that exploitation rates of walleye pollock have been around 10% to 20% of mature stock, relatively low compared with exploitation rates on stressed populations of groundfish in the northwest Atlantic. Based on age-1 survey estimates, pollock recruitment has not declined (Wespestad 1993), so current indications are that the stock is stable and at average abundance levels.

Ecosystem Health Assessment

Trends in individual species and guild biomass in response to fishing and environment support the assessment that most fished populations of fish and crab have been exhibiting cyclic changes in abundance. Changes in other upper-level predators such as marine mammals and birds do not appear to be negatively correlated with fishing removals in the Bering Sea, although the effect of localized prey depletion through fishing activities on these predators is still unknown. No dramatic shifts in species composition have been observed, although reasons for the declines in some species at the beginning of the 1979–1993 period are not clear. The dominance of pollock in the pelagic fish consumer guild may not always have been the case. Prior to the 1960s, there were much larger populations of fur seals, sea lions, whales, and birds. However, because we do not have good time-series data for all these species it is difficult to assess the time trend in species composition of this guild prior to 1979. It is possible that the guild is already a reflection of a dramatic change in species composition prior to the period examined here. Some declines, such as those noted in Greenland turbot and black-legged kittiwakes, were correlated with increases in an exploited population (pollock) and thus could not be interpreted as a negative result of pollock fishing. Furthermore, Greenland turbot recruitment was also correlated with temperature, making it unclear what the actual mechanism controlling Greenland turbot abundance might be. Similarly, the decline in red king crab population size might be due to the natural recovery of the yellowfin sole population after heavy exploitation in the 1960s or it might be due to the possibly high discard mortalities induced by crab fisheries. Unless we know the actual mechanism causing population fluctuations, we have no reliable basis for managing them. Unfortunately, change may be induced by multiple factors simultaneously, and which factor dominates the process may vary across time (Scheffer and Beets 1994). The empirical relations presented here are only the beginnings of moving toward an understanding of important factors affecting population change.

As Rapport et al. (1981) recognize, diagnosing ecosystem distress requires an ability to discern patterns signaling an abnormal change. The first step in learning how to diagnose abnormal change in ecosystems is to separate change induced by external stressors from natural variability in the system. Schindler's (1987) observation that we generally do not know the normal range for most variables due to a limited period of observation is certainly true for the eastern Bering Sea. In the eastern Bering Sea, groundfish and crab biomass data are collected annually. These data need to be supplemented, however, by more detailed information on some components, such as size composition for skate populations, juvenile sea lion movements, and factors affecting survival of crabs. If long-term environmental patterns affect the production and carrying capacity of ecosystems, we also need a more detailed monitoring of primary and secondary production. A model of energy flows in the eastern Bering Sea ecosystem, which is now being parameterized, would also be strengthened by more detailed data on these components. Long-term monitoring programs appear to be the most direct way to establish causal mechanisms responsible for inducing ecosystem change.

References

Austen, D. J., Bayley, P. B., and Menzel, B. M. 1994. Importance of the guild concept to fisheries research and management. Fisheries 19(6):12–20.

Bakkala, R. G. 1993. Structure and Historical Changes in the Groundfish Complex of the Eastern Bering Sea. NOAA Technical Report NMFS 114. Washington, DC: U.S. Department of Commerce.

Bakkala, R. G., and Wakabayashi, K. 1985. Results of the cooperative U.S.-Japan groundfish investigations in the Bering Sea during May–August 1979. Int. N. Pac. Fish. Comm. Bull. 44.

Brodeur, R. D., and Livingston, P. A. 1988. Food habits and diet overlap of various eastern Bering Sea fishes. NOAA Technical Memo NMFS F/NWC-127. Washington, DC: U.S. Department of Commerce.

Chapman, P. M. 1992. Ecosystem health synthesis: Can we get there from here? J. Aquat. Ecosyst. Health 1:69–79.

Connell, J. H., and Sousa, W. P. 1983. On the evidence needed to judge ecological stability or persistence. Am. Nat. 121:789–824.

Costanza, R. 1992. Toward an Operational Definition of Ecosystem Health. In R. Costanza, B. G. Norton, and B. D. Haskell (eds.), Ecosystem Health: New Goals for Environmental Management. Washington, DC: Island Press, pp. 241–256.

Dragoo, D. E., and Dragoo, B. K. 1994. Results of Productivity Monitoring of Kittiwakes and Murres at St. George Island, Alaska in 1993. U.S. Fish and Wildlife Service Report AMNWR 94/06. Homer, Alaska.

Goddard, P., and Zimmerman, M. 1993. Distribution, Abundance, and Biological Characteristics of Groundfish in the Eastern Bering Sea Based on Results of the U.S. Bottom Trawl Survey during June–September 1991. AFSC Processed Rep. 93-15. Seattle: Washington: U.S. Department of Commerce, Natl. Mar. Fish. Serv., Alaska Fish. Sci. Ctr.

Haflinger, K. 1981. A Survey of Benthic Infaunal Communities of the Southeastern Bering Sea Shelf. In D. W. Hood and J. A. Calder (eds.), The Eastern Bering Sea Shelf: Oceanography and Resources, Vol. II. Seattle: University of Washington Press, pp. 1091–1104.

Haflinger, K., and McRoy, C. P. 1983. Yellowfin Sole (*Limanda aspera*) Predation on Three Commercial Crab Species (*Chionoecetes opilio, C. bairdi*, and *Paralithodes camtschatica*) in the Southeastern Bering Sea. Final Report to U.S. Natl. Mar. Fish. Service, Contract 82-ABC-00202. Fairbanks, AK: Institute of Marine Science, University of Alaska.

Hill, M. O. 1973. Diversity and evenness: a unifying notation and its consequences. Ecology 54:427–432.

Ianelli, J. 1993. Greenland Turbot. In Stock Assessment and Fishery Evaluation Report for the Groundfish Resources of the Bering Sea/Aleutian Islands Regions as Projected for 1994. Anchorage, AK: North Pacific Fishery Management Council.

Livingston, P. A. 1993. Importance of predation by groundfish, marine mammals and birds on walleye pollock (*Theragra chalcogramma*) and Pacific herring (*Clupea pallasi*) in the eastern Bering Sea. Mar. Ecol. Prog. Ser. 102:205–215.

Livingston, P. A. 1994. Overview of multispecies interactions involving walleye pollock in the eastern Bering Sea and Gulf of Alaska. Int. Counc. Explor. Sea ICES C.M. 1994/P:1.

Livingston, P. A., Ward, A., Lang, G. M., and Yang, M-S. 1993. Groundfish Food Habits and Predation on Commercially Important Prey Species in the Eastern Bering Sea from 1987 to 1989. NOAA Tech. Memo. NMFS-AFSC-11. Washington, DC: U.S. Department of Commerce.

Niebauer, H. J., and Day, R. H. 1989. Causes of interannual variability in the sea ice cover of the eastern Bering Sea. GeoJournal 18.1:45–49.

Odum, E. P. 1985. Trends expected in stressed ecosystems. BioScience 35:419–422.

Otto, R. S. 1993. Recent Changes in the Species Composition of Bering Sea Fishery Resources. An oral presentation to the National Research Council. Kodiak, AK: National Marine Fisheries Service, Kodiak Laboratory.

Peet, R. K. 1974. The measurement of species diversity. Ann. Rev. Ecol. and Systematics 5:285–307.

Rapport, D. J., Regier, H. A., and Hutchinson, T. C. 1985. Ecosystem behavior under stress. Am. Nat. 125:617–640.

Rapport, D. J., Regier, H. A., and Thorpe, C. 1981. Diagnosis, Prognosis, and Treatment of Ecosystems under Stress. In G. W. Barrett and R. Rosenberg (eds.), Stress Effects on Natural Ecosystems. Chichester: John Wiley and Sons Ltd., pp. 269–280.

Reeves, J. E. 1993. Use of Lower Minimum Size Limits to Reduce Discards in the Bristol Bay Red King Crab (*Paralithodes camtschaticus*) Fishery. NOAA Tech. Memo. NMFS-AFSC-20. Washington, DC: U.S. Department of Commerce.

Scheffer, M., and Beets, J. 1994. Ecological models and the pitfalls of causality. Hydrobiologia 275/276:115–124.

Schindler, D. W. 1987. Detecting ecosystem responses to anthropogenic stress. Can. J. Fish. Aquat. Sci. 44(Suppl. 1):6–25.

Sherman, K. 1993. Emerging Theoretical Basis for Monitoring the Changing States (Health) of Large Marine Ecosystems. NOAA Tech. Memo. NMFS-F/NEC-100. Washington, DC: U.S. Department of Commerce.

Solow, A. R. 1994. Detecting change in the composition of a multispecies community. Biometrics 50:556–565.

Springer, A. M. 1992. A review: walleye pollock in the North Pacific—how much difference do they really make? Fisheries Oceanography 1:80–96.

Suchanek, T. H. 1994. Temperate coastal marine communities: Biodiversity and threats. Amer. Zool. 34:100–114.

Thompson, G., and Methot, R. 1993. Pacific Cod. In Stock Assessment and Fishery Evaluation Report for the Groundfish Resources of the Bering Sea/Aleutian Islands Regions as Projected for 1994. Anchorage, AK: North Pacific Fishery Management Council.

Walters, G. E., and McPhail, M. J. 1982. An Atlas of Demersal fish and Invertebrate Community Structure in the Eastern Bering Sea: Part I, 1978–81. NOAA Tech. Memo. NMFS F/NWC-35. Washington, DC: U.S. Department of Commerce.

Wespestad, V. 1993. Walleye Pollock. In Stock Assessment and Fishery Evaluation Report for the Groundfish Resources of the Bering Sea/Aleutian Islands Regions as Projected for 1994. Anchorage, AK: North Pacific Fishery Management Council.

Wilderbuer, T., and Walters, G. 1993. Rock Sole. In Stock Assessment and Fishery Evaluation Report for the Groundfish Resources of the Bering Sea/Aleutian Islands Regions as Projected for 1994. Anchorage, AK: North Pacific Fishery Management Council.

Wooster, W. S., and Hollowed, A. B. 1995. Decadal Scale Variations in the Subarctic Pacific: A. Winter Ocean Conditions. In R. J. Beamish (ed.), Climate Change and Northern Fish Populations. Canadian Special Publication of Fisheries and Aquatic Sciences 121.

York, A. E. 1991. Sea Surface Temperatures and Their Relationship to the Survival of Juvenile Male Northern Fur Seals from the Pribilof Islands. In F. Trillmich and K. A. Ono (eds.), Pinnipeds and El Niño. Berlin: Springer-Verlag, Berlin, pp. 94–106.

6

Living Marine Resources of the Yellow Sea Ecosystem in Korean Waters: Status and Perspectives

Chang Ik Zhang and Suam Kim

Abstract

The status of the living marine resources of the Yellow Sea ecosystem is reviewed. The major fisheries are at an extremely low level compared with three decades ago and are no longer economically sustainable due mainly to overexploitation of the stocks. In addition, pollution from oil spills and heavy metals in the coastal waters of Korea has increased steadily, causing serious damage to the natural living resources. Land reclamation has increased significantly along the western coast of the Korean peninsula, also harming the fisheries. However, there has been no coordinated research to evaluate the condition of the ecosystem. Many scientists have expressed the need for multidisciplinary research on the Yellow Sea. This led to the formation of the holistic ecosystem monitoring project covering the entire Yellow Sea. The project consists of four major modules: productivity, living marine resources, ecosystem health, and ecosystem management. The Living Marine Resources module has four major research fields: fish biology, distribution and abundance, population dynamics, and ecosystem structure. Field surveys such as bottom trawls, plankton net tows, hydroacoustic measurements, and grab samples will be conducted to obtain biological samples. Also, some commercial data will be used to study population dynamics. Details of the research items for each field are listed and methodologies for sampling and analyses are suggested.

Among the 49 large marine ecosystems (LMEs) in the world ocean, the Yellow Sea LME (Figure 6-1) has been one of the most significantly affected by humans. Fish are the main resource in the Yellow Sea; about 200 species are found. Of these, 45% are warm-water forms, 46% warm-temperate forms, and 9% cold-temperate forms (Zhang et al. 1988). The populations can be divided into two groups: resident species near coastal waters and migratory species. The nearshore species include skates, scaled sardine, spotted sardine, and pomfret. These species are mainly found in bays, estuaries, and around islands during spring through fall, and move to deeper water in winter. The migratory species, which are small yellow croaker, hairtail, Pacific herring, chub mackerel, Spanish mackerel, and filefish, have distinct seasonal movements; some migrate out of

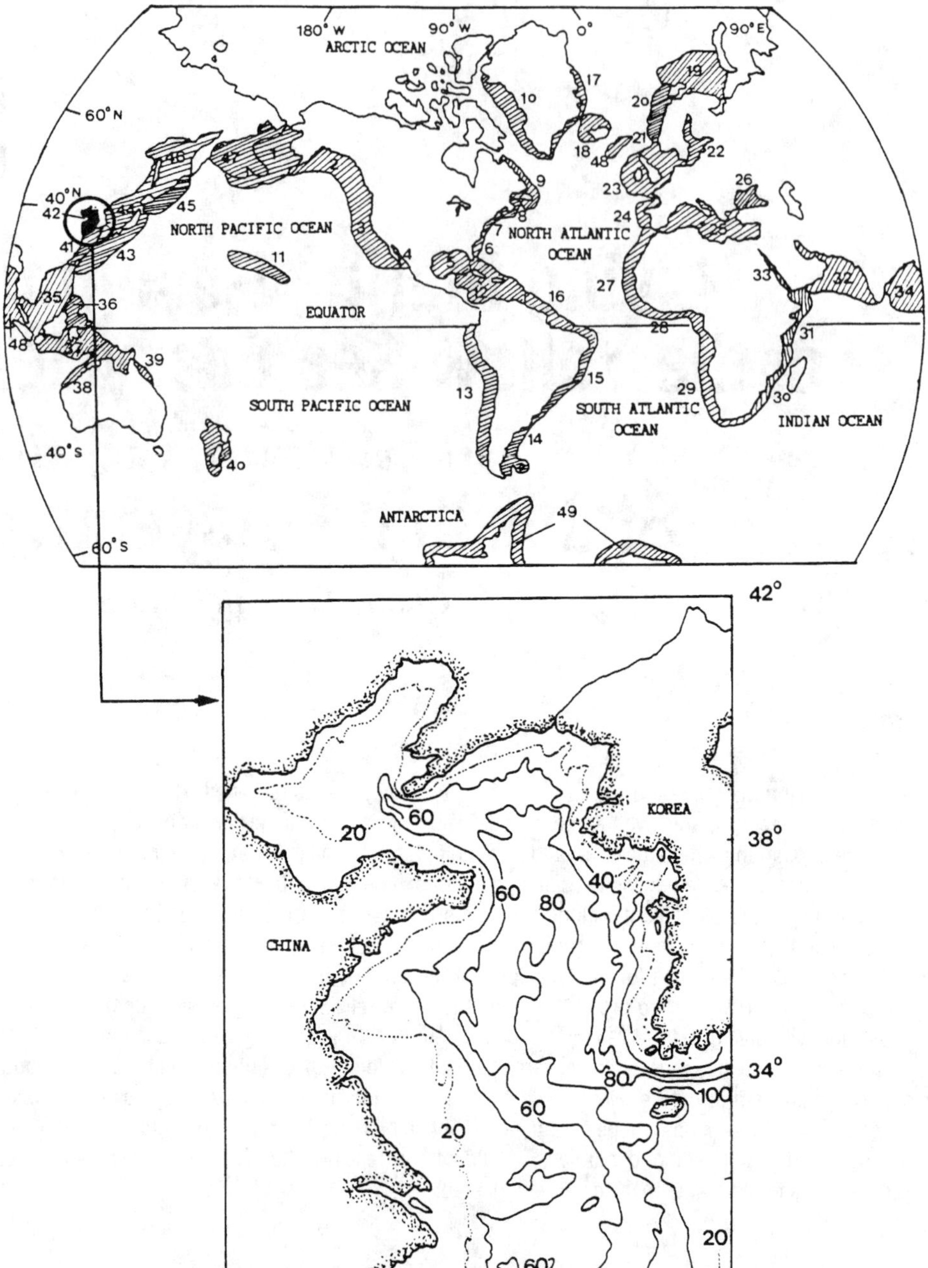

Figure 6-1. Boundaries of 49 large marine ecosystems (above) and bottom topography of the Yellow Sea (below).

the Yellow Sea to the East China Sea in winter. The habitats of these two groups often overlap during spawning and wintering periods.

The commercial utilization of living marine resources in the Yellow Sea dates back several centuries (Park 1976, 1978). With the introduction of bottom trawl vessels in the early twentieth century, many commercial fish stocks began to be substantially exploited by Korea, China, and Japan. Some economically important species, such as red seabream, declined in abundance in the 1920s and 1930s (Xio 1960). By the mid-1960s nearly all the major stocks had been overfished due to a great increase in fishing effort. Since then, the composition of the fish catch has also greatly changed (Xia 1978; Liu 1979; Chikuni 1985; Kang 1987; Zhang and Yoo 1993).

Changes in the Demersal Fish Community

Two large-scale bottom trawl surveys on demersal fish have been conducted in the Korean waters of the Yellow Sea and East China Sea. One survey looked at the distribution of demersal fishes in summer 1967 (Kim et al. 1970), and the other looked at the distribution and relative abundance of demersal fishes in March, May, July, and November 1980/81 (FRDA 1983).

Based on these surveys, ecological characteristics such as changes in biodiversity, species composition, and abundance of major stocks have been examined. For demersal fishes, 134 species of 68 families were identified in 1967, whereas 51 species of 38 families were identified in 1980/81. Thus, there has been a 62% reduction in the number of species from 1967 to 1980/81.

Two interesting phenomena were detected in the species compositions. Tables 6-1 and 6-2 show the rank of the most abundant fish from the 1967 and 1980/81 bottom trawl surveys, respectively. Shifts in the major species are obvious. Among the top 15 abundant species in the

Table 6-1. Rank order of the 20 most abundant fish from the 1967 summer bottom trawl survey

	Taxon			Proportion of total	Cumulative proportion
Rank	Scientific name	(Family)	English name	catch (%)	(%)
1	*Raja kenojei*	(Rajidae)	Skate ray	11.8	11.8
2	*Areliscus trigrammus*	(Cynoglossidae)	Sole	9.4	21.2
3	*Tanakius kitaharae*	(Pleuronectidae)	Flounder	7.2	28.4
4	*Eopsetta grigorjewi*	(Pleuronectidae)	Spotted halibut	5.4	33.8
5	*Lophiomus setigerus*	(Lophiidae)	Sea-devil	5.3	39.1
6	*Helicolenus hilgendorfi*	(Scorpaenidae)	Rosefish	5.3	44.4
7	*Collichys niveatus*	(Sciaenidae)	Corvenias	5.2	49.6
8	*Liparis tanakai*	(Liparidae)	Snailfish	4.3	53.9
9	*Platyrhina sinensis*	(Platyrhinidae)	Ray	3.9	57.8
10	*Branchiostegus japonicus*	(Branchiostegidae)	Tilefish	3.8	61.6
11	*Lepidotrigla microptera*	(Triglidae)	Red robin	3.2	64.8
12	*Raja porosa*	(Rajidae)	Shark ray	2.7	67.5
13	*Pampus argenteus*	(Pampidae)	Pomfret	2.4	69.9
14	*Pseudosciaena polyactis*	(Sciaenidae)	Small yellow croaker	2.2	72.1
15	*Scyliorushinus torazame*	(Scyliorhinidae)	Cat shark	1.9	74.0
16	*Taius tumifrons*	(Sparidae)	Yellow pogy	1.9	75.9
17	*Fugu pardalis*	(Tetraodontidae)	Puffer	1.8	77.7
18	*Trichiurus leptulus*	(Trichiuridae)	Hairtail	1.7	79.4
19	*Fugu niphobles*	(Tetraodontidae)	Puffer	1.6	81.0
20	*Pseudohombus cinnamoneus*	(Bothidae)	Flatfish	1.5	82.5

Table 6-2. Rank of the 15 most abundant fish from the 1980/1981 summer bottom trawl survey

Rank	Taxon: Scientific name	Taxon: (Family)	Taxon: English name	Proportion of total catch (%)	Cumulative proportion (%)
1	*Navodon modestus*	(Aluteridae)	Filefish	23.3	23.3
2	*Trichiurus leptulus*	(Trichiuridae)	Hairtail	15.7	39.0
3	*Raja kenojei*	(Rajidae)	Skate ray	14.0	53.0
4	*Zeus japonicus*	(Zeidae)	John dory	6.6	59.6
5	*Zoarces gillii*	(Zoarcidae)	Eelpout	3.9	63.5
6	*Doderleinia berycoides*	(Serranidae)	Sea bass	3.8	67.3
7	*Pampus argenteus*	(Pampidae)	Pomfret	2.7	70.0
8	*Eopsetta grigorjewi*	(Pleuronectidae)	Spotted halibut	2.1	72.1
9	*Heterodontus japonicus*	(Heterodontidae)	Bullhead shark	1.9	74.0
10	*Lophiomus setigerus*	(Lophiidae)	Sea-devil	1.5	75.5
11	*Liparis anakai*	(Liparidae)	Snailfish	1.4	76.9
12	*Chrysophrys major*	(Sparidae)	Red seabream	1.1	78.0
13	*Branchiostegus japonicus*	(Branchiostegidae)	Tilefish	1.0	79.0
14	*Sebastes inermis*	(Scorpaenidae)	Rockfish	1.0	80.0
15	*Collichthys fragilis*	(Sciaenidae)	Corvenias	1.0	81.0

1980/81 survey, 8 species were not found in the list of 20 abundant species in the 1967 survey. Skate ray, which was first in 1967, was replaced by filefish in 1980/81. This change is remarkable in that filefish was not ranked in the top 20 abundant species in the mid-1960s. Also, hairtail, which was ranked 18th in 1967, became 2nd in 1980/81. It is worth noting that the first three top-ranked species (*Raja kenojei*, *Areliscus trigrammus*, and *Tanakius kitanarae*) in 1967 did not appear in 1980/81. The second phenomenon is that dominant species became more distinct, in that only three species made up 50% of the total biomass in 1980/81, whereas eight species did in 1967.

In the relative stock abundance of major fishes, overall catch per unit effort (CPUE) in 1980/81 was 14.8 kg/haul, whereas the CPUE in 1967 was 26.0 kg/haul, indicating a 43% decline in relative abundance. The CPUE of small yellow croaker, one of the most economically important species, also declined during the 13-year period, from 0.4 kg/haul in 1967 to 0.11 kg/haul in 1980/81.

Biomass Yield and Relative Abundance

The proportion of the annual catch by each nation has changed substantially in the Yellow and East China seas during the past 20 years (Table 6-3). The total catch has increased continuously. In 1970 Japan was the biggest fishing country, with 1.7 million metric tons (44% of the total). China caught approximately 1.6 million mt, and Korea caught 0.6 million mt in 1970. However, Korea's catch increased in 1980, accounting for 24% of the total catch. On the other hand, Japanese activities were reduced during the same period. As a result, the Korean and Chinese proportions of the catch increased about 10% in 20 years, while the Japanese proportion declined about 20%, so that China became the biggest fishing country in the Yellow Sea in the early 1980s.

Table 6-4 shows the annual catch of demersal fishes by the three nations in the Yellow and East China seas for 1970–1991 (Tokimura 1994). The overall catch increased twofold during the two decades. The proportion of demersal fish to the total catch was 33% in 1970, and increased to 40% and 42% in 1980 and 1990, respectively.

Table 6-3. Annual catches (in millions of metric tons) by nation in the Yellow Sea and East China Sea in 1970, 1980, and 1990

	Korea		China		Japan		Total	
Year	Catch	Percentage	Catch	Percentage	Catch	Percentage	Catch	Percentage
1970	0.555	14.6	1.564	41.3	1.670	44.1	3.789	100
1980	1.125	24.3	1.930	41.8	1.567	33.9	4.622	100
1990	1.368	23.0	3.159	53.2	1.410	23.8	5.937	100

Data from FRDA (1994).

Korean fishermen have been fishing in the East China Sea for the last three decades (Figure 6-2). In the 1960s the fishing grounds were within the southeastern Yellow Sea. However, the fishermen extended their fishing grounds toward the East China Sea in the 1970s, and even further in recent years. This expansion was attributed to the decline in the major stocks in the Yellow Sea as well as the increased use of motorized fishing vessels (Zhang et al. 1988).

The Yellow Sea has been one of the major fishing areas for Korea, and it has produced about one-third of the total Korean biomass yield. Figure 6-3 shows Korea's total annual catch and its Yellow Sea catch for 1960–1992. The two catches showed very similar trends, continuously increasing until 1986 and then declining. The current annual catch is nearly at that of the mid-1970s.

Table 6-4. Annual catches of demersal fishes by Korea, China, and Japan in the Yellow Sea and East China Sea, 1970–1991

	Annual Catch (10,000 metric tons)			
Year	China	Korea	Japan	Total
1970	69	28	30	127
1971	76	29	28	133
1972	88	36	24	148
1973	95	42	24	161
1974	101	53	24	178
1975	100	54	23	177
1976	103	51	20	174
1977	104	47	22	173
1978	111	54	21	186
1979	96	63	22	181
1980	98	68	21	187
1981	98	68	19	185
1982	105	65	17	187
1983	104	67	17	188
1984	112	65	15	192
1985	114	69	13	196
1986	122	81	13	216
1987	127	62	12	201
1988	147	64	10	221
1989	160	62	9	231
1990	172	71	8	251
1991	191	55	8	254

Data from Tokimura (1994).

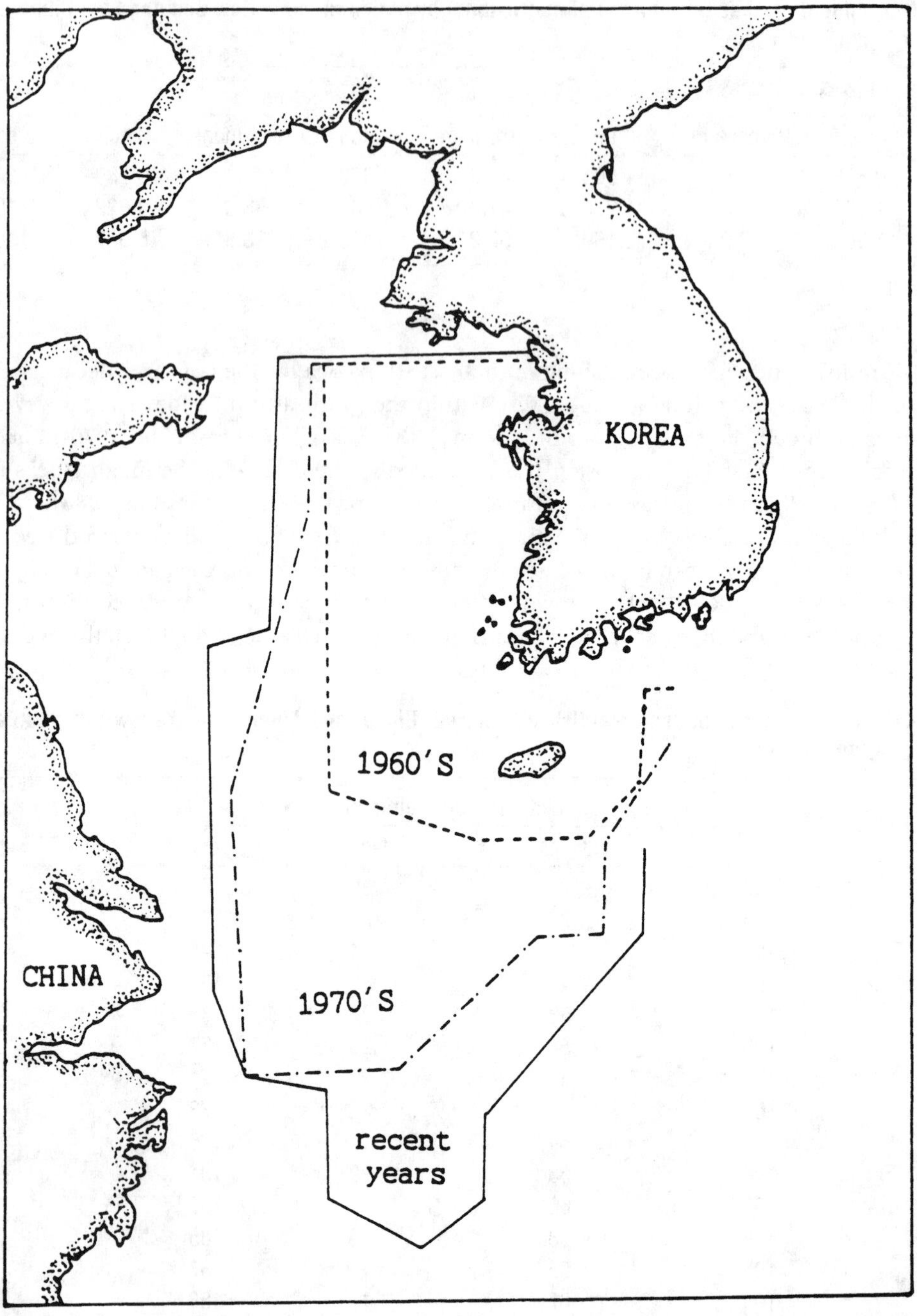

Figure 6-2. Changes in fishing grounds by Korean fishermen in the Yellow Sea and the East China Sea (FRDA 1994).

Figure 6-4 shows annual catch and CPUE of the Yellow Sea Korean fisheries. While annual catch increased continuously until 1986 and declined afterward, annual CPUE showed a different trend, fluctuating between 1960 and 1974 and continuously declining afterward. There was a 12-year time lag between the peaks of catch and the relative abundance. The current annual catch is at the 30-year average level, while the CPUE is far below the average level.

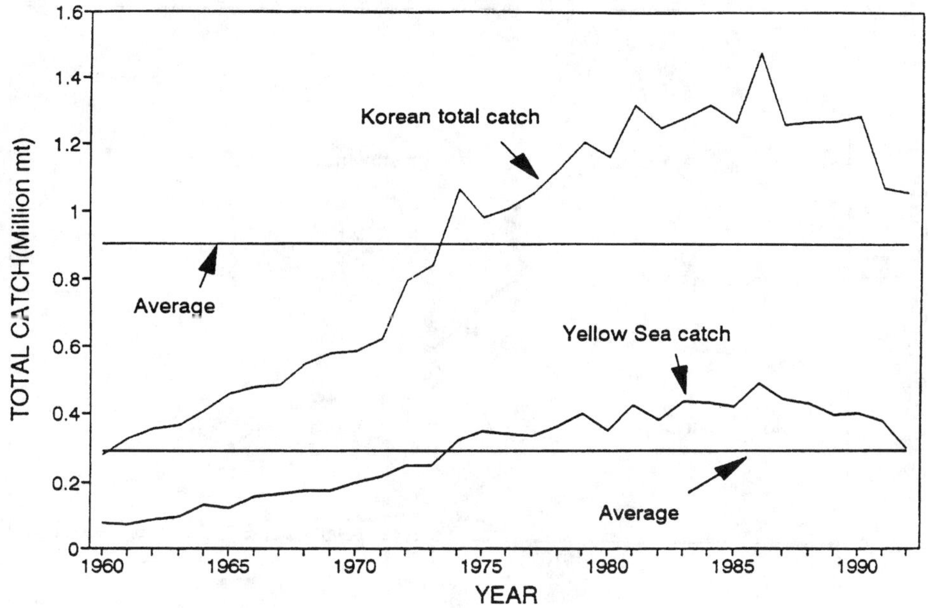

Figure 6-3. Total catches in the Korean adjacent seas and in the Yellow Sea by year for 1960–1992.

The CPUEs of three major fisheries showed a declining trend in the Yellow Sea (Figure 6-5). The large stow net fishery and the large-sized trawl fishery tended to increase, and the mid-sized trawl fishery was stable from the late 1960s to mid-1970s. However, all three fisheries declined dramatically from 1975 until 1980 and then declined more moderately, showing that the relative abundance of the recent years was less than one-tenth that of the 1960s.

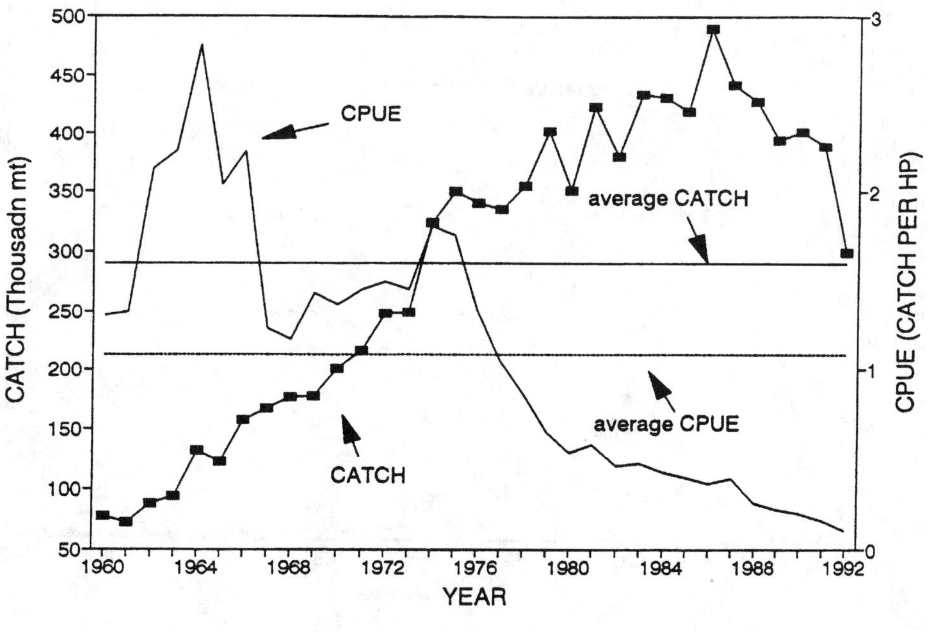

Figure 6-4. Annual catch and CPUE by Korean fisheries in the Yellow Sea for 1960–1992.

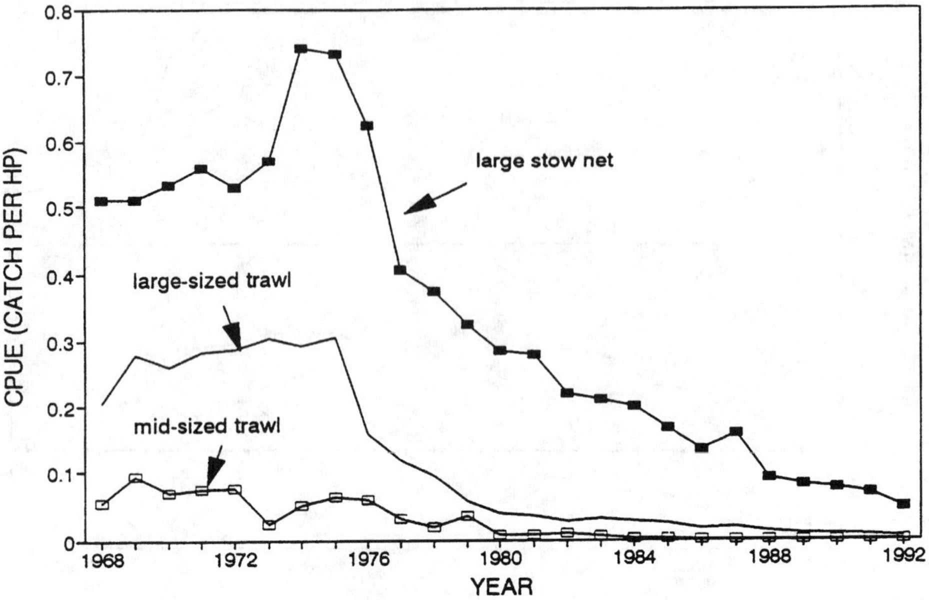

Figure 6-5. CPUE of the large-sized trawl, the medium-sized trawl, and the large stow net fisheries in Korean waters of the Yellow Sea for 1968–1992.

The CPUEs of the two most important species, small yellow croaker and hairtail, also showed a declining trend in the Yellow Sea (Figure 6-6). The CPUE of small yellow croaker started declining in 1964, whereas that of hairtail started declining 10 years later, in 1974. The magnitude of the decline was remarkable.

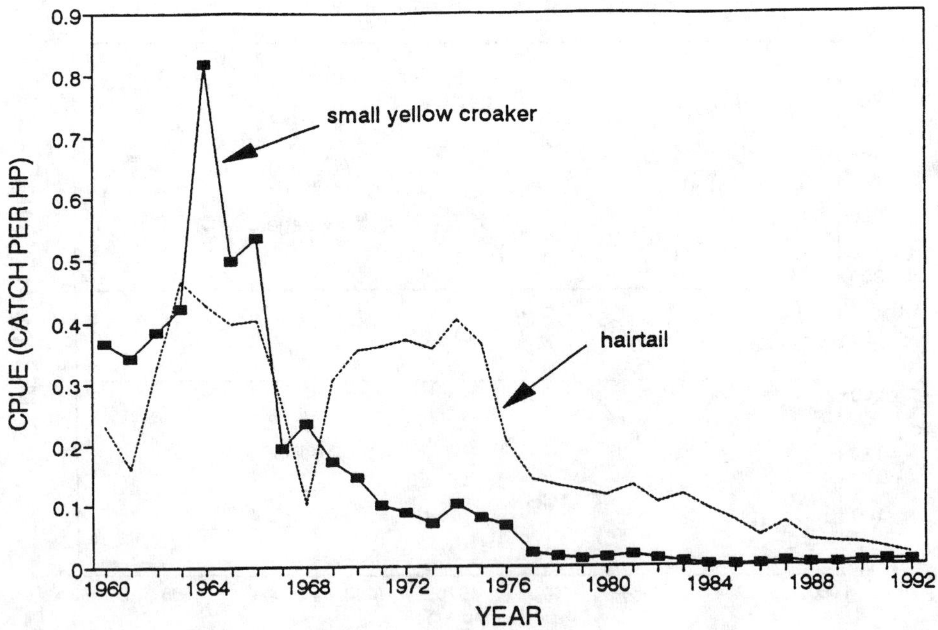

Figure 6-6. CPUE of small yellow croaker and hairtail in Korean waters of the Yellow Sea for 1960–1992.

Catch Composition and Species Dominance

Using official Korean catch statistics for the Yellow Sea, catch composition and species dominance were examined. Changes in species or species group compositions in the catch were substantial. Figure 6-7 shows catch composition by species group for the 1960s and 1980s. The dominant group in the 1960s was Sciaenidae, which accounted for more than 50% of the catch. The second largest group was Trichiuridae, followed by Clupeidae and Scombridae. However, Aluteridae became dominant with 32% during the 1980s, while Trichiuridae remained second. Sciaenidae was less than 10% of the total. Figure 6-8 shows catch composition by species for the same periods. The predominant species in the 1960s was small yellow croaker. The second most abundant species was hairtail, followed by anchovy, skate ray, pomfret, and flounder. In the

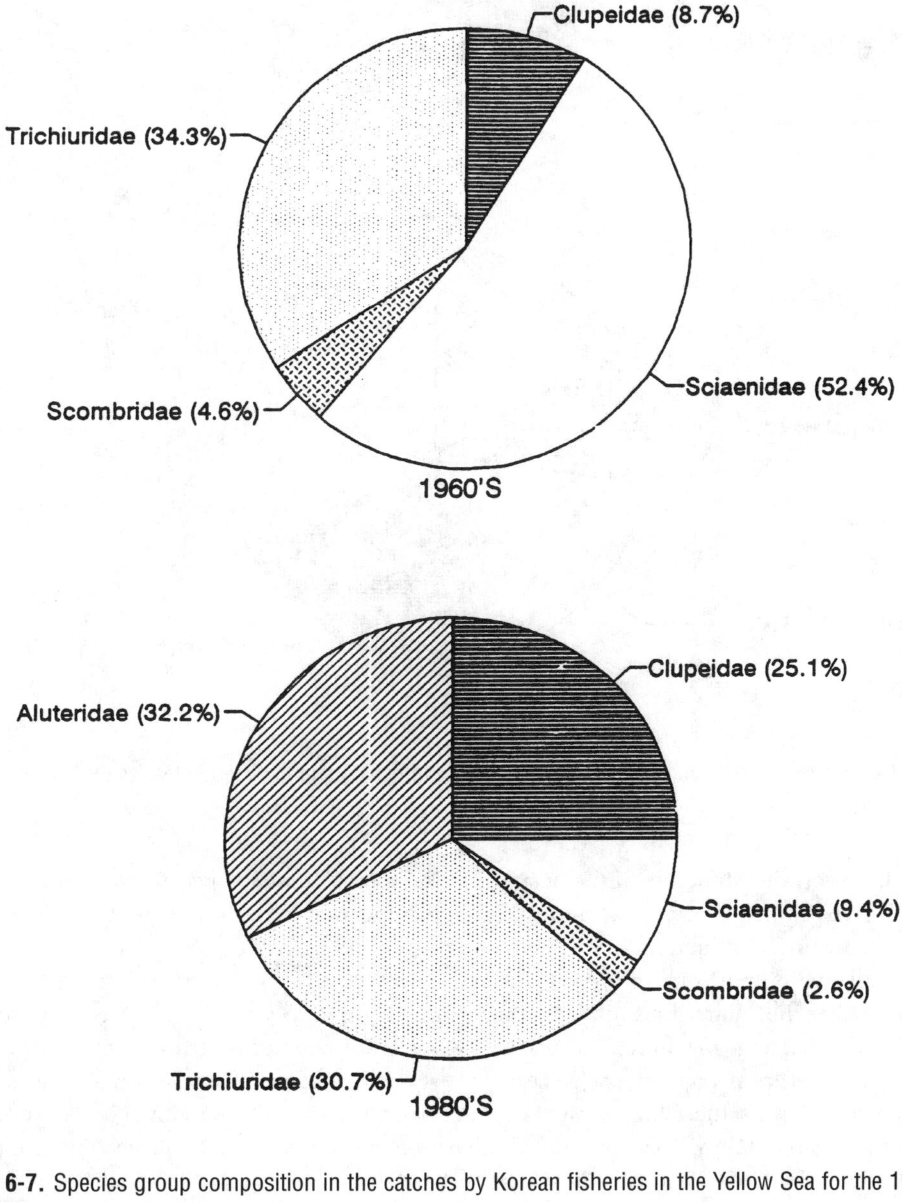

Figure 6-7. Species group composition in the catches by Korean fisheries in the Yellow Sea for the 1960s and 1980s.

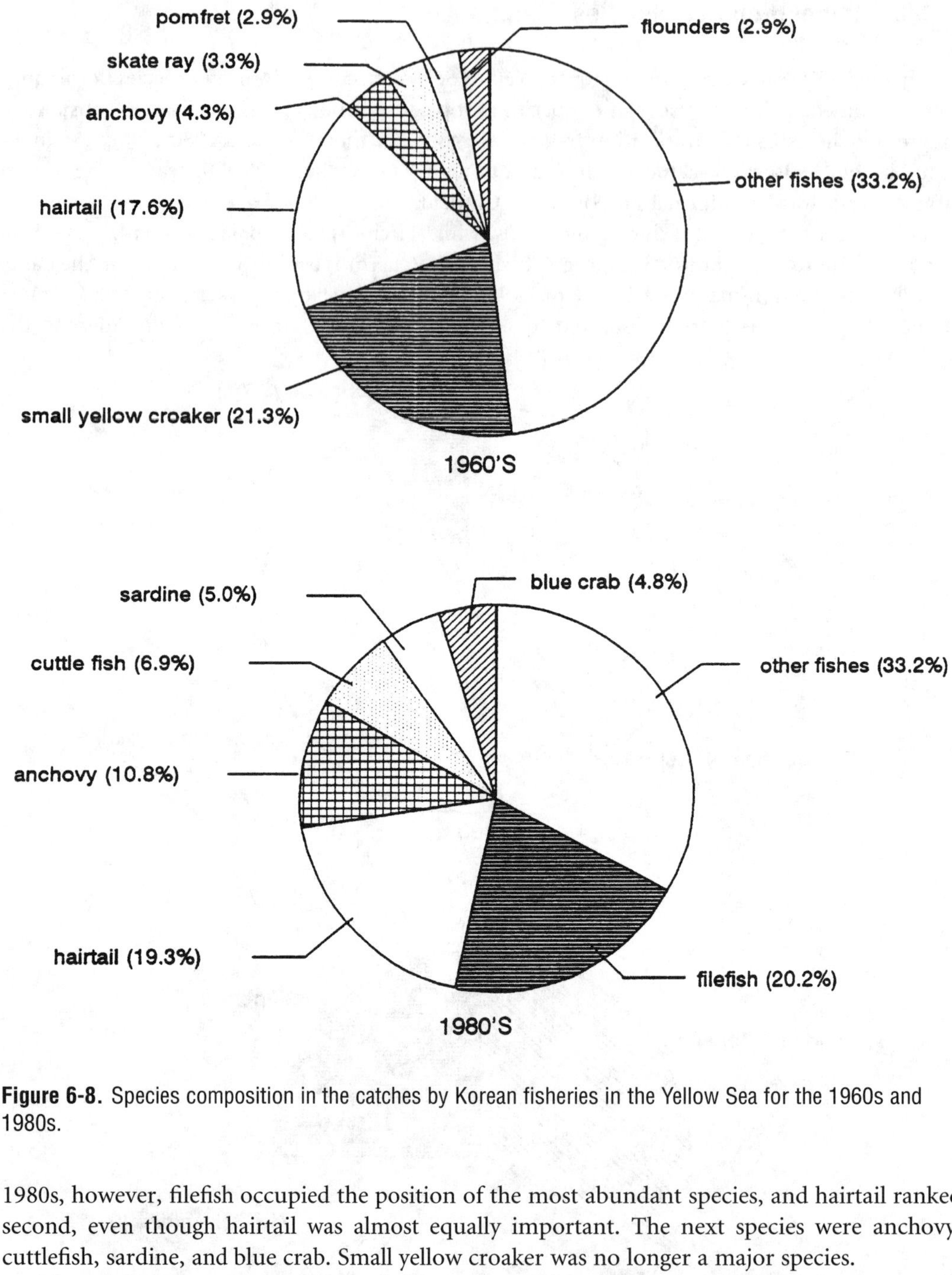

Figure 6-8. Species composition in the catches by Korean fisheries in the Yellow Sea for the 1960s and 1980s.

1980s, however, filefish occupied the position of the most abundant species, and hairtail ranked second, even though hairtail was almost equally important. The next species were anchovy, cuttlefish, sardine, and blue crab. Small yellow croaker was no longer a major species.

The relative proportions and trends in catches of three major groups of fish species—small yellow croaker and hairtail group (demersal), anchovy and sardine (small-sized pelagic), and mackerel and herring (medium-sized pelagic)—were compared to determine whether the trend between catches from the eastern (Korean waters) and western (Chinese waters) parts of the Yellow Sea were the same. Chinese catch data are from Tang (1989). As shown in Figure 6-9, the trends of the small yellow croaker and hairtail group and the mackerel and herring group were quite different between the two areas, whereas the trend of the anchovy and sardine group was

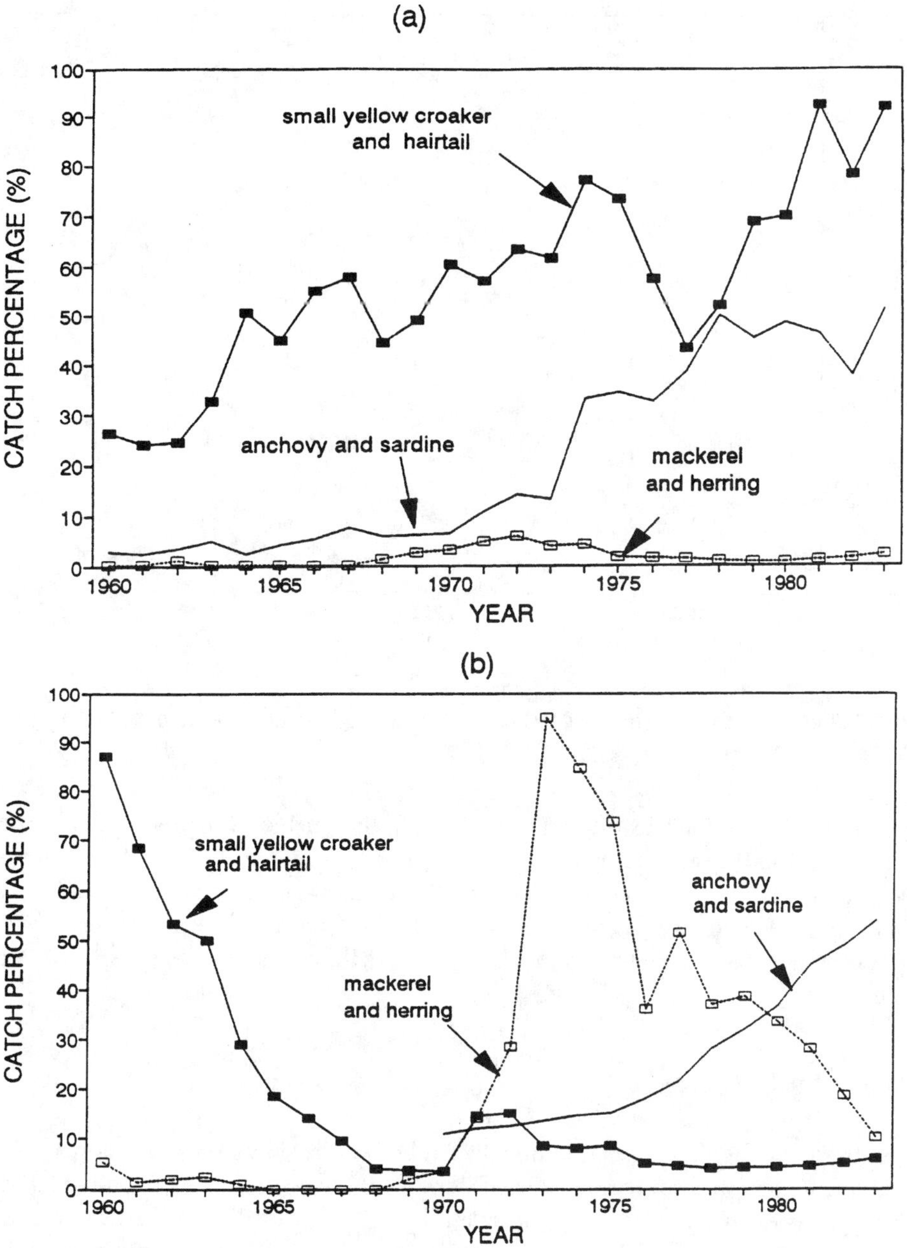

Figure 6-9. Comparisons of trends and proportions in catches for demersal (small yellow croaker and hairtail), small-sized pelagic (anchovy and sardine), and medium-sized pelagic (mackerel and herring) fish of the (a) Korean and (b) Chinese (Tang 1989) waters in the Yellow Sea.

very similar, showing an increasing trend since 1970. The relative magnitude in catch among the three groups was different in the two areas.

Figure 6-10 compares the catch trends for small yellow croaker and hairtail, the two most important species in Korean waters of the Yellow Sea, between 1960 and 1992. In the 1960s the magnitude and trend for both species were similar. However, they began to show an inverse

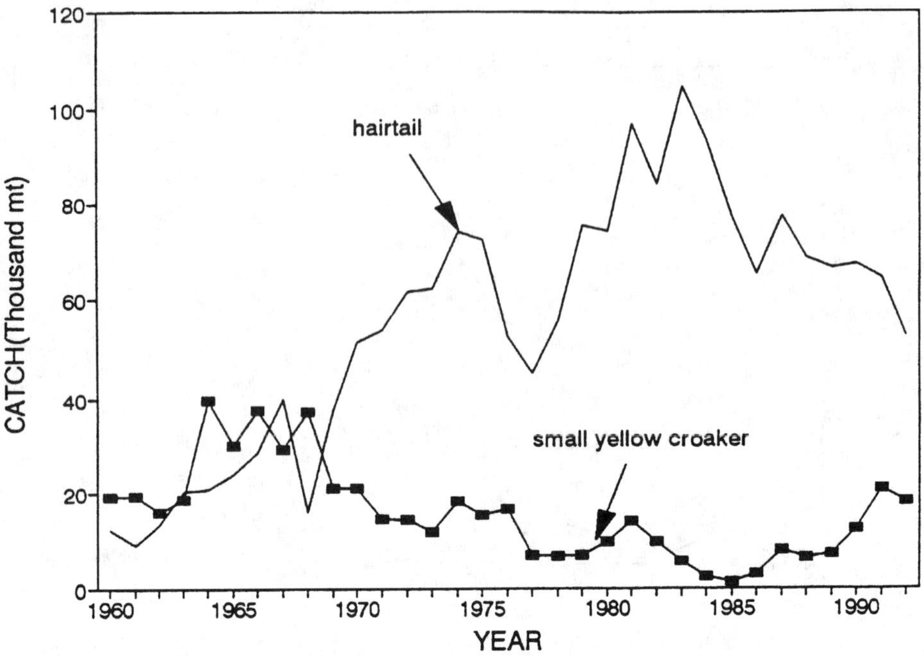

Figure 6-10. Trend in catches of small yellow croaker and hairtail in Korean waters of the Yellow Sea for 1960–1992.

relationship starting in the late 1960s. From 1967 on, the catches of the two species showed an opposite trend, and the magnitude of the catches was quite different. During the 1970s the catch of small yellow croaker dropped to less than one-third that of the late 1960s, whereas the catch of hairtail increased to about twice that of the 1960s. The difference between the catches of the two species was remarkable during the 1980s. In recent years the catches of the two species seemed to be getting closer.

Directions for Research

The Korean side of the Yellow Sea traditionally has been utilized by Korean fishermen. However, few assessments of fish resources have been done. Only two surveys, 1967 and 1980/81, were performed to assess the composition and relative abundance of demersal species. As we have shown above, however, there was a big change in community composition since the mid-1980s, thus requiring new information to diagnose current stock conditions. On the other hand, one of the hot issues in marine science is the understanding of mechanistic processes such as trophodynamics and the interaction between organisms and abiotic environments. Accordingly, the change in the fish community in an ecosystem is viewed as caused by these elements. This concept led to the formation of the Yellow Sea holistic ecosystem research program in Korea.

The objectives of the Living Marine Resources (LMR) research, which is one of the research modules in the Korean Yellow Sea Large Marine Ecosystem (YSLME) project (Figure 6-11), are

- to establish assessment and long-term monitoring programs of LMR with regard to species composition, biomass, biodiversity, trophic interactions, environmental conditions, and stock conditions of major fisheries resources;

- to develop an ecosystem-based management strategy for living resources on the basis of sustainable yield; and
- to provide advice on the implementation of an international fisheries management system.

The LMR module has four components: fish biology, distribution and abundance, population dynamics, and ecosystem structure. Bottom trawls, plankton tows, grab samples, and CTD cast data along with other ancillary samplings will be collected at fixed stations on the Korean side of the Yellow Sea. Sampling locations are designed by the equidistant method and have a uniform distribution with a 30′ × 30′ grid. A total of 88 stations will be covered several times in the first year of the survey (Figure 6-12). Stratified random sampling might be considered in subsequent years, depending on the results of the first-year survey. Also, some hydroacoustic surveys are under consideration, and commercial catch data should be collected and analyzed for population dynamics. A number of fish species that we are interested in have distinct seasonal movements and frequently migrate out of the Korean waters to the Chinese waters of the Yellow Sea and East China Sea. Therefore, the survey should cover the entire habitat area to reveal the

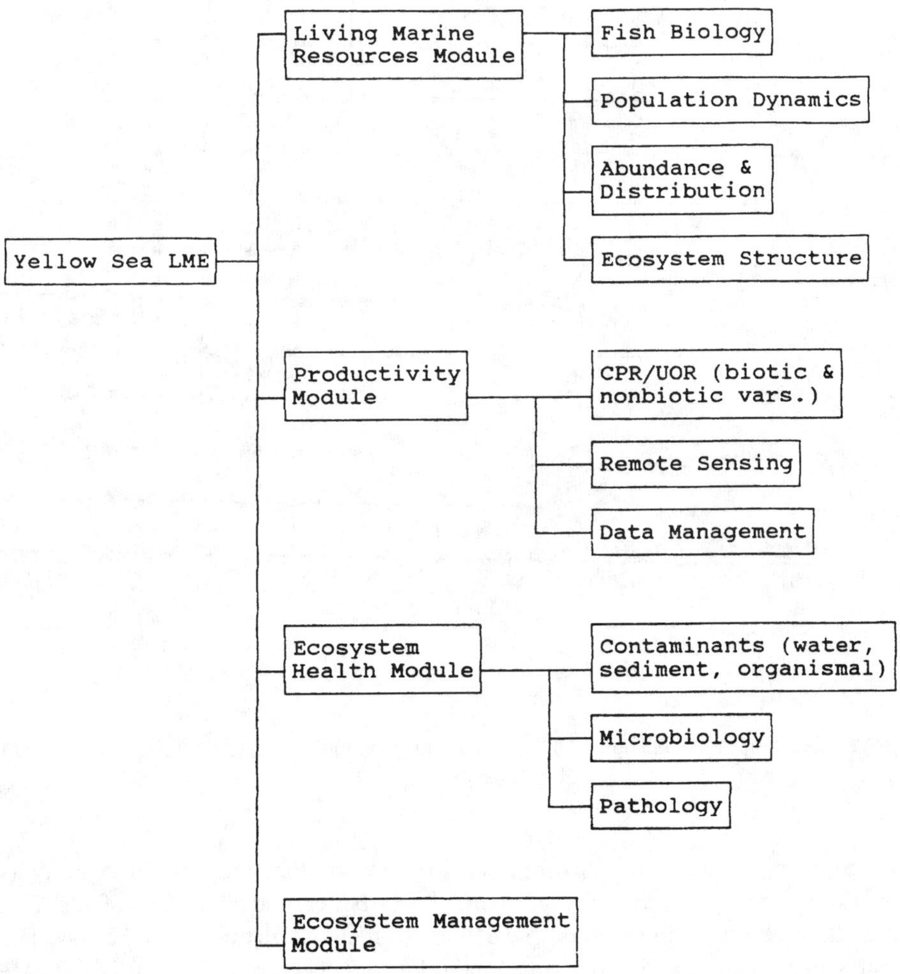

Figure 6-11. Structure of the Korean Yellow Sea Large Marine Ecosystem (YSLME) project.

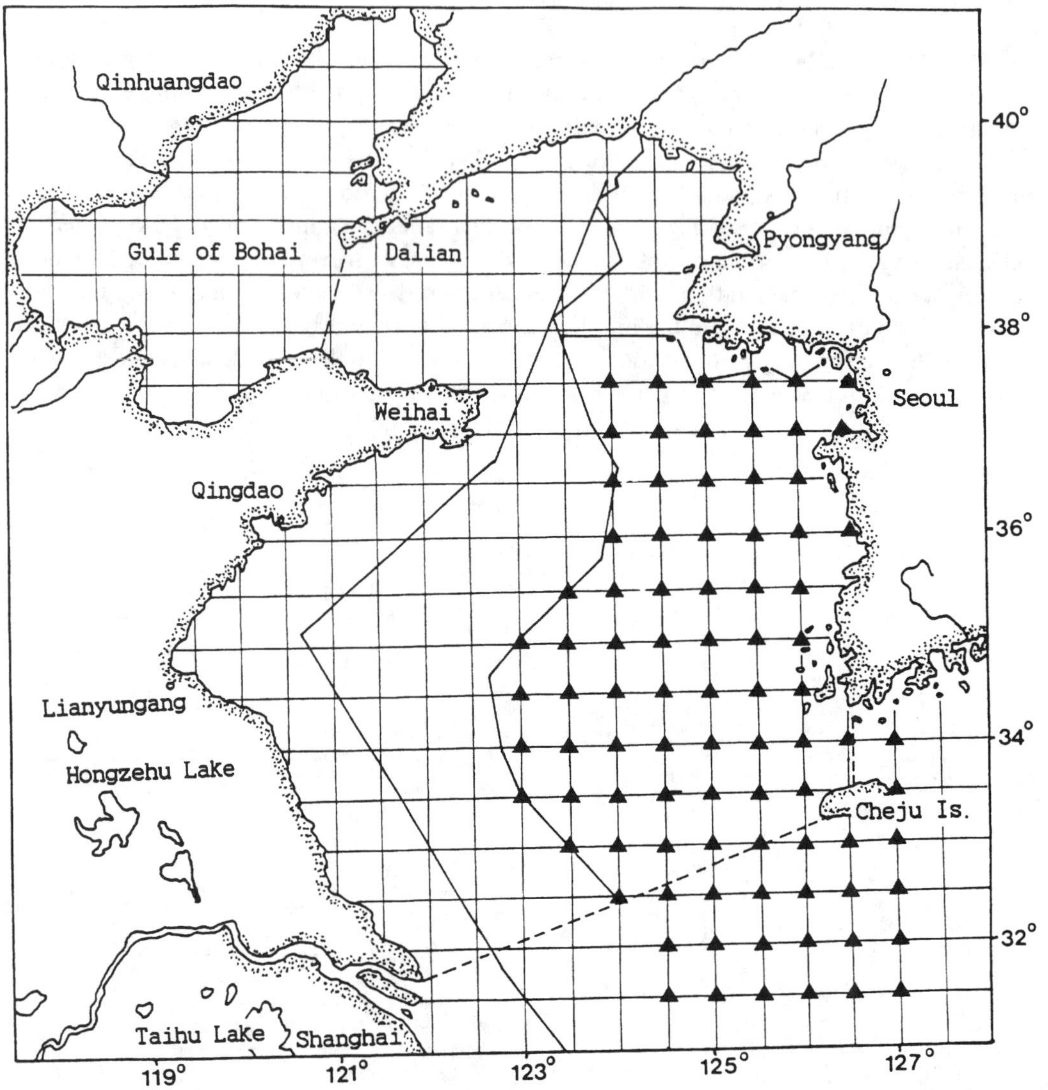

Figure 6-12. Map showing sampling stations for the Living Marine Resources module of the Korean Yellow Sea Large Marine Ecosystem (YSLME) project.

distribution and abundance of those species. This means that joint research by two coastal nations (Korea and China) is essential to develop a better understanding of the Yellow Sea ecosystem. Two objectives in the joint research are to standardize the data to be collected and to have simultaneous surveys using the same methodologies. The research items in the LMR module are tabulated in Table 6-5.

Table 6-5. Details of research items for the Living Marine Resources module of the Korean Yellow Sea Large Marine Ecosystem (YSLME) project

Fish biology	Distribution abundance	Population dynamics	Ecosystem structure
A. Age and growth	**A. Distribution pattern**	**A. Population parameters**	**A. Phyto- and zooplankton**
Age determinations	Seasonal distribution	Growth parameters	Species composition
Age–length keys	Eggs and larval distribution	Instantaneous mortality rates	Distribution
Age compositions	**B. Abundance**	Time-series estimate of recruits	Abundance
B. Maturation and spawning	Density of eggs and larvae	**B. Management parameters**	**B. Ichthyoplankton**
Gonadosomatic index (GSI)	Density of adult fish	Optimal age and/or size at capture (T_r)	Species composition
Fecundity	Species compositions	Optimal fishing intensities (F_{max}, $F_{0.1}$)	Spawning and nursery areas
Spawning ground and season	Length and weight composition	Maximum sustainable yields (MSY)	Abundance
Length–weight relationship	Age composition	Fishing effort at MSY (F_{MSY})	Biomass of adult fish by EPM
C. Mortalities	Sex ratio	Acceptable biological catch (ABC)	**C. Benthos**
Survival rate	Abundance and biomass		Species composition
Instantaneous rates of natural and			Distribution
fishing mortality			Abundance
D. Feeding and ecological			Biomass of major fishes
interactions			**D. Ecosystem diversity and stability**
Stomach contents analysis			Ecological dominance
Predator–prey interaction studies			Diversity
			Stability and resilience

References

Chikuni, S. 1985. The Fish Resources of the Northwest Pacific. FAO Fisheries Tech. Pap. 266. Rome: FAO.

FRDA. 1983. Survey of the Demersal Fisheries Resources in Korean Waters. Technical Report No. 59. Pusan, Korea: National Fisheries Research and Development Agency.

FRDA. 1994. Status of the Korean Yellow Sea fishery. Unpublished manuscript. Pusan, Korea: National Fisheries Research and Development Agency.

Kang, Y. J. 1987. A rational approach for management of fisheries resources in coastal and offshore. Unpublished manuscript.

Kim, Y. S., Kim, Y. M., and Choo, W. I. 1970. A Study on Distribution of Demersal Fishes Near Korean Waters in Summer. Report of Fisheries Resources, No. 8. Pusan, Korea: National Fisheries Research and Development Agency.

Liu, X. 1979. Status of fishery resources in the Bohai and Yellow Seas (in Chinese). Fishery Resources Paper 26:1–17.

Park, K. B. 1976. History of the Korean Herring Fisheries. Thesis Collection of the National Fisheries, University of Pusan, 17:1–37, Korea.

Park, K. B. 1978. History of the Korean Sardine Fisheries. Thesis Collection of the National Fisheries, University of Pusan, 21:21–59, Korea.

Tang, Q. 1989. Changes in the Biomass of the Yellow Sea Ecosystem. In K. Sherman and L. M. Alexander (eds.), Biomass Yields and Geography of Large Marine Ecosystems. AAAS Selected Symposium 111. Boulder, CO: Westview Press.

Tokimura, M. 1994. Status and problems in stock assessment of demersal fisheries resources in the Yellow and East China seas (in Japanese). Jap. Fish. Res. Con., Bull. West Jap. Demersal Fish. 22: 41–57.

Xia, S. 1978. An analysis of changes in fisheries resources of the Bohai Sea, Yellow Sea, and East China Sea (in Chinese). Marine Fisheries Resources Paper 25:1–13.

Xio, S. 1960. Fisheries of the Bohai Sea, Yellow Sea, and East China Sea (in Chinese). Marine Fisheries Resources Paper 2:73–94.

Zhang, C. I., and Yoo, S. 1993. Current Status of the Yellow Sea LME and Suggestions for Future Research. Paper presented at the International LME Symposium and Workshop on the Status and Future of Large Marine Ecosystems of the Indian Ocean, Mombasa, Kenya, 28 March–2 April 1993.

Zhang, C. I., Kim, J. M., and Huh, H. T. 1988. Current Status of Fisheries Resources and Their Rational Management in the Yellow and East China Seas. In Y. B. Go, (ed.), Proceedings of Symposium on Development of Marine Resources and International Cooperation in the Yellow Sea and East China Sea, Cheju, Korea, 19–22 October 1988, pp. 71–96.

7

Ecology and Variability of Economically Important Pelagic Fishes in the Yellow Sea and Bohai Sea

Qisheng Tang and Xianshi Jin

Abstract

The distribution patterns, feeding, reproduction, early life history, growth, population structure, yield, stock size fluctuations, and ecology of seven pelagic fishes of the Yellow and Bohai Seas are analyzed based on surveys from 1981 to 1990. The biomass of small pelagic fish has increased dramatically. Species such as Japanese anchovy, half-fin anchovy, spotted sardine, and scaled sardine have great potential for fisheries, with a total yield of about 0.6 million tonnes. Relatively large-sized species such as Spanish mackerel and chub mackerel have been overfished, and the stock of Yellow Sea herring has collapsed. Species interactions, particularly prey–predator, as well as the influences of fishery and environmental changes may be responsible for such a succession.

In the Yellow and Bohai Seas, changes in several aspects of the fishery, such as species composition of the catch, fluctuations in catch rates, and yield, have been documented over the years. The importance of pelagic fish has been increasing in the Chinese fisheries since the late 1950s, although there have been variations in species composition. The number of pelagic species is considerably less than demersal species in the seas (Zhu et al. 1963; Zhao et al. 1990). In recent years, most commercially important stocks, such as small yellow croaker, hairtail, and Spanish mackerel, have been overfished. The biomass of commercial populations declined by about 40% from the early 1960s through the 1980s (Tang 1990). Meanwhile, the populations of some small pelagic fishes have increased dramatically, such as the Japanese anchovy and sardines, which are small and of low commercial value. They are also vulnerable to changes in environmental factors. Natural conditions may have a large effect on the long-term fluctuations in stock abundance, whereas fishing mortality may play a smaller part in causing changes in stock abundance in pelagic species of the northwest Pacific (Chikuni 1985).

This chapter reviews and discusses the life history, ecology, and variability of resources of seven economically important pelagic fish species in the Yellow and Bohai Seas.

Materials and Methods

The Bohai Sea data used in this study were obtained from surveys (1982–1983) with a pair of bottom trawl vessels (200 hp each) using a trawl net with 1660 mesh × 6.3 cm, with the cod-end of 20 mm and an opening of 5–6 m (Deng 1988). The survey area is shown in Figure 7-1. The Yellow Sea surveys in 1981 and 1985–1988 used two types of survey vessels: a pair of trawlers (600 hp each) using a net of 550 mesh × 30 cm and a cod-end with a 51-mm mesh size in 1981 and the spring surveys, and the R/V *Bei Dou* using a net of 450 mesh × 17 cm, with a 100-mm mesh size cod-end and a 20-mm mesh size cod-end liner for the surveys in summer, fall, and winter. The survey areas (Figure 7-1) varied with season based on previous knowledge of fish distribution.

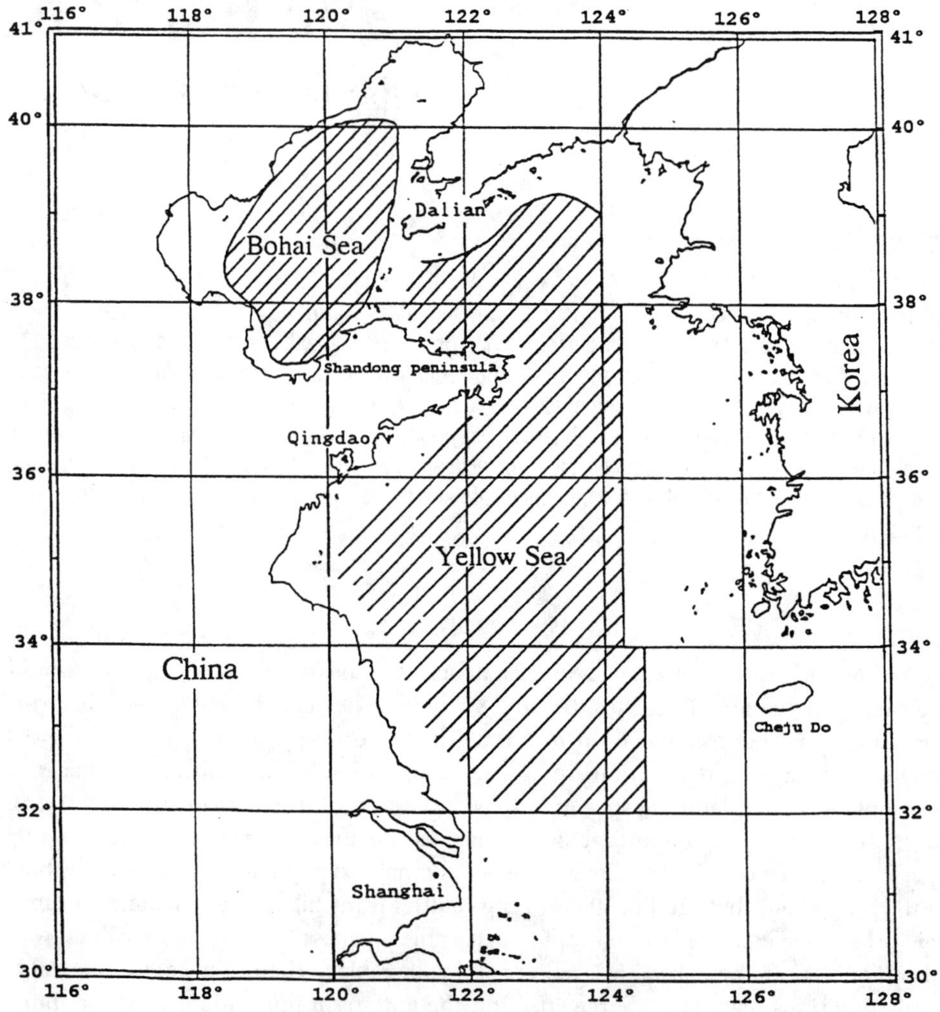

Figure 7-1. Survey area in the Bohai Sea during 1982–1983, and the Yellow Sea in 1986.

Survey data from 1959 were also used in this study (Lin 1965), together with Chinese catch statistics. Surveys on Japanese anchovy from 1984 to 1992 by the R/V *Bei Dou* were used as a reference (Zhu and Iversen 1990).

In addition, 902 scales of half-fin anchovy, 656 scales of spotted sardine, and 469 scales of scaled sardine were read for age determination. The von Bertalanffy growth equations of these species were used.

Results and Discussion

During the 1986 surveys, 21 pelagic fish species were recorded in the Yellow Sea (Table 7-1), of which 7 species belonged to Engraulidae, 5 species to Clupeidae, and the rest were included in seven families (Zhu et al. 1963). Seven economically and ecologically important species, including Japanese anchovy, half-fin anchovy, spotted sardine, scaled sardine, Pacific herring, chub mackerel, and Spanish mackerel, have been studied and are reported on below (Table 7-1).

Distribution and Migration

Japanese anchovy is a small inshore pelagic species that is widely distributed in the Bohai Sea, Yellow Sea, and East China Sea. It migrates seasonally with changes in sea surface temperature. The optimum temperature ranges from 10° to 13°C, and the anchovy is not usually found in

Table 7-1. Pelagic species of the Yellow Sea caught during the bottom trawl surveys in 1986

Common name	Scientific name
Pacific herring	*Clupea pallasi* (Cuvier & Valenciennes)
True sardine	*Sardinops melanosticta* (Schlegel)
Chinese herring	*Ilisha elongata* (Bennett)
Scaled sardine	*Harengula zunasi* (Bleeker)
Spotted sardine	*Clupanodon punctatus* (Temmink & Schlegel)
Japanese anchovy	*Engraulis japonicus* (Temmink & Schlegel)
Anchovy	*Thrissa setirostris* (Broussonet)
Anchovy	*Thrissa mystax* (Bloch & Schneider)
Anchovy	*Thrissa Kammalensis* (Bleeker)
Half-fin anchovy	*Setipinna taty* (Cuvier & Valenciennes)
Long-tailed anchovy	*Coilia mystus* (Linnaeus)
Anchovy	*Coilia ectenes* (Jordan & Seale)
Stinnycheek lanternfish	*Myctophum pterctum* (Alcock)
Blind tasselfish	*Eleutheronema Tetradactylum* (Shaw)
Kuweh	*Atropus atropus* (Bloch & Schneider)
Horse mackerel	*Trachurus japonicus* (Temmink & Schlegel)
Black butterfish	*Formio niger* (Bloch)
Pacific sandlance	*Ammodytes personatus* (Girard)
Chub mackerel	*Pneumatophorus japonicus* (Houttuyn)
Spanish mackerel	*Scomberomorus niphonius* (Cuvier & Valenciennes)
Butterfish	*Stromateoides argenteus* (Euphrasen)

waters below 7°C (Iversen et al. 1993). In November and December, the densest area of distribution is in the northern and central parts of the Yellow Sea. During the winter time the Japanese anchovy migrates to the southeast Yellow Sea and north East China Sea. In spring, as water temperatures increase and the gonads develop, the Japanese anchovy migrates into shallow coastal waters for spawning (Li 1987); it then disperses for feeding and moves into deeper, cooler southern waters. After November, there are very few anchovy left in the Bohai Sea. Dense schools are observed in the southeast Yellow Sea and north East China Sea. It also migrates diurnally depending on light and water temperature. It is usually found in dense schools near the bottom during the daytime and is scattered in the upper to surface layers at night (Zhu and Iversen 1990).

Half-fin anchovy is a small pelagic fish found in the inshore waters. In winter, half-fin anchovy is mainly found on the west of Cheju Do (southeast Yellow Sea), where the bottom-water temperature is 10–14°C and salinity is 33–34 ppt. After winter, when the water temperature increases, it starts a spawning migration toward shallow coastal waters and is found throughout the Yellow and Bohai Seas. Spawning season is from May to June in the Bohai Sea (Deng et al. 1988), and slightly earlier in the Yellow Sea. The estuaries are the major spawning grounds. The stock in the East China Sea spawns from May to July, at a surface water temperature of 15–26°C and salinity of 14–34 ppt (Zhao et al. 1990). During summer the feeding stock is mainly found in the shallow coastal waters of the Bohai Sea and the southern Yellow Sea. In autumn the half-fin anchovy is found in dense schools and begins to move southward for wintering, disappearing from the Bohai Sea and the northern Yellow Sea in December.

The spotted sardine is a small pelagic fish that is widely distributed in inshore waters and estuaries along China's coast. The stock of the Yellow and Bohai Seas winters mainly in the central part of the Sea (34°–36°N, 123°–125°E) from January to March when the bottom-water temperature is 8–11°C and salinity is 32–33.5 ppt. In late March the major stock starts a spawning migration northward and reaches the waters around the Shandong Peninsula in mid-April; part of the stock migrates to bays in the Bohai Sea in late April. The main spawning period is from early May to late June. After spawning, the adults leave the coastal waters for feeding. The 0-group fish feed in the coastal waters. In November both adults and juveniles gradually leave the feeding areas and start their winter migration.

Scaled sardine is another small pelagic fish that is widely distributed in the Bohai, Yellow, and East China Seas. The stock of the Yellow and Bohai Seas winters in the southeastern part of the Yellow Sea (around Cheju Do) during January to March (Zhao et al. 1990). It starts its spawning migration northwestward in batches in March, and arrives at waters off China's coast by late March to mid-May. The stock spawns during the period of May to July, then disperses for feeding in the coastal waters. In November, all schools are mixed in the central and south of the Yellow Sea, and together they migrate southward and return to the wintering ground in January.

Pacific herring in the Yellow Sea is a local stock that only inhabits the central to northern parts of the Yellow Sea (north of 34°N) (Ye et al. 1980). The Yellow Sea stock, found only in small quantities by surveys in the 1950s, was widely found in the Yellow Sea in the late 1960s and onward. The wintering ground is in the central, deep water. In February the adults migrate toward coastal waters off the Shandong Peninsula for spawning, and then migrate into the central and northern Yellow Sea for feeding. In autumn and winter the stock's distribution gradually decreases; it occurs primarily in the central part of the Yellow Sea in winter (Tang 1991).

Chub mackerel is a warm-water fish that has two wintering grounds, one southeast of Cheju Do and one in the central and eastern parts of the East China Sea (Zhu et al. 1982; Wang and Zhu 1983; Zhao et al. 1990). The two stocks (Chikuni 1985) start their spawning migration along 123°30′E from the southern wintering ground in late March or early April; one month later they again migrate along 32°30′–33°31′N from the Cheju Do wintering ground to the Yellow Sea.

Spawning mainly occurs south of Shandong Peninsula from May to June. The feeding period is from July to September and occurs in the deep waters of the central and northern Yellow Sea. As the water temperatures decrease, the population migrates southward for winter along 124°00′–125°00′E. However, the young populations (including the 0-group) feed mostly in the coastal waters off Korea from September to November (Wang 1991).

Spanish mackerel are widely distributed in the northwest Pacific Ocean. The Bohai, Yellow, and East China Seas are the major distribution areas of dense schools. There are two wintering stocks, one in the southeastern Yellow Sea and one offshore in the East China Sea (28°00′–31°20′N and 123°40′–125°30′E, respectively). The stock in the East China Sea spawns from April to May mainly in the Fujian (south China) coastal waters and south of the Shandong Peninsula; the other stock spawns from May to June primarily in the Bohai Sea and the northern Yellow Sea (Wei 1991). After spawning, Spanish mackerel feed in the coastal waters. The distribution of this stock is strongly influenced by water temperature; it migrates southward with decreasing temperature. Until November the main population is distributed in the central and southern Yellow Sea; it returns to the wintering grounds in December.

The pelagic fishes in the Yellow and Bohai Seas are all migratory species. They spawn and feed mainly along the Chinese coast in shallow water and winter in the central and southern Yellow Sea and northern East China Sea. The waters around the Cheju Do are the major wintering ground.

Food and Feeding

The diets of the seven pelagic fishes, as determined from analyses of stomach contents, are shown in Figure 7-2. They feed mainly on zooplankton (copepods, euphausiids, mysids and chaetognaths, and *Acetes chinesis*).

There are some food variations between juveniles (0-group) and adults. The diet of juvenile Japanese anchovy is dominated by copepoda (55% by weight); *Sagitta*, *Lamellibranchia* larvae, and *Coscinodiscus* are also found (Chen 1978; Zhu 1991). As the Japanese anchovy grows, the dominant prey gradually shifts to *Sagitta crassa*, *Calanus pacificus*, *Themisto gracilipes*, *Euphausia pacifica*, and some algae.

Wei and Jiang (1992) reported that the food composition of the half-fin anchovy in the Yellow Sea mainly consists of *Acanthomysis longirostris* (65.9% by weight). Deng et al. (1988) reported that the dominant prey of the half-fin anchovy in the Bohai Sea is *Acetes chinensis* (65.0% by weight). This indicates that the half-fin anchovy diet is not strict and is dependent on food availability.

The stomach contents of spotted sardine in the Yellow Sea were only examined qualitatively and mainly consisted of plankton such as *Coscinodisus*, *Chaetoceros* sp., *Rhizosolemia*, *Oithon* sp., *Harpacticus*, and *Acartia bifilosa* (Wei and Jiang 1992). The major feeding season is in summer and autumn. Data for the spotted sardine in the Bohai Sea are not available.

Scaled sardine mainly feed on zooplankton; *Calanus sinicus* is the primary component. There are some differences in diet between the stocks of the Yellow and Bohai Seas. In addition to *Calanus sinicus*, the Bohai Sea stock feeds on *Acetes chinensis*, Mysidae, and *Sagitta crassa*. The Yellow Sea stock feeds on *Megalopa*, *Euphausia pacifica*, and *Metapenaeopsis dalei* (Deng et al. 1988; Wei and Jiang 1992).

The diet of the Yellow Sea herring is relatively uniform. *Euphausia pacifica* constitutes more than 99% of the total stomach contents by weight (Tang 1980; Wei and Jiang 1992). The feeding period is mainly from April to August; during the rest of the year they feed little or not at all.

Many studies on the stomach contents of chub mackerel show that the dominant prey consists of Japanese anchovy, *Euphausia pacifica*, and Loligo (Zhu 1959; Lin and Yang 1980; Wei and Jiang 1992). The proportion of Japanese anchovy in the diet seems to have increased in recent years.

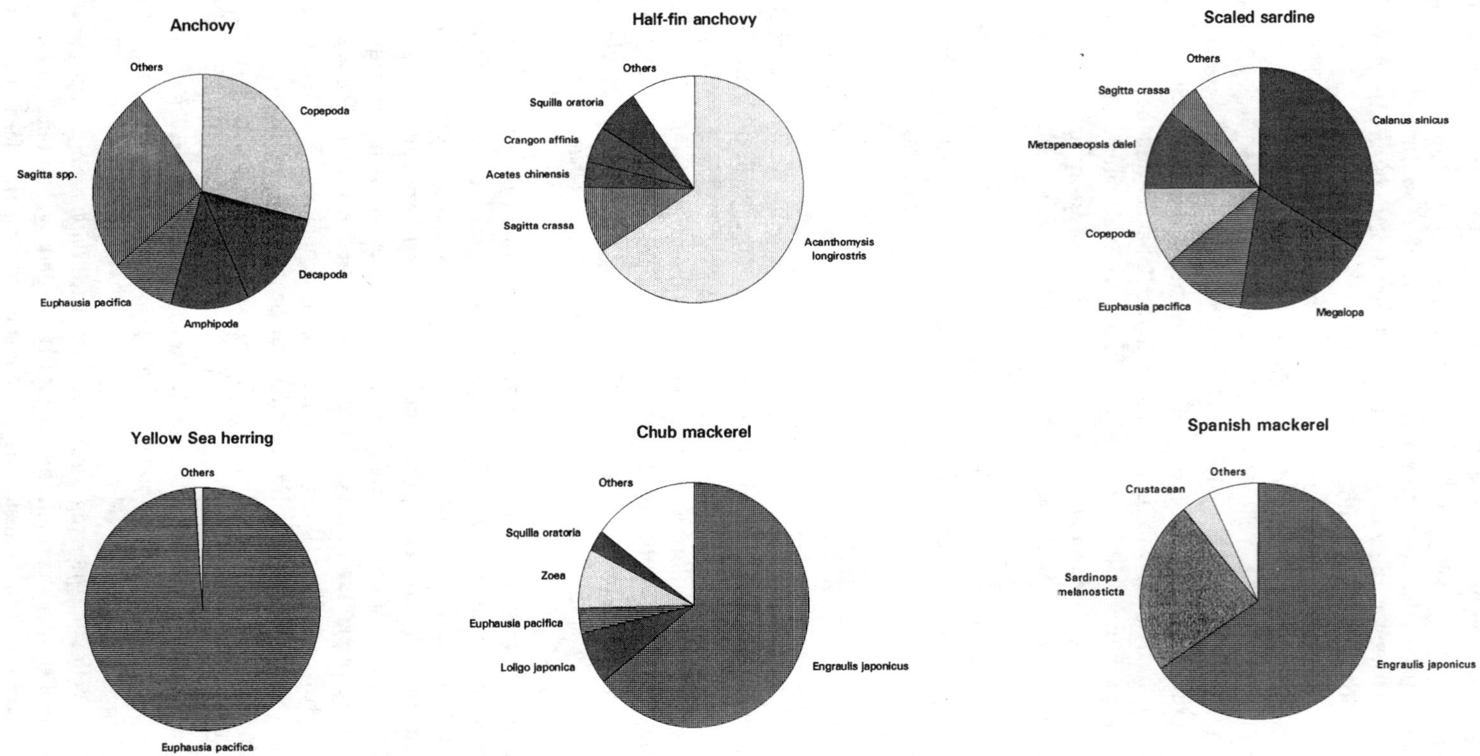

Figure 7-2. Food composition of six pelagic fishes in the Yellow Sea during the surveys in 1985–1986.

Spanish mackerel is believed to have strong piscivorous habits. Small pelagic fish such as Japanese anchovy and sardine account for about 90% of the stomach contents by weight (Tang and Ye 1990). The Japanese anchovy makes up the larger portion and has been named "food of Spanish mackerel" by Chinese fishermen. The size of the prey increases with the growth of the Spanish mackerel, but the prey composition does not vary significantly.

In the Yellow and Bohai Seas, the small pelagic fishes such as anchovies, sardines, and Pacific herring are at a low trophic level and feed mainly on zooplankton (Figure 7-2). There may be competition for food and space among them. Chen et al. (1991) reported that in 1973 to 1974 the biomass of *Euphausia pacifica* increased by 160% in the central Yellow Sea when the herring stock was reduced by half (see the section on Stock Size and Yield Fluctuations). During the surveys from 1981 onward, the biomass of *Euphausia pacifica* decreased sharply to a level even less than that in 1973. This may be due to consumption by chub mackerel, other small pelagic fishes, and demersal fishes. All small pelagic fishes are also preyed upon by the large pelagic fishes (i.e., Spanish mackerel and chub mackerel) and many demersal piscivorous fishes. The two relatively large pelagic fish, Spanish mackerel and chub mackerel, are taken by fishermen.

Reproduction

In general, spawning areas of many species in the Yellow and Bohai Seas are found in the bays of the Bohai Sea and in the coastal waters of Shandong Peninsula. The spawning season is primarily in spring, mainly from May to June (Zhang et al. 1983; Zhao et al. 1990; Liu et al. 1990; Deng and Zhao 1991), although different species may vary slightly. The development and distribution of eggs and larvae of fishes in the Yellow and Bohai Seas have been described based on many experiments and investigations (Zhang et al. 1983). Data on various reproductive characteristics and environmental factors of seven species are given in Table 7-2.

The nearshore banks and bays along the Chinese coast are spawning areas. There are three major spawning areas of the Japanese anchovy: bays in the Bohai Sea, nearshore banks off the south of Shandong Peninsula, and off the Zhejiang coast (southeast China) (Zhang et al. 1983; Zhu and Iversen 1990).

According to the surveys from 1984 to 1988 by the R/V *Bei Dou* (Zhu and Iversen 1990), the spawning season of Japanese anchovy is from May to October, and the peak spawning period is from mid-May to late June, with the optimum temperature being 14–18°C in the Yellow Sea. The 1- to 2-year-old fish dominate the spawning stock. Ovulation can be of multipeak or continual type, and occurs once a year; that is, ovulation occurs as the egg is ripe, without an obvious resting stage in between (Li 1987).

Based on investigations of half-fin anchovy, their spawning areas are mainly in the estuaries and bays of the Yellow and Bohai Seas. Turbid waters with temperatures of 9–13°C in the Yellow and Bohai Seas (Liu et al. 1990) and 15–26°C in the East China Sea (Zhao et al. 1990) are the most favorable.

The spawning period of spotted sardine differs between stocks; it is from July to August in the South and East China Seas, and from April to June in the Yellow and Bohai Seas. Batch spawning is conducted in the shallow coastal bays and estuaries with rather low salinity (18–25 ppt) (Zhang et al. 1983; Zhao et al. 1990). Male fish reach maturity earlier than females. The 2-year-old fish have an absolute fecundity at a mean of 70,000 to 90,000 eggs (Zhao et al. 1990).

The scaled sardine starts spawning in May in the Yellow and Bohai Seas, and one month earlier in the East China Sea, at a relatively high water temperature (19–21°C). The absolute fecundity is low, ranging from 3,500 to 5,500 eggs that have a very small diameter (0.3–0.9 mm) (Zhao et al. 1990; Zhang et al. 1983).

Table 7-2. Data on reproductive biology and environmental factors for seven pelagic fishes of the Yellow Sea

	Species						
	Japanese anchovy	Half-fin anchovy	Spotted sardine	Scaled sardine	Pacific herring	Chub mackerel	Spanish mackerel
Egg diameter (mm)	0.55–1.57	1.40–1.51	1.28–1.6	0.3–0.9	1.42–1.54	0.93–1.15	1.52–1.63
Length at hatching (mm)	3.2	?	4.0–4.4	3.3–3.8	5.24–6.28	2.70	4.27–4.96
Absolute fecundity (× 1,000 eggs/female)	0.6–13.6	0.13–1.48	35–125	3.5–5.5	19.3–78.1	234–861	280–1,100
Age at first maturity (years)	1	1–3	1–2	1–2	1–3	1–2	1–2
Length at first maturity (cm)	6	9.5	13.5	11	16.5–27	26	35,142
Ovulation pattern (time/year)	1	1	batch	l	1	batch	batch
Maximum recorded age (years)	4	6	5	5	9	10	6
Peak spawning season	May–June	Apr.–June	May–June	May–July	Mar.–April	May–July	May–June
Temperature of spawning (°C)	12–22	9–26	14.5–18.5	19–21	0–5	12–25	9–20
Salinity of spawning (%)	28–31	14–34	18–25	?	30±	29–34.5	28–31
Depth of spawning (m)	>25	20±	1–15	20±	3–7	15–50	15–30

The Yellow Sea herring starts spawning in February in very shallow waters (3–7 m) along the coast and bays around the Shandong Peninsula. A small group spawns on the banks off the Liaodong Peninsula and off west Korea. The main spawning season is from March to April, and a water temperature of 0–5°C and a salinity of around 30 ppt are required (Zhang et al. 1983). Tang (1980) found that the egg development of Yellow Sea herring exhibits a definite synchronism, and an individual's absolute fecundity increases linearly with increasing net body weight. The herring eggs stick together and adhere to reefs, algae, and other substances. The hatching time decreases with increased temperature from about 12–14 days at 5.5–10°C to 7–8 days at 15–20°C (Jiang and Chen 1981; Zhang et al. 1983).

The chub mackerel stock in the Yellow Sea spawn around the Shandong Peninsula from May to July. A small portion sometimes enters the Bohai Sea for spawning. The temperature of the spawning areas is relatively high, with an optimum of 12–17°C, and a high salinity is needed (Zhang et al. 1983; Wang 1991). The fecundity of chub mackerel is quite high (Table 7-2) and also increases linearly with net body weight (Liu et al. 1990).

The Yellow Sea stock of Spanish mackerel spawns mostly in the bays of the Bohai Sea, although some spawn along the coast of the Shandong Peninsula. The major spawning season is from May to June, depending on the water temperature. The temperature of spawning areas differs greatly between northern and southern waters, ranging from 9–13°C in the Bohai Sea and 11–20°C in the southern East China Sea, whereas the salinity is similar at 28–31 ppt (Zhang et al. 1983; Wei 1991).

Data on the seven species indicate that all start spawning from an age of about 1 year and have a relatively short life span (Table 7-2). Maturation at an earlier age has been observed in some species. Fecundity in small pelagic species is lower than in larger species. The eggs are released pelagically, except for Yellow Sea herring.

Age and Growth

The development of larvae and postlarvae fishes in China's coastal waters has been summarized based on many observations by Zhang et al. (1983). The von Bertalanffy growth equations for seven pelagic species are given in Table 7-3. The growth equations for half-fin anchovy, spotted sardine, and scaled sardine are based on the survey data from 1982 to 1986.

The length of the Japanese anchovy at hatching ranges from 2.6 to 2.9 mm (Zhao et al. 1990). Zhang et al. (1983) reported that the body length is 3.2 mm after hatching, and at a body length of 18.5 mm (length at metamorphosis) it looks like an adult fish. Five months after spawning the total length reaches 60–90 mm. The growth rate is fast at 1-group and decreases with age (Zhu

Table 7-3. Growth equations of seven pelagic fishes in the Yellow and Bohai Seas

Species	Growth equation	Reference
Japanese anchovy	$L_t = 163\ (1 - e^{-0.8(t+0.2)})$	Zhu and Iversen 1990
Half-fin anchovy	$L_t = 192\ (1 - e^{-0.55(t+1.55)})$	
Spotted sardine	$L_t = 200\ (1 - e^{-0.81(t+0.36)})$	
Scaled sardine	$L_t = 167\ (1 - e^{-0.49(t+1.31)})$	
Yellow Sea herring	$L_t = 305\ (1 - e^{-0.66(t+0.198)})$	Tang 1980
Chub mackerel	$L_t = 425\ (1 - e^{-0.53(t+0.8)})$	Chen et al. 1991
Spanish mackerel	$L_t = 709\ (1 - e^{-0.53(t+0.70)})$	Wei 1991

The equations for half-fin anchovy, spotted sardine, and scaled sardine were fitted based on data from 1982–1986.

and Iversen 1990). At 3 months the fork length of half-fin anchovy is about 30 mm, and at 6 months it is around 75 mm. One-year-old fish reach about 100 mm.

Spotted sardine are 2.9–4.4 mm after hatching, 4.0–4.2 mm in 2 days, and about 5.8 mm in 4 days. In 15 days the spotted sardine is more than 10 mm long, and by October it has grown to 120–130 mm.

The scaled sardine is 3.3–3.8 mm in length after hatching, and reaches 30–45 mm at 1.5 months and weighs 1–2 grams. By mid-October, the 0-group fish increase to 75–98 mm, weighing 4.0–8.5 grams.

According to observations on artificial hatching and embryonic development of Yellow Sea herring (Jiang and Chen 1981), the body lengths of newly hatched larvae are 5.2–6.8 mm. After 4 to 5 days it reaches 7.2–7.8 mm, and in 12 to 13 days reaches 9.9–11.2 mm. The growth rate is fastest during summer, accounting for 43% of total yearly increase, and decreases during autumn (Tang 1991).

The length of newly hatched larvae of chub mackerel is 2.7–3.0 mm, and 3.8–4.0 mm after 4 days (Wang 1991). The 0-group fish grow very quickly: 120–140 mm fish were caught in August. By October the chub mackerel is 200 mm long and weighs about 100 grams.

At hatching the Spanish mackerel measures 4.3–5.0 mm (Zhao et al. 1990). Its growth is the fastest recorded among pelagic fishes in the Yellow and Bohai Seas. 0-group fish can reach 250–300 mm and weigh 200–400 grams at spawning. Female fish grow faster than males, with an average increase in length of 26.1% in the second year after spawning, compared with 21.9% for males (Wei 1991).

There are large differences in size between the seven species (Tables 7-2 and 7-3). The Spanish mackerel is the largest species and the Japanese anchovy the smallest. Growth rates during the first year vary significantly, with smaller species growing more slowly. In addition, the lengths at hatching are not consistent with adult fish size or egg diameter (Table 7-2).

Population Structure

During the 1986 surveys, the population consisted mainly of 0- to 5-year-old fish, with 1- to 2 year-old fish being the major component. Figure 7-3 shows the fork length composition of seven pelagic fishes in the Yellow Sea during the spring and autumn surveys.

The length distribution of the Japanese anchovy shows a unimodal distribution at the range of 6–16 cm, with a mean of 11.9 cm. The 7- to 13-cm length group dominates (92.4% by number) the catch in spring (Figure 7-3). This indicates that the spawning stock depends highly on 1- to 2-year-old fish. The length distribution in autumn was rather scattered, without an obvious dominant length group. The fork length ranged from 6 to 15 cm, with a mean of 10.7 cm.

The length distribution of half-fin anchovy is also unimodal (Figure 7-3), especially in spring. The 14- to 17-cm length group dominated (80.9% by number), with a mean of 15.2 cm. It is similar to the spring length distribution of the Japanese anchovy. The spawning stock mainly consisted of 1- to 2-year-old fish. In the spring 1981 survey along the Jiangsu coast, the composition of the spawning population was 1- to 6-year-old fish, with the 3-year-old fish dominating (Zhao et al. 1990). With the juveniles recruited into the feeding stock, the proportion of the 14- to 17-cm length group was reduced to 76.6%.

The length composition of the spotted sardine consisted of 0- to 5-year-old fish, and 1- to 2-year-old fish were the major year-classes (Figure 7-3). The fork length distribution was 11–21 cm, with a mean of 14.9 cm in spring and 17.7 cm in autumn. This was different from the other pelagic fish, since juvenile fish were not caught during the autumn survey. It is likely that the juvenile fish were in spawning areas in nearshore shallow waters (bays and estuaries) and thus

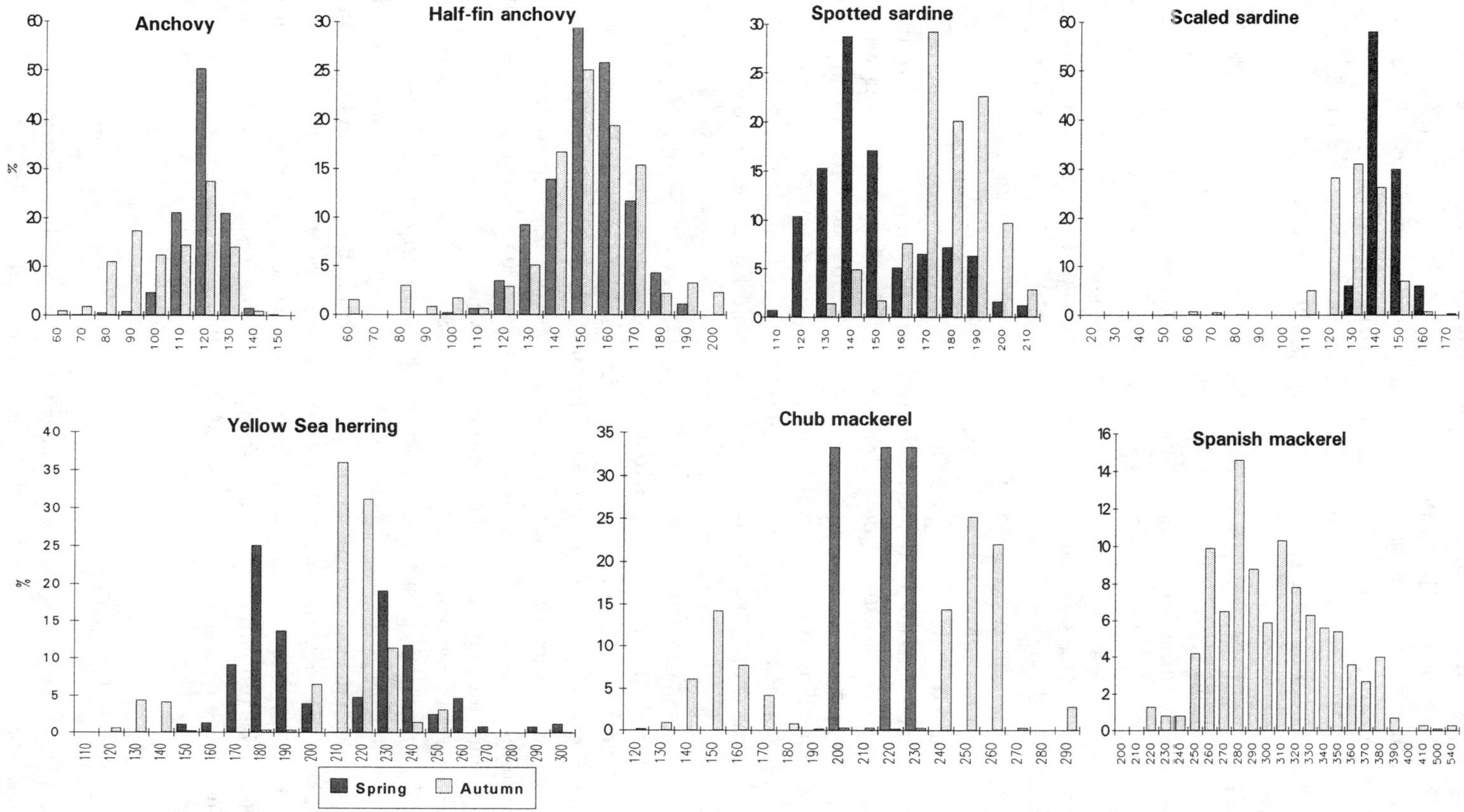

Figure 7-3. Fork length (mm) distribution of seven pelagic fishes in the Yellow Sea during the surveys in 1986.

were not caught. The dominant length group of the spring catch was 13 to 15 cm, accounting for 61.0%, and 17 to 19 cm in autumn, accounting for 71.9%.

The length composition of the scaled sardine in the spring catch was very narrow, ranging from 13 to 16 cm, with a mean of 14.4 cm. The 14- to 15-cm fish accounted for 88.0% in number. In autumn the lengths ranged from 2 to 17 cm, with a mean of 13 cm, and the dominant length group (12–14 cm) accounted for 85.6%. The 0-group fish accounted for 1.4% of the autumn catch (Figure 7-3).

The fork length of the spawning stock of the Yellow Sea herring ranged from 13 to 30 cm, with a mean of 20.8 cm. There were two peaks in the length distribution, one at 18–19 cm and one at 23–24 cm, accounting for 38.8% and 30.9% and corresponding to 1- and 2-year-old fish, respectively (Figure 7-3). However, when the stock was very abundant during the period from 1970 to 1974, the mean age of the catch was 2.5 years old and the fork length and body weight averaged 22.8 cm and 134 grams, respectively. Four-year and older fish accounted for 15.6%, and decreased to 1.5% in the period from 1975 to 1982 (Liu et al. 1990). This demonstrates that the spawning stock at present depends mainly on the 2-year age group. In autumn, the length composition of herring was dominated by the 21- to 22-cm group, accounting for 67.2% by number, and 9.3% of the catch was 0-group fish.

Chub mackerel has a relatively long life span, with a maximum recorded age of 10 years. The population structure was very simple and distinct (Figure 7-3). The data for spring are not reliable because only three specimens were caught. However, the feeding stock in autumn of 1986 shows two peaks, corresponding to 0-group and 1-group fish. The composition of the spawning stock has varied since the 1950s (Liu et al. 1990): The 4- to 5-year-old fish dominated (92.8%) in 1953, and in 1963 decreased to 25%. Meanwhile, the 2-year-old fish increased to 25%, and further increased to 37.8% in 1973. The majority of spawners in 1982 were 3 years old, accounting for 59.1%. The 5 years and older spawners have been very rare since 1968. At present, the spawning stock is highly dependent on 1- and 2-year-old fish.

Spanish mackerel is the largest fishable pelagic fish in the Yellow and Bohai Seas. It was not caught during the spring bottom trawl surveys of 1986. As described previously, the Spanish mackerel's occurrence in the Yellow Sea is highly dependent on the water temperature. The length composition in the autumn catch (Figure 7-3) ranged from 20 to 54 cm in fork length without an obviously dominant length group. Zero-group fish accounted for 99.7% of the catch.

Figure 7-4 shows the variation of mean fork length of Spanish mackerel in the Chinese catches since 1950. The mean lengths have decreased since the mid-1960s, indicating an increasing proportion of young fish. It should be mentioned that the 1-group fish accounted for only 1–7% during the period from 1952 to 1969, increased to 10–27% in the 1970s, and again increased to about 36–56% in the beginning of the 1980s (Wei 1991). The increase in mean length in the mid-1980s was mainly due to a change in fishing method, namely, fast trawling with a larger mesh size. Therefore, the proportion of older fish increased but resulted in the reduction of the spawning stock. As a consequence, the reproductive capacity of the stock may be reduced to a lower level.

Stock Size and Yield Fluctuations

Biomass estimates of stock size of pelagic fishes were not available before 1985. However, the relative abundance index of the Yellow Sea herring was estimated for 1972–1977 (Ye et al. 1980). The first systematic surveys of pelagic fishes, especially Japanese anchovy, were carried out in 1985 (Zhu and Iversen 1990). Estimates based on bottom trawl surveys in 1985 and 1986 are also available. The catch statistics of Chinese fisheries indicate a rapid decline of most bottom fish and

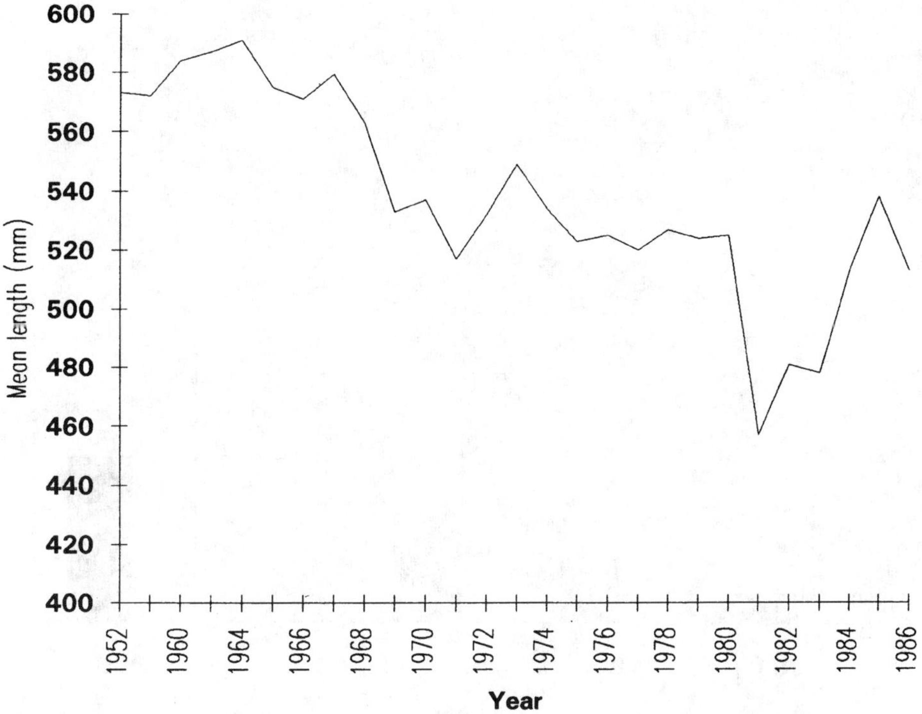

Figure 7-4. Mean length variations of Spanish mackerel in the Chinese catch with year.

some large pelagic fish, and an increase of some small pelagic fish. Figure 7-5 presents the relative proportions of the seven pelagic species in Chinese catches since 1953.

The Japanese anchovy is an undesired species in Chinese fisheries because of its low market value. Only postlarval and juvenile fish have been used by some coastal artisanal fishermen for dried food. In the 1960s, the catch was very small, about 750 to 1,000 mt in the northern Yellow Sea. In recent years, a dramatic increase of abundance occurred in the Yellow and East China Seas—about 3 million mt has been estimated by acoustic surveys, and a potential annual catch

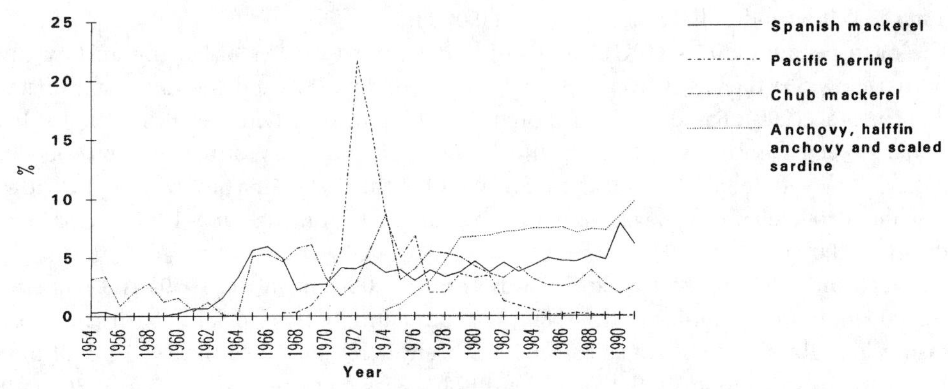

Figure 7-5. Relative proportion variation of major pelagic fishes in the Chinese catch (1953–1992).

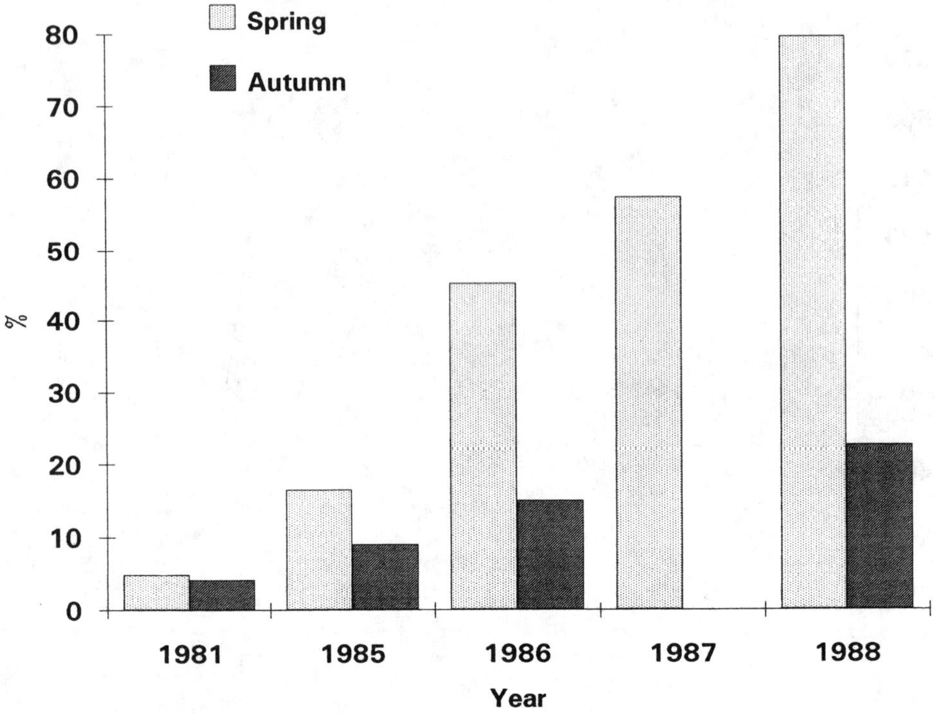

Figure 7-6. Percentage of anchovy in the catches by weight during the surveys of 1981–1988 in the Yellow Sea (no survey made in autumn 1987).

was on the order of half a million mt (Iversen et al. 1993). The stock size has continued to increase in recent years to more than 4 million mt in 1992–1993. Figure 7-6 shows the percentage of Japanese anchovy in the catches during the bottom trawl surveys of 1981 to 1988. Its proportion in the catches increased dramatically, especially in the spring, from 4.8% in 1981 to 79.5% in 1988. The annual catch in China has varied between 30,000 and 150,000 mt during 1989 to 1992.

The half-fin anchovy, spotted sardine, and scaled sardine stocks are also abundant, yet the catches have been relatively small. Their biomass has been increasing in recent years. The total catch in the Yellow and Bohai Seas was about 60,000 mt.

The catch per unit effort (CPUE) index of bottom trawl catches of half-fin anchovy in the southern Yellow Sea increased from 100 in 1971 to 367 in 1982, and the total catch increased sixfold (Liu et al. 1990). Based on the inshore trial fishing in 1981 (Zhao et al. 1990), the density in the Jiangsu coastal water was 1.011 mt/nm^2. The abundance was estimated at 60,000–70,000 mt in the Yellow and Bohai Seas, and about 86,000 mt in the East China Sea, with a total allowable catch of about 100,000 mt. However, the Chinese annual catch during 1980–1983 ranged between 70,000 and 90,000 mt.

Although the catch of spotted sardine was around 1,000 mt in the 1950s and increased to about 5,000 mt in the beginning of the 1980s, the abundance of this stock has been stable or even increasing. The abundance of scaled sardine has been increasing since the 1970s in all the seas along China. The catch index in the Yellow and Bohai Seas increased from 100 in 1976 to 401 in 1981 (Ye 1982), and the catch has been about 22,000 mt in recent years. These species have potential to be further exploited by about 80,000 to 105,000 mt in the Yellow and Bohai Seas.

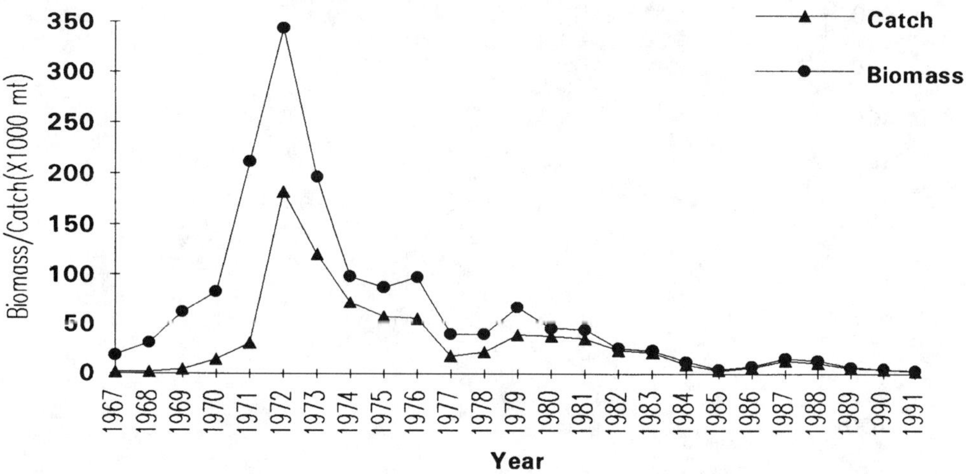

Figure 7-7. Biomass (from virtual population analysis) in Chinese catch variations of Yellow Sea herring in the Yellow Sea (1968–1991).

The Yellow Sea herring fishery has experienced two peaks (about 1900 and 1938) followed by a period of little or no catch (Tang 1990). Since 1967, many demersal stocks in the Yellow Sea have been overfished or depleted. Recovery and outburst of the Yellow Sea herring seem to be associated with depletion of other stocks. The catch of Yellow Sea herring increased rapidly to a peak (more than 180,000 mt) in 1972, and decreased sharply thereafter (Figure 7-7). Tang (1981, 1987) demonstrated that although the fluctuations in recruitment of the Yellow Sea herring have been very large and have had a direct effect on the fishable stock, there is no strong relationship between spawning stock and recruitment. Environmental conditions such as rainfall, wind, and daylight are the major factors affecting the fluctuations in recruitment, and long-term changes in biomass may be correlated with a 36-year cycle of wetness oscillation in eastern China (Tang 1990). Undoubtedly, high fishing pressure speeds up the depletion. During a period of declining stock size, the spawning stock must be conserved to ensure that the reproductive potential is maintained at a satisfactory level. The abundance of adult fish is now reduced to about 1,000 to 3,000 mt (Liu et al. 1990), and the catch has ranged from 1,000 to 2,000 mt in recent years in China.

The Yellow Sea chub mackerel, which had similar fluctuations, experienced two peaks during the periods of 1951–1955 and 1973–1980 (Wang 1991). Ye (1982) noted that the fluctuations in biomass seem to follow an 18-year cycle. The stock in the Yellow Sea is now in a declining period. The spring catch declined from 45,000 mt in 1974 to less than 1,000 mt in 1984. Wang (1991) believes that the Yellow Sea chub mackerel is only a small branch of the northwestern Pacific population, and that because the areas of distribution and spawning are highly restricted it is easily affected by unfavorable environmental factors and fishing. The fluctuations are likely to follow a 22- to 24-year cycle of sunspots (Wang 1991). Fishing is also an important factor in the fluctuation of fisheries resources. There is high fishing pressure on the immature chub mackerel in the wintering grounds, leading to a reduction of recruitment and thus affecting the fishery in the following years. Figure 7-8 shows that in China the CPUE by purse seining in spring decreased sharply in the 1950s and has been at a low level since then.

Spanish mackerel is mainly caught by gill netting in China and is a bycatch in other fisheries. The annual catch increased rapidly in the 1960s and ranged from 20,000 to 40,000 mt (Figure 7-9). The 2-year-old fish dominated the catch (about 75%), and the total mortality was about 65% (Wei

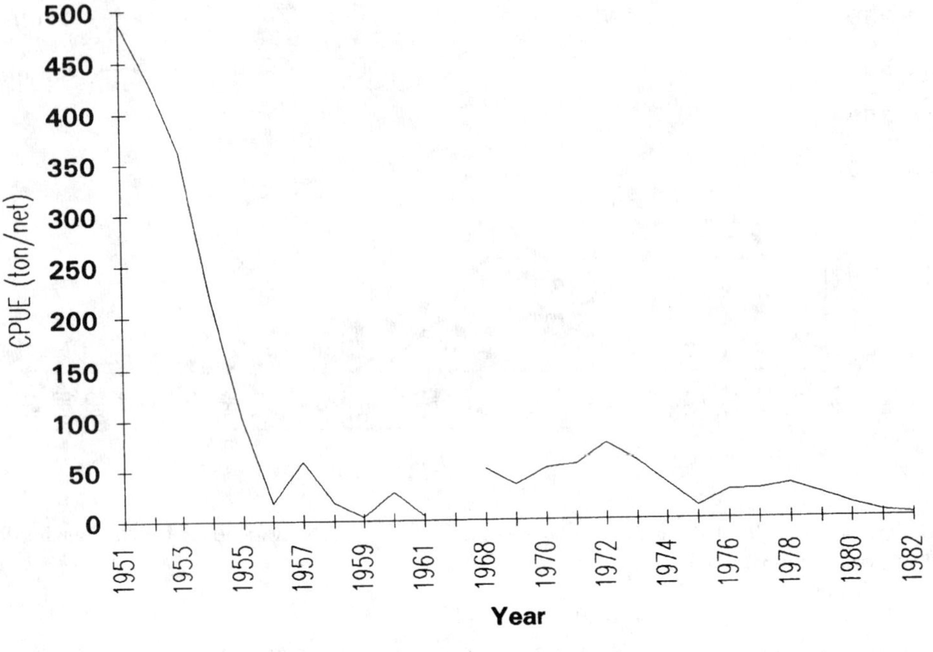

Figure 7-8. CPUE variations of chub mackerel by the Chinese purse seine fishery in spring, 1951–1982.

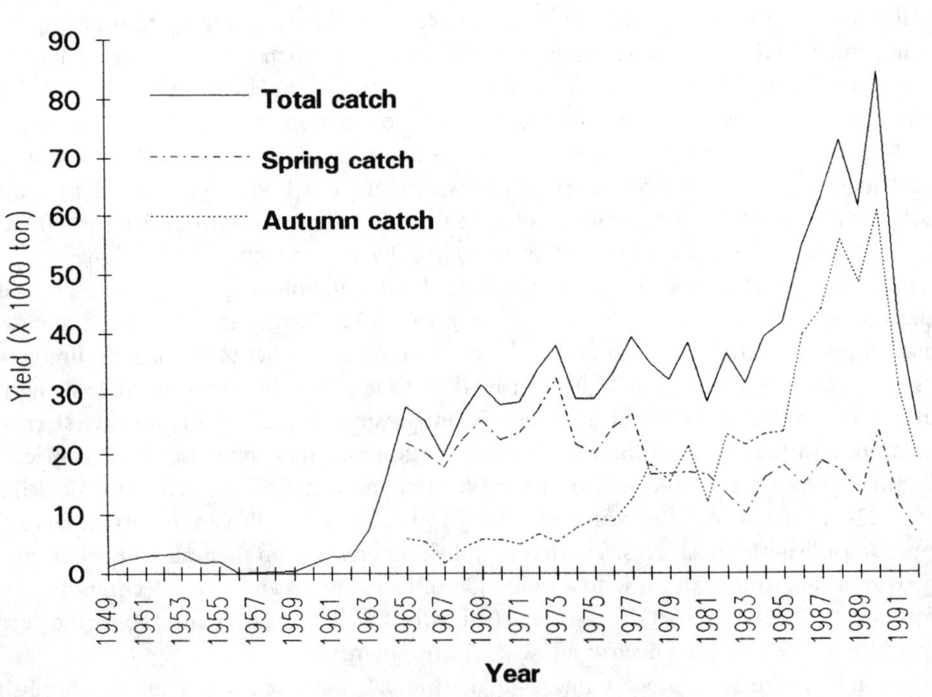

Figure 7-9. Catch variations of Spanish mackerel in China during the period 1949–1992.

1991). By the mid-1970s the catch was mainly taken from the spawning stock in spring. Since the late 1970s, the increase in fishing effort and the large quantity of 0-group fish caught by motorized bottom trawlers in autumn have resulted in a decline of the spring catch and an increase in the summer and autumn catch (Figure 7-9). Therefore, the proportion of older fish decreased and that of younger fish increased in the catches. Thus, the fishery was mainly supported by the younger fish. However, as mentioned previously, the abundance of old spawners has decreased sharply due to the introduction of fast trawling with a larger mesh size. As a result, the catch has declined considerably in recent years.

In the 1950s and 1960s, bottom fish were the major target species in China's fisheries, and their catches increased rapidly. Figure 7-10 shows the catch composition during two bottom trawl surveys. In 1959, small yellow croaker, *Pseudosciaena polvactis*, dominated the spring and autumn catches (44.1% and 20.9% by weight, respectively), whereas the pelagic fish accounted for only 0.1% and 2%, respectively. However, in 1986 pelagic fish dominated the catches (50.7% and 46.3% for spring and autumn, respectively). The stock of Japanese anchovy has increased strikingly and may, together with other small pelagic fishes, take the place of and utilize the surplus food left by Yellow Sea herring and other depleted stocks in the Yellow Sea.

The fishery resources of the Yellow Sea have changed considerably. Chikuni (1985) pointed out that changes in stock abundance in each of the major pelagic resources around Japan have not taken place independently but have been associated with complementary changes in other stocks. Changes in species composition and fluctuations in abundance result from complicated interactions of biological and environmental factors. MacCall (1983) examined the variability of pelagic fish stocks off California and concluded that (1) recruitment fluctuations are particularly important for short-lived fishes, (2) species replacement seems to occur, but is environmentally mediated, and (3) changes in the natural mortality of adults may be an important cause of fluctuations in abundance. Daan (1980) reviewed fisheries data that indicated marked changes in species composition around the world and pointed out that very few reported situations can be conclusively shown to exemplify a replacement phenomenon.

In the Yellow Sea, when the herring stock recovered, the biomass of many demersal fish such as the small yellow croaker, large yellow croaker (*Pseudosciaena crocea*), hairtail (*Trichiurus haumela*), and Pacific cod (*Gadus macrocephalus*) were depleted. The outburst of Yellow Sea herring lasted several years, then collapsed in the late 1970s. The biomass of small pelagic fishes has increased gradually since then. A decline in Spanish mackerel biomass has occurred in the catch of recent years due to the reduced spawning stock. The stock of Japanese anchovy has increased sharply in recent years. It is not possible to simply conclude that herring is replaced by anchovy, similar to the sardine–anchovy interaction off California (Daan 1980), since the long-term change in herring biomass in the Yellow Sea is cyclic (Tang 1990) and there is no strong food competition between the two species, although the distribution of anchovy has covered the distributional area of herring in the Yellow Sea. The depletion of demersal piscivorous fish stocks might be the most important cause for the increase of small plankton feeders.

Conclusion

In the Yellow and Bohai Seas the importance of pelagic fishes has been increasing since the 1950s. The abundance and catches of small-sized pelagic species such as Japanese anchovy, half-fin anchovy, spotted sardine, and scaled sardine have increased dramatically, whereas the relatively larger-sized species such as Spanish mackerel, chub mackerel, and Pacific herring have been depleted or overexploited in recent years.

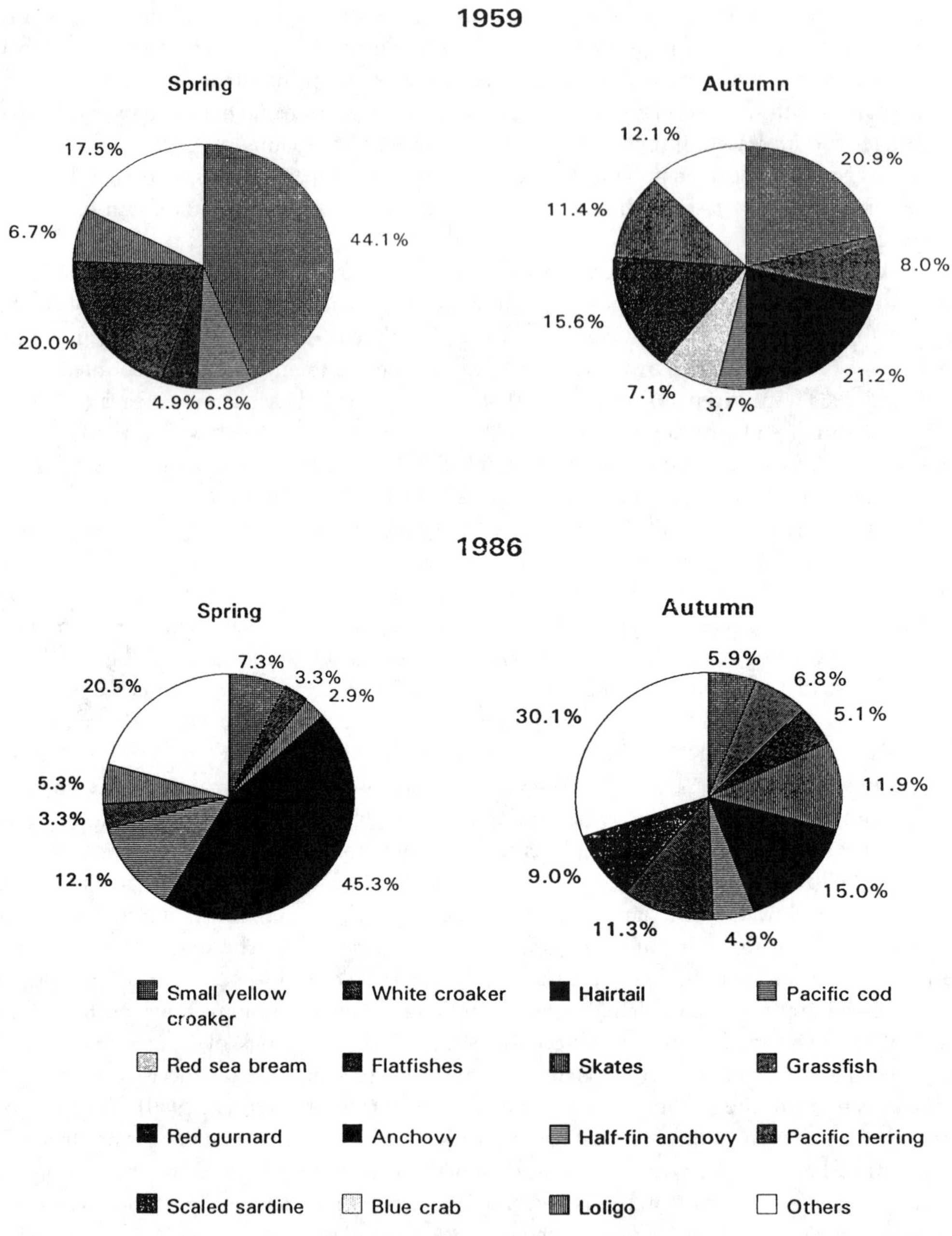

Figure 7-10. Catch composition during bottom trawl surveys in 1959 and 1986.

Compared with their abundance, the catch of small pelagic species is relatively small. This is in contrast with the decline of many species that are larger in size and have a higher commercial value. The Japanese anchovy, in particular, is believed to be the most abundant species in the Yellow Sea, with a potential catch of about half a million tonnes per year. However, the annual catch in China has been quite low in recent years. The CPUE of the half-fin anchovy and scaled sardine has also increased three- to fourfold since the 1970s, but the total catch is still low. These small pelagic species are believed to be underexploited and have great potential in the Yellow and Bohai Seas fisheries.

Small pelagic fishes feed mainly on zooplankton, and are in turn major prey for most of the larger demersal and pelagic fish. It should be noted that the Japanese anchovy plays a very important role as prey, and is eaten by most piscivorous fishes in the Yellow and Bohai Seas. These predators have been mostly depleted or overfished since the early 1960s and the predation pressure has therefore been reduced. The depletion of the Pacific herring stock has provided surplus food and space for small pelagic fishes. Environmental conditions may also be favorable for the small pelagic fishes. These events appear to be related and have caused a succession in the Yellow Sea ecosystem. Further investigations on multispecies interactions and the entire ecosystem are needed.

ACKNOWLEDGMENTS

We are grateful to Dr. A. Johannessen, University of Bergen, and Dr. E. Bakken, Institute of Marine Research, Norway, for their valuable comments on the manuscript. Thanks are also given to our colleagues for helping to collect and analyze the survey data.

References

Chen, J. 1978. Study on feeding behaviour of Japanese anchovy in the northern Yellow Sea. Reports of Liaoning Marine Fisheries Research Institute 43:1–11.

Chen, G. (ed.) 1991. Marine Fishery Environment in China. Zhejiang, China: Zhejiang Science and Technology Press.

Chikuni, S. 1985. The Fish Resources of the Northwest Pacific. FAO Fish. Tech. Pap. 266.

Daan, N. 1980. A Review of Replacement of Depleted Stocks by Other Species and the Mechanisms Underlying Such Replacement. In A. Saville (ed.), The Assessment and Management of Pelagic Fish Stocks. Rapp. P.-v. Réun. Cons. Int. Explor. Mer 177, pp. 405–421.

Deng, J. 1988. Ecological bases of marine ranching and management in the Bohai Sea. Marine Fisheries Research 9:1–10.

Deng, J., and Zhao, Q. (eds.) 1991. Marine Fisheries Biology. China: Agriculture Press.

Deng, J., Meng, T., Ren, S., Qiu, X., and Zhu, J. 1988. Species composition, abundance and distribution of fishes in the Bohai Sea. Marine Fisheries Research 9:11–89.

Iversen, S. A., Zhu, D., Johannessen, A., and Toresen, R. 1993. Stock size, distribution and biology of anchovy in the Yellow Sea and East China Sea. Fish. Res. 16:147–163.

Jiang, Y., and Chen, J. 1981. Preliminary observations on the artificial hatching and embryonic development of Huanghai herring. Acta Oceanologica Sinica 3(3):477–486.

Li, F. 1987. Study on the behaviour of reproduction of the anchovy (*Engraulis japonicus*) in the middle and southern part of the Yellow Sea. Marine Fisheries Research 8:41–50.

Lin, F. 1965. Seasonal changes of bottom fish distribution and catch composition in the Yellow Sea. Marine Fisheries Research 9:56–86.

Lin, J., and Yang, J. 1980. Feeding behaviour of juvenile chub mackerel in the waters off Yantai, Weihai and Qingdao. Marine Fisheries Research 1:1–15.

Liu, X. (ed.). 1990. Investigation and Division of the Yellow Sea and Bohai Sea Fishery Resources. Beijing, China: Ocean Press.

MacCall, A. D. 1983. Variability of pelagic fish stocks off California. FAO Fish. Rep. (291) 2:101–112.

Tang, Q. 1980. Studies on the maturation, fecundity and growth characteristics of Yellow Sea herring, *Clupea harengus pallasi* (Valenciennes). Marine Fisheries Research 1:59–76.

Tang, Q. 1981. A preliminary study on the causes of fluctuations in year class size of Pacific herring in the Yellow Sea. Transac. Oceanol. Limnol. 2:37–45.

Tang, Q. 1987. Estimation of fishing mortality and abundance of Pacific herring in the Yellow Sea by cohort analysis (VPA). Acta Oceanol. Sinica 6(1):132–141.

Tang, Q. 1990. The effects of long-term physical and biological perturbations of the contemporary biomass yields of the Yellow Sea ecosystem. Paper for the International Conference on the Large Marine Ecosystem (LME) Concept and Its Application to Regional Marine Resource Management 1–6 October 1990, Monaco.

Tang, Q. 1991. Yellow Sea Herring. In J. Deng and Q. Zhao (eds.), Marine Fisheries Biology. China: Agriculture Press, pp. 296–356.

Tang, Q. and Ye, M. (eds.). 1990. Exploitation and Conservation of Fishery Resources along Shandong Coast. China: Agriculture Press.

Wang, W. 1991. Chub Mackerel. In J. Deng and Q. Zhao (eds.), Marine Fisheries Biology. China: Agriculture Press, pp. 413–452

Wang, W., and Zhu, D. 1983. Studies on the fisheries biology of mackerel (*Pneumatophorus japonicus* Houttuyn) in the Yellow Sea. II. Studies on the relationship between the distributional behaviour of mackerel and the environment in the Yellow Sea and Bohai Sea. Marine Fisheries Research 6:59–77.

Wang, W., Lu, Z., Yan, Y., and Chen, C. 1983. The biological characteristics of chub mackerel of the fishing ground off the middle and the south of Fujian during the spring fishing season. Marine Fishery 5(2):51–54.

Wei, S. 1991. Spanish Mackerel. In J. Deng and Q. Zhao (eds.), Marine Fisheries Biology. China: Agriculture Press, pp. 357–412.

Wei, S., and Jiang, W. 1992. Study on food web of fishes in the Yellow Sea. Oceanologica et Limnologia Sinica 23(2):182–192.

Ye, M. 1982. Prospective analyses of Yellow Sea chub mackerel resource. Mar. Fish. Sci. and Tech. 1:8–13.

Ye, C., Tang, Q., and Qin, Y. 1980. The Huanghai herring and their fisheries. J. Fisheries of China 4(4):339–352.

Zhang, Y. (ed.). 1983. Inshore Fish Egg and Larvae off China. Shanghai, China: Shanghai Science and Technology Press.

Zhao, Q. (ed.). 1990. Marine Fishery Resources of China. Zhejiang, China: Zhejiang Science and Technology Press.

Zhu, Y. 1959. Reference Materials of Major Marine Fisheries Biology. Proceedings of the Second Workshop of the West Pacific Fisheries Research Committee. Beijing, China: Science Press, pp. 106–143.

Zhu, Y. (ed.). 1963. Ichthyology of the East China Sea. Beijing, China: Science Press.

Zhu, D. 1991. Japanese Anchovy. In J. Deng and Q. Zhao (eds.), Marine Fisheries Biology. China: Agriculture Press, pp. 453–484.

Zhu, D., and Iversen, S. A. (eds.). 1990. Anchovy and Other Fish Resources in the Yellow Sea and East China Sea. November 1984–April 1988, Fisheries Investigations by R/V "Bei Dou." Marine Fisheries Research 11, pp. 1–143.

Zhu, D., Wang, W., Zhang, G., and Li, F. 1982. Studies on the fisheries biology of mackerel (*Pneumatophorus japonicus* Houttuyn) in the Yellow Sea. I. On the migratory and distributional pattern of mackerel in the Yellow Sea and Bohai Sea. Marine Fisheries Research 4:17–30.

8

The Sea of Japan Large Marine Ecosystem

Makoto Terazaki

Abstract

The Sea of Japan is a semi-isolated marginal sea connected to the Pacific by four shallow straits. The only important current flowing into this sea today is the warm Tsushima Current entering through the southern strait. The area of the Sea of Japan is about 978,000 km^2 and its volume is 1,713,000 km^3. The plankton of the Sea of Japan vary considerably both qualitatively and quantitatively. During both winter and summer, phytoplankton abundance is low in the warm-water and offshore cold-water areas, and relatively high in the intermediate temperature region. The mean primary production rates are about 0.21 gC m^{-2} day^{-1} in winter and 0.08 gC m^{-2} day^{-1} in summer. Maximum zooplankton biomass occurs in spring (April–May) and the minimum biomass in winter. Mean annual biomass in cold-water, warm-water, and coastal-water regions were 95 mg m^{-3}, 53 mg m^{-3}, and 57 mg m^{-3}, respectively. High biomass, above 100 mg m^{-3}, was found in the northern region above approximately 40°N. The zooplankton community showed low variance in taxonomic groups and species. Copepods were generally the most abundant group, followed by euphausiids, chaetognaths, amphipods, and mysids. These five groups accounted for over 99% of the total biomass. The Japanese sardine is an abundant pelagic fish in the Sea of Japan and constitutes more than 70% of the total catch. Scad, mackerel, yellowtail, and the common Japanese squid are also major catches. The catch of Japanese sardine attained a maximum (1,169,971 tons) in 1989 after 10 years of poor catches. Long-term fluctuations in the sardine population have been accompanied by noticeable geographic shifts in spawning and nursery grounds.

The area of the Sea of Japan is about 978,000 km^2; its volume is 1,713,000 km^3. It is a semi-isolated marginal sea with an average depth of 1,350 m and a maximum depth of approximately 3,700 m in the northern basin. It is connected to the Sea of Okhotsk, the North Pacific, and the East China Sea through four shallow straits: Tartarskiy Strait (15 m), Soya Strait (55 m), Tsugaru Strait (130 m), and Tsushima Strait (130 m). The only important current flowing into the Sea of Japan today is the Tsushima Current, a branch of the warm Kuroshio Current that enters through the Tsushima Strait between Kyushu and Korea and flows out to the Pacific through both the Tsugaru and Soya Straits (Figure 8-1). The shelf zone forms only about 20% of the total area of the Sea, and the bathyal zone some 40%; the area of the deep-sea floor is also about 40%.

Japanese geologists have studied the paleoenvironmental changes in the Sea of Japan during the last 85,000 years from the sedimentary record in a piston core from the Oki Ridge. According to their reports (Kato 1979; Masuzawa and Kitano 1983; Oba et al. 1991), five distinct changes

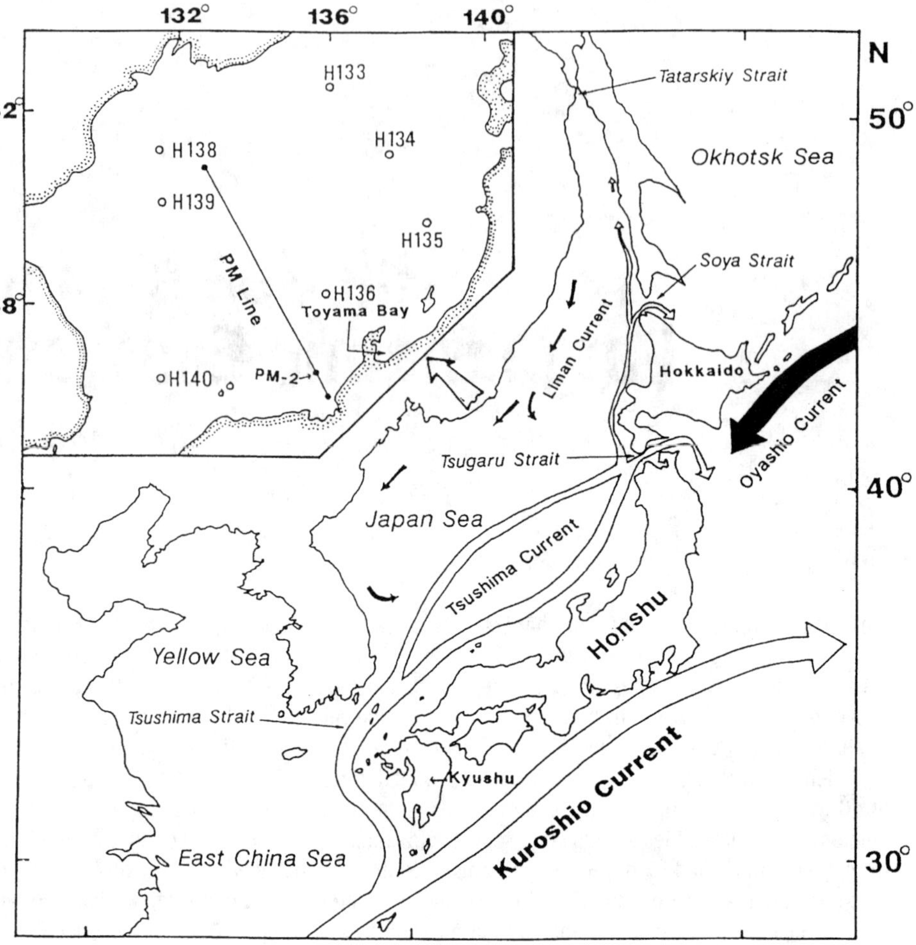

Figure 8-1. Schematic current map of the seas around Japan, and the main sampling stations for zooplankton and micronekton in the Sea of Japan.

have occurred. Between 85,000 and 27,000 years ago, the warm Tsushima Current did not flow into the Sea of Japan, and cold surface water conditions prevailed. Between 27,000 and 20,000 years ago, freshwater input to the Sea of Japan, probably from the Huang Ho River in China, stratified the water column, and severely anoxic conditions eliminated most of the benthic fauna. Between 20,000 and 10,000 years ago, the cold Oyashio Current flowed into the Sea through the Tsugaru Strait, reestablishing deep-water ventilation. Between 10,000 and 8,000 years ago, the foraminiferan compensation level gradually rose to a depth of less than 1,000 m, and bottom conditions changed from anoxic to oxygenated. At 8,000 years ago, the warm Tsushima Current began flowing into the Sea of Japan through the Tsushima Strait, establishing the modern oceanographic regime that has existed for the last 8,000 years.

Zenkevitch (1963) reviewed the flora and fauna of the Sea of Japan. Four hundred and fifty different kinds of plankton have been identified in the Sea of Japan, among them 306 diatoms and 133 peridinians. In contrast to the northern part of the Sea of Okhotsk, the phytoplankton of the Sea of Japan have two maximum blooms, a spring (March–April) diatom bloom and an autumn (September–October) peridinian bloom. The plankton of the Sea of Japan change considerably with depth both qualitatively and quantitatively. The number of bottom-living species decreases with increasing depth. Only 53 species of macrobenthos are known at depths of 1,000 to 2,000 m, 25 species for 2,000 to 3,000 m, and only 5 below 3,000 m (21 in the Sea of Okhotsk). Nishimura (1965, 1968, 1969) outlined the zoogeographical aspects of the Sea of Japan. Tropical to arctic animal populations are distributed in the Sea, some temporarily and others permanently, according to their bioclimatic characteristics. Generally, the tropical–subtropical populations occupy the thin upper layer, whereas the arctic–subarctic populations occupy the lower layer; the boundary between these two layers is very distinct. Seasonal variability is marked, and distinct seasonal patterns exist. In summer, tropical–subtropical populations enter the Sea and spread northward through the upper layer over an extensive area. Many neritic organisms thrive in the northernmost region of the Sea. In winter, arctic–subarctic populations, having risen to the surface layer, drift south in large numbers to the southern coasts of the Sea as or after the tropical–subtropical populations retreat from the north. Compared with the widely spread tropical surface populations mentioned above, the subtropical to warm-temperate mid-water populations are mostly confined to the southeastern half of the Sea of Japan. The northern limit of their distribution is rather sharply defined.

Hydrographic Conditions

The scale of the warm Tsushima Current is small compared with the Kuroshio Current. It is a current with a width of about 60 km, a thickness of less than 200 m, speeds of 0.3–1.7 knots, and a volume transport of $2.0–3.5 \times 10^6$ tons s^{-1}. The flow of the warm current through the Tsushima Strait attains its peak in September. The warm current in the Sea of Japan proper is estimated to be at its strongest during the period from September to October. Because the stream belts of the warm current, especially the offshore one, are free from the stress of any uniform drift currents, they begin to meander on account of the increased lateral shear imposed by a high velocity. The development of meandering brings about several cyclonic and anticyclonic eddies of large scale along either side of the stream belts of the warm current. The formation of several cold-water and warm-water areas along the main stream of the Tsushima Current, observed clearly in late summer to autumn, is supposedly due to such eddies. Generally, there are two types of paths of the Tsushima Current, as shown in Figure 8-2. In type A, two or three stream belts flow nearly straight to the northeast, whereas type B exhibits large-scale meanders. The current alternates at 6-year intervals (Japan Sea National Fisheries Research Institute 1991). When the type A pattern was prevalent (Figure 8-3) an increased water temperature was observed at a depth of 50 m. These eddy areas may trap the pelagic animal populations that have been carried northward by the offshore branch of the warm current and form good fishing grounds.

In spite of the shallowness of the straits connecting the Sea of Japan with the ocean, the isolation of its deep waters is only relative since in winter the deep waters are well aerated as a result of the sinking of cooled surface waters along its slope. It differs from the adjacent parts of the ocean and from the Okhotsk and Bering Seas by its lower temperature and by the absence of an oxygen deficiency in the middle layer. The cold intermediate layer is also absent from the Sea of Japan (Table 8-1).

The Sea of Japan may be divided into two distinct parts according to the position of its temperature gradients. The 0°C isotherm is designated as the boundary. The temperature of the northwestern part falls sharply in winter, and in the Tartarskiy Strait ice is present from November until April, sometimes reaching great thicknesses in the northern part. In the southern part,

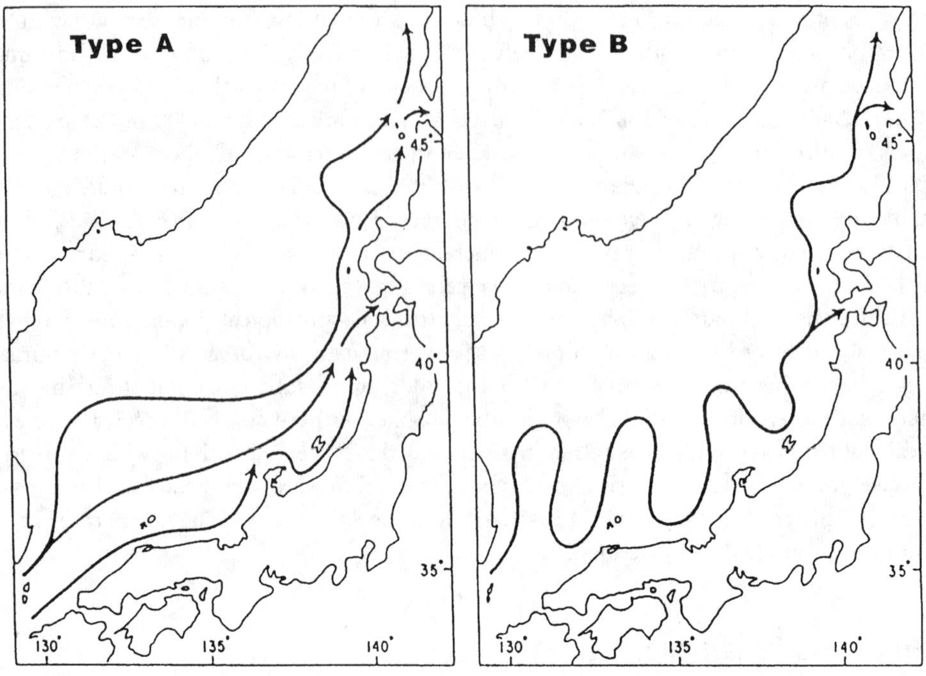

Figure 8-2. Flow path types of the warm Tsushima Current.

the seasonal temperature fluctuations are as high as 14°C, and in the northern part up to 20°C (Figure 8-4).

Imai et al. (1990) estimated the mean concentrations of phosphate, nitrate, and nitrite on a PM section for the four seasons using data from 1972 through 1988. The total amount of nutrients in the Tsushima Current water is less than 0.5% of the total amount in the Sea of Japan. Biomass in the water column is maintained through the nutrients supplied by active mixing with nutrient-

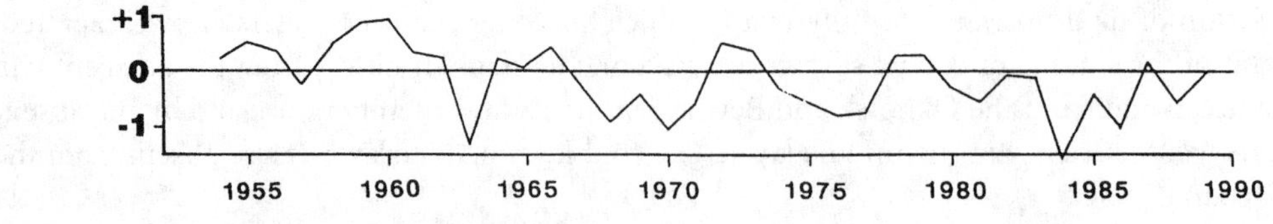

Figure 8-3. Time series of flow types of the warm Tsushima Current in March–October (upper), and annual fluctuations of mean water temperature at 50 m depth in the warm-water region off Honshu (bottom) (after Sea of Japan National Fisheries Research Institute 1991).

Table 8-1. Vertical distribution of temperature, salinity, and oxygen in the Sea of Japan (41°08′N, 137°21′E; August 1970) and Pacific Ocean (40°50′N, 145°00′E; September 1989)

	Temperature (°C)		Salinity (PSU)		Oxygen (ml liter^{-1})	
Depth (m)	Sea of Japan	Pacific	Sea of Japan	Pacific	Sea of Japan	Pacific
0	20.90	17.60	33.62	32.77	4.87	5.91
50	1.88	3.90	34.04	33.30	7.40	6.81
100	0.84	3.10	34.04	33.40	6.46	6.34
200	0.42	1.46	34.08	33.39	6.04	5.47
500	0.37	2.90	34.09	33.93	6.48	1.90
1,000	0.17	2.82	34.07	34.37	5.50	0.84
2,000	0.16	2.38	34.08	34.62	5.26	1.92
3,000	0.21	1.61	34.09	34.68	5.32	2.99

rich subsurface water in winter, as well as through the inflow of nutrient-rich water through the Tsushima Strait (Figure 8-5).

Phytoplankton

In the Sea of Japan a subsurface chlorophyll maximum (SCM) is sometimes found at deep layers, as is the case with many other seas. An SCM was found in almost all springs and in some winters, but was rare in both summer and autumn. The chlorophyll maximum corresponded to depths where the nitrate and phosphate concentrations were notably elevated (Nagata and Kitani 1987). According to Kano et al. (1984), during both winter and summer, phytoplankton abundance is low in the warm-water and offshore cold-water areas, and relatively high in the intermediate-temperature region. The mean primary production rates in the Sea of Japan are about 0.21 gC m^{-2} day^{-1} in winter and 0.08 gC m^{-2} day^{-1} in summer (Figure 8-6).

Ohwada and Ogawa (1966) studied the horizontal distribution of diatoms on the basis of data obtained by the Maizuru Marine Observatory, Japan Meteorological Agency, from 1955 to 1965. Diatoms are abundant in the warm Tsushima Current and near the boundary line, occurring in numbers up to 10^5 per liter, but are less prevalent in the offshore cold waters. In summer, the high diatom density area moves northward, still within the warm Tsushima Current. In the offshore cold waters the population is still smaller. In autumn, the number of diatom cells is also greater in the warm Tsushima Current than in the offshore cold waters. In winter, the same tendency as those in the previous seasons is seen (Figure 8-7). During summer, *Chaetoceros affinis, C. curvisetus, C. didymus* v. *anglica, C. lorenzianus, Leptocylindrus danicus,* and *L. minimus* are abundant in the neritic warm waters. *Dactylisolen mediterraneus, Rhizosolenia alata* f. *gracillima, R. calcar avis, R. styliformis* v. *latissima,* and *Climacodium biconcavum* dominate in the warm waters of the offing. These warm-water species spread to the western part of Hokkaido. In the cold waters, *Chaetoceros atlanticus, C. concavicornis, R. hebetata* f. *semispina,* and *Parafavella pacifica* dominate. In winter *Chaetoceros debilis, C. socialis,* and *Skeletonema coastatum* are the dominant species in the neritic warm waters. *Eucampia zoodiacus, Lauderia borealis,* and *R. stolterfothii* are present in large numbers in the warm waters of the offing. In the cold-water masses, *C. atlanticus* and *C. concavicornis* are still dominant (Ohwada and Ogawa 1966).

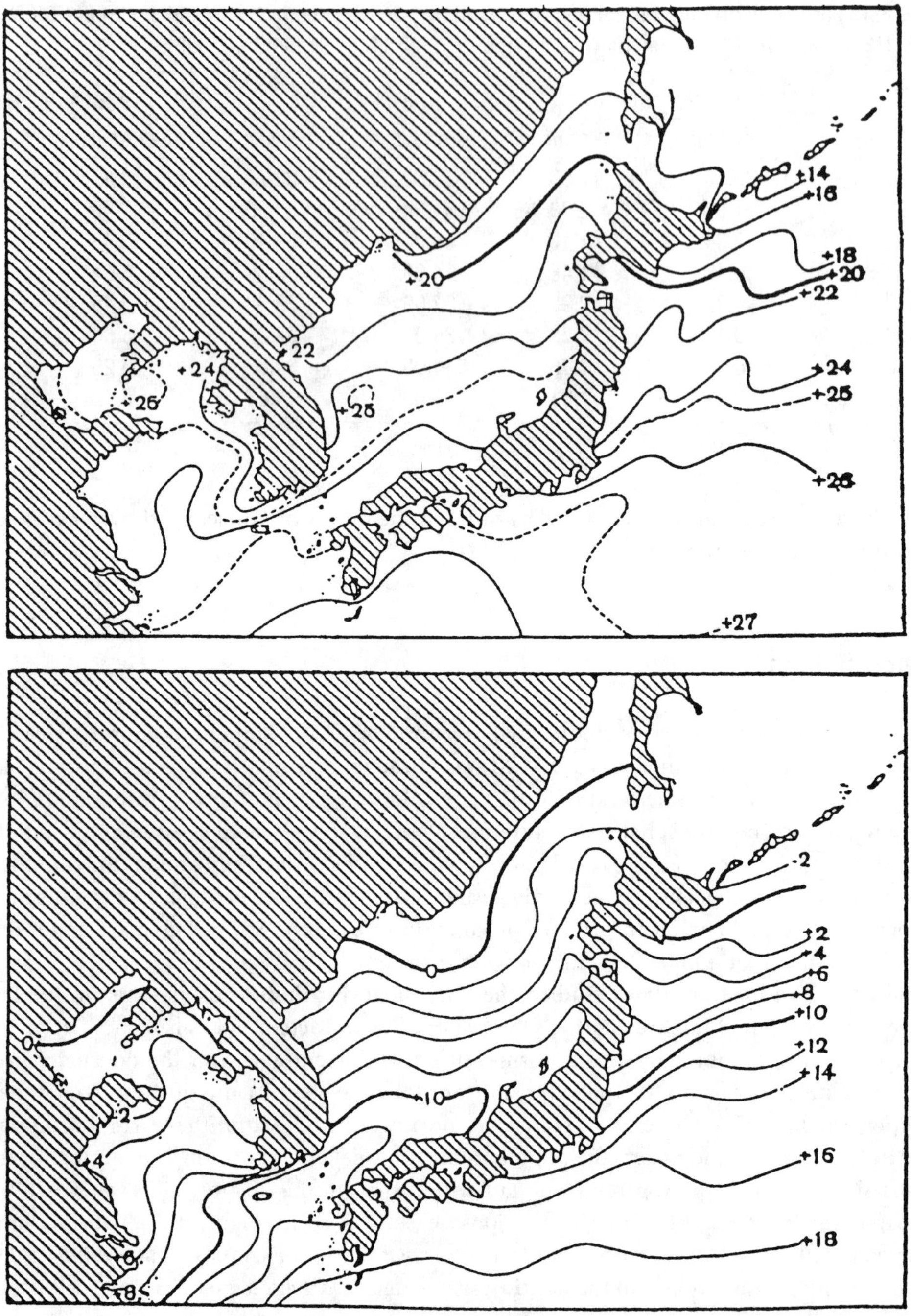

Figure 8-4. Surface temperatures of waters in the Sea of Japan in summer (upper) and winter (bottom) (after Zenkevitch 1963).

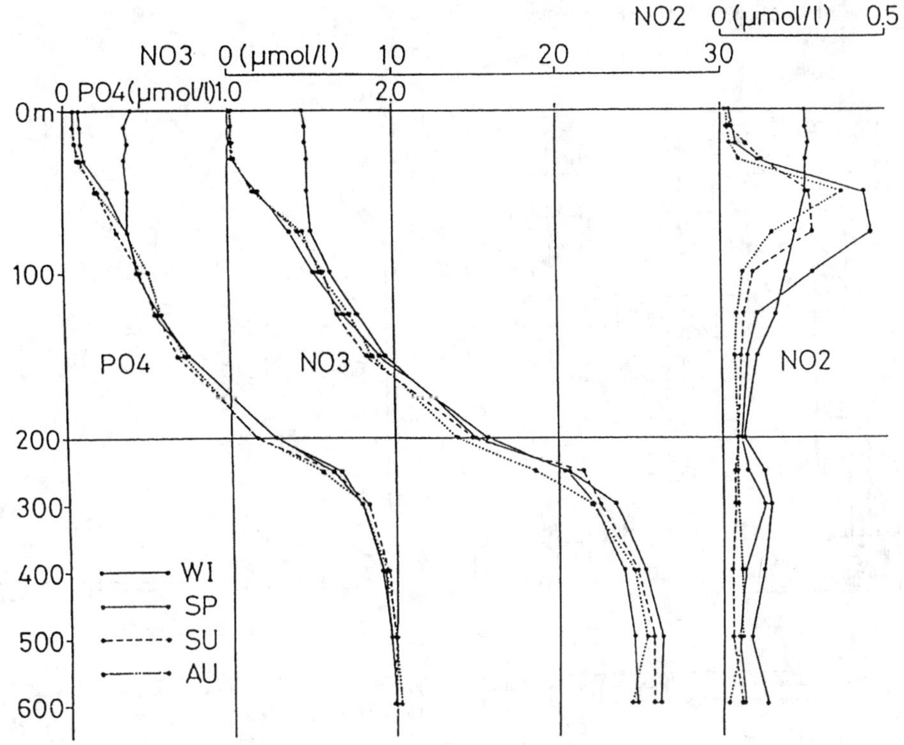

Figure 8-5. Mean vertical profile of nutrients in the four seasons at Station PM-2 (after Imai et al. 1990).

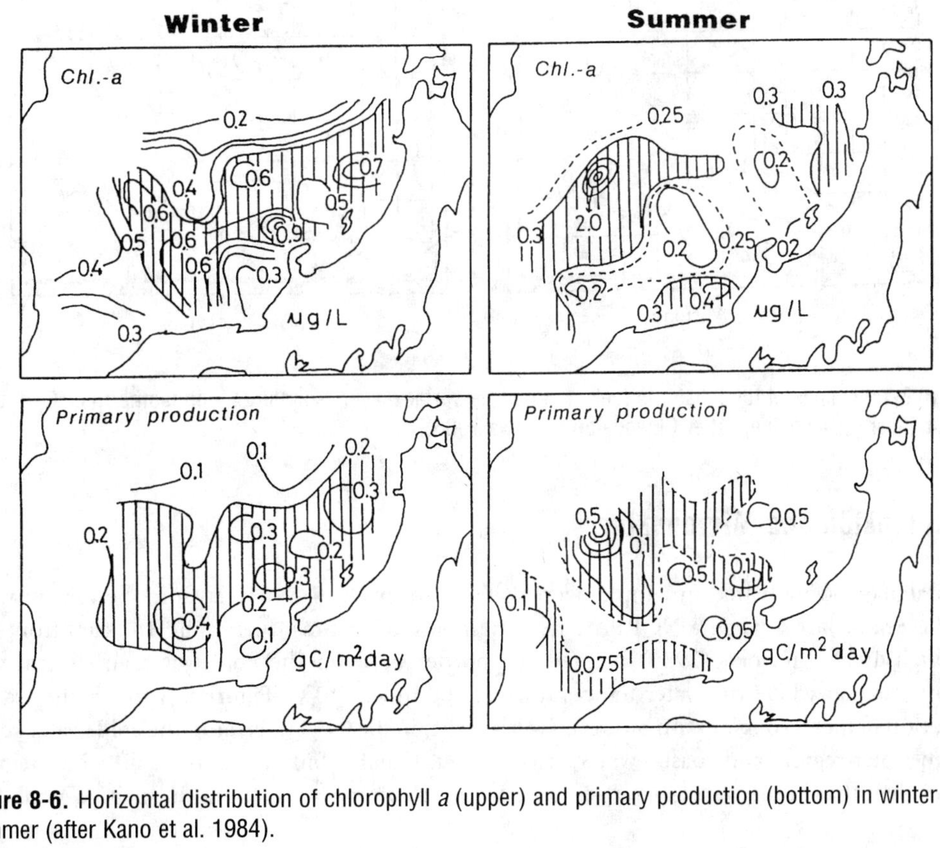

Figure 8-6. Horizontal distribution of chlorophyll *a* (upper) and primary production (bottom) in winter and summer (after Kano et al. 1984).

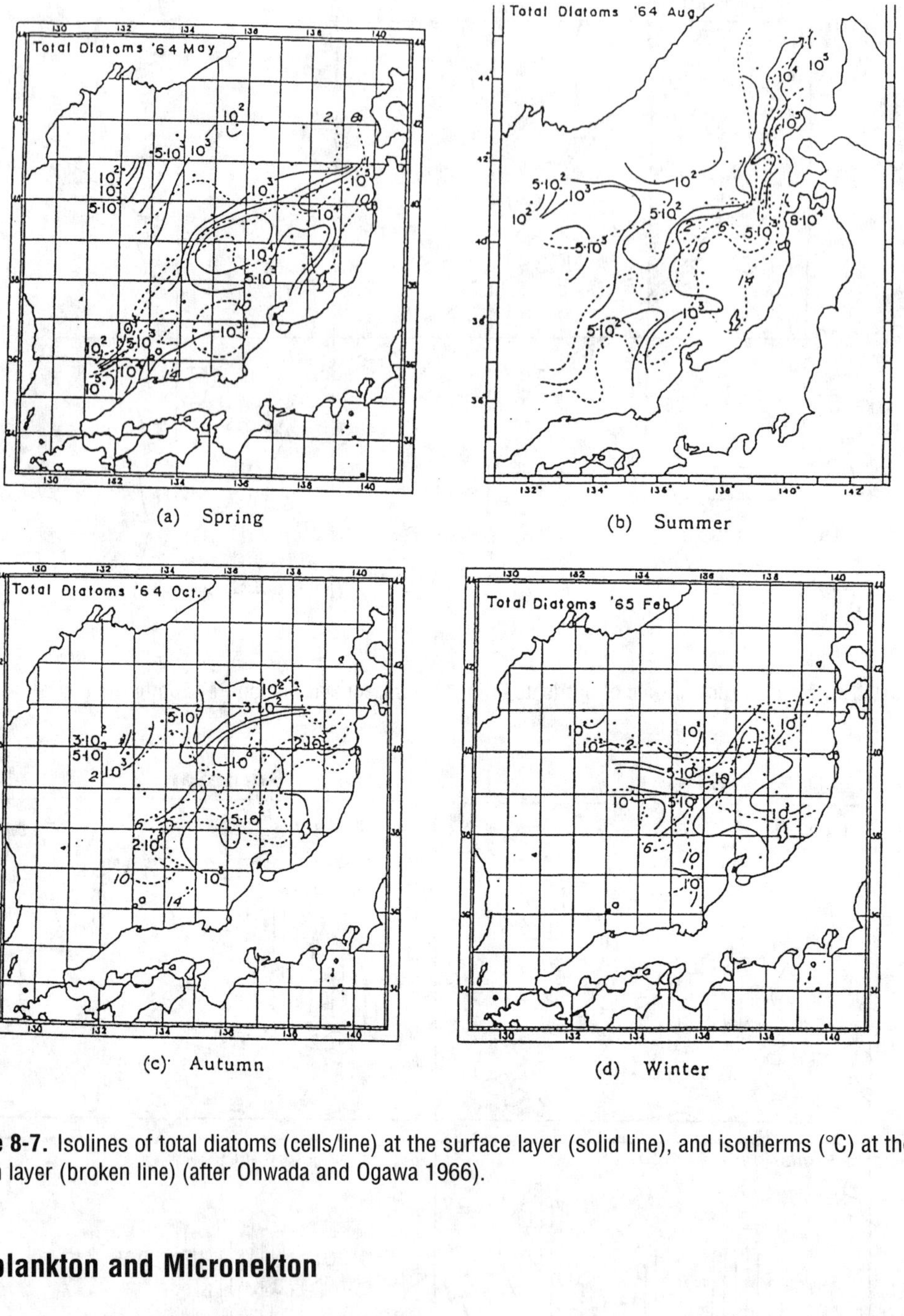

Figure 8-7. Isolines of total diatoms (cells/line) at the surface layer (solid line), and isotherms (°C) at the 100 m layer (broken line) (after Ohwada and Ogawa 1966).

Zooplankton and Micronekton

A total of 6,536 zooplankton samples were collected from the 0–150 m layer in the various waters of the Sea of Japan using a NORPAC net (mesh size 0.35 mm) during the 18 years from 1965 through 1982. Morioka (1985) defined the border between the cold-water and warm-water regions as that where the water temperature at 100 m was 10°C. Figure 8-8 shows the seasonal variation in mean zooplankton biomass (wet weight) in the 0–50 m layer in the cold-water region, warm-water region, and coastal waters off Honshu. Usually, biomass in the cold-water region is larger than that in the warm-water region. Maximum biomass was found in spring (April–May)

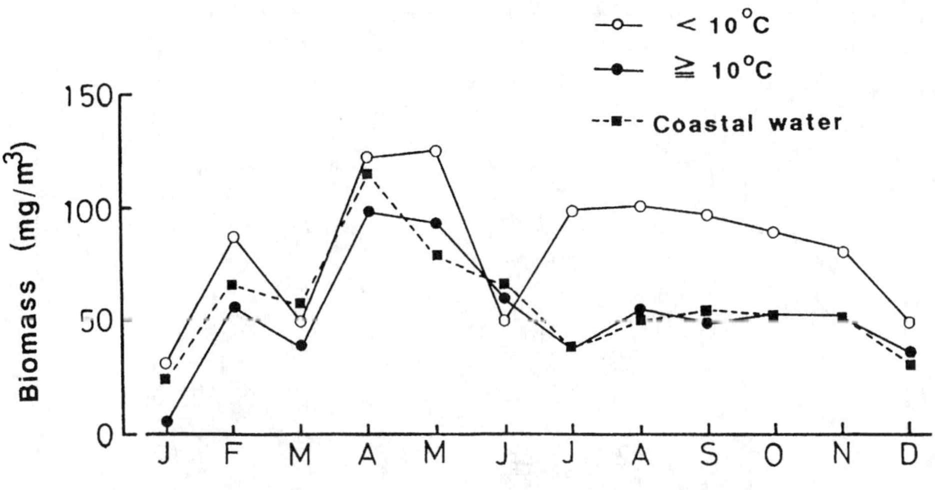

Figure 8-8. Seasonal changes in mean zooplankton biomass (wet weight) in the cold-water (<10°C), warm-water (>10°C), and coastal-water regions in the Sea of Japan (1965–1982) (after Morioka 1985).

and minimum biomass in winter (January) throughout all regions. Mean annual biomass in the cold-water, warm-water, and coastal-water regions was 95 mg m^{-3}, 53 mg m^{-3}, and 57 mg m^{-3}, respectively (Morioka 1985). According to the horizontal distribution of mean annual zooplankton biomass in the 0–150 m layer in the Sea of Japan, zooplankton biomass greater than 100 mg m^{-3} is common in the northern region at latitudes above approximately 40°N (Figure 8-9) (Morioka 1985). During the R/V *Hakuho Maru* KH-70-4 cruise, a 160-cm ORI net was towed obliquely in the 0–1,000 m layer at seven stations to determine the biomass and distribution of meso- and bathypelagic plankton and micronekton. The total biomass of oblique tow samples was large (70–100 mg m^{-3}) at the northern stations, whereas it was small (about 40 mg m^{-3}) at the three southern stations (Figure 8-10) (Marumo et al. 1972). These values are much higher than those (3–10 mg m^{-3}) obtained in the Kuroshio water. The biomass at the northern stations is similar to that obtained in the Oyashio water of the northern North Pacific.

The zooplankton community showed low variance in taxonomic groups and species. Copepods were generally the most abundant group, followed by euphausiids, chaetognaths, amphipods, and mysids. These five groups generally accounted for over 99% of the total biomass. Dominant species in the Sea of Japan are listed below.

Copepods: *Neocalanus cristatus, N. plumchrus, N. flemingeri*
Euphausiids: *Euphausia pacifica, Thysanoessa longipes*
Amphipods: *Themisto iaponica, Primno macropa, Cyphocaris challengeri*
Mysid: *Meterythrops microphathalma*
Chaetognatha: *Sagitta elegans*

In the case of chaetognaths, 17 species belonging to three genera were collected from the Sea of Japan: *Sagitta crassa, S. elegans, S. enflata, S. ferox, S. hexaptera, S. minima, S. nagae, S. neglecta, S. pacifica, S. pseudoserratodentata, S. regularis, S. robusta, S. decipiens, S. lyra, Pterosagitta draco, Krohnitta pacifica,* and *K. subtilis* (Terazaki 1993a). The most abundant species was *Sagitta elegans,* which represented 93.9% to 100% of the individuals collected from the 0 to 1,000 m layer (Figure 8-11).

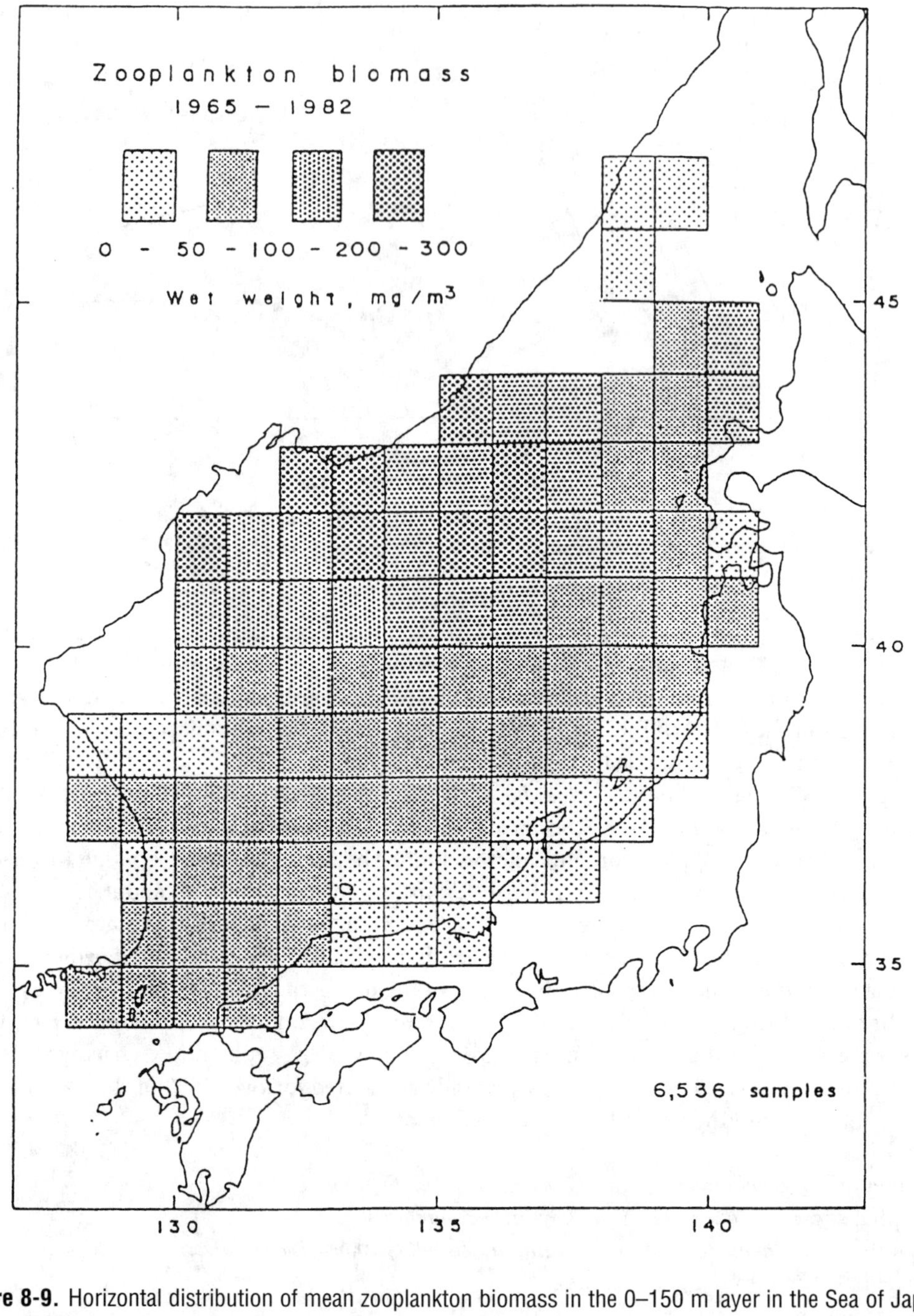

Figure 8-9. Horizontal distribution of mean zooplankton biomass in the 0–150 m layer in the Sea of Japan (1965–1982) (after Morioka 1985).

The vertical distribution of zooplankton biomass was similar day and night. Usually, the largest biomass was found between the surface and 100 m in depth, where a conspicuous thermocline existed. In this area of maximum biomass, all the dominant zooplankton groups mentioned above, except mysids, were present. The biomass decreased greatly with increasing depth (Figure 8-12) (Marumo et al. 1972). Generally, zooplankton biomass was low below 200 m

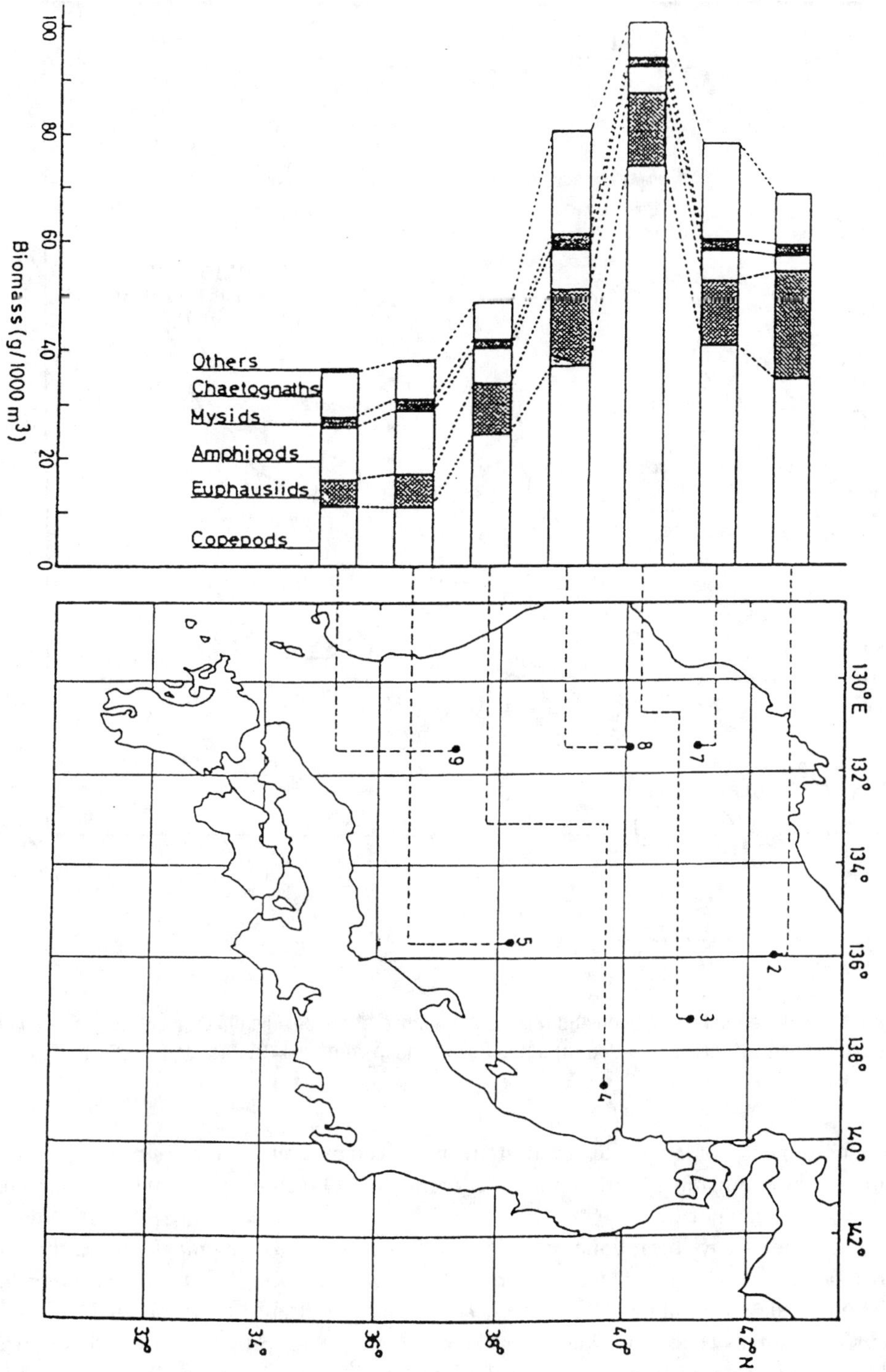

Figure 8-10. Zooplankton biomass and community in the upper 1,000 m of the Sea of Japan, summer 1970 (after Marumo et al. 1972).

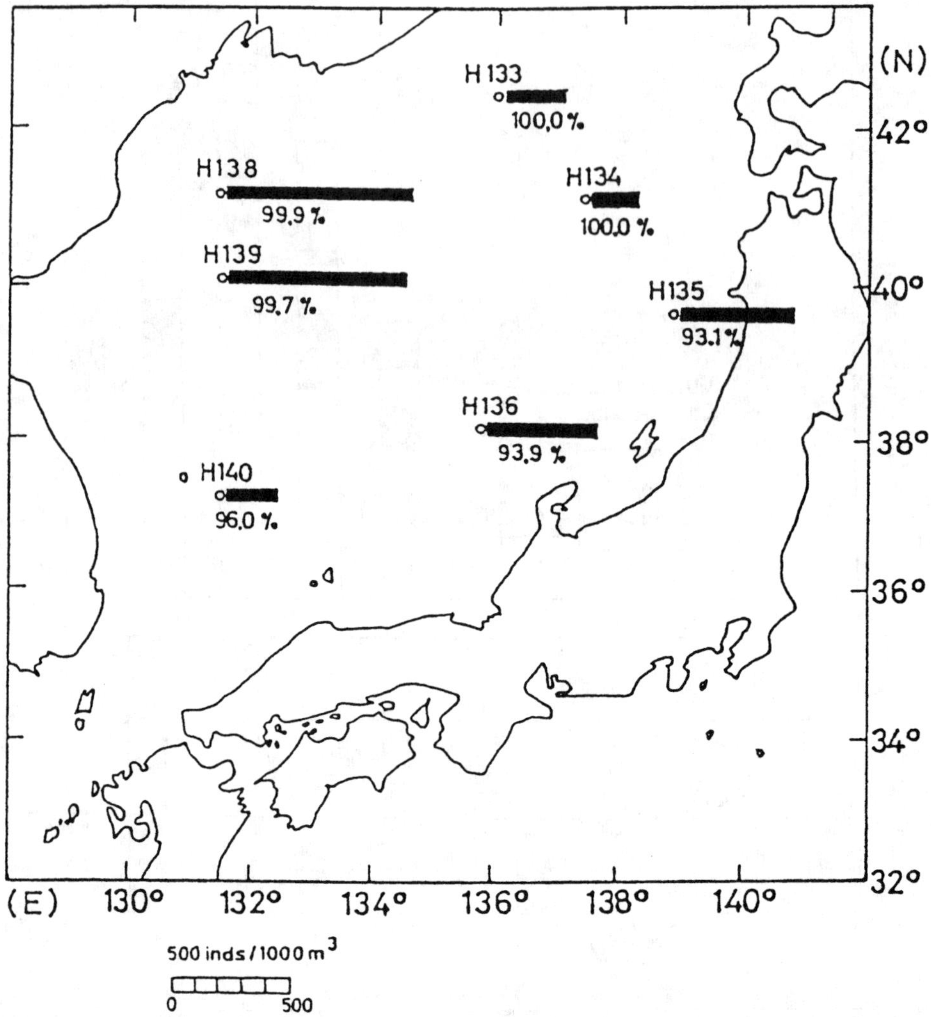

Figure 8-11. Horizontal distribution and relative abundance of the pelagic chaetognath *Sagitta elegans*, collected with an ORI net (0 to 1,000 m) in the Sea of Japan, summer 1970 (after Terazaki 1993a).

throughout the Sea of Japan. However, in the upper 200 m layer there was a distinct tendency for biomass to be comparatively low in the water of the warm Tsushima Current but remarkably high in the water to the north.

Many studies have been done on the vertical distribution of *Sagitta elegans* in the northern North Pacific Ocean (Bieri 1959; Alvariño 1965; Kotori 1972; King 1979; Sullivan 1980; Terazaki and Miller 1986; Terazaki 1992), and *S. elegans* is a typical epipelagic species in the subarctic waters. Individuals collected in the waters of the Sea of Japan proper, below a depth of 200 m, included *S. elegans* and three species of Polychaeta: *Polycrirrus medusa, Chaetozone setosa,* and *Anobothrus gracilis,* which are distributed in the epipelagic layer of the Pacific (Uschakov 1955). Nishimura (1968) reported that many fishes that are epipelagic in the Pacific Ocean were distributed in the deep waters of the Sea of Japan. A large herbivorous copepod, *Neocalanus cristatus,* has been reported to occur as deep as 1,600 m (Miller et al. 1984) and 2,000 m (Sekiguchi 1975)

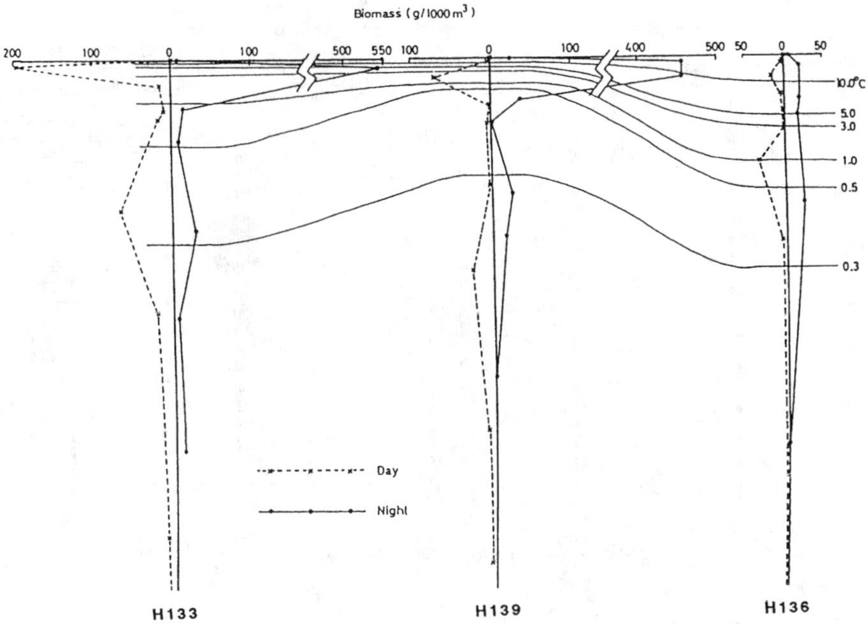

Figure 8-12. Vertical distribution of zooplankton biomass in the upper 1,000 m of the Sea of Japan, summer 1970 (after Marumo et al. 1972).

in the northern North Pacific, whereas the maximum sampling depth containing *N. cristatus* in the Sea of Japan was 3,850 m (Terazaki and Wada 1988).

Between 20,000 and 10,000 years ago, the cold Oyashio Current flowed into the Sea of Japan through its northern straits. The epipelagic chaetognath *Sagitta elegans* was probably introduced into the Sea of Japan during this period, but the shallow straits prevented the invasion of deep-sea chaetognaths. All other chaetognath species were probably introduced into this Sea by warm currents within the last 10,000 years and inhabit the epipelagic layer. The size and number of grasping spines and teeth of the *S. elegans* in the Sea of Japan are larger than those from the Pacific. They also have developed intestinal tissue containing oil droplets similar to meso- and bathypelagic species. These features are adaptations to the deep-water environment. The vertical dispersal of *S. elegans* is facilitated by the absence of competitors such as *Eukrohnia hamata* and *E. fowleri* (Terazaki 1993a) (Figure 8-13).

The production of six crustacean species—*Calanus pacificus, Neocalanus plumchrus, Paracalanus parvus, Pseudocalanus minutus, Metridia lucens,* and *Euphausia pacifica*—in the coastal areas of Toyama Bay was calculated by two methods, age-class analysis and Ikeda's physiological method (Ikeda 1974), using the plankton samples obtained in the spring of 1978. The mean daily production/biomass ratio for the six species is estimated at 0.05. The daily production of zooplankton is estimated at approximately 0.5 mg dry weight or 2.5 mg wet weight m^{-3}, and the mean biomass of the six crustacean species over 50 days was 9.8 mg dry weight or 49.1 mg wet weight m^{-3} (Morioka 1981).

Some reports exist that describe the life history of zooplankton and micronekton in the Sea of Japan (Table 8-2). The pattern and timing of life history events for *Neocalanus plumchrus* are similar to those found in the Georgia Strait; the timing in the two more neritic sites differs in the same way as that in the more oceanic Gulf of Alaska. The fifth copepodite (C5) is the only resting

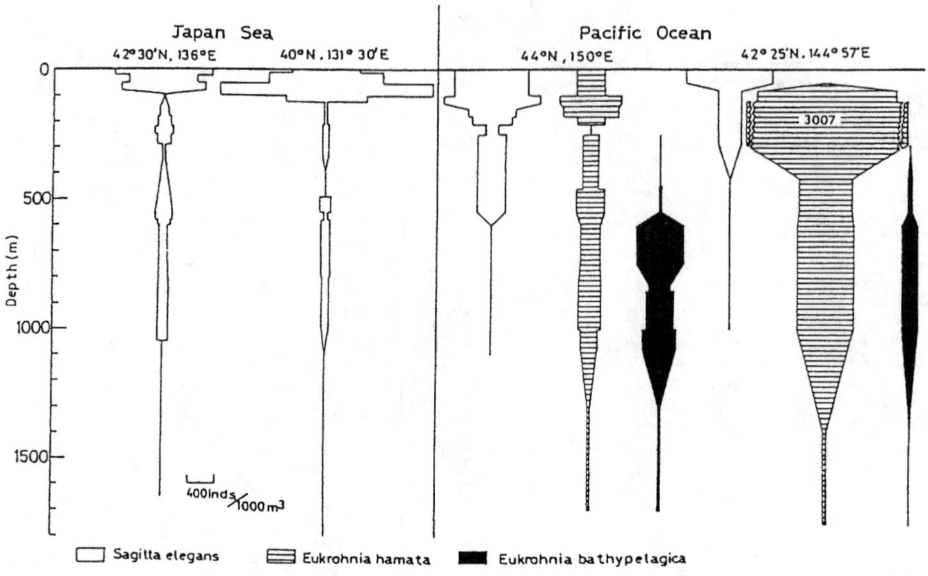

Figure 8-13. Vertical distribution of three dominant chaetognaths in the Sea of Japan and Pacific Ocean.

stage. It matures in winter and spawns at depth without feeding. The young develop to the C5 stage by April, fatten, and return to the resting phase at depth. *N. flemingeri* has two resting stages in the Sea of Japan: the mated female with a quiescent ovary, and the fourth copepodite stage. Fifth copepodites were never observed at depth in a resting phase. Some or all of the population

Table 8-2. Spawning season, number of generations, and estimated life span of macrozooplankton and micronekton in the Sea of Japan

Species	Spawning season	Number of generations	Life span (months)	Reference
Copepoda				
Neocalanus flemingeri	winter	2	24	Miller and Terazaki (1989)
Chaetognatha				
Sagitta elegans	spring, summer	2	10–12	Terazaki (1993b)
Euphausiid				
Euphausia pacifica	spring, summer	1	< 21	Iguchi et al. (1993)
Ostracoda				
Conchoecia pseudodiscophora	whole year	2	30	Ikeda and Imamura (1992)
Amphipoda				
Themisto japonica	whole year	3	?	Sugisaki and Ikeda (unpublished)
Appendicularian				
Oikopleura labradoriensis	spring, summer–fall	2	?	Shiga (unpublished)
Fish micronekton				
Maurolicus japonicus	summer–fall	1	20–21	Ikeda (unpublished)

in the Sea of Japan probably have a 2-year life cycle (Miller and Terazaki 1989). At least some young spawned in a given year develop to C4, rest until the following spring, develop to C5, descend, mature, mate, and rest as females with quiescent ovaries (Figure 8-14).

The proportion of gravid females present indicates that reproduction of the mesopelagic ostracod *Conchoecia pseudodiscophora* in Toyama Bay continues throughout the year, but peaks in April–July. Eggs and instars I and II were distributed below 500 m, whereas the more advanced instars were most abundant in the 350 to 500 m stratum during both day and night. Based on the abundance peaks of each instar in the time-series samples, development times were estimated to be

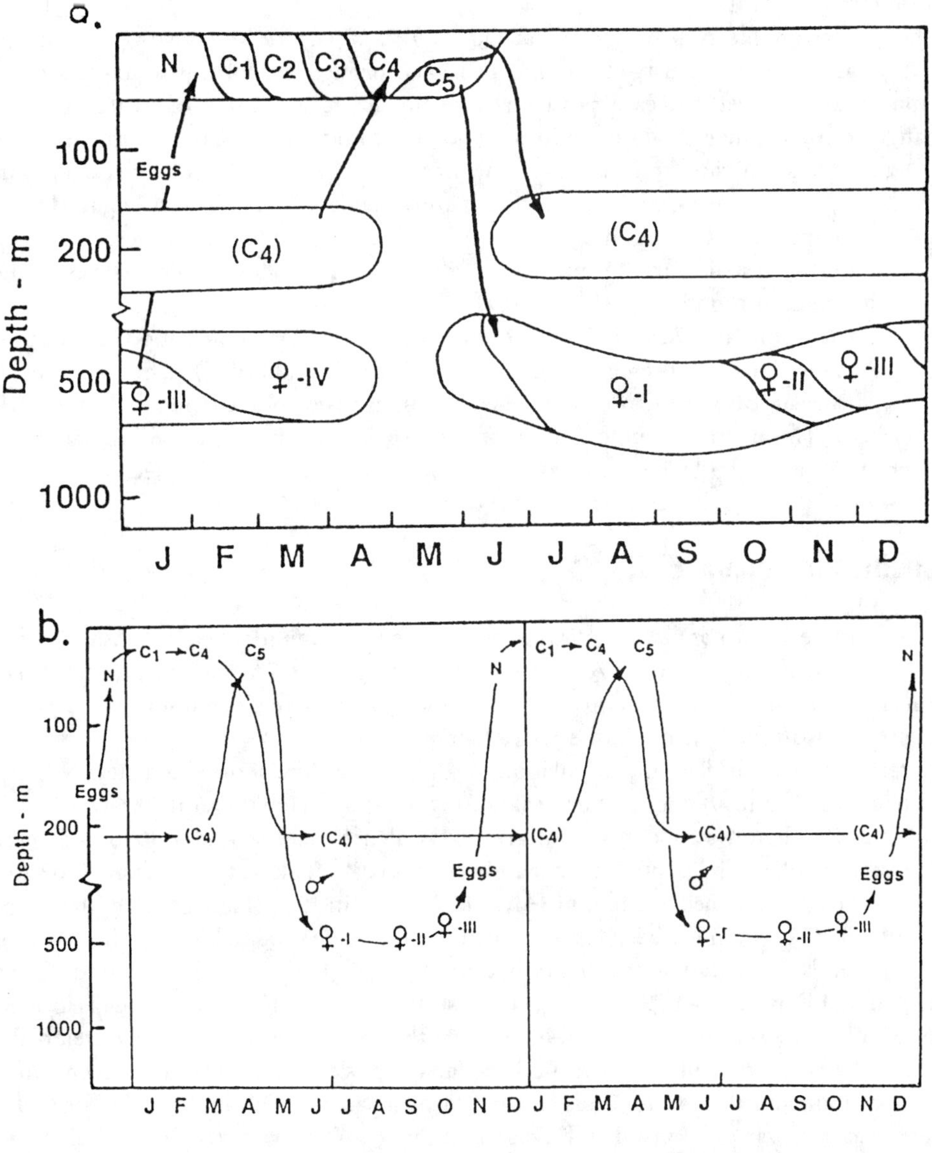

Figure 8-14. Hypothetical life history pattern for *Neocalanus flemingeri* in the Sea of Japan. (C_4) designates fourth copepodite in resting phase. a. Part of each year's egg cohort completes development in 1 year, part in 2 years. b. All individuals require 2 years to complete development. There are nearly separate odd-year and even-year cohorts (after Miller and Terazaki 1989).

2.5, 4, 3, 7, and 11 months for instars III, IV, V, VI, and VII, respectively. Thus, a total of 30 months is required for newly spawned eggs to hatch and reach adulthood (Ikeda and Imamura 1992).

The seasonal variation and life history of the pelagic chaetognath *Sagitta elegans* from Toyama Bay were investigated by doing a series of 0–500 m vertical hauls using a NORPAC net from 1 February 1990 to 30 January 1991. There was no remarkable variation in abundance throughout the year, although the greatest number of individuals was collected in August. Juveniles occurred mostly in spring and summer, from late March to August. Adults (Stage 3) occurred in all seasons except summer. The body length of adults ranged from 26 to 30 mm. There were two principal spawning periods, one in March to May and one in August. Life spans of both cohorts were 10 to 12 months (Terazaki 1993b).

The growth and life cycle of the euphausiid *Euphausia pacifica* were investigated in Toyama Bay. The eggs and larvae occurred from mid-February through late June. After rapid growth (3.1 mm/month) of the larvae and early postlarvae, the modal size (<10 mm) of postlarvae remained unchanged from summer to winter, indicating no growth of *E. pacifica* during this season. After this long growth stagnation from summer to winter, *E. pacifica* grew rapidly once again through the spring phytoplankton bloom of the next year and spawned. *E. pacifica* is thought to have only one generation per year. The minimum maturity size and the maximum size were 11 mm and 22 mm, respectively, for males, and 12 mm and 23 mm, respectively, for females. The estimated life span was less than 21 months (Iguchi et al. 1993).

The pelagic amphipod *Themisto iaponica* from Toyama Bay has a year-round breeding season, and three generations are present (Sugisaki and Ikeda, unpublished). The appendicularian *Oikopleura labradoriensis* in the northern Sea of Japan has two co-occurring generations (Shiga, unpublished). The main spawning season of the fish micronekton *Maurolicus japonicus* was summer to autumn and the life span was 20 to 21 months (Ikeda, unpublished).

Fluctuation of Pelagic Fish

Japanese sardine (*Sardinops melanostica*) is one of the most abundant pelagic fish in the Sea of Japan and contributes more than 70% of the total catch. Scad (*Trachurus japonicus*), mackerel (*Scomber japonicus*), yellowtail (*Seriola quingueradiata*), and Japanese common squid (*Todarodes pacificus*) are also major catches in the Sea of Japan.

Judging from catch data, the population of Japanese sardine showed a generally increasing trend around 1920 and attained its first peak during the 1930s. However, catches were low during World War II, when the fishery was interrupted. Catches increased after 1946 and reached a second peak in 1956. The population abruptly decreased to its lowest level from 1962 to 1971, then recovered and attained a maximum (1,169,971 tons) in 1989 (Figure 8-15). This long-term fluctuation in population has been accompanied by noticeable geographic shifts in spawning and nursery grounds. In the case of the first peak during the 1930s, the spawning ground was located to the south of Kyushu where the southernmost sector of the distributional range is located, and the population size stayed at its peak level. During this period, eggs and larvae developed in the Pacific, and a large-scale migration around the Japanese Islands took place. In recent times, the major spawning and nursery grounds of the Sea of Japan population has been located in the western coastal waters off Honshu and Kyushu (Figure 8-16). However, no relationship has been found between high catches of sardine and the large-scale meandering of the warm Tsushima Current (type B in Figure 8-2).

Annual fluctuations of the anchovy (*Engraulis japonica*) population in the Sea of Japan are significant. Catch data indicate that the population showed an increase beginning in 1952 and

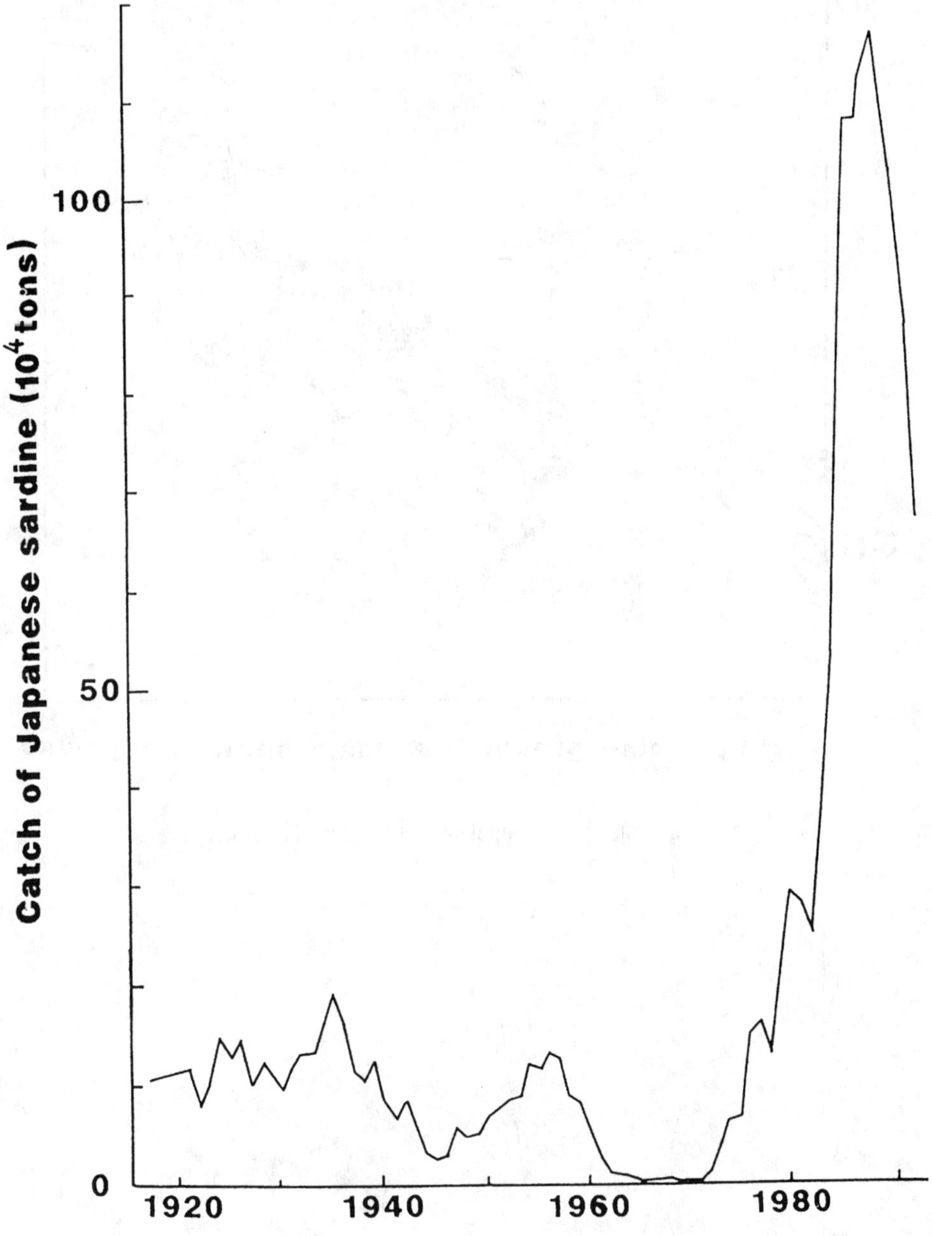

Figure 8-15. Annual fluctuation of sardine catch in the Sea of Japan.

attained a maximum in 1968, whereupon it decreased abruptly. On the other hand, the population of round herring (*Etrumeus terex*) showed a peak during the 1950s and decreased to its lowest level between 1960 and 1972. Both populations have been increasing gradually since 1980 (Figure 8-17).

The major fishing method for yellowtail in the Sea of Japan is the set net, followed by the purse seine and gill net. Catches of yellowtail ranged from 10,000 tons to 20,000 tons during the 17-year period between 1952 and 1968 and then increased. Judging from catch data, the population

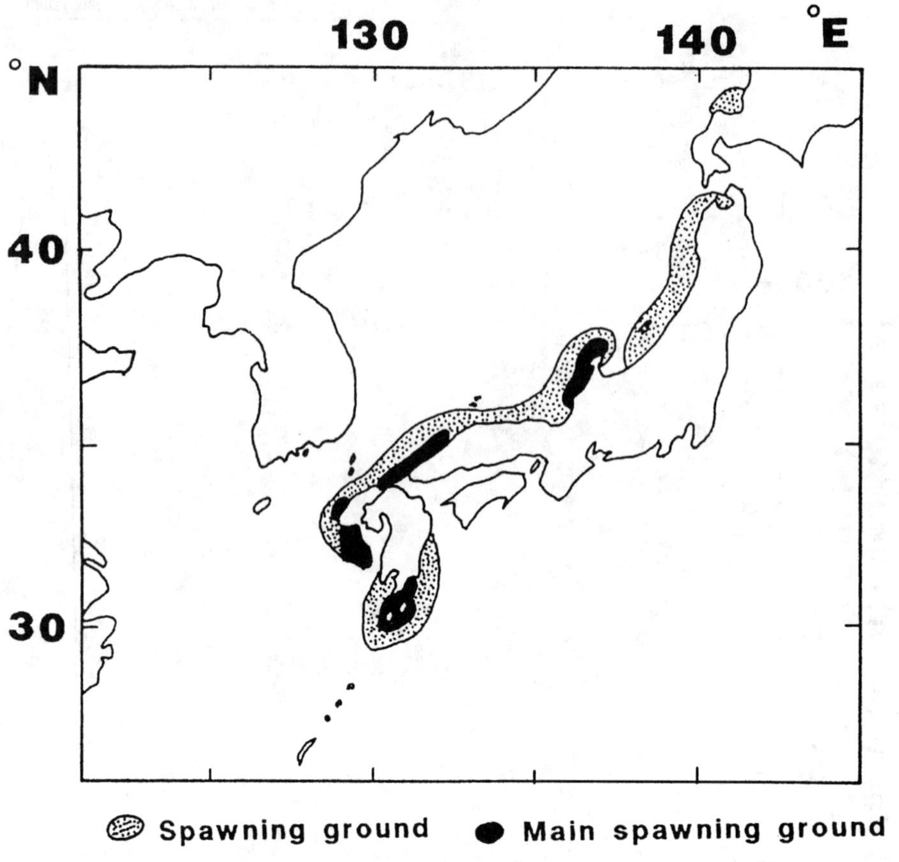

Figure 8-16. Spawning grounds of the sardine population of the Sea of Japan.

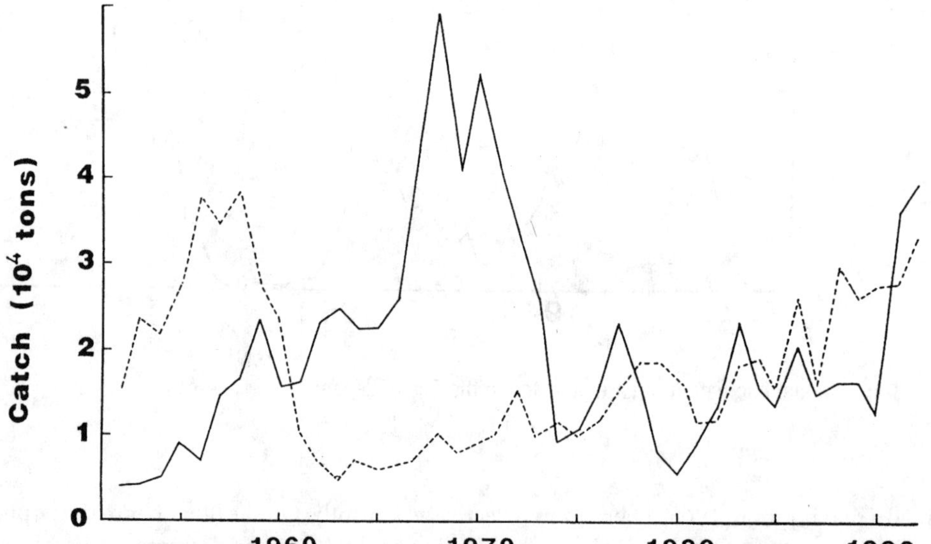

Figure 8-17. Annual fluctuation of anchovy and round herring catches in the Sea of Japan.

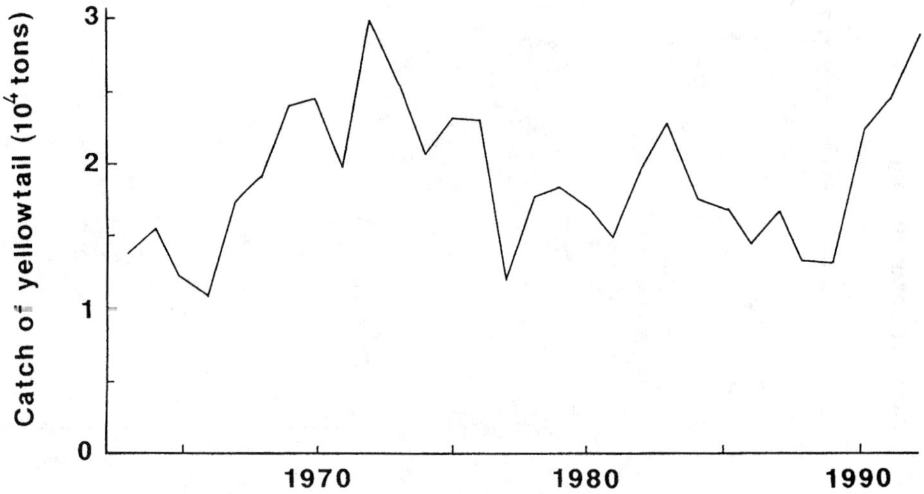

Figure 8-18. Annual fluctuation of yellowtail catch in the Sea of Japan.

attained a maximum (30,000 tons) in 1972; large catches above 20,000 tons have been recorded in the 1990s (Figure 8-18).

Scad is a migratory fish living in warm-water regions. The main spawning grounds of scad are in the East China Sea and the coastal waters of Kyushu, where the scad migrate into the Sea of Japan through the Tsushima Strait. Their fishing grounds are located in the southern Sea of Japan, and catches in the cold water region have been very poor. Catch data indicate that the population of scad increased beginning in 1955 and attained a maximum in 1960 (286,000 tons). Second and third peaks were recorded in 1964 and in 1966. Poor catches (below 40,000 tons) continued during the 16 years from 1971. It is quite possible that the recruitment of young scad from the East China Sea and other waters was very low in this period. Recently, catches have been increasing gradually (Figure 8-19).

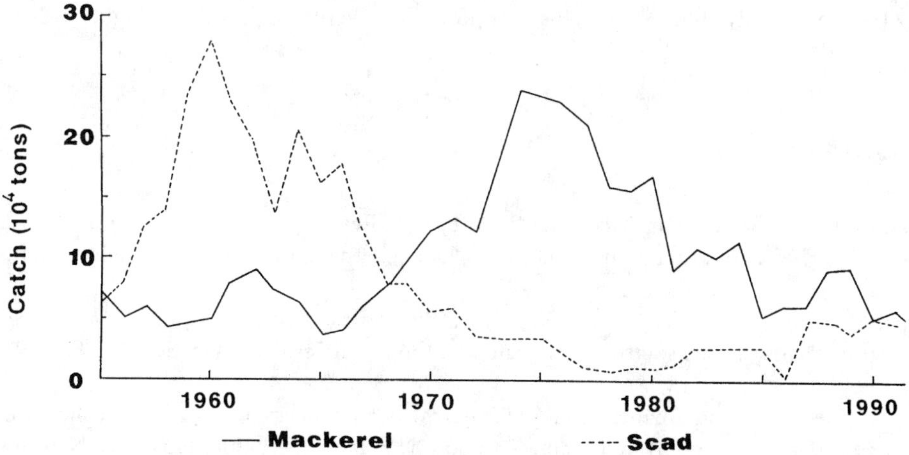

Figure 8-19. Annual fluctuation of mackerel and scad catches in the Sea of Japan.

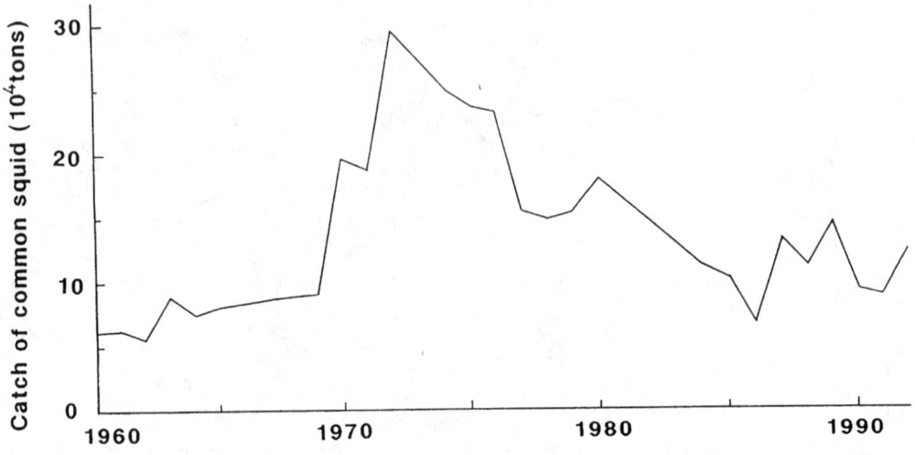

Figure 8-20. Annual fluctuation of common squid catch in the Sea of Japan.

The main fishing grounds for mackerel are also located in the southern Sea of Japan. Annual catch fluctuations are in striking contrast to that of scad. The population attained a maximum in 1975 (226,000 tons). When the warm Tsushima Current meandered (type B in Figure 8-2), many cold or warm eddies occurred along either side of the stream belts. Usually, good fishing grounds for mackerel are formed near eddies (Japan Sea National Fisheries Research Institute 1991), and there was a strong relation between high catches of mackerel and the meandering of the warm Tsushima Current (Figure 8-19).

The spawning grounds of the common squid are in the East China Sea and the coastal waters of southern Honshu and Kyushu. Common squid migrate northward in spring–summer and southward in autumn–winter (Japan Sea National Fisheries Research Institute 1991). Hamabe and Shimizu (1969) determined three distinct groups (summer, autumn, and winter) by analyzing growth, maturity, and migration patterns within the Sea of Japan. According to Kasahara (1991), the largest number of squid mature in autumn, the next largest in winter, and the least in summer. The population of common squid increased abruptly during the 1960s and attained a maximum in 1972 (approximately 300,000 tons), whereupon it has decreased gradually (Figure 8-20).

References

Alvariño, A. 1965. Chaetognaths. Oceanogr. Mar. Biol. A. Rev. 3:115–194.

Bieri, R. 1959. The distribution of the planktonic Chaetognatha in the Pacific and their relationship to the water masses. Limnol. Oceanogr. 4:1–28.

Hamabe, M., and Shimizu, T. 1969. Ecological studies on the common squid, *Todarodes Pacificus* Steenstrup, mainly in the southwestern waters of the Japan Sea. Bull. Jap. Sea Reg. Fish. Res. Lab. 16:13–55.

Iguchi, N., Ikeda, T., and Imamura, A. 1993. Growth and life cycle of a euphausiid crustacean (*Euphausia Pacifica* Hansen) in Toyama Bay, southern Japan Sea. Bull. Japan Sea Natl. Fish Res. Inst. 43:69–81.

Ikeda, T. 1974. Nutrition ecology of marine zooplankton. Mem. Fac. Fish., Hokkaido Univ. 22:1–97.

Ikeda, T., and Imamura, A. 1992. Population structure and life cycle of the mesopelagic ostracod *Conchoecia pseudodiscophora* in Toyama Bay, southern Japan Sea. Mar. Biol. 113:595–601.

Imai, M., Ebara, S., Okimura, H., and Kadono, K. 1990. On the nutrients in the Tsushima Warm Current water (in Japanese). Umi to Sora 66:93–111.

Japan Sea National Fisheries Research Institute. 1991. The Latest Trend of Fisheries Resources in the Japan Sea (in Japanese).

Kano, Y., Baba, N., and Ebara, S. 1984. Chlorophyll a and primary production in the Japan Sea. Oceanogr. Mag. 34:31–39.

Kasahara, S. 1991. The Japan Sea (in Japanese). In K. Naus, T. Okutani, and M. Ogura (ed.), Ika. Tokyo: Seizandoushoten, pp. 143–158.

Kato, M. 1979. Age assigned to dredged siltstone samples and piston core samples. Cruise Rep. Geol. Surv. Japan 13:70–72.

King, K. R. 1979. The life history and vertical distribution of the chaetognath *Sagitta elegans* in Dabob Bay, Washington. J. Plankton Res. 1:153–167.

Kotori, M. 1972. Vertical Distribution of Chaetognaths in the Northern North Pacific Ocean and Bering Sea. In A. Y. Takenouti (ed.), Biological Oceanography of the Northern North Pacific Ocean. Tokyo: Shoten, pp. 291–308.

Marumo, R., Nemoto, T., Omori, M., Aizawa, Y., Terazaki, M., Araki, M., Kawaguchi, K., Komaki, Y., and Honda, T. 1972. Distribution Collected with Norpac Net, ORI Net and IKMT. In R. Marumo (ed.), Preliminary Report of the Hakuho Maru Cruise Kh-70-4 (IBP Cruise). Tokyo: Ocean Research Institute, University of Tokyo, pp. 16–22.

Masuzawa, T., and Kitano, Y. 1983. Sulfate reduction and sulfide deposition in deep-sea sediments from the southwestern Japan Sea. J. Oceanogr. Soc. Japan 39:251–258.

Miller, C. B., and Terazaki, M. 1989. The life history of *Neocalanus flemingeri* and *Neocalanus plumchrus* in the Sea of Japan. Bull. Plankton Soc. Japan 36:27–41.

Miller, C. B., Frost, B. W., Batchelder, H. P., Clemons, M. J., and Conway, R. E. 1984. Life history of large, grazing copepods in a subarctic ocean gyre: *Neocalanus cristatus,* and *Eucalanus bungii* in the Northeast Pacific. Prog. Oceanogr. 13:201–243.

Morioka, Y. 1981. Zooplankton production in the Toyama Bay in March–May, 1978. Bull. Jap. Sea Reg. Fish. Res. Lab. 32:57–64.

Morioka, Y. 1985. Distribution of zooplankton in the Japan Sea. Bull. Jap. Soc. Fish. Oceanogr. 47/48:63–66.

Nagata, H., and Kitani, K. 1987. Vertical distribution of chlorophyll a along the PM line in the Japan Sea. Bull. Jap. Sea Reg. Fish. Res. Lab. 37:13–19.

Nishimura, S. 1965. The zoogeographical aspect of the Japan Sea, Part II. Publs. Seto Mar. Biol. Lab. 13:81–101.

Nishimura, S. 1968. The zoogeographical aspect of the Japan Sea, Part IV. Publs. Seto Mar. Biol. Lab. 15:329–352.

Nishimura, S. 1969. The zoogeographical aspect of the Japan Sea, Part V. Publs. Seto Mar. Biol. Lab. 17:67–142.

Oba, T., Kato, M., Kitazato, H., Koizumi, I., Omura, A., Sakai, T., and Takayama, T. 1991. Paleoenvironmental changes in the Japan Sea during the last 85,000 years. Paleoceanography 6:499–518.

Ohwada, M., and Ogawa, F. 1966. Plankton in the Japan Sea. Oceanogr. Mag. 18:39–42.

Sekiguchi, H. 1975. Seasonal and ontogenetic vertical migrations of some common copepods in the northern region of the North Pacific. Bull. Fac. Mie Univ. 2:29–38.

Sullivan, B. K. 1980. In situ feeding behavior of *Sagitta elegans* and *Eukrohnia hamata* (Chaetognatha) in relation to the vertical distribution and abundance of prey at ocean station P. Limnol. Oceanogr. 25:317–326.

Terazaki, M. 1992. Horizontal and vertical distribution of chaetognaths in a Kuroshio Warm-Core Ring. Deep-Sea Res. 39(Suppl. 1):231–245.

Terazaki, M. 1993a. Deep sea adaptation of epipelagic chaetognatha *Sagitta elegans* in the Japan Sea. Mar. Ecol. Progr. Ser. 98:79–88.

Terazaki, M. 1993b. Seasonal variation and life history of the pelagic chaetognatha *Sagitta elegans* Verrill in Toyama Bay, southern Japan Sea. J. Plankton Res. 15:703–714.

Terazaki, M., and Miller, C. B. 1986. Life history and vertical distribution of pelagic chaetognaths at ocean station P in the subarctic Pacific. Deep-Sea Res. 33:323–337.

Terazaki, M., and Wada, M. 1988. Occurrence of large numbers of carcasses of the large, grazing copepod *Calanus cristatus* from the Japan Sea. Mar. Biol. 97:177–183.

Uschakov, P. V. 1955. Polychaetous Annelides of the Far Eastern Seas of the USSR. Moscow and Leningrad: Izd. AN. USSR.

Zenkevitch, L. A. 1963. Biology of the Seas of the USSR. London: Allen and Unwin.

9

Changes in the Biomass of the East China Sea Ecosystem

Ya-Qu Chen and Xin-Qiang Shen

The East China Sea (ECS) is located in the northwest Pacific Ocean and covers the area from the Taiwan Strait north to the Yellow Sea, east toward the Japanese mainland and Island, and west to mainland China. The ECS is a vast, shallow sea of 770,000 km^2 that includes the Zhoushan, Wentai, and Changjiang estuaries and several important offshore fishing grounds. The East China Sea ecosystem is one of the most intensively exploited large marine ecosystems in the world. During the past 30 years, large-scale changes in biomass yields have occurred, largely through the influence of heavy fishing mortality, resulting in a shift from a fishery based mainly on highly valued demersal species to fast-growing, smaller species such as shrimp and cephalopods.

Climatology and Physical Oceanography

Climatology

The most obvious climatic characteristic over the East China Sea is the alternation of the winter and summer monsoons. The northern wind usually prevails from September to May, whereas the southern wind is dominant from June to August. The yearly average wind velocity is 6.6–8.8 ms^{-1} along the western coast of the East China Sea. The average wind speed is greatest ($>8.0\ ms^{-1}$) over the Taiwan Strait. Because of the effect of cold air, the wind speed increases to 7–8 ms^{-1} over the coast and is above 10 ms^{-1} over the Taiwan Strait in autumn and winter. The mean wind velocity is lower (about 6–7 ms^{-1}) in summer. Winds exceed gale force about 62 to 88 days a year over the East China Sea and 120 days a year over the Taiwan Strait (Marine Fishery Resources 1987).

Temperature

The annual mean air temperature over the East China Sea is about 15–20°C; the temperature distribution is high in the southern part and low in the northern part. The minimum temperature occurs in winter and ranges from 7° to 21°C. The maximum temperature occurs in summer and ranges from 26° to 29°C. Spring is a period of increasing temperature, ranging from 12° to 20°C. Autumn is a period of declining temperature, ranging between 9° and 24°C.

Rainfall

Rainfall is high in the East China Sea, usually about 1,000 mm yr^{-1}. The highest rainfall is in the middle coast of Zhejiang (about 1,200 mm yr^{-1}). The precipitation is mainly concentrated from April to September (70% of the annual precipitation), with May and June the months of peak rainfall.

Typhoons

The weather systems affecting the East China Sea are mainly cold waves, typhoons, and cyclones. The cold waves have an important influence on the East China Sea from September to May, with the major influence of the northern wind acting from November to February. The cold waves affecting the East China Sea influence the northern part more than the southern part. Typhoons may intrude into the East China Sea from May to November, but they pass through principally in July and August (usually five events per season).

Cyclones

The cyclone is one of the more common weather systems over the East China Sea. The cyclones affecting the ECS are mainly of the Donghai and Changjiang types. On average, 24 Donghai cyclones are produced each year. They mainly occur in winter–spring and are usually produced on the sea surface of the northeastern area off Taiwan and directly influence the East China Sea. There are, on average, 15 Changjiang cyclones each year. They mainly occur in spring and summer. After a Changjiang cyclone moves out to sea, it exerts influence on the northern part of the East China Sea and the southern Yellow Sea (Marine Fishery Resources 1987).

Currents

The currents in the East China Sea are mainly composed of the Kuroshio Current and the coastal current. The Kuroshio Current moves to the northeast. A coastal current flows southward along the eastern shore of China, constituting a counterclockwise circulation system (Figure 9-1).

After entering the East China Sea from the south, the Kuroshio Current produces a branch, the Taiwan Warm Current, in the northeast part of the ECS off Taiwan Island. The branch moves to the north along the eastern coast. The major flow moves to the northeast along the continental slope, where it again produces a branch, the Tsushima Warm Current, which flows to the north along the western coast of Kyushu. In the southwestern area of Kyushu off Japan, the major flow turns to the east and comes back to the Pacific Ocean through the Totara Strait. The Tsushima Warm Current is divided into two parts in the southeastern area off Cheju Island. The major portion passes the Korea Strait and enters the Sea of Japan, and the other part, the Yellow Sea Warm Current, moves to the northwest and enters the Yellow Sea. In addition, the Kuroshio Current produces a branch near Bass Strait, the West Branch of Kuroshio, which flows into the Taiwan Strait.

When the Kuroshio Current moves along the continental slope, the position of the mean current axis is between the 200 m and 1,000 m isobaths. The temperature of the surface current axis is usually close to 16°C at the 200 m layer. The mean flux of the Kuroshio Current in the East

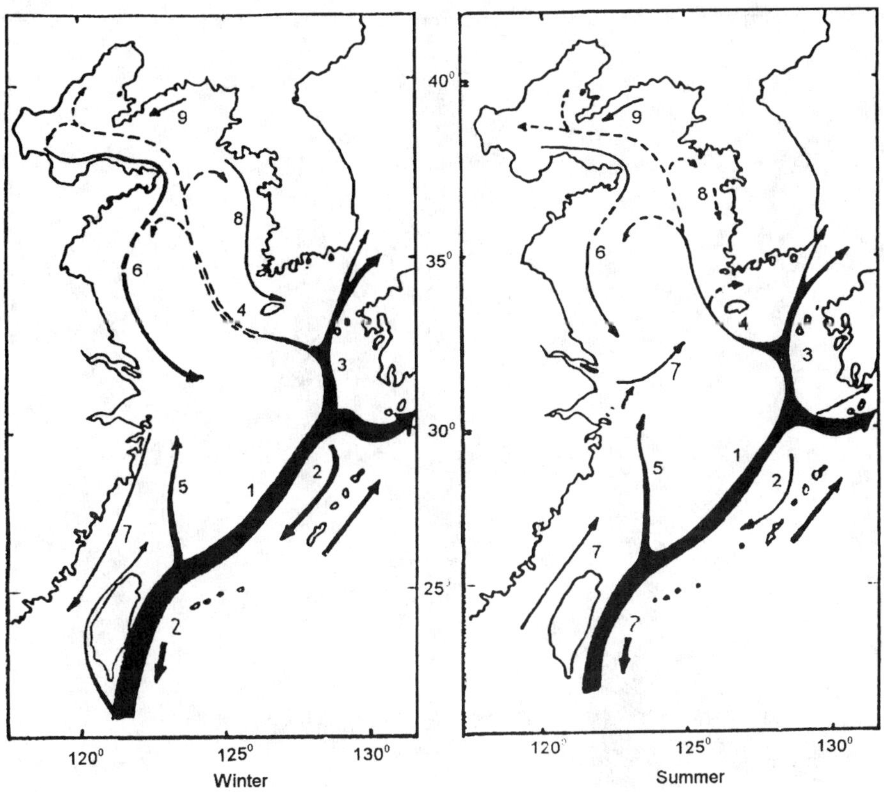

1. Kuroshio
2. Kuroshio (anticurrent)
3. Tsushima Current
4. Yellow Sea Warm Current
5. Taiwan Warm Current
6. Yellow Sea Coastal Current
7. East China Sea Coastal Current
8. West Korean Coastal Current
9. Liaodong Coastal Current

Figure 9-1. East China Sea currents in winter (left) and summer (right).

China Sea is about 35×10^6 m^3 s^{-1}, which is equal to over 1,000 times the mean annual runoff of the Changjiang River. The main flow of the Kuroshio undergoes a periodic change on a 7- to 9-year cycle. The velocity is about 2 to 3 knots and the current width is narrower. The area occupied by the Kuroshio is characterized by high temperature and salinity, with little variation in either. The surface of the Taiwan Warm Current is easily influenced by monsoons, and the flow direction at the middle and lower layers is perennially to the north. The velocity and flux of the Tsushima Warm Current are greater in summer than in winter, but the Yellow Sea Warm Current is stronger in winter than in summer, when the flow takes a major change in direction (Dou et al. 1979).

The East China Sea Coastal Current is formed by the combined waters of the Changjiang River, Qiangjiang River, and other rivers. It is easily influenced by monsoons and its direction and velocity have a clearly seasonal variation. In summer, it flows northeast near Cheju Island in the years when the runoff is great. In winter, the flow is to the south along the shore. The northern area of the East China Sea is influenced by the Yellow Sea Coastal Current in winter.

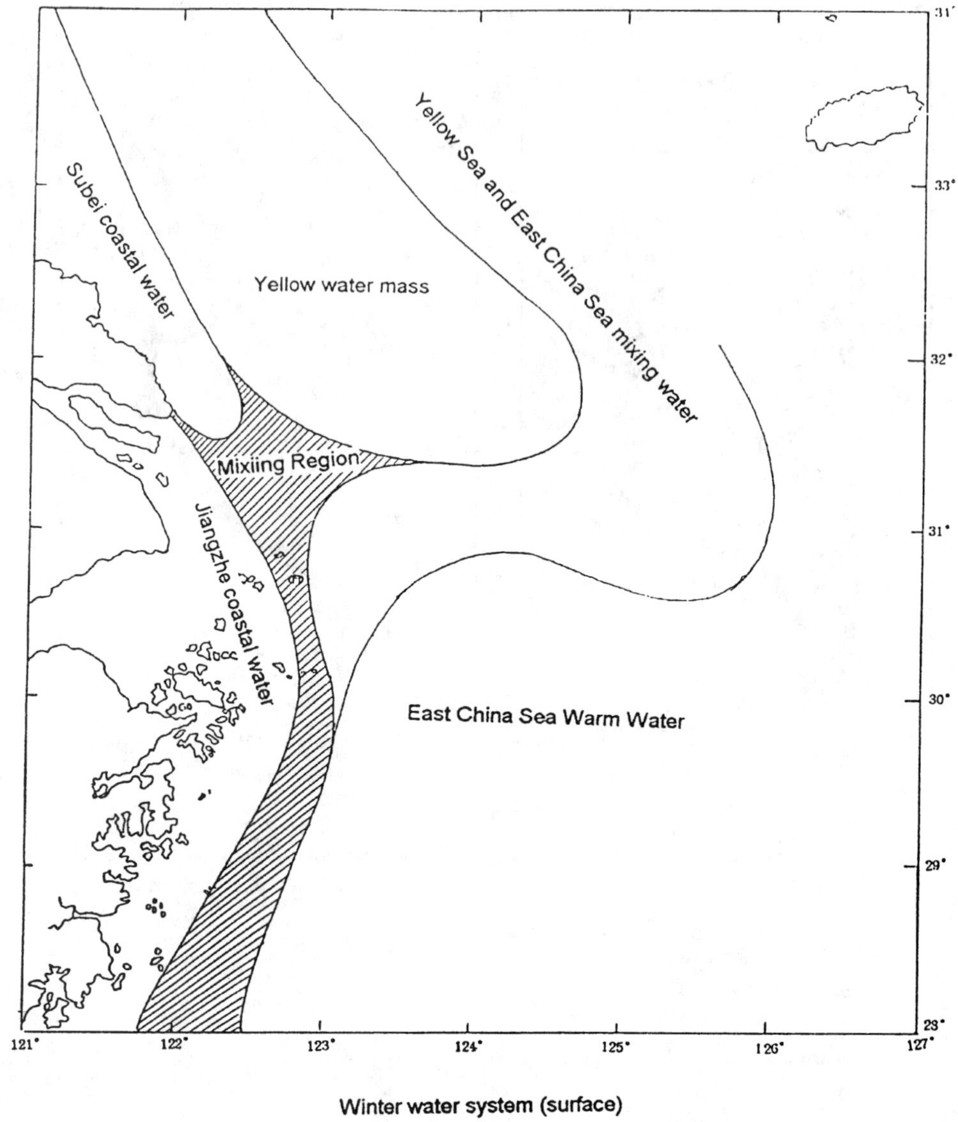

Figure 9-2. Water masses of the East China Sea in (A) winter

Water Masses

The water masses in the East China Sea are mainly composed of Kuroshio water, coastal water, and mixed water formed by the mixing of Kuroshio and coastal water masses (Figure 9-2). Kuroshio water is primarily found in the Kuroshio area of the East China Sea. It can be divided into four temperature and salinity regimes (Yang 1984a). The first is ECS surface water with high temperature (mean 25.0°C) and high salinity (mean 34.3‰). It occurs between 0 and 100 m and is easily influenced by meteorological conditions and adjacent water masses. The hydrological elements change with the season. The second is ECS subsurface water with relatively low temperature (mean 19.5°C) and high salinity (mean 34.8‰). It occurs between 100 and 400 m. In summer, it is upwelled along the slope of the sea bottom and expands west. The third is the ECS

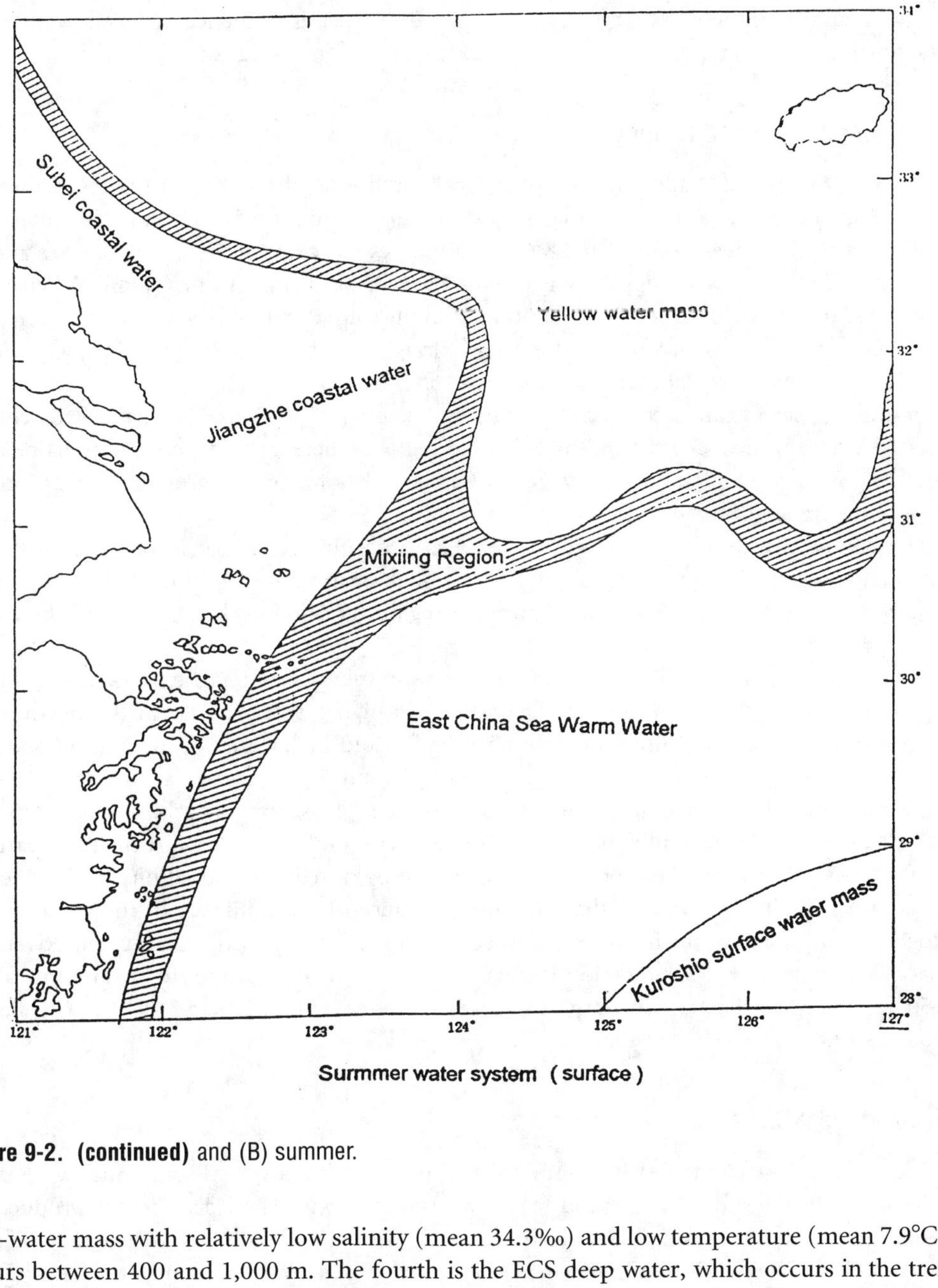

B Surmmer water system (surface)

Figure 9-2. (continued) and (B) summer.

mid-water mass with relatively low salinity (mean 34.3‰) and low temperature (mean 7.9°C). It occurs between 400 and 1,000 m. The fourth is the ECS deep water, which occurs in the trench (below 1,000 m). The temperature is lower and the salinity higher than the mid-water layer (Yang 1984b).

Coastal water is formed by mixing of river runoff and seawater. Its characteristics are mainly influenced by runoff. Coastal water mainly occurs in the shallow area of the western part of the East China Sea and expands to the northeast in summer. The salinity is low (<31.0‰), and the temperature clearly changes with the season. The annual range is 16–23°C at the surface and 8–22°C at the bottom (Dou et al. 1979).

Mixed water is formed by modification of East China Sea water as it spreads to the north. It is located between the Yellow Sea and the East China Sea, with intermediate temperature and salinity. In addition, the northern area of the East China Sea is influenced by Yellow Sea water masses in autumn–winter and by the Yellow Sea cold water in spring–summer.

Water Temperature and Salinity

In response to seasonal changes in meteorological conditions, the surface water temperature of the East China Sea is at a maximum in August and at a minimum in February. In August, the mean temperature ranges from 24° to 28°C along the western coast of the East China Sea and can be over 28°C in the southern region. In February, the range is from 5° to 14°C along the coast and can be 20°C in the southern region. The isothermal lines are distributed southwest to northeast. The greatest annual range of temperature occurs along the western coast (13–22°C), whereas the annual range in the Kuroshio region is small (6–11°C).

In summer, stratification commonly occurs in the upper 20 m. Below 20 m, the water column is relatively homogeneous except in the Kuroshio region, where a permanent thermocline exists below 100 m. In winter, the temperature shows vertical homogeneity from surface to bottom except in the trench of the southeast.

The salinity distribution in the East China Sea is dominated by Kuroshio water with high salinity and coastal water with low salinity. The annual mean salinity is about 33.0‰ in the East China Sea, higher than 34.0‰ in the Kuroshio region, and lower than 31.0‰ in the coastal region.

The salinity is generally higher in winter than in summer. In winter, the trend of isohalines is roughly from southwest to northeast. The salinity is higher than 33.0‰ in the northeastern region, lower than 32.0‰ in the northwestern region, and higher than 34.0‰ in the southern region. In summer, the minimum value of salinity (about 5.0‰) occurs in the estuary of the Changjiang River. The horizontal gradient is largest there. The salinity distribution between 30°N and 34°N, west of 125°E, is influenced by coastal waters with low salinity flowing northeast, but the affected depth is limited to above 10 m. It should be noted that the salinity in the Zhejiang coastal area is higher in summer than in winter because of the influence of the coastal water. Haloclines commonly occur in summer, especially in the estuary of the Changjiang River. The surface salinity in the Kuroshio region is usually higher than 34.0‰, and the maximum value is at 150 m (>34.5‰). The salinity in the Taiwan Strait ranges from 32‰ to 34‰ (Marine Fishery Resources 1987).

Tides and Tidal Currents

The tide of the East China Sea is mainly formed by spreading of the Pacific tidal wave under the influence of the shore outline and sea bottom topography. The tides directly produced by celestial gravitation in this area are relatively small (Dou et al. 1979). It is generally believed that the tide is irregularly semidiurnal in the east between Taiwan Island and Cheju Island and regularly semidiurnal in the western region. Tidal difference is greater in the west than in the east. The maximum tidal difference can reach 8.9 m in the Ganpu of the Hangzhou Bay.

The characteristics of the tidal currents are similar to those of the tides: an irregular semidiurnal tidal current in the east and a regular semidiurnal tidal current in the west. The velocity is larger in the west than in the east. The strongest tidal current appears in Hangzhou Bay and near Zhoushan Archipelago. In addition, the current is tidal offshore and rotary in the open sea.

Pelagic Ecosystem Analysis

Chlorophyll *a* and Primary Production

Based on data obtained during the period from August 1984 to May 1985, the mean value of chlorophyll *a* west of 127°E in the East China Sea is 0.709 mg m^{-3} yr^{-1}. The mean values are 0.959, 0.782, 0.574, and 0.250 mg m^{-3} yr^{-1} in spring, summer, autumn, and winter, respectively. The mean value of primary production is 602 mgC m^{-2} d^{-1}.

Phytoplankton

SPECIES COMPOSITION. A total of 209 species of pelagic phytoplankton have been identified, which includes 168 diatomic species, 32 dinoflagellate species, 2 chlorophyta species, and 1 cyanacea species.

SEASONAL VARIATION IN ABUNDANCE. The mean annual value is 11.2×10^6 indv m^{-3}, the highest of any area in the China Sea. It is much higher than that of the Bohai Sea or the Yellow Sea. The range of variation is also highest in the East China Sea. Two peaks appear in the annual cycle. The highest peak appears in summer (July to September), and a secondary peak appears in spring (April).

QUANTITATIVE DISTRIBUTION. The highest value of phytoplankton biomass in the East China Sea is found west of 124°E. Values are high in the Changjiang estuarine waters and coastal waters of Zhejiang and Fujian Provinces. The regions with high values are the Changjiang Estuary fishing ground, Zhoushan fishing ground, and Yushan fishing ground. In general, the occurrence of phytoplankton peaks coincides with the feeding periods of young, economically valuable fish.

KEY SPECIES. *Nitzchia fusiformis*, *Coscinodiscus* spp., *Skeletonema costatum*, and *Rhizosolenia* spp. are key species. *Nitzchia fusiformis* is a eurythermal and euryhaline species distributed mainly in the Changjiang Estuary, Qiangtonjiang Estuary, and the convergence zone of different water masses in the coastal water of Zhejiang Province; its abundance peaks in summer at $6{,}300 \times 10^3$ indv m^{-3}.

Coscinodiscus species appear during the whole year in the East China Sea. The dominant species of this genus are *C. thorii* and *C. excentricus.* They are distributed mainly in estuarine areas and the convergence zones of coastal waters.

Skeletonema costatum is mainly distributed in the Changjiang Estuary and the coastal waters of Zhejiang Province. Its highest abundance is 5.35×10^6 indv m^{-3} in August. A serious red tide occurs from May to July in Sanmen Bay and Louqing Bay (Zhejiang Province and in northern Meichow Bay and Dayu Harbor, Fujian Province). *Skeletonema* is an estuarine and coastal species, tolerant of a wide range of temperature (5–8°C) and salinity (7–30‰). The apparent optima are 24–30°C and 20–30‰. It is also abundant in organically polluted water (COD >3 mg m^{-3}) and may be a local indicator species for eutrophication.

Zooplankton

About 600 species have been identified, including pelagic crustacea (e.g., Euphausiacea, Copepoda, Mysidacea, Amphipoda, cnidarians, and chaetognaths).

SEASONAL VARIATION OF ZOOPLANKTON BIOMASS. The plankton biomass in the East China Sea (west of lat. 127°E) is high compared with the Bohai Sea, Yellow Sea, and South China Sea. Its mean biomass is 99.8 mg m^{-3}, with high values appearing in the period from April to November. The peak is in August (188 mg m^{-3}); in winter the biomass is much lower. Fluctuations in zooplankton biomass are controlled by water temperature, runoff, and phytoplankton.

QUANTITATIVE DISTRIBUTION OF ZOOPLANKTON. The quantitative distribution of zooplankton is controlled by the different water masses, such as coastal water and the cold-water mass of the Yellow Sea and the Kuroshio. Most high zooplankton density zones are located at convergence zones. The quantitative distribution of zooplankton changes with the season.

Biomass values are lowest in winter (December to February). The area with the highest value is not obvious. The distribution trend is toward higher values in the south. The biomass is lower than 100 mg m^{-3}, except for some fishing grounds such as the Yushan and Wentai fishing grounds.

In spring (March to May), biomass increases sharply after March with increasing water temperature and runoff. Biomass exceeds 100 mg m^{-3} in most waters and fishing grounds such as Yushan, Zhoushan, and Changjiang. During this period, some key species reproduce very quickly, such as *Calanus sinicus* and *Euphausia pacifica.* The area of high biomass moves from the southern zone to the northern zone under the influence of warm currents such as the Kuroshio and Taiwan Currents. The convergence zone is located at the Changjiang Estuary fishing ground. The biomass value in the Zhoushan fishing ground is between 100 and 250 mg m^{-3}. In May the high value zones are the coastal waters of Zhejiang Province, the Changjiang Estuary, and off the Changjiang fishing ground, with biomass ranging between 250 and 500 mg m^{-3}. Biomass could reach 500 to 1,000 mg m^{-3} in convergence areas, but in the Kuroshio area biomass remains low.

In summer (June to August), the distribution trend is similar to that in spring. The high biomass zones remain in the Changjiang Estuary, off the Changjiang fishing ground, and the coastal water of Zhejiang Province. After July, *Calanus sinicus* is reduced in abundance due to increased water temperatures, but *Sagitta nagae* increases, so the total biomass remains stable. In August, in the estuarine area of both the Changjiang River and Quingtan River, biomass values increase because of the large number of reproducing *Labidocera euchaeta.*

In autumn (September to November), biomass values start to be reduced in the Changjiang River estuarine area. The high biomass zone diminishes and disappears in October.

KEY SPECIES. *Calanus sinicus, Labidocera euchaeta, Tortanus vermiculus, Undinula vulgaris, Euchaeta concinna, Euphausia pacifica, Pseudoeuphausia sinica, Sagitta enflata,* and *Sagitta nagae* are the key species of zooplankton in the East China Sea.

Calanus sinicus is one of the most important zooplankton in the China Sea from the Bohai Sea to the South China Sea. Its center of distribution (highest concentration) is located in the Yellow Sea and northwestern part of the East China Sea. It is a dominant species in the East China Sea and is distributed on both sides of southern Japan at 40°N, west of 124°E. The average abundance is 60 indv m^{-3}. Based on survey data, the species produces three generations a year in the Zhoushan and Changjiang fishing areas offshore of Hangzhou Bay. April is the most important period for reproduction. Its temperature range is from 9.5° to 19.5°C, but 11.5–17°C may be the optimal temperature for this species. After July, its quantity is reduced in the east until September due to increasing temperatures and predation by large numbers of fish and other predators. In January and February, it is rare. The highest concentrations are located in the Zhoushan fishing area, where in May the maximum density is about 250–500 indv m^{-3}. High concentrations are also found in the Changjiang fishing area, where it is an important prey species for fish. Concentrations remain high in this fishing area for about 3 months in summer. *Calanus sinicus* is very important for supporting the food base of fish predators in the Zhoushan and

Changjiang fishing areas and in Hangzhou Bay. In November, the species is distributed in waters of salinity of 31.0–33.5‰, but sometimes the species can be found in the brackish water of the estuarine areas and Hangzhou Bay.

Labidocera euchaeta inhabits coastal and estuarine areas where the salinity is less than 30‰. It is mainly distributed in Hangzhou Bay and the Xiangshang and Changjiang estuarine areas. The species is used as an indicator of Changjiang River runoff. Although the species can be found in the East China Sea year round, the peak abundance appears in July and August, with a second peak in May or November. During winter the species is rare. The highest concentrations are located in the estuaries of Changjiang and Hangzhou Bays. In July, the density is about 150 indv m^{-3}. In August, its quantity increases to 944 indv m^{-3} in the Xiangshang area. This species is very sensitive to salinity and has the characteristics of a coastal and low-salinity species.

Tortanus vermiculus is an estuarine species that is dominant in Hangzhou Bay, the Changjiang Estuary, and in the shallow waters of the Zhoushan fishing area. It is found in these areas year round, but its quantity changes temporally and spatially. In the Changjiang Estuary, its abundance in winter and early spring is on average 90 indv m^{-3} in March. In the estuarine area of the southern branch of Changjiang River, its density is high (400–3,000 indv m^{-3}). It is found in Hangzhou Bay as a year round inhabitant. Its average density is 84, 376, 160, and 145 indv m^{-3} in spring (March), summer (July), autumn (September), and winter (January), respectively. It is very rare in Xiangshang, where *Tortanus derjuyini* is always dominant. *T. vermiculus* can be used as an indicator species of runoff in the Zhoushan fishing area and in estuarine waters.

Undinula vulgaris is a tropical species and is an important prey species for fish in the East China Sea. It is distributed mainly in the warm currents where the temperature is over 15°C and salinity is over 33‰. In May it is distributed near 29°N. In August it spreads to near 33°N off the Dasha fishing ground. After November its abundance is reduced greatly and it occurs in the southern area near 29°N.

Euchaeta concinna is a key warm-water, tropical species that is distributed mainly in the South China Sea and the East China Sea. In winter, it occurs mainly south of 32°N. It is rare in this area except for the Yushan fishing ground, where its abundance level is about 6–8 indv m^{-3}. Beginning in May, it is carried to the north with the warm current. In August it increases greatly (5–17 indv m^{-3}) to form a high concentration zone through the entire East China Sea.

Euphausia pacifica is a temperate species that occurs in the North Pacific from January to April, north of 29°N and east of 122°E, where salinity is over 32.5‰. During the period from May to July, it is associated with the cold water of the Yellow Sea. In summer, it almost disappears in the East China Sea, then, after November, extends south to 29°N again. It increases in abundance with increasing advection of cold water, reaching a peak in December. After January its abundance decreases.

Pseudoeuphausia sinica is a local species distributed in the East China Sea mainly in the coastal waters of Jiangsu and Zhejiang Provinces, the estuarine area of Changjiang, and Hangzhou Bay. There is a close relationship between its distribution and salinity and temperature. In general, it occurs mainly in waters where the salinity range is between 31‰ and 33‰ and temperature is between 5° and 29°C. Its abundance is very low from January to April. The peak appears in late summer (September), with high values in waters of the Changjiang estuarine area (15–25 indv m^{-3}). In December, abundance levels decline to below 10 indv m^{-3}.

Sagitta enflata is a key species of chaetognath in the East China Sea and is mainly distributed off Taiwan in the Kuroshio Warm Current where the salinity is higher than 32‰. In summer, it occurs in most areas in the East China Sea where temperature is higher than 26°C and salinity is higher than 33‰. During August and September, high concentrations (100 indv m^{-3}) appear off the Changjiang River estuarine area. The highest value reaches 390 indv m^{-3} wherever fishing grounds of large and small yellow croaker, hairtail, and others are located. During October to December, its abundance declines sharply, and in winter its occurrence is rare.

Sagitta nagae is a temperate and coastal species and is less abundant than the chaetognath *S. enflata*. Seasonal variability of *S. nagae* in the East China Sea is similar to that of *S. enflata*. In January and February, abundance levels are between 5 and 10 indv m^{-3}. In April, abundance increases sharply in the Changjiang estuarine area and along the coastal waters of Zhejiang Province, where there is a convergence zone from May to July. After August its abundance is reduced and is lowest in winter. The relationship between temperature and abundance is significant, with the most favorable temperature being between 14° and 24°C.

Communities of Plankton in the East China Sea and South Yellow Sea

The distribution of plankton in these two seas is characterized by three communities, each with its dominant species and each occupying a definite water body.

The first community of temperate species occurs in the central area of the southern Yellow Sea and extends southward in the autumn and winter. The second community is restricted to along the Jiangsu and Zhejiang Provinces, where the water is characterized by lower salinity and temperature. These species belong to the neritic group. The third community of tropical and subtropical species chiefly inhabits the offshore area of the southeastern East China Sea, where the water is warm and highly saline. There is a transitional zone where the temperate species dominate in winter and spring and the tropical and subtropical species dominate in summer and autumn.

Red Tides in the East China Sea

The East China Sea receives a large amount of runoff from the many rivers draining mainland China, including the Changjiang (Yangtze) River, one of the longest rivers in the world. Its average annual runoff of almost 10^{12} m^3 per year delivers abundant nutrients into a low-nutrient area and affects the composition, distribution, and dynamics of the phytoplankton population. Eastern China is an important economic area whose rapid economic development and growing population have led to large increases in waste water and sewage discharge into the Changjiang River and East China Sea. This has led in recent years to serious eutrophication problems and frequent red tides that are harmful to the ecosystem, fisheries resources, public health, and especially the mariculture in this region.

The occurrence of toxic blooms in recent years has attracted the attention and concern of scientists, fisheries managers, and public health agencies. The high N and P levels of recent years may play a role in red tides. The occurrence of major red tide events has increased in frequency. In 1990 a record number occurred, causing significant economic losses in localized areas. Major red tides are largely confined to the summer months of June to October. But in Xiangshang Gulf, red tides occurred in winter (December to February). The red tides in July 1990 were fairly widespread for a short period in the Zhoushan Archipelago and coincided with the 25‰ isohaline. This isohaline also coincided roughly with a decrease in suspended sediments in several different water bodies, including the turbid northern Jiangsu/Zhejiang coastal water, East China Sea warm water, Kuroshio subsurface water, and Yellow Sea cold water, all of which influence the offshore mixed water mass in different ways and at different times of the year.

Fisheries Resources: Variability and Species Shifts

A total of about 694 species have been identified in the East China Sea. About 200 species (including members of cephalopoda and crustacea) are commercially harvested. The catch of most species is relatively small. The annual catch of key species that are generally considered of commercial importance is shown in Table 9-1.

The annual catch of 25 of these species exceeds 10,000 metric tons (mt). Hairtail (*Trichiurus haumela*) is the most important commercially fished demersal species in the East China Sea, where the catch is 0.4–0.5 million metric tons (mmt) annually. The catch of filefish is 0.3–0.4 mmt; the catch of scad and shrimps, 0.2–0.3 mmt; the catch of small yellow croaker, large yellow croaker, and shrimp mackerel, 0.1–0.2 mmt; and the catch of cuttlefish, cephalopods, and jellyfish, 0.05–0.1 mmt. The demersal species are the major component of the ecosystem and account for 60–85% of the annual total catch.

Chinese fishing activity in the East China Sea has been carried out for a long period, but developed slowly until the 1950s. In 1992, there were more than 90,000 motorized fishing boats supplying 3.85 million kW, with the catch reaching 2.78 mmt. Commercial utilization of fisheries resources was also carried out by Korean, Japanese, and Taiwanese fishing boats.

Due to overexploitation and pollution, the dominant species of the commercial fisheries and major fish stocks such as large yellow croaker, small yellow croaker, jellyfish, and other species have declined sharply. On the other hand, the catch of some low-valued stocks such as filefish, shrimps, and crabs has increased rapidly (Figure 9-3).

Table 9-1. Key species in the fisheries of the East China Sea

Common name	Scientific name
Demersal species	
Hairtail	*Trichiurus haumela*
Small yellow croaker	*Pseudosciaena polyactis*
Large yellow croaker	*Pseudosciaena croacer*
Flatfish	*Pleuronectidae* (mostly *Cleisthenes herzensterini*)
Filefish	*Havodon septentrionalis*
Pomfret	*Pampus argenteus*
Cuttlefish	*Sepiella maindroni*
Crab	*Portunus trituberculatus*
Shrimps	Mostly *Parapenaeopsis hardwickii*
Palaemon gravieri	*Trachy penaeus curvirostris*
Pelagic species	
Spanish mackerel	*Scomberomorus niphonius*
Cuttlefish	*Stromateoides argenteus*
Chinese herring	*Ilisha elongata*
Jellyfish	*Rhopelema hispidim*
Anchovy	*Setipinna*
Mackerel	*Pneumatophorus japonicus*
Scad	*Decapterus maruadsi*
Herring	*Etrumeus micropus*
Sardinella	*Sardinella aurita*

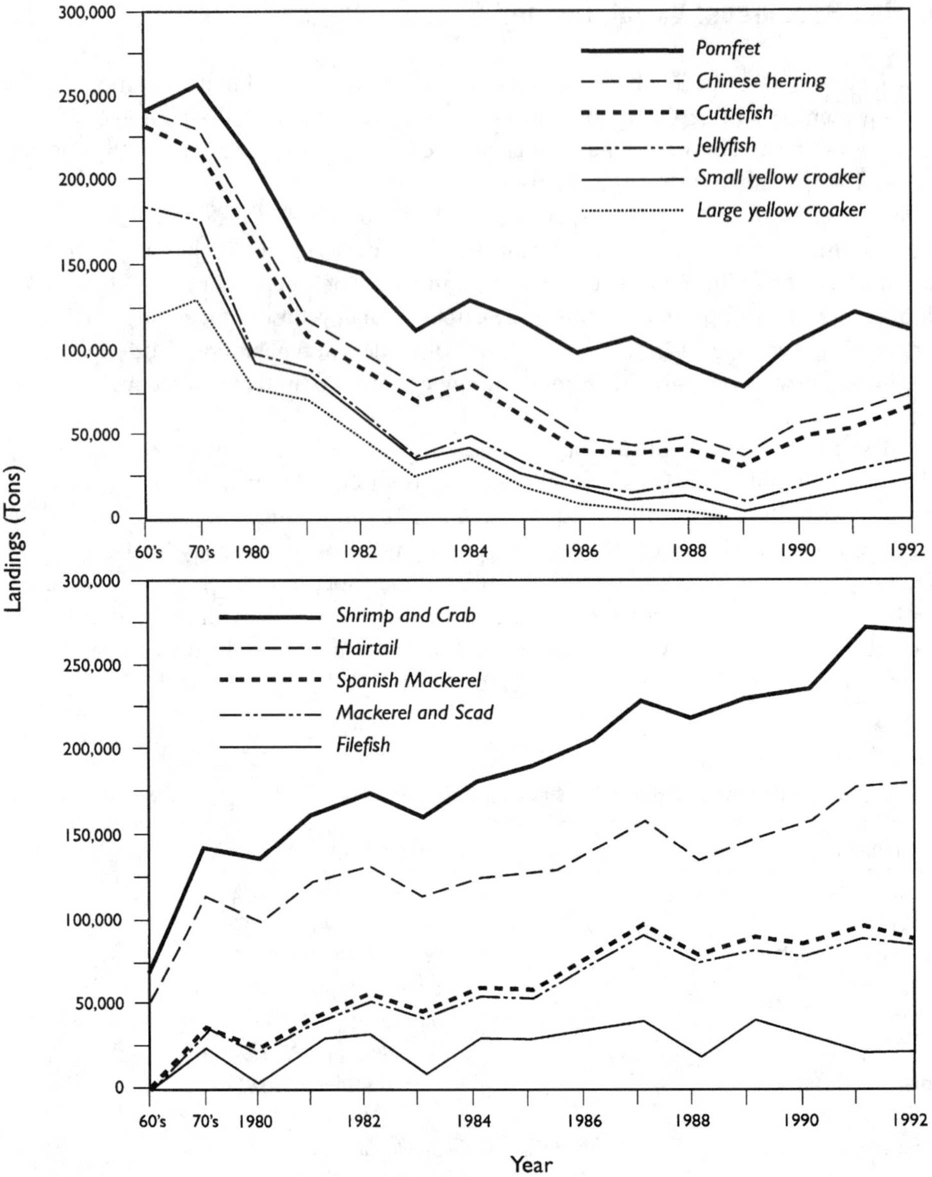

Figure 9-3. East China Sea landings, illustrating change over time (1960s to 1992) in the quantity of species caught.

Table 9-2. Catch by net type (metric tons) in 1992

Marine fishing catch	Trawl	Purse	Draft	Hocu	Set
2,781,750	1,084,273	242,593	180,989	30,830	68,338

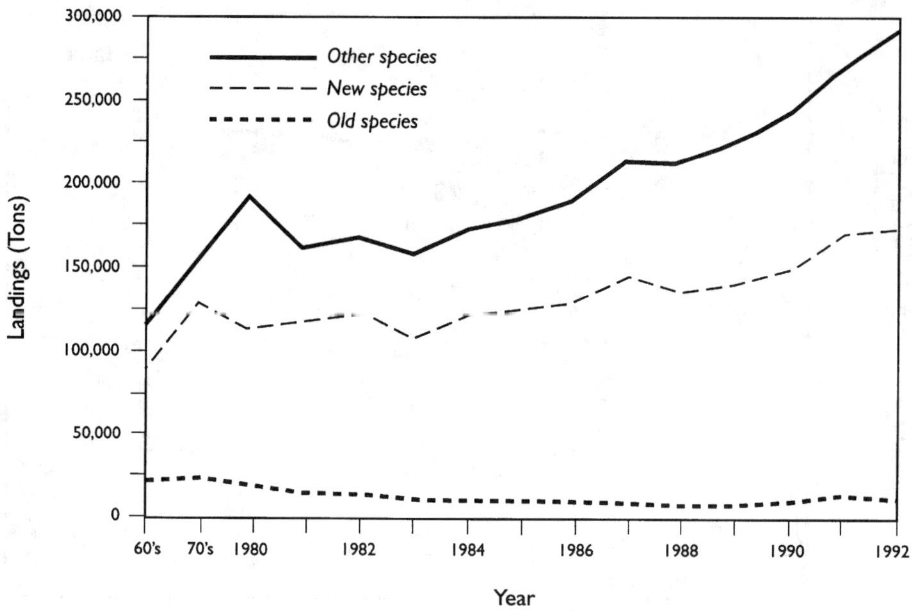

Figure 9-4. East China Sea fishery: old species, new species, and other species.

In 1992, the total catch of aquatic production in the East China Sea was 2.78 mmt. Of the total catch, 2.12 mmt were fish, 0.526 mmt were shrimps and crabs, 4,477 mt were jellyfish, 111,518 mt were shellfish, and 1,253 mt were algae. Catches by net type are shown in Table 9-2.

Status of Landings

Total landings have increased by over 300% (Figure 9-4), from about 893,000 mt in the 1960s to 2,780,000 mt in 1992. The landings of older, traditionally fished, more valuable species have fallen sharply. Landings of less valuable species have increased and more than compensated for the drop in landings of the old species. An increasing amount of other species was also landed, rising from 123,000 mt in the 1960s to 265,000 mt in 1992. All the old and new categories are used for human consumption, whereas large amounts of "other" species are used for feeding cultured fish or poultry.

Trends for total marine fish catch in the East China Sea (Table 9-3) increased gradually over the last 40 years. In the 1950s, the mean value of the total catch was 557,000 mt. In the 1990s it

Table 9-3. Mean catch for the East China Sea, by decade

Decade	Mean catch (10,000 mt)	Increasing rate
1950s	55.7	
1960s	89.4	+60.50
1970s	129.8	+45.18
1980s	172.1	+32.59
1990s (1990–1992)	254.3	+47.76

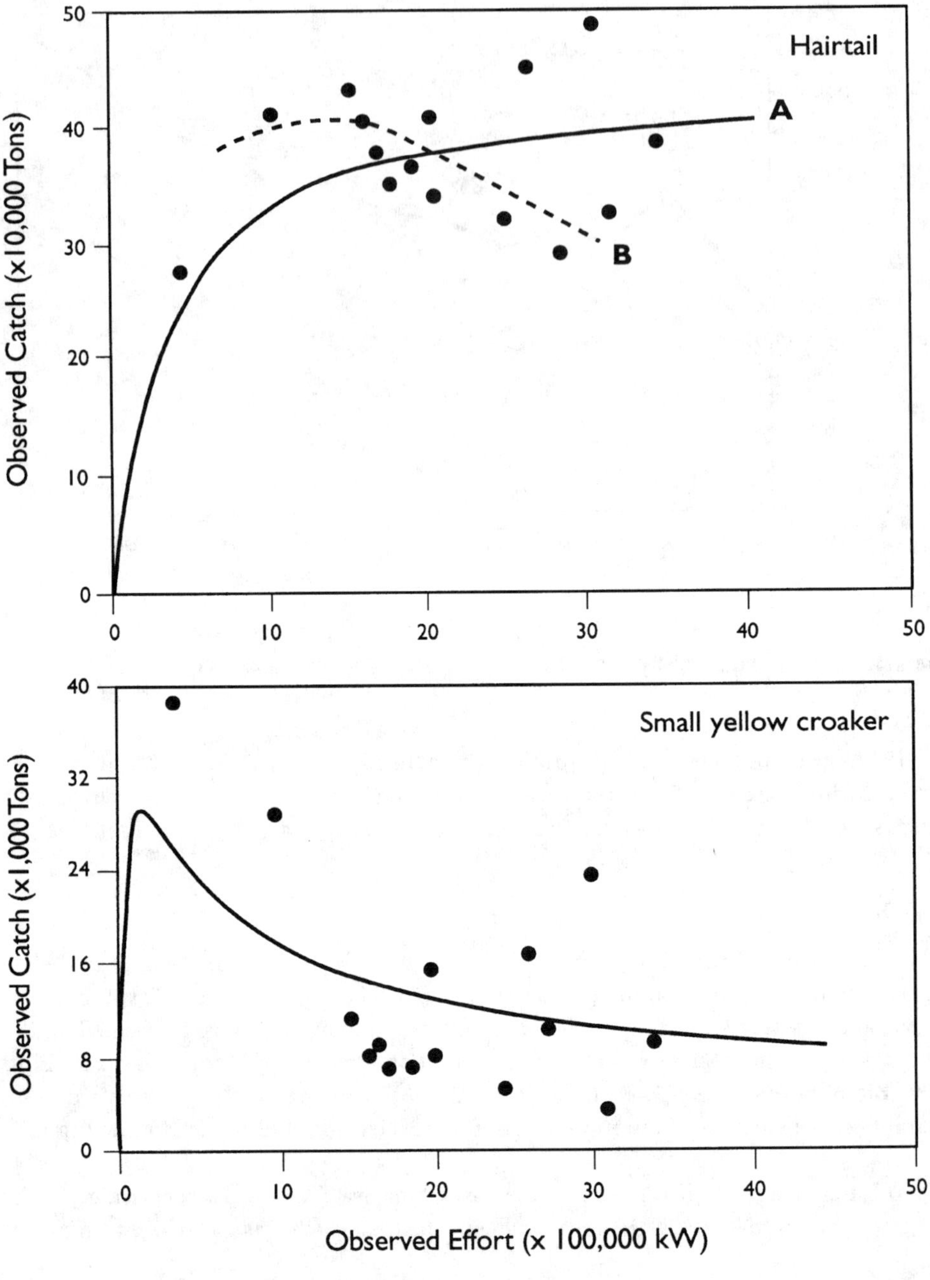

Figure 9-5. Total catch and effort for hairtail and small yellow croaker in the East China Sea.

reached 2,540,000 mt. The maximum rate of increase was 61% in the 1960s, after which it increased more slowly. In the 1990s, it reached 48% due to economic system reforms and changes in fishing operations that enabled the use of more kinds of fishing gear.

Figure 9-5 shows assessments of two important stocks. The *Trichiurus haumela* stock shows that sustainable (equilibrium) yield has leveled off if all the data are used in fitting the yield curve (Line A). The variation around the fitted line is likely due to fluctuating abundance, accessibility, and recruitment. Alternatively, a line has been fitted by eye (Line B) that shows that there was a decline in landings in the 1980s, followed by an increase in landings due to an attempt to manage the fishery.

Biological Assessment Using Effort and Catch

An assessment of East China Sea stocks as a whole (i.e., all species combined) and of the major stocks was carried out using a variety of Fox and Schaefer models. Figure 9-6 shows the results of the assessment for the East China Sea, treating the total landings as a single stock. The best fit obtained (Line A) shows a clearly increasing trend in estimated equilibrium yield as effort is increasing. No estimate of maximum sustainable yield (MSY) was possible because the increasing trend was so masked. A more conservative estimate was obtained by means of a line fitted by eye (Line B), and is used in the following bioeconomic assessment because it provides a more conservative and perhaps a more realistic estimate of stock condition.

CATCH PER UNIT EFFORT. The catch per unit effort (CPUE) of motorized fishing boats in the East China Sea declined except in 1961 and 1963. During 1957–1967, it was about 1.5 mt, but declined to about 1.0 mt during 1968–1978. After 1970, it continued to decline to 0.7–0.44 mt and was reduced to only a third of the catch obtained before 1967.

SPECIES COMPOSITION AND SIZE SHIFTS. Before the 1980s, the key dominant species in the East China Sea were hairtail, large yellow croaker, small yellow croaker, Chinese herring, pomfret, cuttlefish, and jellyfish. After 1980, the species composition changed gradually. The above old

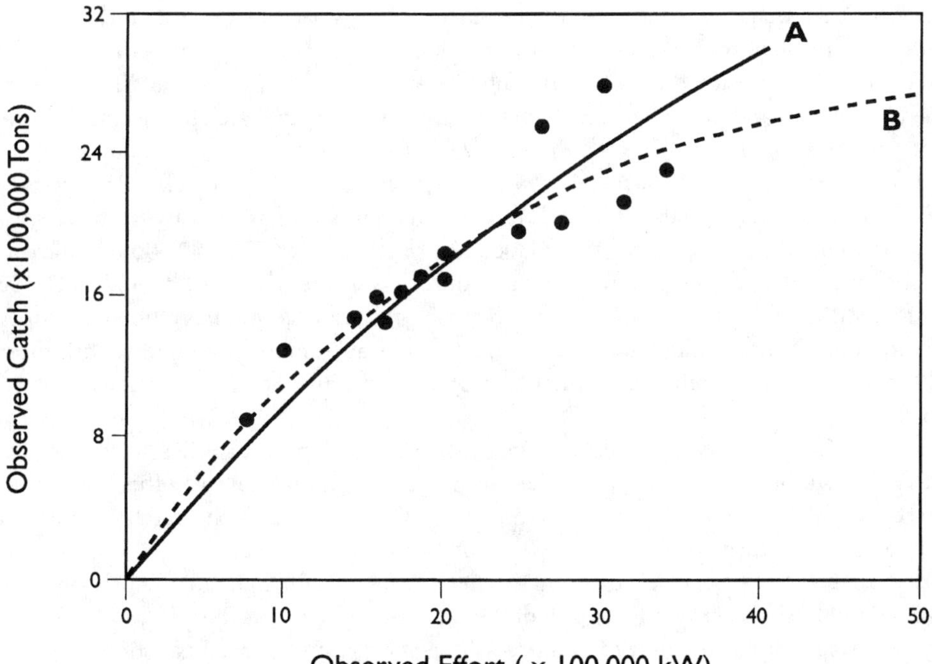

Figure 9-6. Total catch and effort in the East China Sea fisheries.

species declined sharply to 46–49% of the catches during 1990–1992. Also, the fish caught were smaller in size, younger, lower in value, and sexually mature at an earlier age.

KEY SPECIES. Hairtail (*Trichiurus haumela*) is the most commercially important demersal species in the East China Sea. Since the 1950s, the hairtail catch has been increasing sharply. For example, in 1956 the catch was only 100,000 mt, but in 1974 the catch ranged between 350,000 mt and 500,000 mt, representing about 35% of the total catch of marine fish in the East China Sea. In 1992, the catch reached about 492,000 mt.

Large yellow croaker (*Pseudosciena crocea*) is one of the most important commercially fished demersal species in the East China Sea. The catch in the ECS accounts for about 98% of the total catch of the whole China Sea. Before the 1970s, its fishing seasons and fishing grounds were very obvious. During the 1950s to the 1970s, the catch fluctuated between 86,000 and 190,000 mt; after 1974 the catch declined sharply. At the end of the 1980s and early 1990s, the catch had declined to 2,000 mt, only about 1% of the catch in 1974. In 1992, the catch remained at a very low level (2,000 mt). This is an example of the overexploitation of a commercially important species.

The spawning population of this species occurs in shallow, coastal water. Its main spawning grounds are located in the Luisi fishing ground in the southern part of Jiangsu Province, and the shallow waters of the Daiqu fishing ground in the northern part of Zhejiang Province.

The variability of the population of small yellow croaker (*Pseudosciaena polyactis*) was similar to *P. crocea*. During the 1950s and early 1960s, it was abundant, with the catch peaking at 100,000 mt in 1957. Since the end of the 1960s, its distributional range has also become smaller. In the early 1980s, the catch was reduced to less than 10,000 mt. On the other hand, its biological characteristics also have changed. Croaker now grows quickly and reaches sexual maturity early. For example, in 1959, when the length at maturity was 140–160 mm, the percentage of sexually mature fish at age 1 was less than 5%, but in 1981, the length at maturity decreased to 120–140 mm and the percentage of sexually mature fish increased to 40%. The growth rate also increased.

The catch of filefilsh (*Navodon septentrinalis*) increased rapidly beginning in 1974, reaching 192,700 mt in 1976 and exceeding 200,000 mt after the end of the 1980s. In 1987 and 1989, the catches reached 226,900 mt and 215,300 mt, respectively. In recent years the catch has declined sharply; in 1993 it was 13,000 mt. In 1994, it was reduced to 2,000 mt, indicating that the resource is depleted.

Resource exploitation of mackerel (*Pneumatophorus japonicus*) and scad (*Decapterus maruadsi*) began in the mid-1960s. The catch of mackerel and scad has increased rapidly since the 1970s, and in the early 1990s exceeded 300,000 mt. In 1992, it reached 339,000 mt (the catch of *Pneumatophorus* was 126,000 mt). Based on a resources assessment, the stock could reach 500,000 mt, with an MSY of 280,000 mt. Therefore, the mackerel resource still has a potential for future exploitation in the East China Sea. The scad resource is also increasing rapidly, with most fish caught in Fujian Province and the southern part of the East China Sea. It seems that scad fishery is still viable.

Pomfret (*Pampus argenteus* and *P. cinereus*) is also an important species economically. Due to the decline in the catch of yellow croaker, fishing efforts shifted toward other species such as pomfret (caught by nets such as the "setting net," trawl, etc.) and the catch of pomfret has increased since the early 1980s. In 1987, the total catch reached 62,500 mt—the highest ever recorded. In recent years, it reached about 50,000 mt, which is still high, but the fish are small and 1- or 2-year-old fish compose a large part of the catch. It appears that this fish resource is near to being overexploited; it is necessary to reduce fishing efforts on this species as soon as possible.

Chinese herring (*Ilisha elongata*) is one of the traditional commercially caught species in the ECS, mainly caught by drift net. Since 1964, the catch increased rapidly and in 1974 reached a maximum of 26,500 mt. After 1976, the catch has fluctuated around 10,000 mt and the percentage of young fish is increasing (about 30–40%). The resource is fully utilized and is being overexploited. Fishing pressure should be reduced and the resource conserved.

In the 1950s, cuttlefish (*Sopiella maindroni*) was one of four dominant species in the ECS not exploited fully. The average catch per year was 24,000 mt. During 1960 to the early 1970s, the mean catch was about 5,000 mt with small fluctuations. After the mid-1970s, the mean catch declined. Meanwhile, fishing effort was increasing sharply, particularly in the 1960s. At that time, fishing boats not only caught the wintering population but also the spawning population in summer. During the 1978–1980 period, the fishing effort was strengthened and harvested a correspondingly high catch in a short period, but then the resource decreased sharply. In recent years the catch has remained low at 20,000–30,000 mt, which indicates that the resource is fully exploited and should be protected.

Jellyfish (*Rhopelema hispidum*) in the East China Sea is distributed mainly in coastal water from the southern part of Jiangsu Province to Zhejiang Province and is caught by setting net. Before 1975, the resource was much more abundant than it is now, with a catch over 10,000 mt per year. In some years, such as 1960, 1966, 1970, and 1973, the catch exceeded 50,000 mt; the average catch was 26,600 mt per year. Since 1975, the catch has declined sharply. The average catch was less than 20% of previous catches. Currently, the catch is about 1,000 mt per year. Owing to capture of young jellyfish (200 mm) since 1985, resource recovery will prove very difficult.

Shrimp resources are abundant in the ECS and are comprised of about 100 species. Before the 1970s fishermen caught small shrimp by setting net in coastal waters. They rarely exploited large shrimp in this area. Since the end of the 1970s, harvests have increased. In 1966, shrimp began to be exploited by motor trawl fishing boats off Zhoushan Island off the coast in Zhejiang Province. Recently, the annual catch of shrimp has been increasing rapidly in this area, and in 1992 the maximum catch reached 290,000 mt. The key species are *Parapenaeopsis hardwickii, Trachypenaeus curvirostris, Solenocera sinensis, Palaemon gravieri,* and *Acetes chinensis.* Presently, the areas of distribution are declining gradually, and smaller shrimp are appearing in shallow water. This means the resource has been overexploited in shallow water, but still provides a potential for a fishery further seaward.

Crab resources are abundant in the ECS. The key species of crab is *Portunus tribuberculatus,* which also has been a traditionally fished species. It occurs throughout the East China Sea. Fishing boats use both drift and trawl nets. During the 1980s, the catch was about 50,000 mt and constituted about 5% of the total catch in the ECS. In Zhejiang Province, the catch has remained at about 25,000 mt per year since the 1980s. In 1992, the maximum catch reached 91,000 mt, but in recent years the catch has declined.

Analysis of Trends in Resource Dynamics

According to surveys by Chinese and Japanese scientific groups, about 5.50 to 6.00 mmt of fisheries resources exist in the East China Sea, of which about 2.75 to 2.90 mmt could be caught. China, Japan, North and South Korea, and Taiwan have caught 4.00 mmt per year to date, resulting in the overexploitation of these fisheries resources. Species structure and population abundance have changed dramatically. The catch composition of individual species is moving toward a lower age, smaller size, and earlier sexual maturity. Clearly this situation is the result of overfishing. Exploitation levels are different for different species, and although the fishery resources in coastal waters are overexploited, there is potential for further exploitation in the offshore and pelagic areas. Exploitation of demersal fisheries resources is excessive in the ECS, but the potential for pelagic, mid-water fisheries still exists. Traditional species are overexploited. Based on species analysis, the fisheries resources could be divided into five classes:

1. Overexploited fisheries resources that have declined sharply: Large and small yellow croaker, jellyfish, *Platycephalus indicus, Sepiella maindroni,* and *Etrumeus micropus.*
2. Overexploited resources that have clearly declined: *Parapenaeopsis hardwickii, Trachypenaeus curvirostris, Solenocera crassicornis, Stromateoides argenteus, Trichiurus hummela, Ilisha elongata,* and *Portunus trituberculatus.*
3. Fully exploited species: *Muraenesox cinereus, Decapterus maruadsi, Argyrosomus argentatus,* and *Sawara niphonia.*
4. Potential for exploitation: *Pneumatophorus japonicus, Brachiostegus japonicus, Formioniger, Priacanthusma crocanthus, Ommastrephes slooni pacificus, Loligo* spp., and cephalopoda.
5. Not fully exploited: *Auxis tapeinosoma, Negalapis cordyla, Anchoviella commersonii, Engraulis japonicus,* and large or medium-sized shrimps in the deep-sea waters.

Based on quantitative analysis of catch data, the species composition has been changing since the 1950s. In the 1970s the dominant species were large and small yellow croaker (*Pseudosciena crocea* and *P. polyactis*), *Trichiurus haumela,* and *Sepiella maindroni*—the so-called old dominant species. In the 1980s, *Trichiurus haumela, Pneumatophorus japonicus,* shrimps and crab, and filefish (*Navodon septentrinalis*) were the "new" dominant species. In the 1990s, the dominant species and groups fished changed to *Trichiurus haumela, Pneumatophorus japonicus, Decapterus maruadsi,* shrimp, and cephalopods. These kinds of marine organisms, which are low on the food chain, have a short life cycle, and fast growth rates, are key species for fishing.

If we neglect the marine ecological environment of the East China Sea or do not treat pollution effectively while exploiting and utilizing petroleum resources in the area, pollution will become a hazard to the environment, resources, and production of the fisheries, as well as to human health.

Summary and Conclusions

The ecological characteristics and environmental conditions, as well as the biomass of fisheries resources in the East China Sea large marine ecosystem, were described and discussed. A significant change has occurred in the biomass of fisheries resources of this large marine ecosystem during the past few decades. For various reasons there has been overexploitation of some fisheries resources in the East China Sea. On the other hand, it is recognized that variations in environmental conditions (both natural and anthropogenic) can also harm marine organisms and fisheries resources, especially in the coastal and estuarine waters where the important nursery, fishing, and feeding grounds are located. However, it is not clear if the effects of these environmental conditions on the fish populations were secondary to the major impact of overexploitation.

Right now, we must take measures to recover depleted fisheries resources and improve ecological and environmental conditions. Suggestions for improving the conditions of the fish stocks follow.

1. It is necessary to reduce fishing efforts from the current high levels because of the depletion of fish stocks from overexploitation. We must suggest to all countries in the region which species in the ECS to reduce fishing effort for and limit the number of fishing vessels with access to the ecosystem. In particular, we should reduce the fishing effort for some bottom species and traditional species such as large and small croaker, cuttlefish, and jellyfish.

2. It is necessary to carry out a large-scale marine ecosystem study. The East China Sea is an important fishing area. It is clear that a large change has already occurred in both catch composition and ecological diversity (namely, reduction). The study should assess biological, physical, and chemical oceanography, pollution, and fisheries resources. To ensure accurate results in the study, advanced methods, instruments, and facilities must be used. It would be useful if a funding source such as the World Bank could be approached to provide needed support for this research program.
3. All governments should pay more attention to fisheries management. Appropriate laws and regulations to protect fishing grounds and fisheries resources are needed.
4. It is necessary to establish an international cooperative organization to limit fishing effort and facilitate coordination in the assessment and management of fisheries resources. China, Japan, North and South Korea, Hong Kong, and other adjacent countries and areas should be included.
5. It is necessary to develop a fisheries resources enhancement program. An international cooperative program should be initiated to restore fisheries resources and increase scientific exchange among cooperating countries.

References

Dou, Z-X., et al. 1979. Chinese Natural Geography, Marine Geography (in Chinese). Science Press, pp. 92–149.

Marine Fishery Resources Investigation and Division in the Area of the East China Sea (in Chinese). 1987. East China Normal University Press, pp. 24–94.

Yang Tianhong, 1984a. A Preliminary Investigation of the Kuroshio Water-mass in the East China Sea, Studia Marine Sinica, pp. 179–200 (in Chinese).

Yang Tianhong, 1984b. The Statistical Characteristics of the Temperature and Salinity of the Kuroshio Water-mass in the East China Sea, Studia Marine Sinica, 21 pp. 165–178 (in Chinese).

10

Changing State and Health of the Gulf of Thailand Large Marine Ecosystem

T. Piyakarnchana

Abstract

Environmental quality, marine pollution, and resources exploitation are examined to determine the changing state and health of the Gulf of Thailand ecosystem. Natural phenomena such as the monsoon seasons and the intrusion of seawater from the South China Sea are believed to be the major cause of change in the system. However, changes in land use along the coast are also responsible. Mangrove forests are being destroyed to make way for industrial (maricultural), residential, and tourist developments, resulting in the rapid deterioration of the coastal zone. Since 1992, the Thai government has amended several key environmental laws in an effort to improve environmental quality and protect the mangrove forests. Several recommendations are proposed to hasten this process. These include (1) the incorporation of environmental practices into all marine resource development activities, (2) the strict enforcement of the Conservation Laws and Ministerial Regulations, and (3) the implementation of the "Fishing Right" principle.

Gulf of Thailand

The Gulf of Thailand, formerly the Gulf of Siam, covers an area of approximately 350,000 km^2 and is the northwestern portion of the South China Sea. The eastern Gulf is bordered by the coasts of Vietnam, Kampuchea, and Thailand. The northern and western parts of the Gulf are bordered by Thailand and Malaysia. The Gulf is relatively shallow, with a maximum depth of approximately 80 m. The tropical monsoons play a significant role in the changing conditions within the Gulf and the surrounding land.

The Gulf of Thailand might be classified as a semi-enclosed bay due to the existence of two sills at the southernmost end of the Gulf. One sill runs from the Camau Peninsula to the southwest for about 100 km at a depth of less than 25 m; the other runs from Kota Bharue northeast for 150 km at a depth of less than 50 m. A channel about 67 m deep connects the deepest part of the Gulf with the South China Sea.

The Gulf can be geographically divided into an inner sector and outer sector. The line that divides these two sectors on land is approximately from Rayong Province to Prachuab-Kirikhan Province. The inner and outer sectors of the Gulf have an area of 9,323 km^2 and 340,672 km^2, respectively.

The bottom sediments of the Gulf are composed largely of mud and a mixture of mud, sand, and shell fragments. There are about 23 rivers from Thailand's coastline that drain large amounts of freshwater into the Gulf.

Thailand's coastline is approximately 2,600 km. The coastline along the inner part of the Gulf is made up of mud flats, sandy beaches, and rocky shores. The outer sector is mainly sandy beaches.

Piyakarnchana (1987), in a study of yield dynamics as an index of biomass shifts in the Gulf of Thailand ecosystems, compiled data on seawater quality, changes in plankton biomass, finfish catch, and yield of other economically important marine organisms. The study found that although there was a remarkable decrease in the catch of demersal fishes and omnivorous crustaceans, certain species, especially those that were classified as plankton feeders, increased. It was suggested that overfishing and increasing amounts of pollutants in the Gulf were among the major causes for this state of affairs. This study further investigates how these changing environmental conditions are related to the health of the Gulf ecosystem.

Ecosystem Change and Health Indices

Five indices of ecosystem health were accepted at the NOAA Workshop convened in 1993: diversity, stability, yield, productivity, and resilience (Sherman 1994). The concept of "ecosystem health" aims to use these indices to assess the changing state of an ecosystem.

Due to insufficient studies in the Gulf of Thailand on the criteria suggested at the 1993 NOAA Workshop, the author chose to test the changing state and health of the Gulf ecosystem by looking at marine environmental quality, marine pollution, and resource exploitation in the Gulf.

Marine Environmental Quality

Environmental quality in the inner sector of the Gulf clearly shows seasonal variations. For example, in the wet season (July to December), lower-salinity surface water from the northern Gulf is forced to extend farther south to about the middle of the Gulf. This comparatively low-salinity water is also associated with low dissolved oxygen, both characteristics that usually occur in the bottom water of the inner Gulf.

Among the trace metals found, only zinc is influenced by the amount of freshwater discharged. The other trace metals, such as lead, cobalt, cadmium, and mercury, do not appear to be influenced by the runoff of the five major rivers.

Environmental parameters in the inner Gulf have been measured continuously from 1973 to 1986. This is not so for the outer Gulf. However, some parameters, such as water temperature and plant nutrients, have been measured in both areas and can be used for comparison.

There was no significant change in water temperature in the outer Gulf during 1978–1994. Salinity changes were probably caused by the intrusion of higher-salinity water from the South China Sea. Concentrations of nitrate and phosphate are higher during the wet season than the dry season. In contrast, silicate concentrations are higher in the dry season, especially at Prachuab-Kirikhan and Chumporn Provinces, which are located on the western side of the Gulf. Selected parameters in the outer Gulf from 1978 to 1991 are shown in Table 10-1.

Marine Pollution

The water quality at some famous tourist beaches in Thailand was assessed in 1992 by the Pollution Control Department, Ministry of Science Technology and Environment (Pollution

Table 10-1. Some selected ecological parameters in the outer part of the Gulf of Thailand

Station and years	Temp (°C)	S %	O_2 (ml/l)	pH	NO_3 (M)	NO_2 (M)	PO_4 (M)	Si_2O_3 (M)
Prachaub-Chumporn, 1978–1991								
Dry season	27.9–30.5	32.2–35.8	4.3–5.1	8.0–8.3	nd–3.3	nd–0.5	nd–0.4	4.0–38.2
Wet season	27.8–30.2	31.0–34.1	3.8–4.9	8.1–8.2	< 0.1–2.5	nd–2.5	0.3–8.7	1.3–14.2
Total range	27.8–30.5	31.0–35.8	3.8–5.1	8.0–8.3	nd–4.0	nd–2.5	nd–8.7	1.3–38.2
Chumporn Surathani, 1980–1991								
Dry season	28.7–31.1	32.6–33.8	3.4–5.1	7.9–8.3	0.1–4.6	<0.1–0.9	0.1–1.9	—
Wet season	28.1–32.0	31.6–33.9	3.2–5.0	7.9–8.2	nd–15.5	nd–3.7	nd–1.5	17.4–55.0
Total range	28.1–32.0	31.6–33.9	3.2–5.0	7.9–8.3	nd–15.5	nd–3.7	nd–1.5	—
Surathani-Songkla, 1984–1991								
Dry season	26.5–27.7	31.5–32.4	—	—	—	—	—	—
Wet season	—	—	—	—	0.8–5.7	nd–0.4	0.1–3.6	7.0–22.9

Control Department 1992). This report showed that the water quality along the Gulf of Thailand and the Andaman Sea was in fair condition. Coliform bacteria counts were below 1,000 MPN/100 ml, which is suitable for swimming and other water sports. The report also mentioned that the water quality was very good in some places, such as Koh Samet in Rayong Province (on the eastern side of the Gulf). Here coliform bacteria counts were less than 10 MPN/100 ml. However, other beaches, such as Bang Saan Beach (east coast of the Gulf) and Cha-am and Hau-Hin Beaches (west side of the Gulf), were classified as only fair, although still suitable for swimming. Coliform bacteria counts in these places were higher than 1,000 MPN/100 ml.

Petroleum hydrocarbon contamination is also reported in the Gulf of Thailand. However, oil spills in Thailand's waters have caused no serious damage to the marine environment. An issue of the Annual Report on the Status of the Pollution in Thailand by the Pollution Control Department, Ministry of Science Technology and Environment (1992), reported an oil spill of 60,000 liters in the Chao Phraya River near the Bangkok Port in April 1992. The spill was quickly controlled by the Oil Industry Environment Safety Group (IESG,) which is a nongovernmental organization in Thailand.

A large-scale oil spill was reported by the Pollution Control Department Publication in the Malacca Strait on 20 September 1992. This incident actually occurred in international waters, but Thailand was alerted by Malaysia because of the possible movement of the oil slick into the Tarutao Island National Park. Fortunately, a change in the wind direction caused the oil to travel toward the open sea, causing no damage to the Tarutao Island National Park.

On 31 July 1994 and 28 October 1994, large oil slicks were reported on the east coast of the Gulf. Only the July spill was considered serious; however, no environmental damage was reported.

Petroleum pollution in the Gulf has been studied and reported by Wattayakorn (National Research Council 1994). She found hydrocarbon concentrations in the range of 0.65 to 8.30 μg/l in the inner Gulf waters, and between 0.07 and 6.50 μg/l in the outer Gulf. Sediments in the inner and outer Gulf contained 0.70–62.0 μg/g and 0.03–8.30 μg/g, respectively.

Resource Exploitation

CHANGING LAND USE AND LAND COVER. Land use and land cover changes seem to be the most important factors contributing to changes in the Gulf of Thailand. Two areas are of particular importance: hinterland forest conversion and mangrove forest conversion.

Thailand is located in the tropical monsoon region of Southeast Asia. In 1961, about 53% of the total land area, or about 273,629,000 hectares, was forested. In 1991, due to human activities, only 27% remains forested. These forested lands were transformed to rice fields, cash-crop plantations, fruit orchards, grass lands, and residential areas.

In order to increase the productivity of agricultural lands, the use of fertilizers has also greatly increased. The average total amount of nitrogen fertilizer used from 1989 to 1992 was 529,858 tons (Table 10-2). These fertilizers eventually reach the Gulf via land runoff, causing phytoplankton blooms and deterioration of the water quality, especially in the inner Gulf region (Piyakarnchana and Tamiyavanich 1979).

Extensive mangrove forests are found along the coastline of the Gulf of Thailand. They play a significant role in the environmental quality and the economy of the region. They also serve as a buffer zone between the land, the rivers, and the sea, protecting the shore from erosion. They provide habitat and nursing grounds for numerous marine organisms.

In 1961, mangrove forests covered 367,000 hectares, existing in 24 provinces on both sides of the Gulf coast and on the Andaman Sea coast. In 1991, however, at least three provinces reported having lost all their mangrove forests. The total area of the mangrove forests was reduced to 173,600 hectares. These mangrove-covered lands were cleared for various reasons (Table 10-3).

Table 10-2. Imported nitrogen fertilizer (in tons) recorded by a chemical fertilizer regulatory agency

Fertilizer type	1989	1990	1991	1992	Average
Nitrate ammonium type					
Ammonium sulfate	106,267	119,243	97,453	99,237	105,550
Ammonium phosphate	108,914	102,563	85,176	115,222	102,969
Urea	211,340	190,523	150,989	169,189	180,510
Nitrate					
Calcium nitrate	920	450	600	945	729
Other nitrogen	4,804	3,705	6,043	2,486	4,295
Other complex	108,537	162,912	147,243	124,674	135,841
Total nitrogen fertilizer	540,781	579,396	487,504	511,752	529,858

Source: Ministry of Agriculture and Cooperatives

Sixty-four percent of these areas were used for mariculture, mostly for shrimp ponds. Only 3% and 6% were exploited for tin mining and for salt pans, respectively. The remaining 26% were used for miscellaneous purposes, such as residential buildings, industrial developments, and government buildings.

Mariculture, especially shrimp farming, is considered to be the use that has the most detrimental effect on the environmental quality in the coastal zone of Thailand. Tookavinas and Arunlertaree (1993) estimated the amount of waste released into the sea by one crop from the shrimp ponds (Table 10-4). Ammonia is released from these ponds and is toxic to aquatic organisms in high concentrations. The other waste components stimulate phytoplankton growth and result in blooms, thus reducing water clarity and dissolved oxygen (DO) in bay areas. Songsangjinda et al. (1993) reported that nutrient loading from the effluent of the shrimp culture ponds, which contains high concentrations of organic nitrogen, ammonia, nitrate, and nitrite and low levels of orthophosphate, might cause the phytoplankton blooms in the more open water on the western side of the Gulf. They also demonstrated that the elevated nutrient concentrations in

Table 10-3. Major types of activities and areas of utilization of mangrove forest land in Thailand in 1991

Activities	Area of mangrove land (ha)	Percentage
Mariculture (shrimp farms)	110,243.2	64.3
Tin mining	54,506.5	3.2
Salt farms	10,560.0	6.2
Other activities	45,202.3	26.3
Total	220,512.0	100.0

Table 10-4. Estimated amount of wastes (in tons) released in one crop from the shrimp ponds

Parameter	January	February	March	April	Total
NO_2-N	0.06	0.12	0.16	0.18	0.63
NH_3-N	0.54	1.03	1.32	1.48	5.28
PO_4	1.06	2.02	2.59	2.91	10.36
BOD_5	20.54	39.10	50.17	56.24	200.18

Source: Tookavinas, S., and Arunlertaree, C., 1993. Quantity and Quality of Waste Water from Intensive Marine Shrimp Farming at Kung Krabaen Bay: Scope to Study the Impact on the Bay and Coastal Area. In Proceedings of the Eighth National Seminar on Mangrove Ecology, Sustainable Mangrove Resources Management, Vol. 1. National Research Council of Thailand.

the more open coastal areas were caused by effluent that came from the estuary where the shrimp culture ponds were located.

EUTROPHICATION AND HARMFUL ALGAL BLOOMS. The first scientific paper on red tides in the Gulf of Thailand was by Chareonpol (1957). The two species of marine algae reported at that time were *Trichodesmium erythraeum* and *Noctiluca scintillans.* It was not until 1973 that scientists became aware of the harmful effects of these red tides in Thai waters (Piyakarnchana and Tamiyavanich 1979). With the increasing input of untreated wastes from the hinterlands and the cities into the Gulf through the rivers, such deteriorated situations are predicted to become more and more serious. Piyakarnchana and Tamiyavanich also reported frequent occurrences of the red tides in the estuarine regions of the large rivers in the inner Gulf.

The algae species found in the Gulf as red tides differed from place to place and by time of year. Besides *Trichodesmium erythraeum* and *Noctiluca scintillans,* other dinoflagellate species such as *Ceratium furca* were also found to occur in the red tides. The water color caused by this species was yellow-brown at the Lang-Suan River mouth and reddish-brown at Prachuab-Kirikhan Bay. *Ceratium furca* was also found to be a dominant species in the red tides at the Tachin River mouth, in the northern part of the Gulf. During occurrences of heavy algal blooms, marine organisms, especially those living on the bottom, are affected by anoxic conditions caused by very low concentrations of dissolved oxygen.

In recent years the occurrence of red tides has moved southward along the eastern coast of the Gulf. Such spreading patterns were observed with the help of small-plane pilots. They reported red-tide occurrences to the Marine Environmental Watching Unit of NRTC (Thai Net). The results are shown in Figure 10-1 and Table 10-5. There were ten sightings of red tides on the east coast and two on the west coast in 1994. The area includes the Chao Phraya River mouth to around Koh Phai. These findings confirmed the report by Bunpapong and Piyakarnchana (1987) that freshwater discharged to the inner Gulf environment by the five major rivers in the northern part influences algal blooms.

In May 1973, the small fishing village of Prachuab-Kirikhan Province reported 63 cases of illness, including one death, caused by eating green mussels (*Perna viridis*). The symptoms were similar to paralytic shellfish poisoning (PSP). The outbreak was suspected to be caused by a species of dinoflagellate, *Protogonyaulax tamarense* (Fukuyo et al. 1989). However, this species was later found to consist of both toxic and nontoxic strains in the Gulf of Thailand (Kodama et al. 1987; Pholpunthin et al. 1990). The other suspected PSP-toxic dinoflagellate, *Alexandrium cohorticular,* is also found in the Gulf. Green mussels fed with cultures of *A. cohorticular* cells were highly toxic to lab mice (Wisessang et al. 1991). However, no correlation has been found between the occurrence of these two toxic species and the blooms of red tide in the Gulf. Piyakarnchana et al. (1990) argued that tidal influence may affect cell accumulation of *A. cohorticular* in the green mussel beds in the inner part of the Gulf.

In the shallower areas of the upper part of the inner Gulf, another species of dinoflagellate, *Dinophysis caudata,* was reported to cause diarrhea and vomiting to those who ate the blood shell (*Pholas orientalis*). This shell was found to feed on *Dinophysis caudata* and become a bloody red color.

MARINE RESOURCE EXPLOITATION. Piyakarnchana (1987), in his work on yield dynamics as an index of biomass shifts in the Gulf of Thailand ecosystem, reported the total fish catch, the average catching effort (kg h^{-1}), and the yearly decrease or increase (kg h^{-1}) of the demersal fishes. The results indicated a yearly fluctuation of the total catch, and showed that the average catch in the Gulf region decreased drastically from 1967 to 1983.

In the present study, the total fish production in the Gulf from 1985–1991 is examined (Table 10-6). The results show that the total fish catch varies from 1,718,222 metric tons in 1985 to

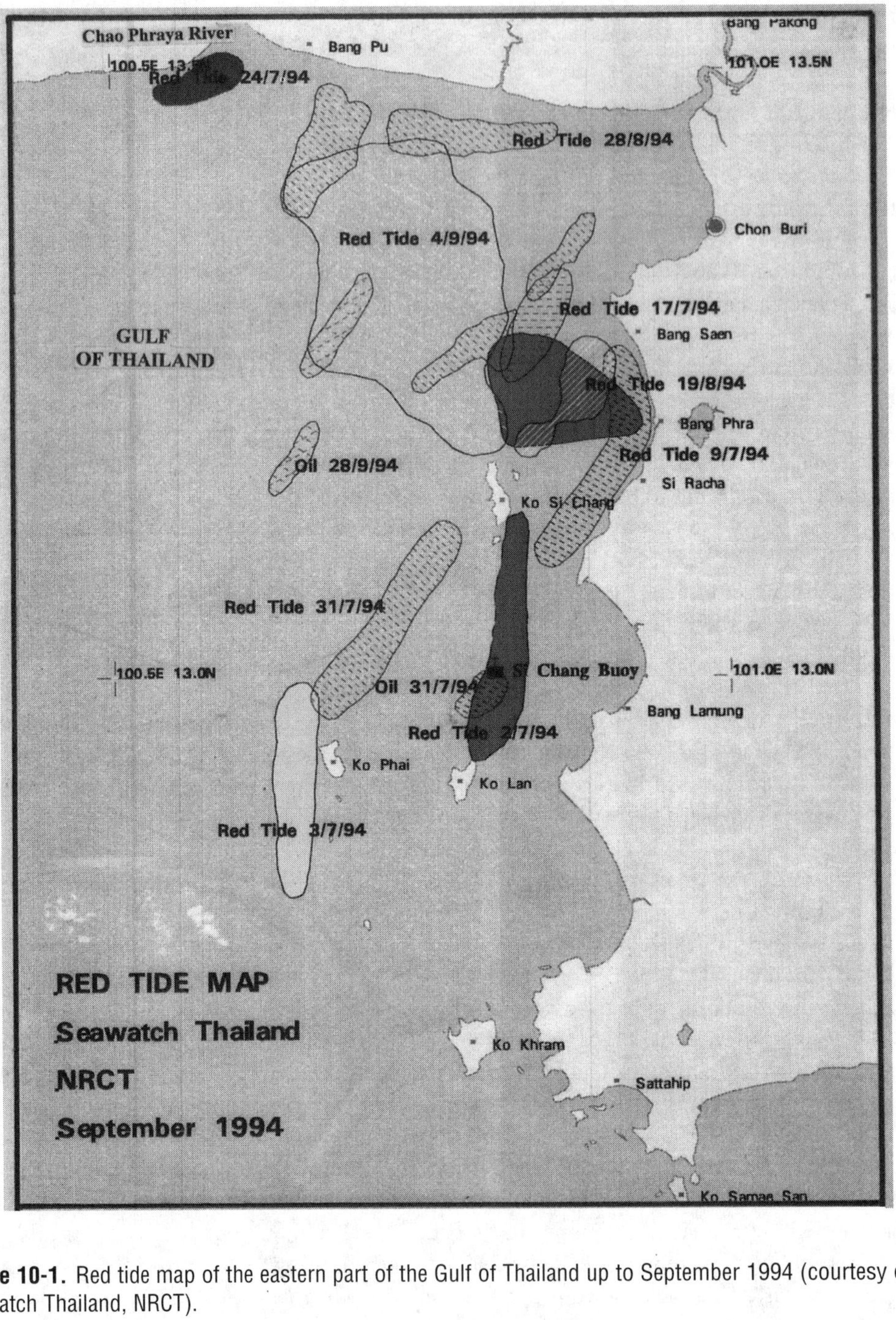

Figure 10-1. Red tide map of the eastern part of the Gulf of Thailand up to September 1994 (courtesy of Seawatch Thailand, NRCT).

1,820,687 metric tons in 1991. It is interesting to observe that the total catch increased to 2.1 million metric tons in 1986–1987.

Piumsomboon (1994), on the other hand, demonstrated the importance of small artisanal fishermen. This author reported that in 1988 the total catch in the Gulf was 2,709,000 metric tons and that small artisanal fishermen caught approximately 352,184 metric tons of this (Table 10-7). Thus, the yearly fish catch from the Gulf during that period was about twice the estimated

Table 10-5. Number of red tides observed on both sides of the inner part of the Gulf of Thailand during July to September 1994

Date	East coast of the Gulf	West coast of the Gulf
02/07/94	1	—
09/07/94	1	—
16/07/94	1	1
17/07/94	1	—
24/07/94	1	—
31/07/94	1	—
14/08/94	1	1
19/08/94	1	—
28/08/94	1	—
04/09/94	1	—

potential stock of 1,150,000 metric tons (Piyakarnchana 1987). The increased catch over the estimated potential stock might partly be due to fishing in international waters and to fishing in cooperation with neighboring countries.

To examine the possible recovery of the fishing stocks in the Gulf region, the total catches of several economically important species were examined (Piyakarnchana 1987). Chayakul (1994) reported on the fisheries status and the maximum sustainable yield (MSY) of several marine species (Table 10-8). The results confirmed the findings by Piyakarnchana (1987) that the fish stocks in the Gulf are still being overexploited. Only one species, *Rostrelliger kanakurta*, which was

Table 10-6. Total fisheries productions (metric tons) in the Gulf of Thailand and the Andaman Sea from 1985 to 1991

Year	Gulf of Thailand	Andaman Sea	Total
1985	1,718,222	339,529	2,057,751
1986	1,964,090	388,144	2,352,204
1987	2,174,942	266,987	2,601,929
1988	2,108,092	399,027	2,466,125
1989	1,963,657	406,890	2,370,548
1990	1,923,163	439,055	2,362,218
1991	1,820,687	657,920	2,478,607

Source: National Research Council, 1994. Collected Papers in the Fifth Marine Science National Seminar on the Status of Thailand Sea and Future Tendency, 22–24 August 1994 (with appendix). National Research Council of Thailand.

Table 10-7. Small-scale fisheries production (in tons) in the Gulf of Thailand from 1977 to 1988

Year	Total marine fisheries	Small-scale fisheries	Percentage
1977	2,067,533	203,900	9.9
1986	2,352,204	388,352	16.5
1988	2,709,000	352,184	13.0

Source: Adapted from Piumsomboon, S. 1994. Social possibility and economical aspect of introducing fishing right for the purpose of coastal fisheries management in Thailand. Thai Fisheries Gazette 17(2):161–168.

Table 10-8. Fisheries status and maximum sustainable yield (MSY) of some economically important marine species in the Gulf of Thailand

Fish species	Present status	MSY (tons)	Concluding remarks
All economically important species	Total catches increased above the MSY since 1976.	1,150,000	Overexploitation from 1976 to present.
Rostrelliger brachysoma	Total catch was at the MSY since 1984. There were signs of decreases during 1990–1991.	104,611	Continue this catching effort.
R. kanakurta	Catches lower than the production capacities.	53,008	Able to increase the catch effort by 20%.
Decapterus morudsi	Reached maximum catches since 1977.	140,000	Overexploitation after 1977.
Penaeus spp.	Decreasing catches.	22,000	Overexploitation since 1982.
Squids (*Sepia* spp., *Loligo* spp., *Octopus* spp.)	Decreasing catches. Total catches in 1991 were 138,000 tons.	106,000	Overexploitation since 1978–1979.

Source: National Research Council, 1994. Collected Papers in the Fifth Marine Science National Seminar on the Status of Thailand Sea and Future Tendency, 22–24 August 1994 (with appendix). National Research Council of Thailand.

found to have increased from 1983, continued to increase; it is suggested that this species is capable of an additional 20% increase in yield (Chayakul 1994). Chayakul (1994) also suggests that the catch effort for the other species of the same genus, *Rostrelliger brachysoma*, could be maintained at the present rate. The recovery of *R. kanakurta* is a direct result of strict enforcement of the Fisheries Regulations, which prohibit the catch of this species during their spawning season.

Conclusion and Recommendations

Three basic criteria are used to evaluate the changing state and the health of the Gulf of Thailand ecosystem: environmental quality, marine pollution, and resources exploitation.

Environmental quality, especially for the major part of the Gulf, can be rated as in good to fair condition. There are, however, some locations of the Gulf where the water quality is lower than acceptable standards by the Ministry of Science Technology and Environment. These areas are mostly located in the inner Gulf region, especially at the mouths of the five large rivers, the popular tourist spots along the coast, and near certain islands.

The factors contributing to the deterioration of the water quality in the Gulf are the increasing inputs of nutrients such as nitrate, phosphate, and silicate from the land and from mariculture ponds. These nutrients come from the increased use of fertilizers in agriculture, the mariculture industry, and from household sewage.

The transformation of large areas of mangrove forests to prawn farming ponds has resulted in significant environmental damage to the Gulf. Although, at present, the shrimp farmers claim to receive larger profits from shrimp production than from the traditional use of the mangrove forests, it is believed that this will be true only in the short run and that shrimp farmers will suffer in the long run.

The close link between humans, the land, the rivers, and the sea can be clearly demonstrated from the study of eutrophication effects and toxic algal blooms in the Gulf of Thailand. Many cities have no sewage treatment and discharge directly into the Gulf, thus severely affecting the marine organisms and the mangrove forests. Clearing of mangrove forests further disturbs the ecological balance in the Gulf. This situation will become more severe if destruction of the mangrove forests continues at its present rate.

The government of Thailand has tried several measures in an attempt to protect the environment and restore it to its original state. For example, since 1992 the Fisheries Acts, the Wildlife Conservation Law, and the Promotion of the National Environmental Quality Act have been amended. The concepts of sustainable use and development were integrated into these laws.

It is strongly recommended that the Conservation Laws be strictly enforced. The government has implemented several well-known methods of fisheries management that have been successful elsewhere in the world, such as building artificial reefs, limiting areas for bottom trawling within 3 kilometers from the coast, controlling the use of certain fishing gears during spawning periods of certain species, and so on. These methods of fisheries management should be continued and expanded to other species.

There has been a proposal to introduce the "Fishing Right" concept into the small artisanal fishing community (Piumsomboon 1994). This concept is extremely important with respect to the low- and medium-income fishermen who are at present the most important sector of the population when considering environmental impacts in the Gulf of Thailand. The government, as well as the general public, should give full support to this matter.

ACKNOWLEDGMENTS

The author wishes to thank Ms. Sangchan Limjirakan (Institute of Environmental Research) for her assistance in preparation of the manuscript, and Dr. Darasri Dowreang (Chief of the Data Analysis sub-division, Remote Sensing Division, National Research Council of Thailand (NRCT)) for providing valuable information and the red tide maps.

References

Bunpapong, M., and Piyakarnchana, T. 1987. Some Trends in the Upper Gulf Water Quality. Proceedings of the Fourth Seminar on the Water Quality and the Quality of Living Resources in Thai Waters, 7–9 July 1987. National Research Council of Thailand, pp. 270–283.

Chareonpol, S. 1957. Preliminary Study of Discolouration of Seawater in the Gulf of Thailand. Hydrographic Department, Royal Thai Navy.

Chayakul, R. 1994. Status of Resources and Marine Fisheries in Thai-Waters. Fifth National Seminar on Status of Thai Marine Waters and Its Future Trends, 22–24 August 1994. National Research Council of Thailand, SB-7-1 to SB-7-35.

Department of Pollution Control, Ministry of Science Technology and Environment. 1992. Report on the Pollution Status in Thailand in 1992.

Fukuyo, Y., Yoshida, K., Ogata, T., Ishimaru, T., Kodama, M., Pholpunthin, P., Wisessang, S., Panichyakarn, V., and Piyakarnchana, T. 1989. Suspected Causative Dinoflagellates of Paralytic Shellfish Poisoning in the Gulf of Thailand. In T. Okaichi, D. M. Anderson, and T. Nemoto (eds.), Red Tides: Biology, Environmental Science and Toxicology. New York: Elsevier, pp. 403–406.

Kodama, M., Ogata, T., Fukuyo, Y., Ishimaru, T., Pholpunthin, P., Wisessang, S., Saitanu, K., Panichyakarn, V., and Piyakarnchana, T. 1987. Non-toxic strains *Protogonyaulax tamarensis* (Lebour) Taylor (Dinophyceae) in the Gulf of Thailand. Nippon Suisan Gakkaishi 53(8):1491.

National Research Council of Thailand. 1994. Collected Papers in the Fifth Marine Science National Seminar on the Status of Thailand Sea and Future Tendency, 22–24 August 1994 (with appendix). National Research Council of Thailand.

Pholpunthin, P., Wisessang, S., Ogata, T., Fukuyo, Y., Kodama, M., Ishimaru, T., and Piyakarnchana, T. 1990. Coexistence of Toxic and Non-toxic Strains of *Alexandrium tamarense* in the Gulf of Thailand. The Second Asian Fisheries Forum 1990. Asian Fisheries Society, pp. 403–406.

Piumsomboon, S. 1994. Social possibility and economical aspect of introducing fishing right for the purpose of coastal fisheries management in Thailand. Thai Fisheries Gazette 17(2):161–168.

Piyakarnchana, T. 1987. Yield Dynamics as an Index of Biomass Shifts in the Gulf of Thailand Ecosystems. In K. Sherman and L. Alexander (eds.), Biomass Yields and Geography of Large Marine Ecosystems. Boulder, CO: Westview Press, pp. 95–142.

Piyakarnchana, T., and Tamiyavanich, S. 1979. Increase in red water phenomena in the upper portion in the Gulf of Thailand. J. Aquatic Animal Diseases 2(4):207–215.

Piyakarnchana, T., Bunpapong, M., Pholpunthin, P., Wisessang, S., and Wattayakorn, G. 1990. Tidal Influence on the Distribution and Population Density of a Toxic Dinoflagellate, *Alexandrium cohorticular* (Balech) Balech, in the Upper Part of the Gulf of Thailand. The Second Asian Fisheries Forum 1990. Asian Fisheries Society, pp. 899–902.

Sherman, K. 1994. Sustainability, biomass yields, and health of coastal ecosystems: An ecological perspective. Mar. Eco. Prog. Ser. 112(3):227–301.

Songsangjinda, P., Tunvilai, D., and Chaiyakam, K. 1993. Water Quality of a Coastal Area from Pakphanang District, Nakornsrithammarat Province of Muang District, Songkla Province in 1990–1991. In Proceedings of the Eighth National Seminar on Mangrove Ecology, Sustainable Mangrove Resources Management, Vol. 1. National Research Council of Thailand.

Tookavinas, S., and Arunlertaree, C. 1993. Quantity and Quality of Waste Water from Intensive Marine Shrimp Farming at Kung Krabaen Bay: Scope to Study the Impact on the Bay and Coastal Area. In Proceedings of the Eighth National Seminar on Mangrove Ecology, Sustainable Mangrove Resources Management, Vol. 1. National Research Council of Thailand.

Wisessang, S., Ogata, T., Kodama, M., Fukuyo, Y., Ishimaru, T., Saitanu, K., Yongvanich, T., and Piyakarnchana, T. 1991. Accumulation of paralytic shellfish toxins by green mussel *Perna viridis* by feeding on cultured cells of *Alexandrium cohorticular* isolated from the Gulf of Thailand. Nippon Sui San Gakkaishi 57(1):127–131.

11

Effects of Physical and Biological Changes on the Biomass Yield of the Humboldt Current Ecosystem*

Jurgen Alheit and Patricio Bernal

Introduction

The Humboldt Current ecosystem is one of the major upwelling systems of the world. It is dominated by four pelagic schooling fish species: anchovy, *Engraulis ringens;* sardine, *Sardinops sagax;* horse mackerel, *Trachurus murphyi;* and mackerel, *Scomber japonicus.* Over the last three decades, fisheries in the Humboldt Current have experienced drastic changes in yields and species dominance. The most spectacular case, which is well documented in many fisheries textbooks, is that of the Peruvian anchovy fishery, once the largest fishery in the world with 12.5 million tons landed in 1970 (Tsukayama 1983). According to Castillo and Mendo (1987), the real catch amounted to 15 million tons in 1970, taking into consideration that underreporting of the real catches was frequent.

The pelagic fish stocks of the Humboldt Current are exposed to three principal factors governing their strong fluctuations: (1) drastic short-term perturbations of their environment; typical for this ecosystem [El Niño–Southern Oscillation (ENSO)]; (2) fishing pressure; and (3) subtle long-term changes in the environment.

The nature and consequences of ENSO events have been described by physical oceanographers and biologists in numerous articles (for key references, see Arntz and Tarazona 1989). The Peruvian anchovy fishery has been analyzed recently from an ecosystem perspective in two books edited by Pauly and Tsukayama (1987) and Pauly et al. (1989).

Descriptions of the fluctuations of the pelagic fisheries in the Humboldt Current have been somewhat limited, as they often concentrated on a single national fishery. Furthermore, the impact of subtle long-term changes in the environment on fish stocks and fisheries has rarely been taken into account. Our objectives are (1) to view fluctuations of pelagic fisheries and their environment simultaneously for the entire Humboldt Current including fisheries of Ecuador, Peru, and Chile, and (2) to point out possible impacts of environmental changes other than El Niño.

*Reprinted with permission from Sherman, K., Alexander, L. M., and Gold, B., eds. Large Marine Ecosystems: Stress, Mitigation, and Sustainability. Washington, DC: AAAS Press, 1993.

Physical Characteristics

Currents

In common with the other large eastern boundary currents, the Humboldt Current is characterized by the principally equatorward flow of cold, low-salinity waters. However, this general description simplifies a rather complex set of flows and counterflows that occur semipermanently between the coast and 1,000 km offshore.

The oceanic sector is dominated by a sluggish, wide flow toward the equator with characteristic velocities of about 0.04 ms^{-1} coinciding with the boundary of the anticyclonic gyre of the Southeast Pacific and which has been recognized as the Chile–Peru Oceanic Current (Robles et al. 1980). Toward the coast, a counterflow to the south, the Peruvian Oceanic Countercurrent (Wyrtki 1967) has been identified at 79°W off Peru and between 76–77°W, 500 km off the Chilean coast. This counterflow has characteristic velocities of about 0.06 ms^{-1}.

The faster Humboldt Current is located somewhat closer to the coast, between 300–400 km offshore, with a core at about 200 m depth. The Humboldt Current often reaches velocities of over 0.18 ms^{-1}. Between the core axis of the Humboldt Current and the coast, three other branches can occasionally be distinguished. They are relatively unstable and show clear seasonal fluctuations. The more permanent branch is a counterflow to the south with a velocity core at around 100 m depth off Peru and at 200 m depth off Chile. This counterflow is poor in oxygen ($<$1 ml/l) and rich in nutrients. It corresponds to the Gunther Current or Peru–Chile Undercurrent described by several authors (Brockmann et al. 1980; Robles et al. 1980). This obvious heterogeneity of flows demonstrates that seasonality and mesoscale processes are dominant in this eastern boundary current system. Satellite imagery (Espinoza et al. 1983; Fonseca and Farias 1987) has shown a rich spectra of plumes, eddies, filaments, and other transient structures that occur regularly in this region and which point to intense horizontal mixing. Given the methods applied to study circulation in this region (mainly through geostrophy), some of the discrete flows identified in the past could in fact correspond to aspects of these complex mesoscale phenomena. Silva and Fonseca (1983) propose a more parsimonious interpretation of the circulation patterns off northern Chile, which is characterized by three main flows that fluctuate between seasons.

Upwelling and Other Localized Processes

Coastal upwelling has been identified as the most important and relevant process in eastern boundary currents, presumably being almost exclusively responsible for the extremely high levels of organic production. In Peru, coastal upwelling reaches a maximum during winter and a minimum during summer. In northern Chile (18–30°S), it peaks during spring, and off central Chile (30–38°S) during late spring and summer. This temporal progression of coastal upwelling from the north to the south is due to the displacement of the subtropical center of high atmospheric pressure that intensifies and moves southward as summer progresses. Therefore, off Peru, upwelling occurs all year round, whereas, in central Chile, it is restricted to spring and summer.

The upwelling paradigm has overshadowed the importance of other phenomena that control distribution and abundance of organisms and explain the location of spawning and feeding centers of fish and, thus, of fishing grounds. The eastern boundary of the Pacific has a comparatively narrow shelf off Peru and an extremely narrow shelf off Chile. However, wherever a significant shelf is present, it is associated with important mesoscale structures resulting from the interaction of circulation with bottom topography and the coastline. Frontal structures develop at the coastal upwelling centers as the pycnocline rises to the surface close to the coast and "breaks" the surface in a zone where hyperbaroclinic conditions are established (Mooers et al. 1978). When upwelled waters are not modified fast enough by mixing, they will necessarily sink at the frontal boundary and so retain nutrients, plankton, and detritus within the upwelled plume. These phenomena have been observed along the entire coast of the Humboldt Current. They play an important role for the maintenance of the spawning grounds of the coastal neritic sardine and anchovy and the feeding grounds of the more oceanic horse mackerel.

Eastern boundary currents have received the close attention of the international oceanographic community after the recognition of the global character of ENSO. ENSO causes a clear oceanographic signal along both eastern boundaries of the Pacific. The ENSO events of 1957–1958, 1972, and 1982–1983 showed a remarkable bi-hemispheric manifestation (Bernal 1981; Wooster and Fluharty 1985). When analyzing the long-term variability of the Humboldt Current system, the ENSO signal will, without doubt, appear as one of the dominant features. However, despite the interannual periodicity of its appearance, its duration is rather short (about a year). For the ecosystem, its evolution and dynamics, ENSOs have to be viewed, therefore, as a typical perturbation, i.e., it is closer to a delta function than to a step function.

Biomass and Catch Data

The acquisition of reliable data on the total biomass of the four dominant pelagic species for the entire period of their fisheries in the Humboldt Current, which started in the early 1950s, proved to be a problem. Catch data are available; however, they are often based on serious underreports of the true catches (Menz and French 1982; Castillo and Mendo 1987). The official catch data, as published by FAO and the fisheries departments of the three countries involved, are used here. Although aware of the problem of underreporting, we believe that the official statistics are the best database available at present, and we assume that they reflect major trends in the fluctuations of fish stocks after the fishery reached a certain level. We are also aware of other factors having a strong influence on trends in catch data, such as improved fishing methods, changes in mesh size, and changes in vulnerability of fish caused by oceanographic factors.

Total Catches of Pelagic Species

The four dominant pelagic species are considered in this analysis. Other clupeiform pelagic species that might be of local importance on the northern and southern edge of the Humboldt Current are omitted from these studies.

Pelagic fisheries in the Humboldt Current began in the early 1950s, reached a peak in the early 1970s, and showed strong fluctuations during the remainder of the 1970s and the 1980s (Figure 11-1). The total pelagic yield was almost exclusively influenced by the rise of the anchovy fishery in the 1950s–1960s, its collapse in 1972, and the subsequent rise of the fishery for sardine, horse mackerel, and mackerel. The usual explanation for these fluctuations is that a combination of extensive overfishing and the strong ENSO in 1972–1973 caused the collapse of the Peruvian anchovy, thus allowing the other three species to expand in an environment where the stocks of the dominant species were substantially reduced. In the following, catch data of anchovy and sardine stocks along the Ecuadorian, Peruvian, and Chilean coasts are analyzed stock by stock in order to identify and compare major population trends.

Anchovy

Three anchovy stocks have been identified in the Humboldt Current ecosystem: (1) the northern and central stock in Peru, (2) the stock in southern Peru and northern Chile, and (3) the stock of Talcahuano in Chile (Serra 1983) (Figure 11-2). Anchovy was the dominant species in these three regions until about 1972. The stock in northern and central Peru, which supported the famous

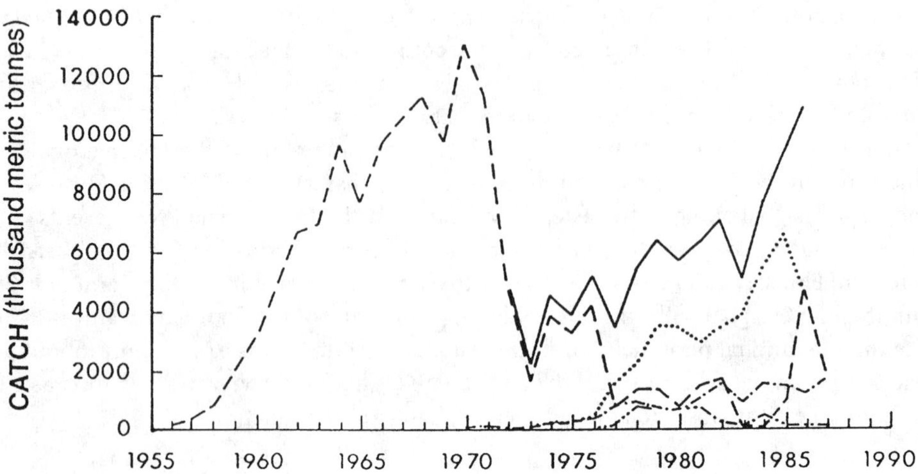

Figure 11-1. Catches (10^3) of pelagic fish in the Humboldt Current ecosystem. Pelagics, total (——); anchovy (– – –); sardine (· · ·); horse mackerel (–·–·–); mackerel (–··–··–).

Peruvian anchovy fishery in the 1960s–1970s, is by far the largest and most important stock. Analyses of fish scale abundance from sediment cores show that anchovy was always the dominant species in this area (DeVries and Pearcy 1982). From 1960 until 1972, its mean biomass was on average 15 million tons (Zuta et al. 1983; Tsukayama 1983). Biomass peaked in 1967 at about 22–23 million tons. The stock decreased dramatically to 9 million tons at the beginning of 1972 and further to 4.1 million tons at the end of 1972 (Zuta et al. 1983; Tsukayama 1983). Within a year, the stock lost about 5 million tons due to excessive fishing. The comparatively strong ENSO of 1972 concentrated the anchovy near the coast where there were still some cold upwelling plumes. This made the anchovy extremely vulnerable to fishing. In 1972, 4.4 million tons of anchovy were caught within 3 months (Tsukayama 1983). Similar trends in anchovy biomass were shown by a monthly VPA-based time series that included predatory effects on anchovy, developed by Pauly and Palomares (1989). According to their estimates, the anchovy biomass peaked in 1970 with 23 million tons.

Anchovy catches peaked in 1970 with 10.9 million tons for the northern and central stock (Figure 11-3). The catches fell dramatically during the ENSO in 1972, remained between 1–6 million tons until 1982, and decreased to an extremely low level during the ENSO of 1982–1983, and particularly thereafter in 1984. However, in 1985 the stock recovered and catches rose to 3 million tons in 1986.

Several interesting aspects of this sequence in catch and biomass data should be noted:

1. According to Zuta et al. (1983) and Tsukayama (1983), the biomass of Peruvian anchovy was 22–23 million tons in 1967. It dropped rather rapidly to about 12 million tons in 1969. According to Pauly and Palomares (1989), the biomass peaked in 1970 with 23 million tons and dropped in 1971 to 9–14 million tons.
2. The recruitment collapse of 1971 occurred *before* the onset of the 1972–1973 ENSO (Mendelsohn and Mendo 1987; Mendelsohn 1989).
3. In August/September 1981, the spawning biomass of the northern/central stock was estimated at 1.2 million tons (95% confidence limits: ±58%) (Santander et al. 1984) by applying the Egg Production Method (Lasker 1985; Alheit 1985).
4. In 1982, the recruitment from the 1981 year-class was very poor (Zuta et al. 1983) and 1.2 million tons of anchovy from the northern/central stock were caught (Figure 11-3). It is,

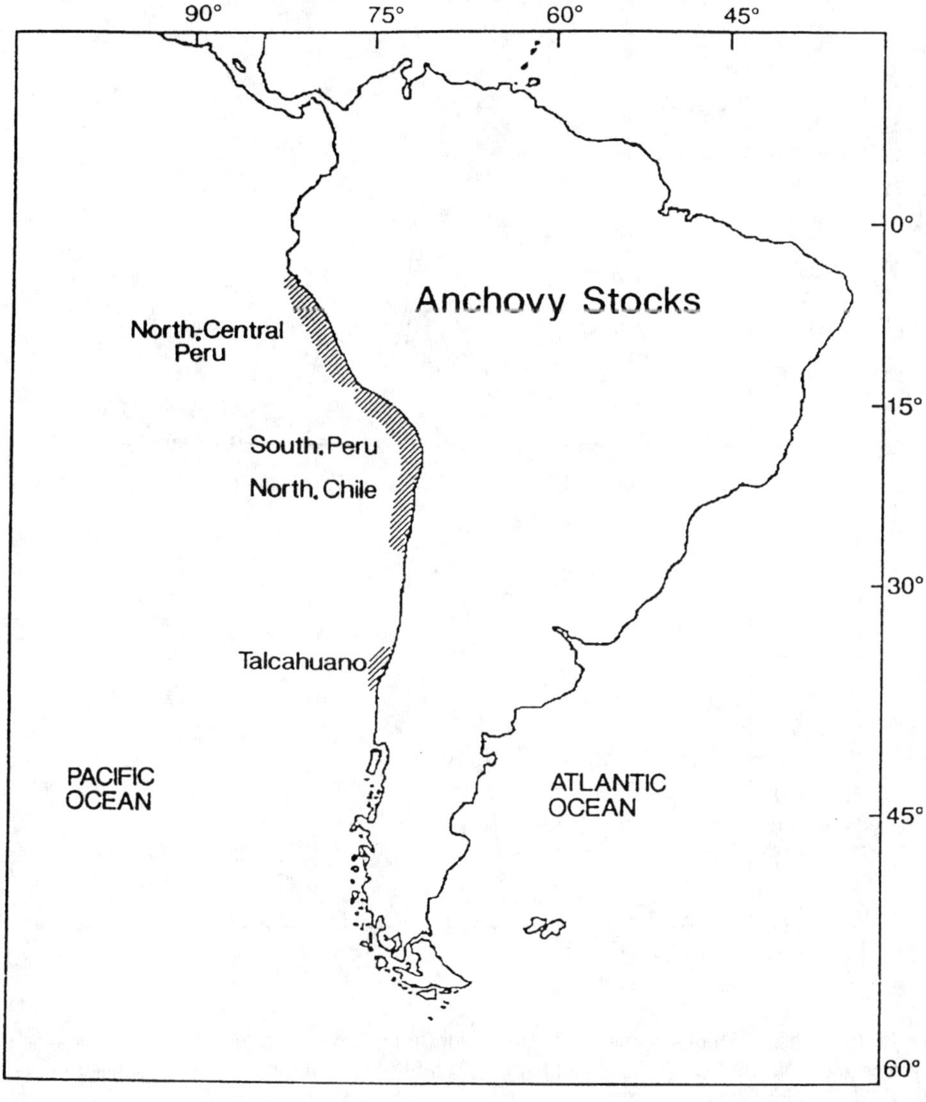

Figure 11-2. Distribution area of anchovy stocks in the Humboldt Current ecosystem.

therefore, concluded that at the onset of the 1982–1983 ENSO, the biomass of this stock must have been extremely low, very likely far below 1 million tons, due to excessive fishing pressure and poor recruitment. Tsukayama and Santander (1987) report that in 1983 Peruvian anchovy had the lowest biomass ever recorded.

5. This extremely low stock virtually exploded to several million tons in 1985 and 1986 (probably well above 6 million tons, according to Csirke [1988]) when 670,000 and 3 million tons, respectively, were caught (Figure 11-3). It is not known where this excellent recruitment originated. Very few anchovies were caught during the ENSO event. Arntz and Tarazona (1989) showed that some anchovies withdrew to deeper water (>100 m) to survive. However, Barber et al. (1985) reported that phytoplankton concentrations at these depths were too low to support larger numbers of anchovy for any length of time. An alternative explanation might be that the recruitment of the northern/central stock off Peru came from the stock off southern Peru and northern Chile, where the effects of the 1982–1983 ENSO were less dramatic.

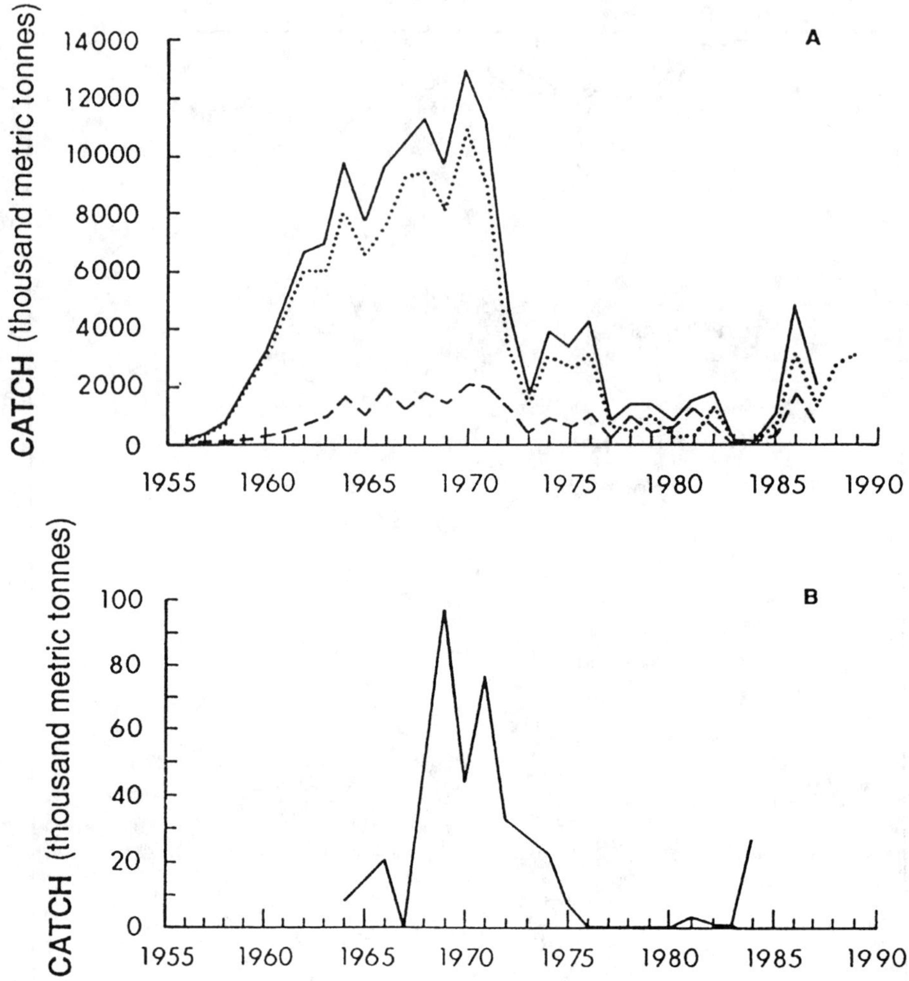

Figure 11-3. Catches (10^3) of anchovy in the Humboldt Current ecosystem. A. Anchovy, total (——); northern/central stock off Peru (· · · ·); southern stock off Peru and northern stock off Chile (– – – –). B. Stock off Talcahuano in Chile.

Whatever caused this surprising recovery of the northern/central stock, it is obvious that the Peruvian anchovy can increase its population size rapidly and drastically once the environmental conditions are favorable. Although the anchovy population is reduced by the effect of ENSO or of "warm" years, ENSO does not seem to have such a long-lasting impact on the anchovy, nor are its effects necessarily as drastic and deleterious in the long term as previously assumed.

Comparing the annual catches of the three anchovy stocks in the Humboldt Current, the almost simultaneous development from 1956 to 1980 is striking (Figure 11-3). Catches rose steadily each year from 1956 on and peaked initially in 1964. They dropped simultaneously in 1965 (ENSO). All three stocks supported high catch levels from 1966 to 1971, and catches of the two large stocks peaked in 1970. Catches decreased again simultaneously in 1972 (ENSO) and entered a period of strong fluctuations until the 1980s. However, catches of the stock off southern Peru and northern Chile remained relatively high and did not show the dramatic, almost steady decline of the stock off northern/central Peru. Catches of all three stocks declined to extreme lows in response to the ENSO event in 1982–1983, but rose again considerably in 1986 (Talcahuano data not included). The anchovy landings in northern Chile even reached a historic record in 1986

(1.25 million tons), when a fully developed fishery switched its target species from sardine to anchovy and benefited from two sudden strong recruitment pulses in 1984 and 1987.

Sardine

Four sardine stocks have been identified in the Humboldt Current ecosystem (Serra and Tsukayama 1988): (1) the stock off Ecuador and northern/central Peru, (2) the stock off southern Peru and northern Chile, (3) the stock off Coquimbo in Chile, and (4) the stock off Talcahuano in Chile (Figure 11-4).

Sardine spawning (Zuta et al. 1983) and catches were insignificant (Serra 1983) during the 1950s–1960s. From 1964 to 1971, the only distinct spawning areas were in northern Peru and in northern Chile (Bernal et al. 1983). After 1971, the sardine expanded to the northern and southern extremities of both refuge areas. Sardine spawning off Peru from 1966–1968 was poor

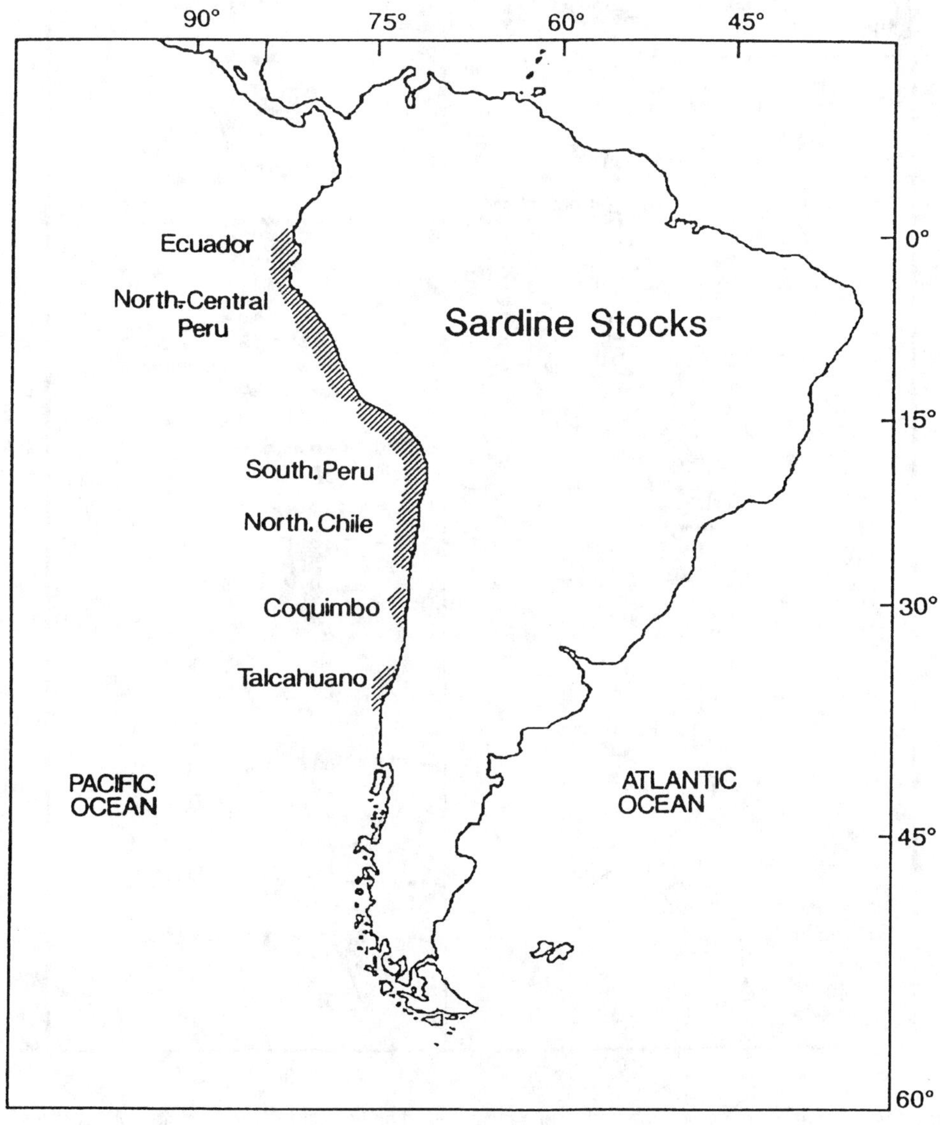

Figure 11-4. Distribution area of sardine stocks in the Humboldt Current ecosystem.

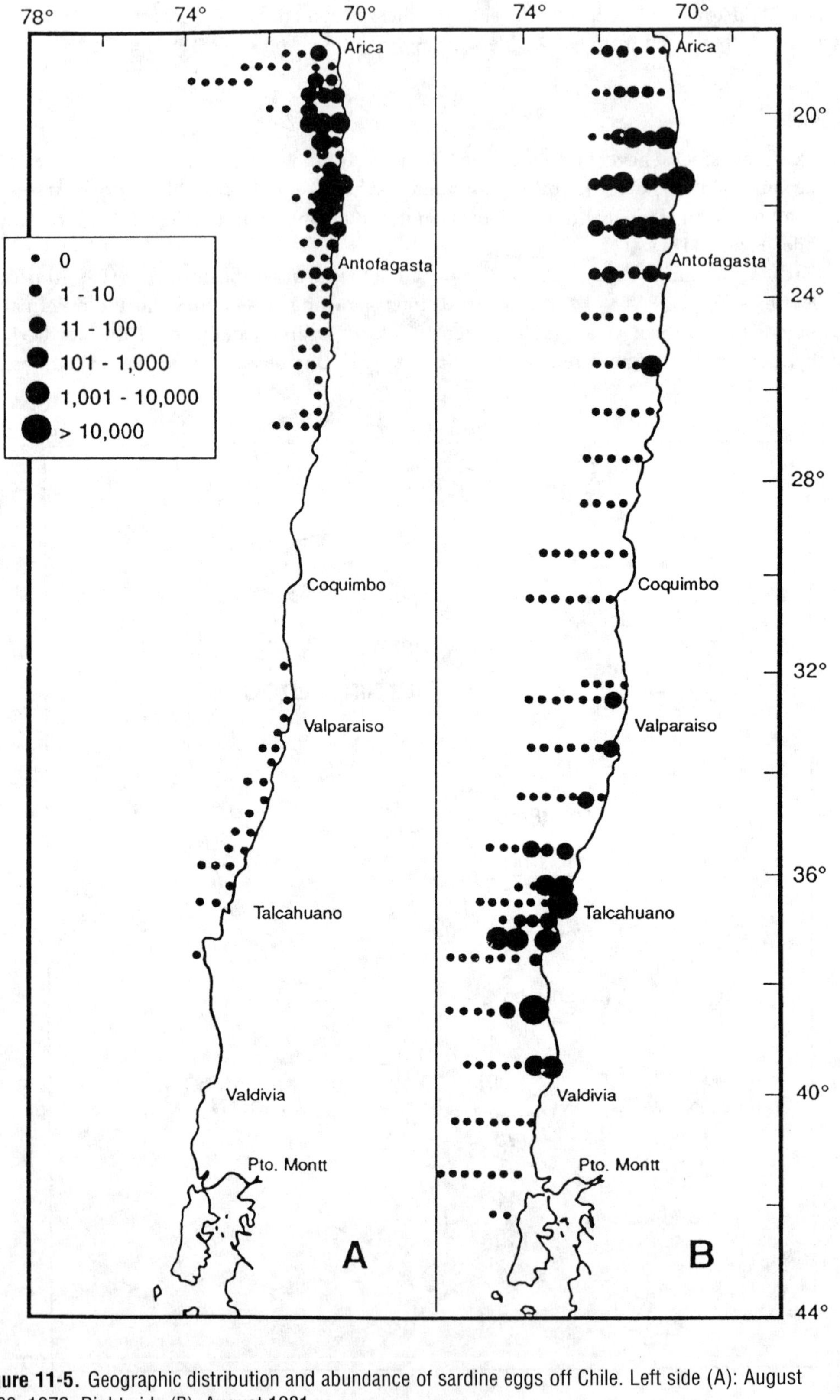

Figure 11-5. Geographic distribution and abundance of sardine eggs off Chile. Left side (A): August 1963–1972. Right side (B): August 1981.

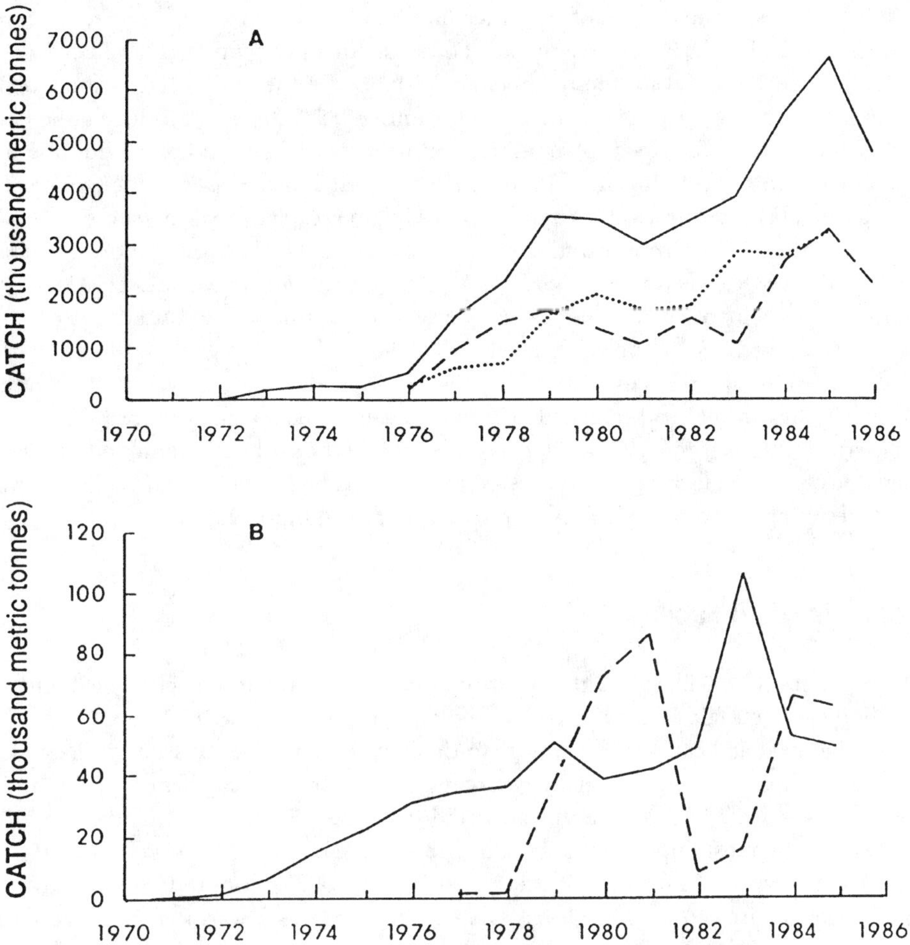

Figure 11-6. Catches (10^3) of sardine in the Humboldt Current ecosystem. A. Sardine, total (——); stock off Ecuador and northern/central Peru (– – – –); stock off southern Peru and northern Chile (· · · ·). B. Stock off Coquimbo (——); stock off Talcahuano (– – – –).

and limited to the region between 6–10°S (Zuta et al. 1983). After 1969, an increase in spawning was observed. After the ENSO of 1972–1973, sardine spawning increased strongly, and the spawning area expanded considerably, although the new spawning areas did not overlap with those previously occupied by the anchovy. From 1976–1980, spawning increased further (Zuta et al. 1983). From 1964–1973, no eggs and larvae were observed in Chile south of 25°S. However, a new spawning area off Talcahuano was established subsequently (Serra 1983; Bernal et al. 1983) (Figure 11-5).

Sardine spawning increased and the geographic distribution of spawning expanded during warm years, e.g., 1969 (Santander and Flores 1983). This phenomenon appears to be related to the more frequent coastal advance of the subtropical surface waters (Santander and Flores 1983; Tsukayama 1983).

An increase in the relative abundance of sardine and horse mackerel was observed in 1970 (Subsecretaria 1983). Catches of sardines in the Humboldt Current ecosystem increased after the ENSO of 1972–1973 when anchovy biomass and landings dropped dramatically (Figure 11-6). Off Coquimbo in 1973, sardine catches increased significantly and, in 1974, the sardine became the dominant species. Anchovies had never been caught in the region off Coquimbo (Mendez 1987).

Significant increases in the catches of the other three sardine stocks were observed several years after the 1972–1973 ENSO: (1) 1976 in the stock off Ecuador and northern/central Peru, (2) in the stock off southern Peru and northern Chile, and (3) 1978 in the stock off Talcahuano (Figure 11-6). Catches of the two large stocks rose steadily until 1985. Interestingly, during the ENSO in 1983, the landings from the stocks off southern Peru/northern Chile and off Coquimbo peaked, whereas the stock off northern/central Peru/Ecuador was at a low. This was due to (1) extensive southward migrations of the sardines apparently trying to escape the warm waters coming from the north, and (2) sardines concentrating close to the coast, thus becoming more vulnerable to fishing (Mendez 1987; Tsukayama and Santander 1987). According to Mendez (1987), in 1983, it was no longer possible to separate the sardine stocks from northern and southern Peru. In 1983, during the strongest ENSO event of this century, Chile landed the highest sardine catches ever recorded, whereas total catches of pelagic fish in Peru dropped considerably. Two years afterward in 1985, when the sardines had returned northward again, Ecuador had a record year of sardine catches (Serra and Tsukayama 1988). There is evidence that the sardines caught off northern Chile had migrated considerable distances southward from Peru. Their fat content was reduced by more than 40%, resulting in a low-quality, lower-priced fish meal (Romo 1985).

Impact of Physical Factors

ENSO is commonly considered as the overriding physical factor influencing yield and species composition of the pelagic catches in the Humboldt Current. The year 1972 saw a dramatic drop in anchovy biomass and catches in response to the ENSO event. The subsequent ENSO in 1976 caused another drastic decrease in anchovy catches. Simultaneously, catches of sardine and horse mackerel surpassed the anchovy landings for the first time (Figure 11-1), and after 1977, the sardine became the most important species in the fishery. The strong ENSO event of 1982–1983 reduced the anchovy to a historic low, with catches around 100,000 tons in 1983. Anchovy catches have subsequently risen to medium levels. Whenever there was a warm year, including ENSO years, the anchovy catches dropped and sardine catches increased.

Clearly, El Niño has extremely adverse effects on the anchovy, reducing the survival of all life history stages and increasing its vulnerability to the fishery, whereas it has a positive effect on sardines. Although ENSO events are relatively frequent (DeVries and Pearcy 1982), the anchovy has always recovered. Therefore, it seems unlikely that ENSO events in the 1970s are the sole cause of the decrease of the anchovy populations and the shift in species dominance in the pelagic fish community of the Humboldt Current. The combined effect of the 1972–1973 ENSO, along with heavy fishing pressure, has often been suggested as an explanation for the decline of the anchovy in the 1970s (Csirke 1988; Muck 1989b). However, the fast recovery of the anchovy after the strong ENSO in 1982–1983 proves otherwise. Most likely, first-order, direct effects of ENSOs on the anchovy and/or sardine are not sufficient to explain the observed changes since ENSOs affect all species in the ecosystem. Therefore, the outcome in terms of species composition at one or more trophic levels must be influenced by a complex set of interspecific second-order interactions and direct effects (Pianka 1987). These more subtle ecological effects are still poorly understood.

It has been demonstrated here that there are a number of changes in the pelagic fish community that began prior to the ENSO of 1972–1973: (1) the recruitment collapse of anchovy in 1971, before the onset of ENSO (Mendelsohn and Mendo 1987; Mendelsohn 1989); (2) an increase in sardine spawning since 1969 (Zuta et al. 1983); and (3) an increase in the relative abundance of sardine and horse mackerel in 1970 (Subsecretaria 1983).

In recruitment studies of the northern/central anchovy stock off Peru, Mendelsohn (1989) defined "recruitment" as the population estimates of the smallest class (length range = 3.75–4.25 cm) in the fishery (Pauly and Palomares 1989). Mendelsohn's estimates increase from about 1963 to 1969–1970, after which they drop dramatically. Recruitment was fairly constant until 1969, peaked conspicuously in 1969, and decreased after 1971 to a new low equilibrium (Mendelsohn 1989). The 1960s may have been a period of exceptionally and consistently high recruitment (Pauly and Palomares 1989).

Mendelsohn (1989) shows a number of time series of environmental variables that he considers in "all likelihood" as surrogates for the actual physical processes that affect the anchovy. Patterns in two of these data sets match the biological trends presented here: (1) a higher mean level of oceanic offshore transport in the 1970s–1980s than in the 1960s, which, in addition, appears to have been a decade of decreased variance in transport; and (2) a slight upward trend in sea surface temperature (SST) off Peru from 1971 onward.

In Chile, the increase of the sardine stock after the ENSO of 1972–1973 is matched by the upward trend of SST off Africa (Figure 11-7). This is in turn reflected by the general warming of the coastal zone off northern Chile, which lasted several years.

According to Muck (1989a), SST variability reflects the dynamics of other physical properties such as temperature, upwelling, turbulence, depth of mixed layer, etc. He studied the standard deviation of SST from simultaneous monthly measurements of nine shore stations along the Peruvian coast. He reported obvious changes in the anomalies of the SST standard deviations after 1968, when the trend switched from mainly negative to positive. The anomalies of the standard deviations of SST determined for the month of October (spawning peak of Peruvian anchovy) were negative or moderately positive until 1968 and from then on mainly positive until 1983. The eight most positive years were in the period from 1969–1978 (Figure 11-7).

Other groups of organisms have also undergone fundamental changes in their population sizes that appear to have started in the late 1960s. Bernal et al. (1983) demonstrated a decrease in zooplankton biomass off northern Chile beginning in 1969 (Figure 11-7). A time series of zooplankton biomass off Peru from 1964–1986 shows the same trend (Carrasco and Lozano 1989). The high peaks of zooplankton volume before 1971 were never again reached in subsequent years (Figure 11-7).

The strongest evidence for biological changes occurring in the Humboldt Current prior to 1971 is presented by Loeb and Rojas (1988). They studied the interannual variation of ichthyoplankton composition and abundance relations off northern Chile from 1964 to 1983 and observed a marked shift in the relative abundance of larval fish of the nonfished mesopelagic species from 1969 to 1970. The abundance of mesopelagic species is not influenced directly by fishing activities and might therefore, in contrast to anchovy and sardine, be a good indicator of environmental changes. Loeb and Rojas (1988) suggested that a subtle, but large-scale (low-frequency), environmental change occurred in the 1969–1970 period. This suggestion is supported by several long-term physical databases from Chile and Peru. From 1960 to 1972, the sea-surface atmospheric pressure anomalies that occurred off northern Chile were predominantly negative, particularly in 1969. After 1972, positive anomalies were predominant (Figure 11-7). Off Peru, the wind-driven turbulent mixing index of surface waters generally increased during and after the ENSO of 1972–1973 (Figure 11-7). Loeb and Rojas (1988) assumed that increased nearshore influence of the subtropical surface waters might be responsible for the increased larval survival and recruitment of sardines, as previously suggested by Santander and Flores (1983). Loeb and Rojas (1988) concluded that an atmospherically driven oceanic circulation change, beginning in the late 1960s and possibly involving onshore presence of subtropical and/or oceanic waters, might be responsible for the shifts observed in zooplankton and larval and adult fish occurring in the late 1960s. They suggested that the fluctuations in anchovy and sardine populations are primarily governed by low-frequency hydrographic events augmented secondarily by fisheries.

Important changes in the Humboldt Current ecosystem, such as the collapse of the anchovy and the rise of the sardine after the ENSO of 1972–1973, have been associated by many authors with the occurrence of El Niño. We are suggesting that there are other more subtle trends in the

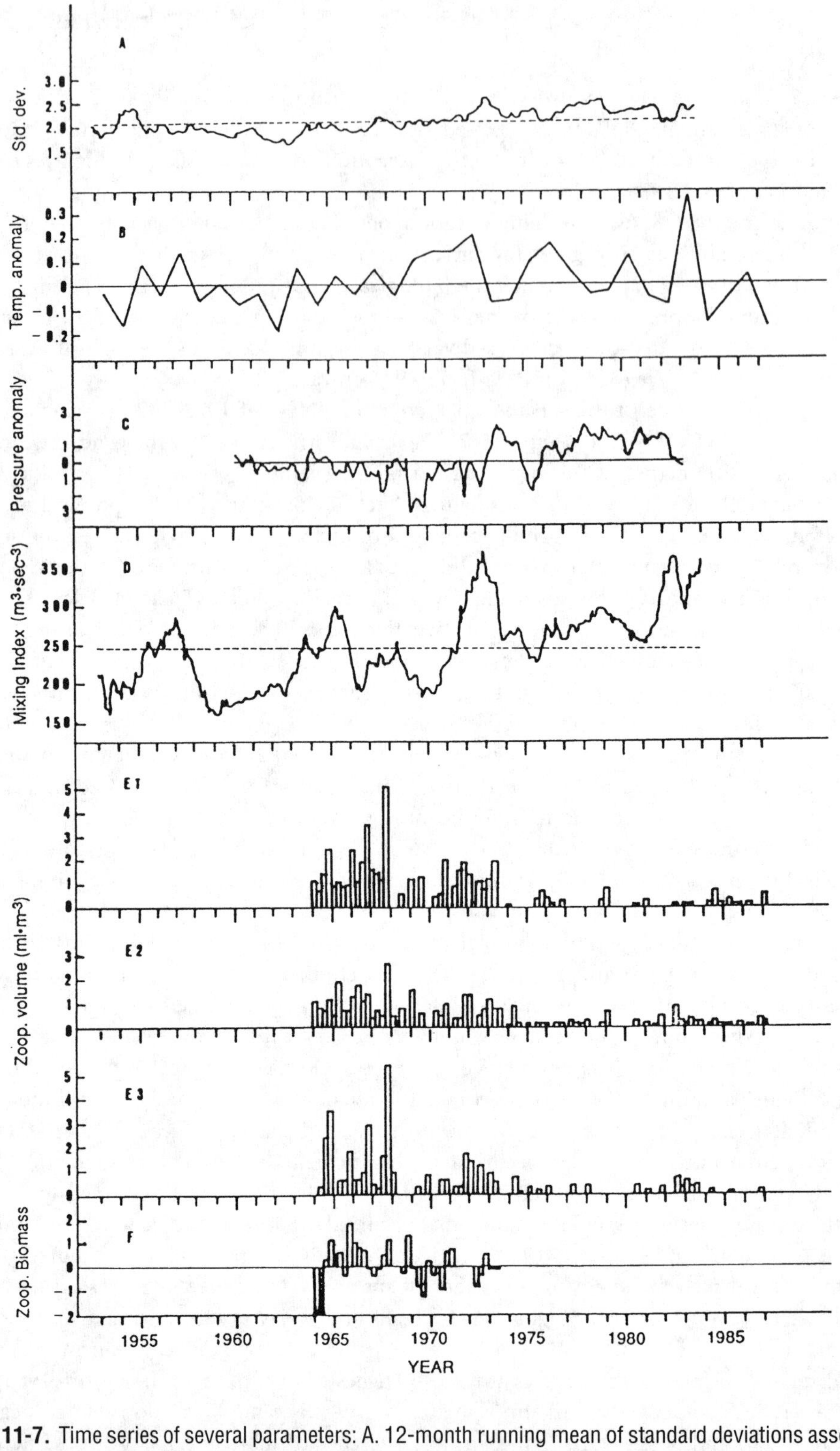

Figure 11-7. Time series of several parameters: A. 12-month running mean of standard deviations associated with mean monthly sea surface temperature values off Peru (5°–15° S), 1953–1984 (after Loeb and Rojas 1988). Dotted line: long-term mean values. B. Anomalies of mean annual standard deviation of sea surface temperature from nine coastal stations of Peru (4°–17° S), 1953–1987 (after Muck 1989a). C. Sea-surface atmospheric pressure anomalies off Arica, Chile (18° S), 1960–1982 (after Loeb and Rojas 1988). D. Surface layer turbulent mixing index values off Peru (5°–15° S), 1953–1984 (after Bakun 1987). E. Zooplankton volume off Peru, 1964–1987 (after Carrasco and Lozano 1989). Areas: E1 (3°30′–5°59′ S); E2 (6°–13°58′ S); E3 (14°–18°30′ S). F. Zooplankton biomass (standardized units) off Chile, 1964–1973 (after Bernal et al. 1983).

physical oceanography of the system that might play a similar or greater role in the changes described for the Humboldt Current. These trends are of lower frequency than the occurrence of the ENSO, and therefore modify local conditions for longer periods of time. Such trends have often been overlooked because they are difficult to observe and extract from the noisy record of environmental variables. The frequency domain in time and space of these low-frequency changes is almost inaccessible in regions where geophysical time series are short, sporadic, or of low quality. In addition, we also suggest that the changes occur through direct and indirect effects among the several species inhabiting the system. Indirect effects include exploitative competition, predation, apparent competition, indirect mutualism, and other processes, including second- and third-order interactions (Pianka 1987).

In spite of the remarkable interest of and research effort applied by the international oceanographic community to the physical processes governing the Humboldt Current ecosystem, there are still important gaps in our knowledge. For example, all data available indicate that during ENSO years global atmospheric forcing of currents dominates. During non-ENSO years, local forcing could play a more important role. The mechanisms of local forcing in the Humboldt Current are not well understood and definitely require more attention in the future (E. Hagen, personal communication). In the studies of global forcing, particular attention was usually given to the impact of equatorial currents on the Humboldt Current and the Peruvian Undercurrent. However, the impact of changes in the southern currents (e.g., Circumpolar Current and West Wind Drift) on the Humboldt Current has been greatly neglected so far (E. Hagen, personal communication). Basinwide atmospheric forcing and its interannual variability must play a role in controlling these large-scale, low-frequency phenomena. Lluch-Belda et al. (1989) reached similar conclusions when comparing worldwide fluctuations of sardine and anchovy stocks. They suggested that long-term fluctuations of these species seem to be dominated by environmental variations that can give rise to large and prolonged changes in biomass and yield, thus causing persistent "regimes" of high and low abundance of the species. The apparent coherence in trends of various sardine stocks on an oceanic scale leads the authors to conclude that the mechanisms responsible for long-term changes in stock sizes, which are not yet clear, are operating also on oceanic scales.

The Role of Predation

The impact of predation on all life history stages of the anchovy has received considerable attention recently.

Egg mortality rates of Peruvian anchovies and sardines can be extremely high. Smith et al. (1989) showed that the instantaneous rates of daily egg mortality (Z) in August/September 1981 were, depending on the region, between 0.90 and 2.14 for anchovies and between 1.28 and 3.88 for sardines off Peru. At times, anchovies feed heavily on their own eggs (de Ciechomski 1967; Hunter and Kimbrell 1980; MacCall 1981; Alheit 1987; Valdes et al. 1987). A considerable part of the high egg mortality in anchovy is due to cannibalism. Twenty-two percent of the total egg mortality at peak spawning time of Peruvian anchovy in August/September 1981 was caused by cannibalism (Alheit 1987); 28% in Californian anchovy (Hunter and Kimbrell 1980; MacCall 1981); and up to 70% in South African anchovy, in an area of particularly high adult density (Valdes et al. 1987).

Whereas in August/September 1981, 76.4% of the Peruvian anchovy had anchovy eggs in their stomachs (Alheit 1987), similar studies in February 1985, the period of the secondary spawning peak, revealed that only few anchovies had fed on their eggs (Alheit, unpublished). This is initially surprising, as the anchovy biomass in 1981 was 1.2 million tons (Santander et al. 1984) and around 3 million tons in February 1985 (Pauly and Palomares 1989) and one might expect lower cannibalism rates at lower biomass values operating as a compensatory mechanism. However, anchovy eggs were hardly encountered farther offshore than 30 miles from the coast in 1981 (Santander et al. 1984), but found up to 90 miles offshore in 1985. Population density appears to be the decisive factor influencing egg cannibalism rates, rather than biomass. In addition, anchovies tend to concentrate near the coast (Jordan 1971) during peak spawning in August/September.

These data are interesting in view of the findings of Mendelsohn and Mendo (1987), who analyzed seasonal recruitment success of the Peruvian anchovy from 1953 to 1981. They found that the annual recruitment peak stems from the secondary spawning peak in January/February, whereas the peak spawning in August/September leads only to few recruits.

Pauly (1987) suggested that egg mortality due to cannibalism is a strong density-dependent mechanism controlling recruitment of the Peruvian anchovy. Smith et al. (1989) discussed thoroughly whether or not interannual differences in the rate of egg cannibalism are sufficient to explain differences in recruitment, and they concluded that this would be unlikely.

In addition to anchovy, other filtering fishes, such as sardines; juveniles of other species; and invertebrates prey heavily on anchovy eggs. Predation by euphausiids on anchovy eggs off California accounts for 28% of the total egg mortality (Theilacker 1988), and the mean number of anchovy eggs in sardine stomachs (per unit of fish weight) was more than double that in anchovy stomachs in Peruvian waters (Alheit 1987). Clearly, the impact of predation on eggs and larvae of anchovies and sardines needs further study.

Until recently, it was assumed that the guano birds were the major predators of anchovy in the Humboldt Current (e.g., Jordan 1967; Schaefer 1970; Murphy 1972; Furness 1982). However, Muck (1989a, b) recently showed that predation by horse mackerel, mackerel, and hake far exceeds the consumption of anchovy by birds. He assumed that predation of horse mackerel on the anchovy of the central/northern stock throughout the 1950s and in the mid-1970s surpassed the anchovy catches of the fishery (Muck and Sanchez 1987). Since horse mackerel and mackerel are usually distributed farther offshore than anchovy, predation is particularly heavy in years of higher SST, which brings the horse mackerel closer to the shore and leads to increasing spatial overlap of both species. According to Muck (1989b), coastal SST appears to be an appropriate parameter for quantification of the distribution overlap between both species, and he suggested that SST-mediated distribution patterns of anchovy and horse mackerel largely determine the dynamics of the anchovy stocks. Consequently, he proposed a flexible exploitation scheme for Peru according to which anchovy should be fished particularly when its biomass is high and SST is low and horse mackerel should be exploited heavily when SST is high.

Conclusion

After consideration of the impact of several physical and biological factors on the dynamics and yield of pelagic fish in the Humboldt Current, the question arises over what management advice could be given. The situation is complicated by the fact that most of the stocks are shared by two countries. The alongshore migrations of the pelagic species, which are governed by hydrographic conditions, often lead to unusually high concentrations of fish in one country, which might cause excessive (over) fishing by this country much to the dismay of the other country. In addition, certain environmental situations, such as ENSO, that have adverse effects on the fishery in one country may increase yields in another country, as demonstrated by the decrease of anchovy catches in Peru and the simultaneous increase of sardine landings in Chile in 1982–1983. Furthermore, during the period when the yields of pelagic fish steadily declined in Peru in the 1970s and early 1980s, total catches of pelagic fish in the northern fishery of Chile, as well as in the southern fishery off Talcahuano, more than doubled. Interestingly, up to 1989, total pelagic yield in the Humboldt Current had almost reattained the high values of the early 1970s with the difference that the fishery had changed from a monospecific to a multispecies fishery.

Because of the overriding importance of physical factors for biomass yields in the Humboldt Current, it seems to be of utmost necessity to establish an environmental monitoring system in which all three countries, Ecuador, Peru, and Chile, actively participate in a cooperative manner. This might be established under the leadership of the Comision Permanente del Pacifico Sur (CPPS) with co-sponsorship of United Nations agencies such as IOC, FAO, or UNEP. A strong environmental database might allow future predictions of changes in the dynamics of the fishes in the Humboldt Current ecosystem and could contribute considerably to a rational fisheries management system for the region.

References

Alheit, J. 1985. Egg Production Method for Spawning Biomass Estimates of Anchovies and Sardines. ICES C.M. 1985/H:41.

Alheit, J. 1987. Egg cannibalism versus egg predation: Their significance in achovies. S. Afr. J. Mar. Sci. 5:467–470.

Alheit J. Unpublished data report. Warnemunde, Germany: Institute for Baltic Research.

Arntz, W., and Tarazona, J. 1989. Effects of El Niño 1982–83 on Benthos, Fish and Fisheries off the South American Pacific Coast. In P. W. Glynn (ed.), Global Ecological Consequences of the 1982/83 El Niño–Southern Oscillation. Elsevier Oceanography Series, pp. 323–360.

Bakun, A. 1987. Monthly Variability in the Ocean Habitat off Peru as Deduced from Maritime Observations, 1953–1984. In D. Pauly and I. Tsukayama (eds.), The Peruvian Anchoveta and Its Upwelling Ecosystem: Three Decades of Change. ICLARM Studies and Reviews 15, pp. 46–74.

Barber, R. T., Chavez, F. P., and Kogelschatz, J. E. 1985. Biological effects of El Niño. Efectos biologicos de El Niño. Com. Perm. Pacif. Sur Bol. ERFEN. 14:3–29.

Bernal, P. 1981. A review of the low-frequency response of the pelagic ecosystem in the California Current. Calif. Coop. Oceanic Fish. Invest. Rep. 22:49–62.

Bernal, P. A., Robles, F. L., and Rojas, O. 1983. Variabilidad fisica y biologica en la region meridional del sistema de corrientes Chile-Peru. FAO Fish. Rep. 291:683–711.

Brockmann, C., Fahrbach, E., Huyer, H., and Smith, R. L. 1980. The poleward undercurrent along the Peru coast: 5–15. Deep-Sea Res. 27A:847–656.

Carrasco, S., and Lozano, O. 1989. Seasonal and Long-term Variations of Zooplankton Volumes in the Peruvian Sea, 1964–1987. In D. Pauly, P. Muck, J. Mendo, and I. Tsukayama (eds.), The Peruvian Upwelling Ecosystem: Dynamics and Interactions. ICLARM Conference Proceedings 18, pp. 82–85.

Castillo, S., and Mendo, J. 1987. Estimation of Unregistered Peruvian Anchoveta (*Engraulis ringens*) in Official Catch Statistics, 1951 to 1982. In D. Pauly and I. Tsukayama (eds.), The Peruvian Anchoveta and Its Upwelling Ecosystem: Three Decades of Change. ICLARM Studies and Reviews 15, pp. 109–116.

de Ciechomski, J. D. 1967. Investigations of food and feeding habits of larvae and juveniles of the Argentine anchovy *Engraulis anchovita*. Calif. Coop. Oceanic Fish. Invest. Rep. 11:72–81.

Csirke, J. 1988. Small Shoaling Pelagic Fish Stocks. In J. A. Gulland (ed.), Fish Population Dynamics. Chichester: Wiley Interscience, pp. 271–302.

DeVries, T. J., and Pearcy, W. G. 1982. Fish debris in sediments of the upwelling zone off central Peru: A late Quaternary record. Deep-Sea Res. 29:87–109.

Espinoza, F. R., Neshyba, S., and Mexicano, Z. 1983. Surface Water Motion off Chile Revealed in Satellite Images of Surface Chlorophyll and Temperature. In P. M. Arana (ed.), Proceedings of the International Conference on Marine Resources of the Pacific. Universidad Catolica de Valparaiso, pp. 41–57.

Fonseca, T. R., and Farias, M. 1987. Estudio del proceso de surgencia en la costa chilena utilizando percepcion remota. Invest. Pesq. (Chile) 34:33–46.

Furness, R. W. 1982. Competition between fisheries and sea bird communities. Adv. Mar. Biol. 20: 225–307.

Hagen, E. 1990. Verbal communication on 17/11/90. Institute for Baltic Research, Warnemunde, Germany.

Hunter, J. R., and Kimbrell, C. A. 1980. Egg cannibalism in the northern anchovy, *Engraulis mordax*. Fish. Bull., U.S. 78:811–816.

Jordan, R. 1967. The predation of Guano birds on the peruvian anchoveta (*Engraulis ringens* Jenyns). Calif. Coop. Oceanic Fish. Invest. Rep. 11:105–109.

Jordan, R. 1971. Distribution of anchoveta (*Engraulis ringens*) in relation to the environment. Invest. Pesq. 35:113–126.

Lasker, R. 1985. An Egg Production Method for Estimating Spawning Biomass of Pelagic Fish: Application to the Northern Anchovy, *Engraulis mordax*. NOAA Technical Rep. NMFS 36.

Lluch-Belda, D., Crawford, R. J. M., Kawasaki, T., MacCall, A. D., Parrish, R. H., Schwartzlose, R. A., and Smith, P. E. 1989. World-wide fluctuations of sardine and anchovy stocks: The regime problem. S. Afr. J. Mar. Sci. 8:195–205.

Loeb, V. J., and Rojas, O. 1988. Interannual variation of ichthyoplankton composition and abundance relations off northern Chile, 1964–83. Fish. Bull., U.S. 86:1–24.

MacCall, A. D. 1981. The consequences of cannibalism in the stock-recruitment relationship of planktivorous pelagic fishes such as *Engraulis*. IOC Workshop Rep. 28:201–220.

Mendelsohn, R. 1989. Reanalysis of Recruitment Estimates of the Peruvian Anchoveta in Relationship to Other Population Parameters and the Surrounding Environment. In D. Pauly, P. Muck, J. Mendo, and I. Tsukayama (eds.), The Peruvian Upwelling Ecosystem: Dynamics and Interactions. ICLARM Conference Proceedings 18, pp. 364–385.

Mendelsohn, R., and Mendo, J. 1987. Exploratory Analysis of Anchoveta Recruitment off Peru and Related Environmental Series. In D. Pauly and I. Tsukayama (eds.), The Peruvian Anchoveta and Its Upwelling Ecosystem: Three Decades of Change. ICLARM Studies and Reviews 15, pp. 294–306.

Mendez, R. 1987. Cambios bioticos y efectos sobre los recursos pesqueros y pesquerias en Chile. Rev. Com. Perm. Pacifico Sur 16:7–96.

Menz, A., and French, S. 1982. The Fishery for Small Pelagic Fishes off the Coast of Ecuador: Its Development and Investigation. In J. C. Castilla (ed.), Segundo Seminario Taller. Bases biologicas para el uso y manejo de recursos naturales renovables: Recursos biologicos marinos. Monogr. Biol. 2:2–17.

Mooers, C. N. K., Flagg, C. N., Boicourt, W. C. 1978. Prograde and Retrograde Fronts. In M. J. Bowman and W. E. Esaias (eds.), Oceanic Fronts in Coastal Processes. Berlin: Springer-Verlag.

Muck, P. 1989a. Relationships between Anchoveta Spawning Strategies and the Spatial Variability of Sea Surface Temperature off Peru. In D. Pauly, P. Muck, J. Mendo, and I. Tsukayama (eds.), The Peruvian Upwelling Ecosystem: Dynamics and Interactions. ICLARM Conference Proceedings 18, pp. 168–173.

Muck, P. 1989b. Major Trends in the Pelagic Ecosystem off Peru and Their Implications for Management. In D. Pauly, P. Muck, J. Mendo, and I. Tsukayama (eds.), The Peruvian Upwelling Ecosystem: Dynamics and Interactions. ICLARM Conference Proceedings 18, pp. 386–403.

Muck, P., and Sanchez, G. 1987. The Importance of Mackerel and Horse Mackerel Predation for the Peruvian Anchoveta Stock (A Population and Feeding Model). In D. Pauly and I. Tsukayama (eds.), The Peruvian Anchoveta and Its Upwelling Ecosystem: Three Decades of Change. ICLARM Studies and Reviews 15, pp. 276–293.

Murphy, G. I. 1972. Fisheries in upwelling regions with special reference to Peruvian waters. Geoforum 11:63–71.

Pauly, D. 1987. Managing the Peruvian Upwelling Ecosystem: A Synthesis. In D. Pauly and I. Tsukayama (eds.), The Peruvian Anchoveta and Its Upwelling Ecosystem: Three Decades of Change. ICLARM Studies and Reviews 15, pp. 325–342.

Pauly, D., and Tsukayama, I. (eds.). 1987. The Peruvian Anchoveta and Its Upwelling Ecosystem: Three Decades of Change. ICLARM Studies and Reviews 15. Instituto del Mar del Peru (IMARPE), Callao, Peru; Deutsche Gesellschaft fur Technische Zusammenarbeit (GTZ), GmbH, Eschborn, Federal Republic of Germany; and International Center for Living Aquatic Resources Management (ICLARM), Manila, Philippines.

Pauly, D., and Palomares, J. L. 1989. New Estimates of Monthly Biomass, Recruitment, and Related Statistics of Anchoveta (*Engraulis ringens*) off Peru (4–14° S), 1953–1985. In D. Pauly, P. Muck, J. Mendo, and I. Tsukayama (eds.), The Peruvian Upwelling Ecosystem: Dynamics and Interactions. ICLARM Conference Proceedings 18, pp. 189–206.

Pauly, D., Muck, P., Mendo, J., and Tsukayama, I. (eds.). 1989. The Peruvian Upwelling Ecosystem: Dynamics and Interactions. ICLARM Conference Proceedings 18. Instituto del Mar del Peru (IMARPE), Callao, Peru; Deutsche Gesellschaft fur Technische Zusammenarbeit (GTZ), GmbH, Eschborn, Federal Republic of Germany; and International Center for Living Aquatic Resources Management (ICLARM), Manila, Philippines.

Pianka, E. R. 1987. The subtlety, complexity and importance of population interactions when more than two species are involved. Revista Chilena de Historia Natural 60:351–361.

Robles, F. L., Alarcon, E., and Ulloa, A. 1980. Water Masses in the Northern Chilean Zone and Their Variations in the Cold Period (1967) and Warm Periods (1969, 1871–73). Proceedings of the Workshop on the Phenomenon Known as "El Niño." UNESCO, pp 83–174.

Romo, D. 1985. Composicion quimica de la harina de pescado chilena durante el fenomeno el Nino 1982–83. Invest. Pesq. (Chile) 32:141–151.

Santander, H., and Flores, R. 1983. Los desoves y distribucion larval de cuatro especies pelagicas y sus relaciones con las variaciones del ambiente marino frente al Peru. FAO Fish. Rep. 291:835–867.

Santander, H., Alheit, J., and Smith, P. E. 1984. Estimacion de la biomasa de la poblacion desovante de anchoveta peruana, *Engraulis ringens*, en 1981 por aplicacion del "Metodo de Produccion de Huevos." Bol. Inst. Mar. Peru, Callao 8:208–250.

Schaefer, M. B. 1970. Men, birds, and anchovies in the Peru current—dynamic interactions. Trans. Am. Fish Soc. 99:461–467.

Serra, J. R. 1983. Changes in the abundance of pelagic resources along the Chilean coast. FAO Fish. Rep. 291:255–284.

Serra, R., and Tsukayama, I. 1988. Sinopsis de datos biologicos y pesqueros de la sardina *Sardinops sagax* (Jenyns, 1842) en el Pacifico suroriental. FAO Sinopsis sobre la Pesca 13.

Silva, N., and Fonesca, T. H. 1983. Geostrophic Component of the Ocean Flow off Northern Chile. In P. M. Arana (ed.), Proceedings of the International Conference on Marine Resources of the Pacific. Universidad Catolica de Valparaiso, pp. 59–70.

Smith, P. E., Santander, H., and Alheit, J. 1989. Comparison of the mortality rates of Pacific sardine, *Sardinops sagax*, and Peruvian anchovy, *Engraulis ringens*, eggs off Peru. Fish. Bull., U.S. 87:497–508.

Subsecretaria de Pesca. 1983. Recursos pelagicos y pesquerias en Chile. Rev. Com. Perm. Pacifico Sur 13:5–23.

Theilacker, G. H. 1988. Euphausiid Predation on Larval Anchovy at Two Contrasting Sites off California Determined with an ELISPOT Immunoassay. In C. M. Yentsch, F. C. Mague, and P. K. Horan (eds.), Immunochemical Approaches to Coastal, Estuarine, and Oceanographic Questions. Berlin: Springer-Verlag, pp. 304–311.

Tsukayama, I. 1983. Recursos pelagicos y sus pesquerias en Peru. Rev. Com. Perm. Pacifico Sur 13:25–63.

Tsukayama, I., and Santander, H. 1987. Cambios bioticos y efectos sobre los recursos pesqueros y las pesquerias en Peru. Rev. Com. Perm. Pacifico Sur 16:97–166.

Valdes, E. S., Shelton, P. A., Armstrong, J. J., and Field, J. G. 1987. Cannibalism in South Africa anchovy: Egg mortality and egg consumption rates. S. Afr. J. Mar. Sci. 5:613–622.

Wooster, W. W., and Fluharty, D. L. 1985. El Niño North: Niño effects in the eastern subarctic Pacific Ocean. Seattle: Washington Sea Grant Program Publication.

Wyrtki, K. 1967. Circulation and water masses in the eastern equatorial Pacific Ocean. Int. J. Oceanol. Limnol. 1:117–147.

Zuta, S., Tsukayama, I., and Villaneuva, R. 1983. El ambiente marino y las fluctuaciones de las principales poblaciones pelagicas de la costa peruana. FAO Fish. Rep. 291:179–253.

12

The Pacific Central American Coastal LME

Andrew Bakun, Jorge Csirke, Daniel Lluch-Belda, and Rafael Steer-Ruiz

Abstract

It is argued that the zone along the tropical Pacific Coast of the Americas located between the California Current LME and the Humboldt Current LME in itself constitutes a separate, distinct large marine ecosystem. The name "Pacific Central American Coastal LME" is proposed. Several of the transboundary issues related to the sharing of this LME coastline by nine separate nation-states are introduced. Various aspects are briefly discussed: the physical setting, seasonal ocean current patterns, upwelling mechanisms, the oxygen minimum layer, fisheries, threatened species, and habitat issues. Some brief socioeconomic background is also provided. Finally, some views on the outlook and needs for the future are offered.

Introduction

Two of the major coastal upwelling regions of the world, the California Current LME (MacCall 1986) and the Humboldt Current LME (Alheit and Bernal 1993) are situated on the Pacific Coasts of North and South America, respectively. However, between the strong tropical temperature fronts which lie off central Mexico and off Ecuador, marking the equatorward limits of these productive subtropical systems, lies a much less studied region which is characterized by recirculating coastal currents and much milder spatial temperature gradients and which possesses the distinct bathymetry, hydrography, productivity, and trophically dependent populations that characterize a discrete large marine ecosystem (Sherman and Alexander 1996). This coastal LME (Figure 12-1) extends along the coast roughly from Cabo Corrientes (about 20.4°N latitude, near the entrance to the largely enclosed Gulf of California LME) southwestwardly along the continental coastline to the vicinity of the equator. We propose that it be recognized as an important component of the LME structure of the world's oceans, and be accorded the designation Pacific Central American Coastal (PCAC) LME.

The PCAC LME stands out as being distinct from the two better known neighboring upwelling LMEs in a number of respects. The temperatures of the upper ocean layers are of course very much warmer, with large-scale monthly mean ocean surface temperatures remaining above 26°C throughout the year over the bulk of the region. Accordingly, the PCAC LME supports a characteristic tropical fauna which is distinctly different from the predominantly temperate fauna of the California and Humboldt systems. Moreover, in contrast to the normally very dry desert-

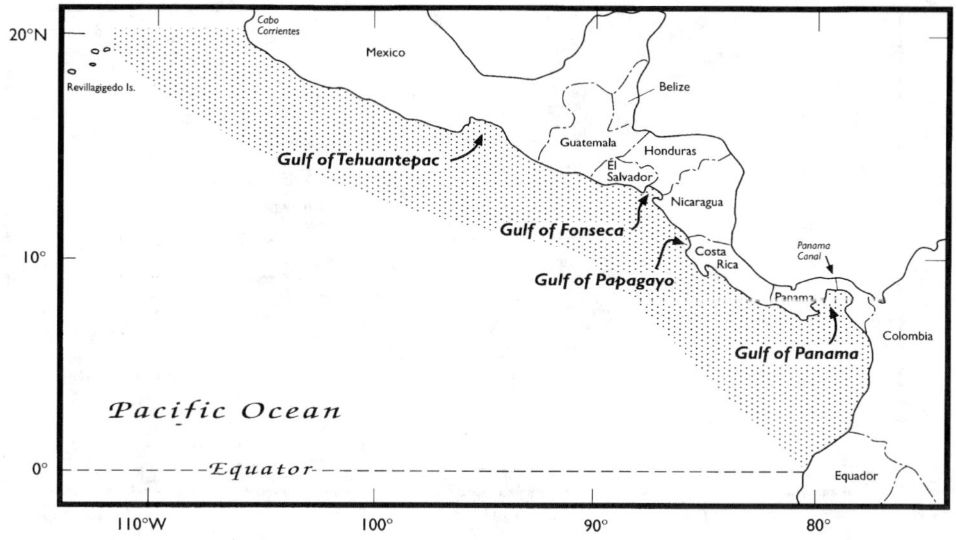

Figure 12-1. Map of the region showing major geographical features discussed in this chapter. Stippled area indicates the area encompassed by the PCAC LME.

like coastlands to the north and to the south, the coastlands of the PCAC LME grade from savanna in the north, to tropical rain forest along much of the Isthmus, to the coastal slopes of Colombia in the far south wherein lie the rainiest areas in all the Americas, surpassing 13 meters mean annual rainfall at some locations on the coastal mountain slopes. The ocean surface currents are not predominantly equatorward as in the upwelling LMEs, but form a seasonally varying complex of gyres, poleward coastal flows, and the recirculation patterns caused by the barrier to the equatorial current flows imposed by the continental boundary. In addition, the predominant enrichment that occurs is not via the "classic" eastern ocean upwelling produced by offshore Ekman transport which is driven by equatorward alongshore wind stress, but occurs through several other mechanisms, including (1) equatorial upwelling, (2) "open ocean" upwelling driven by wind stress curl, (3) as part of the mechanism underlying the Costa Rica Dome structure, or (4) as remarkable episodic "downwind" coastal upwellings forced by winds blowing across the Central American Isthmus from the Caribbean side. In contrast to the strong, coherent equatorward wind patterns of the upwelling LMEs, much of the PCAC LME lies within the sweep of the seasonal movements of the Intertropical Convergence Zone (ITCZ) (Figure 12-2), which is the transocean belt along which the air masses of the Northeast and Southeast trade wind systems flow against one another, producing the band of very weak horizontal winds known as the doldrums. The resulting zone of rising air along the ITCZ is characterized by progressions of spectacular tropical thunderstorm towers and associated torrential rain squalls which "punctuate" the typically smooth sea and calm wind conditions of the doldrums zone.

Nine nation-states share the coastline of the PCAC LME (Figure 12-1). Among these are some of the poorest countries of the Americas. Within these countries, some of the most impoverished segments of the population, driven by economic conditions or political strife, have migrated to the coastal zone, where they manage to glean a livelihood from subsistence fishing and farming. This is also a region of remarkable natural beauty. Accordingly, coastal development activities related to tourism, mariculture activities to produce luxury food items, sport fishing operations, etc., are beginning to compete with and to displace the subsistence-level population.

This is also a region of particularly rich natural ecological structure which may be likewise threatened by coastal development and marine-oriented activities. Very conspicuous factors are

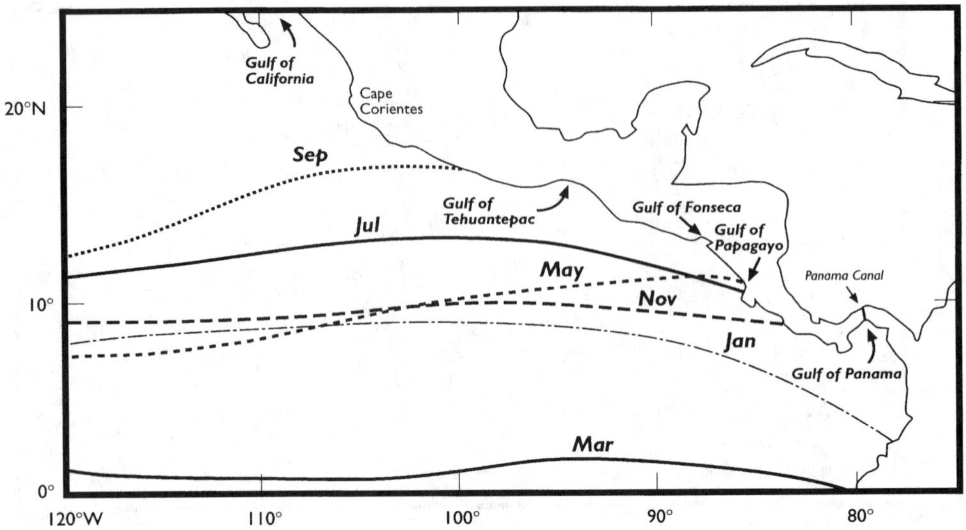

Figure 12-2. Mean monthly positions of the Intertropical Convergence Zone (ITCZ) (extracted from Wyrtki 1965b).

the location of the Panama Canal within this LME and the orientation of the coastlines with respect to maritime routes from the Canal to North and South America and to Asia. These have resulted in heavily trafficked shipping lanes, with attendant dangers of debris and substance spills, paralleling essentially the entire length of PCAC LME coastline.

The Physical Setting

From a geological standpoint, this LME lies at the leading edge of the westward drifting tectonic plate complex which underlies the American continental land masses. Accordingly, a strong crustal subduction zone exists immediately offshore stretching along the entire length of PCAC LME coastline. Consequently the continental slope is very steep, reaching extreme ocean depths very near the continental coast. One of the world's major deep-sea trenches, the Middle American Trench, parallels the coastline only about 100 km offshore. This near-coastal trench reaches depths greater than 4,000 meters all the way from Costa Rica to the northern extremity of the LME, reaching its deepest known depth, 6,662 meters (21,910 feet), off Guatemala. A somewhat shallower deep-sea trench, known as the Colombia Trench, also parallels the coast in the far southern portion of the LME. Because of crustal subduction processes, the entire LME coast lies within a zone of frequent earthquakes.

As commonly occurs at the leading edge of drifting continents, compression and folding of the continental crust has pushed up high coastal mountain ranges. A spectacular string of active volcanoes stretches along the Isthmus from Guatemala to Costa Rica. Another active volcano group exists in Central Mexico. And yet another group exists in Colombia and Ecuador. This close juxtaposition of high coastal mountains with the very deep near-coastal ocean results in one of the steepest land-sea interfaces on earth.

Accordingly, the continental shelves of the LME tend to be very narrow. North of the Gulf of Tehuantepec they are almost nonexistent. From the Gulf of Tehuantepec to the Gulf of Papagayo they widen, but are still narrower than 100 km. South of the Gulf of Papagayo they once again narrow, and except for the area of shelf within the Panama Bight, shelf widths of less than 50 km are the rule essentially to the southern limit of the LME.

Current Patterns

This region lies at the eastern limit of the great trans-Pacific equatorial surface current system, which consists of the North and South Equatorial Currents which flow westward in the same direction as the overlying trade winds, and the Equatorial Countercurrent which flows eastward in the region between them where the trade winds slacken in the doldrums zone associated with the ITCZ.

On approaching the ocean eastern boundary, a large part of the flow of Equatorial Countercurrent veers to the north to merge with the North Equatorial Current around a feature in the ocean density structure called the Costa Rica Dome. This dome is a permanent feature, quite consistently located near lat. 9°N, long. 89°W. Where the anti-clockwise flow around the dome becomes parallel to the coast, it is called the Costa Rica Coastal Current. The other part of the Equatorial Countercurrent flow veers to the south in a clockwise sense to merge into the South Equatorial Current. Between this latter gyre feature and the Colombian Coast, a current pattern tends to form in the opposite rotational sense yielding northward flow along the Colombia coast.

The flow of the North Equatorial Current is also fed from inflow from the California Current which flows southeastwardly along the Mexican coast. The extension of this flow into the area of the PCAC LME varies seasonally, generally extending all the way to the Gulf of Tehuantepec from February to April (Figure 12-3a). However, during the remainder of the year the northwestwardly flowing Costa Rica Coastal Current tends to continue along the coast nearly to the entrance to the Gulf of California (Figure 12-3b,c).

The entire surface current pattern near the eastern edge of the tropical North Pacific Ocean is quite variable seasonally. Wyrtki (1965a, 1965b) has recognized three dominant seasonal patterns (Figure 12-3). From about February to April (Figure 12-3a), when the ITCZ is near its most southerly position (Figure 12-2), the Equatorial Countercurrent is essentially absent from the vicinity of the LME. The two eddy features off Central America (one being around the Costa Rica Dome) are very pronounced. The California Current is in the northwest boundary region of the LME area and supplies most of the water for the North Equatorial Current.

From May to July (Figure 12-3b), the ITCZ is positioned near lat. 10°N (Figure 12-2), allowing the Equatorial Countercurrent to develop, sending most of its water north into the Costa Rica Coastal Current which during this period flows along the coast almost to the entrance of the Gulf of California. The California Current is strong, but swings offshore before entering the PCAC LME.

In August and September, the ITCZ moves to its farthest northward position (Figure 12-2) and then falls rapidly back southward. Through the period from August to December (Figure 12-3c), the California Current influence is weaker, but the Equatorial Countercurrent is strong and well developed, again sending most of its water around the Costa Rica Dome into the Costa Rica Coastal Current which flows strongly northward along the coast of Mexico.

Upwelling Processes

As in all highly stratified tropical ocean systems, a mechanism to transfer nutrients trapped in deeper waters upward to the illuminated surface layers where photosynthesis can occur is crucial to the productivity of the ecosystem. However, in the region containing the PCAC LME, classical

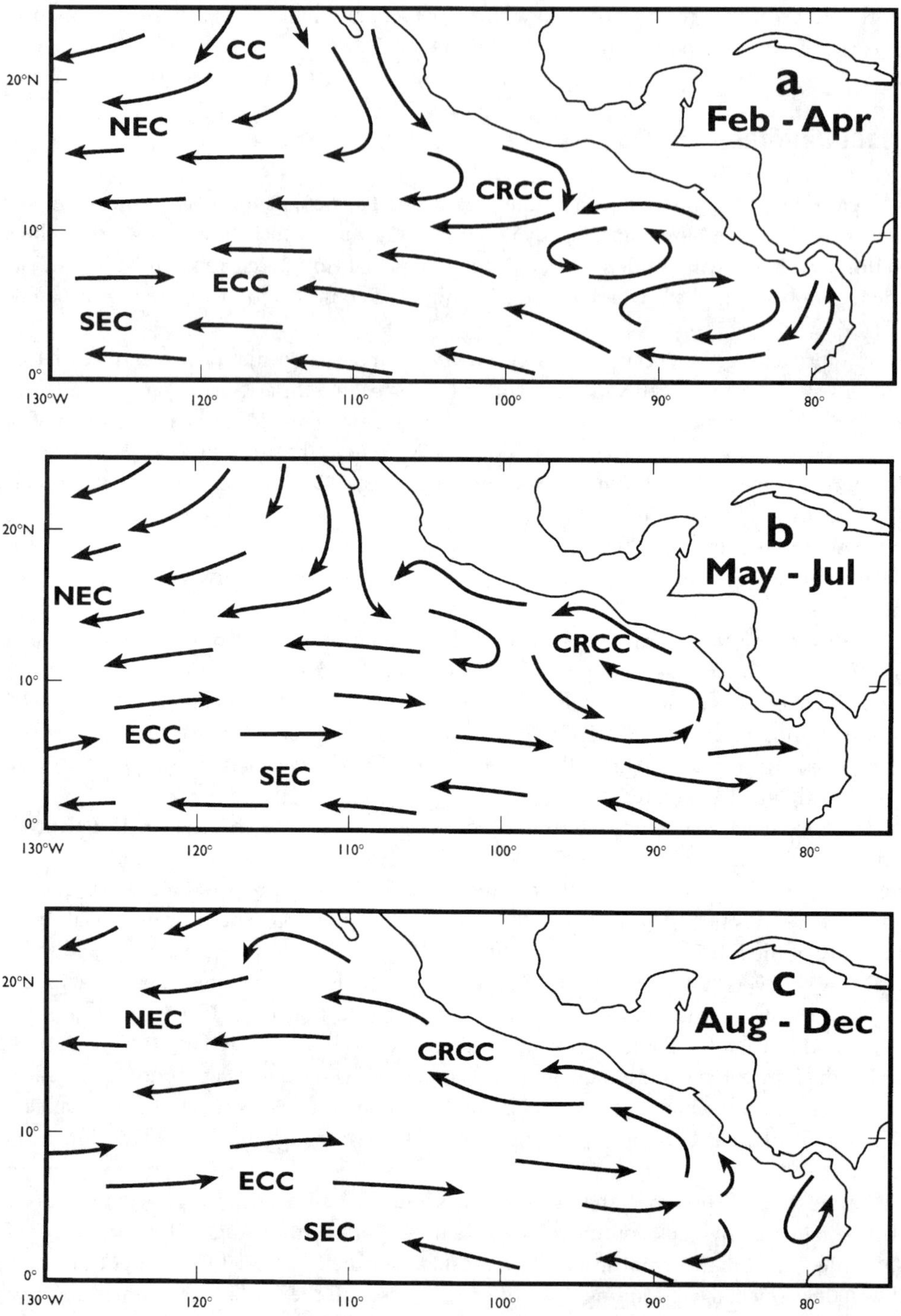

Figure 12-3. Principal seasonal ocean surface circulation patterns (redrawn from Badan-Dangon et al. (1989), originally from Wyrtki (1965a). (a) February to April. (b) May to July. (c) August to December. (NEC: North Equatorial Current; SEC: North Equatorial Current; ECC: Equatorial Countercurrent; CRC: Costa Rica Coastal Current; CC: California Current).

"Ekman" coastal upwelling, driven by winds blowing parallel to the coast in the equatorward sense, does not seem to be as important as it is in many other LMEs. However, several other upwelling mechanisms appear to be important sources of enrichment for biological activity.

One of these mechanisms, acting near the southern limit of the LME, is equatorial upwelling driven by the SE trade winds in the equatorial zone (Figure 12-4a). Because of such upwelling, the equatorial zone constitutes a belt of high nutrient input and high primary productivity that extends across most of the breadth of the Pacific Ocean (FAO 1981). On the eastern side of the Pacific, this equatorial enrichment is greatly augmented by enriched waters being advected directly from the extremely rich Peruvian upwelling system of the Humboldt Current LME.

Another important source of upwelling and nutrient enrichment for the region appears to be the Costa Rica Dome itself. Within the Dome structure the strong tropical thermocline reaches to within 10 meters of the sea surface. At the sea surface, the dome area appears as a zone of slightly reduced temperature, higher salinity and dissolved nutrients, and reduced oxygen saturation (Wyrtki 1964) which are indications of upwelling. This appears to be largely induced by the curvature (Figure 12-5) of the flow pattern (Wyrtki 1964), although Hofmann et al. (1981) suggest Ekman pumping by wind stress curl as a contributing mechanism. Such "open ocean upwelling" (Smith 1968; Bakun 1996) due to Ekman pumping via cyclonic wind stress curl would tend to occur not only in the Costa Rica Dome but all along the cyclonic wind shear pattern (Figure 12-4b) that exists on the northern edge of the ITCZ-doldrums zone that sweeps seasonally through the region (Figure 12-2).

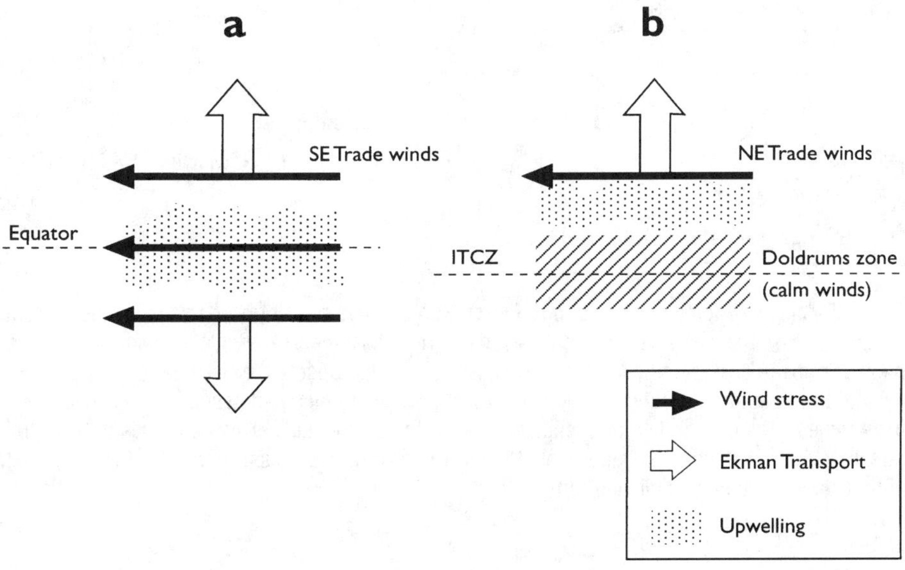

Figure 12-4. (a) Surface Ekman drift is directed to the right of the wind in the northern hemisphere and to the left of the wind in the southern hemisphere (e.g., Bakun 1996). Accordingly, north of the equator the Ekman transport associated with the westward trade wind circulation is northward and south of the equator it is southward. Since transport is directed away from the equator in both hemispheres, there is flow divergence, resulting loss of surface water and corresponding upwelling of deeper waters in the vicinity of the equator. (b) A substantial gradient exists between the calm wind conditions of the doldrums zone which exists in the vicinity of the ITCZ and the strong westward-directed stress of the Northeast Trade Wind zone to the north. (Mathematically, this constitutes a situation of "cyclonic wind stress curl"; see further explanation in Bakun 1996.) Accordingly, a zone of Ekman divergence, loss of surface water, and resulting upwelling exists in the area situated between the minimal Ekman transport of the doldrums zone and the strong northward Ekman transport of the trade wind zone.

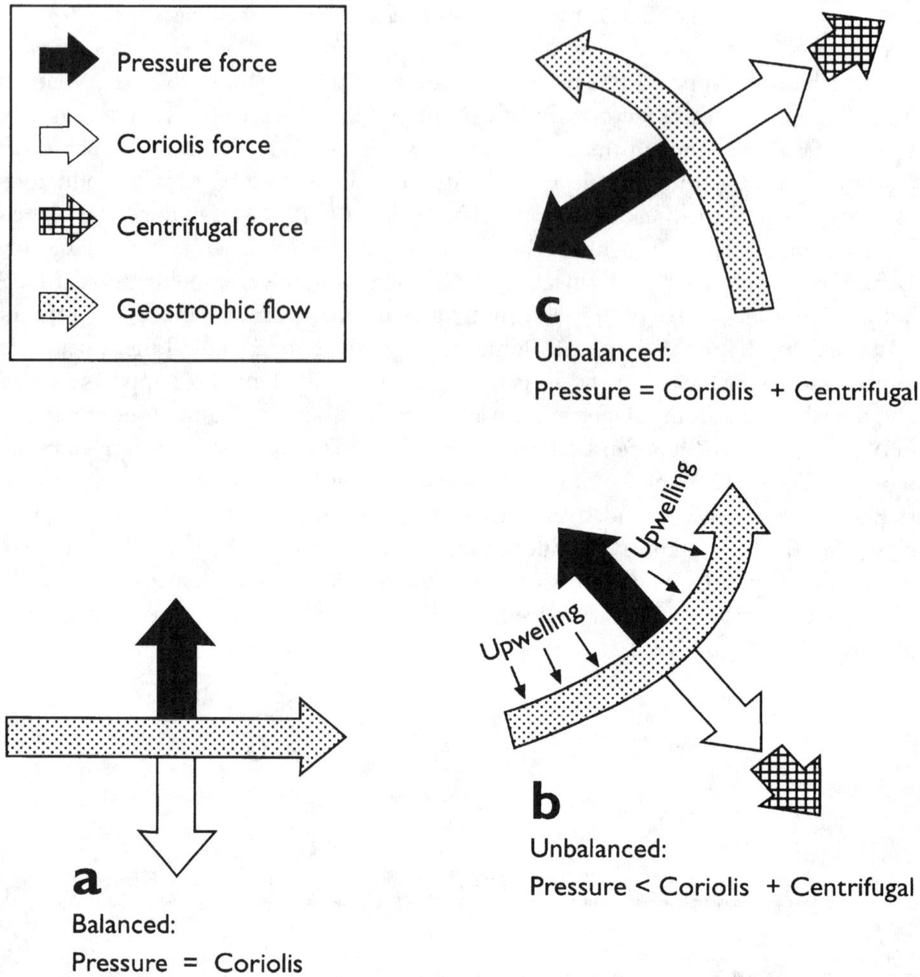

Figure 12-5. Balances of forces involved in the formation of a cyclonic loop in a geostrophic current. (a) In the unperturbed flow, the pressure and Coriolis forces are in balance. (b) As cyclonic curvature develops, a component of outwardly-directed centrifugal force is added. This upsets the balance of forces and causes surface water to move radially outward, with the result that divergence and associated upwelling occurs on the concave side of the flow and convergence occurs on the convex side. This continues until the resulting cross-stream sea surface slope has augmented the inwardly-directed pressure force sufficiently for (c) an equilibrium balance to be reestablished (after Bakun, 1996.)

Strong cyclonic shear zones in the ocean flow pattern, such as exist between the North Equatorial Current and the Equatorial Countercurrent flows, will tend to promote cyclonic eddy formation. This represents another source of upwelling. Intense cyclonic frontal eddies of this type act like moving "nutrient fountains" (Figure 12-6) being advected downstream in the underlying current flow (Bakun 1996).

Among the most distinctive oceanographic features of the PCAC LME are coastal upwelling plumes that often extend offshore from three locations corresponding to major gaps in the mountainous topography of the region. The three locations are within the Gulfs of Tehuantepec, Papagayo, and Panama, respectively (Figure 12-7). McCreary et al. (1989) recount the mechanism. During the winter, high pressure systems, typically lasting from three to ten days, develop in the Gulf of Mexico. The barrier imposed by the intervening high topography results in

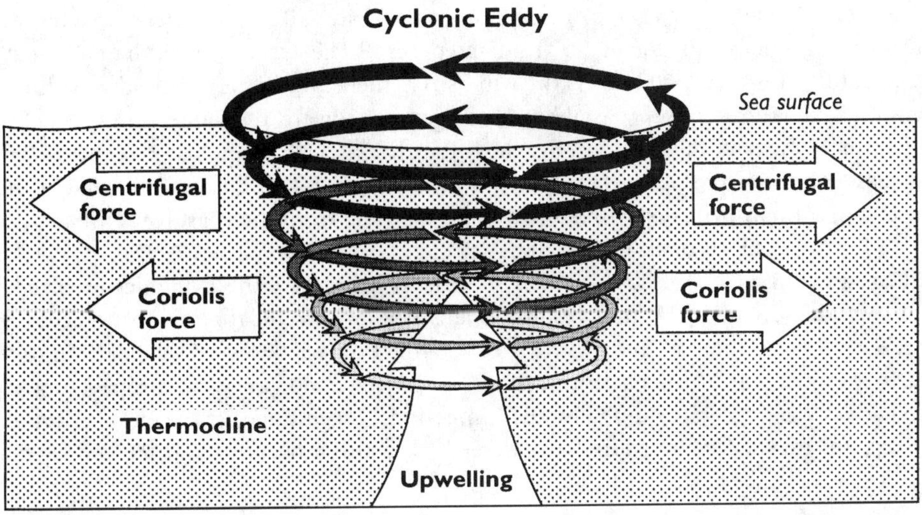

Figure 12-6. Upwelling occurring in a cyclonic eddy. Centrifugal and Coriolis forces due to the water motions are directed outward from the vortex center. These are counteracted by inwardly-directed pressure forces due to upwelling of deeper waters of greater density within the vortex core (after Bakun, 1996).

substantially higher pressures in the lower atmosphere on the Gulf of Mexico side than on the Pacific side of the Isthmus. Because of the relatively low latitudes, the geostropic constraint is reduced. Intense jets of wind blow directly down the pressure gradient through the mountain passes and exert strong offshore-directed stress on the sea surface.

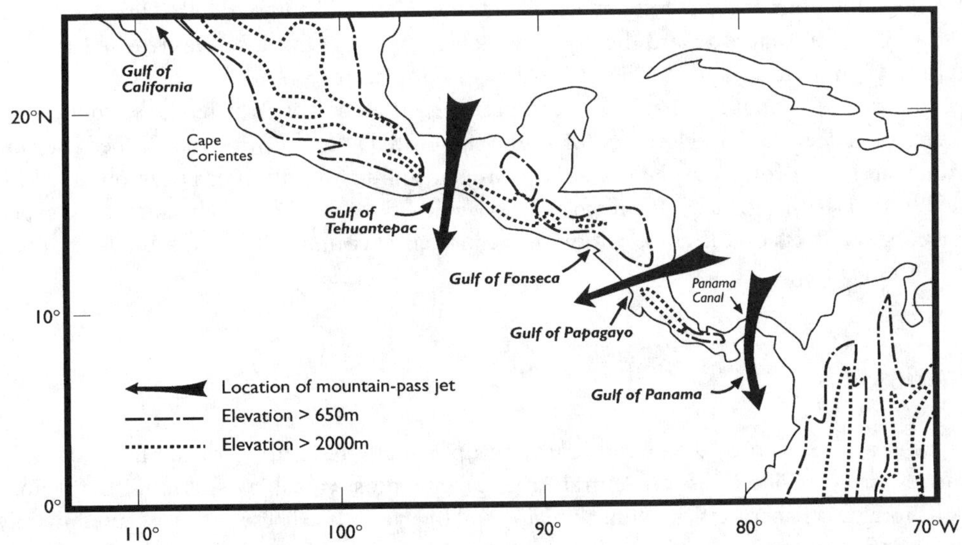

Figure 12-7. A map showing the location of the three mountain pass jets. Light and dark shaded areas indicate where the elevation of the topography is greater than 650 and 2000 m, respectively. The jets occur during the winter when high-pressure systems develop over the Gulf of Mexico. Air flows rapidly through the mountain passes down the pressure gradients, exerting intense offshore-directed wind stress on the sea surface, resulting in rapid offshore transport of surface waters and coastal upwelling which spreads offshore in elongated plumes.

Roden (1961) has studied the phenomenon in the Gulf of Tehuantepec, where the wind jet episodes are called locally by the name "Tehuantepecers." The jets have a width of about 200 km, and extend offshore some 500 km. As the wind strengthens, water moves rapidly offshore along the wind jet axis. In response, coastal currents converge toward the jet axis providing a source for the offshore current; these alongshore flows may have speeds as high as 100 cm/s. Upwelling at the coast can lower the sea surface temperature by some 10°C. Eddies generated at the edge of the jet current may drift slowly westward (~13 km/day) from the coast (Stumpf and Legeckis 1977) and may be still recognizable as far as 2,500 km offshore (Hansen 1987).

McCreary et al. (1989) used computer models to investigate the dynamics involved. Their studies provide several interesting results. The offshore current appears to be ageostrophic. After the wind event there is an opposite ageostrophic flow back toward the coast, immediately acting to readjust the ocean back toward its initial state; no poleward propagating coastal-trapped waves were generated in the model simulations, suggesting that the entire effect remains very localized, except for the eddies that were shed in the process and which propagate westward.

Biology

The Oxygen Minimum Layer

An extensive oxygen-minimum layer exists in this LME. It is more than 1,200 m thick off Mexico and Central America, becoming thinner toward the equator (Wyrtki 1965a). From about 9°N to 16°N latitudes, its upper boundary is shallower than 50 m. Wyrtki attributes the low oxygen layer to sluggish water movement due to the isolation of this region from direct involvement in the major subtropical gyre circulations. From two surveys of the shelf area between the Gulf of Tehuantepec and Gulf of Papagayo during 1987, Bianchi (1991) reports dissolved oxygen levels of 1 ml/l on the floor of the continental shelf between 50 m and 150 m depth, with a tendency to occur in shallower waters toward the north (in September 1987 this level was reached at 75 m off the Gulf of Tehuantepec and at 125 m near the Gulf of Papagayo, while by November/December it had ascended to about 50 m and 75 m, respectively). Below these depths the dissolved oxygen concentration decreased and levels of 0.5 ml/l were located at the continental shelf edge or upper continental slope throughout the year. No hydrogen sulfide formation has been observed in the area (Wyrtki 1965a). However, the dissolved oxygen levels reach low enough values to be expected to have major effects on the composition of the biological community, to inhibit vertical migrations of some pelagic species, etc.

Fisheries

Villegas et al. (1998) provide a comprehensive up-to-date description of the fisheries and fish resources of the Pacific Coast of Central America. The most valuable fisheries of the region are the offshore tuna fisheries (yellowfin, skipjack, and bigeye tuna). These are prosecuted mainly by long range fleets from Korea, Mexico, Japan, USA, Venezuela, and several other countries.

By far the most important fishery for the coastal countries of the PCAC LME is on Penaeid shrimps. This is a fishery for export of an extremely high-priced product, mainly to the USA, and represents a significant source of foreign exchange for all the coastal countries in the area. Most of the shrimp stocks in the region are considered to be overexploited. Vital shrimp nursery grounds are associated with mangroves which are being destroyed in the region at a rapid rate, and with estuaries which are increasingly subject to pollution.

Several hundred species of demersal fishes are commonly taken as bycatch in the shrimp fisheries. Many of these are excellent food fish of potential economic value but do not sustain major commercial fisheries in the area. Nonetheless, even though they are not targeted, their effective levels of exploitation might well be very high due to the heavy fishing pressure exerted on their early life stages by the industrial shrimp fisheries. Accordingly, the shrimp fishery probably acts in a major way to inhibit the establishment of other fisheries for coastal demersals that could potentially benefit many thousands of artisanal fishers in the region.

Significant fisheries on small coastal pelagic fishes, notably Central Pacific anchoveta (Cetengraulis mysticus) and Pacific thread herring (Ophistonema spp.), have existed in the Gulf of Panama since the 1950s, having been established as a source of bait for the offshore eastern tropical Pacific tuna fisheries. Currently, most of the production is used for fish meal and oil. More recently a smaller fishery on Ophistonema was established in Costa Rica and lasted for more than a decade. This fishery supplied a small local cannery, which ceased its operations due to economic factors.

There are artisanal shark fisheries operating in El Salvador and Guatemala, and sharks are also caught as bycatch in other countries. Sharks have low reproductive rates and are known to be vulnerable to population collapse under exploitation by fisheries. There is a serious need for assessment of current population trends in exploited shark species of this region.

Shark fisheries are but one example of a serious inadequacy of fishery statistics in most of the countries of the region. This makes it extremely difficult to confidently assess the states of the various fishery resource stocks of the area.

Ecosystem Issues

The crucial nature of transboundary issues in this region are epitomized by the situation of the Gulf of Fonseca (location shown in Figure 12-1) which is an important nursery grounds for several valuable shrimp species. This relatively small water body, being only a little more than 50 km in its largest linear extent, contains some 22% of the entire mangrove area of the Pacific coasts of all the states of Central America proper (Guatemala to Panama). The shoreline of this small gulf itself encompasses coastlands of three separate nation states, El Salvador, Honduras, and Nicaragua. Pollution arising from the operations of extensive banana plantations is a serious problem and destruction of mangrove forests is rampant.

Threatened species are an issue in the region. Sea turtles are a particular concern, although protected beaches for turtle spawning exist in some countries. Some species of sharks also may be threatened. Generally, the steep coastal topography acts in connection with the coastwise gradation of rainfall and other tropical climatic and ocean conditions to produce a unique complex of local coastal environments of both terrestrial and marine types (notably the Gulf of Fonseca and other even smaller, partially enclosed areas). Threats to these finely structured ecosystem segments from coastal development, the heavy ship traffic that exists along this coast, expanding tourism, and competition for fishery resources clearly pose important concerns for biodiversity preservation.

Brief Socioeconomic Synopsis

In the past twenty years, Central America has been torn by tragic civil wars in Nicaragua, El Salvador, and Honduras. Currently the situation is much improved. All Central American nations now are independent republics with stable governments. Their economies are diverse and developing. Since 1991, Mexico, Guatemala, Honduras, Nicaragua, and Costa Rica have integrated their economies in a Central American Common Market and have joined in forming the Organization of Central American States (CAE) which promotes social integration to assist development and fight poverty. All the countries of the PCAC LME region are members of the Organization of American States (OAS).

The main commercial products of the region are derived from agriculture (mainly bananas, coffee, cane sugar, and cotton), from forestry (pine and tropical hardwoods), and from manufacturing (cement, clothing, textiles). Oil and derived products are exported from Colombia and Mexico. There are also significant mining activities, notably gold in Nicaragua and Colombia, silver and lead in Honduras, and nickel and copper in other countries. Development of tourism has become a priority activity. For example, tourism recently represented some 28.2% of foreign exchange earnings for Costa Rica, and 20.4% for the Central American region as a whole (OMT 1994).

In the coastal zone, tourism, fisheries, aquaculture, and agriculture are the main economic activities. Shipping and industrial activities also have significant impact. Within Central America (i.e., not including Mexico and Colombia) aquaculture occupies an area of 25,900 ha. In 1994, 23,000 tons of shrimp were produced (PRADEPESCA 1995), 70% being from Honduras and Panama, with the Gulf of Fonseca by itself representing 52% of the regional production. At least 90% of the shrimp farms have been constructed on former mangrove or salt pond areas.

A large proportion of the coastal population lives in rural areas where the standard of living and the availability of basic services (health, education, water, sewage, etc.) are greatly reduced. Serious problems include pollution of coastal waters, lack of facilities for proper solid waste disposal, deficiencies in basic infrastructure, a lack of integrated development planning that may exacerbate natural hazards, problems related to mining of sand beaches and corals and to harvest of wood from coastal wetlands, and social and cultural displacement of native and ethnic entities.

Marine Ecosystem Outlook and Needs

Among the most pressing needs for this region are development of much more extensive resources and infrastructure for data collection and monitoring of ecosystem changes. The fishery catch statistics are generally poor and unreliable. For example, most landing reports contain no species identification. On the "habitat" side, there is a pressing need for reliable statistics for and mapping of mangrove areas and for initiation of a system of stress mitigation and mangrove restoration. There is a need for strengthening of local administrations, which are poorly equipped and staffed to do proper fisheries monitoring and management. Well-founded analyses of basic issues need to be generated. For example, analyses need to be produced of potential sources of conflict between coastal rural development (including artisanal fisheries) and other land/sea uses (urban development, industrial development, tourism, sport fishing). Also, there is a need for bioeconomic analyses of shrimp fisheries interactions (through bycatch) with other potential coastal fisheries. In order to promote the economic well-being of the local inhabitants, the great potential for tourism that exists due to the rich natural attributes of the region should be, and will be, developed. But this development needs to proceed carefully and wisely in order to avoid destroying its unique and fragile basis.

With these caveats in mind, there appears to be room for development of limited (but still of local importance) sport and/or small-scale commercial fisheries for mid-size pelagics and other "oceanic" species (Coryphaena, Caranx, Decapterus, bill fishes, swordfish, sharks, tunas). But this should be approached with a careful view toward the potential sources of conflicts between artisanal and sport fishing referred to above. The same appears to be true for coastal fish stocks (Centropomus, Sciaenids, Haemulopsis, Lutjanus, Epinephelus, alosas, etc.). But again, this should be done with careful consideration given to potential sources of conflicts between the industrial shrimp fishery and artisanal and sport fishing activities. There seems also to be a limited potential for development of deep-water shrimp fisheries on species (mostly Galatheidae) known to be abundant off parts of Central America.

Finally, there needs to be much work done on strengthening awareness of the importance of preservation of ecosystem integrity among the local people and governments, and also in putting into place effective mechanisms (Sherman 1996) to translate such awareness and concern into appropriate and effective actions.

References

Alheit, J. and Bernal, P. 1993. Effects of physical and biological changes on the biomass yield of the Humboldt Current ecosystem. In K. Sherman, L. M. Alexander, and B. Gold (eds.), Large Marine Ecosystems—Stress, Mitigation, and Sustainability. American Association for the Advancement of Science pp. 53-68.

Badan-Dangon, A., Robles, J.M. and Garcia, J. 1989. Poleward flows off Mexico's Pacific Coast. In S.J. Neshyba, C.N.K. Mooers, R.L. Smith and R.T. Barber (eds.), Poleward Flows Along Eastern Ocean Boundaries. Springer-Verlag, pp. 176-201.

Bakun, A. 1996. Patterns in the Ocean: Ocean Processes and Marine Population Dynamics. University of California Sea Grant, San Diego, California, USA, in cooperation with Centro de Investigaciones Biológicas de Noroeste, La Paz, Baja California Sur, Mexico. 323 pp.

Bianchi, G. 1991. Demersal Assemblages of the continental shelf and slope edge between the Gulf of Tehuantepec (Mexico) and the Gulf of Papagayo (Costa Rica). Mar. Ecol. Prog. Ser. 73: 121-140.

FAO, 1981. Atlas of the Living Resources of the Seas. Food and Agriculture Organization of the United Nations. Rome. 23 pp. 69 maps.

Hansen, D.V. 1987. History and status of the NOAA EPOC5 program. In E.I.G. Katz and J.M. Witte (eds.), Further Progress in Equatorial Oceanography. Nova University Press.

Hofmann, E.E., A.J. Busalacchi and J.J. O'Brien. 1981. Wind generation of the Costa Rica Dome. Science 214: 552–554.

MacCall, A.D. 1986. Changes in the biomass of the California Current ecosystem. In K. Sherman and L.M. Alexander (eds.). Variability and Management of large Marine Ecosystems. AAAS Selected Symposium 99. Washington, D.C.: American Association for the Advancement of Science, pp. 33-54.

McCreary Jr., J.P., Hyong, S.L. and Enfield, D.B. 1989. The response of the coastal ocean to strong offshore winds: With application to circulations on the Gulfs of Tehuantepec and Papagayo. J. Mar. Res. 47: 81-109.

OMT. 1994. Tendencias del Mercado Turístico: Américas, 1980-1992. Comisión de la Organización Mundial del Turismo para las Américas. Documento preparado para la vigésima sexta reunión. San Salvador, 24 de mayo 1994.

Roden, G.I. 1961. On the wind-driven circulation in the Gulf of Tehuantepec and its effect upon surface temperatures. Geofis. Int. 1: 55-72.

PRADEPESCA, 1995. Situación Actual y Perspectivas del Cultivo de Camarón en el Istmo Centroamericano. Resumen del III Simposio Centroamericano sobre Camarón Cultivado (ANDAH-FPX), Honduras. PRADEPESCA. Panamá.

Sherman, K. 1996. Achieving regional cooperation in the management of marine ecosystems: The use of the large marine ecosystem approach. Ocean and Coastal Management 29: 165-185.

Sherman, K. and Alexander, L. (eds.) 1986. Variability and Management of Large Marine Ecosystems. AAAS Selected Symposium 99. Boulder: Westview Press, 317 pp.

Smith, R.L. 1968. Upwelling. Oceanogr. Mar. Biol. Ann. Rev. 6: 11-46.

Stumph, H.G., and Legeckis, R.V. 1977. Satellite observations of mesoscale eddy dynamics in the eastern tropical Pacific ocean. J. Phys. Oceanogr. 7: 648-658.

Villegas, L., Csirke, J. and Arosemena, D. 1998. Los recursos pesqueros del literal Pacifico de America Central. FAO Fisheries Technical Paper. 149 pp. (in press).
Wyrtki, K. 1964. Upwelling in the Costa Rica Dome. Fish. Bull. US 63: 355-372.
Wyrtki, K. 1965a. Summary of the physical oceanography of the Eastern Pacific Ocean. Institute of Marine Resources, University of California, San Diego, Ref. 65-10. 78 pp.
Wyrtki, K. 1965b. Surface Currents of the Eastern Tropical Pacific Ocean. Inter-Amer. Trop. Tuna Comm., Bull. 9(5): 271-304.

III

Methodological Studies

13

Differences between Boreal and Temperate Multispecies Fisheries Systems: Form and Implications for Multispecies Assessment and Management

Jake Rice

Abstract

The International Council for the Exploration of the Sea (ICES) Multispecies Assessment Working Group has conducted extensive analyses of feeding and growth patterns of fish predators in marine ecosystems around the North Atlantic. I review some key results to provide a basis for looking at the properties of marine ecosystems of the Pacific and at approaches to assessment and management of those systems.

These analyses have demonstrated empirically that a number of features are similar among the various boreal seas and similar among the temperate seas, but differ between the two types of ecosystems. These features include diet diversity, the ability of predators to compensate for fluctuations in prey stocks, patterns of interannual variation in growth, and spatial heterogeneity in diets.

The empirical differences in fish predator feeding and growth mean that models appropriate for multispecies assessment and management of boreal systems will have some features that differ from models appropriate for multispecies assessment and management of temperate systems, such as how predator ration is handled, how spatial effects are handled, and the role of environmental forcing. The nature and implications of these differences are also addressed. I suggest ways in which knowledge from the rigorously empirical studies of the Multispecies Assessment Working Group may be transferred to assessment and management problems in the marine ecosystems of the Pacific.

It is commonplace to see statements that single-species fish stock assessment and management are doomed because they do not provide an adequate foundation for sustainability or for achievement of biological, economic, or social objectives in fisheries. Some critics focus concern on physical oceanographic influences on fish distribution and population dynamics; others focus on trophic interactions among species. Some critics call for "full ecosystem management." For example, the vision statement for the U.S. NOAA Science Review on Build Sustainable Fisheries declared that "NOAA will integrate its operational management activities in an ecosystem-wide approach . . . and undertake a more proactive effort in understanding and maintaining biodiversity" (NOAA 1994). The Fisheries Resource Conservation Council in the Canadian Atlantic stated, "A mandate for conservation requires treatment of the fishery in its full context, as a component of the broad ecosystem. In this report, efforts have been made to move in this direction . . . These are clearly only initial moves toward an ecosystem approach . . . **Nevertheless, the Council believes progress can be made in developing more complete ecosystem-based approaches, and is pursuing this further**" (FRCC 1993, bold type in original). The announcement for this symposium highlights that the concern is shared in the Pacific jurisdictions as well.

The need for fish stock assessment and management to take a wide perspective is not a new insight. Around the North Atlantic, where fisheries assessment and management organizations have longer histories than around the Pacific, both ICES and the Canadian Atlantic Fisheries Scientific Advisory Council had formed, respectively, the Multispecies Assessment Working Group and the Marine Environment and Ecosystems Subcommittee by 1980. The Northwest Atlantic Fisheries Organization had a different structure, but always devoted a portion of each assessment meeting to presentations and discussions of environmental and trophic influences on stocks (NAFO 1980).

If there is so much history on assessment and management in an ecosystem context, why do recent national policy statements suggesting the need to do such work represent a change from present practice? If assessment scientists and managers are receptive to the ecosystem context, why are so many single-species assessments still being done? The answer is that although the concept of ecosystem management is attractive, it has been very difficult to make operational. Why is this so?

Basically, it is hard to apply the type of advice that comes from many "ecosystem approaches" to assessment. As a rule of thumb, once the uncertainty associated with an estimate of stock status or advised harvest exceeds around 20% of the estimate, the usefulness of that advice to a fisheries manager decreases. Two eminent ecosystem modelers who independently have given some attention to fisheries ecosystem problems informally commented to me that they found fisheries a difficult field in which to work. They were used to concluding they had an adequate model "if I got all the signs right in my relationships" or "if the relative parameter estimates are within a couple orders of magnitude of what we think are reasonable." The gap is large between such ecosystem models and scientific advice that fisheries managers can actually use.

Moreover, we already know the first-order answers to major ecosystem management questions. Without further investment in ecosystem science and modeling, we can already advise any marine ecosystem with developed fisheries that long-term sustainability would be served if exploitation rates were lowered, and that heavy fishing of lower trophic levels is living dangerously. However, the worldwide overcapitalization of fleets and unfavorable economics of most fisheries make it difficult to reduce exploitation rates effectively without compelling scientific justifications for such actions.

In a field where progress generally has been disappointing, the ICES Multispecies Assessment Working Group can claim 15 years of progress in making multispecies assessment and advice operational. Consistent with the applied focus of the Multispecies Assessment Working Group, its work has focused on estimating predation mortality and analyzing variation of predator growth or size at age. Analyses of stomach contents data were important to the investigation of both processes.

In this chapter I will review a few of the areas where they have made significant progress on broadening the context for assessment and advice to fisheries managers. I will also highlight a number of generalizations that may be valuable in transferring the progress in the North Atlantic to Pacific stocks and systems.

Multispecies Assessment Working Group Approach

A key tool of placing stock assessments into a multispecies context has been the multispecies virtual population analysis (MSVPA). The programs are based on the work of Andersen and Ursin (1977) and Helgason and Gislason (1979). Reviews are available in Sparre (1991). A key point is that MSVPA adds extensive stomach contents data to the usual catch-age and survey data used in single-species assessments. Quarterly "suitability" parameters are estimated from the population and stomach contents data to link predators to prey, adjusting for abundances of both. MSVPA then provides not just the usual numbers and fishing mortalities at age for each species but also empirical estimates of predation mortality caused by each age of each predator in the model on each age of each prey in the model. The estimates of numbers and mortality by species and age are produced iteratively, and the full assemblage of interacting predators and prey converge numerically. Changes in the predation mortality component of natural mortality over years reflect both changes in predator and prey biomasses and changes in abundance of other prey eaten by the MSVPA predators.

Most of the Working Group's actual MSVPA results are specific to the North Sea and will not be reviewed here. Two important generalizations do come directly from MSVPA analyses, however. The first is that the results of annual population estimates in a single-species and a multispecies context do not differ in any important (or statistically significant) way (ICES 1991) (Table 13-1). The short-term changes in natural mortality are small compared with other changes in population dynamics. The MSVPA needs to be run and tuned two or three times a decade to output new values for natural mortality. As long as the natural mortality values used in single-species assessments are then updated, single-species virtual population analyses (VPAs) provide satisfactory estimates of numbers at age for both the predators and the prey.

The second important finding is that MSVPA is a predictive tool. Stomach data sets adequate to parameterize the full model were collected in both 1981 and 1991. The Working Group conducted a large number of tests on the comparability of suitabilities estimated with each data set and on the ability of MSVPA with one year's data to predict diets observed in the other year. Although sizes of some of the predator and prey populations changed greatly, the Working Group concluded that "[t]he suitabilities do vary between runs with different data sets. The changes are

Table 13-1. Ratios of estimated numbers at age from multispecies VPAs and single-species VPAs from 1990 Working Group reports, for six key species in the North Sea (values are ratios of means over 1983–1988)

	Species					
Age	Cod	Whiting	Saithe	Haddock	Herring	Norway pout
1	1.10	1.25	0.95	1.45	0.94	2.06
2	1.04	0.99	0.96	1.09	1.18	0.97
3	1.06	1.01	0.97	1.06	1.10	1.26
4	1.06	1.03	0.98	0.99	1.26	—
5	1.05	1.02	0.99	1.04	0.96	—
6	1.06	1.03	0.99	1.05	1.01	—
7	1.06	1.07	0.97	1.10	0.95	—
8	1.06	1.00	1.01	1.21	0.90	—

Table 13-2. Results of fitting linear models with various terms to suitabilities of major MSVPA predators

Species	Fit index	Grand mean	Prey species nested under year	Predator species nested under year	Species	Size	Residual
Cod	SumSQ	120.2	49.0	11.3	5.0	0.9	53.9
	df	1,518	29	1	28	4	1,445
	% var		41	9	4	1	
Whiting	SumSQ	96.4	42.1	11.9	7.9	2.2	32.4
	df	962	28	1	25	4	903
	% var		44	12	8	2	
Saithe	SumSQ	90.8	69.4	0.7	3.4	1.5	15.8
	df	748	25	1	19	4	698
	% var		76	1	4	2	
Mackerel	SumSQ	119.2	60.1	5.7	4.3	1.2	47.8
	df	614	17	1	13	4	578
	% var		50	5	4	1	
Haddock	SumSQ	118.3	40.8	29.5	11.8	0.4	35.8
	df	557	18	1	12	4	521
	% var		34	25	10	0	

Note: Suitabilities are estimated with all North Sea stomach data. Terms are added from left to right, each successively reducing the total sum of squares.

small, however. The forecasting properties of MSVPA appear generally robust to the observed level of changes in suitabilities. There are no models available which deal with changes in suitabilities better than MSVPA does" (ICES 1994).

As an example of what is meant by a "small" change, a model with predator and prey species main effects and a variant of the Andersen-Ursin size preference function can account for between 50% (mackerel) and 76% (saithe) of the variance in the over 2,000 separate suitability estimates (Table 13-2). Adding terms for year captures only an additional 4% to 8% of the variance.

This stability is important because it allows MSVPA to be used to evaluate mid-term and long-term consequences of management actions, such as changes in mesh size or effort (Figure 13-1). These forecasts do differ between single-species and multispecies analyses, sometimes by large amounts (ICES 1991). It is in the selection of management strategies, not in the annual advisory processes, that ecosystem interactions will become important.

Comparative Analyses

Overview

The ICES Multispecies Assessment Working Group does much more than just run MSVPA programs. The core data bases for MSVPA of the North Sea (size and numbers at age for eight major predators and six major fish prey, and stomach contents data for ages 2 onward of all eight predators) contain a wealth of information on species interactions. Moreover, these data sets have been augmented by data sets on Atlantic cod (*Gadus morhua*) numbers, sizes, and diets at age for other ecosystems around the North Atlantic. These data sets have been subjected to extensive analyses to improve our insights into the multispecies interactions of these ecosystems and facilitate development of better assessment and management models of these systems.

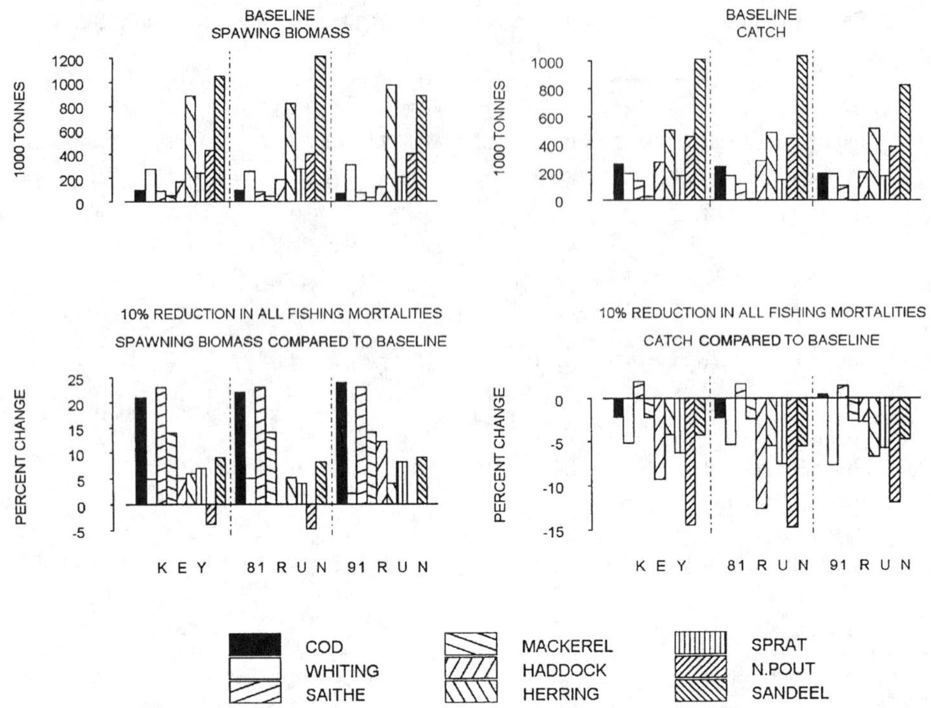

Figure 13-1. Illustration of application of multispecies VPA to evaluate consequences of management actions in the North Sea fisheries. Top panels present MSVPA estimates for 1994 of spawning biomass (left) and catch (right) at effort levels reported in 1993 assessments. Lower panels present changes in MSVPA estimates (relative to top panels) of spawning biomass and catch if baseline effort levels were reduced by 10%. For each panel, KEY represents results using all stomach data, 81 RUN represents results using only stomach data from 1981, and 91 RUN represents results using only stomach data from 1991.

Growth Increments of Cod in Boreal Systems

The core data here were length-at-age data for cod from the Barents Sea, Iceland, Greenland, and the Newfoundland-Labrador Shelf, generally spanning a decade or more (ICES 1991) (Table 13-3). Environmental variables included estimates of cod biomass (for density-dependent effects), biomass of capelin (and in some cases other prey), and water temperature for most of these cod stocks. The North Sea data were not used because of absence of capelin in the North Sea and because the water temperature regime overlapped very little with the regimes of the boreal systems.

In any comparative analysis of data sets from different places, times, and species, there is a possibility of artifacts due to (sometimes small) differences in data collection methods. This is true in these analyses, particularly for the capelin data, but a number of measures were taken to reduce such inconsistencies (ICES 1991). Moreover, a variety of models, differentially affected by irregularities in the data, were tested in most analyses. Effects that appear consistently across analyses and systems are robust to details of sampling.

Because length-at-age data may not be independent over years of individual cohorts, standardized growth increments as well as lengths at age by cohort and year were analyzed. A variety of linear models were fit to the growth increment data, treating the year, the cohort, and the environmental variables in different ways. Key models included a standard general linear model (GLM) with terms for cohort, year, and environmental variables; the Millar-Myers (1990) modified von Bertalanffy curve [where $L_{\text{infinity}} = L_0 + b \times$ (environmental factors)]; and the Horbowy and Swinder (1989) modification of the Andersen-Ursin functional growth increment model.

Table 13-3. Years of data coverage for analyses of cod growth increments and lengths at age from various Atlantic marine systems

	Barents Sea	Greenland Sea	Iceland	Nfld.-Lab. Shelf	New England	North Sea
Cod lengths	1979–1990	1976–1990	1965–1989	1976–1990	1977–1990	1972–1990
Cod biomass	1979–1990	1975–1989	1978–1990	1962–1989	1964–1989	1972–1990
Cod ages present (yr)	3–15	3–12+	3–9	3–13	3–12	2–12
Ages analyzed (yr)	3–8	5–8	4–9	4–9	4–8	2–7
Bottom temperatures	1921–1989	1973–1989	1972–1990	1946–1989	N.A.[a]	N.A.[a]
Capelin biomass	1973–1979	N.A.[a]	1978–1990	1972–1989	N.P.[b]	N.P.[b]

Note: Because key environmental variables were missing, many analyses of lengths were not conducted for New England and North Sea cod stocks.

[a]N.A. means data were not available at the time the Working Group conducted the analyses.

[b]N.P. means capelin are not present in noteworthy amounts in the system.

Table 13-4. Results of GLM analyses of growth increments for cod in four Atlantic ecosystems

	System			
Term	Barents Sea	Greenland Sea[a]	Iceland Sea	Newfndlnd.-Lab. Shelf[b]
Cohort (sig.; r^2)	–; 0.12	–; 0.25	–; 0.15	–; 0.10
Age	–; 0.00	–; 0.13	**; 0.09	***; 0.31
Year	***; 0.67	***; 0.72	***; 0.77	***; 0.24
Age + Year	–	**	***	***
	***; 0.68	***; 0.81	***; 0.84	***; 0.73
Age + Cohort	–	–	*	***
	–; 0.13	–; 0.47	–; 0.47	–; 0.45
Year + Cohort	***	***	***	***
	–; 0.72	***; 0.97	*; 0.92	***; 0.65
Age + Cohort	–	–	–	***
+ Year	–	**	–	–
	***; 0.73	***; 0.97	***; 0.93	***; 0.78

– means $0.10 < p$; *$0.05 < P < 0.10$; ** $0.01 < P < 0.05$; ***$P < 0.01$.
[a]Most analyses split by East and West Greenland.
[b]Many analyses split by NAFO Divisions 2J, 3K, and 3L.

The analyses revealed many informative details, but also some important generalizations. The year effect is significant in almost all analyses in all models, whereas the cohort effect is significant less often and is much smaller than the year effect (Table 13-4). Model results suggest that cod have limited ability to compensate for fluctuations in capelin biomass. The dominance of the year effects suggests that some strong environmental signals should be present; however, no environmental relationships were consistent across systems or models within systems. Depending on the details of the model and the system, inverse biomass (density dependence), temperature, capelin biomass, and per capita capelin biomass (prey per predator) were occasionally significant terms. Moreover, year effects remained after accounting for cod growth variations due to prey, temperature, and cod stock fluctuations. These effects were much stronger than the variance explained by the specific environmental variables (Table 13-5).

Table 13-5. Significance of environmental variables in stepwise GLM analyses of cod growth data

	System			
Term	Barents Sea	Greenland Sea[a]	Iceland	Nfld.-.Lab Shelf[b]
1/Biomass	***; ;***	d; ;d	d; ;d	***; ;**
Cohort biomass	***; ;***		d; ;d	d; ;d
Capelin biomass	***; ;***		**; ;*	***; ;***
Capelin biomass/ Cod biomass	***; ;***		d; ;d	*; ;d
Bottom temperature	***; ;***	d; ;d	***; ;***	*; ;*
Cohort	d; ;d	d; ;d	d; ;d	d; ;d
Age	d; ;–	d; ;–	d; ;–	***; ;–
Error df	50; ;50	17; ;–	30; ;41	69; ;80
R^2	0.54; ;0.61	0.37; ;–	0.47; ;0.38	0.52; ;0.29

Note: First entry of each cell is for growth increment; second entry is for standardized growth increment. *d* means term was deleted in a backward stepping procedure.
Negative sign: $0.10 < p$; *$0.05 < P < 0.10$; **$0.01 < P < 0.05$; ***$P < 0.01$.
[a]Most analyses split by East and West Greenland.
[b]Many analyses split by NAFO Divisions 2J, 3K, and 3L.

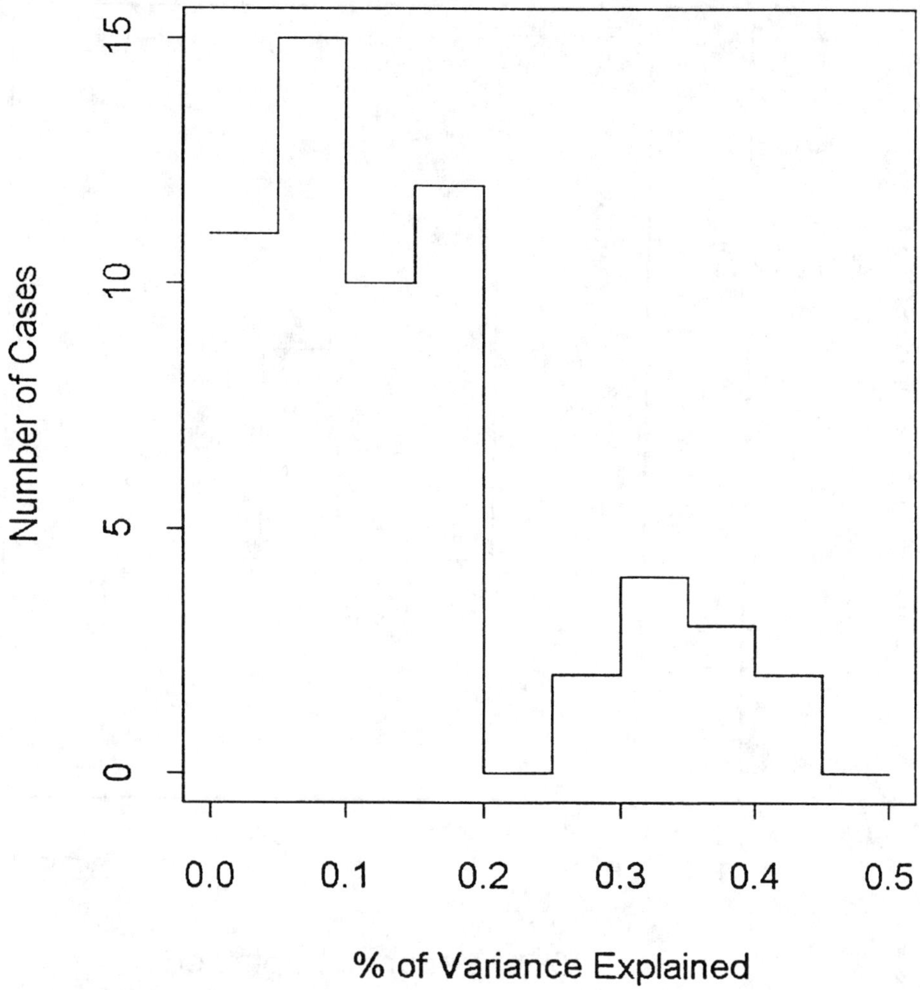

Figure 13-2. Ability of diverse general linear models to account for variance in length increments of cod, across four ecosystems. Sets of cases contributing to histograms are not independent, due to inclusion of subsets of terms in different models fit to the same data, so statistical tests of distributions are inappropriate.

In the exploration of proper formulations for the basic models (bulked or unbulked data, log transformed or untransformed, etc.), even complex models rarely captured as much as 20% of the variance in the data (Figure 13-2). In all cases where r^2 exceeded 20%, predator size was the influential effect (Figure 13-3).

The Working Group concluded, "Either these data sets do not contain adequate descriptions of these [environmental] variables as they relate to cod growth, or there are other, yet unexplained factors influencing measured cod growth in the Boreal Atlantic systems" (ICES 1991).

Stomach Contents

The Working Group had extensive stomach data sets from six circum-Atlantic systems: New England–Gulf of Maine, Newfoundland Shelf, Iceland, Barents Sea, North Sea, and Baltic Sea. Analyses focused on data from cod, which are a dominant predator in all six systems. There were

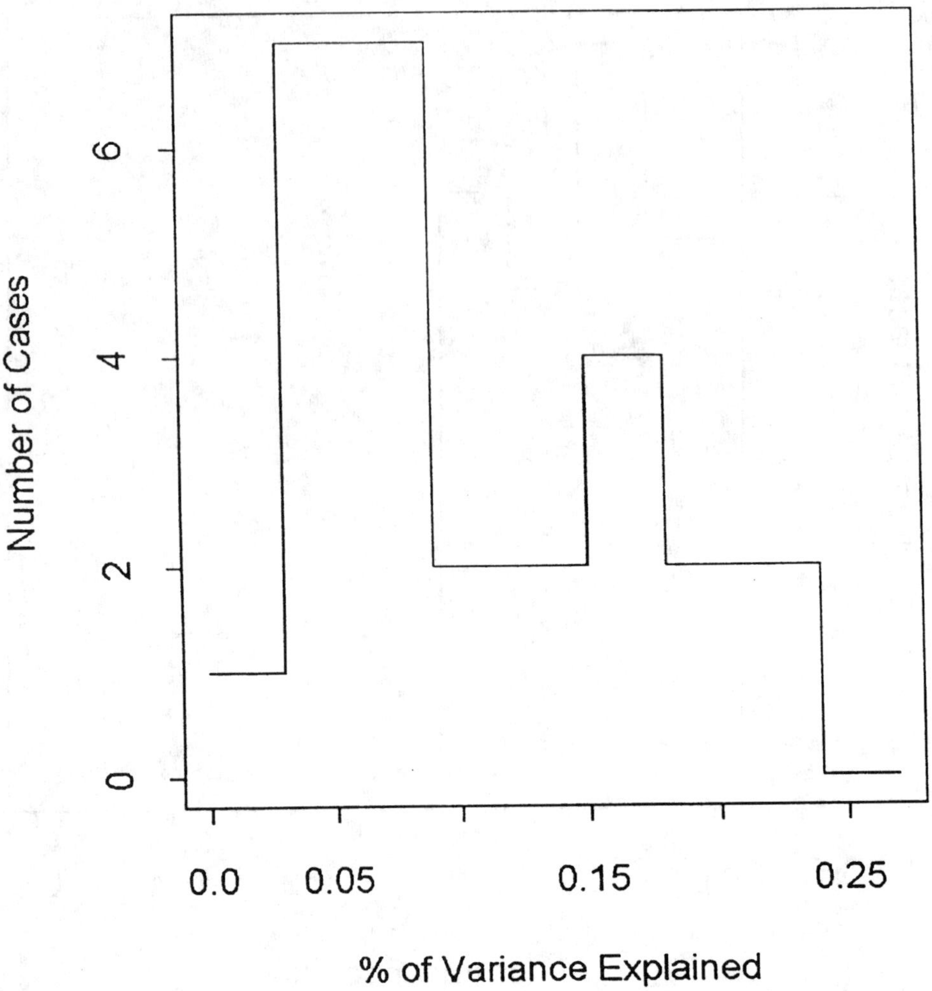

Figure 13-3. Ability of size terms to account for variance in length increments of cod, across four ecosystems. Cases are a subset of models presented in Figure 13-2, with similar lack of independence among sets of cases.

a number of differences in data collection, such as seasonal coverage, years included, bulked versus single stomach processing, and length stratification. Effects of a number of these factors are reported in ICES (1992) but will not be treated here.

The number of stomachs appears large (Table 13-6), but results of running MSVPA (ICES 1994) with independent stomach data sets in 1981 and 1991 demonstrate that this level of sampling is necessary. With each data set, the full suite of quarterly suitabilities (each age of each prey for each age of each predator) was estimated. These were differenced and plotted as a function of the mean sample size used to estimate the particular combination of predator, prey, age, and quarter. The results (Figure 13-4) show no interannual bias, but there is noteworthy scatter, particularly for predator-prey-age-quarter combinations with fewer than about 400 stomachs. These results indicate that stomach sampling programs can provide the basis for stable and unbiased descriptions of diets. However, for results to be quantitatively useful, sampling programs must reach levels of several hundred stomachs per predator age group and quarter, and more if the predator eats a variety of prey.

Table 13-6. Characteristics of data sets of cod stomach contents, by system

	Trait				
System	Number of samples	Number of stomachs	Mean predator length (cm)	Mean food weight (gm)	Percentage of empty stomachs
North Sea	7,871	41,161	37.48	30.39	13.85
Baltic Sea	8,966	50,737	41.05	13.44	20.55
Barents Sea	35,753	35,753	47.43	21.55	21.75
Iceland	9,364	38,210	54.27	45.40	15.89
Newfoundland and Labrador	26,273	26,273	51.53	33.73	2.95
New England	8,501	8,501	53.75	27.74	15.62
Total	96,728	200,635	45.87	27.70	15.99

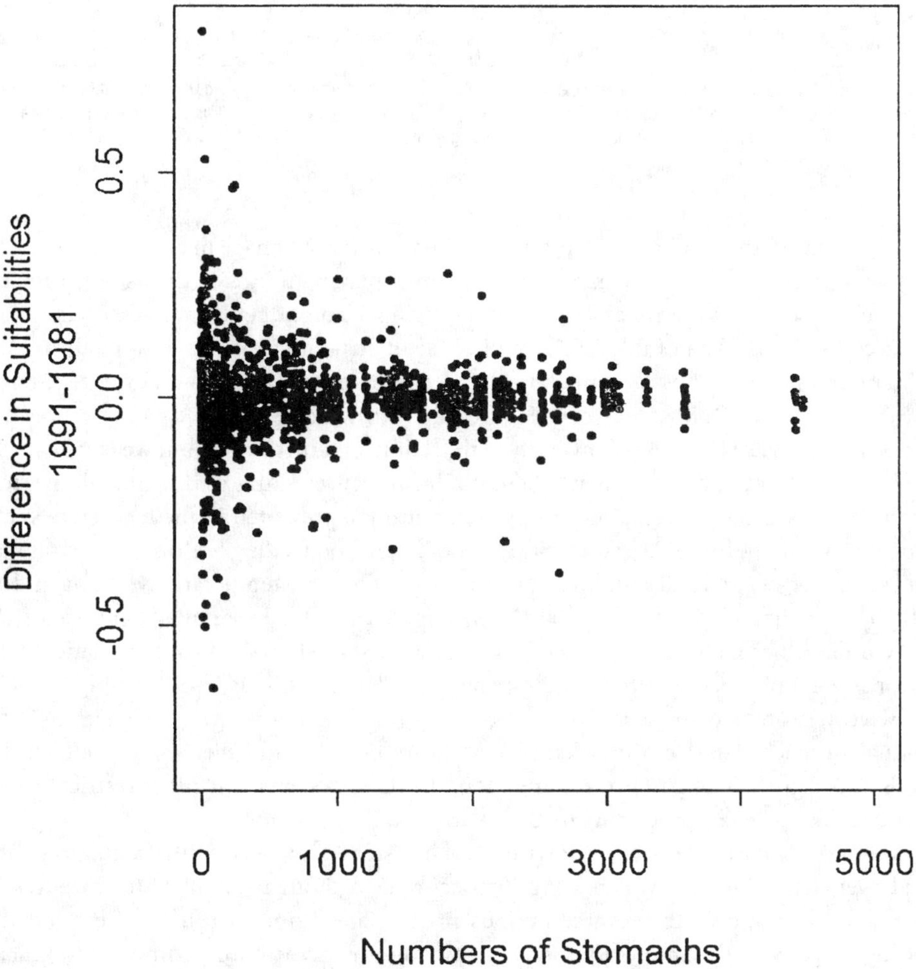

Figure 13-4. Relationship between stability of suitability estimates using single years of data and the number of stomachs examined. The *x*-axis is number of stomachs for a specific combination of predator age, prey age, and quarter; the *y*-axis is suitability estimated from 1991 data minus suitability estimated from 1981 data. Each entry is for a separate combination of predator age, prey age, and quarter.

Table 13-7. Summary of principal component structure for various attributes of cod stomach contents

	System					
Attribute	North Sea	Baltic Sea	Barents Sea	Iceland	Nfld.-Lab.	New England
Capelin	X	X	3; 1	2; −1	3	X
Unident. fish	X	4	3; 1	4	3	3
Other fish	1	−2	4	4	3	2
Crustaceans	−1	2	2	3	−1	−1
Molluscs	3; 2	3	1	2	2	2
Polychaetes	3	−3	1	2	2	4; 3
Echinoderms	2	X	1	2	2	4
Other food	2	1	1; 2	4	4	3
Depth	X	X	−3; 1	1	1	−2; 1
Temperature	X	X	3	1	1	X
CPUE	4	X	4	2	−4	X
Length	1	−1	−2; 4	−3	−2	1
Number of eigenvalues > 1.0	3	3	4	4	4	4

PCAs were performed on data from each system separately. Where PC loading for an attribute exceeded 0.30, the number of the component is entered in the table; a negative sign indicates negative loading. Multiple entries in a column mean that the attribute loaded on more than one component. X denotes that the variable was missing in the data set.

Compared with the general linear models, ordination analyses have many statistical inadequacies. However, they are a more convenient tool for displaying trends in data sets. In the case of the diet data, several patterns emerged from principal component analyses (PCAs) of each system's cod stomach data (Table 13-7). Similar factor structure was apparent in all six systems. The environmental variables of temperature and depth generally covary as expected, but more important, no diet variables are associated strongly with whichever factor picks up the environmental variables. Size is associated inversely with feeding on crustaceans, but weakly, if at all, with feeding on fish. In each boreal system, there is a factor representing feeding on all invertebrates other than crustacea, but this factor is not systematically represented in the temperate systems.

Looking at the specific PCA scores, the effect of larger cod feeding less on invertebrates covers the full size range of cod in all systems but is stronger in the three temperate (New England, North Sea, Baltic Sea) than in the three boreal (Newfoundland, Iceland, Barents Sea) systems (Figure 13-5). On the other hand, the variance in this relationship shows a stronger relationship with predator size in boreal systems than in temperate ones (Figure 13-6). A relationship between size and piscivory is only apparent in the North Sea. A series of scatter plots (Figure 13-7) for the individual stomach data from Newfoundland shows how the full diet space occupied by cod expands as cod grow. This pattern was present in the other systems, but either restricted amounts of data or bulking before processing made it harder to demonstrate as clearly.

The ordination scores of the factor from each system associated with feeding on fish were plotted over the 15 years in which sampling occurred. Although not all systems were sampled in all years, there appear to be three periods in the data. Prior to 1982, scores tended to be either high or low, and scores of specific systems were highly variable from year to year. From 1982 to the late 1980s, all systems showed "typical" (i.e., factor score near zero) and stable levels of feeding on fish. In the most recent years the three systems with a continuous sampling effort have diverged greatly from the "average" condition of the mid-1980s, but Newfoundland has shown increased feeding on fish, whereas the Baltic and Barents Seas cod are feeding much less on fish.

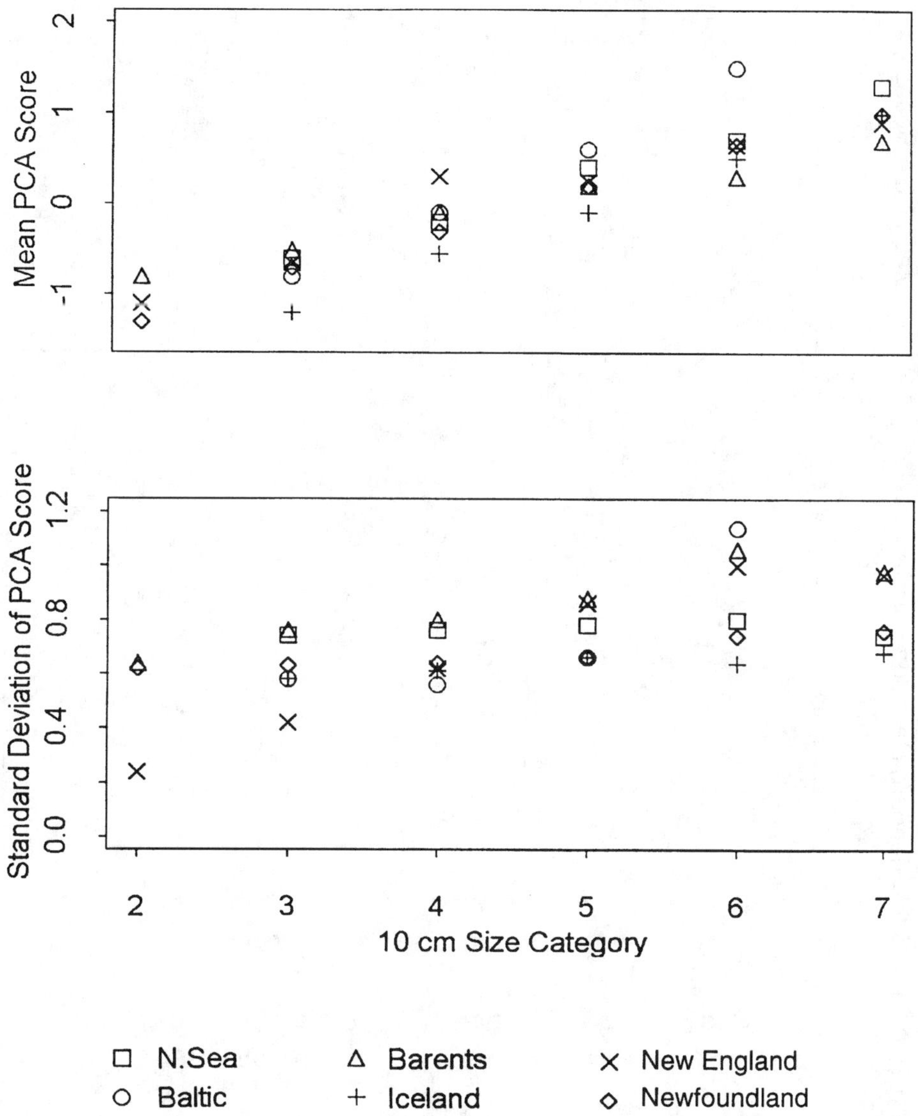

Figure 13-5. Relationship of degree of feeding on crustaceans to length of cod, across six ecosystems. Upper panel represents mean principal component scores for each system; lower panel represents standard deviation of scores.

Applications to Pacific Systems

The results above do not exhaust the ecosystem insights arising from the analyses of the Multispecies Assessment Working Group. Rather, together they give some flavor of the range of empirical investigations required to move to operational multispecies management of fisheries. Equally important, they convey several messages that must be considered as multispecies advice and management of fisheries ecosystems in the Pacific are developed.

The first overall message is positive: It is possible to develop tools and collect sufficient data for empirically based and consistent advice. The second message is also positive, namely, that in the short term, reasonably accurate fisheries advice and management can be developed without including many multispecies complexities. It is in the selection of management strategies, and the

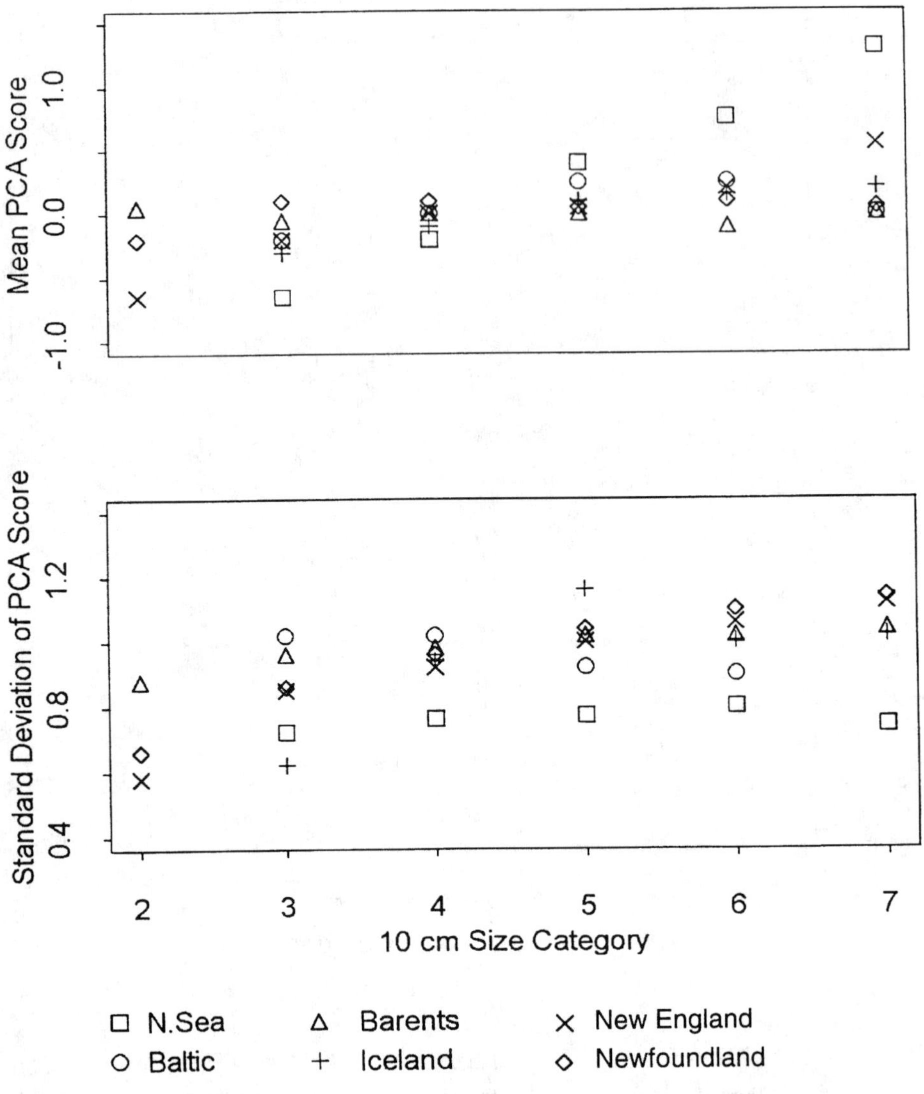

Figure 13-6. Relationship of degree of feeding on fish to length of cod, across six ecosystems. Upper panel represents mean principal component scores for each system; lower panel represents standard deviation of scores.

evaluation of longer-term consequences of management actions, that the ecosystem-level considerations come into play. The third message is also positive: Multispecies assessment tools exist that actually can be applied to a selection of fisheries strategies and management actions, and these tools produce predictions that can be tested and have been verified.

Beyond this, the messages present more challenges. At the multiple-ecosystem scale, the current thoughts about regimes and regime shifts are consistent with the diet analyses. For example, the representation of fish and invertebrates in diets of cod for six systems coincided for several years. However, the systems diverted from the shared pattern at different times, and when they did so they diverged at different rates and in different directions. Hence, a demonstration of a period of similar trajectory among Pacific ecosystems will not be sufficient to know where each will head when a regime shift does occur.

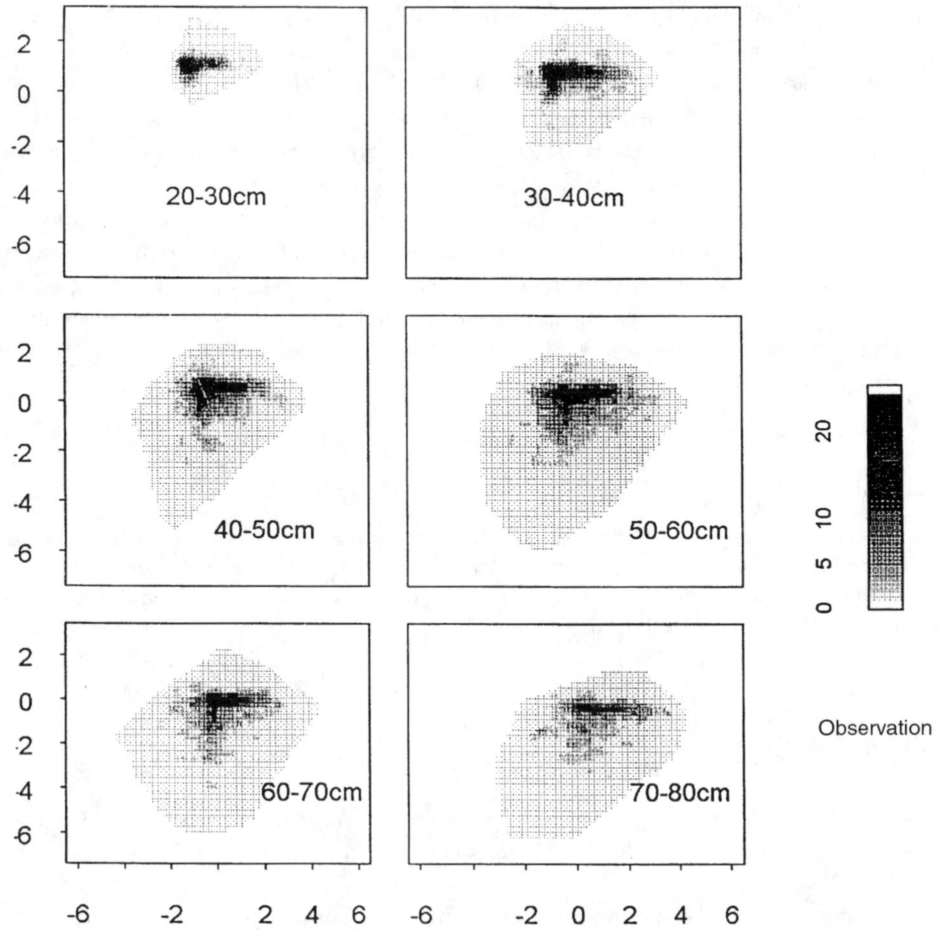

Figure 13-7. Illustration of the predator "space" occupied by Newfoundland cod, based on analysis of all stomach contents data. The *x*-axis is scores of principal component associated with fish in stomachs; the *y*-axis is scores of principal component associated with feeding on crustaceans. Panels present scores for progressive 10-cm size classes of cod from 20–30 cm to 70–80 cm.

Both the cod growth and cod diet data suggest that stocks in boreal systems are going to be more challenging to assess and manage than are stocks in more temperate systems. Year-to-year differences in diets, growth, prey compensation, and ration are all large, and the readily available annual environmental indices account for little of the interannual signals. Both the patterns and the inconsistencies in relationships with covariates are especially large in the boreal systems. That boreal systems show greater variance in PCA scores for the stomach contents data means the predator–prey relationships in these systems will be particularly hard to quantify in ways useful for development of multispecies management advice.

The North Sea stomach data highlight just how variable stomach contents data really are. Whereas the North Sea data were fairly consistent in the multi-ecosystem analyses, the contrast of suitabilities from the 1981 and the 1991 data sets limits the degree of stability. Moreover, the consistent results were due not just to data quality (which was high) but also to data quantity and a decade of refining model formulation and parameterization.

This standard for data and models returns us to why multispecies assessment and management have proven so difficult to make operational. The interest in ecosystem management arose from the unsatisfyingly large uncertainties that are associated with single-species assessment and advice, and the need for new kinds of advice on new kinds of problems. It was logical to infer that much of the uncertainty was due to multispecies interactions. What was not logical was to assume that the inclusion of multispecies considerations would readily reduce that uncertainty or make the new kinds of advice clear and useful to managers. Rather, inclusion of multispecies interactions requires representing additional relationships in the assessment and advisory models and then parameterizing those relationships. Both steps bring with them opportunities for adding model and parameter uncertainty to assessments and advice rather than reducing uncertainty through representing the relationships at all. The Multispecies Assessment Working Group has shown that both parts of the job can be done well, and that neither part can be done easily. Those working on assessment and management of fisheries in the Pacific can build effectively on much of their progress.

ACKNOWLEDGMENTS

Many results I report were produced at meetings of the ICES Multispecies Assessment Working Group. They reflect data and analyses contributed by many laboratories and scientists. Credit for ideas and efforts must be shared with all Working Group participants. However, the author, not the Working Group members or ICES, is responsible for the contents of this manuscript.

References

Andersen, K. P., and Ursin, E. 1977. A multispecies extension to the Beverton and Holt theory of fishing, with accounts of phosphorous circulation and primary production. Medd. Dan. Fisk. -og Havundrs. N.S. 7:319–435.

Fisheries Resource Conservation Council (FRCC). 1993. 1994 Conservation Requirements for Atlantic Groundfish—Report to the Minister of Fisheries and Oceans. Dept. of Supply and Services Document Fs 1-61/1994E.

Helgason, T., and Gislason, H. 1979. VPA Analysis with Species Interaction due to Predation. ICES C.M. 1979/G:52.

Horbowy, J., and Swinder, M. 1989. Growth of Baltic sprat in relation to stock density and food biomass. Rapp. P.-v. Réun. Cons. Int. Explor. Mer. 190:102–104.

International Council for the Exploration of the Sea (ICES). 1991. Report of the Multispecies Assessment Working Group. ICES C.M. Assess:7.

International Council for the Exploration of the Sea (ICES). 1992. Report of the Multispecies Assessment Working Group. Copenhagen, 16–25 June. ICES C.M. 1992/Assess:9.

International Council for the Exploration of the Sea (ICES). 1994. Report of the Multispecies Assessment Working Group. ICES C.M. Assess:9.

Millar, R. B., and Myers, R. A. 1990. Modelling Environmentally Induced Change in Growth for Atlantic Canada Cod Stocks. ICES Doc. C.M. 1990/G:24.

National Oceanographic and Atmospheric Administration (NOAA). 1994. NOAA Science Review—Build Sustainable Fisheries. Washington, DC: U.S. Department of Commerce.

Northwest Atlantic Fisheries Organization (NAFO). 1980. Scientific Council Reports—1983. Dartmouth, Nova Scotia: Northwest Atlantic Fisheries Organization.

Sparre, P. 1991. Introduction to Multispecies Virtual Population Analysis. In N. Daan and M. P. Sissenwine (eds.), Multispecies Models Relevant to Management of Living Resources. ICES Mar. Sci. Symp. 193, pp. 12–21.

14

The Significance and Prediction of Predation in Marine Fisheries

Nicholas J. Bax

Abstract

There is a high level of predation on fish from marine mammals, seabirds, and squid and other invertebrate predators in addition to that from other fish. This predation can be highly variable between years. The variety of predation processes in marine fishery ecosystems complicates the interpretation of particular predation events, but there are examples of predation structuring marine ecosystems. Advantages and drawbacks of the available approaches for understanding and predicting predation's impacts are discussed. Much work remains to be done, and it is stressed that one of the difficulties in understanding predation is in defining the spatial scale of the study and the functional groups that constitute the ecosystem. Several examples are provided of models of the interaction between fur seals and commercial fisheries, where it is demonstrated that it may not be possible to determine the impact of predation by consideration of the direct predation alone; indirect predation may nullify or reverse the direct predation effects. It is concluded that improved understanding of the role of predation in structuring fishery ecosystems will require an ongoing and nondisruptive experimental approach to fisheries management.

Many of the world's marine ecosystems have undergone major shifts in species composition that have had severe consequences on their biological functioning and the derived economic yield. The mechanisms underlying these major shifts are poorly understood because they appear to result from not one but several causes that are effected through an interactive system of species, habitat, and environment (Caddy 1993). The complexity imposed by these interactions severely complicates management measures implemented to reverse these shifts (Wilson et al. 1994).

Management at the level of individual species may not be sufficient to reverse these ecosystem shifts or to prevent major shifts in other ecosystems. Even if a marine ecosystem were characterized by stability and a consistent species composition, it would prove impossible to manage a predator–prey system in that ecosystem by managing the species individually (May et al. 1979). Where ecosystems do not remain stable over time frames of interest to human users, there is little likelihood that management implemented one species at a time will be successful over the long term. Without an understanding of interactions between species and between fisheries in the context of the physical environment and habitat in which these interactions occur, management is forced to concentrate on short-term tactical results to the detriment of medium-term planning and long-term strategic management (Sharp 1992). Because, once initiated, short-term tactical issues appear more pressing than strategic issues, more effort is expended on short-term research and less on strategic issues, creating a vicious cycle known as the tyranny of the urgent.

An important aspect of ecosystem functioning is species interaction through predation. These effects are direct when a predator kills a prey, and indirect in a multitude of ways, including predation on shared prey resources, predation involving intermediate predators, and predation that influences prey behavior and resource availability. Knowledge of predatory interactions is one source of information necessary to address strategic issues in ecosystem management. This chapter reviews the magnitude, importance, and prediction of predation in marine systems with emphasis on those species of direct interest to commercial fisheries. The review concludes with examples of models that predict predation in marine commercial fisheries.

Overview of Predation Processes

That predation in marine systems is a complicated affair has been clear since at least the publication of Hardy's classic diagram of the details involved in the "simple" trophic chain whereby herring eat copepods that in turn eat algae (Cushing 1975) (Figure 14-1). The level of detail in predation interactions rapidly becomes overwhelming, and the potential interactions in even comparatively simple seven-compartment models make prediction highly uncertain (Gomes 1993).

Because it is not possible to look at all predation processes in a fishery ecosystem, it is necessary to concentrate on those of greatest importance, being always aware that those aspects that are excluded may attain greater importance when there is a change in the context in which the predation processes operate (Sharp 1992). At the other extreme from considering all predation interactions is the assumption that the combined predation on a stock imposes a constant rate of mortality and that the stock can be assessed as an individual stock. One step removed from this extreme is to concentrate only on direct (linear) predation interactions; for example, if seals eat

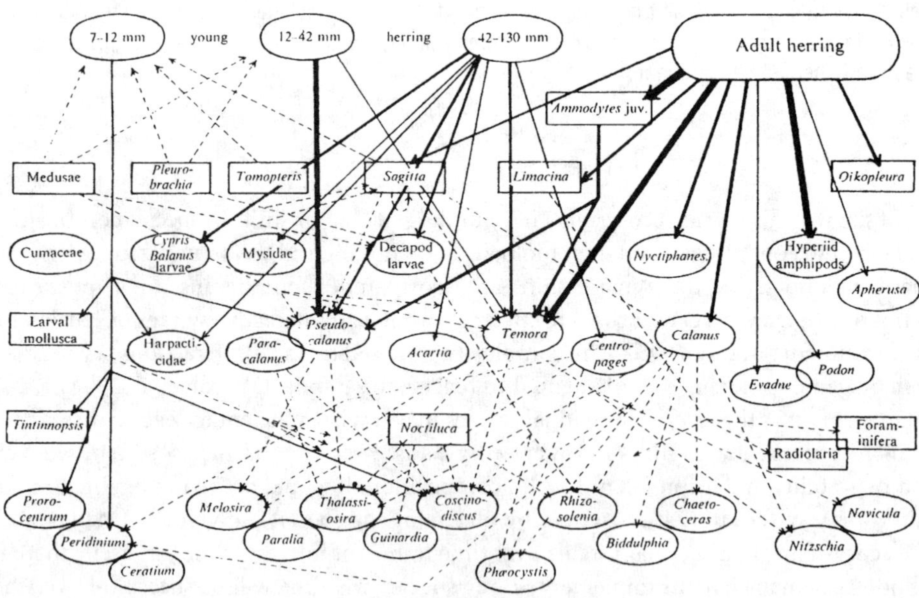

Figure 14-1. The food web representing the food sources that North Sea herring depends on during its life cycle, as described by Hardy in 1924. (Reprinted by permission from Cushing, D. H. 1975. Marine Ecology and Fisheries. Cambridge, England: Cambridge University Press.)

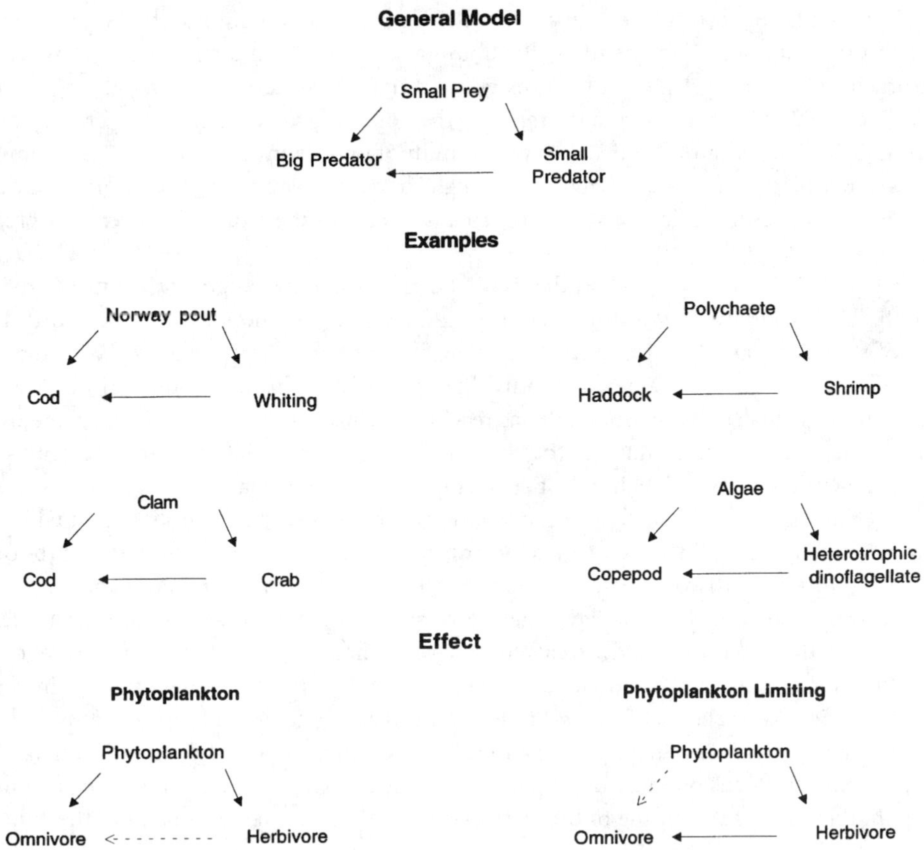

Figure 14-2. Triangular food web (top) proposed by Ursin (1982) as a mechanism of stability in marine ecosystems, some examples of triangular food webs (middle), and an illustration of how the interactions alter to confer stability on a plankton community (redrawn from Ursin 1982 and Jones and Henderson 1987).

cod, then reducing consumption by seals will cause a decreased consumption of cod, and the survival of cod will increase. The next step might be to consider first-order indirect (triangular) interactions that have the potential to reverse the effects of direct interactions (e.g., May et al. 1979; Butterworth et al. 1988) and may confer stability on marine ecosystems (Figure 14-2). When fishing is superimposed on these triangular interactions, the impacts can be either symmetrical or asymmetrical, depending on whether the fishery and the big predator exploit the same prey or not, respectively (Beverton 1985). Symmetrical interactions enable the qualitative assessment of the likely impact of a change in abundance of the big predator (e.g., marine mammals) on the abundance of the commercially exploited fish, whereas even the direction of change in the abundance of the commercial fish following a change in abundance of the big predator is indeterminate for asymmetrical interactions. Beyond this level of abstraction the predation interactions become more complicated and the results dependent on the model specifications. It is at this level that general regulatory patterns are unknown and much work remains to be done, although there have been advances in less speciose freshwater ecosystems (e.g., Kitchell et al. 1994).

Predation does not always have a regulatory effect on a population; the mortality rate caused by predation can be either independent of the prey animals' density (density independent) or dependent on this density (density dependent). Because density-independent processes are not related to abundance of the prey in any consistent manner, they are generally destabilizing. A good example is predation of eggs and larval fish in the planktonic community, where feeding is opportunistic and highly dependent on the abundance and spatial distribution of other planktonic organisms. Density-dependent processes are usually considered to be stabilizing (compensatory) because the rate of predation decreases as the population decreases; however, they can also be destabilizing (depensatory) if the absolute amount of predation stays the same as a population changes. Examples are reduced predation through predator swamping at high prey abundance, and constant catches regardless of stock size from a directed fishery on a consistently aggregating stock.

Much of the focus in predation studies is on the predator and changes in the rate of predation. However, the selectivity of predation requires comparable attention. Size is the most obvious aspect of selectivity, and the size range of prey available to a predator increases with the size of that predator. Similarly, the value of an individual to the future of the population varies with the size of that individual; an individual that has reached maturity and is ready to pass gametes on to the next generation has more value to the continuation of the population than a freshly spawned individual that has a small likelihood of surviving to the size of maturity. This is especially the case for teleost fish, which usually produce thousands of eggs or larvae, only two of which need to survive and successfully spawn to enable continuation of the population. The pronounced selectivity of predation is only just beginning to be integrated into ecological models with the understanding that modeling the "average" response of a cohort may obscure the underlying trend for only the exceptional individuals to survive to maturity (e.g., Beyer and Laurence 1981; West 1988; Rice et al. 1993). Individual differences tend to be stabilizing (at least in models) because individuals or classes of individuals with decreased vulnerability to predators provide a refuge for the species from the predator, and refuges are in general stabilizing (Smith and Mead 1974). The current focus on individual species in marine ecosystems may in the long run impede understanding of the functioning of those ecosystems to a greater extent than has the concept of distinct trophic levels (Platt 1985).

Magnitude of Predation in Marine Ecosystems

Predatory interactions in marine ecosystems are ubiquitous. The following review of the magnitude of predation summarizes only a fraction of the available reports but provides an idea of predation's potential in influencing ecosystem dynamics.

Marine Mammals

Several species of whales and seals are known to be major predators of groundfish, herring, capelin, and squid in the Bering Sea (Lowry and Frost 1985), where they have been estimated to consume approximately 10% of all commercial fish stocks annually, compared with 5% by man and 2% by birds (NMFS 1979; Laevastu and Larkins 1981). Toothed whales in Icelandic waters are estimated to consume about 850,000 tonnes of fish annually (Sigurjónsson and Vikingsson 1992). Common and bottle-nosed dolphins stranded on the coast of northwest Spain contained an average of 210 fish otoliths per stomach, while Risso's dolphin, pilot whales, and a sperm whale contained an average of 14 cephalopod upper and lower beaks per stomach (Gonzàlez et al. 1994). Gadidae, primarily blue whiting, formed the majority (65%) of the fish consumed. Gray seals in the North Sea were estimated to consume 14.3 million 1- to 4-year-old cod, or 4.8% of this size component of the stock, imposing a natural mortality in the range of 0.0034 to 0.1259. Cape fur seals in southern Africa were estimated to eat approximately 1 million tonnes of food per annum, the majority of which was fish, equivalent to between a third and all of the commercial catch in their feeding area (Shaughnessy 1985).

A recent collection of papers on the interactions between marine mammals and fisheries (Beddington et al. 1985) indicates that the amount of commercially valuable fish eaten by marine mammals is similar to or exceeds that taken by the commercial fishery in several areas (South Georgia, Kerguelen Islands, southern Africa, northwest Atlantic, Bering Sea). In other areas (U.S. Pacific Northwest coast, Bristol Bay), marine mammal predation causes a localized depletion of invertebrates that would prevent development of a commercial shellfish fishery.

Seabirds

Seabirds have a high metabolic rate, high energy requirements, and consequently a high rate of predation. For example, it has been calculated that murres consume between 10% and 33% of their body weight daily (Vermeer et al. 1987). Therefore, in the Scotia-Weddell Sea, although the average biomass of seabirds is estimated to be one-third that of seals and one-sixth that of whales, their daily energy requirements are comparable (Schalk et al. 1993). Seabird consumption of fish is high in absolute terms; for example, fish-feeding alcids were estimated to consume 2,675 tonnes of fish per day in the eastern North Pacific, while all other seabirds consumed 866 tonnes per day (Vermeer et al. 1987). Seabird consumption of fish is also high in terms of overall production, accounting for 22% of the fish production off Oregon, USA; 29% of the fish production within the 45-km radius of a colony in the Shetland Isles, UK; 24% of the mean catch off South Africa; 17% of the anchovy off Peru from 1961 to 1965; and 4% to 7% of the adult herring off the Gulf Islands, Canada (Furness 1982). Seabirds also consume large quantities of squid and were estimated by Furness (1994) to consume 40,000 tonnes in the northeastern North Atlantic, 63,000 tonnes in the southeastern North Atlantic, 3,000 tonnes in the Mediterranean, 466,000 tonnes off South Georgia, and 223,000 tonnes off Hawaii.

Northern gannets in Newfoundland waters consume the same pelagic prey species at similar sizes as the commercial fishery in that area. The gannets' estimated harvest of mackerel and squid often exceeds commercial landings (Montevecchi et al. 1988). Between 1977 and 1986 there were four coincidental collapses in the human and avian harvests of squid, and one coincidental collapse in the two harvests for mackerel. Seabirds have been shown to deplete food resources in the waters surrounding their colonies (Furness and Birkhead 1984; Birt et al. 1987), and have been suggested as an indicator of the availability (and by extrapolation, abundance) of fish (Cairns 1987).

Caution is needed before concluding that the occurrence of fish in the diet of seabirds implies predation on those fish. Where commercial fisheries operate, a large proportion of seabirds' diet can come from fish discarded from the commercial fishery. For example, Furness et al. (1988) estimated that 90% of offal and 75% of whole fish discarded from commercial fisheries around the British Isles were consumed by seabirds, representing between 5% and 10% of the landed catch weight for flatfish and gadoid fisheries. However, breeding seabirds, which may have to maximize energy obtained per unit time, may no longer scavenge for offal and discards but instead concentrate on catching shoaling fish such as sand eels (Furness et al. 1988). Their breeding success may be critically dependent on the abundance of suitable prey such as sand eels at this time, as shown by poor breeding success and chick survival in the Shetland Isles in the mid-1980s following declines in the abundance of sand eels.

Squid

Squid are abundant, widespread, rapidly growing piscivores considered to play an important but relatively unknown role as predators in marine ecosystems. Their rapid increase in population biomass and the importance of juvenile and larval fish in their diet suggest that squid could be significant regulators of recruitment in marine fish. Fish occurred in more than 80% of nonempty stomachs of two species of the long-finned squid, *Loligo* sp., in the northeastern Atlantic, from

Scotland to the Azores (Pierce et al. 1994). The most frequently identified prey in northern waters included members of commercial families Gadidae and Clupeidae, whereas scad, *Trachurus picturatus*, were more important in southern waters. Fish were also the most abundant prey (42% by volume; herring 12% by volume) in short-finned squid, *Illex* sp., sampled off Iceland (Jónsson 1980). Zero-group cod occurred in 8% of short-finned squid caught off Newfoundland (Dawe 1992).

Off the northeastern United States, two cephalopod species were estimated to consume between 33,000 and 43,000 tonnes (*Illex illecibrosus*) and between 182,000 and 703,000 tonnes (*Loligo pealei*) of fish per annum. As the majority of this predation in some quarters was on prerecruits of commercial species, including Atlantic cod, Maurer and Bowman (1985) concluded that the squid were a significant source of prerecruit mortality. Radchenko (1992) estimates that the annual production of Gonatidae (60% of the squid biomass) in the Bering Sea, at 9.4 million tonnes, exceeds production of the dominant families of fish. This production provides an estimated 4.2 million tonnes for other predators, while 2.4 million tonnes is eaten by the Gonatidae themselves. One component of the diet of squid is juvenile fish. Longhurst and Pauly (1987) estimate that cephalopods in the Gulf of Thailand consumed 700,000 tonnes per year of prey in the early 1960s and 500,000 tonnes per year in the early 1980s, compared with a total annual fisheries harvest of about 200,000 and 800,000 tonnes in the same periods. Two of the five major squid species in the Benguela ecosystem consume commercial species, with one species' consumption consisting of 10% to 50% commercial fish species during 60% of the year when its distribution overlaps with hake and anchovy. Annual consumption of commercial fish for this species is estimated at between 8,500 and 1.6 million tonnes (Lipínski 1992). The high variability in these figures illustrates the difficulty in obtaining precise estimates of the impacts of squid on fish in marine ecosystems.

Squid are not just predators of fish, but also their prey. Results from the ICES stomach sampling program show squid to be a small but significant component of the diet of 0-group whitefish (Boyle and Pierce 1994). There have been several instances of rapid increases in squid biomass following the onset of commercial fishing (Gulf of Thailand, Mauretania). It is suggested that reduced predation by fish on juvenile squid provides a more plausible explanation for these rapid increases than reduced trophic competition (Longhurst and Pauly 1987).

Invertebrates (Excluding Squid)

Fish eggs and larvae form a major component of the diet for many marine invertebrates. An extensive review by Bailey and Houde (1989) reported that species of gelatinous zooplankton, copepods, chaetognaths, euphausiids, amphipods, and snails have been recorded as predators of fish eggs and/or larvae. The relative abundance of marine invertebrates means that even a small predation rate on fish eggs and larvae can result in a substantial mortality rate. Removal of invertebrate predation can increase survival of fish larvae (cod, herring, capelin, and plaice) to metamorphosis (in enclosures) by over 50% (Oiestad 1983). Medusae have been estimated to consume the equivalent of 57 Pacific herring larvae per day (Arai and Hay 1982), or up to 28% and 60% of larval herring stocks daily in some locations off British Columbia (Purcell et al. 1987). Medusae were estimated to consume 2–5% of the total stock of herring larvae daily in the Kiel Bight (Möller 1980), causing a measurable decline in the size of the larval herring population (Möller 1984).

North Sea plaice eggs and larvae suffer heavy mortality on their way from the spawning grounds to the nursery areas, some of which is attributable to predation by herring, which selectively eat the eggs (Daan et al. 1985), but as they reach coastal waters, the later arrivals are subject to high predation by coelenterates. Somewhere between 1% and 2.5% of the *Pleurobrachia pileus* population in the western Wadden Sea was estimated to have consumed a flatfish larva in the few hours before capture, resulting in a predation of between 140 and 350 larvae consumed per 1,000 m^3 every couple of hours. This would cause a significant mortality when larval abundance is estimated to be 230 larvae per 1,000 m^3 and is thought to effectively end plaice immigration to the nursery area (Veer 1985). Presettlement predation is followed by predation by brown shrimp after settlement, possibly removing between 2% and 14% of the population per day (Veer 1987). Flounder larvae that migrate to the nursery areas later in the year suffer a greater percentage mortality than plaice, causing the ratio of the relative abundance of flounder larvae to plaice larvae to drop from 10:1 before settlement to 1:4 after settlement (Veer and Zijlstra 1982). *P. pileus* was the suspected predator.

Fish

Fish are a major predator of other fish in marine ecosystems and may be a major source of mortality for the smaller size classes. Even if a larva escapes its attacker, its probability of survival is likely reduced. Bailey and Houde (1989) report that predation of fish eggs and larvae by other fish has been recorded since early this century; for example, Bowman (1922, in Bailey and Houde 1989) noted haddock gorged with herring eggs. Johannessen (1980) found an average of 15,000 to 20,000 eggs in the stomachs of cod caught near a herring spawning ground in a Norwegian fjord and concluded that predation could remove 40% to 60% of total egg production in years of low egg deposition. Dragesund and Nakken (1973) concluded that haddock and saithe consume up to 40% of the eggs of spring-spawning herring. Livingston (1991) estimated that Bering Sea yellowfin sole consumed 2×10^9 0-group arrowtooth flounder and 80×10^9 0-group Greenland halibut in 1984, but none in 1985 or 1986. The numbers of arrowtooth flounder consumed were 13 times greater than subsequent recruitment at age 2 (150×10^6), whereas 27,000 times as many Greenland halibut were consumed as subsequently recruited at age 4 (3×10^6). Moksness and Oiestad (1987) report the rapid disappearance of capelin larvae in mesocosms when herring started to school, and suggested that herring predation on capelin might explain the absence of good recruitment of capelin in the Barents Sea when herring are abundant.

Cannibalism is also an important source of egg mortality. Cannibalism by adult northern anchovy accounted for 17% of the egg abundance per day, or 32% of the total egg mortality (Bailey and Houde 1989). Peruvian anchoveta were responsible for consuming 15% of the egg abundance, or 22% of the total egg mortality for this species (Bailey and Houde 1989).

The predation of fish on later life stages of fish is not as large (in terms of numbers per fish) as predation on eggs and larvae, but because it may occur after recruitment levels for that year are determined, its influence on the adult population and commercial fisheries yield may be more direct. An analysis of the sources of fish mortality in six diverse exploited marine ecosystems (Bax 1991) showed that the major source of mortality for fish beyond the larval and early juvenile stages in these ecosystems is predation by the fish themselves (Table 14-1).

Table 14-1. The major sources of fish mortality in six diverse exploited marine ecosystems

	Fish loss (tonnes/km²)				
Ecosystem	Birds	Mammals	Fish	Fishery	Fish/Fishery
Benguela Current	0.3	2.6	56.5	1.6	35
Georges Bank	2.0	5.4	42.5	6.1	7
Balsfjord	0.0	0.0	14.1	1.5	9
E. Bering Sea	0.2	1.5	11.0	1.4	8
North Sea	0.6	0.1	7.0	4.4	2
Barents Sea	0.0	3.0	5.1	1.8	3

Original data sources provided in Bax (1991).

Table 14-2. Applications of the ECOPATH II box model to obtain preliminary estimates of the predation by trophic groups for four continental shelf ecosystems for biomass lost to fish predation and the fishery

			Biomass lost		
Location (reference)[a]	Area (km^2)	Units	Fishery	Fish	Fish/Fishery
Yucatan, Mexico (1)	100,000	g dw/m^2/yr	0.09	2.12	25
Gulf of Mexico (2)		g dw/m^2/yr	2.40	8.24	3
Venezuela (3)	30,000	t/km^2/yr	5.20	32.02	6
Brunei Darussalam (4)	7,396	t/km^2/yr	0.70	15.06	20

[a]All papers appear in Christensen and Pauly (1993). (1) Arreguin-Sanchez et al. 1993; (2) Browder 1993; (3) Mendoza 1993; (4) Silvestre et al. 1993.

In these six marine ecosystems that have been analyzed in detail, the indications are that the predation by fish is 2 to 35 times the loss to commercial fisheries. It needs to be stressed that the major causes of mortality will vary between species and that fishing will have greater impact for larger specimens of commercial species, whereas predation will have greater impact on smaller specimens and most noncommercial species.

The ECOPATH II box model (Christensen and Pauly 1993) was applied to obtain preliminary estimates of the predation by trophic groups for four continental shelf ecosystems for biomass lost to fish predation and the fishery (Table 14-2). Preliminary modeling indicates that predation by fish is between 3 and 25 times as large as the commercial catch for these four ecosystems. (Note that commercial catch figures for the Gulf of Mexico include the large bycatch of the shrimp fishery.)

It should be mentioned that predation can be highly variable between years. For example, Mehl (1989) estimated that from 1984 to 1986 Barents Sea cod consumed 0.4 to 1.2 times their own biomass of capelin, which included an estimated 85% of the numbers of 3- and 4-group capelin in 1985. The consumption of capelin by cod increased threefold between 1984 and 1985, an increase attributed to growth of the strong year-class of cod to a size at which larger capelin could be taken and to a greater overlap between the two species in early 1985. A similar interaction between cod and capelin was observed in Icelandic waters from 1979 to 1984 (Magnússon and Pálsson 1989), where as capelin declined, the proportion of capelin in the cod diet also declined. Because there were insufficient alternative prey resources, the total consumption by cod also declined; only 30% of any reduction in capelin in the diet was replaced by alternate prey. The approximately 50% decline in the biomass of economically important finfish (e.g., cod, haddock, flounders, herring, and mackerel; Murawski et al. 1983) on the U.S. Northeast Continental Shelf ecosystem in the late 1960s and early 1970s due to intensive fishing was followed by an increase in the biomass of lower-valued species (sandlance, dogfish, and skates) and led to alterations in consumption and predation throughout the ecosystem (Overholtz and Nicholas 1979; Powers and Brown 1987; Payne et al. 1990).

How Important Is Predation in Marine Ecosystems?

Given the demonstrated high levels of predation in marine ecosystems, it might be expected that the importance of predation in regulating those systems would be as easily demonstrated. This has not proven to be the case. Much research has been conducted to determine whether starvation (derived from Hjort's 1914 critical period concept) or predation determine year-class strength in commercial fish stocks worldwide, but little evidence has been found to support the critical period concept (May 1974; Bailey and Houde 1989). The general absence of large numbers of starving

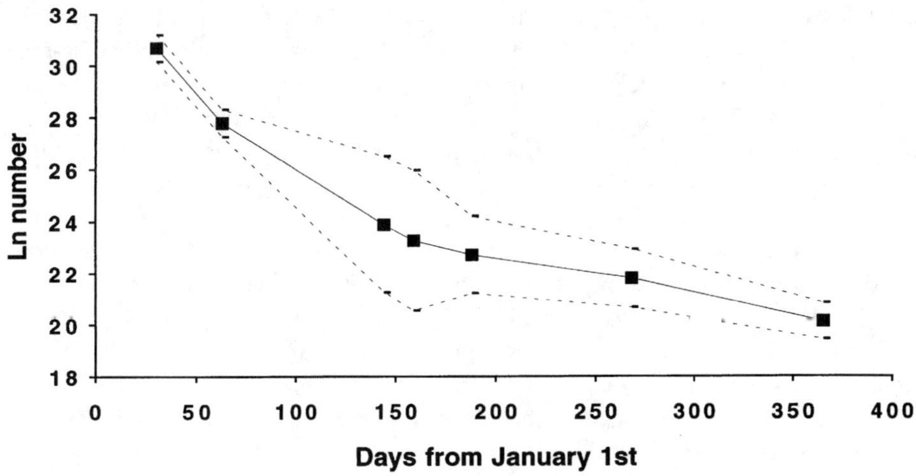

Figure 14-3. Survival curve for North Sea plaice from parents to the end of the first year, showing maximum ratios of year-class strength at certain intermediate ages. The expansion of the ratios at about 150 days expresses the possibility of very high variability (200-fold) during the pelagic larval stage, which is then progressively damped by density-dependent mortality to the observed fourfold range at recruitment (redrawn from Beverton and Iles 1992).

larvae in the sea, the higher mortality for egg and pre-feeding larvae with plentiful yolk reserves than for larvae at yolk absorption (the critical period), and high survival rates of several commercial species raised in enclosures in the absence of predators all indicate that starvation may not be the primary determinant of year-class strength (Bailey and Houde 1989). However, with few exceptions, it has also not been possible to demonstrate that predation determines year-class strength. The effects of predation are intertwined with those of food availability: Starvation leads to an increased susceptibility to predation (Bailey 1984; Rice et al. 1987). The potential exists for some generalist predators (e.g., jellyfish on herring larvae) to affect the prey in several ways: through direct predation and through competition for food, thereby reducing growth rates and increasing the prey's vulnerability to further predation (Purcell et al. 1987). However, at present, there is no unequivocal proof that reduced food supplies causing reduced growth rates lead to increased predation (Leggett and DeBlois 1994).

Although predation on the eggs, larvae, and juveniles of fish is high, it has been difficult to define a relationship between abundance at this stage and recruitment to the mature stock (Sissenwine 1984, though see Hylen and Dragesund 1973; Meyers and Cadigan 1993). One of the difficulties in relating high predation on eggs and larvae to subsequent recruitment is that the eggs and larvae are often only one of several available food sources and predation on them is adventitious. In addition, significant mortality due to predation in a flexible multispecies community can be a rapid process occurring over a restricted period in specific locations as a consequence of independent events, thus making it difficult to define regulatory processes. This is well illustrated by the Japanese flounder, which will move up off the bottom 2 days after settlement unless sufficient food is found (Tanaka et al. 1989). Once off the bottom it again becomes more susceptible to predators, including larger juveniles that settled earlier.

Despite opportunistic feeding behaviors of egg and larval fish predators, the high specificity of predation to time and place and the multiplicity of predators provide many opportunities for population regulation. This is best documented for the southern North Sea plaice, where the overall magnitude of plaice recruitment appears determined during the pelagic egg and larval stage when plaice suffer heavy mortality from selective predation by herring in open waters (Daan et al.

1985), followed by coelenterates in coastal waters (Veer 1986). At this point the coefficient of variation of year-class strength is 62%; it is not until the postsettlement and density-dependent predation by shrimp that the final regulation of year-class strength occurs, reducing the coefficient of variation to 30%, which is similar to that of 2-year-old plaice recruiting to the fishery. Beverton and Iles (1992) concluded that for North Sea plaice, mortality in the pelagic larval stage was density independent and caused high variability in year-class strength, but that postsettlement mortality was density dependent (perhaps operating through three sequential stages) and was sufficient to cause the low year-class variability characteristic of North Sea plaice (Figure 14-3). The conclusion that density-independent interannual variability in population size is determined in the larval stage but that density-dependent mortality in the juvenile stage attenuates this variability has been extended to 13 other North Atlantic demersal species (Meyers and Cadigan 1993).

Despite the important role of predation in regulating recruitment in flatfish populations, predation may not be the general mechanism determining recruitment. Indeed there is unlikely to be a single mechanism for recruitment regulation, but more likely a suite of factors affecting different life history stages that in different years will have their effects enhanced or diminished by environmental variability. Both the synchrony of recruitment over wide geographical areas (Koslow 1984) and the impact of large-scale climatic variability on local recruitment events (Sharp 1992) indicate the importance of the environment in modifying recruitment success.

Marine mammals are major predators of fish, including commercially fished populations, generally concentrating on fully recruited fish. It might be expected that determining the impacts of their predation on the size of fish populations would be relatively straightforward, but here again it has been difficult to demonstrate that predation has a regulatory impact. Even in situations where it has been estimated that predation by marine mammals is equal to or exceeds commercial catches, there is insufficient evidence to conclude that predation by marine mammals drives the changes in fish populations (reviewed by Beverton 1985). Only in an isolated freshwater lake, where the food resources were limited in variety, was there good evidence that seals reduced fish abundance (Power and Gregoire 1978). In this instance, the susceptible species (lake trout) had fewer larger size classes, increased growth rates, increased fecundity, and a reduced age of maturity in the presence of seals, whereas the nonsusceptible species (brook trout) were more abundant in the presence of seals, indicating an indirect effect of mammal predation.

Marine mammals have been shown to regulate populations over several trophic levels in the marine intertidal zone (Simenstad et al. 1978; Duggins 1988; Estes and VanBlaricom 1985). Duggins (1988) showed that removal of sea urchins leads to the development within one year of a kelp association of high biomass and diversity, followed in the second year by lowered biomass and diversity as a single kelp species becomes dominant. A similar trend in kelp biomass and diversity was observed when contrasting Torch Bay, which has no urchin-predating sea otters, with Deer Harbor, where sea otters had been present for less than 2 years, and Surge Bay, where sea otters had been present for 15 years (Figure 14-4).

Population regulation by marine mammals appears to occur (or is detected) primarily where their diet is specialized. Northridge and Beddington (1992) contrast the recognizable impacts of marine mammals having specialized diets with the lack of detectable impact for marine mammals having more generalized feeding habits. Specialized predators such as gray whales, which feed by sucking sediment into their mouths and extracting food items, cause pits that may cover more than 30% of the sea bed in places. Dense tube mats of ampeliscid and other amphipod crustaceans are destroyed in this feeding process and replaced by other scavenging invertebrates (Oliver and Slattery 1985). Similarly, areas of high walrus density in Alaskan waters are associated with relatively low bivalve mollusc densities (Fay et al. 1977, in Northridge and Beddington 1992), and walruses in Bristol Bay appear to consume the surf clam at levels exceeding their maximum sustainable yield (Northridge and Beddington 1992). Conversely, gray seals in the North Sea that feed on a variety of mobile fish species were not considered to be a major factor influencing

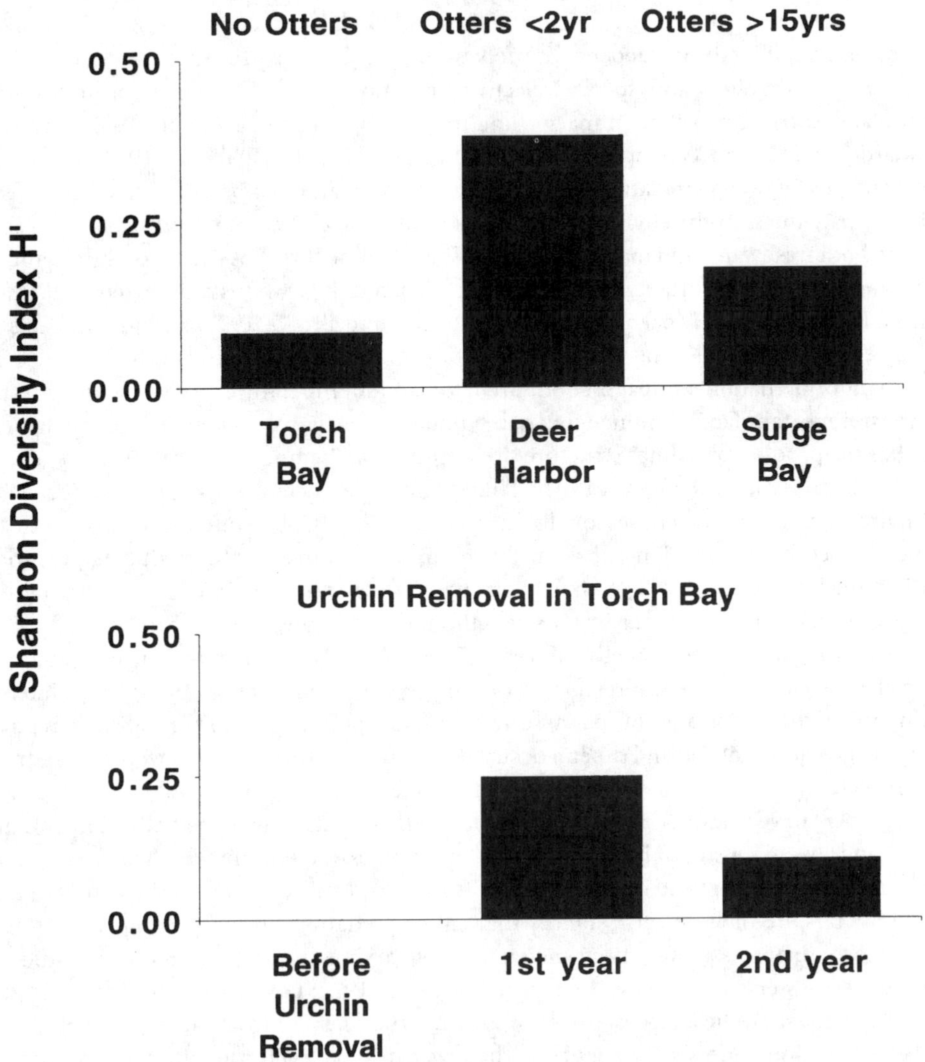

Figure 14-4. Species diversity of kelps in areas with otters present for different time periods (top) and in one experimental area following the removal of sea urchins (bottom). There was no growth of macroscopic algae in removal areas before experiments were begun (redrawn from Duggins 1980).

community dynamics because, in general, the mortality caused by seals was one or two orders of magnitude less than that caused by fishing and it was not restricted to one or two species (Harwood and Croxall 1988). In this instance the alternate viewpoint of fisheries interests was that a cull of the seals was desirable for several reasons, including that annual consumption of commercially exploitable fish by seals in 1976 was approximately 112,000 tonnes and if fishermen were able to catch half of this amount this would represent an increased harvest to the fishery of 15–20 million pounds sterling.

The frequently opportunistic impacts of predation complicate the determination of its impact on ecosystem functioning. Similarly, the interaction of predation with growth and behavior raises the possibility that the indirect effects of predation or its concomitant factors can determine its final impact. The indirect effects of predation are well illustrated in freshwater communities, where predation that changes the habitat occupancy and foraging success of individual fish can lead to decreased individual growth rates (Werner et al. 1983; Werner and Hall 1988; Tonn et al.

1992), decreased overwinter survival (Post and Evans 1989; Tonn et al. 1992), and an increased period of susceptibility to size-dependent predation. Size-selective predation can lead to changed population growth rates, and species-selective predation can lead to a changed community structure and nutrient recycling in marine and freshwater ecosystems (Vanni 1987; Hanson and Chouinard 1992; He and Wright 1992; Hixon and Beets 1993; Kitchell et al. 1994).

Interactions between predation and feeding in freshwater ecosystems are often driven by small-scale patchiness in the environment, a feature often neglected in marine ecosystem studies. However, both freshwater and marine fish will respond to the threat of predation by staying in the habitat where the risk of mortality or the mortality risk per unit growth is minimized (Gilliam 1982; Gilliam and Fraser 1987; Tanaka et al. 1989; Gotceitas and Brown 1993; Hixon and Beets 1993; Kennedy et al. 1994), suggesting that the interactions between predation and habitat are a major determinant of predation's impact. Food, predators, and temperature were concluded to be the three most important factors influencing habitat quality for flatfish, where habitat quality can be defined as the product of habitat structure, food, predators, temperature, salinity, oxygen, water depth, and hydrodynamics (Gibson 1994). Thus, habitat quality in the marine ecosystem will vary with environmental conditions, seasonally and between years. Cold winters in the southern North Sea may restrict the distribution of brown shrimp in plaice nursery areas, reducing predation on recently settled plaice larvae and leading to strong year-classes (Veer 1987); however, there is no single event that determines year-class strength, and the strong plaice year-class of 1981 was associated with a warm winter (Neill et al. 1994). Interactions between predation and habitat occur in both the pelagic and demersal realms. For example, Rose and Leggett (1990), in an interesting examination of the importance of spatial scale on inferences drawn from field surveys, report that although aggregations of cod and capelin occur in the same local area, capelin were concentrated in water temperatures outside the temperature range preferred by cod.

Just as there have been attempts to determine whether starvation or predation control survival of larval and juvenile fish, so have there been attempts to determine whether predation (top-down) or food availability and competition (bottom-up) control the overall dynamics of aquatic ecosystems. Despite divisive early stances that either predation or competition was of greatest importance, both processes have been shown to be of varying importance in different studies and in different ecosystems (see reviews by Hunter and Price 1992; Power 1993; Sherman 1992; Polis 1994). Theoretical studies suggest that top-down processes control the upper trophic levels, whereas production controls abundance in the lower trophic levels, the ultimate productivity of the system, and the number of trophic levels (Leibold 1989). An alternative, and not necessarily contradictory, view developed through extensive observation and manipulation of freshwater lake ecosystems is that potential productivity at all levels is set by nutrient supply, whereas actual productivity is determined by the recycling of nutrients by different populations (Carpenter et al. 1985). Caddy (1993) concluded that in enclosed and semi-enclosed marine ecosystems, nutrient enrichment and fishing acted in synchrony to produce the observed changes in the trophic structure.

A good test of the importance of predation in structuring marine ecosystems is the response to its removal. The hunting to near extinction of sea otters in the North Pacific earlier this century led to increased survival of sea urchins and the loss of kelp beds along large stretches of North Pacific coastline. However, with the cessation of hunting, sea otter populations are now increasing, reducing grazing by urchins and promoting growth of dense kelp beds (Estes and VanBlaricom 1985). The renewed kelp forests provide refuge for kelp-bed fish, providing food for the otters, which increased in abundance above the levels attained when invertebrates were the primary food resource. In this instance the effects of predator removal were reversible; this may not always be the case. For example, an intertidal community on the west coast of South Africa is normally dominated by the rock lobster, which consumes other benthic components, especially whelks, reducing their abundance but increasing the relative abundance of one whelk species

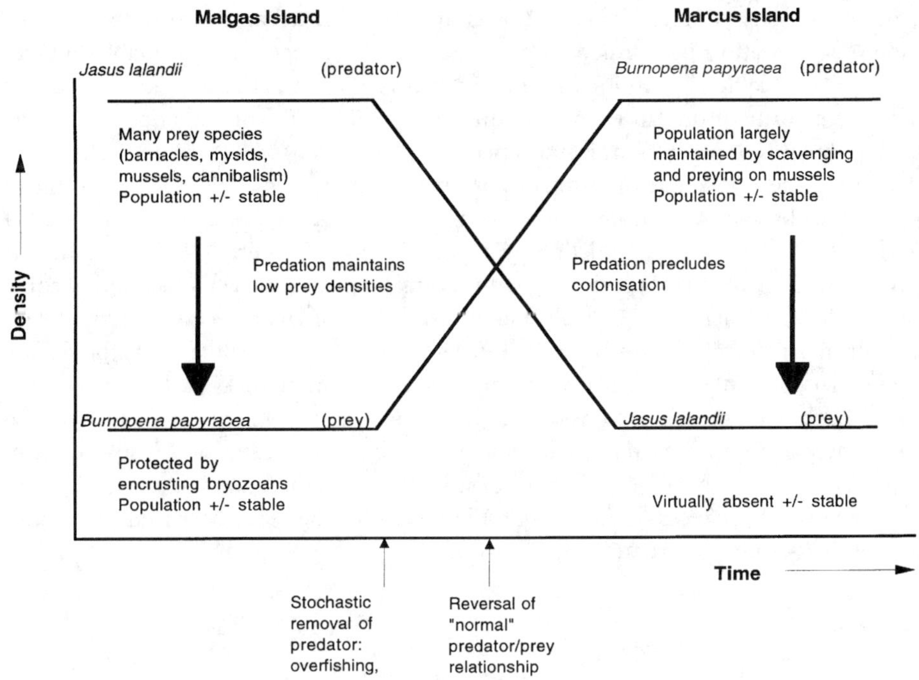

Figure 14-5. Schematic portraying the reversal of roles as predator and prey for rock lobster (*Jasus lalandii*) and the whelk (*Burnopena papyracea*) at Marcus and Malgas Islands. The populations have been monitored and found to be stable for more than 4 years. (Reprinted by permission from Barkai, A., and McQuaid, C. 1988. Predator-prey role reversal in a marine benthic ecosystem. Science 242:62–64.)

covered by an unpalatable bryozoan (Barkai and McQuaid 1988). However, on one island, from which lobsters disappeared for unknown reasons in the last 20 years, there is a high abundance of several whelk species and extensive beds of black mussels, holothurians, and sea urchins not found where rock lobsters are present. This indicates the effects of predator removal on the benthic community, but what is more interesting is that when rock lobsters were reintroduced into the whelk-dominated habitat they did not survive, but were killed on average within 15 minutes of reintroduction by approximately 300 whelks that attached themselves to each lobster. In this instance, the community has entered a second stable state in which the former dominant predator can no longer survive, let alone structure the community (Figure 14-5).

The U.S. Northeast Shelf ecosystem provides a large-scale (in geographic and economic terms) example where the selective removal of higher-valued, demersal predatory fish (e.g., cod, haddock, flounders) through intensive fishing has led to an ecosystem dominated by lower-valued predatory fish (dogfish and skates constituted over 70% of the demersal biomass in 1986 compared with about 25% in 1963) that are remaining despite reduction in fishing effort (Sherman 1992). It is not known whether this ecosystem has reached an alternate stable state that will require concerted management action to correct, or whether it will return to closer to its original condition given time.

Predicting Predation

It is currently not possible to model predation in marine ecosystems explicitly. Many simplifying assumptions have to be made. Which simplifying assumptions can be made while still adequately representing the system of interest is the subject of continuing debate. In most fisheries management analyses, predation is assumed to be constant, which seems unreasonable but has been shown to be a reasonable assumption in some instances (Punt and Hilborn 1994). Even in models in which the majority of predatory interactions are included, the chosen stock and recruitment relationship remains of paramount importance (Shepherd 1988).

There are three immediate questions to ask before modeling a marine system: the level of aggregation of subsystems, the time scale to be used, and the geographic scale to be used (Larkin and Gazey 1982). Hierarchical modeling promotes the use of different scales in a hierarchy of models and provides a method for dealing with several levels of scale while restricting complexity to achievable levels at each scale (O'Neill et al. 1986). It may often be necessary to model one part of the whole interconnected assemblage in sufficient detail to be revealing while the remainder is treated as either constant or with relatively simplistic approximation. As Larkin and Gazey (1982) conclude, "Those with greater pretension will more often fail; those with (lesser) pretension will more often be irrelevant." The danger in modeling marine systems at only one scale is not just that important effects may be omitted but also that the individual units (e.g., stocks, species, or fisheries) chosen may reflect the interests and judgment of the researcher more than the actual structure of the system (Silvert and Dickie 1982).

Available Models

Models of fish predation in marine ecosystems vary from whole ecosystem simulation models (e.g., Andersen and Ursin 1977; Laevastu et al. 1982) and the more simplified box models (e.g., Polovina 1984; Christensen and Pauly 1992) that simulate predation and energy flow through the ecosystem, to highly idealized models derived from theoretical ecology (e.g., May et al. 1979). Intermediate between these are models that provide a retrospective analysis of catch data of a limited number of species while accounting for interspecific predation (e.g., Helgason and Gislason 1979; Pope 1979).

The whole-ecosystem models provide a more complete representation of predation in the ecosystem and can be used to isolate dominant interactions, but require extensive data input and are unable to provide precise estimates. They have the advantage of being able to provide an alternate assessment of marine resource abundance, especially for valuable stocks and areas where abundance estimation by standard techniques is impossible (Laevastu and Larkins 1981; Bax and Eliassen 1990). These models, and the more tractable box models that now incorporate inverse analysis to provide fits to undetermined systems (Christensen and Pauly 1992), are useful in determining which species need to be incorporated or aggregated in reduced and more rigorous models.

Simpler theoretical models provide a means of exploring the interactions between predator and prey. They are unable to capture the complexity of natural systems but provide a valuable means of understanding the relevance of particular ecological processes and mathematical functions to ecosystem functioning (e.g., Yodzis 1994).

Multispecies extensions to retrospective catch models provide the possibility of incorporating interactions between the major species that are either of interest to, or susceptible to, management intervention; considerable effort has been invested in such models. Forward projections from these retrospective analyses using a multispecies extension to age-structured yield per recruit models have provided a clear indication of the potential for predatory interactions influencing standard fisheries prognoses in the North Sea (Figure 14-6) and Irish Sea (Macer and Shepherd 1987; Brander 1988). The ICES multispecies virtual population analysis (MSVPA) is the most researched model, but other approaches have been used to explore interactions of marine mammals with fisheries (e.g., Lleonart et al. 1985; Punt 1994), and several of the available catch-at-age analyses can be configured to examine the implications of direct predation, including cannibalism.

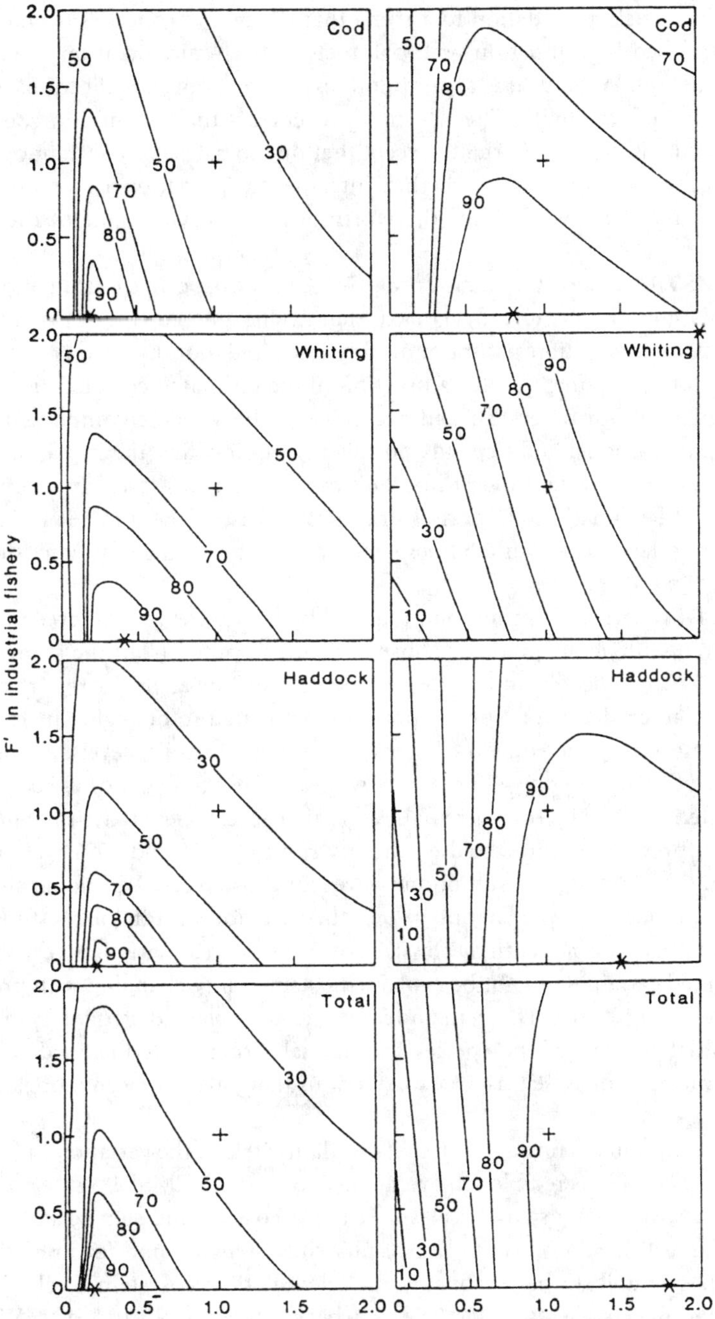

Figure 14-6. Yield per recruit isopleths expressed as percentages of maximum yield per recruit (*) for given levels of fishing mortality in the industrial fishery (targeting small fish) and the human consumption fishery (targeting larger fish) for North Sea cod, whiting, haddock, and the combined species. Isopleth graphs in the left column exclude predation effects; isopleth graphs in the right column include predation effects for these three species. Fishing mortality is expressed relative to current situation (+). (Reprinted by permission from Macer, C. T., and Shepherd, J. G. 1987. The effect of species and fishery interactions on the assessment of some North Sea gadoids. Fish. Res. 5:303–317.)

It is usual in modeling predation to assume that the main properties of interspecific predation can be described by a function with parameters that remain constant over the period of analysis or prediction. Without this assumption, extensive stomach collections would be necessary over all scales of variability. This assumption replaces the assumption common to single-species models, that is, that all mortality except that due to fishing is constant over the period of interest. It is generally not known whether interspecific predation can be described by this assumption, but the ICES MSVPA for the North Sea has provided one of the first tests of its validity.

The ICES MSVPA generalizes single-species VPA (a retrospective examination of catch data) so that natural mortality is no longer a constant value but consists of a fixed component, which represents nonpredation natural mortality and predation by species not explicitly modeled, and a variable component that is a function of the estimated consumption of the modeled fish and the suitability of other fish and age classes. The variable component of mortality is estimated within the model and depends on the assumption that the suitability of a prey item for a predator remains constant over time. This assumption was tested by configuring MSVPA with data from the stomach collections from 1981 (55,166 fish) and 1991 (92,894 fish) in a key run and using the constant suitabilities generated by the model to predict stomach contents in 1981 and 1991.

In Figure 14-7, relative consumption predicted by the model for five predators for 1981 and 1991 is compared with actual predator consumption computed from the stomach contents for the two years. Further details and analyses are provided in an ICES report (1994). Relative consumption by all predators of Norway pout was estimated to be higher in 1991 than in 1981 (Figure 14-7), but was predicted by the key run to remain at similar levels. At the same time the relative consumption of sprat by all predators was estimated to decline between 1981 and 1991, but was predicted by the key run to remain at similar levels. Generally, the figures for relative consumption of prey species predicted by the key run for 1981 and 1991 were more similar to each other than the estimated consumption figures for those years. Additional analyses indicated that the MSVPA model explained about 50% of the variation in suitability estimates for the two years. Thus, suitability was not entirely constant over the 10-year period. A very small proportion of the unexplained variability might be explained by suitability changing with prey biomass, but this was not a strong effect. The remainder of the unexplained variability may result from changes in spatial overlap of the species (no spatial structure is included in the North Sea MSVPA), other aspects of predator–prey interactions that have not been modeled, and sampling variability.

That the assumption of constant suitability explains 50% of the variation in stomach contents over the 10-year period is a credible achievement. The North Sea MSVPA integrates predator and prey interactions over many spatial scales and has a coarse time step relative to many of the interactions that will have occurred. The results suggest that some features of predator–prey interactions may be able to be reasonably modeled at the ecosystem level without explicitly representing their finer-scale attributes. The drawback is that the North Sea MSVPA has required a tremendous amount of field collection, data analysis, and modeling. There are few other marine ecosystems where such a level of effort is feasible.

The models described so far consider the ecosystem as a collection of interacting species. The complex interactions between species in a natural ecosystem cannot be represented in entirety, and in no case are the interactions at the egg, larval, and early juvenile stages modeled, although this may be where year-class strength is determined for many species. Many attributes of species interactions in marine communities can be explained on the basis of size (e.g., Gomes 1993). Approaches that do not start at the species level have been explored with the aim of describing system properties at higher levels. Most effort in this direction has involved particle-size spectrum models (e.g., Sheldon et al. 1977; Platt 1985) in which it is assumed (based on considerable

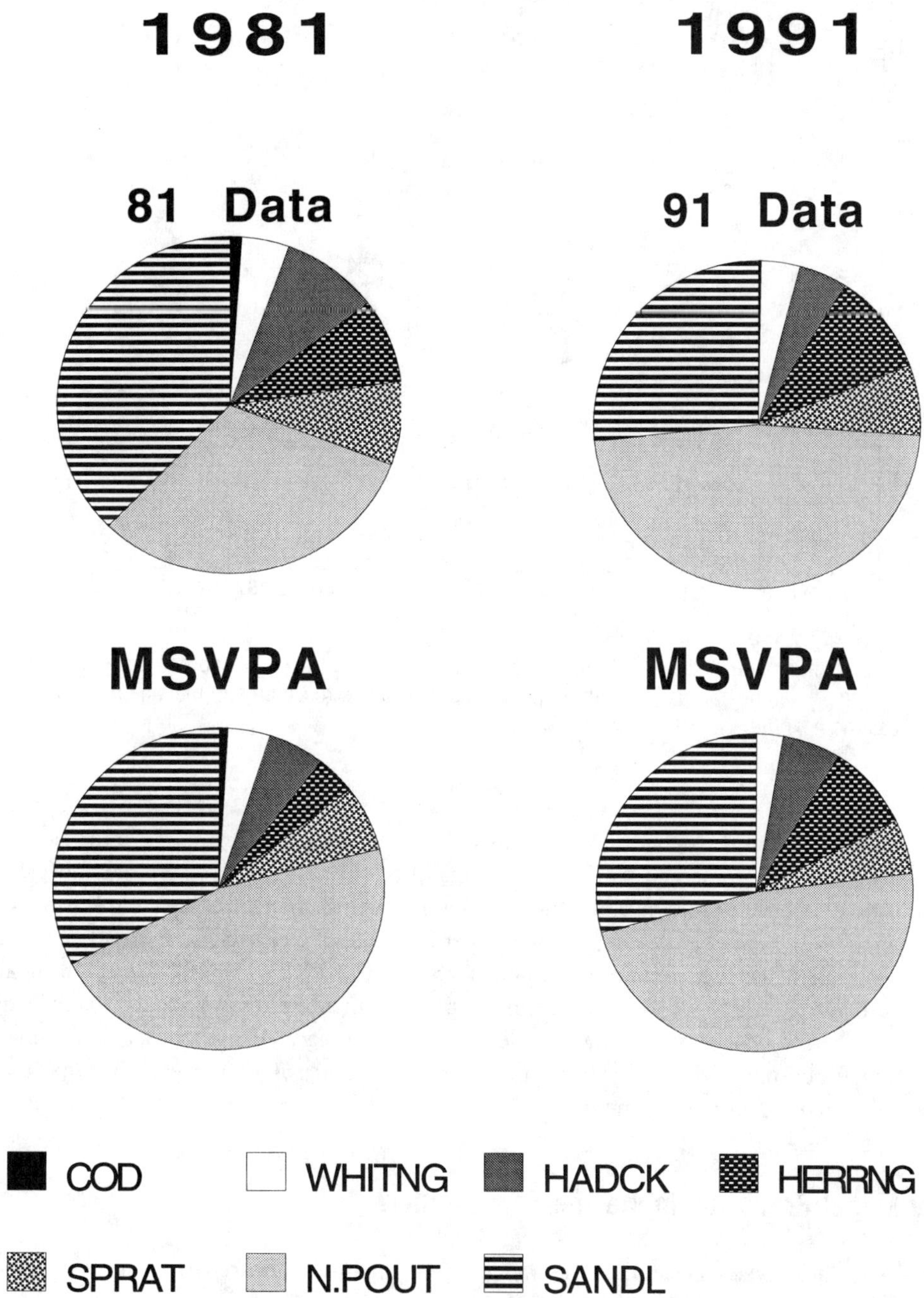

Figure 14-7. Relative consumption by all North Sea MSVPA predators in 1981 and 1991 as estimated from the data collected in those years, compared with that estimated by the MSVPA model key run, which uses all data (redrawn from ICES 1994).

observation) that the frequency distribution of particles is represented by an orderly decline with increasing size. With this assumption the biomass of one sector of the community (large fish, for example) can be estimated from another sector (phytoplankton, for example). This approach was used by Moloney and Field (1985) to estimate potential pelagic fish yields off South Africa. Pope

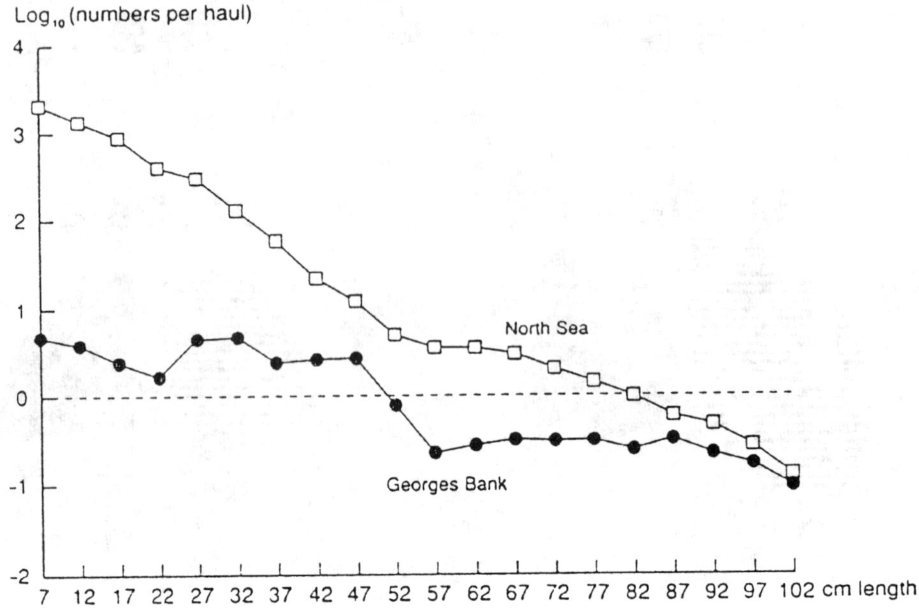

Figure 14-8. Comparison of the total finfish catch at length per trawl haul on Georges Bank and in the North Sea. (Reprinted by permission from Pope, J. 1989. Fisheries research and management for the North Sea: The next hundred years. Dana 8:33–43.)

and Knights (1982) and Ursin (1982) demonstrated that the size spectrum of groundfish taken in a North Sea survey remained uniform between different years and periods of different fish composition. Pope (1989) compares the slopes for North Sea and Georges Bank fish over the range of 10 to 110 cm (Figure 14-8). The steeper slope for the North Sea fish indicates a larger proportion of smaller fish compared with Georges Bank, and is the expected result of the higher fishing pressure in the North Sea. These models provide an alternative approach to assessing and managing marine ecosystems but will need to become integrated with other determinants of ecosystem functioning, especially habitat and trophic guilds, if they are to become of more relevance to ecosystem management.

Why Model Predation in Marine Ecosystems?

Models are "a marvelous crutch to the imagination, expanding awareness of subtle interactions that on first acquaintance are called 'counterintuitive results'" (Larkin 1978).

Given the uncertainties and variability of predation in marine ecosystems, the necessity of modeling it must be questioned. Punt and Hilborn (1994) make the case that in some instances including predation (cannibalism in this instance) in stock assessment models does not improve their performance, but a distinction must be drawn between short-term tactical and medium- to long-term strategic forecasting. In tactical management it may be reasonable to assume that natural mortality remains fixed, especially as year-class strengths and fishable biomass may have already been determined for the next few years. However, over the time periods of interest in strategic management, year-class strengths will not have been determined and the possibility exists that manipulation of the relative abundance of predator and prey species or predator and prey size classes could alter fishable biomass. Two examples will be discussed here: the ICES MSVPA and models of interactions between seal and hake off South Africa.

ICES MSVPA

The ICES MSVPA model has had a major impact on the scientific advice provided to managers. Quite early in its development, it became apparent that the convenient assumption that juvenile fish suffered the same instantaneous mortality rate as adult fish was incorrect. An incorrect schedule of mortality rates would substantially alter the scientific advice provided to managers on mesh selectivity, for example. As Gislason and Helgason (1985) concluded, "a changed exploitation pattern and a higher natural mortality for the younger age-groups will change both the yield per recruit and the mesh-size assessments. It thus seems likely that much of the current advice, which is based on calculations where the same natural mortality has been applied to all ages, is wrong." Since this time, updated mortality rates developed in the MSVPA have been included in single-species assessments for North Sea fish.

A good example of the value of strategic modeling is provided by application of a forecasting model following on from the ICES MSVPA to address the likely impact of an increased mesh size in the North Sea fishery (ICES 1989). A priori, from the perspective of individual species, it was expected that increasing mesh size would increase yield per recruit of the major commercial fish by increasing survival of the younger, smaller fish until the time when their rapid somatic growth declined. However, once multispecies interactions were included it was found that yields would decrease for almost all commercial species, reversing the conclusions of single-species analyses. The ICES (1992) report concluded that the quantitative results from predictions with this multispecies model underscore an earlier warning that "intuition and qualitative models can mislead when used to evaluate indirect effects of fishing . . . Even the directions of responses are indeterminate and magnitudes of responses differ greatly in response to a fixed change in harvest."

Seals and Hake

Diet flexibility in marine mammals makes it very difficult to determine the impact of predators on a prey species or on the ecosystem because decreased predation by marine mammals would affect several species in addition to the commercial species of interest and these other fish species might be competitors or predators (e.g., Northridge and Beddington 1992). Bowen (1985) concluded from a review of the interactions between seals and fisheries in the northwest Atlantic that "we simply cannot say what effect an increase or decrease in the harp seal populations will have on fish catches over the long term" because other predators (whales, seabirds, and cod) make it extremely likely that compensatory, density-dependent natural mortality would occur in the various commercial fish populations. The Royal Commission on Seals and Sealing in Canada (1986) concluded that although there was strong evidence that seals could affect the abundance of fish stocks, it was "not aware of any instance in which known and measured change in the abundance of seals has had a measurable effect on fish catches." Butterworth et al. (1988), investigating the scientific basis for reducing the South African seal population, could find no "proven link between seal population reduction and a consequential increase in yield to any fishery world-wide."

Despite these cautions, there has been interest in the North and Southeast Atlantic in reducing the population of seals to increase returns to the commercial fisheries. The recent report by the Fisheries Resource Conservation Council (FRCC 1994), an independent advisory body set up in 1993 to make formal public recommendations to the Minister of Fisheries and Oceans,

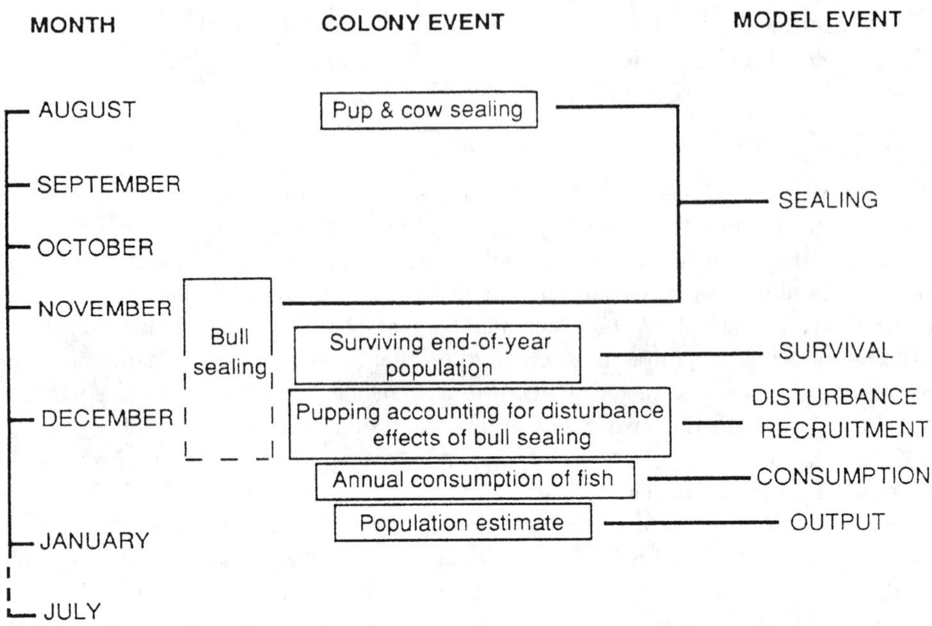

Figure 14-9. Schematic showing the temporal sequence of events at a South African fur seal colony and the corresponding sequence in the simulation model. (Reprinted by permission from Wickens, P. A., Shelton, P. A., David, J. H. M., Field, J. G., Oosthuizen, W. H., Roux, J. P., and Starfield, A. M. 1992. A fur seal simulation model to explore alternative management strategies. Can. J. Fish. Aquati. Sci. 49:1396–1405.)

Canada, reaffirmed the appropriateness of the "ecological approach" and concluded (among other things) that early action should be taken to significantly reduce the populations of seals to help recovery of the groundfish stocks. Interestingly, the Council concluded that good decisions can be made on the basis of common sense, even though the complex interactions operating in the ecosystem are not formally understood. An example of where common sense might not prevail is provided by modeling associated with the hake and seal interaction off southern Africa.

There has been interest in reducing the populations of Cape fur seals, which have tripled since the introduction of legal controls in 1893 and are now estimated to consume an amount of hake comparable to that taken by the fishery, and for which there are no current resource limits to continued expansion (Wickens et al. 1992). A model developed by Wickens et al. (1992) represents a single unstructured population of seals from which (among many other things) the annual consumption of fish is estimated under various harvest strategies. The structure of this model is represented in Figure 14-9. The model was designed to test different management hypotheses against their success in "efficiently: (1) culling to reduce seal population growth rate, (2) culling to reduce fish consumption, and (3) harvesting to maximize number of seals removed over a period of time." Simulation model runs indicated that reduction of fish consumption was achieved under any scenario that reduced seal population growth rate (Figure 14-10). No attempt was made to define the species composition of fish consumed, which is an important distinction from the two following models, where the type of fish consumed is shown to be of importance in predicting the impacts of seals on the commercial fishery.

A second model, developed by Punt (1994), is described as a "minimal realistic" model that incorporates predation on hake by seals, other hake species, and cannibalism only. The structure

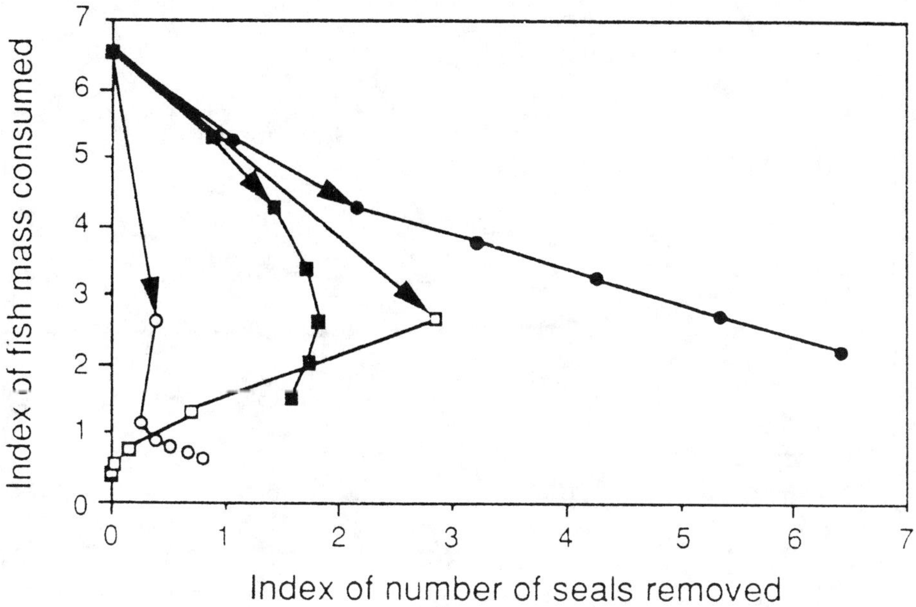

Figure 14-10. Index of fish consumption by the South African fur seal for different levels and different selectivity of seal culling, from a simulation model. Each line represents increases in the cull rate from 0% to 60% in steps of 10% in the direction of the arrows. Culling is simulated for pups (solid squares), cows (open squares), bulls excluding disturbance (solid circles), and bulls including disturbance (open circles). (Reprinted by permission from Wickens, P. A., Shelton, P. A., David, J. H. M., Field, J. G., Oosthuizen, W. H., Roux, J. P., and Starfield, A. M. 1992. A fur seal simulation model to explore alternative management strategies. Can. J. Fish. Aquati. Sci. 49:1396–1405.)

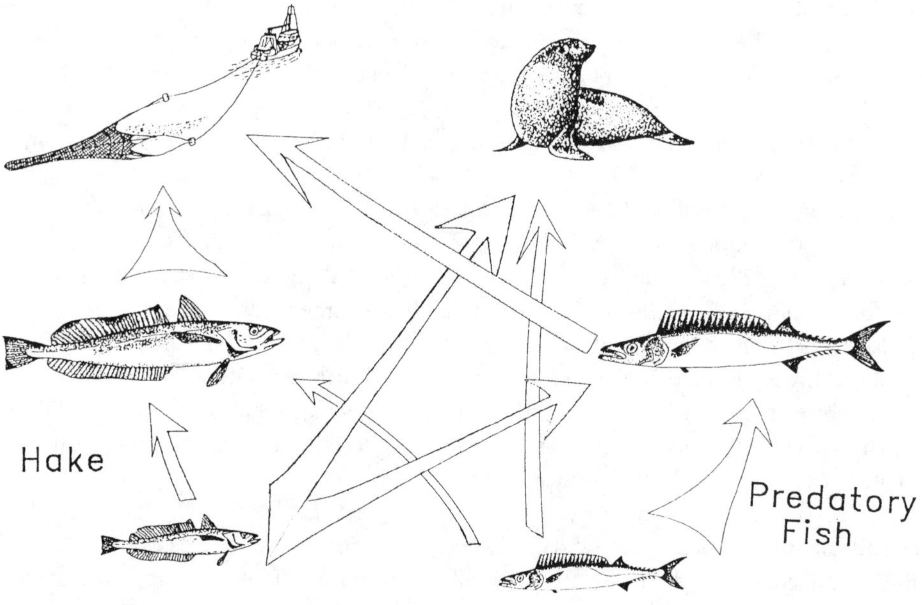

Figure 14-11. Schematic showing the trophic groups and their interactions in the minimal realistic model of Cape fur seal and Cape hake interactions off South Africa. (Reprinted by permission from Punt, A. E. 1994. Data Analysis and Modelling of the Seal-Hake Biological Interaction off the South African West Coast. Report submitted to the Sea Fisheries Research Institute, South Africa.)

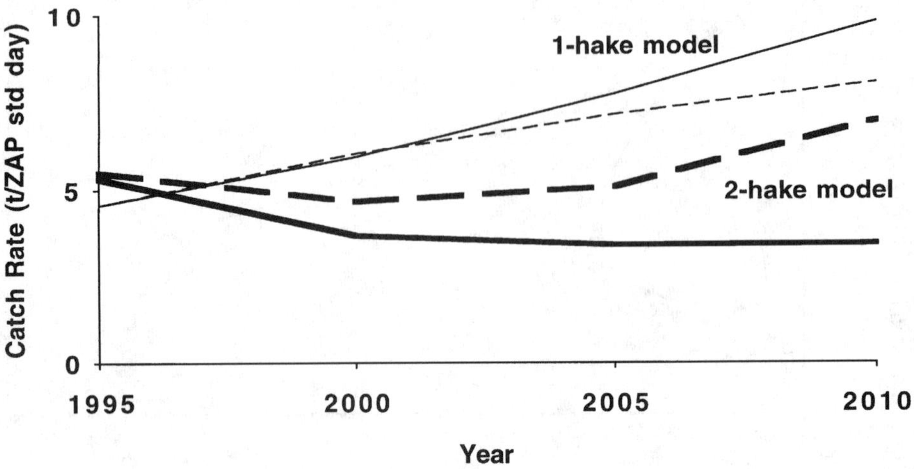

Figure 14-12. Comparison of the effects on future commercial hake catches of increased seal removals (broken lines) against no seal removals (solid lines) for a minimal realistic model of seal and hake interaction configured so that it either aggregates (1-hake model) or separates (2-hake model) the two hake species (redrawn from Punt 1994).

of this model is provided in Figure 14-11 and can be configured to include one or two hake species. When configured to include only one hake species, the model consistently predicts that a cull of seals would lead to slightly increased average catches, catch rates, and profits, balanced by a decreased consumption of hake by seals. However, when configured to include two morphologically different hake species interacting through predation, the opposite result occurs, namely, reduced average catches, catch rates, and profits following a seal culling (Figure 14-12). This result occurs because only one of the two hake species is vulnerable to predation by seals, and adults of this species consume juveniles of the other species. A reduced seal population leads to increased numbers of the predatory hake, causing a decline in abundance of the prey hake and an overall decline in catch rates.

The third model, developed by Bax and Crawford (unpublished data), was based on a simplified ecosystem model of the type developed by Laevastu and Larkins (1981). The structure of this model is provided in Figure 14-13. The model was designed to represent predation and competition by all major species in the ecosystem. A one-time reduction in the population size of seals led directly to reduced seal predation, but this was more than compensated for by an increase in indirect predation on juvenile Cape hake by predatory fish (including piscivorous squid), the abundance of which increased following reduced predation from the seals (Figure 14-14). The lower abundance of juvenile hake led to reduced recruitment to the adult stock and thus to little overall change in the adult hake population.

The potential for indirect interactions to modify direct interactions is clear from the above results, and the importance of considering the effects of additional species on simple predator–prey interactions has been demonstrated. It is also important to consider the ontogenetic development of the species under consideration, since predator–prey interactions may affect adults and juveniles differently. It is not possible to claim that any of the above models correctly predicts the system response to a change in predation. It is possible, however, to say that predicting the results of altered predation is uncertain and that indirect effects have the capacity to reverse the expected direct effects.

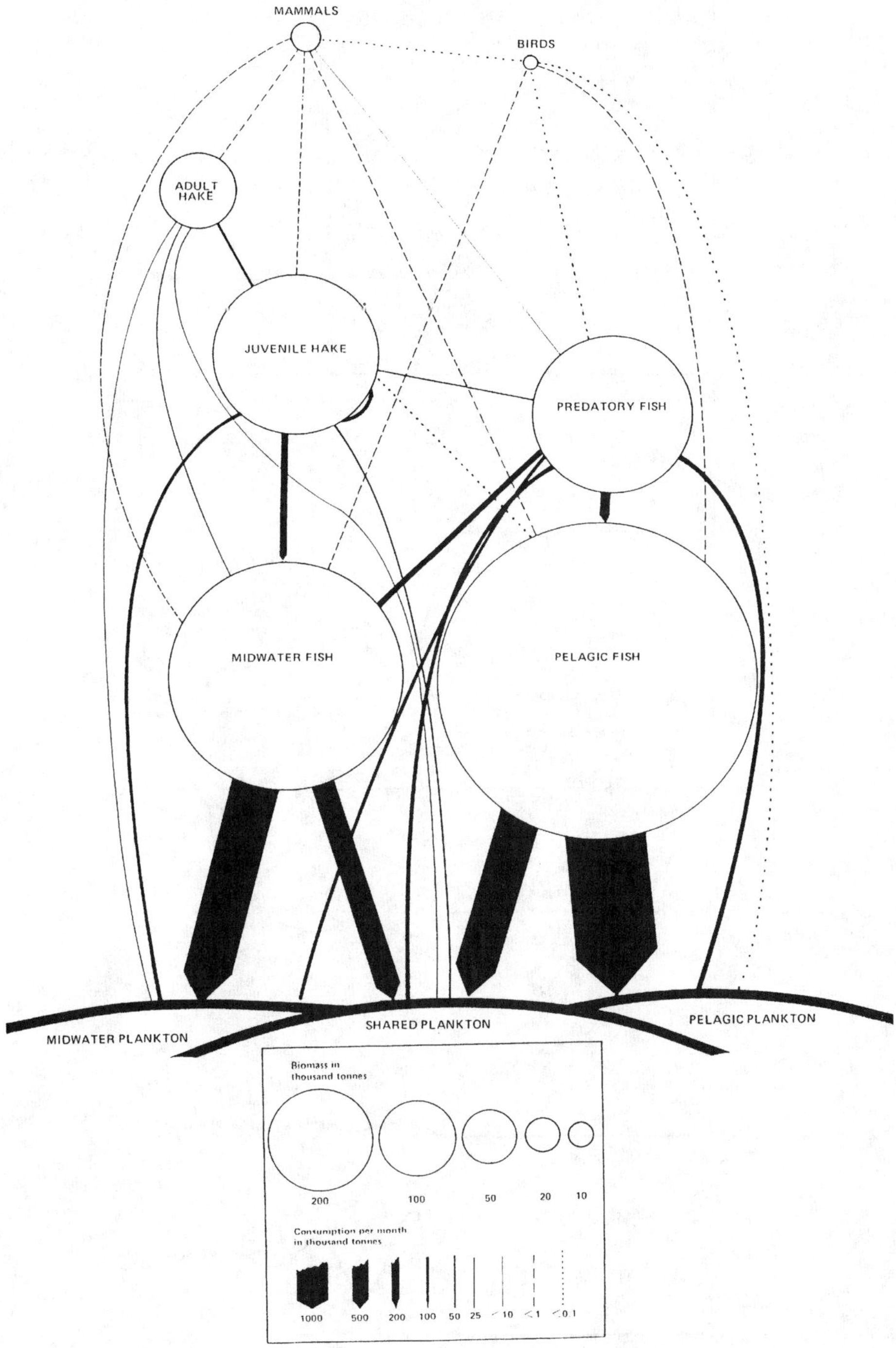

Figure 14-13. Steady-state representation of the main trophic groups in the Benguela Current ecosystem. The area of each group is proportional to its biomass, and the width of each arrow is proportional to the trophic transfer between groups (from Bax and Crawford, unpublished results).

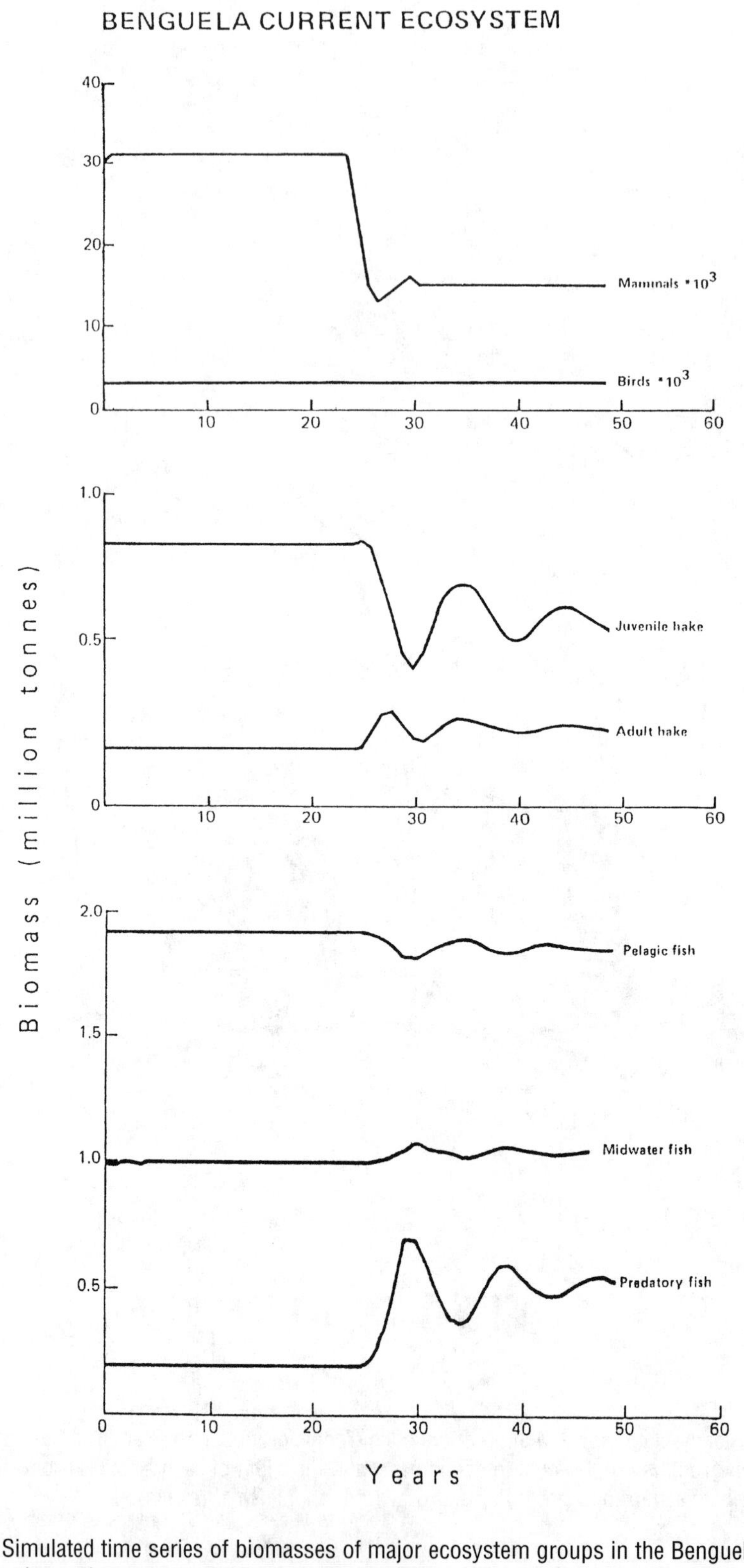

Figure 14-14. Simulated time series of biomasses of major ecosystem groups in the Benguela Current ecosystem following an instantaneous 50% culling of seals (from Bax and Crawford, unpublished results).

Summary and Conclusions

The diversity of predator–prey interactions in aquatic systems is truly amazing. It is a diversity based on small-scale spatial patterns, temporal synchrony, and the individual responses of prey to highly selective predators. At a community level, this diversity is magnified by the wealth of direct and indirect interactions, often represented by transitory encounters. There is a growing consensus among population ecologists and fisheries scientists that it is not possible to predict predation events in marine ecosystems numerically, yet at the same time there is a growing recognition, perhaps demand, that fisheries should be managed in their context as a component of marine ecosystems.

Proven techniques are available for the tactical management of marine fisheries, although data often limit their reliability. It is in strategic management where there is a lack of proven techniques and a concern that numerical prediction may never be viable. Even qualitative prediction is suspect, while predictive models remain unable to represent the small-scale, seemingly insignificant events that may determine community structure. Management by intuition seems doomed to failure given the importance of indirect predator–prey interactions in determining system response to an intervention. Experimental, or adaptive, management (e.g., Walters and Hilborn 1976) has gained adherents in the 15 years since its introduction to fisheries, but there are few examples of its successful introduction; too often it has been used as an excuse to promote short-term, experimental harvesting (Halbert 1993). Adaptive management can only work if it becomes an integrated part of resource management used as an ongoing technique to monitor and direct management interventions and harvesting techniques. Approaches to the application of adaptive management are clearly needed that take advantage of the geographical and spatial diversity of fishery resources and the selectivity of harvest techniques to gather information on marine ecosystems in an ongoing but nondisruptive manner (see Sainsbury 1988). Management techniques that are now used as management endpoints, for example, area closures and mesh size restrictions, need to be considered as experimental tools that can be modified to gain information on the resource without affecting short-term yields. Such approaches will require flexibility from all involved in resource management: the harvesters, managers, scientists, politicians, and public. These approaches may be especially applicable in the most diverse multispecies, multigear fisheries, which are the most difficult to assess and predict.

A second approach that has gained recognition in recent years is the wise management and conservation of the production bases for fishery ecosystems—habitat, prerecruits, and forage fish. Some suggest that this may be the only practical approach if marine ecosystems are characterized by complex dynamics (Wilson et al. 1994). However, a distinction needs to be drawn between complex dynamics and unknown dynamics. For example, Ursin (1982) comments that the concentration on unpredictable short-term variability in the dynamics of marine stocks has diverted attention from the stocks' long-term stability and predictable dynamics. There will be times, however, such as when the level of knowledge of a marine system is considered too low or the costs of obtaining further knowledge on a system are considered too great (see Punt and Butterworth 1991), when the most practical approach might be to conserve the production basis instead of believing that managing species as though they were independent can provide the basis for strategic management.

Just as it is clear that no single model is appropriate for the study of multispecies interactions (Larkin and Gazey 1982; Starfield et al. 1988), it is clear that no single management policy is appropriate for the management of multispecies fisheries. Clear distinction needs to be made between those analytical and management techniques suitable for tactical management and those suitable for strategic management. A single survey or a single adaptive management experiment may meet tactical management needs, but strategic management needs will only be met through an ongoing commitment to exploring the properties of the managed system. New experimental and management techniques are required that utilize the diversity of marine ecosystems to gain information in a nondisruptive manner, while at the same time preserving the systems' productivity. This approach will require the cooperation of resource users, scientists, and managers and provides the best potential for furthering an understanding of the importance of predation in marine ecosystems and the consequences of its possible modification.

References

Andersen, K. P., and Ursin, E. 1977. A multispecies extension to the Beverton and Holt theory of fishing, with accounts of phosphorous circulation and primary production. Medd. Dan. Fisk. Havunders. 7:319–435.

Arai, M. N., and Hay, D. E. 1982. Predation by medusae on Pacific herring (*Clupea harengus pallasi*) larvae. Can. J. Fish. Aquat. Sci. 39:1537–1540.

Bailey, K. M. 1984. Comparison of laboratory rates of predation on five species of marine fish larvae by three planktonic invertebrates: Effects of larval size on vulnerability. Mar. Biol. 79:303–309.

Bailey, K. M. 1994. Predation on juvenile flatfish and recruitment variability. Neth. J. Sea Res. 32:175–189.

Bailey, K. M., and Houde, E. D. 1989. Predation of eggs and larvae of marine fishes and the recruitment problem. Adv. Marine Biol. 25:1–83.

Barkai, A., and McQuaid, C. 1988. Predator-prey role reversal in a marine benthic ecosystem. Science 242:62–64.

Bax, N. J. 1991. A comparison of the fish biomass flow to fish, fisheries, and mammals on six marine ecosystems. Int. Counc. Explor. Mer 193:217–224.

Bax, N. J., and Eliassen, J. E. 1990. Multispecies analysis in Balsfjord, northern Norway: Solution and sensitivity analysis of a simple ecosystem model. J. Cons. Int. Explor. Mer 47:175–204.

Beddington, J. R., Beverton, R. J. H., and Lavigen, D. M. (eds.). 1985. Marine Mammals and Fisheries. London: George Allen and Unwin.

Beverton, R. J .H. 1985. Analysis of Marine Mammal-Fisheries Interactions. In J. R. Beddington, R. J. H. Beverton, and D. M. Lavigne (eds.), Marine Mammals and Fisheries. London: George Allen and Unwin, pp. 3–33.

Beverton, R. J. H., and Iles, T. C. 1992. Mortality rates of 0-group plaice (*Pleuronectes platessa*) and turbot (*Scophthalmus maximum* L.) in European waters. III. Density-dependence of mortality rates of 0-group plaice and some demographic consequences. Neth. J. Sea Res. 29:61–79.

Beyer, J. E., and Laurence, G. C. 1981. Aspects of stochasticity in modelling growth and survival of clupeoid fish larvae. Rapp. P.-v. Réun. Cons. Int. Explor. Mer 178:17–23.

Birt, V. L., Birt, T. P., Goulet, D., Cairns, D. K., and Montevecchi, W. A. 1987. Ashmole's halo: Direct evidence for prey depletion by a seabird. Mar. Ecol. Prog. Ser. 40:205–208.

Bowen, W. D. 1985. Harp Seal Feeding and Interactions with Commercial Fisheries in the North-west Atlantic. In J. R. Beddington, R. J. H. Beverton, and D. M. Lavigne (eds.), Marine Mammals and Fisheries. London: George Allen and Unwin, pp. 135–152.

Boyle, P. R., and Pierce, G. J. 1994. Fishery biology of Northeast Atlantic squid: An overview. Fish. Res. 21:1–15.

Brander, K. 1988. Multispecies Fisheries of the Irish Sea. In J. A. Gulland (ed.), Fish Population Dynamics. John Wiley & Sons, pp. 303–328.

Butterworth, D. S., Duffy, D. C., Best, P. B., and Bergh, M. O. 1988. On the scientific basis for reducing the South African seal population. Suid-Afrikaanse Tydskrif vir Wetenskap. 84:179–188.

Caddy, J. F. 1993. Toward a comparative evaluation of human impacts on fishery ecosystems of enclosed and semi-enclosed seas. Rev. Fish. Sci. 1(1):57–95.

Cairns, D. K. 1987. Seabirds as indicators of marine food supplies. Biol. Oceanogr. 5:261–271.

Carpenter, S. R., Kitchell, J. F., and Hodgson, J. R. 1985. Cascading trophic interactions and lake productivity: Fish predation and herbivory can regulate lake ecosystems. BioScience 35(10): 634–639.

Christensen, V., and Pauly, D. 1992. ECOPATH II—A software for balancing steady-state ecosystem models and calculating network characteristics. Ecological Modelling 61:169–185.

Christensen, V., and Pauly, D. (eds.). 1993. Trophic Models of Aquatic Ecosystems. ICLARM Conference Proceedings 26.

Cushing, D. H. 1975. Marine Ecology and Fisheries. Cambridge, England: Cambridge University Press.

Daan, N., Rijnsdorp, A. D., and van Overbeeke, G. R. 1985. Predation by North Sea herring *Clupea harengus* on eggs of plaice *Pleuronectes platessa* and cod *Gadus morhua*. Trans. Am. Fish. Soc. 114:499–506.

Dawe, E. G. 1992. Predation by Short-Finned Squid on Atlantic Cod and Other Fish at Newfoundland, Canada. ICES C.M. 1992/K:32.

Dragesund, O., and Nakken, O. 1973. Relationship of parent stock size and year-class strength in Norwegian spring spawning herring. Rapp. P.-v. Réun. Cons. Int. Explor. Mer 164:15–29.

Duggins, D. O. 1988. Kelp beds and sea otters: An experimental approach. Ecology 61:447–453.

Estes, J. A., and VanBlaricom, G. R. 1985. Sea-otters and shellfisheries. In J. R. Beddington, R. J. H. Beverton, and D. M. Lavigne (eds.), Marine Mammals and Fisheries. London: George Allen and Unwin, pp. 187–235.

Fisheries Resource Conservation Council (FRCC). 1994. Conservation—Stay the Course. Third Report from the Fisheries Resource Conservation Council to the Minister of Fisheries and Oceans, Canada.

Furness, R. W. 1982. Competition between fisheries and seabird communities. Adv. Marine Biol. 30:225–307.

Furness, R. W. 1994. An estimate of the quantity of squid consumed by seabirds in the eastern North Atlantic and adjoining seas. Fish. Res. 21:165–177.

Furness, R. W., and Birkhead, T. R. 1984. Seabird colony distributions suggest competition for food supplies during the breeding season. Nature 311:655–656.

Furness, R. W., Hudson, A. V., and Ensor, K. 1988. Interactions between Scavenging Seabirds and Commercial Fisheries around the British Isles. In J. Burger (ed.), Seabirds and Other Marine Vertebrates. Competition, Predation and Other Interactions. New York: Columbia University Press, pp. 240–268.

Gibson, R. N. 1994. Impact of habitat quality on the recruitment of juvenile flatfish. Neth. J. Sea Res. 32:191–206.

Gilliam, J. F. 1982. Habitat Use and Competitive Bottlenecks in Size-Structured Fish Populations. Ph.D. dissertation, Michigan State University, East Lansing.

Gilliam, J. F., and Fraser, D. F. 1987. Habitat selection under predation hazard: Test of a model with foraging minnows. Ecology 68:1856–1862.

Gislason, H., and Helgason, T. 1985. Species interaction in assessment of fish stocks with special application to the North Sea. Dana 5:1–44.

Gomes, M. do C. 1993. Predictions under Uncertainty: Fish Assemblages and Food Webs on the Grand Banks of Newfoundland. Newfoundland: Institute for Social and Economic Research, Memorial University, Newfoundland, Canada.

Gonzàlez, A. F., López, A., Guerra, A., and Barreiro, A. 1994. Diets of marine mammals stranded on the northwestern Spanish Atlantic coast with special reference to Cephalopoda. Fish. Res. 21:179–191.

Gotceitas, V., and Brown, J. A. 1993. Substrate selection by juvenile Atlantic cod (*Gadus morhua*): Effects of predation risk. Oecologia 93:31–37

Halbert, C. L. 1993. How adaptive is adaptive management? Implementing adaptive management in Washington State and British Columbia. Rev. Fish. Sci. 1(3):261–283.

Hanson, J. M., and Chouinard, G. A. 1992. Evidence that size-selective mortality affects growth of Atlantic cod (*Gadus morhua* L.) in the Southern Gulf of St Lawrence. J. Fish Biol. 41:31–41.

Harwood, J., and Croxall, J. P. 1988. The assessment of competition between seals and commercial fisheries in the North Sea and the Antarctic. Marine Mammal Sci. 4:13–33.

He, X., and Wright, R. A. 1992. An experimental study of piscivore-planktivore interactions: Population and community responses to predation. Can. J. Fish. Aquat. Sci. 49:1176–1183.

Helgason, T., and Gislason, H. 1979. VPA Analysis with Species Interaction due to Predation. ICES CM 1979/G:52.

Hixon, M. A., and Beets, J. P. 1993. Predation, prey refuges, and the structure of coral reef fish assemblages. Ecol. Monogr. 63:77–101.

Hunter, M. D., and Price, P. W. 1992. Playing chutes and ladders: Heterogeneity and the relative roles of bottom-up and top-down forces in natural communities. Ecology 73(3):724–732.

Hylen, A., and Dragesund, O. 1973. Recruitment of young Arcto-Norwegian cod and haddock in relation to parent stock size. Rapp. P.-v. Réun. Cons. Int. Explor. Mer 164:57–68.

International Council for the Exploration of the Sea (ICES). 1989. Report of the Multispecies Assessment Working Group. ICES. C.M. 1989/Assess:20.

International Council for the Exploration of the Sea (ICES). 1992. Report of the Study Group on Ecosystem Effects of Fishing Activities. ICES C.M. 1992/G:11.

International Council for the Exploration of the Sea (ICES). 1994. Report of the Multispecies Assessment Working Group. ICES C.M. 1994/ Assess:9.

Johannessen, A. 1980. Predation on Herring (*Clupea harengus*) Eggs and Young Larvae. ICES C.M. 1980/H:33 (Mimeo).

Jones, R., and Henderson, E. W. 1987. The Dynamics of Energy Transfer in Marine Food Chains. In A. I. L. Payne, J. A. Gulland, and K. H. Brinks (eds.) The Benguela and Comparable Ecosystems. S. Afr. J. Mar Sci. 5, pp. 447–465.

Jónsson, E. 1980. Biological studies on squid (*Todarodes sagittatus*) in Icelandic waters during the autumn 1979 with notes on its distribution and migration (English summary). Hafrannsoknastofnunin Fjolrit 7.

Kennedy, M., Shave, C. R., Spencer, H. G., and Gray, R. D. 1994. Quantifying the effect of predation risk on foraging bullies: No need to assume an IFD. Ecology 75:2220–2226.

Kitchell, J. F., Eby, L. A., Schindler, D. E., and Wright, R. A. 1994. Predator-prey dynamics in an ecosystem context. J. Fish. Biol. 45(Suppl. A):209–226.

Koslow, J. A. 1984. Recruitment patterns in northwest Atlantic fish stocks. Can. J. Fish Aquat. Sci. 41:1722–1729.

Laevastu, T., and Larkins, H. A. 1981. Marine Fisheries Ecosystem: Its Quantitative Evaluation and Management. Farnham, England: Fishing News Books.

Laevastu, T., Favorite, F., and Larkins, H. A. 1982. Resource Assessment and Evaluation of the Dynamics of the Fisheries Resources in the Northeastern Pacific with Numerical Ecosystem Models. In M. C. Mercer (ed.), Multispecies Approaches to Fisheries Management Advice. Can. Spec. Publ. Fish. Aquat. Sci. 59, pp. 70–81.

Larkin, P. A. 1978. Fisheries management —An essay for ecologists. Anu. Rev. Ecol. Syst. 9:57–74.

Larkin, P. A., and Gazey, W. 1982. Applications of Ecological Simulation Models to Management of Tropical Multispecies Fisheries. In D. Pauly and G. I. Murphy (eds.), ICLARM/CSIRO Workshop on the Theory and Management of Tropical Fisheries. Manilla, Philippines: ICLARM, pp. 123–140.

Leggett, W. C., and Deblois, E. 1994. Recruitment variation in marine fishes: Is it regulated by starvation and predation in the egg and larval stages? Neth. J. Sea Res. 32:119–134.

Leibold, M. A. 1989. Resource edibility and the effects of predators and productivity on the outcome of trophic interactions. Am. Nat. 134:922–949.

Lipínski, M. R. 1992. Cephalopods and the Benguela ecosystem: Trophic relations and impact. S. Afr. J. Mar. Sci. 12:791–802.

Livingston, P.A. 1991. Total Groundfish Consumption of Commercially Important Prey. In P. A. Livingston. Groundfish Food Habits and Predation on Commercially Important Prey Species in the Eastern Bering Sea from 1984 to 1986. NOAA Tech. Memo. NMFS F/NWC-207. Seattle: Alaska Fisheries Science Center.

Lleonart, J., Salar, J., and Macpherson, E. 1985. CVPA, an expanded VPA with cannibalism. Application to a hake population. Fish. Res. 3:61–79.

Longhurst, A. R., and Pauly, D. 1987. Ecology of Tropical Oceans. San Diego: Academic Press.

Lowry, L. F., and Frost, K. J. 1985. Biological Interactions between Marine Mammals and Commercial Fisheries in the Bering Sea. In J. R. Beddington, R. J .H. Beverton, and D. M. Lavigne (eds.), Marine Mammals and Fisheries. London: George Allen and Unwin, pp. 42–61.

Macer, C. T., and Shepherd, J. G. 1987. The effect of species and fishery interactions on the assessment of some North Sea gadoids. Fish. Res. 5:303–317.

Magnússon, K. G., and Pálsson, O. K. 1989. Trophic ecological relationships of Icelandic cod. Rapp. P.-v. Réun. Cons. Int. Explor. Mer 188:206–224.

Maurer, R. O., and Bowman, R. E. 1985. Food consumption of squids (*Illex illecebrosus* and *Loligo pealei*) off the northeastern United States. NAFO Scientific Council Studies 9:117–124.

May, R. C. 1974. Larval Mortality in Marine Fishes and the Critical Period Concept. In J. H. S. Blaxter (ed.), The Early Life History of Fish. Berlin: Springer-Verlag, pp. 3–19.

May, R. M., Beddington, J. R., Clark, C. W., Holt, S. J., and Laws, R. M. 1979. Management of multispecies fisheries. Science 205:267–277.

Mehl, S. 1989. The northeast Arctic cod stock's consumption of commercially exploited prey species in 1984–1986. Rapp. P.-v. Réun. Cons. Int. Explor. Mer 188:185–205.

Meyers, R. A., and Cadigan, N. G. 1993. Density-dependent juvenile mortality in marine demersal fish. Can. J. Fish. Aquat. Sci. 50:1576–1590.

Moksness, E., and Oiestad, V. 1987. Interaction of Norwegian spring-spawning herring larvae (*Clupea harengus*) and Barents Sea capelin larvae (*Mallotus villosus*) in a mesocosm study. J. Cons. Int. Explor. Mer 44:32–42.

Möller, H. 1980. Scyphomedusae as predators and food competitors of larval fish. Meeresforch. Rep. Mar. Res. 28:90–100.

Möller, H. 1984. Reduction of a larval herring population by jellyfish predator. Science 224:621–622.

Moloney, C. L., and Field J. G. 1985. Use of particle-size data to predict potential pelagic-fish yields of some southern African areas. S. Afr. J. Mar. Sci. 3:119–128.

Montevecchi, W. A., Birt, V. L., and Cairns, D. K. 1988. Dietary changes of seabirds associated with local fishery failures. Biol. Oceanogr. 5:153–161.

Murawski, S. A., Lange, A. M., Sissenwine, M. P., and Mayo, R. K. 1983. Definition and analysis of multispecies otter-trawl fisheries of the northeast coast of the United States. J. Cons. Int. Explor. Mer 41:13–27.

National Marine Fisheries Service (NMFS). 1979. Draft Environmental Impact Statement of the Interim Convention on Conservation of North Pacific Fur Seals. Seattle, Washington: U.S. Department of Commerce, NMFS.

Neill, W. H., Miller, J. M., Van der Veer, H. W., and Winemiller, K. O. 1994. Ecophysiology of marine fish recruitment: A conceptual framework for understanding interannual variability. Neth. J. Sea Res. 32:132–152.

Northridge, S., and Beddington, J. 1992. Marine Mammals. In M. J. Crawley (ed.), Natural Enemies: The Population Biology of Predators, Parasites, and Diseases. Boston: Blackwell Scientific Publications.

Oiestad, V. 1983. Predation on Fish Larvae as a Regulatory Force Illustrated in Enclosure Experiments with Large Groups of Larvae. ICES C.M. 1983/L:39.

Oliver, J. S., and Slattery, P. N. 1985. Destruction and opportunity on the sea floor: Effects of gray whale feeding. Ecology 66:1965–1975.

O'Neill, R. V., DeAngelis, D. L., Waide, J. B., and Allen, T. F. H. 1986. A Hierarchical Concept of Ecosystems. Monographs in Population Biology 23. Princeton, NJ: Princeton University Press.

Overholtz, W. J., and Nicholas, J. R. 1979. Apparent feeding by the fin whale, *Balaenoptera physalus*, and humpback whale, *Megoptera novaeangliae*, on the American sand lance, *Ammodytes americanus*, in the Northwest Atlantic. Fish. Bull. U.S. 77:285–287.

Pálsson, Ó. K. 1994. A review of the trophic interactions of cod stocks in the North Atlantic. ICES Mar. Sci. Symp. 198:553–575.

Payne, P. M., Wiley, D. N., Young, S. B., Pittman, S., Clapham, P. J., and Jossi, J. W. 1990. Recent fluctuations in the abundance of baleen whales in the southern Gulf of Maine in relation to changes in selected prey. Fish. Bull. U.S. 88:687–696.

Pierce, G. J., Boyle, P. R., Hastie, L. C., and Begoña Santos, M. 1994. Diet of squid *Loligo forbesi* and *Loligo vulgaris* in the northeast Atlantic. Fish. Res. 21:149–163.

Platt, T. 1985. Structure of the Marine Ecosystem: Its Allometric Basis. In R. E. Ulanowicz and T. Platt (eds.), Ecosystem Theory for Biological Oceanography. Can. Bull. Fish. Aquat. Sci. 213, pp. 55–64.

Polis, G. A. 1994. Food webs, trophic cascades and community structure. Australian J. Ecol. 19:121–136.

Polovina, J. 1984. Model of a coral reef ecosystem. Coral Reefs 3:1–11.

Pope, J. 1979. A Modified Cohort Analysis in which Constant Natural Mortality Is Replaced by Estimates of Predation Levels. ICES CM/H:16.

Pope, J. 1989. Fisheries research and management for the North Sea: The next hundred years. Dana 8:33–43.

Pope, J. G., and Knights, B. J. 1982. Comparison of Length Distributions of Combined Catches of All Demersal Fishes in Surveys in the North Sea and at Faroe Bank. In M. C. Mercer (ed.), Multispecies Approaches to Fisheries Management Advice. Can. Spec. Publ. Fish. Aquat. Soc. 59, pp. 116–118.

Post, J. R., and Evans, D. O. 1989. Size-dependent overwinter mortality of young-of the-year yellow perch (*Perca flavens*): Laboratory, in situ enclosure, and field experiments. Can. J. Fish. Aquat. Sci. 46:1958–1968.

Power, G., and Gregoire, J. 1978. Predation by freshwater seals on the fish community of Lower Seal Lake, Quebec. J. Fish. Res. Board Can. 35:844–850.

Power, M. E. 1993. Top-down and bottom-up forces in food webs: Do plants have primacy? Ecology 73(3):733–746.

Powers, K. D., and Brown, R. G. 1987. Seabirds. In R. H. Backus (ed.), Georges Bank. Cambridge, MA: Cambridge Press, pp. 359–371.

Punt, A. E. 1994. Data Analysis and Modelling of the Seal-Hake Biological Interaction off the South African West Coast. Report submitted to the Sea Fisheries Research Institute, South Africa.

Punt, A. E., and Butterworth, D. S. 1991. On an approach for comparing the implications of alternative fish stock assessments, with application to the stock of Cape hake, *Merluccius* spp. off Northern Namibia. S. Afr. J. Mar. Sci. 10:219–240.

Punt, A. E., and Hilborn, R. 1994. A comparison of fishery models with and without cannibalism with implications for the management of the Cape hake resource off southern Africa. ICES J. Mar. Sci. 51:19–29.

Purcell, J. E., Siferd, T. D., and Marliave, J. B. 1987. Vulnerability of larval herring (*Clupea harengus pallasi*) to capture by the jellyfish, *Aequorea victoria*. Marine Biol. 94:157–162.

Radchenko, V. I. 1992. The role of squids in the pelagic ecosystem of the Bering Sea. Oceanology 32:762–767.

Rice, J. A., Crowder, L. B., and Binkowski, F. B. 1987. Evaluating potential sources of mortality for larval bloater (*Coregonus hoyi*): Starvation and vulnerability to predation. Can. J. Fish. Aquat. Sci. 44:467–472.

Rice, J. A., Miller, T. J., Rose, K. A., Crowder, L. B., Marschall, E. A., Trebitz, A. S., and DeAngelis, D. L. 1993. Growth rate variation and larval survival: Inferences from an individual based size-dependent predation model. Can. J. Fish. Aquat. Sci. 50:133–142.

Rose, G. A., and Leggett, W. C. 1990. The importance of scale to predator-prey spatial correlations: An example of Atlantic fishes. Ecology 71(1):33–43.

Royal Commission on Seals and Sealing in Canada. 1986. Seals and Sealing in Canada. Report of the Royal Commission, Vol 3. Ottawa: Canadian Government Publishing Centre.

Sainsbury, K. J. 1988. The Ecological Basis of Multispecies Fisheries, and Management of a Demersal Fishery in Tropical Australia.In J. A. Gulland (ed.), Fish Population Dynamics, 2nd ed. New York: John Wiley & Sons, pp. 349–382.

Schalk, P. H., Brey, T., Bathmann, U., Arntz, W., Gerdes, D., Dieckmann, G., Ekau, W., Gradinger, R., Plotz, J., Nothig, E., Schnack-Schiel, S. B., Siegel, V., Smetacek, V. S., and van Franeker, J. A. 1993.

Towards a Conceptual Model of the Weddell Sea Ecosystem, Antarctica. In V. Christensen and D. Pauly (eds.), Trophic Models of Aquatic Ecosystems. ICLARM Conference Proceedings 26, pp. 323–337.

Sharp, G. D. 1992. Pieces of the Puzzle: Anecdotes, Time-Series and Insights. In A. I. L. Payne, K. H. Brink, K. H. Mann, and R. Hilborn (eds.), Benguela Trophic Functioning. S. Afr. J. Mar. Sci. 12, pp. 1081–1092.

Shaughnessy, P. D. 1985. Interactions between Fisheries and Cape Fur Seals in Southern Africa. In J. R. Beddington, R. J. H. Beverton, and D. M. Lavigne (eds.), Marine Mammals and Fisheries. London: George Allen and Unwin, pp. 119–134.

Sheldon, R. W., Sutcliffe, W. H., Jr., and Paranjape, M. A. 1977. The structure of the pelagic food chain and the relationship between plankton and fish production. J. Fish. Res. Board Can. 34:2344–2353.

Shepherd, J. G. 1988. An exploratory method for the assessment of multispecies fisheries. J. Cons. Int. Explor. Mer 44:189–199.

Sherman, K. 1992. Monitoring and Assessment of Large Marine Ecosystems: A Global and Regional Perspective. In D. H. McKenzie, D. E. Hyatt, and V. J. McDonald (eds.), Ecological Indicators. Essex, England: Elsevier Press, pp. 1041–1074.

Sigurjónsson, J., and Vikingsson, G. 1992. Investigations on the Ecological Role of Cetaceans in Icelandic and Adjacent Waters. ICES CM 1992/N:24.

Silvert, W., and Dickie, L. M. 1982. Multispecies Interactions between Fish and Fishermen. In M. C. Mercer (ed.), Multispecies Approaches to Fisheries Management Advice. Can. Spec. Publ. Fish. Aquat. Sci. 59, pp. 163–169.

Simenstad, C. A., Estes, J. A., and Kenyon, K. W. 1978. Aleuts, sea otters, and alternate stable-state communities. Science 200:403–411.

Sissenwine, M. P. 1984. The uncertain environment of fishery scientists and managers. Marine Resource Economics 1(1):1–30.

Smith, R. H., and Mead, R. 1974. Age structure and stability in models of predator-prey systems. Theor. Pop. Biol. 6:308–322.

Starfield, A. M., Shelton, P. A., Fields, J. G., Crawford, R. J. M., and Armstrong, M. J. 1988. Note on a modelling schema for renewable resource problems. S. Afr. J. Mar. Sci. 7:299–303.

Tanaka, M., Goto, T., Tomiyama, M., and Sudo, H. 1989. Immigration, settlement and mortality of flounder (*Paralichthys olivaceus*) larvae and juveniles in a nursery ground, Shijiki Bay, Japan. Neth. J. Sea Res. 24(1):57–67.

Tonn, W. M., Paszkowski, C. A., and Holopainen, I. J. 1992. Piscivory and recruitment: Mechanisms structuring prey populations in small lakes. Ecology 73:951–958.

Ursin, E. 1982. Stability and variability in the marine ecosystem. Dana 2:51–67.

Vanni, M. J. 1987. Indirect Effect of Predators on Age-Structured Prey Populations: Planktivorous Fish and Zooplankton. In W. C. Kerfoot and A. Sih (eds.). Predation: Direct and Indirect Effects on Aquatic Communities. Hanover, NH: University Press of New England, pp. 149–160.

Veer, H. W. van der. 1985. Impact of coelenterate predation on larval plaice *Pleuronectes platessa* and flounder *Platichthys flesus* stock in the western Wadden Sea. Mar. Ecol. Prog. Ser. 25:229–238.

Veer, H. W. van der. 1986. Immigration, settlement, and density-dependent mortality of a larval and early post-larval 0-group plaice (*Pleuronectes platessa*) population in the western Wadden Sea. Mar. Ecol. Prog. Ser. 29:223–2236.

Veer, H. W. van der. 1987. Predation by crustaceans on a newly settled 0-group plaice *Pleuronectes platessa* population in the western Wadden Sea. Mar. Ecol. Prog. Ser. 35:203–215.

Veer, H. W. van der, and Zijlstra, J. J. 1982. Predation of Flatfish Larvae by *Pleurobrachia pileus* in Coastal Waters. ICES C.M. 1982/G:16.

Vermeer, K., Sealy, S. G., and Sanger, G. A. 1987. Feeding Ecology of Alcidae in the Eastern North Pacific. In J. P. Croxall (ed.), Seabirds: Feeding, Ecology and Role in Marine Ecosystems. Cambridge, England: Cambridge University Press, pp. 189–227.

Walters, C., and Hilborn, R. 1976. Adaptive control of fishing systems. J. Fish Res. Board Canada. 33:145–159.

Werner, E. E., and Hall, D. J. 1988. Ontogenetic habitat shifts in bluegill: The foraging rate-predation risk trade-off. Ecology 69:1352–1366.

Werner, E. E., Hall, D. J., and Mittlebach, G. G. 1983. An experimental test of the effects of predation risk on habitat use in fish. Ecology 64:1540–1548.

West, L. 1988. Prey selection by the tropical snail *Thais melonges*: A study of interindividual variation. Ecology 69:1839–1854.

Wickens, P. A., Shelton, P. A., David, J. H. M., Field, J. G., Oosthuizen, W. H., Roux, J. P., and Starfield, A. M. 1992. A fur seal simulation model to explore alternative management strategies. Can. J. Fish. Aquati. Sci. 49:1396–1405.

Wilson, J. A., Acheson, J. M., Metcalfe, M., and Kleban, P. 1994. Chaos, complexity and community management of fisheries. Marine Policy 18(4):291–305.

Yodzis, P. 1994. Predator-prey theory and management of multispecies fisheries. Ecological Applications 4(1):51–58.

15

Acoustic Assessment as an Available Technique for Monitoring the Living Resources of Large Marine Ecosystems

Qisheng Tang, Xianyong Zhao, and Xianshi Jin

For proper management of living marine resources, knowledge of the state of the entire ecosystem, such as the size of each individual population at a specific time, is essential. To obtain this knowledge, various methods have been employed; the most frequently used techniques are bottom trawl, egg and larvae methods, and acoustic methods. The bottom trawl survey is an effective method for multispecies studies, but is mainly used for demersal communities. Because survival rates during species' early life history are often not well known, the egg and larvae methods have large uncertainties. Even though little data can be obtained on the uppermost sea surface "blind zone" and the bottom "dead zone," the acoustic method has gained great popularity and has been relatively successful in fisheries research.

A large marine ecosystem (LME) management program requires the entire ecosystem to be sampled in as much detail as possible over a relatively short time period. The quasi-continuous sampling feature coupled with relatively high sampling speed (around 10 knots of vessel speed) makes the acoustic method a promising choice for the implementation of such a LME program. The application of the acoustic technique in monitoring nekton and micronekton resources has been briefly reviewed by Holliday (1993); its potential application for monitoring zooplankton in LMEs was also examined.

Chinese scientists at the Yellow Sea Fisheries Research Institute (YSFRI), Qingdao, China, partly in cooperation with the Norwegian scientists at the Institute of Marine Research (IMR), Bergen, Norway, have conducted a series of acoustic surveys in various areas between 1984 and 1994 (Zhu and Iversen 1990; Iversen et al. 1993; Tang et al. 1995). Results from these surveys are presented in this chapter. The effectiveness of the acoustic assessment as an available technique in monitoring the living resources in LMEs is discussed.

Development of Acoustic Techniques for the Investigation of Living Marine Resources

The most notable pioneer application of underwater acoustics in the study of marine fish was published in 1935 (Sund 1935). This study may be seen as a prototype for the modern application of acoustic techniques for the investigation of living marine resources.

After the Second World War, acoustic instruments developed by the military were modified for civilian use. Echo sounders were used for fish detection. The design of the first analog echo integrator, described by Dragesund and Olsen (1965), was a milestone for acoustic techniques used in fisheries. This instrument made feasible the quantitative assessments of fish resources on a large scale, although at that time the accuracy of the estimates was relatively poor owing to imprecise calibration methods and the lack of sufficient knowledge on the target strength of fish. Further development in high-performance scientific echo sounder systems (such as the SIMRAD EK400 and QD digital echo integrator) and calibration methods (Foote et al. 1987), as well as direct in situ target-strength measurement techniques (such as the dual beam and split beam methods), coupled with better understanding of the limitations of the acoustic method, have led to greatly improved assessments of fish resources.

With the advent of acoustic instruments having large dynamic range and working on frequencies ranging from tens to hundreds of kilohertz, abundance estimates have been extended from nekton species to micronekton and even larger zooplankton. Many of these applications have been reviewed by Holliday (1993).

Acoustic Assessment of Fish Populations, with Special Reference to the Yellow Sea Anchovy (*Engraulis japonicus*)

Estimating fish abundance is probably the most important application of acoustics in fisheries research (MacLennan 1990). Many commercially and trophically important fish stocks have been monitored acoustically. Scientists at the YSFRI and the IMR have conducted a series of acoustic trawl surveys in the Yellow Sea and the East China Sea in the period between 1984 and 1994. The purpose of these investigations has been to map the distribution and to measure the biomass of anchovy and other fish resources in the ecosystems in question. Most of the results from these surveys are given by Zhu and Iversen (1990) and Iversen et al. (1993). The survey results on wintering anchovy will be updated here.

Vessel, Equipment, and Methodology

All surveys were carried out aboard the R/V *Bei Dou*, a 56.2-m oceanographic research vessel capable of conducting acoustic surveys, stern trawls, and plankton sampling. The acoustic data were collected with an echo sounding integration system, namely the SIMRAD EK400/38 kHz echo sounder and the SIMRAD QD digital echo integrator. The transducer was mounted on a blister located at the fore-port side of the hull, 4.5–5.0 m deep in the water depending on the loading condition of the vessel. The output of the integrator was given for each elementary distance sampling unit (EDSU) of 5 nautical miles (nm). An additional SIMRAD EK400/120 kHz echo sounder was run to help discriminate acoustic echo signs resulting from different targets. The overall acoustic system was calibrated prior to each survey according to the procedure described by Foote et al. (1983, 1987).

Identification and sampling of the target echo signs were done either by bottom trawl (18-mm stretched mesh size in the cod-end) or semi-pelagic trawl (22-mm stretched mesh size in the

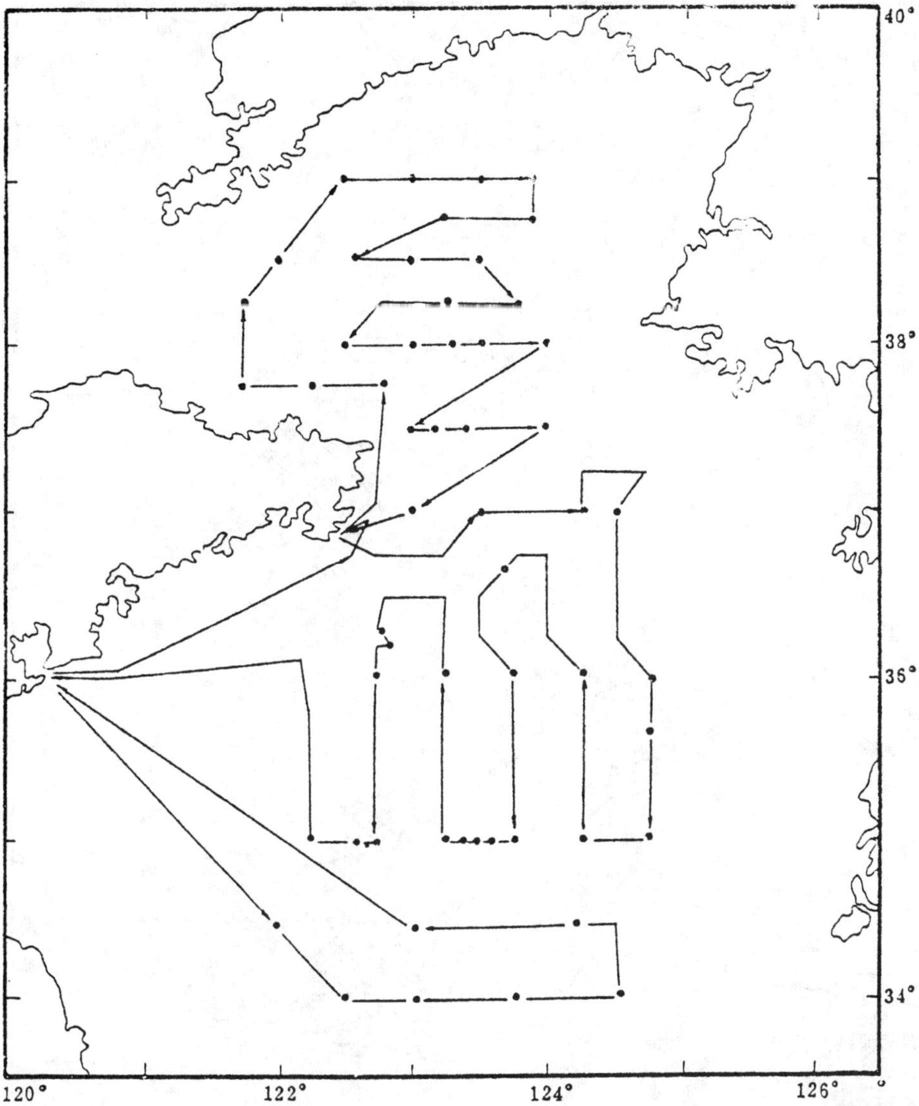

Figure 15-1. Cruise track and sampling stations of Japanese anchovy survey by the R/V *Bei Dou* in the Yellow Sea. (Reprinted by permission from Zhu, D., and Iversen, S. A. 1990). Anchovy and Other Fish Resources in the Yellow Sea and East China Sea, November 1984–April 1988 (in Chinese with English abstract). Mar. Fish. Res. China 11.)

cod-end) depending on the target's vertical distribution. The surveys were conducted along parallel transects in most cases; however, this was subject to minor modifications depending on the weather conditions and fish distribution. The vessel speed was about 12 knots at normal weather conditions. An example of the cruise tracks is shown in Figure 15-1.

Monitoring Geographic Distribution

During an acoustic survey, a large volume of water over a large area can be sampled in a relatively short time in a quasi-continuous manner. This provides an excellent means for monitoring fish distributions and their annual variations.

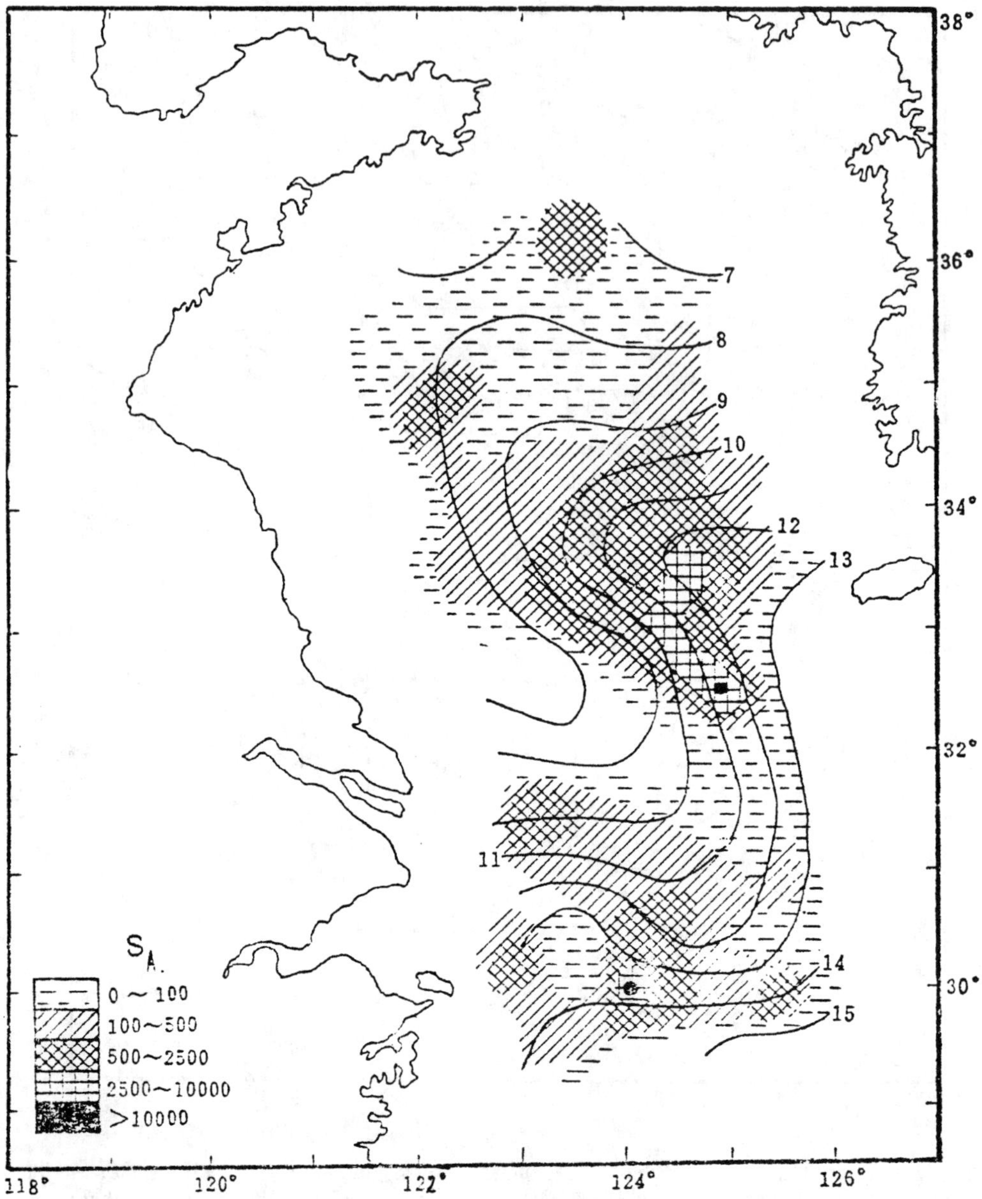

Figure 15-2. Distribution and relative density (S_A) of Japanese anchovy during winter. The contours are surface water temperature. (Reprinted by permission from Zhu, D., and Iversen, S. A. (eds.). 1990. Anchovy and Other Fish Resources in the Yellow Sea and East China Sea, November 1984–April 1988 (in Chinese with English abstract). Mar. Fish. Res. China 11.)

Surveys of the Yellow Sea anchovy indicated that the geographic distribution of the anchovy is strongly affected by changes in the physical environment (Zhu and Iversen 1990). In November and December, anchovy occur mainly in the northern and central parts of the Yellow Sea, with the southern border at about 34°N. The Yellow Sea Warm Current (a branch of the Kuroshio Current) is believed to act as a physical barrier for the distribution of anchovy. With the progressive winter cooling of the sea, the anchovy migrate south and eastward. Figure 15-2 shows a typical distribution pattern of the wintering anchovy.

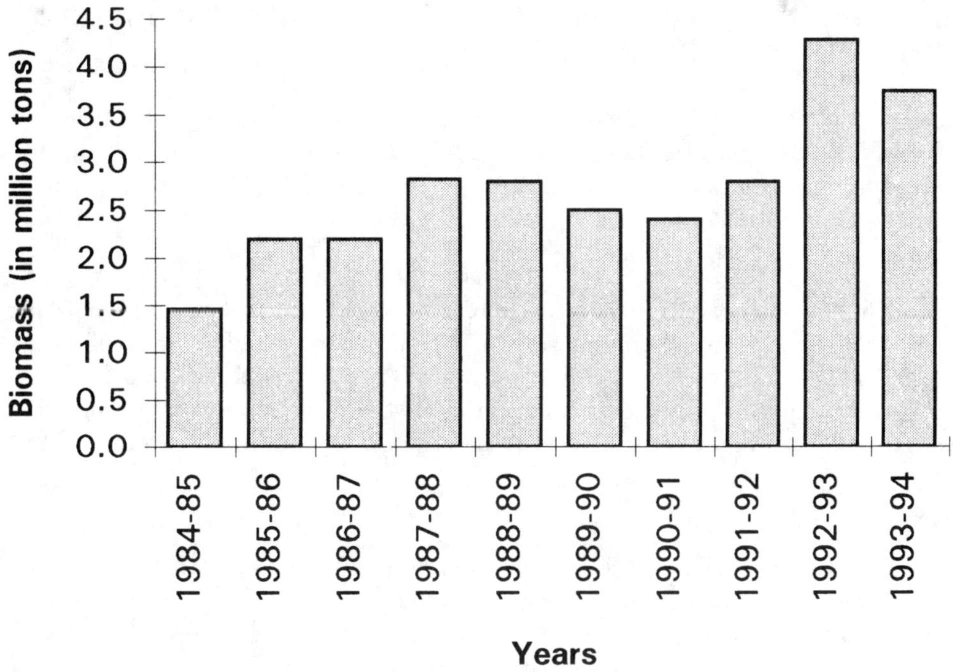

Figure 15-3. Annual variations of Japanese anchovy biomass in the Yellow Sea during winter 1984–1994.

Stock Assessment

Figure 15-3 shows the biomass of anchovy in the Yellow Sea (north of 32°N) from the winter seasons of 1984–1985 to 1993–1994. The estimates were calculated with the subarea method. In the beginning years of the study, the biomass in each surveyed rectangle (0.5 degree of latitude by 0.5 degree of longitude) was estimated based on the average S_A (area backscattering coefficient, m^2/nm^2) value and weight/length information of anchovy in the rectangle. The total biomass in the surveyed area was simply the sum of all the subareas (Iversen et al. 1993). Recently, the subareas were determined by the poststratification method. Five to 16 geographical strata were selected based mainly on echo density distributions. The method of strata determination based on anchovy length information was also exercised. No significant differences in biomass estimates obtained from the methods mentioned above were observed. The target strength (TS) to fish length (L, cm) relation used was $TS = 20 \log L + 72.5$ dB. In general, in situ measurements agreed with this relation (Chen and Zhao 1990).

The highest estimate of anchovy biomass in the Yellow Sea was 4 million mt, obtained from a survey in winter 1992–1993. This high estimate was believed to be due to the strong 1991 year-class and the relatively large area covered by the survey, especially in the west and east. During the survey, very dense concentrations of anchovy were also found near the Korean side, at the southeastern corner of the surveyed area. This has confirmed the concern raised of inadequate coverage of the shallow-water component (Iversen et al. 1993), especially the components along the Korean side. Better knowledge of the distribution and biomass of the Yellow Sea anchovy may be obtained with improved coverage of the shallow inshore areas (maybe with a smaller boat) and better collaboration with Korean scientists under the LME spirit.

Acoustic Assessment of 0-Group Fish: The Case of Age-0 Walleye Pollock (*Theragra chalcogramma*) in the Aleutian Basin

An acoustic mid-water trawl survey on the abundance of walleye pollock in the Aleutian Basin was carried out during the period of 28 June to 2 August 1994. During the survey, large concentrations of 0-group pollock were observed and assessed acoustically (Tang et al. 1996).

Methodology

The research vessel and acoustic instruments were the same as those used for the anchovy surveys. With the calibrated echo sounding integration system, the acoustic data were collected along parallel transects, with a transect space of 22 nm around Bogoslof Island and 44 nm in other areas. The biological samples were collected with a pelagic trawl with mesh sizes ranging from 400 cm at the wings and 2.2 cm at the cod-end. The vertical opening of the net mouth measured 30 to 40 m.

Distribution of Age-0 Pollock

The age-0 pollock were mainly distributed in the northern and eastern parts of the Aleutian Basin, extending from the edge of the eastern Bering Sea shelf to the basin area, and in the northern and eastern parts of the Bering High Seas, with the zero distributional isoline approximately parallel to and about 100 nm from the 200-m bottom contour line. Areas with the densest concentrations were observed at the northeastern corner of the surveyed area (north of 56°–30°N and east of 180°E); the highest S_A value measured approximately 500. Another area with a relatively dense concentration was east of 168°W, around Bogoslof Island, where the S_A value was approximately 100.

During the survey, age-0 pollock performed distinct diurnal vertical migrations. From 0800 to 1400 hours, they were mainly found in the water column from 80- to 120-m depth. From about 1600 to 2000 h, they gradually migrated to 20–60-m water layer. Around midnight (2200 to 0200 h), age-0 pollock were concentrated in the upper 10-m layer and vulnerable to surface plankton sampling gears. At dawn (0300 to 0400 h), they descended rapidly to the depth of 60–90 m and gradually stabilized at about 100 m. At all depths, age-0 pollock showed a strong schooling behavior and concentrated in a layer 10 to 20 m thick.

Biological Information

During the survey, whenever dense concentrations were observed around 100-m depth and shallower, a mid-water trawl was carried out. Age-0 pollock dominated the catch in all cases in the northeastern part of the Aleutian Basin.

The fork length of age-0 pollock sampled by the mid-water trawl around the northeastern part of the Bering High Sea ranged from 29 to 47 mm, with a mean of 40.2 mm; the dominant length ranged from 36 to 44 mm. The individual body weight averaged 0.43 gm. Slight differences were observed in the mean size of fish caught at different locations. The age-0 pollock sampled at 58° to 40°N and 178° to 30°W ranged from 29 to 41 mm, with a mean of 36 mm, whereas samples from a haul at 53° to 40°N and 173° to 47°W ranged from 34 to 47 mm, with a mean of approximately 43 mm.

Discussion and Conclusion

The theory and instrumentation of using underwater acoustics to estimate fish abundance have gained much attention (Forbes and Nakken 1972; Johannesson and Mitson 1983; MacLennan and Simmonds 1992). Acoustic assessments of fish resources have been done both on pelagic species such as Atlantic herring and Peruvian anchoveta, and demersal or semidemersal species such as Arctic cod and Bering Sea walleye pollock. The problems of the surface "blind zone" and the bottom "dead zone" can be potentially or partly solved by the development of the quantitative side-looking sonar system (Misund et al. 1996) and improved understanding on the acoustic sampling and signal processing near the sea bottom (Ona and Mitson 1996).

With its capability of mapping fish distribution and measuring absolute fish density continuously in the entire water column along each transect, the acoustic method appears to be superior to other methods such as the systematic trawling method, at least for fishes performing strong diurnal vertical migrations with patchy distributions. Our experiences on the acoustic surveys of anchovy and other fish resources in the Yellow Sea and the East China Sea have also demonstrated that acoustic techniques will be the most important tool for the monitoring of the fish resources in LMEs.

Acoustic abundance estimations of 0-group fish have been performed since the very beginning of the existence of the echo integration technique (Dragesund and Olsen 1965). However, when acoustic fish abundance estimations are extended to apply to 0-group fish, three main problems arise: (1) the availability of age-0 fish to acoustic detection, (2) identification of acoustic echo signs, and (3) the acoustic scattering properties of age-0 fish.

Availability of Age-0 Fish to Acoustic Detection

This problem may be encountered when the fish is distributed close to the sea surface, where it cannot be detected by down-looking echo sounders mounted on vessels of several hundred tons. An alternative acoustic technique is to use horizontal-searching sonar similar to the one used by Misund et al. (1996).

The nursery grounds of juvenile fish may often extend to shallow inshore waters, where it is difficult for any large boat to operate. To solve this problem, a portable scientific echo sounder, such as the SIMRAD EY500, may be used aboard a ship of appropriate size. In the case of age-0 pollock in the Aleutian Basin, this is not a significant problem because there is only a relatively short time period in which the fish are distributed in the uppermost surface layers.

Identification of Acoustic Echo Signs

The identification of acoustic echo signs is important both for species discrimination and for S_A value allocation, as well as for biological classification. Our experience with age-0 pollock in the Aleutian Basin indicated that mid-water trawls are an effective tool for species identification of juvenile fish inhabiting relatively deep waters (Tang et al., 1996). When a dense scattering layer centered at about 100-m depth was observed for the first time during the acoustic survey in the Aleutian Basin, the mid-water trawl was towed through the layer. This layer consisted mainly of age-0 pollock together with a few other juvenile fish species. Later experience showed that a half-hour tow was adequate to provide samples for identification and size classification of juvenile fishes. However, the results were certainly biased due to the mesh size of the trawl used. Trawls with smaller mesh size may provide improved knowledge of both the species compositions and the size distributions of each species. However, plankton associated with the distribution of juvenile fish can never be sampled by mid-water trawls, thus creating a great barrier for the thorough understanding of the composition of the acoustic targets when mid-water trawls are the only sampling gear.

Acoustic Scattering Properties of 0-Group Fish

For quantitative estimates of a fish resource, it is essential to know the target strength of the fish in question. Knowledge of the effective target strength of fish is relatively poor, especially for juvenile fish. Nevertheless, efforts have been made in this direction (Ona 1994).

When the target strength to fish length relation currently used for walleye pollock, TS = 20 log$L$ + 66 dB (Foote and Traynor 1988), is extrapolated to apply to age-0 pollock in the Aleutian Basin, a rough biomass estimate of 20,000 mt is reached, corresponding to an abundance of 58.2 billion individuals.

Although there are still a number of problems in the application of acoustic techniques for the assessment of 0-groups, the survey on age-0 pollock in the Aleutian Basin has shown that the acoustic method is a promising technique that deserves to be further developed for such surveys.

In addition to fish abundance estimates, the use of acoustic techniques for plankton investigations has gained attention over the last decade (Holliday 1993) and was an important topic in the Fourth International Symposium on Fisheries and Plankton Acoustics held in Aberdeen, Scotland, in June 1995.

Acoustic systems operating at two frequencies (38 and 120 kHz) during the acoustic surveys in the Yellow Sea and the East China Sea have provided us with information about the plankton community (potentially mixed with some micronektons) in the survey area. Acoustic techniques' fast, continuous sampling make them a promising tool for mapping the spatial distribution and for measuring the relative density of the plankton community. With this in mind, the potential value of the acoustic techniques in the monitoring of LMEs should not be ignored.

References

Chen, Y., and Zhao, X. 1990. In situ target strength measurements on anchovy (*Engraulis japonicus*) and Sardine (*Sardinops melanostrictus*). In Proceedings of International Workshop on Marine Acoustics, 26–30 March 1990, Beijing, China. China Ocean Press.

Dragesund, O., and Olsen, S. 1965. On the possibility of estimating year-class strength by measuring echo-abundance of 0-group fish. Fisk Dir. Skr. Ser. Havunders 13:47–75.

Foote, K. G., and Traynor, J. J. 1988. Comparison of walleye pollock target strength estimates determined from in situ measurements and calculations based on swimbladder form. J. Acoust. Soc. Am. 83:9–17.

Foote, K. G., Knudsen, H. P., and Vestnes, G. 1983. Standard calibration of echo sounders and integrators with optimal copper spheres. Fisk Dir. Skr. Ser. Havunders 17:335–346.

Foote, K. G., Knudsen, H. P., Vestnes, G., MacLennan, D. N., and Simmonds, E. J. 1987. Calibration of Acoustic Instruments for Fish Density Estimation: A Practical Guide. Int. Coun. Explor. Sea Coop. Res. Rep. 144.

Forbes, S. T., and Nakken, O. (eds.). 1972. Manual of Methods for Fisheries Resources Survey and Appraisal. Part 2: The Use of Acoustic Instruments for Fish Detection and Abundance Estimation. FAO Man. Fish. Sci.

Holliday, D. V. 1993. Applications of Advanced Acoustic Technology in Large Marine Ecosystem Studies. In K. Sherman, L. M. Alexander and B. Gold (eds.), Stress, Mitigation and Sustainability of Large Marine Ecosystems. Washington, DC: AAAS Press, pp. 301–319.

Iversen, S. A., Zhu, D., Johannessen, A., and Toresen, R. 1993. Stock size, distribution and biology of anchovy in the Yellow Sea and East China Sea. Fish. Res. 16:147–163.

Johannesson, K. A., and Mitson, R. B. 1983. Fisheries Acoustics: A Practical Manual for Biomass Estimation. FAO Fish. Tech. Pap. 240.

MacLennan, D. N. 1990. Acoustical measurement of fish abundance. J. Acoust. Soc. Am. 87:1–15.

MacLennan, D. N., and Simmonds, E. J. 1992. Fisheries Acoustics. London: Chapman & Hall.

Misund, O. A., Aglen, A., Hamre, J., Ona, E., Røttingen, I., Skagen, D., and Valdermarsen, J. W. 1996. Improved mapping of schooling fish near the surface: Comparison of abundance estimates obtained by sonar and echo integration. ICES J. Mar. Sci. 53:383–388.

Ona, E. 1994. Detailed In Situ Target Strength Measurement of 0-Group Cod. ICES CM 1994/B:30.

Ona, E., and Mitson, R. B. 1996. Acoustic sampling and signal processing near the seabed: the deadzone revisited. ICES J. Mar. Sci. 53:677–690.

Sund, O. 1935. Echo sounding in fishery research. Nature. 135:953.

Tang, Q., Jin, X., Li, F., Chen, J., Wang, W., Chen, Y., Zhao, X., and Dai, F. 1996. Summer Distribution and Abundance of Age-0 Walleye Pollock, *Theragra chalcogramma* in the Aleutian Basin. U.S. Dep. Commer. NOAA Tech Rep. NMFS 126:35–45.

Tang, Q., Wang, W., Chen, Y., Li., Jin, X., Zhao, X., Chen, J., and Dai, F. 1995. Stock assessment of walleye pollock in the north Pacific Ocean by acoustic survey (in Chinese with English abstract). J. Fish. China 19(1):8–20.

Zhu, D., and Iversen, S. A. (eds.). 1990. Anchovy and Other Fish Resources in the Yellow Sea and East China Sea, November 1984– April 1988 (in Chinese with English abstract). Mar. Fish. Res. China 11.

16

Shoal Structure of Commercially Important Pelagic Fish in Relation to the Dynamics of Upwelling Marine Ecosystems

Virgilio Arenas and Carlos Robinson

Fishes live in specific or multispecific groups, sometimes structured as a school or, more often, as shoals, according to the benefits each individual gets from the group. As a result, group size, composition, integrity, and behavior may vary widely in relation to ecological conditions. Shoaling is a strategy resulting from eons of effort to maximize the individual's evolutionary fitness. The shoaling fishes of the oceans provide food and important economic and social benefits for the growing human population.

The most important fisheries are located in the most productive ecosystems of the world: upwelling ecosystems. Small pelagic shoaling fishes, such as anchovies, sardines, and herring, are not only important to fishermen but also are an important part of the food chain. Shoaling fishes compete as herbivores by grazing and filtering phytoplankton, or as carnivores by filtering and biting zooplankton and micronekton. Small pelagic fishes play several important roles in the structure and function of the ecological system. They contribute to the processes that facilitate the maintenance and enhancement of primary productivity, either through the "seeding" of new upwelled waters or the cycling and distribution of nutrients. By their shoaling feeding behavior, they graze selectively; by their shoaling predation avoidance behavior, they transport materials and biomass. Small pelagic fishes are key elements for the extended influence of the physically altered systems.

When managing upwelling ecosystems, one should consider that fishing the small pelagic species disturbs the ecological balance. It is better to see fishing as a tool for managing ecological processes. In the perspective of climate change, understanding the role of the shoaling behavior of small pelagic fishes is important. In this paper we present some results of our project off the west coast of Baja California, the goal of which is studying shoaling behavior as a key factor in the dynamics of an upwelling ecosystem and as a tool for fisheries management.

Study Area

The west coast of Baja California is the southern end of one of the best-known oceanographic regions of the world, the California Cooperative Fisheries Investigations (CalCOFI) region. On several occasions during the past 30 years the area has been included in the CalCOFI programs and was considered for stock assessment of sardines or anchovies. The reports contain a great deal of data and are released annually.

A description of the North Pacific Ocean climatology and large-scale circulation is necessary to understand the current system off the west coast of North America. An important review of the California Current system was given by Hickey (1979). An equatorward surface flow along the west coast of the United States and Baja California, Mexico, is fed by the West Wind Drift, the northern limb of the North Pacific Gyre. Beneath this flow and concentrated over the continental slope is a poleward flow. These two currents form what is referred to as the Eastern Boundary Current system, or in this case, the California Current system; similar systems occur in other oceans, for example, the Peru Current system, Benguela Current system, and the Canary Current system. From Vancouver to the tip of Baja California, the nearshore current shows seasonal variation related to that of the alongshore component of wind stress at the coast. The current is northward in winter north of Cape Mendocino, and southward in spring and summer at all locations. The current arrangements in the region can be described as follows: Between the arctic and tropical regions of the North Pacific Ocean there are two water masses, the Subarctic Cyclonic Gyre and the Subtropical Anticyclonic Gyre; the North Pacific Current connects to the North Equatorial Current by way of the wide, sluggish, and complex California Current flowing to the south off California and Mexico.

Methods

Four oceanographic cruises took place on the west coast of Baja California onboard the B/O *El Puma* on the following dates: October 19 to November 3, 1992; May 10 to May 27, 1993; November 28 to December 16, 1993; and a preliminary cruise for methodology testing.

Approximately 50 stations were used. These stations were paired, one inside the neritic zone and the other 10 miles away in the oceanic zone. At each station the following parameters were measured: temperature, salinity, oxygen, phosphate, nitrate, nitrite, ammonia, silicate, phytoplankton at standard depths, and a vertical zooplankton sampling to a maximum depth of 200 m. During the interstation trip, an onboard continuous recording of temperature and salinity was obtained. Also, a continuous ecointegration system using HADAS (Hydroacoustic Data Acquisition System) was running with a transducer of 200 kHz and a Symrad EY 200 ecosonde (Lindem and Houri 1988).

Three areas were selected within the upwelling region for temporal and spatial small-scale sampling (SSS): in the north near the U.S. border, in the middle, and in the south in front of Magdalena Bay. The SSS was designed to study the diel variations within a 10-mile transect. During each transect, three stations 5 miles apart were sampled: one in the neritic zone, another at the shelf break, and a third in the oceanic zone. Each station was sampled six to seven times during the diel cycle. During the October 1992 cruise, 24-hour sampling for ecointegration, temperature, and salinity was performed along a transect almost parallel to the coast. During the December 1993 cruise, three SSS transects in each region were done.

Primary productivity was estimated at standard depths with in situ ^{14}C and oxygen incubations of light and dark bottles. In accord with the information coming from the ecointegration, micronekton trawls were done at the target depths for variable times using an Issacs-Kid net.

Results: The Ecological Scenario

The Physics

An upwelling system is a local perturbation of the normal distribution of a water mass. Due to the wind pattern blowing south, the surface water moves offshore and is replaced by subsurface water that is cool and usually rich in nutrients. During spring and summer, the north and northwesterly winds produce an offshore transport of surface water, and the vertical movement of water produces a cooling of surface water. The upwelling phenomenon in the region varies from some kilometers to hundreds of kilometers along the coast and extends offshore and onshore as an identifiable horizontal perturbation hundreds of kilometers away from the actual site of upwelling.

Water ascends from depths of tens to hundreds of meters at velocities of meters per day. The distribution, extension, amplitude, strength, and duration of the upwelling depends on the responses of local topography, as well as the submarine topography, to the wind field. The structure and dynamics of the ocean also affect the response of the system to the wind effect.

The physical-ecological scenario has to take into account the mesoscale eddylike hydrographic features existing in coastal upwelling regions. Offshore upwelled water is advected equatorward, and in response to wind relaxation or reversal events is advected onshore; during these times frontal gradients intensify (Schwing et al. 1991). The biotic components of the ecosystem may take advantage of these oceanographic processes to enhance development, avoid predators, and position themselves in the optimal conditions.

Over the past several decades, linear empirical analyses of the relationship between local wind effects and recruitment variability of pelagic fish yielded nonsatisfying results. Recently, nonlinear methods of analysis applied to sardines and anchovies resulted in a domed-shaped relationship, or "optimal environmental window," a worthwhile concept to apply in order to understand feeding behaviors (Bakun 1994). The oceanographic features described are linked to the local characteristics of the site (e.g., wind forcing, flow–bathymetry interactions) in a way that leads to the seemingly incoherent arrangement of typical upwelling system features such as the ones we have been following during diel cycles in three locations.

The California Coastal Upwelling ecosystem includes most of the U.S. Pacific coast; however, the west coast of Baja California changes dramatically due to the presence of the Gulf of California, which affects the wind field. The El Niño–Southern Oscillation (ENSO), the very well-known cycles of warming along this coast, takes effect approximately every 5 years or so (2 to 10 years). Some of these events are strong enough to have a massive impact along the eastern boundary of the Pacific Ocean, particularly in the tropical areas. Descriptions of the effects of ENSO events on the upwelling ecosystem have been well documented since 1958. The 1983 ENSO event (Glynn 1989) is regarded as the strongest this century.

This description of the physical scenario should include something about local changes due to global climate change (Arenas et al. 1990). There is much uncertainty about the nature of the global response to elevated CO_2 and other greenhouse gas levels due to lack of interdisciplinary knowledge. Positive and negative feedback mechanisms substantially reduce the confidence with which predictions can be made. First, it is expected that under the enhanced greenhouse climate, the region would be warmed 2–4°C; there would be an increase in precipitation (20%), with an accompanying shift of geographical patterns of cloudiness and rainfall. Storms would be more intense and storm tracks would move poleward. There would be an increase in wind intensity and probably in the next century ENSO events would become more common and intense.

Since a major greenhouse gas is water vapor, we expect the effects would be more evident in subtropical and temperate zones. This relative warming would diminish the temperature gradient and weaken the ocean currents bounding the Pacific Central Gyre.

Our classic wind-driven coastal upwelling ecosystem could be affected in the following way. CO_2 buildup could inhibit the nighttime cooling and enhance the daytime heating of the arid inland valleys. Seasonally, we would expect that in these areas short-range, vigorous wind systems would form. This could lead to the intensification of the alongshore winds and the acceleration of the upwelling circulation. Strong rising trends have been noted in the eastern ocean upwelling systems (Bakun 1992). The coastal system would thus be more mixed and turbulent. Simulation models of the atmospheric circulation support the hypothesis that southward wind stress could increase off northern California. On the west coast of Baja California this pattern may be affected by the presence of the Gulf of California, which could suppress the enhancement effects of inland valleys (Bakun 1990).

The Chemistry

From the point of view of biogeochemistry, the upwelling ecosystem represents a natural replenishment of nutrients consumed at the euphotic zone. In accord with the physics of the event, the upwelling carries a certain type and amount of material that circulates within a cell. These materials are eventually transported outside the system after a number of paths and processes occur.

The concentrations of CO_2, O_2, nitrogen, phosphorus, and silicate change dramatically according to their physics and the degree of coupling to the biological components of the system. Usually more than 40% of the nutrients are recycled within local ecological processes. During a diel cycle in S. Quintin (north of Baja), we detected changes in phosphate concentrations in the upper 100 meters from the detection limit (0.032 μg-at/l) to more than 2.0 μg-at/l in two pulses related to displacement of water masses. Similar changes were seen for ammonia, nitrate, and nitrite. The atomic relation N:P varied in a 24-hour period far from the 15:20 expected. This reflects the biological influence on the ecological scenario that extends beyond the day–night cycle.

The Biology

At the upwelling ecosystem, besides heating, there are two main inputs of energy: primary productivity based on photosynthesis, and the secondary energy that actually drives ecological processes (Margalef 1978). As a consequence of physics and chemistry, the upwelling environment is well suited to reach the level of being the most sustainable ecosystem (Longhurst and Pauly 1987).

On the west coast of Baja, the primary productivity is high and varies locally, temporarily giving a strong local heterogeneity. Under the effects of predation during the diel cycles, phytoplankton populations change from a single-species community of millions per cubic meter to one of a few thousand cells of more than several dozen species. During these cycles the phytoplankton community shows certain coherence. Most of the species with relatively lower abundance were also present more frequently with depths and time; a reason may be that intensive and selective grazing leaves a few cells of several species.

Micronekton also have a very patchy structure and reach a very high biomass. They usually form scattering layers as well as zooplankton aggregations (Smith et al. 1989). According to our hydroacoustic analysis, several groups of animals compose this level of the pelagic environment. Sixteen species of euphausiids were recorded during the seven diel cycles from the December cruise; the species, sizes, and biomass varied from 5% to 96% of the macroplanktonic communities (Gomez et al. 1994). The euphausiaceans *Euphausia eximia,* a crustacean of the family phasiphaidea, *Nyctiphanes simplex,* and a decapod galatheidae, *Pleuroncodes planipes,* known as a red crab, were identified by their target strength in a huge scattering layer in an upwelling region off Magdalena Bay (24°41′N and 112°W). This area is recognized as one of the most productive zones along the southwest coast of Baja California (Robinson et al., submitted for publication). Results showed that organisms with different target strengths were not distributed randomly within the scattering layer. The degree of shoaling (compactness), as well as the abundance of each of the three most abundant species, varied widely according to the depth, time of the day, and location. The slope section showed the highest abundance and shoaling; fishes were 100% in shoals around 75 m depth, and euphausiaceans and red crabs reached only 20% of shoaling. At the oceanic section the three groups were also present, but in much lower densities, and occurred as individuals rather than aggregated (according to the acoustic threshold established).

Populations of small pelagic fishes fluctuated greatly; their remarkable declines and recoveries are famous in the fishery literature. Major changes in populations are often accompanied by alterations in the dominant species in the ecosystem. There are indications that these shifts might be recognized by changes in the planktonic communities before they become obvious in the fishery. Fluctuations seem to be in phase with similar species in distant regions of the world ocean. Japanese and California ecosystems seem to be teleconnected in a Pacific basin-wide atmospheric process (Lluch-Belda et al. 1989). Owing to their great numerical abundance and dense concentrations in coastal areas, the scales of small pelagic fishes deposited in the sediments provide records of profound fluctuations over the last 2,000 years.

Biological data for the northern anchovy taken in Mexican waters are available in a database for the years 1978–1989 (Arenas et al. 1994), and in hard copy up to the present. During 1976–1991, landings in the Port of Ensenada, Mexico, accounted for about 80% of the total harvest. The ENSO event in 1982–1984 profoundly affected the growth rate of juvenile anchovies, whereas the effects on larval growth rate were minimal relative to the previous non–El Niño year (Butler 1989), showing the effects of an environmental perturbation.

Shoaling (Pitcher 1983) is a very prominent characteristic of exploited small pelagic species like sardines and anchovies, for which the small-magnitude event has a low level of importance or ecological relevance.

The Fisheries

Fishing on the west coast of Baja is diverse and intense, and is of economic importance to the region. The main catch is small pelagic fish. Other resources associated with the upwelling ecosystem are microphyte harvesting, sport fishing, and commercial fishing of pelagic fish, demersal fish, and sea-urchins. The two main problems of the fisheries are prediction of availability and multiple use demands coming from industries, social groups, and politicians.

The small pelagic species (sardines and anchovies) had been used mainly for reduction to fish meal. The northern anchovy (*Engraulis mordax*) ranges from the Gulf of California to British Columbia and has a northern, central, southern, and Gulf of California population (Vrooman et al. 1981). The central population is the largest of the four; it moves as a Mexico–U.S. transboundary resource, from 30° to 38°N latitude. This population, or subpopulation, and the southern stocks are the subject of this paper. Until recently the central population supported a significant reduction fishery in both the United States and Mexico. Landings reached a peak of 308,000 metric tons during the 1980–1981 season but then declined to lower levels (Jacobson et al. 1993). During the 1990–1993 season, landings for both countries were less than 12,000 metric tons a year. At the present time the anchovy fishery in Mexico and the United States is dormant. However, in recent years schools of Pacific sardine, once thought extinct, were found to be growing. A biomass stock survey took place in 1994 as a shared effort between Mexico and the United States.

Discussion

The Fish Behavior

Shoaling is a very prominent characteristic of exploited pelagic species like anchovies and sardines (Pitcher 1983). Several studies on shoaling fish have investigated what factors influence variations in shoal structure, abundance, and distribution by focusing on some particular sets of environmentally controlled stimuli, such as obstacles in the water and reaction to the noise of vessels. Other researchers have looked at abundance and distribution of shoals using different methods and objectives (Blaxter and Hunter 1982) and have concluded that the size, form, integrity, and structure of shoals are very diverse and influenced by many factors. Actual shoaling behaviors come from hundreds of selective recruitments adapting to the variability of the environment. Other studies have shown that shoaling behavior has a prominent ecological role. Both actual and theoretical studies have shown that shoaling is a successful survival strategy; these studies recognize that shoaling results from individuals making decisions to cooperate only when it pays to do so (similar rules seem to apply for several other social organizations where hierarchy is absent). Each individual tries to maximize its own success through shoaling, and the behavior contributes to the individual's fitness. These studies indicate that food and predation are key factors to understanding shoaling (Pitcher 1986).

Our studies on the diel vertical and offshore–inshore movements of anchovies off central California (Robinson and Arenas, submitted for publication) show that apparently some key factors apply. Results suggest that local differences in degree of shoaling and abundance are related to the amount of food available; anchovies were more congregated in zones with high concentrations of the euphausiids that they eat. A significant relationship was found among echo intensities of both organisms during the diel cycles at two locations in close proximity that showed no differences in physical characteristics. Anchovies move very rapidly, forming aggregates at depths where they may rest after meals and then migrating to the exposed neritic environment in a risk-taking search for food. Here they form patterns characterized by extended uncompacted shoals. Response to predators should also be reflected by the shoals.

Actually, anchovies, as well as many other fishes, are in constant search for food and are in constant risk of being preyed upon. In their shoal each individual relies on other individuals. Fish move inside the shoal motivated by hunger and fear of predation. Hungrier fish take the front, riskier position (Robinson and Arenas 1993). The shoal will move in search of food regardless of risk or availability. When food is scarce, the shoal may vary in size as well as in composition (Arenas and Robinson, submitted for publication). These are the reasons why Nonacs et al. (1994) concluded that modeling the behavior of the northern anchovy as a schooling predator better mimics the growth rates of wild fish. From the shoaling behavior of individual fish evolves the population-level phenomena that we need to predict to optimize management.

Toward a Sensible Management of Upwelling Ecosystems

Management implies different things to different people. It is generally taken to mean an effort to make the best use of the resource, in this case the ecosystem, including the fish. Management implies a degree of conservation of the resource, and again conservation means something different to each of us. Conservation may mean to harvest the fish at the maximum sustainable yield or to get maximum employment or maximum economic returns. When considering the management of upwelling ecosystems, the same problems apply. To get the best use of the resource, we propose to examine and assess the ecological processes through the disturbances they cause.

A key to understanding the ecology of upwelling ecosystems is to understand the ecology of small pelagic shoaling fishes. It is important to understand how individual fish behave and try to allocate resources and time among activities related to maintenance, growth, and reproduction in the face of a changing milieu and under the pressure of varying exploitation regimes.

Conceptually, fishing can be considered a perturbation event that propagates as a storm wave in the spatial–temporal matrix of the ecological field. In the areas and times where the fishing is actually taking place, it is a local catastrophe that influences other components of the biotic and abiotic environment. Over time, the event is perceived as an anomaly of mean values. For the fish, the magnitude of the effects of fishing depends not only on the local intensity and frequency of fishing, but also on its relative importance due to variable conditions depending on adaptation, evolution, and spatial flexibility. Small pelagic fishes usually live in shoals, within which an individual's survival is determined by the survival of the shoal.

Fisheries management of shoaling pelagic fishes is particularly complicated because their vulnerability to capture tends to increase as population size is depleted by fishing or as populations become more available to fisheries. In such conditions, heavily exploited fisheries are unstable, leading to stock collapse. In Mexico, conventional regulations on fishing effort have been the basis for the management of the depleted sardine stock, with the hope that it will recover fully. The best strategy for the management of this fishery should include early warnings of collapse either caused by overfishing or by historic variation such as those recorded in the sediments (Berger and Herguera 1992) or decadal changes of regime (Lluch-Belda et al. 1989). The challenge is the handling of collapse.

A collapse may proceed as follows: Fishing starts in "marginal environmental windows" (MEWs) as well as in the "optimal environmental windows" (OEWs) because relative population abundance is not proportional to the optimality of the environment. The basic idea is that shoals do not feed or reproduce in marginal environmental windows until the quality of optimal areas decreases due to the effects of increased populations (Arenas 1992). As fishing starts depleting the population, or MEW areas are reduced, the fish increase their relative abundance in OEW areas, the internal mixing of populations is reduced, and fishing and catch ability increases. There can also be pulses of contraction and expansion of the whole population, resulting in variations in biological characteristics such as size, growth rates, size distribution, and reproduction in response to environmental conditions. Paradoxically, a very well-known characteristic of shoaling species is the tendency within members of the same shoal toward a phenotypic, physiological, and behavioral homogeneity; this is a key factor for the success of shoaling benefits. Similarities between members facilitate efficient escape from predators and decrease competition for food (Pitcher 1986). Fishing of shoaling species splits natural aggregations, causing heterogeneity of the remaining fish. Consequently the adaptive ability to survive decreases. Populations under fishing pressure thus may be at a disadvantage with respect to oceanic and biotic fluctuations.

References

Arenas, P. 1992. Spatial Behavior of Fish and Fishermen: The Use of Habitat Selection and Optimal Foraging Theory in Fisheries. Southwest Fisheries Science Center, USA Adm. Rep. LJ-92-09.

Arenas, P., Arenas, V., Mangel, M., and Smith, P. 1990. Global Climate Change, the Mexico and United States Pacific Coast, and the Transboundary Fisheries Resources. UC-MEXUS Critical Issues Conference. San Diego, California.

Arenas, P., Barnes, J., and Jacobson, L. D. 1994. Fishery and Biological Data for Northern Anchovy Taken in Mexican Waters. Southwest Fisheries Science Center, USA Adm. Rep. LJ-94-03.

Arenas, V., and Robinson, C. (Submitted for publication.) Fish Face Fast in Groups.

Bakun, A. 1990. Global climate change and intensification of coastal ocean upwelling. Science 247:198–201.

Bakun, A. 1992. Global greenhouse effects, multidecadal wind trends, and potential impacts on coastal pelagic fish populations. ICES Mar. Sci. Symp. 195:316–325.

Bakun, A. 1994. Climate Change and Marine Populations: Interactions of Physical and Biological Dynamics. Workshop on the Scope, Significance, and Policy Implications of Global Change and the Marine Environment, 14–17 May, 1994, University of Rhode Island.

Berger, W. H., and Herguera, J. C. 1992. Reading the Sedimentary Record of the Ocean's Productivity. In P. G. Falkowski and A. D. Woodhead (eds.), Primary Productivity and Biogeochemical Cycles in the Sea. New York: Plenum Press, pp. 455–486.

Blaxter, J. H. S., and Hunter, J. R. 1982. The biology of the clupeoid fishes. Adv. Mar. Biol. 20:1–223.

Butler, J. L. 1989. Growth during the larval and juveniles stages of the northern anchovy, *Engraulis mordax*, in the California Current during 1980–1984. U.S. Fish. Bull. 87:645–652.

Glynn, P. W. (ed.). 1989. Global Ecological Consequences of the 1982–1983 El Niño–Southern Oscillation. Amsterdam, Holland: Elsevier.

Gomez, J., Arenas, V., and Robinson, C. 1994. Biomass and Structure Population of the Euphausiids on Seven Diel Cycles along the West Coast of Baja California, Mexico. CalCOFI Meeting, 1994.

Hickey, B. M. 1979. The California Current System—Hypotheses and facts. Prog. Oceanogr. 8:191–279.

Jacobson, L. D., Lo, C. H., and Barnes, S. T. 1993. A biomass-based assessment model for northern anchovies, *Engraulis mordax.* U.S. Fish. Bull. 52:711–724.

Lindem, T., and Houri, H. A. 1988. Hydro Acoustic Data Acquisition System HADAS. Oslo, Norway: Department of Physics, University of Oslo.

Lluch-Belda, D., Crawford, R. J., Kawasaki, T., MacCall, A., Parrish, R. H., Schwartlose, R. A., and Smith, P. E. 1989. Worldwide fluctuations of sardine and anchovy stocks: The regime problem. S. Afr. J. Mar. Sci. 8:195–205.

Longhurst, A. R., and Pauly, D. 1987. Ecology of Tropical Oceans. San Diego: Academic Press.

Margalef, R. 1978. What Is an Upwelling System? In R. Boje and M. Tomcak (eds.), Upwelling Regions. Berlin: Springer-Verlag, pp. 12–14.

Nonacs, P., Smith, P., Bouskila, A., and Lutberg, B. 1994. Modeling the behaviour of the northern anchovy, *Engraulis mordax*, as a schooling predator exploiting patchy prey. Deep-Sea Res. II 41:1, 147–169.

Pitcher, T. J. 1983. Heuristic definitions of fish shoaling behaviour. Anim. Behav. 31:611–612.

Pitcher, T. J. 1986. Predators and food are the keys to understanding fish shoals: A review of recent experiments. Naturaliste Can. 113:225–233.

Robinson, C., and Arenas, V. 1993. Responses to Food Rewards and Risk of Predation in Pelagic Schooling Fish. An. Inst. Ciencias del Mar, UNAM, Mexico.

Robinson, C., and Arenas, V. (Submitted for publication.) Diel vertical and offshore-inshore movements of anchovies off central Baja California.

Robinson, C., Gomez, J., and Arenas, V. (Submitted for publication.) On the structure of a scattering layer in Baja California using both net trawl samples and hydroacoustic devices.

Schwing, F. B., Husby, D. M., Garfield, N., and Tracy, D. 1991. Mesoscale oceanic response to wind events off Central California in spring 1989: CTD surveys and AVHRR imagery. CalCOFI Rep. 32:47–64.

Smith, P. E., Ohman, M. D., and Eber, L. E. 1989. Analysis of the patterns of distribution of zooplankton aggregations from an acoustic Doppler Current Profiler. Calif. Coop. Ocean. Invest. 30:88–103.

Vrooman, A., Paloma, P. A., and Zweifel, J. R. 1981. Electrophoretic, morphometric and meristic studies of subpopulations of northern anchovy, *Engraulis mordax.* Calif. Fish and Game 67:39–51.

17

Measurements of the Upper Ocean Structure Using Towed Profiling Systems

J. Aiken, R. Pollard, R. Williams,
G. Griffiths, and Ian Bellan

Abstract

One of the most cost-effective methods for spatially and temporally monitoring the environmental variables necessary for evaluating the status, or health, of a large marine ecosystem is by the use of towed instrumentation. This may take the form of measurements from research ships or from ships-of-opportunity. Two towed instruments that provide these capabilities are the Undulating Oceanographic Recorder/Aquashuttle and the SeaSoar. New data from these highly developed towed systems, collected from various oceans of the world by research and commercial vessels, are presented.

Addressing future strategies for biological oceanography, the National Academy of Sciences made the following recommendation: "As a complement to the development of a satellite system, a towed sensor is needed that can remotely sense in the upper layer of the ocean, where biological activity is most concentrated (i.e., top 100 meters approximately). This sensor should have the capability to measure a major feature of the physical structure (temperature), obtain a measure of phytoplankton (fluorescence), and derive estimates of major components of the herbivores and carnivores (multifrequency acoustics)" (NAS 1982).

All these capabilities are provided by the Undulating Oceanographic Recorder (UOR) (Aiken 1981a, 1985; Aiken and Bellan 1986a, 1990) and SeaSoar (Pollard 1986; Pollard and Pu 1985), which are high-technology towed profiling sensor systems ideally suited to the upper-ocean observation requirements of global programs such as the World Ocean Circulation Experiment (WOCE) and Joint Global Ocean Flux Study (JGOFS), and for monitoring large marine ecosystems (LMEs) (Sherman 1993, 1994). (Chelsea Instruments markets both the UOR, under the trade name Aquashuttle, and SeaSoar.) Batfish (Dessureault 1976; Herman and Dauphinee 1980) and UOR Mark 1 (Bruce and Aiken 1975; Aiken et al. 1977) were the forerunners of these highly developed towed systems. SeaSoar, carrying a precision CTD (conductivity, temperature, depth) sensor, has been developed to be the state of the art in towed-body technology, undulating to geostrophic depths (500 m) and optimized to provide precise measurements of upper-ocean physical structures with a spatial resolution capable of resolving the smallest geostrophic structures (10 km). Used in conjunction with a ship-mounted Acoustic Doppler Current Profiler (ADCP), SeaSoar has the ability to produce three-dimensional ocean surveys and provide data that address the needs of all the WOCE core objectives: global measurements, southern ocean studies, and gyre and upper-ocean dynamics.

In contrast to SeaSoar, the UOR Mark 2 was optimized to be a semiautonomous profiling sensor system for biological oceanography that is suitable for deployment from a high-speed ship-of-opportunity (merchant ship) (Williams et al. 1994; Williams and Lindley 1992; Lindley and Williams 1994). The UOR is similar to the Continuous Plankton Recorder (CPR), although the latter is a fixed-depth (10 m) mechanical plankton sampler, optionally fitted with electronic sensors and a data logger (Williams and Aiken 1990). The CPR has been used in a monthly all-year survey of the North Sea since 1931 (Hardy 1939) and of the North Atlantic Ocean since 1948 (Glover 1967). There is a western North Atlantic counterpart operated by the National Marine Fisheries Service, Narragansett, Rhode Island (USA), and new surveys have been started or planned for the study of LMEs in west Africa and the Yellow Sea.

Although originating from different scientific perspectives, both Aquashuttle and SeaSoar have a common operational philosophy and capability, namely, the ability to exploit, for scientific purposes, the time spent by research ships (or merchant ships, in the case of the Aquashuttle) on passage between stations or ports and to conduct special surveys of large ocean areas at or near full cruising speeds. In doing so, specific research objectives requiring high-resolution, quasi-synoptic measurements or regular seasonal/annual coverage, unobtainable by conventional research ship methods, can be targeted (Williams 1994). CPRs, augmented with self-logging electronic instrument systems and physical and bio-optical sensors, provide a useful platform for the measurement of the bio-optical properties of the surface mixed layer (Williams and Aiken 1990). This towed vehicle and instrumentation, along with the Aquashuttle, was recommended by the 1991 Cornell Workshop (Sherman and Laughlin 1992) as one of the instruments to provide data for monitoring the changing states of LMEs.

Technologically and scientifically, both Aquashuttle and SeaSoar are semiautonomous systems, with intermediate status between conventional oceanographic deployments from stationary research vessels and fully autonomous underwater vehicles (AUVs) such as Autosub (NERC 1992), still under development. In each case, the vehicle is attached by a towing cable to a mother ship and the data from the sensors are recorded in situ or communicated up the cable to a shipboard recorder or computer; in future these data could be relayed via satellite to a laboratory base ashore. Operationally, these systems have a range comparable with that projected for AUVs and can provide measurements of physical and biogeochemical processes and fluxes on basin scales. The Aquashuttle and SeaSoar can cover distances of 300 to 500 km per day from a research ship or up to 900 km per day from a high-speed merchant ship. These scales of measurement, surveyed quasi-synoptically, are complementary to the scales observed in daily satellite imagery of the oceans (wave height, altimetry, sea surface temperature, and water color). Both Aquashuttle and SeaSoar have the added dimension of vertical structure, with depth ranges from 5 to 100 m for the Aquashuttle at 5 to 6 m s^{-1} (10–12 knots) on unfaired towing cable, and 0 to 500 m for SeaSoar at around 4.5 m s^{-1} on a fully faired cable. These capabilities provide a unique synergy with satellite large area observations.

The development of SeaSoar and the Aquashuttle has focused on two main design criteria: maximum depth and undulation amplitude attainable at maximum towing speed, and vehicle stability at top speed. Although the diving forces generated by the vehicle are important, the maximum depths attainable by towed vehicles such as SeaSoar and Aquashuttle are limited principally by the drag of the towing cable, which is proportional to the cable diameter and the square of the towing speed. The weight of the vehicle and the weight of the towing cable assist the maximum target depth but hinder the attainment of the minimum depth target, particularly at low speeds. Thus, to maximize the depth performance it is essential to keep the cable diameter to a minimum (minimizing drag and weight) while operating within a safe working load (normally set at 20% of the breaking strength of the cable). Greater depths can be attained by reducing the cable drag by cable fairings, gaining two additional benefits: first, the load on the cable is less for a given maximum depth and speed condition, so greater operational depths are allowed within the same safe working load of the cable; second, fairings reduce cable "strumming" and vibration, thus extending the fatigue lifetime of the cable. The cable drag factor (dimensionless) for a bare cable is around 1.2, which can be reduced to 0.6 for the best "haired" fairings and to less than 0.1 for hydrodynamically designed fairings such as Flexnose (T.M. Fathom Inc.). Special faired cables are expensive and require special winches at considerable cost. However, the benefits in the depth performance more than justify this expense, extending the maximum depth performance for SeaSoar and Aquashuttle from less than 100 m on a bare cable to 500 m for SeaSoar on a fully faired cable at 4.5 m s^{-1}. The Aquashuttle has been towed only rarely on faired cables, achieving a notable increase in maximum depth; the projected maximum possible (estimated by computer simulation) is somewhat less than that attained by SeaSoar.

Vehicle stability is the other major consideration in the design of towed bodies. A study of the equations of motion shows that the "wingspan" of the diving planes is the major parameter responsible for lateral instabilities, whereas the surface area of the tail fin(s) is the major contributor to the forces restoring yaw and roll perturbations (Laitinen 1967). SeaSoar, with a wingspan of 1.25 m, was originally subject to roll instabilities, though an active roll-control rudder provides satisfactory stable towing performance over the operational range of 4 to 5 m s^{-1}. With the diving planes positioned at the center of towing forces and close to the center of gravity in water, SeaSoar maintains a near-horizontal pitch angle throughout its undulation cycle, with only small amplitude pitch fluctuations. The diving planes are actuated hydraulically, with power provided by an impeller-driven pump at the rear of the vehicle. The depth profile is servo-controlled by a surface electronic control unit via the conductor-cored towing cable, with the minimum and maximum depth targets and the undulation pitch adjustable by the operator at will.

In contrast, the Aquashuttle was designed for high-speed ship-of-opportunity (merchant ship) operations with a top speed in excess of 20 knots (10 m s^{-1}) by minimizing wingspan and maximizing tail fin surface area. The Aquashuttle develops its diving forces by the body itself acting as the diving plane (though fixed wings are fitted for extra lift at speeds less than 12 knots), and it has twin tail fins that double the total surface area. Because of this, the Aquashuttle changes its pitch angle as it undulates through the water column, from approximately 20° nose-down when deep to approximately 10° nose-up when shallow. The body pitch, and hence the depth undulation, is controlled by the tail plane, which is actuated by a microprocessor-controlled electromechanical servo system powered by an electrical alternator and impeller at the rear. The alternator frequency (impeller frequency) also provides a distance reference for the depth/distance program; the undulation pitch is adjustable from 800 m to 4,000 m. The Aquashuttle is normally towed on ordinary steel cable with the depth limits (minimum, maximum) and undulation pitch preprogrammed; the limits can be adjusted during the tow by RS232 link to the servo-mechanism via a conductor-cored cable.

Sensors for Aquashuttle and SeaSoar need to have many of the characteristics of autonomous sensors: small size, low power consumption, reliable rugged construction, and stable electronic sensing and output signals compatible with low-power logging or computer systems. The current suite of sensors carried by the Aquashuttle consists of CTD, chlorophyll fluorescence (Aiken 1981b), water transmission (SeaTech, 0.25-m path length, 660 nm), and multispectral irradiance sensors (Ed (λ), Eu (λ) at 412, 443, 490, 520, 554, and 632 nm, downwelling PAR and upwelling radiance at 683 nm). For measurements of irradiance, consideration needs to be given to the positional attitude of the platform and potential artifacts of the towing method (Aiken and Bellan 1986b, 1990). Additional sensors that have been accommodated on the UOR are as follows: a fluorometer for dissolved organic compounds/material (DOC/DOM); blue and green transmissometers; pH electrode; dissolved oxygen (DO) electrode; nitrite and nitrate analyzers (Williams et al. 1994); bioluminescence detector (Aiken and Kelly 1984); particle counter; optical particle counter (Herman 1988); pump and probe fluorometer and its derivative, the Fast Repetition Rate (FRR) fluorometer; and the continuous plankton sampler. Data are logged in situ on a solid-state logger or communicated up the towing cable for real-time display and processing by computer.

SeaSoar carries a Neil Brown Mark III CTD, a fluorometer or transmissometer, DO electrode, and a suite of irradiance sensors as used on the Aquashuttle. Data are communicated up the cable to a shipboard computer for logging and processing. With electronic miniaturization, many more sensors can be carried simultaneously; a "stretched SeaSoar" with optical plankton counter, acoustic biomass sensors, multispectral transmissometers, and chemical (nutrient) electrodes and analyzers in addition to the standard sensors is being considered. The basic technical specifications of the Aquashuttle and SeaSoar are given in Table 17-1; the towed systems are shown in Figure 17-1.

The ship-of-opportunity methods pioneered by the CPR survey and exploited by the Aquashuttle are equally pertinent today, with scientific issues focused on global problems, ocean–atmosphere interactions, and so forth. Ship-of-opportunity surveys can examine full seasonal changes and interannual variability on longer time scales and are ideally suited for monitoring LMEs (Williams 1994). This is in contrast to research ships, which are usually only available for periods of 30 to 40 days maximum. The Aquashuttle has been deployed in ship-of-opportunity surveys in the western English Channel (Aiken 1981a; Robinson et al. 1986), in the North Sea (Williams and Aiken 1990; Lindley and Williams 1994), and in the North Atlantic Ocean from Iceland to Portugal (1990–1991), coupled with semiautonomous shipboard measurements of the partial pressure of CO_2 (pCO_2); these measurements have been extended to other routes in the North Atlantic as of 1994.

Table 17-1. Technical specifications for UOR and SeaSoar

Item	UOR	SeaSoar
Dimension		
Length	0.99 m	2.0 m (impellor guard to bridle)
Tail height	0.5 m	1.0 m
Body width	0.38 m	0.4 m
Wingspan	0.65 m	1.25 m
Weight (air)	80 kg	150 kg (including hydraulic unit)
Towing specifications		
Speed	7.5 to 26 knots	7 to 10 knots
	3.8 to 13.4 m s^{-1}	3.6 to 5.1 m s^{-1}
Cable diameter	8 mm outer diameter	8 mm outer diameter
Cable length	300–700 m bare	700 m fully faired
Cable load	0.75 tonnes maximum	1.6 tonnes maximum
Depth range	0 to 100 m at 5.5 m s^{-1}	0 to 500 m at 4 m s^{-1}
	0 to 40 m at 10.0 m s^{-1}	
Undulation pitch	800 to 4000 m	2000 to 4000 m
Sensor suite		
Depth (m)	0 to 250 m; 1% resolution	0 to 500 m; 1% resolution
Temperature (°C)	Partial resistance; 0.05°C	(Neil Brown Mark III)
Conductivity (Sal. psu)	Anderra; 0.05 psu	(0.01°C/0.001 psu)
Chlorophyll (mg m^{-3})	Aiken (1981b); 0.01 mg m^{-3}	Chelsea 0.01 mg m^{-3}
Dissolved oxygen	Orbisphere	Beckman

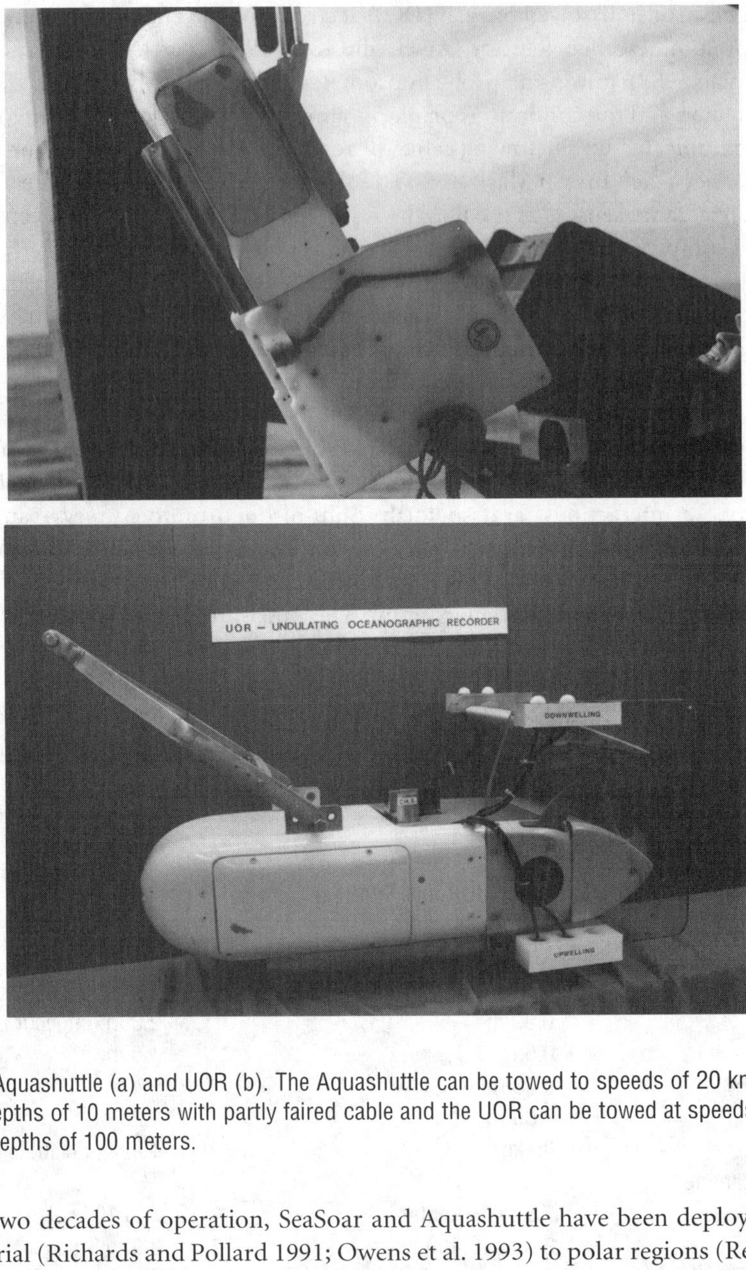

a)

b)

Figure 17-1. Aquashuttle (a) and UOR (b). The Aquashuttle can be towed to speeds of 20 knots and deployed to depths of 10 meters with partly faired cable and the UOR can be towed at speeds up to 10 knots and to depths of 100 meters.

In over two decades of operation, SeaSoar and Aquashuttle have been deployed worldwide from equatorial (Richards and Pollard 1991; Owens et al. 1993) to polar regions (Read et al. 1993; Trees et al. 1992) and in dynamic frontal and current systems such as Agulhas and the Antarctic circumpolar currents (Pollard et al. 1987; Read et al. 1993). Major surveys have included FASINEX (Pollard and Regier 1990, 1992); Vivaldi in the northeast Atlantic and Azores (39°N to 54°N and 12°W to 34°W) in April/May 1991; the JGOFS; the North Atlantic Bloom Experiment in 1989 (Weeks et al. 1993), 1990 (Savidge et al. 1992), and 1991 (Holligan et al. 1993); and the U.S. JGOFS equatorial Pacific surveys, 12°N to 12°S, 140°W, spring and fall 1992 (Aiken et al. 1993, and this paper). The Aquashuttle and SeaSoar have been used in many surveys of the North Sea (e.g., Owens et al. 1991; Williams and Lindley 1992) and the English Channel and southwest approaches (Holligan et al. 1983; Aiken and Kelly 1984; Fasham et al. 1983), and in fish egg and larval studies (Coombs et al. 1981, 1990, 1992). In this paper, new, unpublished data acquired using the Aquashuttle and SeaSoar are described to further demonstrate the applicability of these towed vehicles to surveying large marine ecosystems.

Northeast Atlantic Ocean

Research Vessel Deployments

Aquashuttle measurements of temperature structure, chlorophyll concentration, water reflectance, and water transmission (beam attenuation) were acquired in the Iceland Basin in June 1991 from a research ship, with contemporary AVHRR imagery of sea surface temperature (SST) and visible band reflectance. Extraordinarily high values of reflectance were measured arising from a bloom of *Emiliania huxleyi* and its detached coccoliths (calcite platelets around 1 μm in diameter). At concentrations up to 400,000 lith. ml^{-1}, 30% or 40% of the visible blue-green light (400–550 nm) is reflected back out of the water column (nearly 10 times the normal reflectance), giving the water a characteristic milky-blue color. Coccolithophores hold a special interest in the context of climate change because they affect the air–sea exchange of CO_2, reducing the capacity of the ocean to take up CO_2 by lowering the alkalinity of the water and emitting dimethylsulfide (DMS), which can form cloud condensation nuclei in the atmosphere, changing cloud albedo. Additionally, they influence ocean albedo by intense light scattering in the surface layer. The extreme heterogeneity of the area is illustrated by the NOAA-11 visible band image (atmospherically corrected) for 20 June, shown in Figure 17-2; the transect of the UOR across the bloom from 60°N 20°W to 63°N 20°W on the same day is marked in the figure. Nine successive days of imagery (from 15 to 23 June) showed the dynamic complexity of the water mass occupied by the

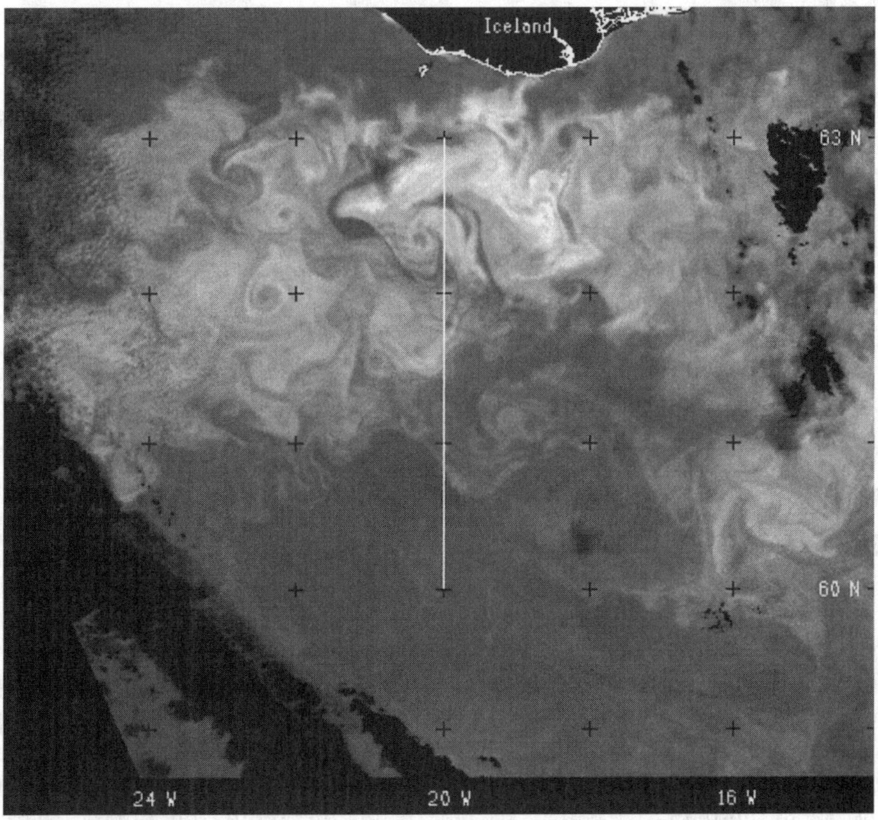

Figure 17-2. A VHRR atmospherically corrected visible band image of the Iceland Basin for 20 June 1991, marked with the track of the UOR from 60° to 63°N, 20°W.

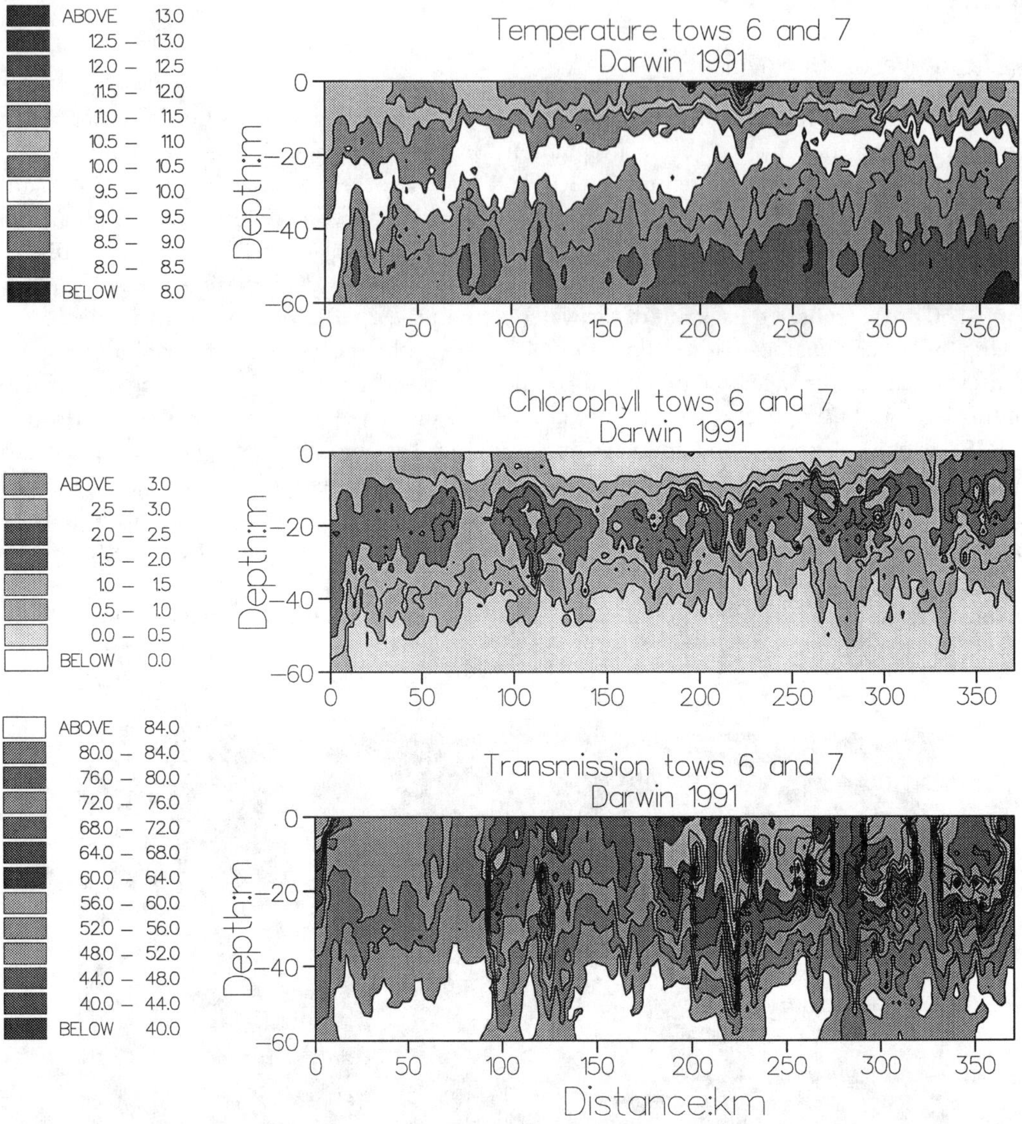

Figure 17-3. UOR vertical sections of temperature, chlorophyll concentration and beam attenuation transmission coefficient (c660) from 60° to 63°N, 20°W from 0500 GMT to 2400 GMT on 20 June 1991.

coccolithophores and the variability along its frontal boundaries. Active entrainment of water from the Iceland Shelf to the north and North Atlantic water to the south can be observed from the imagery, although the frontal boundaries remain sharply defined. These observations show that the coccolithophores occupied a discrete water mass with specific biological and environmental characteristics that favored the bloom production.

The Aquashuttle measurements (see Figure 17-3) showed that the bottom water at 60 m, under the bloom (north of 61°30′N), was colder than farther south (south of 61°), whereas the surface water (0–5 m) was warmer, presumably as a result of the dissipation of the solar radiant energy in a shallow surface layer by multiple scattering with the coccoliths.

The generally uniform beam attenuation coefficient, c660, from the surface to a depth of 20 m was confirmation of the existence of a homogeneous surface layer of biogenic particles. The surface values of c660 from 1.2 to 1.4 m^{-1} in the southern zone were indicative of phytoplankton chlorophyll as the main biogenic constituent of the water; values over 1.4 m^{-1} (and there were a few patches in this southern zone) can be interpreted as indicating moderate concentrations of coccolithophores and coccoliths. Immediately to the north of the frontal zone (70–75 km from the start), the water transmission dropped sharply, with c660 increasing to over 1.6 m^{-1} and substantial patches of water with c660 $>$ 3.0 m^{-1} toward the northern end. These constituted some of the highest open-ocean water beam attenuation coefficients ever recorded.

The uniformly high attenuation coefficient for blue-green light (approximately 0.25 to 0.3 m^{-1}, 400–550 nm) and high reflectance (up to 30% at 440 nm) produced characteristically milky-blue-colored water. The color spectra were nearly uniform from 410 to 520 nm, that is, "white" but with a higher than normal fraction of blue light (Figure 17-4 shows nearly equal values of R443, R490, and R520). Optical depths of only 2 to 3 m were typical. UOR's reflectances (R443, R490, R520), surface layer transmission (532 nm), beam attenuation coefficient c (532), and AVHRR reflectance are shown in Figure 17-4 for the tow from 61.6°N to 63° N, 20°W, 1700 to 2400 GMT on 20 June 1991. Note the high correlation between all the measured quantities (e.g., c (532) = 0.268 R(AVHRR) + 0.952; $R^2 = 0.86$); for further details see Holligan et al. (1993).

Ship-of-Opportunity Deployments

From 1990 to 1992 the Aquashuttle was deployed from the commercial vessel M.V. *Selfoss* (EIMSKIP, The Icelandic Steamship Co. Ltd.) between Reykjavik, Iceland, and Portugal—a distance of over 2,800 km. The ship's speed ranged from 7 to 11 knots, and the Aquashuttle was deployed on long-distance tows undulating to a depth of approximately 40 m. The primary objective of the program was to establish ship-of-opportunity protocols for towing the Aquashuttle from a commercial vessel. Two tows of 350-km and 375-km duration were done between 1 and 3 July 1991, showing the chlorophyll and temperature structure in the upper 40 m. Deployments were carried out on 8-mm cable from the ship's mooring winch through a simple davit over the stern starboard quarter. Over the whole 735-km tow the stability of the mixed layer in the northeast Atlantic appeared to be only vertically perturbed by internal waves. Chlorophyll values were low, but discrete elevated patches occurred throughout Tow 1, close to the Icelandic shelf, and at the edge of Hatton Bank at the end of Tow 1. These tows demonstrated the feasibility of the use of this towed vehicle from ships-of-opportunity, although initially the ships need to be manned, and the equipment operated, by a trained technician to establish protocols with the ship's officers and crew.

North Sea Ship-of-Opportunity Deployments

A full ship-of-opportunity program was initiated in March 1989 in the North Sea using the M/V *Astrea* (NOR-CARGO Ltd.) operating from Grimsby, sailing up the east coast of the United Kingdom to Aberdeen and then on to Stavanger/Bergen (Norway) before returning to Grimsby within the week. The ship was a Ro/Ro ferry of 1,599 gross tonnage with an operating speed between 14 and 17 knots. Because of the high operating speed and the necessity of launching and recovering at these speeds, a 10-mm-diameter steel cable of 140-m length was used and an SWL of 1.2 tons for towing. The ship had been used regularly since 1979 to tow the CPR and was equipped with a simple davit on the starboard quarter. The Aquashuttle was deployed using the mooring capstan through a number of lead blocks through the davit over the stern, and could be controlled by one person. On recovery at these high towing speeds, a small canvas sea anchor

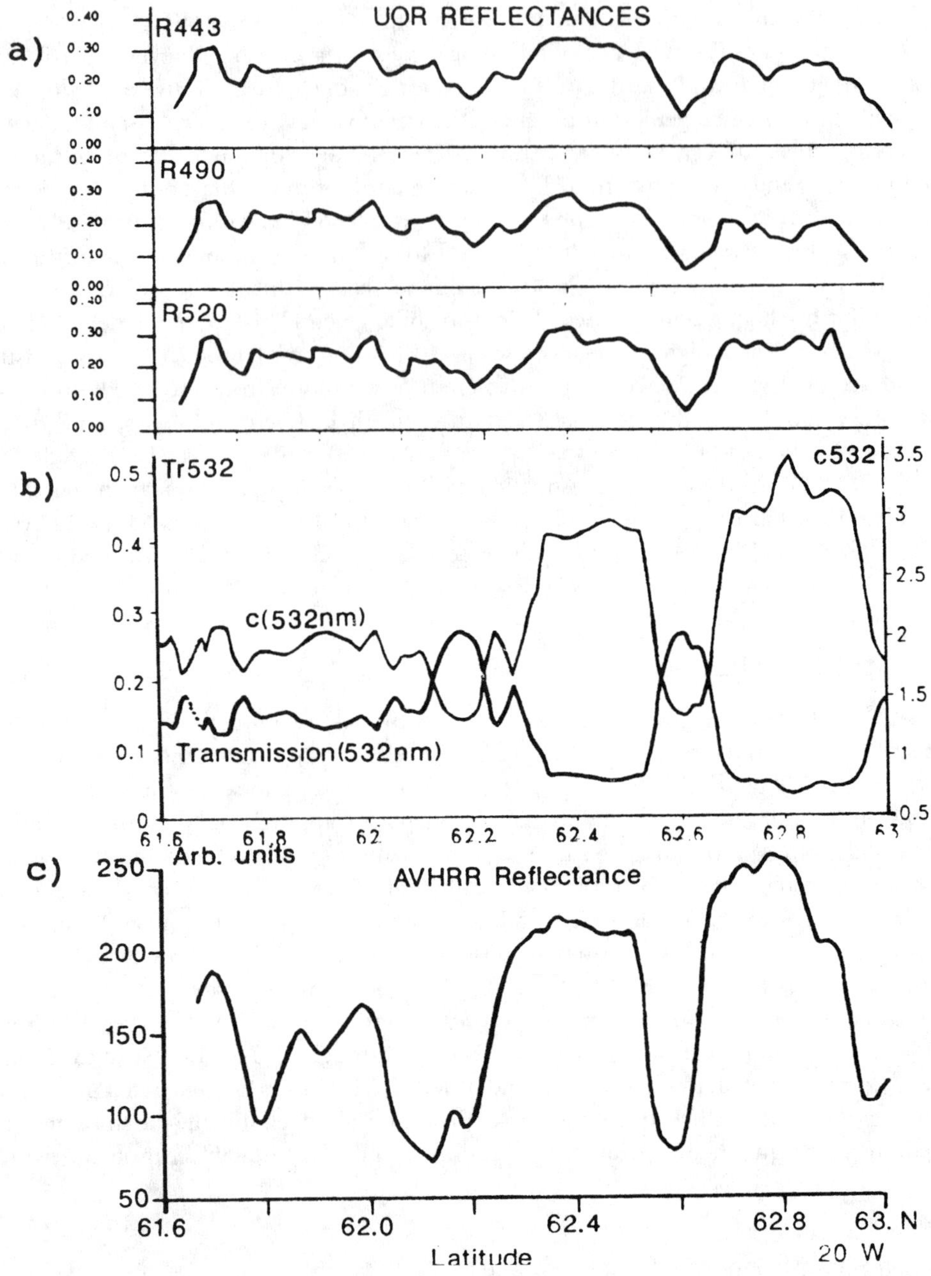

Figure 17-4. UOR bio-optical measurement and satellite reflectances on 20 June 1991 for transects from 61.6°N to 63.0°N, 20°W (1700 GMT to 2400 GMT). (a) UOR reflectances (R520, R490, R443). (b) Surface layer transmission (532 nm) and beam attenuation coefficient c (532 nm). (c) AVHRR reflectance (580 to 680 nm).

attached to 10 m of rope and shackle was slipped down the towing cable when the vehicle was just below the surface. This held the Aquashuttle very stable during recovery up the ship's side. Between March 1989 and August 1991, 29 separate high-speed deployments and recoveries were carried out with little damage and no loss of equipment. The tracks towed in May 1991 are shown

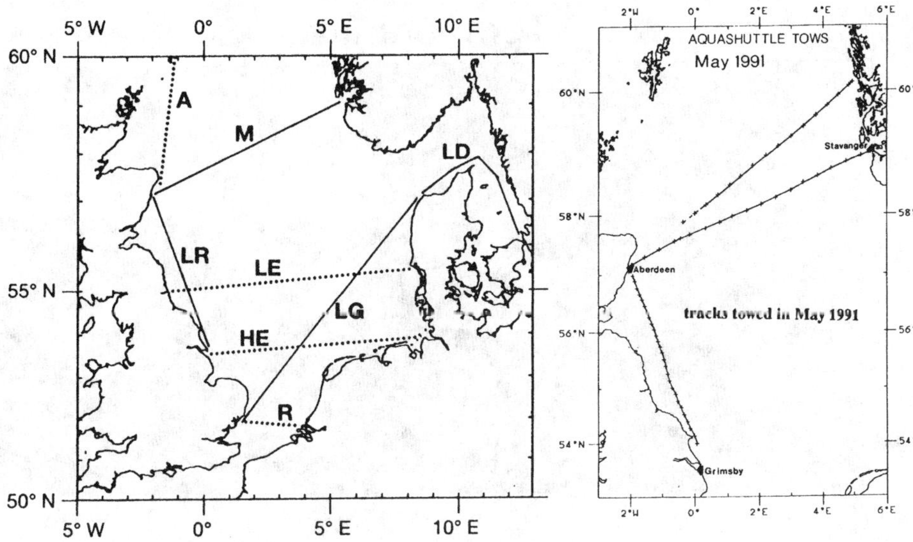

Figure 17-5. Standard routes of the CPR survey of the North Sea in 1989. Continuous lines indicate routes on which instrumented CPRs were deployed (i.e., environmental variables were measured on the same tow). Dotted lines indicate routes where CPRs sampled plankton only. Aquashuttles were towed on the LR and M routes.

in Figure 17-5. The Aquashuttles used in this program were the original Mark 1s. The basic sensor suite and undulation profiles are given in Williams and Lindley (1992).

This ship-of-opportunity program was undertaken because of the growing concern that the blooms of phytoplankton, and especially dinoflagellates, are becoming more prevalent in the North Sea. These blooms are primarily a coastal phenomenon and are in part thought to be caused by anthropogenic inputs from the land, although specific hydrological and meteorological conditions contribute to and are necessary for their formation. The cysts of one particular species, *Alexandrium tamarense*, are known to overwinter in great numbers in the sediments along the east coast and are brought to the surface in spring through turbulent mixing of the water column. This species is responsible for paralytic shellfish poisoning (PSP) in humans when shellfish contaminated by these microorganisms are consumed. If the environmental conditions are favorable, *Alexandrium* can reach bloom proportions, as it did in 1990. It was to research and monitor the basic environmental parameters along the east coast of the United Kingdom that the 3-year Aquashuttle program was instigated.

Examples of the data for a tow taken between Aberdeen and Stavanger (9–10 May 1991) are shown in Figure 17-6. The parameters reported are temperature, chlorophyll, salinity, and turbidity (15-cm path length transmissometer with pulsed infrared LED light source and photodiode detector, Partech Electronics Ltd.). (The contour grids of these plots are based on grids of 200 × 90 points.) The changes in temperature and, more so, in salinity as the Aquashuttle crosses from the shoal waters of the northern North Sea into the northern flowing coastal current, over the deeper water of the Norwegian Rinne, are clearly seen. At the shelf break there is a higher chlorophyll density marked by concentrations of the dinoflagellate *Ceratium* spp. The large volume of dinoflagellate cells gave a strong turbidity signal. A further example is given in a tow taken during 6–7 August 1991 between Grimsby and Aberdeen (Figure 17-7), which was when the strongest thermal stratification was measured along this route. Subsurface chlorophyll maxima associated with this strong thermocline are clearly shown.

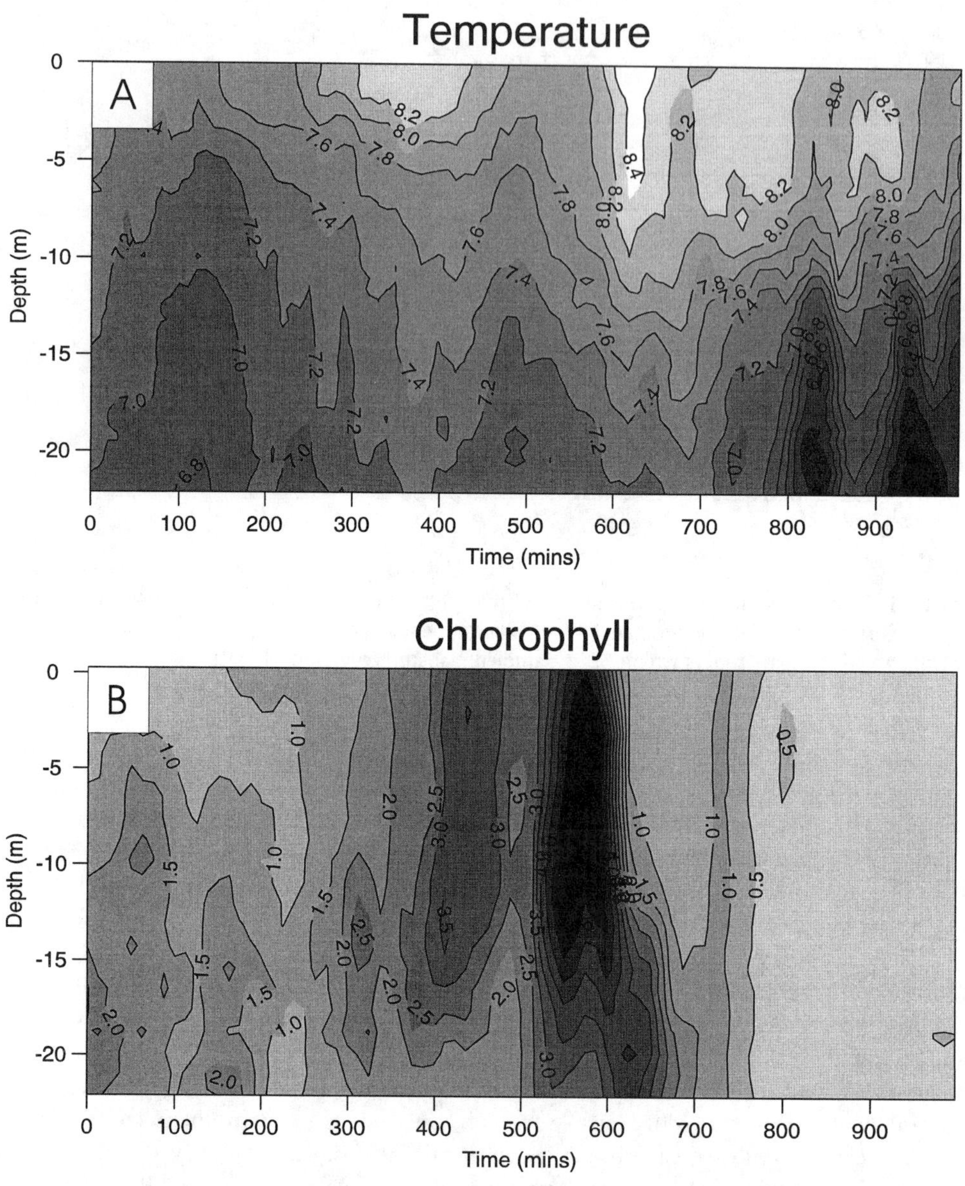

Figure 17-6. Ship-of-opportunity UOR/Aquashuttle tow on M/V ASTREA form Aberdeen to Stravagner (Norway) 9–10 May 1991, showing (a) Temperature, (b) Chlorophyll-a, (c) Salinity, and (d) Turbidity in the upper 20 m.

Equatorial Pacific Ocean, 1992

As reported by Richards and Pollard (1991), SeaSoar had been used in transequatorial surveys at 160°E in 1988. In 1992, SeaSoar and Aquashuttle measurements contributed greatly to the COARE survey and the U.S. JGOFS equatorial Pacific surveys from Hawaii to Tahiti. The Aquashuttle was towed between stations from 12°N to 12°S along 140°W in spring (January–March) and fall (August–September) 1992, providing a description of the vertical structure (surface to 100 m) of the key physical and bio-optical properties: temperature, salinity, chlorophyll concentration, water transmission (beam attenuation), PAR, dissolved oxygen concentra-

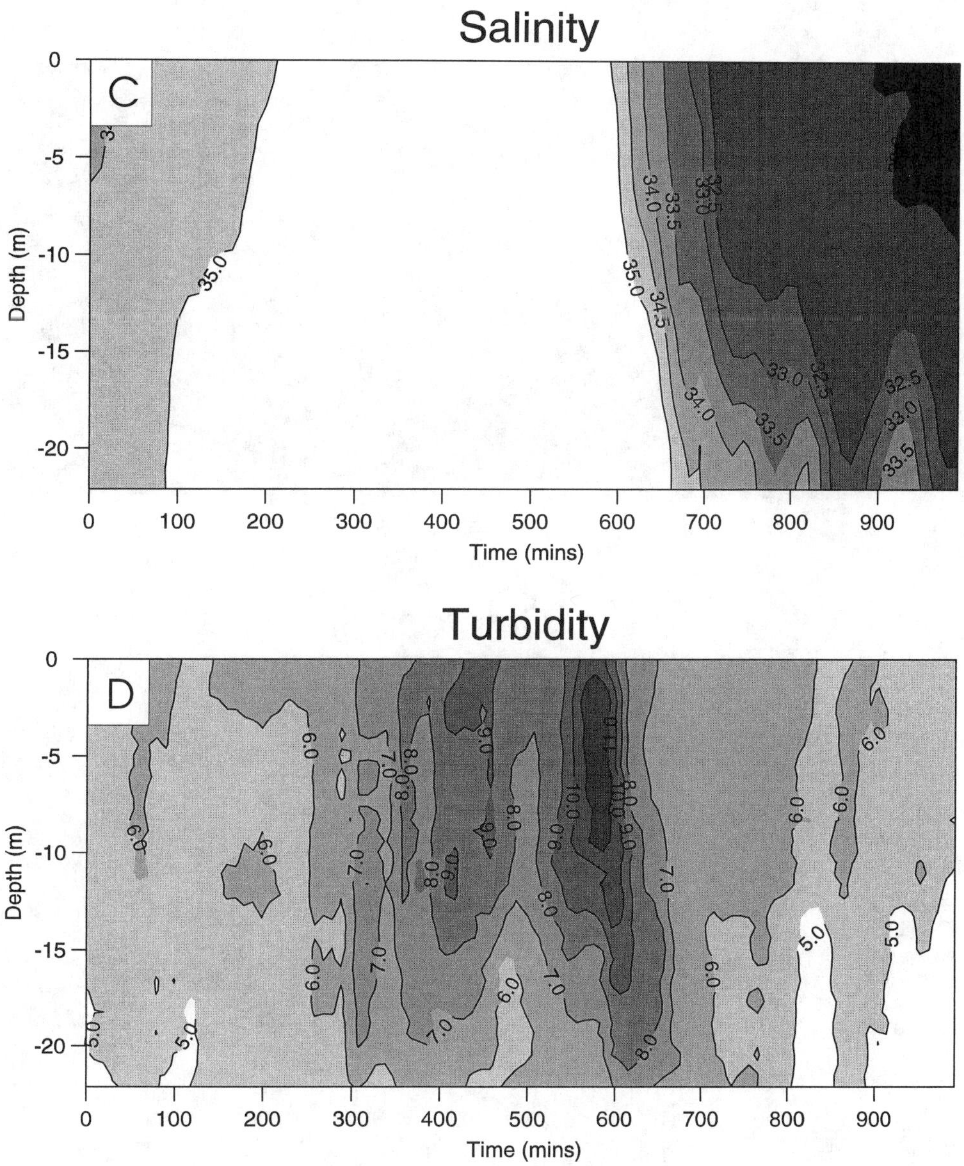

Figure 17-6. (*Continued*)

tion, water attenuation, and reflectance at 412, 443, 490, 510, 555, and 632 nm. The data showed the characteristic water structures, convergences, and divergences associated with the North and South Equatorial Currents, Equatorial Counter Current, and Equatorial Undercurrent.

The spring survey found the El Niño–Southern Oscillation (ENSO) in full spate, with positive surface temperature anomalies of 1.5° to 3°C above the climatological means from 7°N to 12°S. For most of this region, phytoplankton biomass (chlorophyll concentration) was concentrated in the surface layer above the permanent thermocline, at values of 0.3 to 0.4 mg m^{-3}. This was far in excess of the values of 0.05 mg m^{-3} that could be considered typical of the permanently stratified oligotrophic ocean. Only at the extreme ends of the transect, 12° to 9°N and 9° to 12°S, were what would be considered conventional structures observed, with a subsurface chlorophyll maximum up to 0.3 to 0.4 mg m^{-3} in the thermocline, and surface values of 0.05 mg m^{-3}.

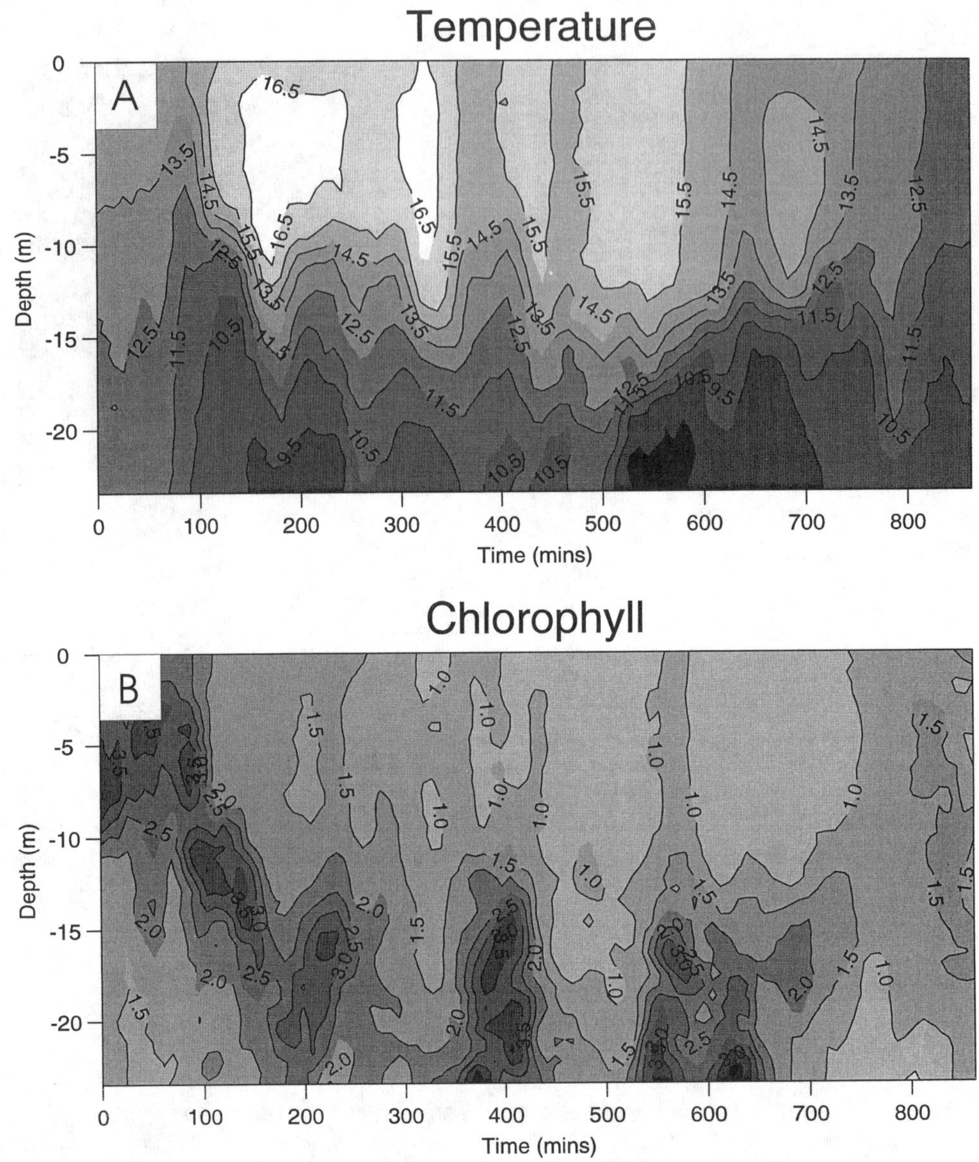

Figure 17-7. Ship-of-opportunity UOR/Aquashuttle tow on M/V ASTREA along the East coast of the United Kingdom from Grimsby to Aberdeen, 6–7 August 1991, showing (a) Temperature, (b) Chlorophyll-a structure in the thermocline.

In August, the section across the equatorial front, between the North Equatorial Current and the Equatorial Counter Current, showed a spectacular phytoplankton accumulation of *Rhizosolenia castracanei*, reaching surface values of 10 to 30 mg m^{-3} in patches—conditions not generally expected in an oligotrophic ocean system.

The Aquashuttle tows (see Figure 17-8), nominally along 140°W, encountered the front at the end of Tow 5 (around 2°N) and the start of Tow 6 (2°N), though the vessel had drifted west by about 140 km while "on station" for 2.5 days at 2°N. A second prominent high-chlorophyll feature was observed midway along Tow 6 at around 1.5°N. The initial front was marked by a 3°C change of surface temperature from >27°C to approximately 24°C in a distance of roughly 25 km. In the

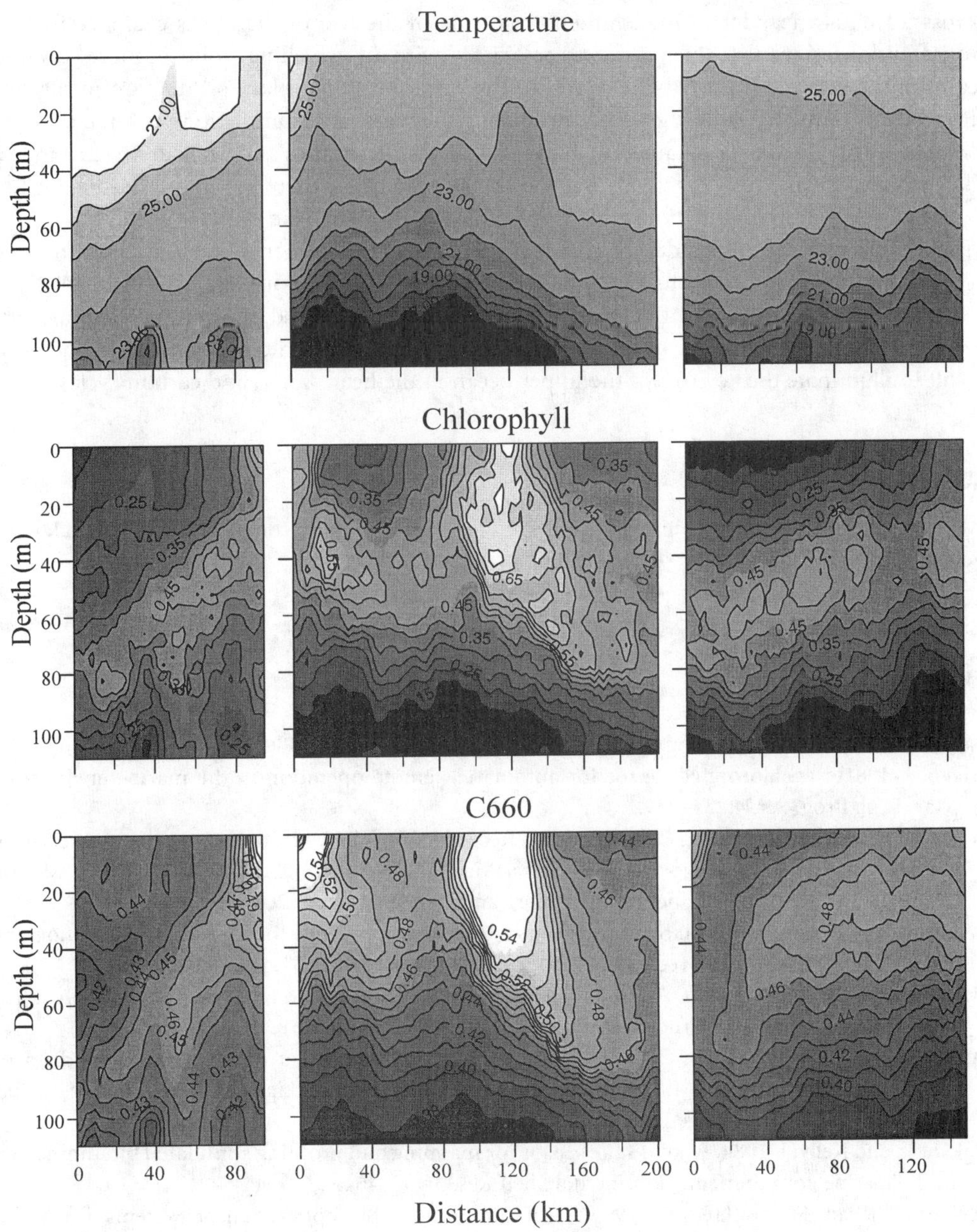

Figure 17-8. UOR vertical sections of Temperature, Chlorophyll concentration and Beam attenuation coefficient (c660) for tows 5, 6, and 7, across the equatorial front, in the Equatorial Pacific Ocean, August 1992.

main patch, measured at the start of Tow 6, chlorophyll concentrations over 0.5 mg m^{-3} were observed to a depth of 50 m. The front at this point exhibited a strong surface convergence, with strong currents to the southwest on the north (warm) side and to the northwest on the south (cold) side. The second feature at around 1.5°N had little surface temperature expression (<0.5°C) but the chlorophyll patch was even larger, with concentrations up to 1 mg m^{-3} to a depth of 50 m extending for 50 to 60 km. It may be that the observation was a tangential section across a frontal meander in the southern cold side of the front. There was evidence from the Aquashuttle sections that this patch was associated with an upwelling, wind-induced, divergent feature bringing cold, nutrient-rich water to the surface that fed plant production in the stable surface layer immediately to the south, producing the massive accumulations observed.

Surveys of recent data obtained with towed UOR/Aquashuttles and SeaSoar systems indicate that it is only in the last few years that the spatial resolution that they attain has begun to be exploited. Used in combination with ship-mounted profilers, the systems yield even greater insights. Over the next decade we anticipate that other instruments, such as high-frequency acoustic profilers, autonomous nutrient samplers, and particle counters (Herman 1988), will reach the stage of development where they can be installed in towed or, eventually, autonomous vehicles. New insights into the physical, biological, and chemical structure of the upper ocean will result to illuminate the key role of the upper ocean in the heat, water, and carbon cycles.

ACKNOWLEDGMENTS

We would like to thank all the captains, officers, crews, and shipping companies who have allowed us onboard and supported us in the deployment of these towed vehicles.

References

Aiken, J. 1981a. The Undulating Oceanographic Recorder Mark 2. J. Plankton Res. 3:551–560.

Aiken, J. 1981b. A chlorophyll sensor for automatic, remote operation in the marine environment. Mar. Ecol. Prog. Ser. 4:235–239.

Aiken, J. 1985. The Undulating Oceanographic Recorder Mark 2. A Multirole Oceanographic Sampler for Mapping and Modelling the Biophysical Marine Environment. In A. Zirino (ed.), Mapping Strategies in Chemical Oceanography. American Chemical Society 209, pp. 315–332.

Aiken, J., and Bellan, I. 1986a. Synoptic Optical Oceanography with the Undulating Oceanographic Recorder. In M. A. Blizard (ed.), Ocean Optics VIII. Proc. SPIE 637, pp. 221–229.

Aiken, J., and Bellan, I. 1986b. A Simple Hemispherical, Logarithmic Light Sensor. In M. A. Blizard (ed.), Ocean Optics VIII. Proc. SPIE 637, pp. 211–216.

Aiken, J., and Bellan, I. 1990. Optical Oceanography: An Assessment of a Towed Method. In P. J. Herring, A. K. Campbell, M. Whitfield, and O. Maddock (eds.), Light and Life in the Sea. Cambridge, England: Cambridge University Press, pp. 39–57.

Aiken, J., and Kelly, J. 1984. A solid state sensor for mapping and profiling stimulated bioluminescence in the marine environment. Continental Shelf Res. 3:455–464.

Aiken, J., Bellan, I., and Trees, C. 1993. Semi-autonomous Bio-optical Sensor Systems. Proceedings ABOOS Symposium. In press.

Aiken, J., Bruce, R. H., and Lindley, J. A. 1977. Ecological investigations with the Undulating Oceanographic Recorder: The hydrography and plankton of the waters adjacent to the Orkney and Shetland Islands. Marine Biol. 39:77–91.

Bruce, R. H., and Aiken, J. 1975. The Undulating Oceanographic Recorder—A new instrument system for sampling plankton and recording physical variables in the euphotic zone from a ship underway. Mar. Biol. 32:85–97.

Coombs, S. H., Aiken, J., and Griffin, T. D. 1990. Aetiology of mackerel spawning west of the British Isles. Meeresforschung 33:52–75.

Coombs, S .H., Aiken, J., and Lockwood, S. J. 1981. A Survey of Mackerel Spawning and Environmental Conditions in the Celtic Sea in May, 1981. ICES CM 1981/H:32 (mimeo).

Coombs, S. H., Nichols, J. H., Conway, D. V. P., Milligan, S., and Halliday, N. C. 1992. Food availability for sprat larvae in the Irish Sea. J. Mar. Biol. Assoc. U.K. 72:821–834.

Dessureault, J.-G. 1976. "Batfish". A depth controllable towed body for collecting oceanographic data. Ocean. Engineering 3:99–111.

Fasham, M. J. R., Holligan, P. M., and Pugh, P. R. 1983. The spatial and temporal development of the spring phytoplankton bloom in the Celtic Sea, April 1979. Progr. Oceanogr. 12:87–145.

Fasham, M. J. R., Platt, T., Irwin, B., and Jones, K. 1985. Factors affecting the spatial pattern of the deep chlorophyll maximum in the region of the Azores front. Progr. Oceanogr. 14:129–165.

Glover, R. S. 1967. The Continuous Plankton Recorder survey of the North Atlantic. Symp. Zool. Soc. Lond. 19:189–210.

Hardy, A. C. 1939. Ecological investigations with the Continuous Plankton Recorder. Object, plan and methods. Hull Bull. Mar. Ecol. 1:1–57.

Herman, A. W. 1988. Simultaneous measurement of zooplankton and light attenuance with a new optical plankton counter. Continental Shelf Res. 8:205–221.

Herman, A. W., and Dauphinee, T. M. 1980. Continuous and rapid profiling of zooplankton with an electronic counter mounted on a "Batfish" vehicle. Deep-Sea Res. 27:79–96.

Holligan, P. M., Fernandez, E., Aiken, J., Balch, W. M., Boyd, P., Burkill, P. H., Finch, M., Groom, S. B., Malin, G., Muller, K., Purdie, D. A., Robinson, C., Trees, C. C., Turner, S. M., van der Wal, P. 1993. A biogeochemical study of the coccolithophore *Emiliania huxleyi*, in the North Atlantic. Global Biogeochemical Cycles. In press.

Holligan, P. M., Viollier, M., Dupouy, C., and Aiken, J. 1983. Satellite studies on the distribution of chlorophyll and dinoflagellate blooms in the western English Channel. Continental Shelf Res. 2:81–96.

Laitinen, P. O. 1967. Cable-Towed Underwater Body Design. Report 1452. San Diego: U.S. Naval Electronics Laboratory.

Lindley, J. A., and Williams, R. 1994. Relating plankton assemblages to environmental variables using instruments towed by ships-of-opportunity. Mar. Ecol. Prog. Ser. 107:245–262.

National Academy of Sciences (NAS). 1982. Two Special Issues in Satellite Oceanography. Ocean Dynamics and Biological Oceanography. Washington, DC: National Academy Press.

Natural Environment Research Council (NERC). 1992. The Natural Environment Research Council Report 1991–1992. NERC Publication Series.

Owens, N. J. P., Burkill, P. H., Mantoura, R. F. C., Woodward, E. M.S., Bellan, I., Aiken, J., Howland, R. J. M., and Llewellyn, C. A. 1993. Size fractionated primary production and nitrogen assimilation in the North western Indian Ocean. Deep-Sea Res. 10:697–709.

Owens, N. J. P., Woodward, E. M. S., Aiken, J., Bellan, I., and Rees, A. P. 1991. Primary production and nitrogen assimilation in the North Sea during July 1987. Netherlands J. Sea Res. 25:143–154.

Pollard, R. T. 1986. Frontal surveys with a towed profiling conductivity/temperature/depth measurement package (SeaSoar). Nature 323:433–435.

Pollard, R. T., and Pu, S. 1985. Structure and ventilation of the upper Atlantic Ocean northeast of the Azores. Prog. Oceanogr. 14:443–462.

Pollard, R. T., and Regier, L. 1990. Large variations in potential vorticity at small spatial scales in the upper ocean. Nature 348:227–229.

Pollard, R. T., and Regier, L. A. 1992. Vorticity and vertical circulation at an ocean front. J. Phys. Oceanogr. 22:609–625.

Pollard, R. T., Read, J. F., and Smithers, J. 1987. CTD and SeaSoar Data from the Agulhas Retroflection Zone. Report 245, Institute of Oceanographic Sciences Deacon Laboratory.

Read, J. F., Allen, J. T., Machin, P., Miller, G. W. J., Morrison, A. I., Pollard, R. T., and Taylor, T. G. 1993. SeaSoar Data Collected on RRS Discovery Cruise 198 (Sterna) across Drake Passage and in the Bellingshausen Sea. Internal Document No. 320, Institute of Oceanographic Sciences Deacon Laboratory.

Richards, K. J., and Pollard, R. T. 1991. Structure of the upper ocean in the western equatorial Pacific. Nature 350(6313):48–50.

Robinson, G. A., Aiken, J., and Hunt, H. G. 1986. Synoptic surveys of the western English Channel. The relationships between plankton and hydrography. J. Mar. Biol. Assoc. U.K. 66:201–218.

Savidge, G., Turner, D. R., Burkill, P. H., Watson, A. J., Angel, M. V., Pingree, R. D., Leach, H., and Richards, K. J. 1992. The BOFS 1990 Spring Bloom Experiment: Temporal evolution and spatial variability of the hydrographic field. Prog. Oceans 29:235–281.

Sherman, K. 1993. Large Marine Ecosystems as Global Units for Marine Resource Management: An Ecological Perspective. In K. Sherman, L. Alexander, and B. Gold (eds.), Large Marine Ecosystems V: Stress, Mitigation and Sustainability. Washington, DC: AAAS Press.

Sherman, K. 1994. Sustainability, biomass yields, and health of coastal ecosystems: An ecological perspective. Mar. Ecol. Prog. Ser. 112:277–301.

Sherman, K., and Laughlin, T. (eds.). 1992. Large Marine Ecosystems Monitoring Workshop Report. NOAA Technical Memo NMFS-F/NEC-93. Washington, DC: U.S. Department of Commerce.

Trees, C. C., Aiken, J., Hirche, H.-J., and Groom, S. B. 1992. Bio-optical variability across the Arctic Front as determined by shipboard and satellite observations. Polar Biol. 12:455–461.

Weeks, A. R., Fasham, M. J. R., Aiken, J., Harbour, D. S., Read, J. F., and Bellan, I. 1993. The spatial and temporal development of the spring bloom during the JGOFS North Atlantic Bloom Experiment, 1989. J. Mar. Biol. Assoc. U.K. 73:253–282.

Williams, R. 1994. Evaluation of New Techniques for Monitoring and Assessing the Health of Large Marine Ecosystems. NATO Advanced Resarch Workshop on Evaluating and Monitoring the Health of Large-Scale Ecosystems, 10–15 October 1993, Montebello, Canada. Springer-Verlag.

Williams, R., and Aiken, J. 1990. Optical Measurements from Underwater Towed Vehicles Deployed from Ships-of-Opportunity in the North Sea. In Proceedings of the Third International Congress on Optical Science and Engineering. Proc. SPIE 1269, pp. 186–194.

Williams, R., and Lindley, J. A. 1992. Aquashuttle deployments in the North Sea. SeaSoar News 3:8–11.

Williams, R., Lindley, J. A., and Bloomer, N. 1994. Nutrient measurements using Aquashuttle. Undulations (CI Ltd. Newsletter) 4:2–3.

18

High-Resolution Biogeochemical Monitoring for Assessing Environmental and Ecological Changes in the Marginal Seas Using Ferry Boats

A. Harashima, R. Tsuda, Y. Tanaka, T. Kimoto, and T. Hagiwara

Abstract

An automated biogeochemical monitoring system using the seawater intake of a ferry boat was developed and deployed to investigate the temporal and spatial variations of marine environmental parameters (dissolved nutrients, fluorescence, temperature, salinity, and pH). Semicontinuous data recording (every 10 seconds during 4 transect cruises per week between Pusan, Korea, and Kobe, Japan) and automated bottle sampling with filtration (24 cruises per year) were performed using the seawater taken continuously from the engine cooling system intake at 6 m depth. Nearly 2 years of operation has clarified the temporal variability on a daily to seasonal time scale. These data will be used to assess the decadal changes in the coastal and marginal seas due to anthropogenic influences. Furthermore, several attended research cruises were performed to measure the distribution of pCO_2, concentration and species compositions of phytoplankton and zooplankton, and phytoplankton particle size spectra by laser technique. In 1993, the Kobe–Pusan Ferry Line was discontinued and we initiated two other ferry monitoring programs in a similar way (Osaka–Seto Inland Sea–Beppu line and the Osaka–Okinawa line) to determine the zonation structure, from estuary to semi-enclosed sea to continental shelf slope to the Kuroshio Subtropical Gyre. The ferry-based monitoring was found to be feasible and applicable to a variety of marine environmental issues, particularly in the marginal seas. Some technical aspects are reported in this paper.

Marginal seas (estuaries, bays, and continental shelf seas) are rich ecosystems that are maintained by the normal cycling of elements such as carbon, nitrogen, and phosphorus. These cycles have been perturbed from preindustrial or preagricultural conditions by anthropogenic activities. The anthropogenic perturbations are, however, far more difficult to assess compared with their counterparts in the atmosphere because they must be extracted from many kinds of natural variations and because the ocean is more heterogeneous than the atmosphere. One urgent need is to set up monitoring schemes and obtain time series of various parameters to evaluate environmental changes. These schemes should provide data for policy makers through the creation and validation of models for the assessment and prediction of environmental change. It is crucial to develop platforms suitable for gathering long-term time-series data with spatial resolution beyond that achievable by typical research vessels.

Time- and space-dependent data are acquired in either of two conceptual ways, Lagrangian or Eulerian observations. The former may be acquired by operating drifting buoys, and the latter from a set of moored stations. Either Lagrangian or Eulerian observations can provide enough data to outline the models in theory, provided the number of Lagrangian buoys are sufficient to cover most of the different water masses in the basin or that the number of Eulerian stations are sufficient to resolve spatial variations and for tracing the displacement or propagation of water masses. These requirements are rarely fulfilled in practice because the ocean is so heterogeneous and extensive that it is difficult to deploy a sufficient number of either Lagrangian or Eulerian apparatuses. In the extreme case of watching from a single mooring point, we cannot determine whether the change is due to a reaction in a water mass or the advection of different water masses past the sensors.

On the other hand, satellites can acquire two-dimensional images and cover the ocean globally on a daily basis. Data from ocean color sensors are effective in mapping phytoplankton distribution (Gordon et al. 1985; Esaias et al. 1986; Hooker and Esaias 1993; Barale and Schlittenhard 1993). Therefore, we made a database of monthly composite images of CZCS data in the northwestern Pacific area (Fukushima et al. 1997). However, satellite data are often compromised by cloud cover. Moreover, the set of satellite data for cloud-free conditions is not appropriate as a time series because it cannot be regarded as random sampling, considering the variability of air–sea heat exchange and photosynthesis and their possible relationships to cloud cover. The satellite data are indirect signals of the in situ physical and biological conditions that should be validated by direct measurements. We made a database of ship-based biogeochemical data for the area around Japan (Harashima and Kikuchi 1990) to retrospectively analyze the relation to the satellite data. However, the possibility of the coincidence of satellite and ship observations in time and space was small because the frequency of regular ship observations was insufficient. Therefore, an in situ platform that can regularly scan the ocean (in this sense similar to satellite missions) is required even if it cannot cover it two-dimensionally.

Regular research vessels are not suitable because we cannot use them exclusively for long-term monitoring programs despite the importance of global environmental issues. It is worth considering the fact that most of the merchant and passenger ships take seawater into their engine cooling systems and that any commercial ship could potentially be an ocean observation platform as a volunteer observation ship or ship-of-opportunity if the owners and their crews are cooperative. Among these vessels, the ferries are of special utility because they frequently sail in fixed routes and their draft depths are stable compared with those of cargo ships or tankers.

Because of the importance of ocean margin processes (Holligan and de Boois 1993; Smetacek et al. 1991; Mackenzie et al. 1991; Wollast et al. 1993), we selected the ferry route between Pusan, Korea, and Kobe, Japan (ship's name: *Dan-Noh*, Osaka International Ferry, Co., Ltd.; Figure 18-1). We monitored temperature, salinity, pH, and in vivo fluorescence (with a Turner-Design fluorometer) semi-continuously (every 10 seconds), and took bottle and filter samples for dissolved nutrients, chlorophyll-a, and pheopigments with automated filtration samplers. These procedures have been described in a previous report (Harashima 1993b).

This report describes the methods not covered in the previous report (Harashima 1995a, and with color plates and numerical data on CD-ROM, 1995b), data processing and export to CD-ROM, the results and remarks on the malfunction of the first implementation of the monitoring system, and improvements to be made were a network for biogeochemical monitoring to be established using ferry boats.

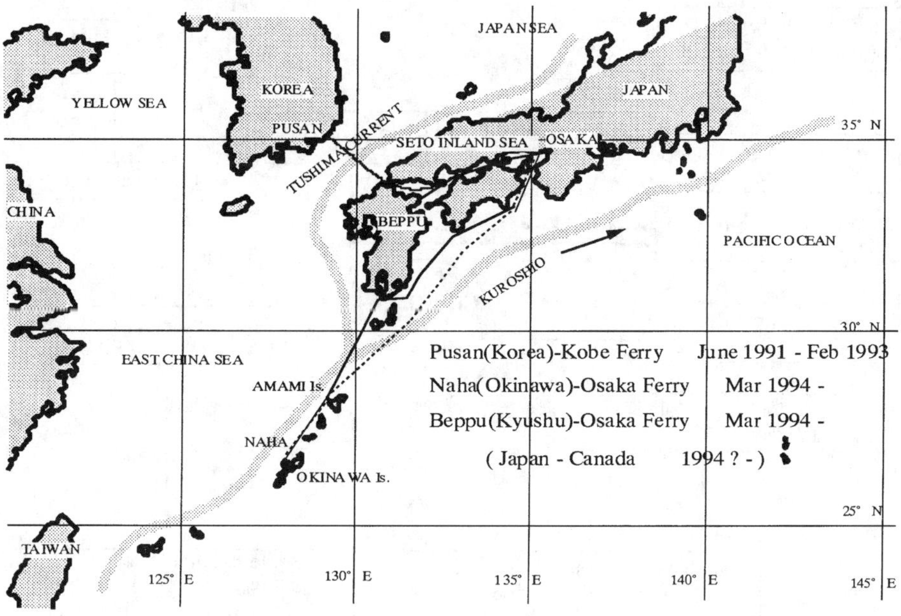

Figure 18-1. Routes of the Japan–Korea ferry, Osaka–Okinawa ferry, and Osaka–Seto Inland Sea–Beppu ferry.

Methods for Unmanned Measurements and Sampling

Figure 18-2 shows the schematic diagram of the monitoring system equipped for the first ferry monitoring program, performed during June 1991 to February 1993. The regular monitoring menu consists of the analysis of temperature (T), salinity (S), pH, and in vivo fluorescence measurements every 10 seconds (hereafter termed continuously sensed data), and the bottles and filters collected automatically every 2 weeks (hereafter called the bottle and filter data).

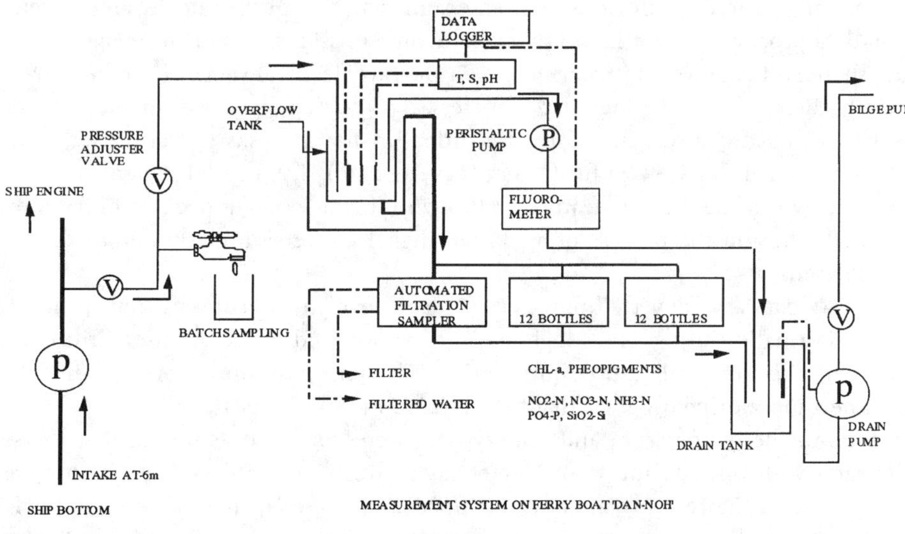

Figure 18-2. Schematic diagram of the monitoring system on the ferry *Dan-Noh* used in the first ferry monitoring program from June 1991. The system has been improved for the following program that began in 1994.

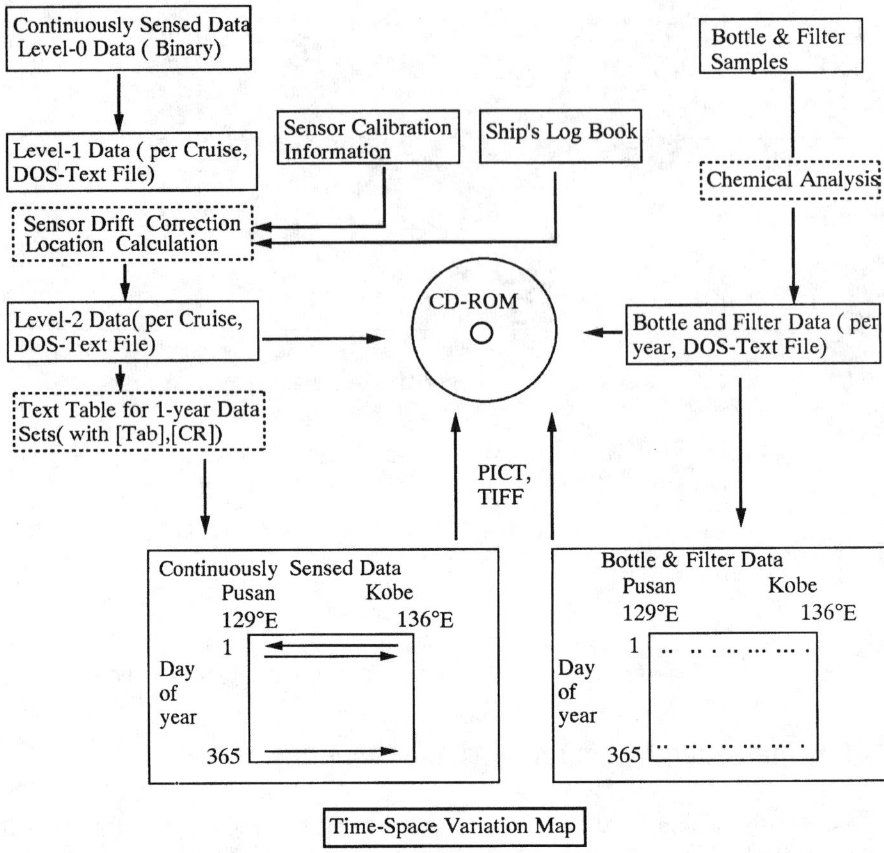

Figure 18-3. Data flow diagram.

The function of a ferry platform is in a sense similar to that of the satellite and therefore the method of data processing is similar to those in satellite missions, as shown in Figure 18-3. The continuously sensed data were transformed from the raw electrical signals into the corresponding physical, chemical, and biological parameters. The analog signals from the sensors were converted to 12-bit digital (binary) form, recorded on floppy disks (Level-0 data), and subsequently converted into DOS text file format (Level-1 data). The Level-1 data format is appropriate for quickly viewing the data features, although it includes slight departures from the true values created either in the process of analog-to-digital conversion or by output drift due to fouling of the sensors.

The Level-1 data are converted into Level-2 data through the computation of the time to location transformation using ship's logbook information and a linear sensor drift correction using pre- and postcruise calibration information. The ship position is obtained directly from GPS in the new monitoring program that began in 1994 (Harashima 1995b).

The drifts of the temperature and salinity data were corrected assuming that the sensors drifted linearly with time during 1-week intervals between calibrations. The fluorometer was calibrated using four fluorescein standards (0, 10, 20, and 30 μg/l), and the values in the Level-2 data are shown as ppb units of equivalent fluorescein concentration. The pH electrodes were calibrated against two freshwater-based NBS scale buffer solutions.

The automated filtration sampling was performed approximately every 2 weeks. On average, 30 bottle and filter samples were taken per cruise. They were then sent to the laboratory for chemical analysis. On 20 May, 1992, the method of nutrient analysis was changed from manual analysis (Strickland and Parsons 1968) to autoanalyzer (Bran Luebbe).

The data for each variable have time (date) and space (position along the ship route) axes, so a two-dimensional representation using computer graphics is more comprehensible. Such plots were found to be useful for checking data quality and for identifying latent malfunction of the system. As shown in Figure 18-3, the Level-2 data of the continuously sensed variables were transformed into text tables in which each line and column corresponded to 1 day and 30 seconds of longitudinal range, respectively. In this process, the original Level-2 data that fell within each 30 seconds' range (approximately 1 km) were gathered and averaged, excluding missing or abnormal data.

Thus, tables of 720 columns (129°E to 135°E: Pusan to Kobe) × 365 lines (1 to 365 days per year) were generated for each variable. These text tables were imported into computer graphics software to generate color sections, shown in Harashima (1995a, Figures 5 to 8). The bottle and filter data were transformed into a similar tabled text and processed to generate the color sections shown in Harashima (1995a, Figures 9 to 15).

The data have been distributed by CD-ROM (Harashima 1995b; Harashima 1997). The CD-ROM contains a text file for showing a synopsis of this monitoring program, the ship's route map, a schematic diagram of the monitoring system, the numerical data files, and the image files saved in both PICT and TIFF formats for use in Macintosh, DOS, and UNIX systems.

Optional Attended Investigations

Attended investigations were performed seven times during the period from 1991 to 1993 with the aim of performing research-oriented experiments and obtaining additional measurements of seawater properties. Another purpose was to check for malfunctions during the cruise and explore ways to develop and improve the automatic sampling system. These additional measurements were performed on samples taken from an auxiliary sampling faucet branching from the seawater supply pipe feeding the automatic sampling system. We attempted to draw water for the additional measurements and the automatic sampling simultaneously.

Zooplankton and phytoplankton samples were analyzed by light microscopy for comparison, and the nanoplankton and picoplankton by fluorescence microscopy. Part of the plankton count data were transformed to values of carbon biomass in each size fraction, with the data for size class and taxonomic classification to the genus level (Furusawa et al. 1993) based on the method of Strathmann (1967).

The phytoplankton biomass concentration with size spectra was also measured using laser. The instrument detects the fluorescence signals of the individual chlorophyll-containing particle in a flow cell excited by the laser light via glass fiber and counts the number of fluorescence events in each size-fraction (Tsuda et al. 1991; Tanaka et al. 1993; Harashima et al. 1997).

The distribution of seawater CO_2 partial pressure was measured on five cruises with a seawater CO_2 analyzer (Kimoto and Harashima 1993) that made use of a newly developed bubble-based equilibrator for better time response and that minimized the original water temperature compared with the traditional shower-based types.

Results

The in vivo fluorescence data revealed the occurrence of spring and autumn blooms and the spatial variations within them. Values were higher in the eastern Seto Inland Sea and lower in the Tsushima Strait, reflecting the overall level of nutrients. However, the data show not a simple correlation but a time-dependent interrelation characterized by uptake, decomposition, and supply from land. Both the fluorescence-derived and directly measured chlorophyll-*a* data indicate a decline of phytoplankton biomass in midsummer. The yearly cycle of phytoplankton blooms was closely connected to nutrient concentrations.

The nutrient concentrations were generally higher in the eastern part of the Seto Inland Sea than in the Tsushima Current area. Concentrations of PO_4-P, NO_2-N, and NO_3-N showed similar variations, that is, decreasing during phytoplankton blooms, becoming depleted in midsummer, and recovering when vertical mixing began to entrain deeper nutrient-rich water. NH_4-N concentrations were low throughout the year except in the area of the Kanmon Strait, where phytoplankton biomass was relatively high and vertical mixing was strong throughout the year. Dissolved Si also varied in parallel with phytoplankton biomass concentration; however, the Si recovery occurred before those of N and P because of summer runoff in the eastern part of the Seto Inland Sea. There, N or P limited the phytoplankton growth, preventing the complete depletion of dissolved Si.

PO_4-P and NO_3-N data for the period during phytoplankton blooms in the Tsushima Strait showed that about 0.1 μM of PO_4-P remained when NO_3-N was completely depleted by phytoplankton uptake; that is, N limitation rather than the other factors presumably terminated the blooms. The depletion of N continued throughout the five months from early summer until vertical mixing began in October. Silicon did not vanish in midsummer and was presumably not the limiting factor for the growth of the phytoplankton, particularly diatoms, although the Si concentration threshold for diatom growth is not clearly defined. The persistence of Si in the mixed layer may be due to the entrainment of East China seawater containing Si of terrestrial origin.

Nitrogen and phosphorus in the eastern Seto Inland Sea were not completely depleted, although they decreased during the phytoplankton blooms in which concentration of chlorophyll-*a* remained near or above 1 μg/l throughout the summer. Nitrogen and phosphorus were probably resupplied by remineralization from sedimentary organic matter that increased with increasing water temperature and from river runoff.

The pH observed in both 1991 and 1992 in the eastern Seto Inland Sea (Figure 18-4) was lower by 0.2 pH units relative to other areas. It was somewhat unusual because, generally speaking, the spatial variations of pH should be small because of the large buffering capacity of the seawater carbonate system. This pH low was not due to rainfall or the consequent characteristically low-pH river runoff because the salinity in this area was not remarkably lower than those of other areas. It was therefore attributed to the mixing of benthic water, in which the decomposition of organic matter lowers pH, into the surface layer.

It was found that phytoplankton biomass was correlated with pH, implying that photosynthesis consumed CO_2 and therefore raised pH. pCO_2 measurements (Kimoto and Harashima 1993), particularly those made during the manned investigation on the March 1992 cruise, supported this explanation. This cruise took place during the initiation phase of a phytoplankton bloom characterized by intense photosynthetic activity. pCO_2 was low and pH was high in an area of high fluorometric chlorophyll.

On the other hand, the data of the July 1992 cruise did not reveal a similar distribution. This cruise took place when a phytoplankton bloom had terminated and therefore pCO_2 was higher in the area where phytoplankton had peaked in early spring than the other area. This contrast may be partly due to the fact that the higher production in the past contributed to the current high pCO_2 via a high rate of respiration or biodecomposition of organic matter. It is also true that higher water temperature increased the pCO_2 as a result of the reduced solubility of gases.

The difference of the biogeochemical states between the pre- and postbloom periods also appeared in the composition of phytoplankton species. Figure 18-5 shows the size fractionation and genus composition observed in the season when the spring bloom had terminated (July

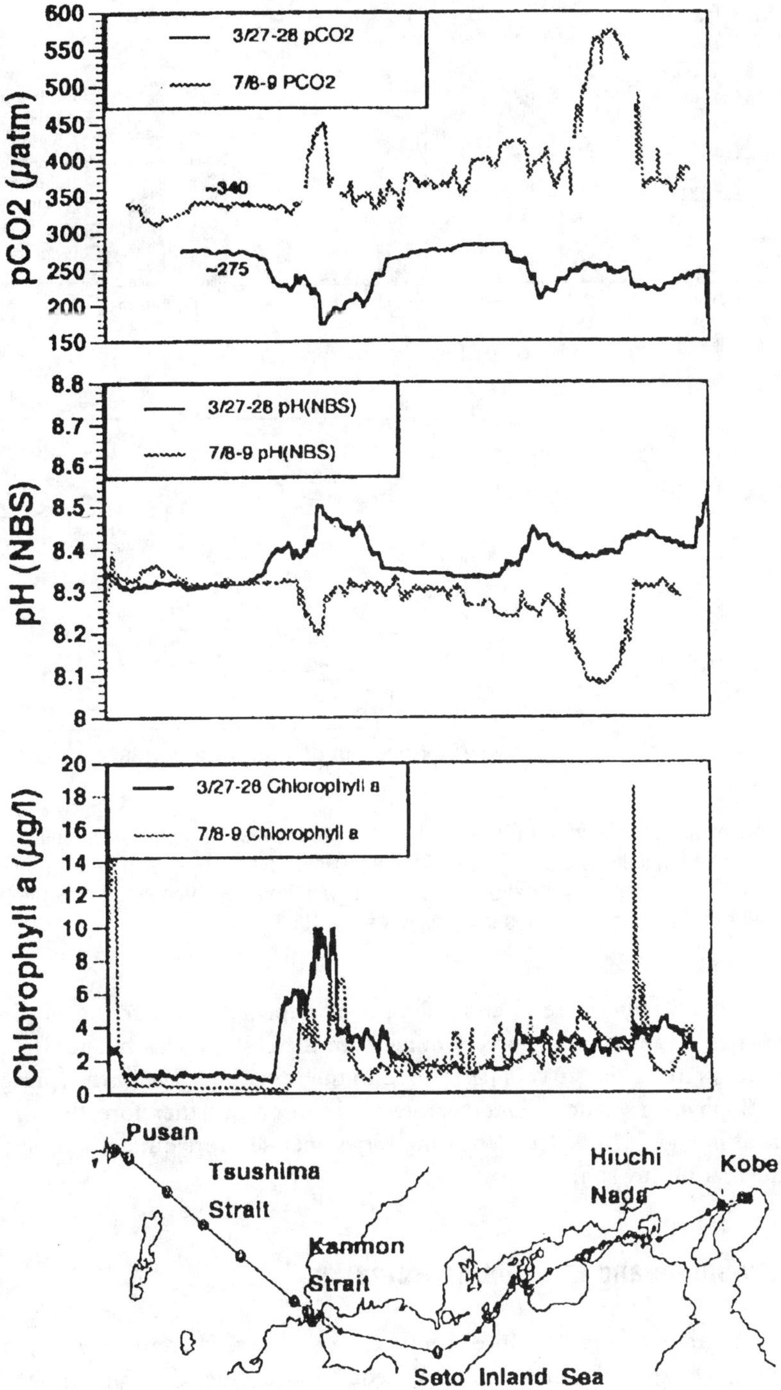

Figure 18-4. Horizontal distribution of pCO_2, pH, and chlorophyll-*a* revealed in two unattended cruises carried out in March (solid curve) and July (dotted curve) of 1992. (Reprinted by permission from Kimoto, K., and Harashima, A. 1993. High Resolution Time/Space Monitoring of the Surface Seawater CO_2 Partial Pressure by Ship-of-Opportunity. Proceedings of the Fourth International CO_2 Conference World Meteorological Organization. Global Atmosphere Watch 89, pp. 88–91.)

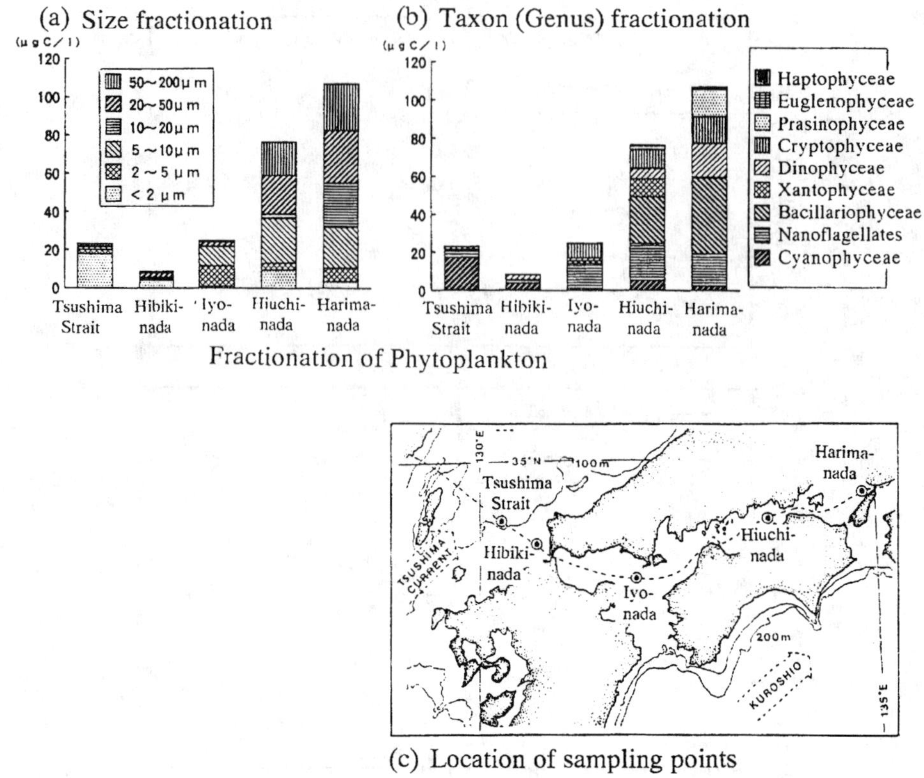

(c) Location of sampling points

Figure 18-5. Horizontal distribution of (a) size fractionation and (b) taxon (genus) composition of phytoplankton observed in July 1992. Locations of the sampling points are shown in (c). Conversion from the cell number to the biomass concentration (carbon unit) was done by the method of Strathmann (1967). (Modified from the original figures in Furusawa et al. 1993.)

1992). In Tsushima Strait, where N and P had been depleted, the fraction of phytoplankton smaller than 2 μm (cyanophyceae) was dominant. In the eastern Seto Inland Sea, a portion of dinophyceae appeared. In the surveys performed in the initiating stage of the spring bloom, the diatom Bacillariophyceae was dominant regardless of location and, therefore, the contrast was not as remarkable as in Figure 18-5. Thus, the time-series analyses were quite important in assessing the state of the marine environment.

Technical Problems and Concluding Remarks

The first ferry program on a Japan–Korea route was a test case. Therefore, some portions of the system worked as we expected, whereas others needed improvements or further development for fruitful results. The selection of parameters to be monitored was found to be reasonable.

Several papers concerning the efficiency of chlorophyll—in various simple or sophisticated forms—as an indicator of marine environmental change have appeared since our ferry monitoring scheme was designed. Falkowski and Wilson (1992) analyzed historical Secchi disk depth data, which depend on the chlorophyll concentration, to investigate whether the phytoplankton biomass has increased in the past 70 years as a result of atmospheric CO_2 increases. On the other hand, the use of a dual-flash fluorometer has been proposed as a sophisticated method to measure the photosynthetic efficiency (Falkowski and Kolber 1993). As far as our automated monitoring program is concerned, the present setup of in vivo fluorescence measurements by a typical fluorometer seems appropriate because laboratory-quality space cannot always be provided aboard volunteer observation ships. However, sophisticated sensors can be installed on the seawater intake system of ferries provided that the sensor system is somewhat robust.

High-pressure liquid chromatography (HPLC) has recently been recommended as a standard method for the analysis of photosynthetic pigments. However, before switching to this new standard, there should be a period during which HPLC and the traditional fluorescent or spectrophotometric chlorophyll-*a* and pheopigment analyses are used in parallel. Such a transition period would permit better utilization of the extensive pigment data set for the assessment of long-term environmental change. It may be necessary to continue cross-checking the results of the HPLC and the fluorescent and spectrophotometric chlorophyll-pheopigment methods for a certain period. In the present configuration of the automated filtration sampler, the seawater quantity for each filtration is around 450 ml, which is insufficient for analysis by HPLC. One possible procedure would be to use in vivo fluorescence for the time–space survey with parallel use of a limited number of HPLC samples on the attended cruises.

The importance of measuring pH has been recognized, and the available methods have been improved (Dickson 1993; Millero 1986; Millero et al. 1993). pH is a suitable variable with which to evaluate the seawater carbonate system. The measurement of any two of the four carbonate system variables (total alkalinity, total inorganic carbon, partial pressure of dissolved carbon dioxide, and pH) permits the calculation of the other two quantities using equilibrium relationships (Dickson and Goyet 1994). The desired pH measurement accuracy for this purpose is 0.002, and the colorimetric technique is expected to fulfill this. The pH electrode sensor installed on the ferry does not fulfill this requirement in terms of absolute accuracy. However, it is suitable for visualizing temporal and spatial patterns to explore related processes as shown in the previous section. Several problems have been found in calibrating the output from the pH electrode. From October to December of 1991, the baseline of the pH signal was anomalously elevated compared with other seasons by about 0.3 pH unit. This difference is attributed to a change in the quality of the pH buffers. It also became apparent that the use of freshwater-based calibration buffers could lead to a substantial offset error due to the difference in ionic strength between the buffers and seawater. In the new program beginning in 1994, a seawater-based buffer solution (Smith 1968) has been used to calibrate the intercept of the pH electrode output after measuring the slope with two freshwater-based buffer solutions.

Apart from the specific procedures for each measurement, the method of getting a seawater supply from the engine cooling pipe to the monitoring system is crucial. To minimize the time taken for the water to reach the measurement system from the entrance of the intake pipe, a high flow rate is needed. This is particularly important in the measurement of dissolved pCO_2 because the carbonate equilibrium depends on the in situ water temperature. On the other hand, leakage of seawater and damages to the flow cells of the measurement system need to be considered. Thus, the pipe pressure was adjusted to around 1 kgW/cm^2, which produced a flow rate of around 10 liters per minute. Setting up an overflow tank with an open surface (see Figure 18-2) in which the sensors are dipped was found to be crucial to solving the pressure/flow rate problem. As we could not find a pressure regulator for seawater, we asked the ship's crew to watch the pressure gauge occasionally.

The ship usually uses MGPS, that is, injection of chlorine gas to prevent biofouling in the seawater pipe. The resulting concentration of chlorine gas is usually very low and seems harmless for the measurement of T, S, pH, and fluorescence. To avoid bias in the specific chemical measurements, unchlorinated seawater may be safe. When we performed the bottle and filter sampling and manned pCO_2 measurements, we asked the ship's crew to suspend MGPS.

Other improvements that began in 1994 are the direct monitoring of ship position using GPS on the bridge, and automatic stoppage of the seawater supply when sensors detect leakage on the floor.

Despite the inconveniences of working aboard nonresearch vessels, the feasibility of biogeochemical monitoring from ferries has been demonstrated. Such monitoring programs can obtain time-series data of directly sensed variables with high spatial resolution unlike those available from satellites or other platforms.

ACKNOWLEDGMENTS

This project owes much to the cooperation of the Osaka International Ferry Co. Ltd. The authors also thank Korea Ocean Research & Development Institute (KORDI) for their collaboration under the Japan–Korea Science and Technology Agreement. This project has been performed by the participants and organizations of the Monitoring Execution Group based on discussions with the Subcommittee for Japan–Korea Ferry Monitoring (1991–1993) and Subcommittee for Continental Shelf Ferry Monitoring (1994–), which were organized by NIES-CGER and chaired by N. Handa of Nagoya University.

References

Barale, V. and Schlittenhard, P.M. 1993. Ocean Colour: Theory and Applications in a Decade of CZCS Experience, Kluwer Academic Publishers.

Dickson, A. G. 1993. The measurement of seawater pH. Marine Chem. 44:131–142.

Dickson, A. G., and Goyet, C. (eds.). 1994. Handbook of Methods for the Analysis of the Various Parameters of the Carbon Dioxide System in Seawater, Version 2. Unpublished manuscripts, Department of the Environment.

Esaias, W. E., Feldman, G. C., McClain, C. R., and Elrod, J. A. 1986. Monthly satellite-derived phytoplankton pigment distribution for the North Atlantic Ocean basin. EOS 67(44):835–837.

Falkowski, P. G., and Kolber, Z. 1993. Estimation of phytoplankton photosynthesis by active fluorescence. ICES Marine Sci. Symp. 197:92–103.

Falkowski, P. G., and Wilson, C. 1992. Phytoplankton productivity in the North Pacific ocean since 1900 and implications for absorption of anthropogenic CO_2. Nature 358:741–743.

Fukushima, H., Toratani, M., Matsumura, S., and Harashima, A. 1997. Time Series of Phytoplankton Bikomass Distribution over the Northwestern Pacific Area by Monthly Composite Images from Nimbus7-CZCS Data, CGER-D015(CD)-'97, Center for Global Environmental Research, National Institute for Environmental Studies.

Furusawa, K., Kurokawa, O., Teishima, H., and Harashima, A. 1993. Phytoplankton Biomass Fractionation in Size and Taxon in Seto Inland Sea and Tsushima Strait (in Japanese). Proceedings of 1993 Spring Assembly of Japan Oceanographic Society, pp. 337–338.

Gordon, H. R., Austin, R. W., Clark, D. K., Hovis, W. A., and Yentsch, C. S. 1985. Ocean Color Measurements. In B. Saltzman (ed.), Satellite Oceanic Remote Sensing. Advances in Geophysics 27, pp. 297–333.

Harashima, A. 1993a. Continuous Ocean Biogeochemical Monitoring by Japan-Korea Ferry Boat for the Validation of Ocean Color Remote Sensing. Proceedings of Environment '93 Symposium on Remote Sensing in Environmental Research and Global Change. Science & Technology. University of Hong Kong. Hong Kong: The Commercial Press (H.K.) Ltd., pp. 92–104.

Harashima, A. 1993b. High Frequency Marine Biogeochemical Monitoring from a Japan-Korea Ferry—1991 Results. Annual Report on Global Environmental Monitoring. Center for Global Environmental Research, National Institute for Environmental Studies, Japan, pp. 33–45.

Harashima, A. 1995a. High Temporal-Spatial Resolution Biogeochemical Monitoring from a Japan-Korea Ferry—1992 to 1993 Results. Monitoring Report on Global Environment. Center for Global Environmental Research, National Institute for Environmental Studies, Japan, pp. 13–50.

Harashima, A. (ed.). 1995b. Collected Data of the Biogeochemical Monitoring by Japan-Korea Ferry Boat. CGER-D007(CD)-'97. Center for Global Environmental Research, National Institute for Environmental Studies, Japan.

Harashima, A. 1997. Collected Data of High Temporal-Spacial Resolution Marine Biogeochemical Monitoring from Ferries in the East Asian Marginal Seas (April 1994–December 1995) CGER-D012(CD)-'95, Center for Global Environmental Research, National Institute for Environmental Studies.

Harashima, A., and Kikuchi, A. 1990. Biogeophysical remote sensing, a ground truth data base and graphics system for the northwestern Pacific Ocean. EOS 71(10):314–315.

Harashima, A., Tsuda, R., Tanaka, Y., Kimoto, T., Tatsuta, H., and Furusawa, K. 1997 Monitoring Algal Blooms and Related Biogeochemical Changes with a Flow-through System Deployed on Ferries in the Adjacent Seas of Japan. In M. Kahru and C. B. Brown (eds.), Monitoring Algal Blooms—New Technique for Detecting Large Environmental Change, Landes Bioscience Publishers and Springer, pp. 85-111.

Holligan, P. M., and de Boois, H. 1993. Land-Ocean Interactions in the Coastal Zone (LOICZ). IGBP Report No. 25. Stockholm: ICSU.

Hooker, S. B., and Esaias, W. E. 1993. An overview of the SeaWiFS project. EOS 74(21):241–246.

Kimoto, K., and Harashima, A. 1993. High Resolution Time/Space Monitoring of the Surface Seawater CO_2 Partial Pressure by Ship-of-Opportunity. Proceedings of the Fourth International CO_2 Conference World Meteorological Organization. Global Atmosphere Watch 89, pp. 88–91.

Mackenzie, F. T., Brewer, J. M. , Charlson, J. J., Hofmann, E. E., Knauer, G. A., Kraft, J. C., Nöthig, E. M., Quack, B., Walsh, J. J., Whitfield, M., and Wollast, R. 1991. What Is the Importance of Ocean Margin Processes in Global Change? In R. F. C. Mantoura, J. M. Martin, and R. Wollast (eds.), Ocean Margin Processes in Global Change. New York: John Wiley and Sons, pp. 433–454.

Mantoura, R. F. C., Martin, J. M., and Wollast, R. (eds.). 1991. Ocean Margin Processes in Global Change. New York: John Wiley and Sons.

Martin, J. H., Coale, K. H., Johnson, J. S., Fitzwater, S. E. 1994. Testing the iron hypothesis in ecosystems of the equatorial Pacific Ocean. Nature 371:123–129.

Millero, F. J. 1986. The pH of estuarine waters. Limnol. Oceanogr. 31:839–847.

Millero, F.J., Zhang, J.-Z., Fiol, S., Sotolongo, S., Roy, R. N., Lee, K., and Mane, S. 1993. The use of buffers to measure the pH of seawater. Marine Chem. 44:143–152.

Milliman, J. D. 1991. Flux and Fate of Fluvial Sediment and Water in Coastal Seas. In R. F. C. Mantoura, J. M. Martin, and R. Wollast (eds.), Ocean Margin Processes in Global Change. New York: John Wiley and Sons, pp. 69–90.

Smetacek, A., Bathmann, U., Nöthig, E. M., and Scharek, R. 1991. Coastal Eutrophication: Causes and Consequences. In R. F. C. Mantoura, J. M. Martin, and R. Wollast (eds.), Ocean Margin Processes in Global Change. New York: John Wiley and Sons, pp. 251–279.

Smith, S. V. 1968. Calcification and organic carbon metabolism as indicated by carbon dioxide. Monogr. Oceanogr. Methodol. 5:469–484.

Strathmann, R. R. 1967. Estimating the organic carbon content of phytoplankton from cell volume or plasma volume. Limnol. Oceanogr. 12:411–418.

Strickland, J. D. H., and Parsons, T. R. 1968. A Practical Handbook of Seawater Analysis J. Fish. Res. Board Canada. Bull. No. 167.

Tanaka, Y., Tsuda, R., and Harashima, A. 1993. In Situ Monitoring and Size Spectra of Phytoplankton via Laser-Induced Fluorescence through an Optical Fiber on Japan- Korea Ferry. Abstracts of the Second Annual Meeting of PICES (North Pacific Marine Science Organization), Seattle, Washington, p. 38.

Tsuda, R., Kakui, Y., and Kumagai, M. 1991. Analysis of phytoplankton size using optical fiber cables with the laser technique. Verh. Internat. Verein. Limnol. 24:1088–1090.

Turner Designs. 1983. Chlorophyll and Pheophytin. Turner Fluorometric Facts Bulletin 101.

Wollast, R., Mackenzie, F. T., and Chou, L. (eds.). 1993. Interactions of C, N, P and S Biogeochemical Cycles and Global Change. New York: Springer-Verlag.

IV

Indicators of Ecosystem Health

19

Nutrients in the Bohai Sea

Yi Cui and Yunli Song

The Bohai Sea has an area of approximately 180,000 km^2, with an average depth of 18 m. It is bordered by land on three sides and has three bays: Laizhou, Bohai, and Liaodong. There are many large and small rivers that discharge nutrient-rich freshwater into the Bohai Sea. It has a high productivity and is the main nursery and feeding ground for many economically valuable fishes and prawn.

Eutrophication results from large quantities of organic matter and nutrients entering the sea. Nitrogen and phosphorus are the main contributors to eutrophication and are responsible for the occurrence of red tides (Thomas 1970; Japanese Aquatic Society 1982). This paper discusses the distribution and variation of nitrogen and phosphorus in the Bohai Sea using survey data from 1992 and 1993, and compares this data with historical data to understand nutrients in the Bohai Sea. This information is necessary to protect the marine environment and aquaculture interests.

Materials and Methods

Surveys were carried out in August and October of 1992 and February and May of 1993 (referred to collectively in the text as the 1992 survey). Analyses performed included dissolved oxygen (DO), pH, NO_2-N, NO_3-N, NH_4-N, PO_4-P, and SiO_3-Si. These data will be compared with a survey done during the same time periods in 1982–1983 (referred to as the 1982 survey).

Results and Discussion

Table 19-1 gives the water quality standards used in China to ensure human and fisheries health. The survey data are listed in Tables 19-2 to 19-5. It can be seen that inorganic nitrogen (IN) and phosphorus (IP) have a clear monthly and annual trend. The annual mean of inorganic nitrogen is less than 100 μg/l for all years; thus, it is in the nutrient-poor category of Table 19-1. The annual mean of inorganic phosphorus was more than 15 μg/l in 1960 and 1982, but in 1992 was less than 15 μg/l (Table 19-2). The 1982 monthly and annual means of inorganic phosphorus were about 30 μg/l, whereas in 1992 they were less than 15 μg/l, except for the February monthly mean (Tables 19-3 and 19-4). The monthly mean of inorganic nitrogen was less than 100 μg/l, except during the winter, which had higher concentrations. The annual mean of nitrate was less than 100 μg/l, with Laizhou Bay having higher concentrations than the other bays.

The data indicate that nitrogen concentrations are increasing, whereas phosphorus is decreasing. This may be attributed to changes in river runoff. For example, the utilization of water from the Yellow, Luanhe, and Haihe Rivers has increased greatly due to increased industrial activity in

Table 19-1. Water quality standards ($\mu g/l$)

Standard	Nutrient-poor	Nutrient-average	Nutrient-rich
Inorganic nitrogen	<100	175	250
Inorganic phosphorus	<15	30	45

Table 19-2. Annual nutrient means in the Bohai Sea ($\mu g/l$)

Year	PO_4-P	NO_3-N	Σ IN
1960	24.78	9.94	36.54
1982	29.11	21.42	42.14
1992	10.22	37.10	49.84

Table 19-3. Monthly nutrient means in the Bohai Sea ($\mu g/l$)

	1982				1992			
Nutrient	Feb.	May	August	Oct.	Feb.	May	August	Oct.
PO_4-P	46.8	39.6		26.6	16.4	10.5	5.9	8.1
NO_3-N	48.3	14.3	8.8	14.6	67.3	24.9	19.7	36.1
Σ IN	69.0	21.6	20.9	29.4	77.1	38.4	35.1	48.3

Table 19-4. Comparison of annual nutrient means in the Bohai Sea waters ($\mu g/l$)

	NO_3-N		PO_4-P	
Bay	1982	1992	1982	1992
Laizhou Bay	45.2	65.5	21.1	6.5
Bohai Bay	20.7	51.2	36.2	8.1
Liaodong Bay	9.1	21.7	30.0	6.8
Bohai Center	13.2	24.4	30.5	11.2

Table 19-5. Inorganic nitrogen and inorganic phosphorus and the N/P ratio in the Bohai Sea waters

Waters	Inorganic nitrogen ($\mu mol/l$)	Inorganic phosphorus ($\mu mol/l$)	N/P ratio
Laizhou Bay	5.98	0.21	28.7
Bohai Bay	5.09	0.26	19.6
Liaodong Bay	2.16	0.22	9.8
Bohai Center	2.56	0.36	7.2
Bohai Sea	3.56	0.33	10.8

the coastal zone. The annual amount of water drawn from the Yellow River has exceeded the amount flowing into the sea (Qian 1993).

Tide gates and water ditches to retain water have been built along the entire length of the Haihe and Luanhe Rivers (China Coastal Survey 1989; Tangshan 1991). During the nonrainy season there is very little water flowing into the sea, and only a small amount of water flows into the sea during the rainy season. Thus, the nutrient concentrations are changing due to lack of dilution. Also, nitrogen concentrations are increasing due to greater use of chemical fertilizers in coastal areas. It is reported that the amount of nitrogen fertilizer draining into the Bohai Sea is 1.7×10^5 tons a year (Zhang 1980). The nitrate concentrations at the mouth of the Yangzi River also appear to be rising. For example, the maximum value of nitrate was 810 μg/l in a joint Sino-American survey in 1983, whereas the maximum value of nitrate in 1963 was only 220 μg/l. This is believed to be due to the increased use of nitrogen fertilizers. In addition, salinity in these areas is also changing; these variations in salinity may be due to a decrease in the amount of runoff. The salinity at the mouth of the Yellow River was 29.37 and 27.75 ppt in August and October of 1982, respectively, whereas during the corresponding period in 1992 the salinity was 30.76 and 29.74 ppt, respectively.

Phosphate decreases seen in the 1992 survey may be linked to a decrease in biomass; excretion of phosphorus by zooplankton is often the most important means of phosphorus regeneration (Riley 1982). The cause of the decrease needs to be examined further.

Chu (1949) demonstrated in an indoor culture experiment that at the lower limits of optimum concentration, inorganic nitrogen and phosphorus utilization by phytoplankton was 80 μg/l and 18 μg/l, respectively. Zou (1983) thought that 15 μg/l of phosphorus was the lowest concentration necessary for reproduction and growth of phytoplankton. Some researchers believe that increased concentrations of nitrogen and phosphorus are the cause of algae reproduction, and that phosphorus is the key factor (Wu 1983). Therefore, the fact that the number of phytoplankton (98×10^4 cell/m^3) in 1992 was lower than in 1982 (322×10^4 cell/m^3) may be due to lower phosphorus concentrations. This relationship needs to be examined further.

The N/P ratio is an important index when assessing eutrophication. The normal seawater N/P ratio is 16:1 (Pilson 1985), and phytoplankton take in nitrogen and phosphorus from seawater at a 16:1 ratio also (Zheng 1986). Thus, if this N/P ratio deviates in either direction, it will restrict phytoplankton growth. The N/P ratios of the Laizhou and Bohai Bays are greater than 16:1. Laizhou Bay is influenced mostly by runoff from the Yellow River. The N/P ratios of the other parts of the Bohai Sea are less than 16:1 (Table 19-5). In view of this, it appears that there is a nitrogen excess in the Laizhou and Bohai Bays, and that in the other waters there is a nitrogen shortage. In addition, the annual means of nitrogen and phosphorus in each bay also show the same results (Table 19-4). For example, the nutrient concentrations in Laizhou and Bohai Bays fluctuate, reaching the lowest limit for maintaining a healthy phytoplankton population. Phosphorus reaches concentrations lower than the optimum for phytoplankton. For the other waters, nitrogen is lower compared with phosphorus, and the N/P ratio is much lower than 16:1. It is suggested that in Laizhou and Bohai Bays the phosphorus is the limiting factor, whereas for the other waters nitrogen may be the limiting factor. This interpretation agrees with Zou (1983). However, the N/P ratio for the whole Bohai Sea is still lower than the Redfield ratio, although the concentration of nitrogen is rising and the N/P ratio of 1992 is 2.4 times greater than in 1982. Thus, inorganic nitrogen is still the limiting factor for phytoplankton growth in the Bohai Sea (Cui 1992). Harvey (1957) showed that nitrogen utilization was 9 times that of phosphorus in culture experiments. This indicates that there is still a depletion of nitrogen. It must be pointed out that the phosphorus content is below the lower limit for phytoplankton survival during all seasons except winter. Thus, phosphorus, to a certain extent, may also play an important role in the overall system.

Conclusion

Nutrient concentrations in the Bohai Sea are still in the mid to low range. Similar results were found by Tian (1983). However, in some areas of the Bohai Sea, such as in Laizhou Bay, nitrogen concentrations are higher, whereas phosphorus concentrations are comparable. Here, the phytoplankton concentrations are higher than elsewhere, indicating favorable conditions for reproduction and development.

An increasing amount of industrial and domestic sewage is being discharged into some densely populated coastal areas due to increased industrialization. As a result, eutrophication is occurring in the area of Tianjin City, located in western Bohai Bay. In the coastal zone near Huanghua, the drainage of prawn ponds has resulted in the occurrence of red tides. These ponds become densely populated with prawn and experience high levels of mortality, especially during the warm season. When the crop has been harvested, these ponds are drained into the coastal waters, resulting in nutrient concentrations far in excess of the purification and dilution capacity of seawater, thus causing eutrophication. Hence, to protect the marine environment and aquaculture interests, special attention should be paid to areas of increased population and places where industries, both land-based and aquaculture, are developing.

References

China Coastal Survey. 1989. Science Press, pp. 215–225.

Chu, S. P. 1949. Experimental studies on the environmental factors influencing the growth of phytoplankton. Cont. Fish. Res. Inst., Dept. Fish., Nat. Univ. Shantung 1:37–52.

Cui, Y. 1992. The study on the relation between phytoplankton and environmental factors in the Bohai Sea. Marine Environ. Sci. 11(3):56–59.

Harvey, H. W. 1957. The Chemistry and Fertility of Seawaters. Cambridge University Press.

Japanese Aquatic Society. 1982. Organisms Index on Eutrophication along Coastal Waters. Hengxing Shehou Press, pp. 1–60.

Pilson, M. E. Q. 1985. Annual cycles of nutrients and chlorophyll in Narragansett Bay, Rhode Island. J. Mar. Res. 43(4):849–873.

Qian, Y. 1993. The Variance on Water and Sand of Yellow River. Chinese Building Material Press, pp. 7–96.

Riley, J. P., et al. 1982. Chemistry and Oceanography, Vol. 2 (Cui Qingchen, trans.). Marine Press, pp. 268–300.

Tangshan Plan Committee. 1991. The Resource of Tangshan. Tianjing Press.

Thomas, W. H. 1970. Effect of ammonia and nitrate concentration on chlorophyll increases in natural tropical Pacific phytoplankton populations. Limnol. Oceanogr. 15(3):386–394.

Tian, J. 1983. The primary study of organic pollution and red tide organisms in the mouth of Yellow River. Marine Environ. Sci. 2(1):46–50.

Wu, X. 1983. Chinese Encyclopedia. Chinese Encyclopedia Press, pp. 102–103.

Zhang, J. 1980. The Primary Study on the Aquatic Production in the Yellow and Bohai Seas. Chinese Aquatic Society.

Zheng, Z. 1986. Proceedings of Ocean Biology and Ecology. Xiamen University Press.

Zou, J. 1983. The primary study of eutrophication and red tide in the Bohai Bay. Marine Environ. Sci. 2(2):45–54.

20

Ecological Carrying Capacity of Semi-enclosed Large Marine Ecosystems

S. M. Konovalov

Implementation of the sustainable development concept (Daily and Ehrlich 1992) should include an ecological assessment of freshwater bodies and seas because large amounts of pollutants accumulate in rivers and are transported into coastal areas, redistributed throughout the adjacent sea, and even penetrate the deep oceans. Atmospheric circulation and hydrodynamics in the open sea and the ocean are also responsible for the redistribution of pollutants, causing high concentrations of pollutants in lakes, reservoirs, estuaries, bays and semi-enclosed seas. This situation is due primarily to poor water exchange compared with rivers and open oceanic waters and secondarily to temperature, salinity, density, and other gradients that contribute to the development of a geochemical barrier (Nowicki and Oviatt 1990). The ever-increasing human population living along the coast greatly contributes to pollution in coastal areas where water exchange is poor.

Of all the large marine ecosystems (LMEs), semi-enclosed seas are subject to the greatest anthropogenic impact due to poor water exchange and high population density on their shores. We have distinguished 25 semi-enclosed LMEs and endeavored to assess the probable extent of human population loading on them and their resistance to anthropogenic impacts. This approach should provide information on which LMEs are displaying critical ecological conditions, which have more or less satisfactory conditions, and which are in intermediate conditions with regard to human-induced eutrophication. This comparative analysis should provide a better understanding of the rate of marine ecosystem transformation due to anthropogenic loading and provide information for managing and restoring ecosystems.

Ecological Carrying Capacity

The ecological carrying capacity (ECC) of a reservoir can be estimated using characteristics of the water cycle, chemical compounds involved in biochemical cycles of autotrophic and heterotrophic organisms, and by studying pathways and cycles of pollutants. The neighboring Azov and Black Seas show remarkably different geomorphology and water mass volume. The Azov Sea is 1,680 times smaller than the Black Sea, with 0.008 km^3 of water per km^2 of sea surface, whereas for the Black Sea this estimate is 0.787 km^3 of water per km^2. Naturally, the greater the water volume per unit sea surface, the slower the deterioration due to anthropogenic impacts, which is to the advantage of the Black Sea.

When there is a large water exchange with other seas or with the ocean, the effects of pollutants on the system are minimized due to transport away from the sea. For example, in the Azov Sea, complete water exchange takes 6.4 years; in the Black Sea it takes about 960 years. Hence, the Azov Sea is more resistant to anthropogenic inputs only because pollutants are rapidly transported out of the system.

Several scenarios can be applied to pollutants that persist in the sea. These scenarios are similar to those recognized for certain types and modes of sedimentation by natural cycling of substances (Shepard 1963). Pollutants dissolved in seawater may be involved in the biological cycle of matter and the formation of biogenic deposits. Pollutants that take part in chemical processes are referred to as hydrogenic deposits. Products of biological and chemical degradation of some pollutants are also attributed to biogenic and hydrogenic deposits (Polikarpov 1981). Polikarpov and Egorov (1981) believe that pollutants are withdrawn mostly through biosedimentation. According to their conclusions, up to 3.25 million tons of lead, 46 thousand tons of mercury, and 380 thousand tons of PCBs are removed annually from the photic layer of the world ocean due to biosedimentation. Geochemical transformation of a number of substances into insoluble compounds, and their deposition in areas that show pronounced geochemical barriers (traps), are more likely to be understood and assessed from these studies (Albani et al. 1989; Nowicki and Oviatt 1990).

Thus, the ecological carrying capacity of the sea is the total of the natural geomorphologic and hydrographic characteristics of water bodies with regard to the removal of pollutants, the decomposition or transformation of pollutants into insoluble compounds, and the geochemical sedimentation cycle. This is also the ability of an ecosystem to accumulate pollutants and perform biogeochemical sedimentation. In other words, ecological carrying capacity is the ability of seas to receive and transfer various amounts of pollutants and to endure disturbances of the natural cycles (mostly, nitrogen and phosphorus compounds) that do not cause irreversible changes in the ecosystem.

The present and the future of marine ecosystems depend on the extent of anthropogenic loading and the rate of pollutant deposition or removal, expressed as the ratio between the level of pollution and the ecological carrying capacity of the water body. When the population load exceeds the ecological carrying capacity, marine ecosystems respond immediately at every trophic level. Therefore, the anthropogenic load should be carefully examined, and point and nonpoint land-based sources and air transfer of pollutants taken into account. The anthropogenic load has been assessed for the North and Baltic Seas (Northern Europe's Seas 1989), but most of the semi-enclosed seas are still items on the agenda for this sort of research, which is costly and requires trained individuals to conduct.

The population density along the coastal areas around the basin of an LME can be used as an indicator of the extent of pollution in the sea. Metabolic rates have been well studied in man (Rast and Lee 1983) and domestic animals (Johnes and O'Sullivan 1989) in relation to the amounts of phosphorus and nitrogen excreted by metabolic products into the environment. For example, 1.0 kg of phosphorus is excreted by man annually. Of this amount, about 0.7 kg P person^{-1} yr^{-1} is from human metabolic waste and 0.3 kg P person^{-1} yr^{-1} is from household laundry and other detergents. Similar correlations have been found between the output of energy, fertilizers, chemical and metallurgic production, and the pollution level per capita. I shall use the index of population density in the comparative analysis that follows.

The response of ecosystems and the sea to pollution can be assessed based on data about the replacement of dominant species at every trophic level, reduction of biodiversity, decrease in average size of organisms in the population, and accelerated maturation of many plants and animals that prefer less polluted aquatic habitats. Considerable data are available on contemporary changes in the conditions of ecosystems, which permit a comparative analysis among LMEs where ecological conditions are badly degraded.

To evaluate population density in the basins of different seas, let us consider their geomorphologic and hydrographic characteristics in order to classify the seas according to their ecological carrying capacity. Then we shall examine correlations between population density and the rate of ecological change of the ecosystem.

Hydrographic Characteristics of Semi-enclosed Large Marine Ecosystems

Surface area and average depth are ecologically significant characteristics. The amount of fallout from airborne pollutants and precipitation over the seas depends on the area of sea surface. Average depth and the dependent value of water volume per km^2 of sea surface are used to estimate the volume of water per unit area that receives pollutants. The carrying capacity of an LME is indicative of the bulk of water that the LME contains and of the potential water volume that could dissipate pollutants.

Now, let us examine 21 LMEs that are semi-enclosed seas (Sherman 1994, 1995) in accordance with the above-stated characteristics. The Seto Inland Sea, the Azov Sea, and the Adriatic Sea have also been included in this discussion. Two groups are easily distinguished: those with an ecological carrying capacity up to 600,000 km^3 and those with an ECC of more than 1,000,000 km^3 (Tables 20-1 and 20-2). In the first group, the lowest values of average depth and water volume are in the Yellow Sea LME, with the Baltic and North Sea LMEs showing somewhat higher estimates. Maximum volumes have been estimated for the West Greenland Shelf and the Black Sea LMEs. In the latter, however, the oxygenated water column makes up only 54,000 km^3, while the rest is a H_2S^- zone where anaerobic life consists of mostly microorganisms (Vinogradov 1989). The oxygenated water volume is regarded as the active capacity in calculating the ECC of the sea. Thus, the Black Sea LME is assumed to have the volume of 54,000 km^3 in line with the North, Baltic, and Yellow Seas. This group also includes the Barents Sea LME, the West Greenland LME, and the Northern Australian Shelf LME, which have much greater values of sea surface area and capacity.

The Gulf of California is of special concern because of its small surface area and great average depth. The Seto Inland Sea is the smallest in water area, and the Azov Sea has the smallest capacity. The size (area) and the water volume of these seas, and hence their ecological carrying capacity, are considerably less than those of the second group. When the anthropogenic load is high, which is typical of all the seas except the Red Sea, the processes of ecosystem transformation are more evident and profound.

The second group is represented by the seas of greater size, depth, and water volume. The lowest average depth and capacity are found in the Okhotsk Sea and the maximum in the Arabian Sea (Table 20-2). Correspondingly, the estimate of ECC is highest in the Arabian Sea and lowest in the Okhotsk, South China, and East China Sea LMEs. Small as these three are, they exceed relevant estimates calculated for the seas of the first group (Table 20-1). One can hypothesize that in the second group dramatic ecological perturbations would take place over smaller water areas adjacent to estuaries of large rivers or to ports. According to such characteristics as water volume per km^3 of sea surface area and the total water volume, all LMEs that are regarded as semi-enclosed seas fall in the range between the two extremes, namely, between the Yellow Sea (small water volume) and the Arabian Sea (large water volume and great depths).

Some hydrodynamic characteristics of the first group should be examined separately. This is because specific modes of water exchange or freshwater inflow into the sea are important for the LMEs with small water volume and low ECC, for these may contribute either to reducing or to increasing the ecological carrying capacity. Two main pathways of water inflow into an LME are land runoff, indicative of river discharge, and rainfall over the LME. Given that river discharge depends to a great extent on the corresponding drainage basin, let us consider the role that the land plays in freshwater discharge into the seas. One must consider the ratio between the drainage

Table 20-1. Some geomorphologic characteristics of semi-enclosed LMEs (first group)

LME[a]		Sea area (1×10^6 km^2) (*x*)	Average depth (m)	Sea volume (1×10^3 km^3) (*y*)	Volume per km^2	References
1 (4)	Gulf of California	177	818	145	0.818	Dobrovolsky and Zalogin 1992
2 (10)	West Greenland Shelf	689	861	593	0.861	
3 (19)	Barents Sea	1,424	230	316	0.230	Bruevich 1948; Skjoldal and Rey 1989
4 (21)	North Sea	575	90	47	0.090	Northern Europe's Seas 1989
5 (22)	Baltic Sea	400	55	22	0.055	Northern Europe's Seas 1989
6 (26)	Black Sea	423	1,272	538 (54)[b]	1.272	Remizova 1992
7 (38)	North Australian Shelf	1,449	260	377	0.260	Dobrovolsky and Zalogin 1992
8 (41)	East China Sea	836	306	258	0.306	Dobrovolsky and Zalogin 1992
9 (42)	Yellow Sea	416	38	17	0.038	Dobrovolsky and Zalogin 1992
10 (33)	Red Sea	460	437	201	0.437	Bogdanova 1966
11 (—)	Azov Sea	38	8	0.32	0.008	Remizova 1992
12 (—)	Seto Inland Sea	22	37.3	0.82	0.037	Goda 1992; Nakanishi et al. 1992
13 (—)	Adriatic Sea	138	254	35	0.254	Bombace 1993; Dederen 1992

[a]Numbers in parentheses correspond to Figure 1-1 in Sherman (1996).

[b]The oxygenated water column; the rest is a H_2S^- zone where anaerobic life consists of mostly microorganisms (Vinogradov 1989).

Table 20-2. Some geomorphologic characteristics of semi-enclosed LMEs (second group)

LME*		Sea area (1×10^6 km^2) (x)	Average depth (m)	Sea volume (1×10^3 km^3) (y)	References
1 (1,47)	Bering Sea	2,315	1,640	3,796	Dobrovolsky and Zalogin 1992
2 (5)	Gulf of Mexico	1,555	1,511	2,332	Darnell and Defenbaugh 1990
3 (12)	Caribbean Sea	2,777	2,429	6,745	Dobrovolsky and Zalogin 1992; Richards and Bohnsack 1992
4 (17,18,20)	Norwegian Sea	1,340	1,735	2,325	Dobrovolsky and Zalogin 1992
5 (25)	Mediterranean Sea	2,505	1,538	3,754	Dobrovolsky and Zalogin 1992; Dederen 1992
6 (32)	Arabian Sea	4,832	3,006	14,523	Dobrovolsky and Zalogin 1992
7 (34)	Bay of Bengal	2,172	2,586	5,616	Dobrovolsky and Zalogin 1992
8 (35)	South China Sea	3,537	1,024	3,623	Pauly and Christensen 1993
9 (36)	Sulu-Celebes Seas	788	1,570	2,050	Dobrovolsky and Zalogin 1992
10 (37)	Indonesian Sea	1,483	2,935	7,440	Dobrovolsky and Zalogin 1992
11 (44)	Sea of Japan	1,062	1,536	1,631	Dobrovolsky and Zalogin 1992
12 (46)	Sea of Okhotsk	1,340	821	1,316	Dobrovolsky and Zalogin 1992

*Numbers in parentheses correspond to Figure 1-1 in Sherman (1996).

area and the area of sea surface (*z*/*x* in Table 20-3). Among the LMEs with a small water volume (first group), this ratio is smallest for the North Australian Shelf and largest for the Azov Sea. The Baltic, Black, and Yellow Seas show similar values. This correlation is important for evaluating the anthropogenic load. The greater the estimate, the more developed the catchment area and the better the potential for reducing the population load. In other words, the more developed the drainage area, the lower the ECC value of the sea.

Another important relationship of both natural (erosion) and anthropogenic impacts on an LME is the ratio between river influx and precipitation over the LME (Table 20-3). It is interesting to compare two small seas, for example, the Azov and the Seto Inland Seas. The shallow Azov Sea has twice the surface area of the Seto Inland Sea and a larger drainage basin, whereas the Seto Inland Sea has a small drainage area but an annual freshwater influx that is remarkably constant. From this, one can deduce that the ecological carrying capacity of the Seto Inland Sea is greater, given that the drainage area is less developed and rainfall more abundant.

Freshwater inflow and water exchange with neighboring water bodies are also important because pollutants are exchanged between them. The ratio between the annual freshwater supply and water volume of the sea is given in Table 20-4. In the Azov Sea in previous years, the annual river flow (when the rivers have not been overused) and the precipitation over the sea made up one-sixth of the total sea volume. Hypothetically, in 6 years, freshwater would replace seawater in the absence of evaporation and exchange with other seas. Similarly, freshwater replaces 1/14 of the seawater volume for the Seto Inland Sea.

Annual inflow of freshwater into the North Sea equals 1/89 of the sea volume, while the total exchange would hypothetically be completed in 3 years (Northern Europe's Seas 1989). The explanation lies in a pronounced exchange between the North Sea and the Atlantic Ocean, and the Norwegian and Baltic Seas. However, the influx of more polluted Baltic water contributes to anthropogenic loading of the sea. Yet a rapid, 3-year exchange promotes a considerable increase in the ECC of the North Sea LME. In the Azov Sea, freshwater input decreased by twofold due to a fourfold reduction in river inflow (11 km^3 per year). The overuse of the sea and a sharp decline of river inflow resulted in a residence time estimated at about 12 years.

A high rate of exchange is apparently inherent to the Yellow Sea, too. The ratio of freshwater input to the total sea volume is 1/20. Water exchange is theoretically completed in 6 years because this LME receives a huge amount of water from the East China Sea. The annual influx of freshwater in the Black Sea constitutes 1/92 of the oxygenated water volume and 1/921 of the total volume of the sea. Hence, the entire cycle of exchange would take less than 1 year. Although the annual share contributed by freshwater inflow to the Adriatic Sea is 1/140 of the total sea volume, the entire exchange would take about 50 years because of the intensive exchange with the Mediterranean Sea through the Ionic Sea.

Table 20-5 gives the geomorphologic and hydrographic characteristics and the totals of the first group. Two semi-enclosed seas of a very low ECC are distinguished, namely, the Baltic and the Yellow Seas. The Azov Sea, not presently included in the list of LMEs (Sherman 1994, 1995), should be included in this group. These three seas, the Gulf of California, and the Seto Inland Sea make up the association that displays a similar ECC if based on the considered characteristics.

The Barents, North, East China, and Red Seas are characterized by higher ECC. However, the determining characteristic is the duration of water exchange (residence time) because of its much greater scale compared with other constituents. It is noteworthy that in the Baltic Sea the rate of exchange is lower than in the other seas.

The seas of the second group are distinguished by a greater average depth and much greater (2.5- to 27-fold) volume compared with others. As Table 20-2 shows, the volume is very small for the Sea of Japan and the Sea of Okhotsk, and very large in the Arabian, Caribbean, and Indonesian Seas. These seas are all characterized by a less developed drainage area as compared with the area of sea surface. More or less developed are the drainage basins of the South China

Table 20-3. Some characteristics of water circulation in semi-enclosed LMEs (first group)

LME	Ratio z/x*	River flow (km^3/y) (r)	Precipitation (km^3/y) (p)	r/p	$r + p$ (km^3/y)	References
Gulf of California	5.3	min. (4)	min.			
Barents Sea	0.9	200	427	0.47	627	
North Sea	1.5	206	320	0.64	526	Heip 1989; Dederen 1992; Tent 1987
Baltic Sea	4.4	436	223	1.96	659	Northern Europe's Seas 1989
Black Sea	5.6	346	238	1.45	584	Remizova 1992; Reshetnikov 1984
North Australian Shelf	0.8	70	1,449	0.05	1,519	
East China Sea	1.0	7,250	1,672	4.34	8,922	
Yellow Sea	4.8	470	380	1.20	850	Milliman 1991
Red Sea	9.0 (?)	min.	min.	—	—	
Azov Sea	16.9	41	14	2.93	55	Bronfman and Chlebnikov 1985
Seto Inland Sea	2.5	36	22	1.64	58	Goda 1992
Adriatic	1.5	153	97	1.58	250	Franco and Michelato 1992; Dorten et al. 1991; Catanni and Corni 1992

*z = drainage area.

Table 20-4. Water exchange characteristics in semi-enclosed LMEs (first group)

LME	River inflow and precipitation (km^3/y) ($r + p$)	Sea volume (1×10^3 km^3/y) (y)	Ratio $y/(r + p)$	Residence time (yr)	References
Gulf of California	min.	145	—	25 (?)	
Barents Sea	627	316	500	4	
North Sea	526	47	89	3	Northern Europe's Seas 1989
Baltic Sea	659	22	33	22.7	Northern Europe's Seas 1989
Black Sea	584	54/538	92/921	1450 (?)	Bogdanova 1969; Remizova 1992
North Australian Shelf	1,519	377	248	20 (?)	
East China Sea	8,922	258	30	10 (?)	
Yellow Sea	850	17	20	15 (?)	Milliman 1991
Red Sea	min.	201		9	Bogdanova 1966
Azov Sea	55	0.32	6	6, 4 (?)	Remizova 1992
Seto Inland Sea	58	0.82	14	10 (?)	Goda 1992
Adriatic Sea	250	35	140	18 (?)	Vollenweider et al. 1992

Table 20-5. Geomorphologic and hydrographic characteristics of semi-enclosed LMEs (first group)

LME	Depth average	Volume of the sea (y)	Ratio z/x	Ratio $y/(r + p)$	Residence time (yr)	Total (sum of previous columns)	Range
Gulf of California	11	8	3	2	2	26	5
West Greenland Shelf	12	13	12	?	—	—	
Barents Sea	7	11	10	4	11	43	11
North Sea	5	6	7	4	12	37	8
Baltic Sea	4	4	5	5	3	24	1
Black Sea	13	7	2	3	1	26	4
North Australian Shelf	6	12	12	5	4	39	9
East China Sea	9	10	9	9	7	44	12
Yellow Sea	3	3	4	10	6	26	3
Red Sea	10	9	11	1	9	40	10
Azov Sea	1	1	1	12	10	25	2
Seto Inland Sea	2	2	6	11	8	29	6
Adriatic Sea	8	5	8	6	5	32	7

Table 20-6. Some characteristics of water circulation in semi-enclosed LMEs (second group)

LME	Ratio z/x	River inflow (km³/y) (r)	Precipitation (km³/y) (p)	r/p	$r + p$ (km³/y)	References
Bering Sea	1.0	400	1,600	0.25	2,000	Ivanenkov 1964
Gulf of Mexico	1.8	1,060	3,000	0.35	4,060	Richards and McGowan 1989
Caribbean Sea	0.4	400 (?)	5,000	0.08	5,400	Dagg et al. 1991
Norwegian Sea	min.	?	?		?	
Mediterranean Sea	0.5	430	1,000	0.43	1,430	
Arabian Sea	0.2	140	1,200	0.12	1,340	
Bay of Bengal	1.8	1,600	1,100	1.45	2,700	Dwivedi 1993
South China Sea	2.0	1,000	7,074	0.14	8,074	
Sulu-Celebes Seas	0.23	50 (?)	2,360	0.02	2,410	
Indonesian Seas	0.6	150 (?)	4,450	0.03	4,600	
Sea of Japan	1.4	350 (?)	1,270	0.28	1,620	
Sea of Okhotsk	1.0	270 (?)	940	0.29	1,210	

Sea, the Gulf of Mexico, and the Bay of Bengal (Table 20-6), which is beneficial for their ECCs. The relatively high ratio between the drainage area and the sea surface for the Sea of Japan is solely due to the fact that two-thirds of the catchment area of the Amur River is included. For the Arabian Sea, the area of sea surface is at times as large as the drainage area, resulting in the largest estimated ECC among seas of the same group. For the Sulu-Celebes Seas the situation is similar (Table 20-6).

Because of the poorly developed catchment areas, the estimate of precipitation falling over the sea by far exceeds that of the river influx. The only exception is the Bay of Bengal, which receives freshwater from rivers as large as the Ganges, the Brahmaputra, and the Irrawaddy. River influx and precipitation contribute only a small portion of the total volume of the seas under consideration (Tables 20-7 and 20-8). If freshwater sources alone were considered in exchange

Table 20-7. Water exchange characteristics in semi-enclosed LMEs (second group)

LME	River inflow and precipitation (km³/y) ($r + p$)	Sea volume (1×10 km³/y) (y)	Ratio $y/(r + p)$	Residence time (yr)	Reference
Bering Sea	2,000	3,796	1,898	38	Ivanenkov 1964; Arsenjev 1967
Gulf of Mexico	4,030	2,332	574	3	Brown et al. 1991
Caribbean Sea	5,400	2,429	450	3	
Norwegian Sea	?	2,325	—		
Mediterranean Sea	1,430	3,754	2,625	94	Dederen 1992
Arabian Sea	1,340	14,523	10,838	8 (?)	
Bay of Bengal	2,700	5,616	2,080	10 (?)	
South China Sea	8,074	3,623	449	15 (?)	
Sulu-Celebes Seas	2,410	2,050	850	25 (?)	
Indonesian Seas	4,600	7,440	1,617	20 (?)	
Sea of Japan	1,620	1,631	1,007	20	
Sea of Okhotsk	1,210	1,316	1,088	10 (?)	

Table 20-8. The semi-enclosed LMEs (second group) arranged according to geomorphologic and hydrographic characteristics

LME	Depth average	Volume of the sea (y)	Ratio z/x	Ratio $y/(r+p)$	Residence time (yr)	Total (sum of previous columns)	Range
Bering Sea	7	8	6	8	2	31	7
Gulf of Mexico	3	5	3	3	11	25	4
Caribbean Sea	9	10	9	2	10	40	9
Norwegian Sea	8	4	12	—	—	—	—
Mediterranean Sea	5	7	8	10	1	31	6
Arabian Sea	12	12	11	11	9	55	11
Bay of Bengal	10	9	2	9	8	38	8
South China Sea	2	6	1	1	6	16	1
Sulu-Celebes Seas	6	3	10	4	3	26	5
Indonesian Seas	11	11	7	7	5	41	10
Sea of Japan	4	2	4	5	4	19	2
Sea of Okhotsk	1	1	5	6	7	20	3

calculations, the total exchange period would be from 0.5 to 11 thousand years. Estimates of the actual rate of exchange are very high because these involve other seas and oceans (the Mediterranean Sea, with the longest residence time of 94 years, being an exception). The lowest relative values known are for the Caribbean Sea and the Gulf of Mexico, due to intense hydrodynamics in the area where the Gulf Stream originates.

High exchange rates and large volumes found in the second group of semi-enclosed LMEs are responsible for much greater ECCs than those of the first group. Modest ecological carrying capacity values have been obtained for the Sea of Japan, the South China Sea, and Okhotsk Sea LMEs, and the greatest values for the Arabian, Caribbean, and the Indonesian Seas. It is quite possible that the Norwegian Shelf has a considerable ECC, but difficulties involved in describing relevant characteristics made us refrain from defining its place among the examined seas.

Thus, with regard to the ECC value, the Yellow, Baltic, Azov, and Black Sea LMEs are at one extreme of the range and the Arabian, Caribbean, and Indonesian Sea LMEs are at the other. The ECCs of the rest of the LMEs are intermediate in value. With this classification in mind, let us compare the estimates of ECC and those of human loading for each sea.

Population Loading of Semi-enclosed Large Marine Ecosystems

Each human living on the planet contributes to the anthropogenic load on marine ecosystems. Population densities in coastal areas are not uniform and hence the population load on them differs from one to another. Metabolic products from people and animals as well as industrial products contribute to the natural cycles. Metabolic rates of humans and domestic animals only depend slightly on the level of economic development and consumption of energy and hydrocarbons; steel and chemical production is much more dependent. Recent statistics enable us to easily obtain transitory coefficients to relate the consumption of various resources to the extent that the cycle is disturbed in different countries. That is why the population density, described as the number of people settled in the drainage area per unit area or volume of the sea, is an important integral characteristic of anthropogenic loading on the LME. It enables the estimation of loadings as they relate to the economic development of a country.

The world ocean contains about a billion km^3 of water (Northern Europe's Seas 1989). Humans inhabiting its drainage area number 5.66 billion. Thus, the population density is about 5.7 individuals per km^3 of the world ocean on the average. Uneven distribution of the human population in coastal areas results in a disturbance in the balance between the ECC and the population density load (PL) and, ultimately, in an ecological disaster of a different scale. High population density is unusual for the semi-enclosed LMEs of low ECC.

Table 20-9 presents population density per unit of sea area for the first group of LMEs. The Yellow Sea LME has the greatest density, whereas estimates for the Seto Inland Sea, the East China Sea, and the Azov Sea are an order of magnitude lower. The North, Baltic, Black, and Adriatic Seas have population densities 3 to 4 times lower. Population density is low for terrestrial areas adjacent to the Gulf of California, the Red Sea, and the North Australian Shelf, and extremely low for the West Greenland Shelf and the Barents Sea, compared with the average obtained for the world ocean.

The morphology of the LME may also influence the actual human load on the water body, due to the fact that the water column per unit area may vary considerably among LMEs. For example, in the first group of semi-enclosed LMEs the range of variation in depth is very large—160-fold, the average depth being 8 m for the Azov Sea and 1,272 m for the Black Sea. Therefore, human population per unit of sea volume is a more reliable index describing population load on the LME. In the first group, the maximum estimate—109,375 humans per km^3—has

Table 20-9. Population load on semi-enclosed LMEs (first group)

LME	Volume of the sea ($1 \times 10\ km^3$)	Population in drainage area (mln.)	Population per km^2 of the sea	Population per km^3 of the sea	Range	Reference
Gulf of California	145	8	45	55	11	
West Greenland Shelf	593	0.05	0.07	0.09	13	
Barents Sea	316	3.5	2.5	11	12	
North Sea	47	210	365	4,468	4	Northern Europe's Seas 1989
Baltic Sea	22	75	205	3,472	5	Wulff et al. 1990
Black Sea	51	140	330	2,593	6	
North Australian Shelf	377	30 (?)	21	80	9	Kort and Salnikov 1982
East China Sea	258	400	1,307	1,550	7	
Yellow Sea	17	600	15,789	35,294	3	Kort and Salnikov 1981
Red Sea	201	15	33	75	10	Kort and Salnikov 1982
Azov Sea	0.32	35	921	109,375	1	
Seto Inland Sea	0.82	29	1,318	38,366	2	Goda 1992; Nakanishi et al. 1992
Adriatic Sea	35	32	232	914	8	Giordani and Hammond 1992; Marchetti et al. 1989

been calculated for the Azov Sea. Population density for the Yellow and Seto Inland Seas is one-third that of the Azov Sea and is an order of magnitude lower for the North, Baltic, Black, and Adriatic Seas. Low population densities have been reported for the Gulf of California, the North Australian Shelf, and the Red Sea, and an extraordinarily low density was reported for the West Greenland Shelf. Thus, the semi-enclosed LMEs of the first group can be divided into five groups depending on the number of humans per unit of sea volume. The maximum value exceeds the average estimated for the world ocean.

Semi-enclosed LMEs of the second group show quite low values for population density. The greatest estimates, 414 persons per km^2 and 160 persons per km^3 (Table 20-10), are for the Bay of Bengal, whereas those for the Bering, Norwegian, and Okhotsk Seas are lower than the average obtained for the world ocean. The Mediterranean and the South China Seas also have densely settled basins. For the rest of the seas, population density is moderate (equivalent to 15 to 45 humans per km^3). The Bay of Bengal has the maximum population load. It decreases slightly for the Mediterranean and South China Seas and is 2 to 4 times lower for most of the LMEs included in the second group.

When the semi-enclosed LMEs are arranged according to decreasing population density per unit of sea volume, the top ten are the Azov, Yellow, Seto Inland, North, Baltic, Black, South China, Adriatic, and Mediterranean Seas LMEs and the Bay of Bengal LME. It should be noted that of these, four seas are part of the Mediterranean Sea system and six are in Europe. The rest of the LMEs are in Southeast Asia, where about half the human population lives.

The relationship between the level of ECC and the population load (PL) was examined. The lowest estimates of the two indicate that low ECC would match high PL. Conversely, high estimates are characteristic of high ECC combined with low PL. The case when both ECC and PL are low demonstrates a balance. In the first group of seas, a disbalance, that is, low ECC and high PL, is obvious in the Azov and Yellow Seas. The low ECC of the Baltic Sea is accompanied by a decreased PL. For the Seto Inland Sea both characteristics are high, and the Black Sea shows low ECC with a pronounced PL.

Combining the two indices allows a more accurate comparison among the LMEs. Thus, the Azov, Yellow, Baltic, Black, and Seto Inland Seas are the five seas of the first group whose ECC and PL are not in balance. Similarly, in the second group the five seas are the South China Sea, the Sea of Japan, the Gulf of Mexico, and the Mediterranean and Sulu-Celebes Seas. Arrangement of the LMEs according to the anthropogenic loading adds the Bay of Bengal and the South China Sea of the second group. Of the 25 seas, the most favorable is the West Greenland Shelf LME, where ECC is high and PL lowest. This is followed by the Norwegian, Bering, Okhotsk, and Barents Seas, for their basins are scarcely settled. However, use of the seas for nuclear weapons trials or burying radioactive waste could markedly change the present classification.

Ecological Transformation in Semi-enclosed Large Marine Ecosystems

Human loadings of LMEs include the impact of chemical pollutants and eutrophication on the water body. Chemical pollutants (heavy metals, hydrocarbons, pesticides, PCBs, and other synthetic chemicals) are known to suppress the physiological activity of cellular membranes, cells, and tissues of marine organisms. This leads to increases in mortality of larvae, perturbations in development, retarded growth, disturbed sexual development, and a variety of genetic defects in chromosomes and genes (Dorigan and Harrison 1987).

Ultimately, the increased content of these pollutants in the environment may result in lower population numbers and poorer viability of surviving organisms, and the extinction of some species. Primarily, this refers to upper trophic levels. Bacteria, microalgae, protozoans, and the smaller aquatic animals with a short life span are capable of surviving due to mutations that

Table 20-10. Population load on semi-enclosed LMEs (second group)

LME	Volume of the sea (1×10 km^3)	Population in drainage area (mln.)	Population per km^2 of the sea	Population per km^3 of the sea	Range	Reference
Bering Sea	3,796	0.5	0.22	0.13	12	
Gulf of Mexico	2,332	100	64	43	4	Lewis 1989
Caribbean Sea	6,745	100	36	15	9	Kort and Salnikov 1984
Norwegian Sea	2,325	2.5	0.5	0.2	11	
Mediterranean Sea	3,754	350	140	93	2	Pastor 1988
Arabian Sea	14,523	280	58	19	7	Myers 1993; Kort and Salnikov 1982
Bay of Bengal	5,616	900	414	160	1	Mathur 1992; Milliman et al. 1989
South China Sea	6,323	500	141	79	3	Gomez 1988
Sulu-Celebes Seas	2,050	55	70	27	6	
Indonesian Seas	7,440	120	13	16	8	Gomez 1988
Sea of Japan	1,631	60	56	37	5	Okamura 1989
Sea of Okhotsk	1,316	4	3	3.0	10	

mitigate the effects of pollutants in the following generations. High reproduction rates and, hence, abundance of these organisms and high genetic turnover are the factors that make positive mutations dominant and protect these organisms against toxins.

According to Vollenweider (1992), "eutrophication is the process of enrichment of waters with plant nutrients, primarily nitrogen and phosphorus." This process can involve replacement of dominant microalgal species and changes in the dynamics of seasonal blooms. Microalgal blooms can be harmful. Macrophytes can be forced from the sublittoral region because of a decreased photic layer. Moreover, oxidation of organic matter left after phytoplankton blooms induces an oxygen deficiency and therefore suppresses the viability of macrophytes.

Degradation of benthic communities is mostly a result of oxygen deficiency in the near-bottom water layer subject to intense inputs of organic matter from the bloom. Some phytoplankton have been found to produce toxins that suppress the vital activity of both benthic and pelagic communities, including fish (Viviani 1992; Underdal et al. 1989). However, pelagic communities become extinct not solely because of toxins but also from oxygen deficiencies. The thick soup of microalgae can kill fish by clogging their gills. Thus, on the one hand eutrophication involves enhanced reproduction of some groups and species of phytoplankton and, on the other hand, hampers normal functioning of the marine ecosystem.

The extent to which the balance between ECC and PL is disturbed determines the severity of the impact from chemical pollutants and excessive nutrients. The disturbed balance will manifest itself by disrupting the functioning of the ecosystem. Therefore, examining the semi-enclosed LMEs with very low ECC can contribute to forecasting changes and resulting conditions in seas with greater ECC. In turn, general trends in the dynamics of marine ecosystem degradation can be revealed and described through studies of the disturbances in how the ecosystem functions for a series of LMEs where increasing ECC is combined with decreasing population loading.

Intense anthropogenic impact can be responsible for reduced biodiversity, dominant species shifts, altered energy flows, a dramatically reduced photic layer, degradation of macrophytic, benthic, and pelagic communities, and decreased productivity of upper trophic levels. Some species vanish because of a declining body size and a reduced life span. Ecosystem stress and dysfunction can generate conditions favorable for colonizers that start reproducing actively owing to a feeding preference for the lowest trophic level, including newly dominant phytoplankton and zooplankton species.

Let us focus our attention on the most significant disturbances in the functioning of different LMEs. Ecosystem stress is characterized, at first, by a reduction in the numbers and biomass of dominant autotrophic and heterotrophic species in the upper trophic levels (Table 20-11). The next step is the decreasing range of populations of the dominant species, with some species vanishing completely. In the East China and Mediterranean Seas and in the Bay of Bengal—where the discrepancy between ECC and population load is most pronounced—these processes take place on a local level, whereas on the whole the ecosystem is not affected. The available data are too few to make judgments about the Seto Inland Sea.

Among the rest of the LMEs examined, the most threatened are the Azov and Black Seas. Apparently, critical conditions are developing in the Yellow Sea also. Population numbers and ranges are decreasing in the North and Baltic Seas, but extinction of species is not as pronounced as in the Azov and Black Seas. For the Adriatic Sea, the situation is similar, although there have been recent improvements (Marchetti 1992).

Another stage in the transformation of ecosystems is the degradation of macrophytes and of benthic and pelagic communities (Table 20-12). These large-scale phenomena are usually provoked by eutrophication and dumping. Intense oxidation of organic substances causes hypoxic and anoxic conditions that permit only anaerobes to survive. In the Black Sea, the anoxic near-bottom environment was found to extend for tens of thousands of kilometers (Zaitsev et al. 1987). Elmgren (1989) reported broken food chains with links missing for the Baltic Sea LME

Table 20-11. Transformation of ecosystems: Area, species numbers and biomass, and biodiversity (decreasing)

	Numbers, biomass (decreasing)		Area (decreasing)		Biodiversity (decreasing)	
LME	Minimum	Maximum	Minimum	Maximum	Minimum	Maximum
Azov Sea	–	+	–	+	–	+
Seto Inland Sea	?	–	?	–	?	–
Yellow Sea	–	+	–	+	+ (?)	–
North Sea	+	–	+	–	+	–
Baltic Sea	–	+	–	+	+	–
Black Sea	–	+	–	+	–	+
Adriatic Sea	–	+	–	+	+	–
East China Sea	+	–	+	–	+	–
Bay of Bengal	+	–	+	–	+	–
Mediterranean Sea	+	–	+	–	+	–

because of an oxygen deficiency in the benthic layer over an area as vast as 100,000 km^2. Bladder wrack (*Fucus vesiculosus*) vanished over 3,500 km^2 in the Gulf of Bothnia (Rapport 1989). In coastal waters of the North Sea, plankton blooms are an occasional catastrophe (Underdal et al. 1989) causing mass mortality of sea urchins, starfish, molluscs, and fishes, and, ultimately, the degradation of near-shore benthic communities over hundreds of kilometers down to 30 m depth.

In the Adriatic Sea, new biocenoses are found well adapted to occasionally adverse conditions (Crema et al. 1991). Though Tang (1993) showed a marked increase of primary production in the Bohai Sea, no data are available that would provide evidence of a catastrophic plankton bloom in the Yellow Sea. Perhaps the effects of eutrophication are considerably mitigated by intensively cultivated *Laminaria japonica* and mariculture of other macrophytes. Regrettably, no data are available on hypoxic and anoxic conditions in the Seto Inland Sea, the Bay of Bengal, or the Mediterranean Sea.

A decreasing photic layer is convincing evidence for eutrophication in LMEs. Reproducing intensively, microalgae form a thick covering that does not allow light to penetrate downward. As a result, in the Black Sea LME macrophytes are vanishing. Their proportion of the total autotro-

Table 20-12. Transformation of ecosystems: Degradation of phytobenthic, benthic, and pelagic communities

LME	Phytobenthic	Benthic	Pelagic
Azov Sea	Yes	Yes	Yes
Seto Inland Sea	?	?	?
Yellow Sea	Yes	Yes	Yes
North Sea	?	Yes	Yes
Baltic Sea	Yes	Yes	Yes
Black Sea	Yes	Yes	Yes
Adriatic Sea	Yes	Yes	Yes
East China Sea	No	No	No
Bay of Bengal	No	No	No
Mediterranean Sea	No	No	No

Table 20-13. Transformation of ecosystems: Decreasing transparency, replacement of dominant species, and impact on energy flow

LME	Decreasing transparency	Replacement of dominant species	Transformation of energy flow
Azov Sea	Yes	Yes	Yes
Seto Inland Sea	Yes	?	?
Yellow Sea	Yes (?)	Yes	Yes
North Sea	Yes (?)	Yes	Yes
Baltic Sea	Yes	Yes	Yes
Black Sea	Yes	Yes	Yes
Adriatic Sea	Yes	Yes	Yes
East China Sea	No (?)	No	No
Bay of Bengal	No (?)	No	No
Mediterranean Sea	No	No	No

phic organisms is now negligible. For example, formerly huge fields of *Phillophora* spp. in the northwestern Black Sea decreased from 10 mt to 0.5 mt and less (Zaitsev et al. 1987). Reduction of water transparency by 1.5 to 2 times in the offshore zone and 5 to 10 times in coastal waters (Zaitsev et al. 1987) is responsible for the *Phillophora* field elimination. The shallowness of the Azov Sea neutralizes the effect from the reduced photic layer. In the Baltic Sea, bladder wrack is becoming extinct (Baden et al. 1990; Cederwall and Elmgren 1990).

In the North, Baltic (Cederwall and Elmgren 1990), and Adriatic Seas (Marchetti et al. 1989; Krstulovic and Solic 1990), the photic layer shows a tendency to decrease. In the Yellow and Seto Inland Seas the photic layer is not as deep as in the past. Happily, in the East China and Mediterranean Seas and the Bay of Bengal the photic layer persists within the previously reported ranges.

Most obvious are shifts in species composition. Transformations in food webs and energy flows are observed in the first seven of the LMEs given in Table 20-13, and in the Azov and Black Seas, where there are occasional outbreaks of jellyfishes, *Aurelia aurita* and *Rhizostoma pulma.* Twenty years ago their biomass in the Black Sea amounted to 300–500 mt (Shushkina and Arnautov 1987) and 13.5 mt in the Azov Sea (Studenikina et al. 1991). In the late 1980s the ctenophore *Mnemiopsis leidyi* was introduced from Chesapeake Bay in the ballast water of a ship. *M. leidyi* was found to prevail in the Azov Sea's zooplankton beginning in 1989–1990, with its biomass reaching more than 4 mt (Studenikina et al. 1991). In the Black Sea, the biomass estimates were 400 mt for *M. leidyi* and 60 mt for *A. aurita* and *R. pulma* in 1990 (Vinogradov et al. 1992). Transformations on such a large scale in the structure of the energy flow have been recorded in no other seas but the Azov and Black Seas.

Ecological stress is possible in the Yellow Sea; however, the large-scale cultivation of macrophytes is a beneficial factor alleviating the situation. At the same time, transformation of ichthyocenoses and profound changes in the structure of catches have been attributed to overexploitation (Tang 1989, 1993).

The improved conditions of the Baltic and Seto Inland Seas owing to improvements in sewage treatment and technology lead to the conclusion that these seas are ecologically safe compared with the Azov and Black Seas. An indicator of this is an observed increase in productivity of fish per unit area of the Baltic Sea (Elmgren 1989; Hansson and Rudstan 1990). Despite a fast rate of exchange, the ecological situation is similar in the North Sea, although the biocenoses of many coastal areas require some level of recovery. In recent years the Adriatic Sea has shown signs of a dramatically deteriorating ecological state. Its northern part may become as afflicted as the Black Sea unless eutrophication is reduced.

Conclusion

Comparing the ecological carrying capacity and population density (population load) in the drainage areas of semi-enclosed LMEs per km^3 of sea allows a quantitative means of identifying the most threatened marine ecosystems. The situation in which the ECC is extremely low and the population load is high is indicative of ecosystem stress. This implies a decreasing population range and biomass of dominant species, reduced biodiversity, degradation of phytobenthic and benthic communities, replacement of dominant species, and reduced and transformed food chains and energy flows in the ecosystem.

By using these characteristics, the comparative analysis of the relationship between ECC and PL has shown that ECC and PL are correlated, contributing to our knowledge about the actual ecological states of the semi-enclosed LMEs. The greatest discrepancy between the two indices has been found in the Azov Sea, the Seto Inland Sea, and the Yellow Sea. The survey of the biological conditions of marine ecosystems permitted corrections to be made and has shown that in addition to the Azov Sea, the Black Sea is also subject to severe ecosystem stress. This may be due to the high river inflow and a high population density in the drainage area of the northwestern part of the Black Sea. Uneven distributions of human population loadings may result. Formerly this area was known for high biodiversity and productivity at every trophic level. The degradation of the northwestern Black Sea ecosystem is thought to have affected the ecological state of the entire Black Sea.

With regard to ecology, the Yellow and the Adriatic Seas are similar to the Azov and Black Seas. As with the Black Sea, population load on these seas is distributed rather unevenly. The impact is maximum in the Bohai Sea (part of the Yellow Sea) where the Hwang Ho and the Liao Ho empty and exchange with other seas is limited, and in the northern part of the Adriatic Sea where the Po River enters. Indeed, stress and degradation spreading from local subsystems are a potential danger to the entire sea.

The several hot ecological areas that have been identified for the Baltic Sea are the Gulfs of Riga, Finland, and Bothnia. There the ecological stress is most pronounced. On the whole, however, the situation is not that bad and even tends to improve in the North Sea and Baltic Sea LMEs. Ecological conditions in the rest of the LMEs are relatively good, although local disturbances may occasionally affect the structure and functioning of the ecosystems.

At present, 73% of the total population of the planet has settled around the drainage basins of the semi-enclosed LMEs (Table 20-14). Constituting just 6% of the total volume of the world ocean, these seas require a special management approach. It should be noted that the total volume of the first group of semi-enclosed LMEs contributes but 0.2% to the total volume of the world ocean, although 29.6% of the total population inhabits their basins. The combined volume of the second group makes up 5.8% of the world ocean. The drainage basins of these seas contain 43.4% of the total population.

To ensure sustainable development in the drainage basins of semi-enclosed LMEs, countries must be willing to cooperate to mitigate the stresses on the systems. The Baltic countries are

Table 20-14. Population density of the semi-enclosed LMEs and the world ocean

	Volume (mln km)	Percentage	Population density (billion)	Percentage
World ocean, total	1,000	100	5.66	100
First group of the semi-enclosed LMEs	2.066	0.2	1.677	29.6
Second group of the semi-enclosed LMEs	57.851	5.8	2.472	43.4
World ocean without semi-enclosed LMEs	940	94	1.511	27

showing the way to a most fruitful and rewarding cooperation. Along with the long-term coordination of scientific research in this field, a financial system has been developed to support management of ecologically advanced technologies that save natural resources and recycle waste (Nordic Programme for Environment 1989; Sweden 1990; Marine Pollution 1990; Wulff and Niemi 1992).

The European Community is undertaking similar work on the North Sea (Howarth 1988; Kasoulides 1988). The measures aimed at protecting the Mediterranean Sea include ratifying the Convention for the Protection of the Mediterranean Sea against pollution, and a series of protocols on dumping and reducing land-based sources of pollution. The most rewarding measure is the data exchange that has been organized by CIESM. Governments of Black Sea countries have ratified a convention for the protection of the Black Sea against pollution and developed an Action Plan. Still, intergovernmental activity is not yet as vigorous as that in the Baltic region.

In closing, special emphasis should be placed on the vital importance of international cooperation in promoting and implementing the concept of sustainable development and protection of the ecosystems of semi-enclosed seas.

ACKNOWLEDGMENTS

I sincerely thank Dr. K. Sherman for sharing a deep interest in the problem. His friendly concern and support with relevant references were a powerful stimulus to perform the reported work devoted to LMEs of the world ocean.

References

Albani, A. D., Rickwood, P. C., Favero, V. M., and Serandrei Barbero, R. 1989. The geochemical anomalies in sediments on the shelf near the Lagoon of Venice, Italy. Marine Pollution Bull. 20 (9):438–442.

Arsenjev, V. S. 1967. Currents and Water Masses of the Bering Sea (in Russian). Moscow: Nauka Publ.

Baden, S. P., Loo, L., Pihl, L., and Rosenberg, R. 1990. Effects of eutrophication on benthic communities including fish: Swedish West Coast. Ambio 19 (3):113–122.

Bogdanova, A. K. 1966. Hydrological characteristics of the Red Sea and Gulf of the Aden according to the observations of the R/V "Academician Kovalevsky" (in Russian). Oceanology Res. 15:45–68.

Bogdanova, A. K. 1969. Hydrology of Bosporus and Adjacent Water of the Black Sea. In V. A. Vodjanitski (ed.), Water Exchange through Bosporus and Influence on Hydrological Regime of the Black Sea. Kiev: Naukova Dumka Publ., pp. 5–121.

Bombace, G. 1993. Ecological and Fishing Features of the Adriatic Sea. In K. Sherman, L. M. Alexander, and B. D. Gold (eds.), Large Marine Ecosystems: Stress, Mitigation, and Sustainability. Washington, DC: AAAS Press, pp. 119–136.

Bronfman, A. M., and Chlebnikov, E. P. 1985. The Azov Sea. The Backgrounds of Restoration (in Russian). Leningrad: Hydrometeoizdat.

Brown, B. E., Browder, J. A., Powers, J., and Goodyear, C. D. 1991. Biomass, Yield Models, and Management Strategies for the Gulf of Mexico Ecosystem. In K. Sherman, L. M. Alexander, and B. D. Gold (eds.), Food Chains, Yields, Models, and Management of Large Marine Ecosystems. Boulder, CO: Westview Press, pp. 125–164.

Bruevich, S. V. 1948. The essay of hydrochemistry of the Barents Sea (in Russian). Proceed. Institute Oceanogr. 10(22):80–116.

Catanni, O., and Corni, M. G. 1992. The Role of Zooplankton in Eutrophication, with Special Reference to the Northern Adriatic Sea. In R. A. Vollenweider, R. Marchetti, and R. Viviani (eds.), Marine Coastal Eutrophication. Amsterdam: Elsevier, pp. 137–158.

Cederwall, H., and Elmgren, R. 1990. Biological effects of eutrophication in the Baltic Sea, particularly the coastal zone. Ambio 19 (3):109–112.

Crema, R., Castelli, A., and Prevedelli, D. 1991. Long term eutrophication effects on macrofaunal communities in Northern Adriatic Sea. Marine Pollution Bull. 22(10):503–508.

Dagg, M., Grimes, C., Lohrenz, S., McKee, B., Twilley, R., and Wiseman, W. 1991. Continental Shelf Food Chains of the Northern Gulf of Mexico. In K. Sherman, L. M. Alexander, and B. D. Gold (eds.), Food Chains, Yields, Models, and Management of Large Marine Ecosystems. Boulder, CO: Westview Press, pp. 67–106.

Daily, G. C., and Ehrlich, P. R. 1992. Population, sustainability, and Earth's carrying capacity. BioScience 42(10):761–771.

Darnell, R. M., and Defenbaugh, R. E. 1990. Gulf of Mexico: Environmental overview and history of environmental research. Amer. Zool. 30(1):36.

Dederen, L. H. T. 1992. Marine Eutrophication in Europe: Similarities and Regional Differences in Appearance. In R. A. Vollenweider, R. Marchetti, and R. Viviani (eds.), Marine Coastal Eutrophication. Amsterdam: Elsevier, pp. 663–672.

Dobrovolsky, A. D., and Zalogin, B. S. 1992. Regional Oceanology (in Russian). Moscow: Moscow University Press.

Dorigan, J. V., and Harrison, F. L. (eds.). 1987. Physiological Responses of Marine Organisms to Environmental Stresses. Washington, DC: U.S. Department of Energy.

Dorten, W. S., Elbaz-Poulichet, F., Mart, L. R., and Martin, J. M. 1991. Reassessment on the river input of trace metals into the Mediterranean Sea. Ambio 20(1):26.

Dwivedi, S. N. 1993. Long-Term Variability in the Food Chains, Biomass Yields, and Oceanography of the Bay of Bengal Ecosystem. In K. Sherman, L. M. Alexander, and B. D. Gold (eds.), Large Marine Ecosystems: Stress, Mitigation, and Sustainability. Washington, DC: AAAS Press, pp. 43–52.

Elmgren, R. 1989. Man's impact on the ecosystem of the Baltic Sea: Energy flows today and at the turn of the century. Ambio 18(6):326–332.

Franco, P., and Michelato, A. 1992. Northern Adriatic Sea. In R. A. Vollenweider, R. Marchetti, and R. Viviani (eds.), Marine Coastal Eutrophication. Amsterdam: Elsevier, pp. 35–62.

Giordani, P., and Hammond, D. E. 1992. Benthic Fluxes and Nutrients. In R. A. Vollenweider, R. Marchetti, and R. Viviani (eds), Marine Coastal Eutrophication. Amsterdam: Elsevier.

Goda, T. 1992. Eutrophication and Its Remedies in the Seto Inland Sea with Respect to the Characteristics of Bed Deposition. In R. A. Vollenweider, R. Marchetti, and R. Viviani (eds.), Marine Coastal Eutrophication. Amsterdam: Elsevier, pp. 1217–1238.

Gomez, E. D. 1988. Overview of environmental problems in the East Asian seas region. Ambio 17(3): 166–169.

Hansson, S., and Rudstan, L. G. 1990. Eutrophication and Baltic fish communities. Ambio 19(3): 123–125.

Heip, C. 1989. The ecology of the estuaries of Rhine, Meuse and Scheldt in the Netherlands. Scientia Marina 53(23):457–463.

Howarth, M. J. 1988. The (NERC) North Sea Community project, 1987–1992. Adv. Underwater Tech. 16:247–260.

Ivanenkov, V. N. 1964. Hydrochemistry of the Bering Sea (in Russian). Moscow: Nauka Publ.

Johnes, P. J., and O'Sullivan, P. E. 1989. The natural history of Slapto Ley. Natural reserve XVIII. Nitrogen and phosphorus losses from the catchment: An export coefficient approach. Field Studies 7(2):285–309.

Kasoulides, G. C. 1988. London North Sea Conference. Marine Pollution Bull. 19(3):97–99.

Kort, V. G., and Salnikov, S. S. (eds.). 1981. The Pacific Ocean (in Russian). Leningrad: Nauka Publ.

Kort, V. G., and Salnikov, S. S. (eds.). 1982. The Indian Ocean (in Russian). Leningrad: Nauka Publ.

Kort, V. G., and Salnikov, S. S. (eds.). 1984. The Atlantic Ocean (in Russian). Leningrad: Nauka Publ.

Krstulovic, N., and Solic, M. 1990. Long-term study of heterotrophic bacteria as indicators of eutrophication of the open-middle Adriatic Sea. Estuarine, Coastal and Shelf Science 30(6):611–617.

Lewis, J. 1989. Changing the fate of the Gulf. EPA Journal 15(5):46–47.

Marchetti, R. 1992. The Problems of the Emilia Romagna Coastal Waters: Facts and Interpretations. In R. A. Vollenweider, R. Marchetti, and R. Viviani (eds.), Marine Coastal Eutrophication. Amsterdam: Elsevier, pp. 21–33.

Marchetti, R., Provini, A., and Crosa, G. 1989. Nutrient load carried by the River Po into the Adriatic Sea, 1968–87. Marine Pollution Bull. 20(4):168–172.

Marine Pollution '90. 1990. Action Programme. Stockholm: Swedish Environmental Protection Agency.

Mathur, H. S. 1992. Population Growth and Food Resource Development in India: The Needs for a Second Green Revolution. Washington, DC: International Geographic Congress, pp. 398–399.

Milliman, J. D. 1991. Flux and Fate of Fluvial Sediment and Water in Coastal Seas. In R. F. C. Mantoura, J. M. Martin, and R. Wallast (eds.), Ocean Margin Processes in Global Change. New York: John Wiley & Sons, pp. 69–89.

Milliman, J. D., Broadus, J. M., and Gable, F. 1989. Environmental and economic implications of rising sea level and subsiding deltas: The Nile and Bengal examples. Ambio 18(6):340–345.

Myers, N. 1993. Population post-Rio. People and Planet 2(1):31.

Nakanishi, H., Ukita, M., Sekine, M., Fukugawa, M., and Murakami, S. 1992. Eutrophication Control in the Seto Inland Sea. In R. A. Vollenweider, R. Marchetti, and R. Viviani (eds.), Marine Coastal Eutrophication. Amsterdam: Elsevier, pp. 1239–1256.

Nordic Programme for the Environment. 1989. Copenhagen: Nordic Council of Ministers.

Northern Europe's Seas. Northern Europe's Environment. 1989. Stockholm: Norstedts Tryckeri AB.

Nowicki, B. L., and Oviatt, C. A. 1990. Are estuaries traps for anthropogenic nutrients? Evidence from estuarine mesocosms. Marine Ecol. Progr. Ser. 66(12):131–146.

Okamura, K. 1989. Future Ocean Space Utilization in Japan. EEZ Resources: Technol. Assess. Conference, Honolulu, Hawaii, pp. 22–26.

Pastor, X. 1988. The Mediterranean deserves better than this. Naturope 60:15–20.

Pauly, D., and Christensen, V. 1993. Stratified Models of Large Marine Ecosystems: A General Approach and an Application to the South China Sea. In K. Sherman, L. M. Alexander, and B. D. Gold (eds.) Large Marine Ecosystems: Stress, Mitigation, and Sustainability. Washington, DC: AAAS Press, pp. 148–174.

Polikarpov, G. G. 1981. Ecological background for management of the hydrosphere under anthropogenic load (in Russian). J. Hydrobiol. 17(6):39.

Polikarpov, G. G., and Egorov, V. N. 1981. Capability of marine ecosystems to neutralize radioactive and chemical pollutants (in Ukrainian). News of Academy of Sciences (Ukraine) 2:73–84.

Rapport, D. J. 1989. Symptoms of pathology in the Gulf of Bothnia (Baltic Sea): Ecosystem response to stress from human activity. Biological J. Linnean Soc. 37(12):33–49.

Rast, W., and Lee, G. F. 1983. Nutrient loading estimates for lakes. J. Environ. Engr. Div. ASCE 109(3):502–517.

Remizova, S. S. 1992. Peculiarities of the hydrological regime of the Black and Azov Seas for the last seven thousand years (in Russian). Water Resources 4:115–123.

Reshetnikov, V. I. 1984. River inflow to the Black Sea from Turkey drainage area (in Russian). Meteorol. Hydrol. 11:114–117.

Richards, W. J., and Bohnsack, J. A. 1992. The Caribbean Sea: A Large Marine Ecosystem in Crisis. In K. Sherman, L. M. Alexander, and B. D. Gold (eds.), Large Marine Ecosystems: Patterns, Processes, and Yields. Washington, DC: AAAS Press, pp. 44–53.

Richards, W. J., and McGowan, M. F. 1989. Biological Productivity in the Gulf of Mexico: Identifying the Causes of Variability in Fisheries. In K. Sherman, and L. M. Alexander (eds.), Biomass Yield and Geography of Large Marine Ecosystems. Boulder, CO: Westview Press, pp. 287–326.

Shepard, F. P. 1963. Submarine Geology. New York: Harper and Row.

Sherman, K. 1995. Sustainability, biomass yields, and health of coastal ecosystems: an ecological perspective. Mar. Ecol. Prog. Ser. 112: 277–301.

Sherman, J. 1994. Achieving regional cooperation in the management of marine ecosystems: the use of the large marine ecosystem approach. Ocean & Coastal Management 29(1–3): 165–185. Elsevier.

Shushkina, E. A., and Arnautov, G. N. 1987. *Medusae Aurelia* and Black Sea Plankton in May 1984 (in Russian). In M. E. Vinogradov and M. V. Flint (eds.), Present State of the Black Sea Ecosystem. Moscow: Nauka Publ., pp. 186–196.

Skjoldal, H. R., and Rey, F. 1989. Pelagic Production and Variability of the Barents Sea Ecosystem. In K. Sherman and L. M. Alexander (eds.), Biomass Yields and Geography of Large Marine Ecosystems. Boulder, CO: Westview Press, pp. 241–268.

Studenikina, E. I, Volovik, S. P., Mirsoyan, I. A., and Luts, G. I. 1991. *Mnemiopsis leidyi* in the Azov Sea (in Russian). Oceanology 31(6):981–985.

Sweden. 1990. A living environment. Main proposals. Swedish Government Bill 1990/91. Riksdagen, Stockholm.

Tang, Q. 1989. Changes in the Biomass of the Yellow Sea Ecosystem. In K. Sherman and L. M. Alexander (eds.), Biomass Yields and Geography of Large Marine Ecosystems. Boulder, CO: Westview Press, pp. 7–35.

Tang, Q. 1993. Effects of Long-Term Physical and Biological Perturbations on the Contemporary Biomass Yields of the Yellow Sea Ecosystem. In K. Sherman, L. M. Alexander, and B. D. Gold (eds.), Large Marine Ecosystems: Stress, Mitigation, and Sustainability. Washington, DC: AAAS Press, pp. 79–93.

Tent, L. 1987. Contaminated sediments in the Elbe estuary: Ecological and economic problems for the Port of Hamburg. Hydrobiologia 149(2):189–199.

Underdal, B., Skulberg, O. M., Dahl, E., and Aune, T. 1989. Disastrous bloom of *Chrysochromulina polylepis* (Prymnesiophyceae) in Norwegian coastal waters 1988—Mortality in marine biota. Ambio 18(5):265–268.

Vinogradov, M. E. 1989. The vertical zonality of pelagic ecosystem in the Black Sea (in Russian). Reports Acad. Sci. 306(3):717–721.

Vinogradov, M. E., Sapozhnikov, V. V., and Shushkina, E. A. 1992. The Black Sea Ecosystem (in Russian). Moscow: Nauka Publ.

Viviani, R. 1992. Eutrophication, Marine Biotoxins, Human Health. In R. A. Vollenweider, R. Marchetti, and R. Viviani (eds.), Marine Coastal Eutrophication. Amsterdam: Elsevier, pp. 631–662.

Vollenweider, R. A. 1992. Coastal Marine Eutrophication: Principles and Control. In R. A. Vollenweider, R. Marchetti, and R. Viviani (eds.), Marine Coastal Eutrophication. Amsterdam: Elsevier, p. 120.

Vollenweider, R .A., Rinaldi, A., and Montanari, G. 1992. Eutrophication, Structure and Dynamics of a Marine Coastal System: Results of Ten Year Monitoring along the Emilia-Romagna Coast (Northwest Adriatic Sea). In R. A. Vollenweider, R. Marchetti, and R. Viviani (eds.), Marine Coastal Eutrophication. Amsterdam: Elsevier, pp. 63–106.

Wulff, F., and Niemi, A. 1992. Priorities for the restoration of the Baltic Sea—A scientific perspective. Ambio 21(2):193–195.

Wulff, F., Stigebrandt, A., and Rahm, L. 1990. Nutrient dynamics of the Baltic Sea. Ambio 19(3):126–133.

Zaitsev, Y. P., Garkavaya, G. P., Nesterova, D. A., Polishchuk, L. N., and Tsokur, A. G. 1987. Present State of the North Western Black Sea Ecosystem under Anthropogenic Impact. In M. E. Vinigradov and M. V. Flint (eds.), Present State of the Black Sea Ecosystem. Moscow: Nauka Publ., pp. 216–229.

21

Alteration of Fatty Acid Composition of Coastal Diatoms under Cadmium Stress

Jae Sam Yang and Wonho Yih

Abstract

The fatty acid composition of eight coastal diatom clones was determined using gas chromatography. The diatoms, *Skeletonema costatum, Chaetoceros* spp., *Cylindrotheca closterium*, and *Thalassiosira* spp., which are dominant species in Korean waters, were isolated from four different areas: the neritic and coastal Yellow Sea, a tidal pool, and Jinhae Bay. The diatoms were kept in BWM media, and the effects of acute cadmium toxicity on the composition of fatty acids were investigated. The major fatty acid components were 14:0, 16:0, 16:1, 18:1, and 18:2. *Thalassiosira* spp. were distinguished by a high content of 16:1, whereas other species showed a high content of both 14:0 and 16:1. Interspecific differences in the fatty acid composition of a control group were noted. A sharp decline in polyunsaturated fatty acids (PUFA) was found for most of the diatoms tested in media with cadmium. The diatoms from polluted waters such as Jinhae Bay maintained a stable composition of PUFA under cadmium stress, as compared with those from unpolluted waters. PUFA composition is suggested as a possible indicator for evaluating ecosystem health.

The lipid composition in phytoplankton and the adaptation of cell membranes to stressed environments have been frequently studied (Fisher and Schwarzenbach 1978; Gillan et al. 1983; Henderson et al. 1990). Algal lipids are associated with membranes and energy systems; for example, polar lipids, important constituents of algal membranes, are present in diatoms as phospholipids, sulfolipids, and galactosyl lipids. When fatty acids are esterified as nonpolar triacylglyceride, they can be important energy storage products in some cells. Such structural and functional roles of fatty acids in cells suggest that fatty acids might also be a useful indicator of adaptive strategies to stressed environments. Morris and Culkin (1976) found that most investigators have examined algal fatty acids as biogeochemical tracers or as assessment tools for their nutritional importance to herbivores. Here, the alteration of fatty acid composition in coastal diatoms under toxic metal stress was investigated. The implications of these results are discussed with reference to some physiological processes of fatty acids. The fatty acid composition of eight diatoms from four habitats was studied, and the differences in fatty acid composition among the clones are discussed.

Table 21-1. Species, IC-50, and test concentrations of cadmium for the toxicity experiment

Species	Clones	$IC\text{-}50_{fby}$*	Test concentration (ppm)
Chaetoceros spp.	GS-12	0.0028	0.05
	J-21	1.920	1.00
	Br-16	ND	1.00
Skeletonema costatum	NC-37	0.975	1.00
	Br-4	ND	1.00
	YS-4	ND	1.00
Thalassiosira spp.	NC-29	1.265	1.00
Cylindrotheca closterium	YS-26	ND	1.00

*$IC\text{-}50_{fby}$ is inhibition concentration by 50% for final biomass yield.

Methodology

Culture

Experiments were conducted with cultures of seven species of centric diatoms (*Thalassiosira* spp., *Chaetoceros* spp., and *Skeletonema costatum*) and one pennate diatom (*Cylindrotheca closterium*). In all experiments, cells were grown in BWM media using 2-liter autoclavable containers (Nalgene Co.). Cultures were maintained at 20±1°C under continuous light from cool-white fluorescent lamps (5,800 lux).

The experiments were performed under three different conditions. One group (Control) was incubated in a media without cadmium. A second group (Cd-1X) was grown in unaltered media for 3 days, then Cd was added at IC-50 concentration and incubated for the next 24 hours. The third group (Cd-2X) was incubated in media with IC-50 concentration of Cd for 4 to 5 days. The culture experiment was initiated with an inoculation (1 ml) and terminated when the harvested volume of the cellular mass reached at least 2 ml. For the culturing of all species, batch cultures were done until the end of the exponential growth. Table 21-1 lists the species, clones, and IC-50 and test concentrations of Cd (ppm) for the toxicity experiment.

Analyses

Two liters of culture media and diatom cells were centrifuged to make a pellet. Twice, chloroform:methanol (2:1, V/V) containing BHT (5 mg/100 ml) was added to the samples and kept for 12 hours in a dark, cool place. The samples were transferred to a borosilicate glass tube and vacuum-dried at 40°C. Next, 2 ml of BF-methanol were added to the samples; they were then placed in a brass reaction tube in boiling water for 30 minutes. After cooling, 4 ml of hexane containing BHT (5 mg/100 ml) were added to the tube before 1 ml of 5 N NaOH was slowly added to the hexane. The hexane fraction was transferred to another container, and again hexane was added to the tube. The two hexane fractions were combined and then treated with anhydrous sodium sulfate to remove moisture, and concentrated to 0.1 ml. Two μl of the hexane fraction were used to determine the fatty acid composition by gas chromatography. For the first step of identification, the retention times of peaks on the chromatograms were compared with the standard materials (Sigma Chemical Co.), but a gas chromatograph/mass spectrometer (Hewlett Packard Model 5890A-5970) was used for the final confirmation of a fatty acid. The operating conditions of the gas chromatograph are described in Table 21-2.

Table 21-2. Operating conditions of Hewlett-Packard 5890 II gas chromatograph

Column	Supelco Wax 10, 30 m × 0.53 mm I.D.
Carrier gas	Nitrogen
Split ratio	1/20
Column temperature	Initial: 180°C; Final: 240°C
	Gradient: 5°C / min.
Injection port temperature	250°C
Detector temperature	280°C
Detector	FID

Results and Discussion

Fatty Acid Composition of the Control Group

Fifteen fatty acids were identified from the lipid extracts. The fatty acids common to all the clones were 14:0, 14:1, 16:0, 16:1, 18:0, 18:1, 18:2, 18:3, 20:4, 20:5, 22:1, and 22:6. The major components were 14:0, 16:0, 16:1, 18:1, and 18:2. *Thalassiosira* spp. were distinguished by a high content of 16:1, whereas other species showed high contents of both 14:0 and 16:1.

Clones of the same species originating from different locations were compared. Table 21-3 compares the fatty acid composition of the control group of clones collected from different areas. Each species showed their own species-specific composition of fatty acids. For example, the two *Skeletonema costatum* clones, which were collected from a tidal pool and the Yellow Sea, showed the identical fatty acid composition. Diatoms, as a group, have a characteristic fatty acid composition (Chuecas and Riley 1969; Shaw 1966). However, centric diatoms exhibit differences from pennate diatoms (DeMort et al. 1972). The interspecific difference in the fatty acid composition among the three centric diatoms in this work was much more evident than their intraspecific variability, in spite of their contrasting natural habitats (Table 21-3).

The fatty acid content of any species may vary widely in response to environmental conditions. Ackman et al. (1964) have shown that the fatty acid composition of *Skeletonema costatum* changes with the age of the culture. Numerous studies have shown that low-nitrogen media cause an accumulation of fat in the cell (Shaw 1966). Shaw also demonstrated that fat content increased at high light intensities.

Table 21-3. Major fatty acids of the diatom clones in the control group

Species	Clones	Location and habitat	Major fatty acids
Chaetoceros spp.	GS-12	Yellow Sea, neritic	14:0, 16:1, 16:0, 18:1
	J-21	Jinhae Bay, eutrophicated	14:0, 16:1, 16:0, 18:1
	Br-16	Boryung Coast, coastal	14:0, 16:1, 16:0, 18:2
Skeletonema costatum	NC-37	Kunsan Coast, tidal pool	14:0, 16:1, 16:0, 18:1
	Br-4	Boryung Coast, coastal	14:0, 16:1, 16:0, 18:2
	YS-4	Yellow Sea, neritic	14:0, 16:1, 16:0, 18:1
Thalassiosira spp.	NC-29	Kunsan Coast, tidal pool	16:1, 16:0, 14:0, 22:6
Cylindrotheca closterium	YS-26	Yellow Sea, neritic	16:1, 16:0, 14:0, 18:1

Note: "Yellow Sea" represents unpolluted neritic Yellow Sea water, "Jinhae Bay" represents highly eutrophicated water with some metal pollution, "Boryung" represents typical Yellow Sea coastal water, and "tidal pool" is an anonymous tidal pool near Kunsan.

Table 21-4. Reduction in number of polyunsaturated fatty acids (PUFAs) due to cadmium stress

			Number of PUFAs	
Habitat	Clones	Species	Without Cd	With Cd
Eutrophicated (Jinhae Bay)	J-21	*Chaetoceros* spp.	5	0
Tidal pool	NC-29	*Thalassiosira* spp.	5	4
	NC-37	*Skeletonema costatum*	5	1
Neritic and coastal (Yellow Sea)	GS-12	*Chaetoceros* spp.	5	3
	Br-16	*Skeletonema costatum*	5	3
	Br-4	*Skeletonema costatum*	5	1
	YS-4	*Skeletonema costatum*	5	5
	YS-26	*Cylindrotheca closterium*	5	5

Alteration of Fatty Acid Composition under Cadmium Stress

A sharp decline in the number of polyunsaturated fatty acids (PUFAs) was found in most of the species tested in media containing IC-50 equivalents of Cd (Table 21-4).

Stability of the PUFAs also varied with and without Cd present. A clone from the heavily polluted area J-21 (Jinhae Bay) maintained a more stable PUFA composition under Cd stress than those from other, unpolluted waters. For example, all of the PUFAs of YS-26 and YS-4 were transformed to saturated or single unsaturated fatty acids. A similar pattern of PUFA transformation was found in the clones from the tidal pools and the neritic waters. Hence, this study suggests that fatty acid composition and PUFA transformation in diatoms may be possible indicators of coastal ecosystem health.

References

Ackman, R. G., Jangaard, P. M., Hoyle, R. J., and Brockerhoff, H. 1964. Origin of marine fatty acids. I. Analyses of the fatty acids produced by the diatom *Skeletonema costatum.* J. Fish. Res. Bd. Can. 21:747–56.

Chuecas, L., and Riley, J. P. 1969. Component fatty acids of the total lipids of some marine phytoplankton. J. Mar. Biol. Assoc. U.K. 49:97–116.

DeMort, C. L., Lowry, R., Tinsley, I., and Phinney, H. K. 1972. The biochemical analysis of some estuarine phytoplankton species. I. Fatty acid composition. J. Phycol. 8:211–216.

Fisher, N. S., and Schwarzenbach, R. P. 1978. Fatty acid dynamics in *Thalassiosira pseudonana* (Bacillariophyceae): Implications for physiological ecology. J. Phycol. 14(2):143–150.

Gillan, F. T., Fisher, N. S., and Johns, R. B. 1983. The effects of copper and zinc on the fatty acids and carotenoids in the marine diatom *Asterionella japonica.* Botanica Marina 26:255–257.

Henderson, R. J., Mackinlay, E. E., Hodgson, P., and Harwood, J. L. 1990. Differential effects of the substituted pyridazinone herbicide sandoz 9785 on lipid composition and biosynthesis in photosynthetic and non-photosynthetic marine microalgae. J. Exp. Bot. 41(227):729–736.

Morris, R. J., and Culkin, F. 1976. Marine lipids: Analytical techniques and fatty acid ester analysis. Oceanogr. Mar. Biol. Ann. Rev. 14:391–433.

Shaw, R. 1966. Polyunsaturated fatty acids of microorganisms. Advan. Lipid Res. 4:114–174.

22

Gills and Skin as Bioindicators of Water Pollution in Fish from Freshwater and a Marine Habitat

Sjoerd Wendelaar Bonga, Rob Lock, Li Jie, Chen Minshan, Chen Bijuan, and Deng Jingyao

Abstract

The delicate epithelia covering the gills and skin of fish are easily damaged by water pollutants. This causes disturbance of ion homeostasis of the blood and is associated with reduced growth and reproduction and increased mortality. We have compared the effects of cadmium (50 μgCd/l) in freshwater on the gills and skin of the freshwater tilapia, *Oreochromis mossambicus*, with the effects of the severely polluted Jiaozhou Bay (Yellow Sea) on this tissue of the greenling, *Hexagrammos otakii*. The effects were very similar and included necrosis and apoptosis of pavement cells of gills and skin and of the branchial chloride cells and, in both tissues, an increased secretion rate of pavement cells on the filaments and mucous cells. Furthermore, widening of intercellular spaces and infiltration in these spaces by leukocytes, including lymphocytes, macrophages, and granulocytes, were observed. We conclude that the changes occurring in gills and skin are good indicators of environmental pollutants in both freshwater and saltwater fish. The changes are indicators of disturbed ion regulation and are likely associated with reduced growth and reproduction. Thus, when observed in natural waters, they represent a serious threat to the survival of the fish population and possibly also to the freshwater and marine ecosystems concerned.

Pollution of freshwater and saltwater affects natural fish populations by reducing growth and reproduction. This will impair survival, and thus may result in reduction or elimination of natural fish populations (Dethlefsen and Tiews 1985; Wendelaar Bonga and Lock 1992).

Gills represent a large part of the body surface of fish, and their delicate structure is very sensitive to water pollutants such as heavy metals or organic chemicals such as pesticides. Not surprisingly, many structural alterations have been described in the gills of fish exposed to pollutants under experimental or natural conditions. These include cell degeneration, edematous swelling, and rupture of the epithelium covering the gills (Mallatt 1985). These effects seriously hamper the normal physiological functions of the gills, which include respiration and ion regulation. In particular, inhibited ion uptake and diffusive loss of ions resulting in low plasma ion levels have been frequently reported for freshwater fish; in saltwater fish, reduced branchial extrusion of ions leading to elevated plasma ion levels have been reported (Eddy 1981; Evans 1987; Wendelaar Bonga and Lock 1992).

Unlike the branchial epithelium, the skin surface has received little attention, although this tissue is in intimate contact with the environment. Our studies have shown that the skin responds as rapidly and profoundly to environmental pollutants as the branchial epithelium. The skin of fish, in particular the covering epithelium, is a metabolically very active border tissue consisting of different cell types. The filament cells are the dominant cell type. In addition, mucous cells, sensory cells, and an occasional chloride cell are present (Whitear 1986; Iger and Wendelaar Bonga 1994). The skin surface, similar to the surface of the branchial epithelium, is covered by a mucous layer that serves many functions, including protection against mechanical and chemical influences, and has antimicrobial properties. Environmental factors such as water acidification, elevated copper concentrations, or rapid temperature increases affect the structure of this epithelium, leading to degeneration of different cell types, increased mucous secretion, and infiltration of the skin by leukocytes (Iger and Wendelaar Bonga 1994; Iger et al. 1994a, 1994b).

Thus, the epithelia of gills and skin represent excellent bioindicators of water quality. The water quality of the Yellow Sea is becoming increasingly polluted by a variety of chemicals. In order to evaluate the effects of the degraded water quality on commercially important fish species, we studied the effects of cadmium exposure on some osmoregulatory parameters and on the ultrastructure of gills and skin of the tilapia, *Oreochromis mossambicus*, and compared these results with the ultrastructure of gills and skin of the greenling, *Hexagrammos otakii*, caught in two areas in the Yellow Sea near Qingdao.

Materials and Methods

Male tilapia (*Oreochromis mossambicus*), an African cichlid species weighing 10–20 g, were obtained from laboratory stock and were kept in tap water (for ionic composition see Wendelaar Bonga and Van der Meij 1989). Groups of eight fish were exposed to cadmium in concentrations of 1, 10, 50, 100, and 1,000 μgCd/l for 2, 4, 6, 14, and 35 days. Blood was collected from the caudal arteries; blood plasma osmolarity was determined in a micro-osmometer, and plasma Na^+, Cl^+, and total Ca were determined by atomic absorption spectrophotometry. Small pieces of gill filaments and the ventral skin covering the lower jaws from fish exposed for 10 days to 50 μgCd/l were dissected and prepared for electron microscopy as described by Wendelaar Bonga and Van der Meij (1989). Na^+/K^+-ATPase activity of the gills was determined according to Wendelaar Bonga et al. (1990). Greenlings of about 250 g body weight were collected by netting in Jiaozhou Bay near Qingdao (Figure 22-1), a heavily polluted part of the Yellow Sea. The water in this area contains high concentrations of heavy metals and organic pollutants. Metal concentrations determined at the sampling sites are given in Table 22-1. Gills and skin tissues were dissected and prepared for electron microscopy as described above.

Results

Electron Microscopy of Gills and Skin of Tilapia

THE GILLS. The gills consist of gill filaments with respiratory lamellae projecting from these filaments. The filament epithelium is multilayered and consists of cells containing many microfilaments, the filament cells, and an upper layer of pavement cells, typically with an outer cell

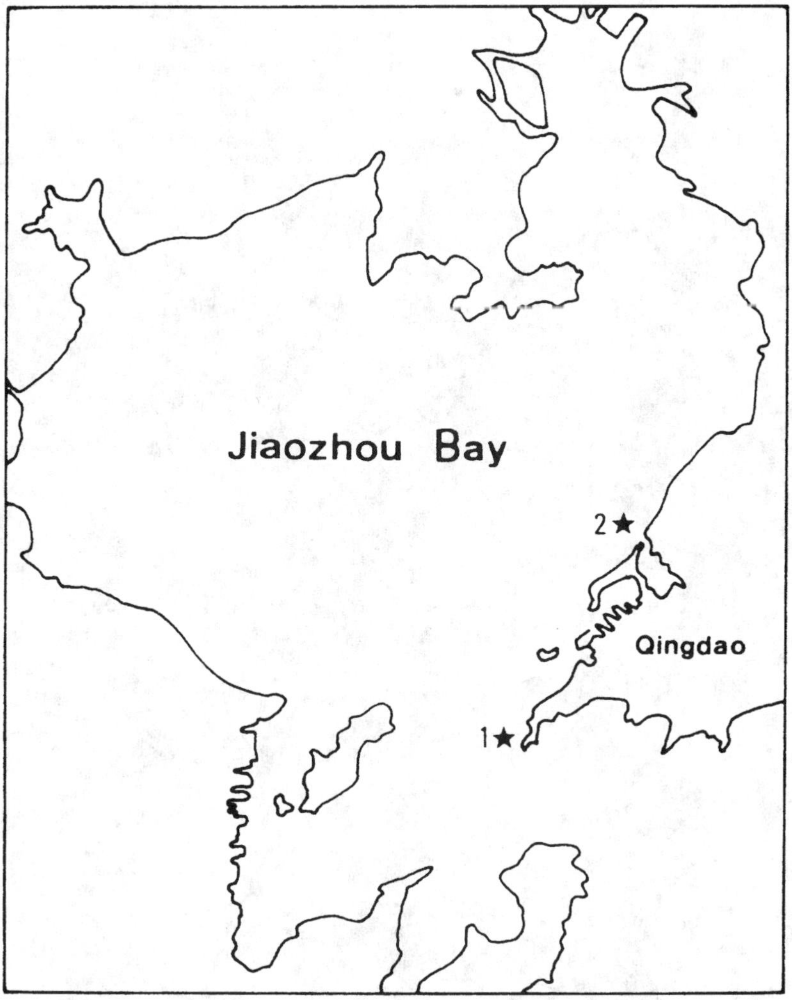

Figure 22-1. Sampling localities for greenling in the Yellow Sea near Qingdao, China: (1) at the entrance of Jiaozhou Bay, (2) estuary of Haibo River.

membrane folded in microridges. In between these cells chloride cells are found. The lamellae are covered by one layer of pavement cells, also called respiratory cells (Figure 22-2).

Examination of the gills of tilapia exposed for 10 days to 50 μgCd/l revealed substantial structural damage of the branchial epithelium. Necrosis and apoptosis, two types of cellular degeneration, were observed for pavement cells and chloride cells. Necrosis represents the acci-

Table 22-1. Concentrations of heavy metals (μg/l) in seawater at two locations in Jiaozhou Bay

Locality	Zinc	Cadmium	Lead	Copper
1[a]	0.59	0.34	5.85	0.32
2[b]	2.46	0.61	13.10	0.26

[a]Location 1 is the sea area around the entrance of Jiaozhou Bay (see Figure 22-1).
[b]Location 2 is the sea area around the estuary of Haibo River (see Figure 22-1).

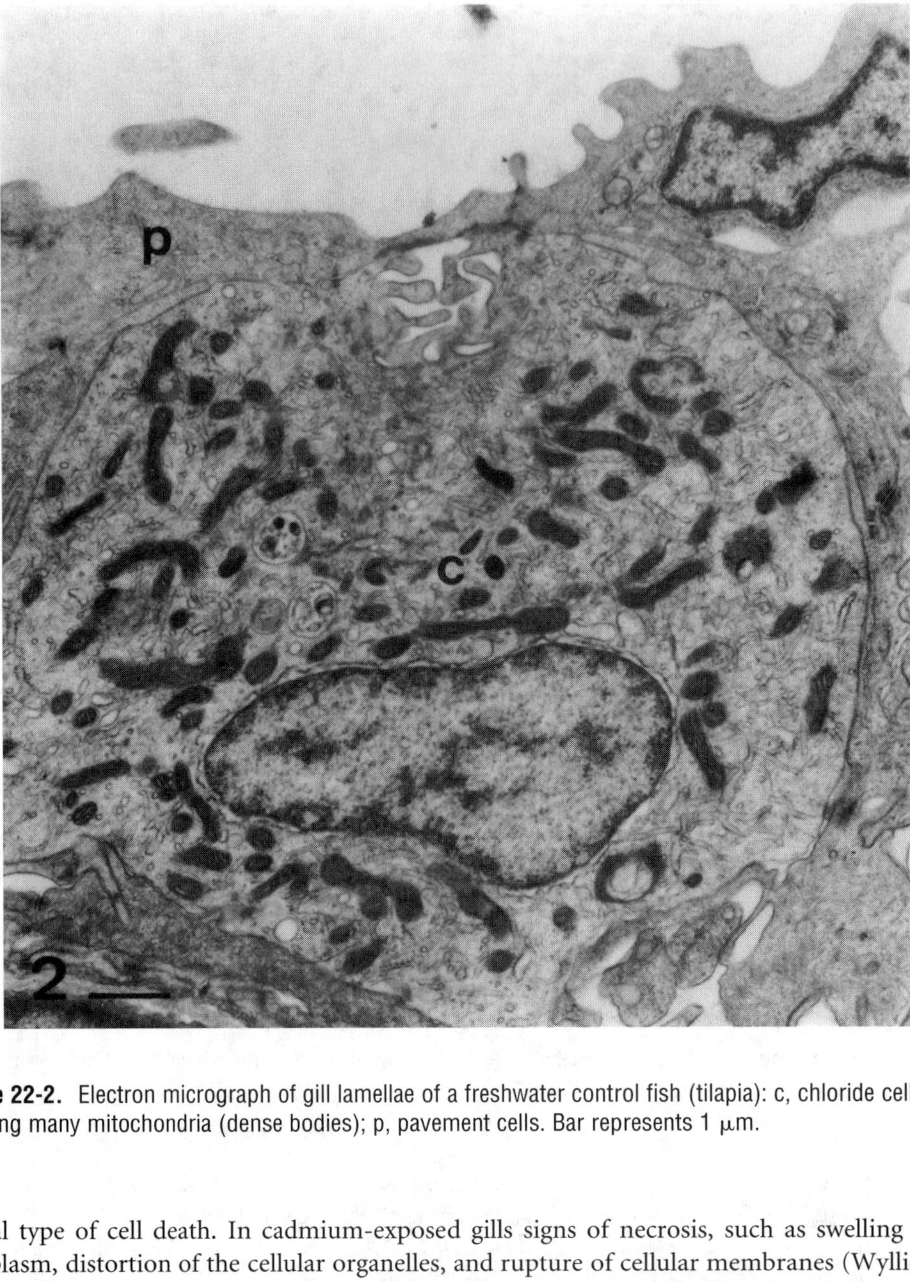

Figure 22-2. Electron micrograph of gill lamellae of a freshwater control fish (tilapia): c, chloride cell, showing many mitochondria (dense bodies); p, pavement cells. Bar represents 1 μm.

dental type of cell death. In cadmium-exposed gills signs of necrosis, such as swelling of the cytoplasm, distortion of the cellular organelles, and rupture of cellular membranes (Wyllie et al. 1980), were commonly found in both the pavement cells and the chloride cells. Apoptosis, representing physiologically controlled cell death (Wyllie et al. 1980), was also common. It is characterized by densification caused by cell shrinkage of the nuclei, the mitochondria, and the cytoplasmic matrix. Both apoptotic pavement cells and chloride cells were observed. Increase of intercellular spaces, indicating epithelial swelling, was also commonly found. In these spaces leukocytes were frequently present, in particular, lymphocytes and macrophages. These phenomena were rarely observed in the gills of the control group (Figure 22-3).

THE SKIN. The skin epithelium is mainly composed of filament cells, with an upper layer differentiated in pavement cells (Figure 22-4). In control fish some mucous cells and an occasional leukocyte were observed. In the cadmium-exposed fish, necrosis and apoptosis of the pavement cells were common, as were enlargement of intercellular spaces and the occurrence of leukocytes

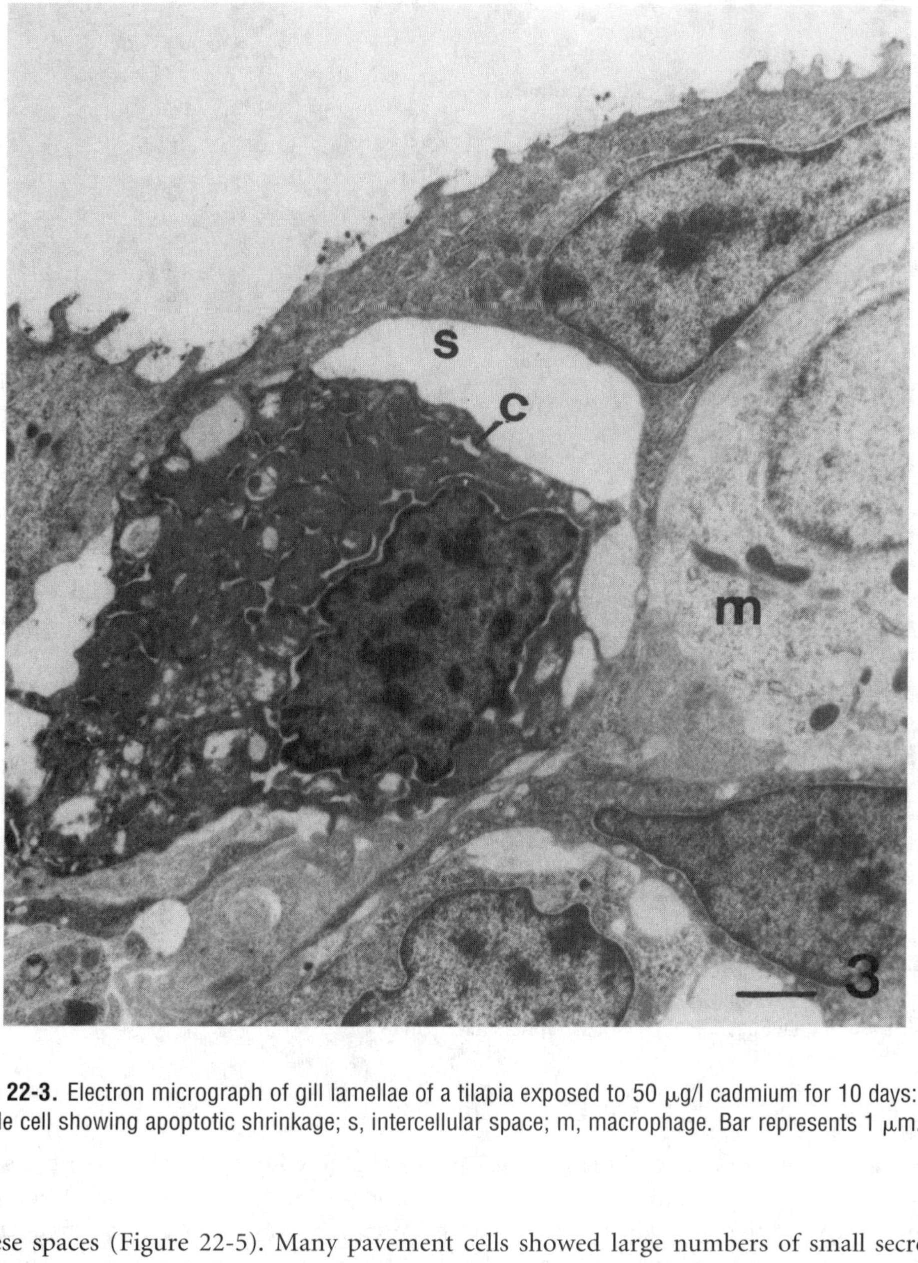

Figure 22-3. Electron micrograph of gill lamellae of a tilapia exposed to 50 μg/l cadmium for 10 days: c, chloride cell showing apoptotic shrinkage; s, intercellular space; m, macrophage. Bar represents 1 μm.

in these spaces (Figure 22-5). Many pavement cells showed large numbers of small secretory granules that released their contents by exocytosis at the surface. These phenomena were rarely observed in the control group.

PLASMA OSMOLARITY AND ELECTROLYTE LEVELS OF TILAPIA. Figure 22-6 shows the results of exposure of tilapia to different concentrations of cadmium. Whereas at 1 μgCd/l no difference was observed with the control levels, higher concentrations induced a concentration-dependent reduction in plasma osmolarity that was corrected after 14 days of exposure at 10 μgCd/l but led to a progressive reduction at higher concentrations and to rapid death within 8 days at the highest concentration used (1,000 μgCd/l). Analysis of plasma ion levels revealed that the drop in osmolarity may be caused by reduction of plasma Na^+, Cl^-, and Ca levels (Table 22-2). The specific activity of Na^+/K^+-ATPase, a key enzyme in Na^+ transport across the gills, was reduced at exposure levels, inducing substantial ion losses (Table 22-1).

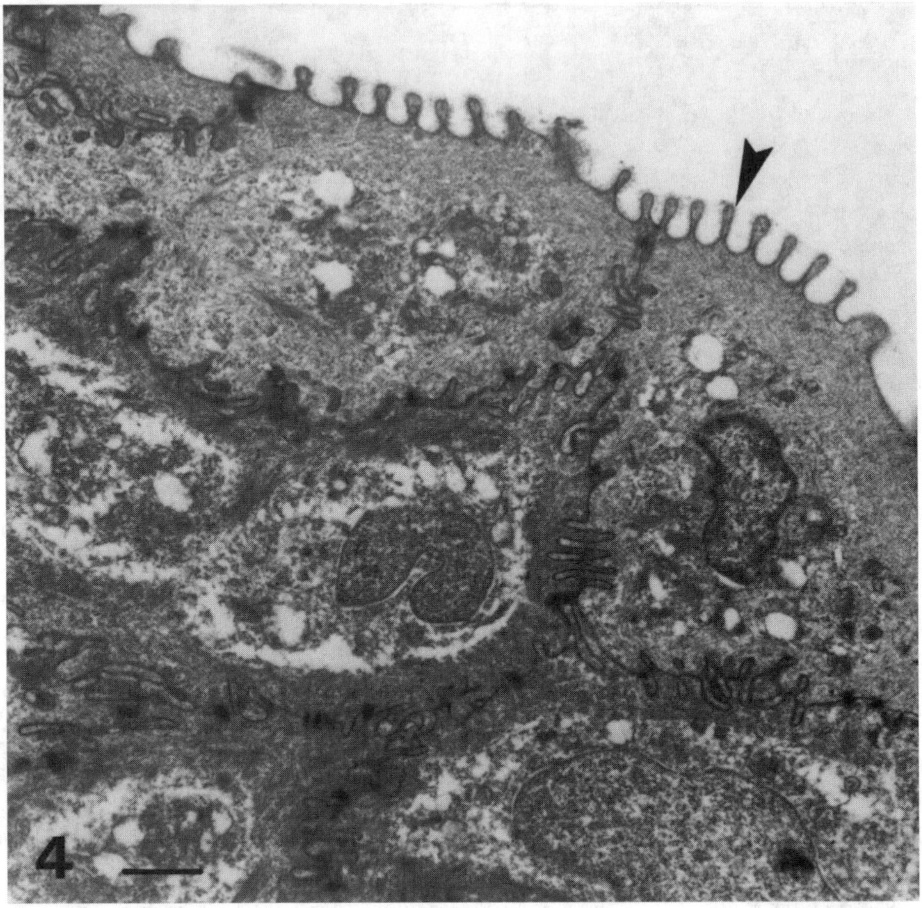

Figure 22-4. Electron micrograph of skin of tilapia (control), showing filament cells and (upper layer) pavement cells. Arrow: apical membrane of pavement cells folded into microridges, the typical surface specialization of intact pavement cells. Bar represents 1 μm.

Electron Microscopy of Gills and Skin of Greenling

THE GILLS. The gills of fish collected from Jiaozhou Bay showed the general structure described for the gills of teleosts, including tilapia (Wendelaar Bonga et al. 1990). The epithelium forming the upper layer of the gill filaments and the respiratory lamellae showed a high frequency of necrotic and apoptotic pavement cells and chloride cells (Figure 22-7). The cytoplasmic spaces between these cells were enlarged and many leukocytes were present in these spaces, in particular, lymphocytes and neutrophilic granulocytes. The blood lacunae of the lamellae were often found to be distorted (Figure 22-8).

THE SKIN. The general ultrastructure of the skin conformed to the general pattern of teleost skin (Whitear 1986) and did not substantially deviate from that of tilapia. The skin of the fish from Jiaozhou Bay was characterized by many apoptotic and necrotic pavement cells (Figure 22-9). Also, some mucous cells occasionally showed necrosis or apoptosis. The cytoplasm of normal pavement cells frequently showed large amounts of secretory granules (Figure 22-10). The intercellular spaces were widened in many places, and leukocytes were commonly observed in these spaces. Many mucous cells were found. The fish obtained from both localities in Jiaozhou Bay were affected to a similar extent.

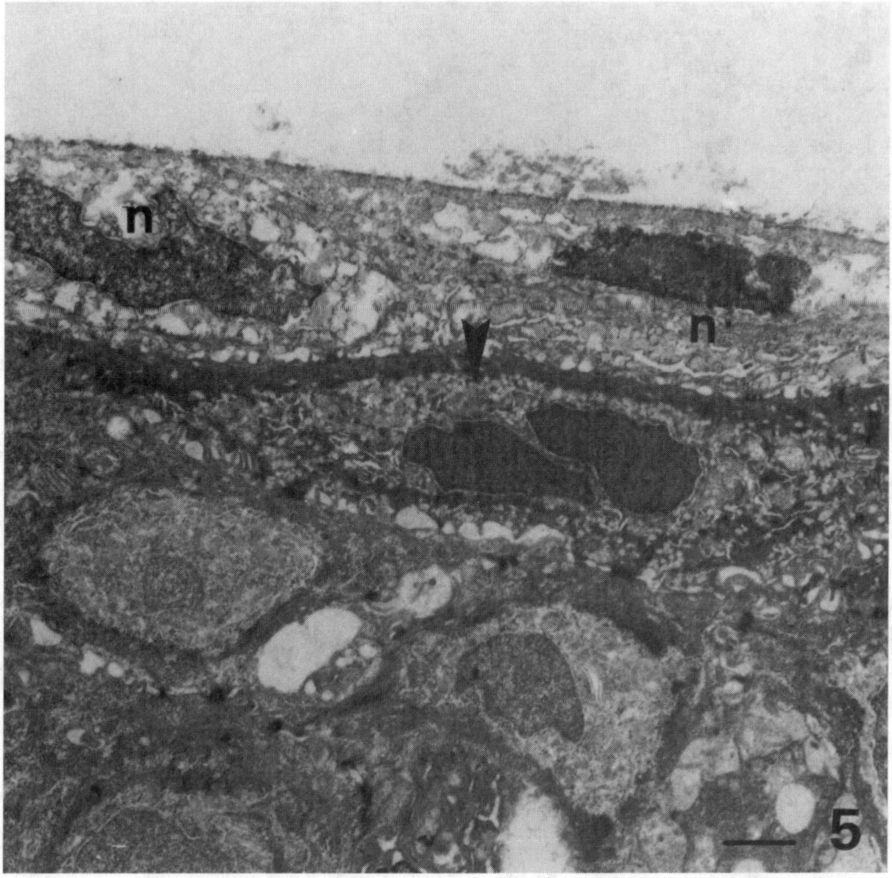

Figure 22-5. Electron micrograph of skin of tilapia exposed to 50 μgCd/l for 10 days. The upper layer of pavement cells is necrotic (showing loss of microridges and cytoplasmic swelling). Arrow: filament cell showing many small secretory vesicles typical for the filament cells of stressed fish; s, intercellular swelling. Bar represents 1 μm.

Discussion

The effects of cadmium at a concentration of 50 μg/l on the structure of gills and skin are similar to those reported earlier for tilapia at 10 μg/l (Pratap et al. 1989), although the damage to these tissues was much more serious than reported for the lower concentration. In particular, the chloride cells were affected, showing degeneration by necrosis and apoptosis. Necrosis reflects accidental cell death caused by substantial and immediate damage to the cells. Apoptosis also reflects mortal damage but to an extent that enables the cells to commit suicide, a physiologically controlled process (Wyllie et al. 1980). Nondegenerating pavement cells showed stimulated secretion of small secretory granules, which have been shown to contain peroxidase activity and may have an antimicrobial function (Iger and Wendelaar Bonga 1994). The widening of the intercellular spaces reflects loss of structural, and possibly also functional, integrity of the epithelia. The appearance of leukocytes in these spaces indicates migration of these cells out of the blood space. Their accumulation in the branchial area, especially of the lymphocytes and granulocytes and in the skin, may be a response to antigens penetrating these tissues, probably as a result of increased permeability caused by the toxicant. The macrophages are most likely involved in the removal of the remnants of degenerated cells.

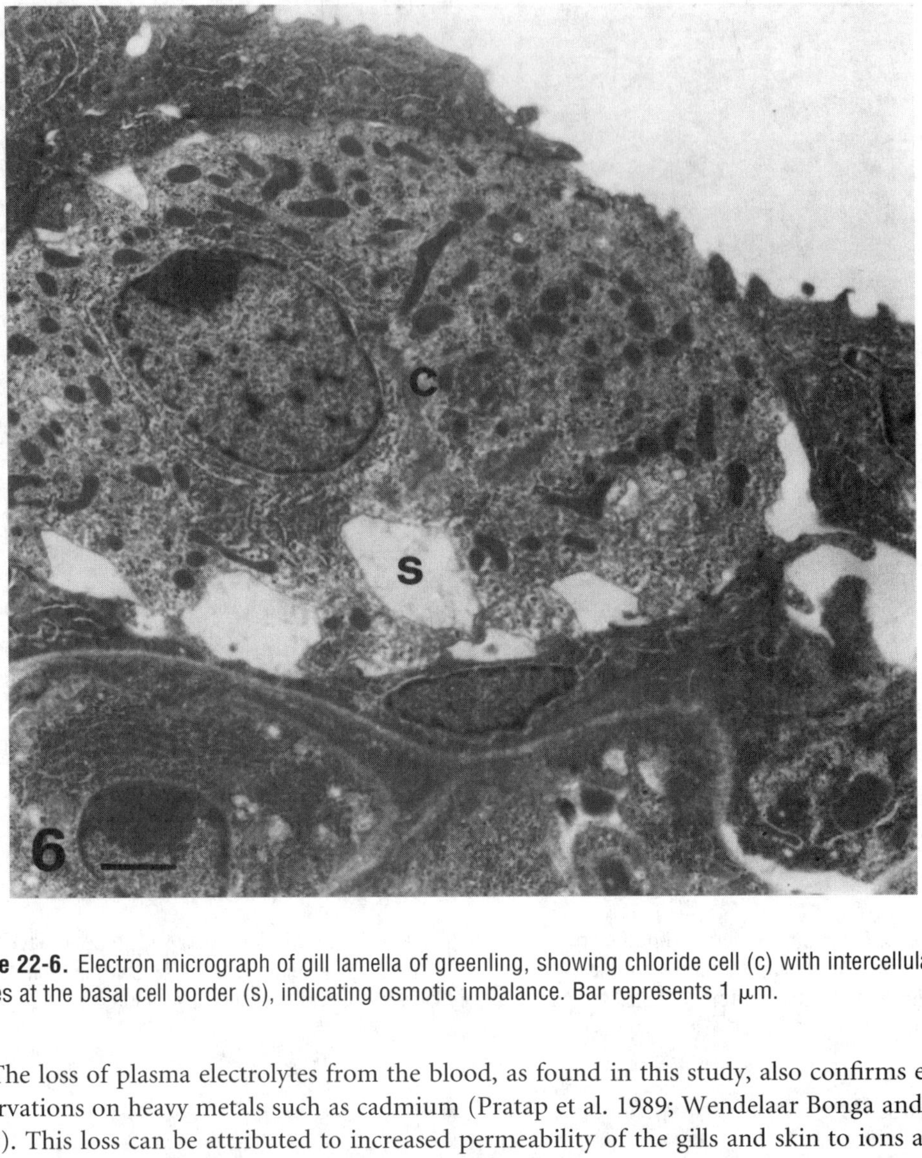

Figure 22-6. Electron micrograph of gill lamella of greenling, showing chloride cell (c) with intercellular spaces at the basal cell border (s), indicating osmotic imbalance. Bar represents 1 μm.

The loss of plasma electrolytes from the blood, as found in this study, also confirms earlier observations on heavy metals such as cadmium (Pratap et al. 1989; Wendelaar Bonga and Lock 1992). This loss can be attributed to increased permeability of the gills and skin to ions and to reduced ion-transport capacity of the chloride cells in the gills, the principal sites of active ion uptake in freshwater fish (Wendelaar Bonga and Lock 1992). This conclusion is supported by the

Table 22-2. Specific Na^+/K^+-ATPase activity ($\mu molP_i.h^{-1}$ per mg protein of the gill filament) and of Na^+, Ca, and Cl^- concentrations (in mmol/l) in the blood plasma of tilapia exposed for 6 days to 50 μg Cd/l in freshwater

	50 μg Cd/l		Controls	
Na^+/K^+-ATPase activity	35.2 ± 8.1*	(6)	58.1 ± 13.4	(7)
Plasma Na^+	135 ± 7*	(8)	148 ± 6	(8)
Plasma Ca	3.15 ± 0.16**	(7)	268 ± 0.13	(8)
Plasma Cl^-	126 ± 8*	(7)	107 ± 6	(7)

Note: Means ± S.E.M.; numbers of fish indicated in parentheses.
*$P<0.05$
**$P<0.01$, Student's t-test

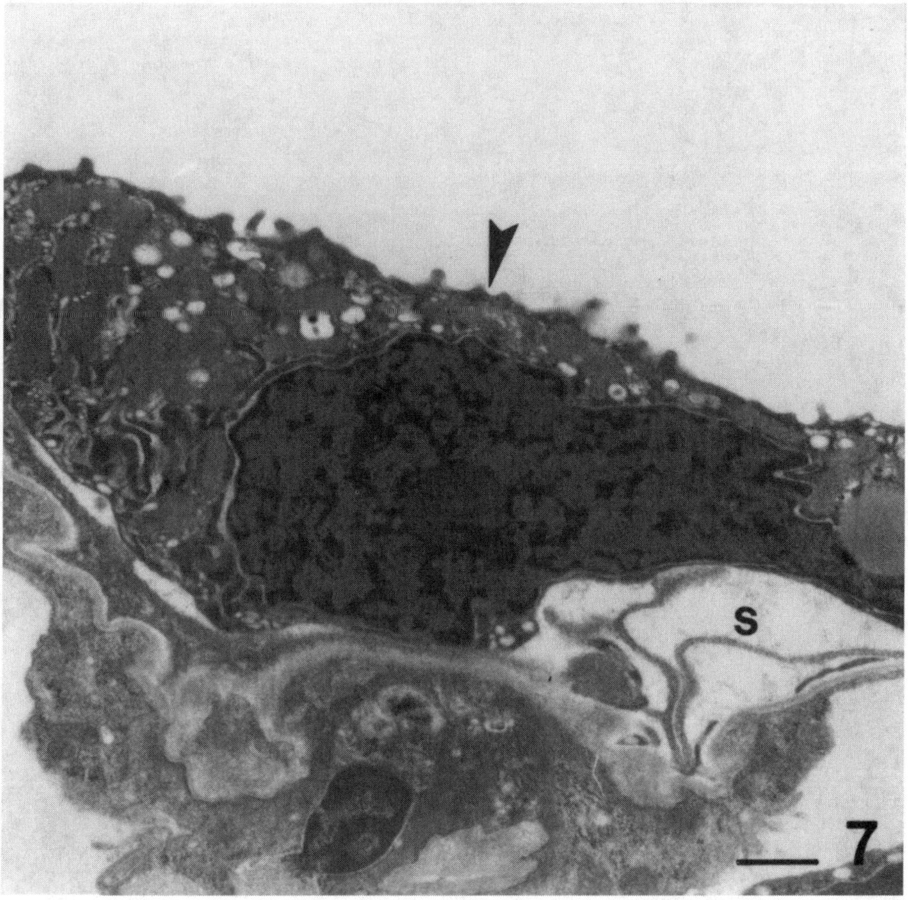

Figure 22-7. Electron micrograph of gill lamella of greenling, showing an apoptotic chloride cell (c), large intercellular spaces, and a disrupted endothelial layer (e) lining the lamellar blood space. Bar represents 1 μm.

present data on tilapia, showing a high number of degenerating chloride cells and a reduced Na^+/K^+-ATPase activity in the gills. The chloride cells are the main location of this Na transport-facilitating enzyme. This enzyme activity is often reduced in polluted water (Wendelaar Bonga and Lock 1992). The structural damage of the gills and skin of the greenlings obtained from the Yellow Sea was qualitatively similar to that of the cadmium-exposed tilapia. In both species, necrosis and apoptosis of the pavement cells and widened intercellular spaces containing many leukocytes were observed. From a quantitative point of view the condition of the skin and gills of the greenlings was at least as poor as that of tilapia gills exposed to 50 μgCd/l.

Several conclusions can be drawn from these observations. First, the effects of polluted seawater on the skin and gills of marine fish are apparently comparable to a high extent to the effects of different pollutants in freshwater fish. Our studies on freshwater tilapia, carp, and trout have already shown that the effects of pollutants and disturbing factors such as heavy metals, a rapid increase in temperature, or acidified water containing aluminum, are all very similar (Iger and Wendelaar Bonga 1994; Iger et al. 1994a, 1994b; Pratap et al. 1989) and comparable to the effects described in this paper for cadmium-exposed tilapia. The present data on greenlings indicate that pollutants have similar effects on a marine species. Apparently, the structural changes described in this study represent nonspecific effects of pollutants on gills and skin of

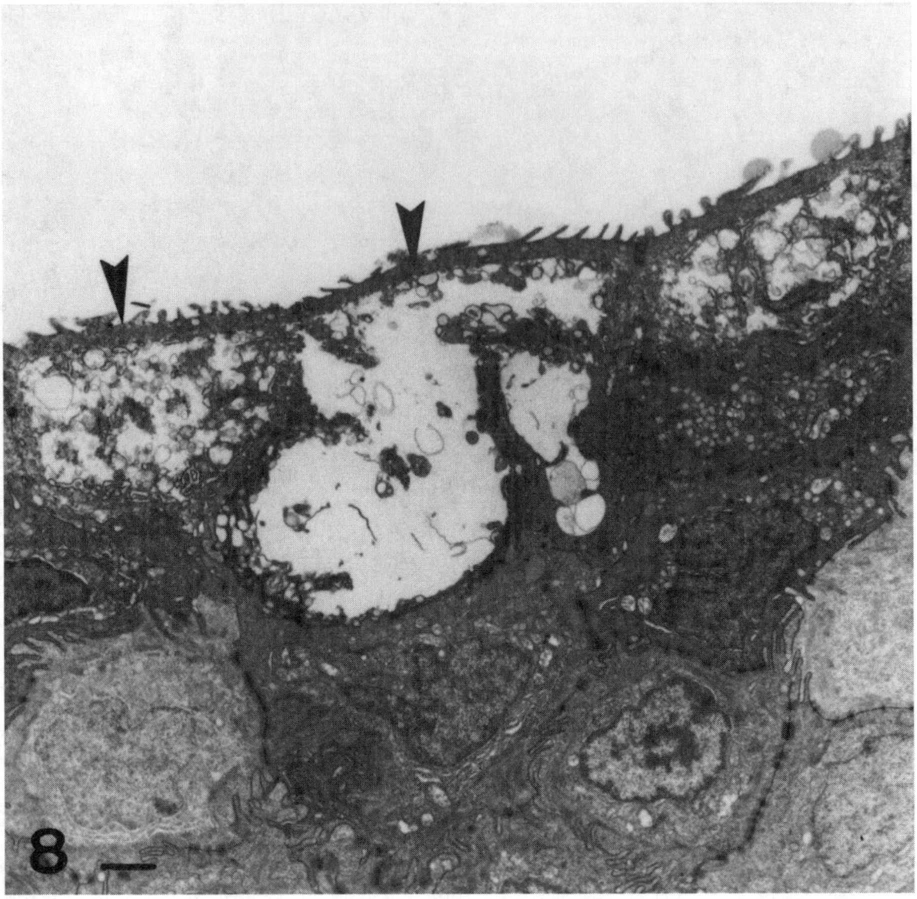

Figure 22-8. Electron micrograph of skin of greenling, showing necrotic pavement cells with substantial cytoplasmic swelling. Bar represents 1 μm.

fishes irrespective of their habitat. This implies that the observed effects are good indicators for water pollution. Second, the severe degree of tissue damage in the greenling was comparable to that of a highly toxic concentration of cadmium in tilapia. This indicates that the degree of pollution of Jiaozhou Bay at the sampling localities is indeed very serious. Cadmium concentrations of 0.34 and 0.61 μg/l were measured in these areas. This is well below the concentration of 50 μgCd/l used in our tilapia experiment. The structural effects on the gills and skin of these tilapia were qualitatively and quantitatively similar to those observed in the greenlings, which may have been caused by the presence of many other toxic pollutants in the Jiaozhou Bay area. In addition to cadmium, zinc, lead, and copper (Table 22-1), substantial concentrations of organic toxicants have been reported in Jiaozhou Bay. Most likely, the effects of these different pollutants are partially or completely additive (see, for example, Pelgrom et al. 1995 for the interaction of cadmium and copper in fish).

Finally, the structural damage observed in the present study has been associated in tilapia and carp with substantially reduced rates of growth and reproduction (our unpublished results). In nature, this will result in reduction and eventually disappearance of fish populations. There is reason for concern about the water quality in Jiaozhou Bay. The present data show that the structural changes in gills and skin represent excellent biomarkers for the analysis of pollution in marine habitats, as has been demonstrated earlier for freshwater habitats.

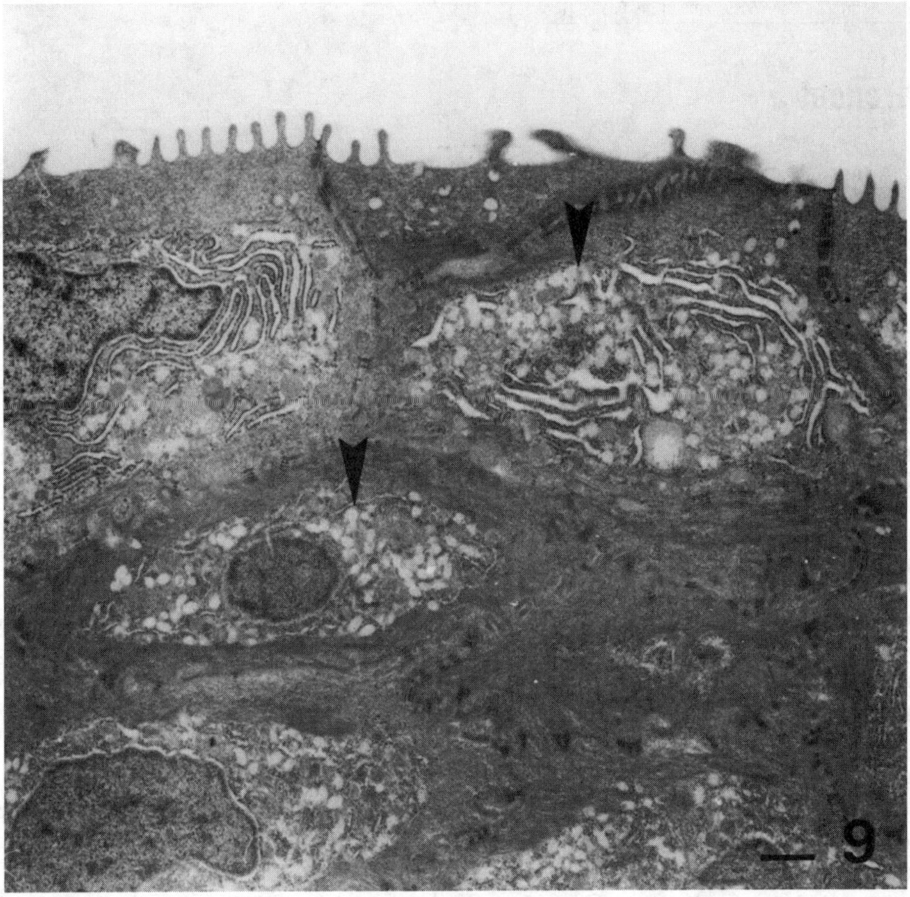

Figure 22-9. Electron micrograph of skin of greenling, showing an intact area of the skin. Arrows: upper filament cells showing many secretory granules. Bar represents 1 μm.

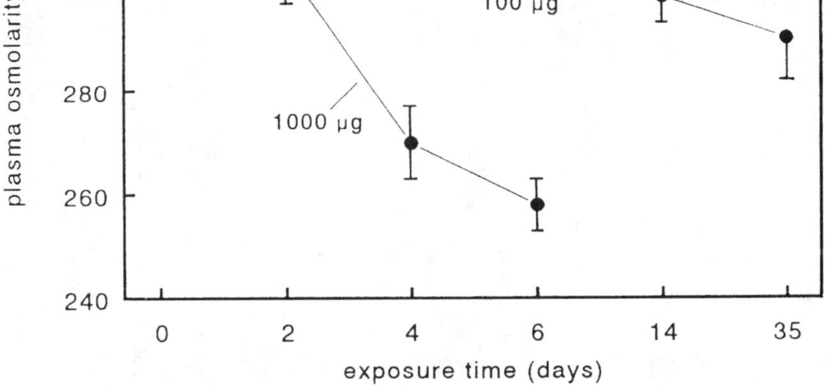

Figure 22-10. Plasma osmolarity of freshwater tilapia after exposure to indicated concentrations of waterborne cadmium; means ± S.E. of eight fish. At 1,000 μgCd/l all fish died within 8 days.

References

Dethlefsen, V., and Tiews, K. 1985. Review on the effects of pollution on marine fish life and fisheries in the North Sea. J. Appl. Ichthyol. 3:97–118.

Eddy, F. B. 1981. Effects of Stress on Osmotic and Ionic Regulation in Fish. In A. D. Pickering (ed.), Stress in Fish. London: Academic Press, pp. 77–102.

Evans, D. H. 1987. The fish gill: Site of action and model for toxic effects of environmental pollutants. Environ. Health Perspect. 71:74–58.

Iger, Y., and Wendelaar Bonga, S. E. 1994. Cellular responses of the skin of carp (*Cyprinus carpio*) exposed to acidified water. Cell Tiss. Res. 275:481–492.

Iger, Y., Jenner, H. A., and Wendelaar Bonga, S. E. 1994a. Cellular responses in the skin of the trout (*Oncorhynchus mykiss*) exposed to temperature elevation. J. Fish Biol. 44(3):921–935.

Iger, Y., Lock, R. A. C., Jenner, H. A., and Wendelaar Bonga, S. E. 1994b. Cellular responses in the skin of carp (*Cyprinus carpio*) exposed to copper. Aquatic Toxicol. 29:49–64.

Mallatt, J. 1985. Fish gill structural changes induced by toxicants and other irritants: A statistical review. Can. J. Fish. Aquat. Sci. 42:630–648.

Pelgrom, S. M. G. J., Lamers, L .P .M., Garritsen, J .A. M., Pels, B. M., Lock, R. A. C., Balm, P. H. M. and Wendelaar Bonga, S. E. 1995. Interactions between copper and cadmium during single and combined exposure in juvenile tilapia *Oreochromis mossambicus*: Influence of feeding condition on whole body metal accumulation and the effect of the metals on tissue water and ion content. Comp. Biochem. Physical C. Pharmacol. Toxicol. Endocrinol. 111(2):227–235.

Pratap, H. B., Fu, H., Lock, R. A. C., and Wendelaar Bonga, S. E. 1989. Effect of waterborne and dietary cadmium on plasma ions of the teleost *Oreochromis mossambicus* in relation to water calcium levels. Arch. Environ. Contam. Toxicol. 18(14):568–575.

Wendelaar Bonga, S. E., and Lock, R. A. C. 1992. Toxicants and osmoregulation. Neth. J. Zool. 42:478–493.

Wendelaar Bonga, S. E., and Van der Meij, J. C. A. 1989. Degeneration and death, by apoptosis and necrosis, of the pavement and chloride cells in the gills of the teleost *Oreochromis mossambicus.* Cell Tissue Res. 255:235–243.

Wendelaar Bonga, S. E., Flik, G., Balm, P. H. M., and Van der Meij, J. C. A. 1990. The ultrastructure of chloride cells in the gills of the teleost *Oreochromis mossambicus* during exposure to acidified water. Cell Tiss. Res. 259(3):575–585.

Whitear, M. 1986. The Skin of Fishes Including Cyclostomes—Epidermis. In J. Bereiter Hahn, A. G. Matotsy, and K. S. Richards (eds.), Biology of the Integument. 2. Vertebrates. Heidelberg: Springer, pp. 8–38.

Wyllie, A. H., Kerr, J. F. R., and Curry, A. R. 1980. Cell death: The significance of apoptosis. Int. Rev. Cytol. 68:251–306.

23

Pollution in the Yellow Sea Large Marine Ecosystem: Monitoring, Research, and Ecological Effects

Jun She

In recent decades, the Yellow Sea Large Marine Ecosystem (YSLME) has been affected by pollution, overfishing, and coastal and ocean engineering. Some anomalous phenomena have occurred and can be looked at in two different time scales. On a scale of several days to several months, organisms die in certain areas due to the input of large amounts of wastewater or due to an alteration in the food chain caused by eutrophication. On the interannual or interdecadal scale, there is a sharp decline in quality and quantity of harvested resources and the disappearance of certain species. Within a dynamic system, changes of this kind mean that the state of the entire system may be approaching an irreversible condition, at which point the resources in the YSLME are in critical danger.

Of the various environmental stressors, such as pollution, overfishing, and natural variation, pollution can cause the disappearance of a species in a biome quite rapidly. Because of this, pollution is of special importance in a marine ecosystem.

In the first section, a summary of the previous monitoring programs in the YSLME is given. Sources and concentrations of pollutants entering the YSLME and its coastal areas through the air, water, and sediment are described in the second section. In the third section the ecological effects of pollution in the YSLME are assessed. The concept of "self-purification" is reviewed in the next section. Finally, the mechanisms and effects of pollution are summarized, questions are raised, and suggestions for monitoring and studying pollution in the YSLME are given.

Previous Studies on Pollution and the Environmental Status of the Yellow Sea LME

Two kinds of studies have been done on pollution and the environmental status of the YSLME. The first includes routine observation from weather stations, meteorological–ocean stations, environmental monitoring stations, and biological monitoring stations. There are 18 weather stations, 20 meteorological–ocean stations, and several environmental monitoring stations in the coastal areas of the YSLME, along with several regular ship transections and mooring buoys in the Yellow Sea. These stations provide a large database on a wide variety of environmental parameters.

The second kind of study is one that has a special purpose and a finite time frame, such as the following:

- The Nationwide Industrial Pollution Sources Investigation, initiated in 1985.
- A large-scale comprehensive offshore oceanographic study carried out in 1958–1959, which can be used as the baseline study for the YSLME. This study investigated the plankton community, the benthic community, hydrometeorology, hydrochemistry, geology, and so forth.
- The Yellow Sea Comprehensive Investigation on Pollution in 1982.
- The Nationwide Coastal Zone and Marine Resources Investigation carried out between 1982 and 1985. Areas investigated included supralittoral vegetation; intertidal organisms; phytoplankton; bottom-dwelling fish, shrimp, crabs, and shellfish; as well as various physical and chemical parameters.
- The current Nationwide Marine Zoning, Island, and Harbour Investigations, which include an extensive survey of the biological resources. This study provides a large amount of data for developing rational use and management plans for the coastal areas of the Yellow Sea.

In all these investigations, pollution and the environmental state were monitored in the region of interest. Although a large amount of information has been gathered on the distribution of pollution sources, inputs of contaminants to the sea, and water and sediment pollution in coastal areas (isobaths deeper than 20 m), more is needed. However, the existing information does provide an important basis for assessing the long-term response of the YSLME to pollution.

The Pollution Features in the Yellow Sea Large Marine Ecosystem

The YSLME is a semi-enclosed shelf sea. Affected by the economic development in China, the YSLME has some unique pollution characteristics.

Distribution and Sources of Principal Pollutants

Statistics show that 1,700 million tons of wastewater flow into the YSLME per year. Of this, 84% is industrial wastewater and the rest is domestic sewage. Table 23-1 shows the inputs to the Yellow Sea and the Bohai Sea for 12 kinds of contaminants. The main pollutants include organics, petroleum, phenoxide, chloride, and inorganic nitrogen and metals. Of the metals, zinc, arsenic, and chromium account for the largest proportion.

Table 23-1. River inputs of pollutants (tons/year) to the Yellow Sea and Bohai Sea in 1989

	Oils	COD	Hg	Cd	Pb	Zn	As	Cr	Cu	Phenoxide	Chloride	Inorganic nitrogen
Bohai Sea	21,178	544,598	9.9	26.5	65	432	119	99	82	644	52	25,636
Yellow Sea	11,210	383,850	5.5	0.7	130	1,156	217	245	111	928	358	19,120

Table 23-2. Loading and retention rates in the Bohai Sea

	Oil		Inorganic nitrogen		Active phosphate		COD	
Year	Loading	Rate	Loading	Rate	Loading	Rate	Loading	Rate
1985	0.023	3.3	0.014	5	0.018	55	0.74	0
1986	0.015	1.4	0.036	11	0.007	7	0.78	0
1987	0.019	1	0.031	4	0.011	31	0.83	0
1988	0.012	—	0.116	65	0.008	20	1.02	1
1989	0.021	8	0.072	21	0.010	19	1.05	—
1990	0.030	20	0.070	31	0.003	6	0.98	—
1991	0.014	2	0.123	45	0.012	33	1.31	1
Mean	0.2		0.010		0.95		1	

Units: density = mg/l, rate = %.

Generally speaking, the Bohai Sea receives twice as much pollution as the Yellow Sea. But for some inputs such as phenoxide, chloride, and metals, the Yellow Sea receives twice as much as the Bohai Sea, probably because of the local industries.

In Tables 23-2 and 23-3, the average loadings and retention rates of oil, COD, inorganic nitrogen, and active phosphate during 1985–1991 in the Yellow and Bohai Seas are presented. Although there may be problems of representativeness and consistency in the data, Tables 23-2 and 23-3 still give a good indication of the water pollution problems. On average, the retention rate of oil, inorganic nitrogen, and active phosphate is about 20%, whereas the retention rate of COD is less. It is uncertain whether the concentration of a certain pollutant in the water is proportional to the pollutant's inputs to the sea.

The most serious source of pollution comes from the rivers that discharge into estuaries and harbors. The main pollutants in several heavily polluted bays are given in Table 23-4.

In recent years, many studies have been carried out on the distribution of pollutants in the shallow sea areas in the YSLME. These include studies on the distribution of mercury, copper, and zinc in the sediments of the Bohai Sea (Zheng and Zhang 1985; Zhang and Liu 1986), the loading of offshore areas of China (Liu et al. 1988), the fluxes of metals in the sediment of the North East Sea of China, and pollutant loadings in estuaries (Sun et al. 1984; Liu et al. 1988; Shao and Wang 1992).

Table 23-3. Loading and retention rates in the Yellow Sea

	Oil		Inorganic nitrogen		Active phosphate		COD	
Year	Loading	Rate	Loading	Rate	Loading	Rate	Loading	Rate
1985	0.029	18	0.021	6	0.005	11	0.71	1
1986	0.030	33	0.065	29	0.005	14	0.68	0
1987	0.016	6	0.049	18	0.006	22	0.72	0
1988	0.031	8	0.004	—	0.001	—	0.88	0
1989	0.028	17	0.070	22	0.018	21	1.39	9
1990	0.043	20	0.064	32	0.004	13	0.98	—
1991	0.020	5	0.102	33	0.011	15	0.99	1
Mean	0.2		0.015		1.0			

Units: density = mg/l, rate = %.

Table 23-4. Pollutants in heavily polluted bays

Dalian Bay	Jinzhou Bay	Bohai Bay	Liazhou Bay	Jiaozhou Bay	Haizhou Bay
Petroleum	Heavy metals	Organics	Organics	Chromium	Mercury
Organics		Petroleum	Petroleum	Calcium	
Arsenic		Phenoxides	Phenoxides		
			Heavy metals		

The following conclusions were reached:

1. Pollution loadings are greatest near the coast and decrease with distance from the coast.
2. Surface sediments contain large amounts of toxic metals. For example, the concentration of mercury in sediments in the Bohai Sea is 3,000 to 7,000 times that in the water. This concurs with measurements shown by Bronfman et al. (1976). The result is that the pollutant flux is from the sediment to the water column. In Jiaozhou Bay this is believed to be a significant source of pollution.
3. Some pollutants are in the surface microlayer of water, such as lead, petroleum, etc. A study of trace metal loadings in aerosols in the Kuroshio area indicated that lead from aerosols enters the ocean through the air–sea interface (Chen et al. 1992).
4. Monitoring of organic chloride pesticides in 1982 indicated that these pesticides are present in the water and in surface sediments in the YSLME.

Overall, although pollutant loadings are increasing yearly in the shallow water areas, they still do not exceed acceptable levels; however, bottom sediment concentrations are at seriously high levels and are affecting the bottom biota.

Characteristics of Air Pollution and the Air–Sea Fluxes of Pollutants

Many papers have been published on air pollution in coastal cities of the YSLME. It has been concluded that air pollution has increased significantly. For example, the visibility in Beijing decreased from 60 km to 10 km between 1970 and 1979. In the winter the probability of visibility being lower than 4 km increased by a factor of 10 (Su et al. 1986). Aerosol studies indicate that the maximum transport of aerosols occurs at 1,000 to 1,500 meters (Wang et al. 1984).

Monitoring has shown that offshore areas are less polluted than coastal areas. A joint China–Japan investigation in Kuroshio (Chen et al. 1992) collected and analyzed 2 years of aerosol data. The results are as follows:

1. The main pollutants in aerosols (cadmium, lead, and copper) are consistent with those released by land-based sources.
2. Solubility is greatest in summer, and Cd is more soluble than Pd, which is more soluble than Cu.
3. Aerosol loadings are highest in the winter.
4. Flux studies show that sulfate and nitrate are deposited as solids. Of the dissolved metals, copper has the highest flux to the sea, at a rate of 10^{-1} $\mu g/m^2d$. Lead and cadmium have a rate of 10^{-1} $\mu g/m^2d$.

It is clear that our knowledge of air pollution and air–sea fluxes in the YSLME is very limited. However, research indicates that the flux from the air cannot be ignored, especially in enclosed or semi-enclosed shallow seas, where the flux of some pollutants from the atmosphere may be greater than that from the rivers (Hoogweg et al. 1991).

Ecological Effects of Pollution in the Yellow Sea Large Marine Ecosystem

Pollution in the YSLME may be responsible for a series of ecological phenomena, including the disappearance of certain species, the increasing occurrence of red tides, and increasing concentrations of pollutants in organisms.

The Disappearance of Species

Due to high concentrations of pollutants, many species have suffered and some have disappeared. This is particularly true in spawning and nursery ground areas. Some commercial fish, shrimp, crab, and shellfish have been seriously affected. In Dalian Bay, some localized areas are barren of organisms and fishes. In Bohai Bay, Laizhou Bay, and Jiaozhou Bay, there have been events in which large numbers of fish, shrimp, crab, and shellfish have died.

High sediment concentrations in some areas have harmed the benthic communities, and the number of benthic species has decreased significantly. In Jiaozhou Bay, 147 macrobenthic species were collected in 1963–1964, 30 in 1975, and only 17 in 1980–1981 (Wu et al. 1992).

Increased Occurrence of Red Tides

Although red tides have occurred naturally since early times, they were rare. Since the 1970s, the frequency of red tides has gradually increased, the range of these tides has expanded, and their duration has lengthened. In 1989, the red tides were continuous in the Huanghua area for about 70 days.

There are three principal causes of red tides. These are eutrophication, increasing metal loadings from pollution and from mariculture, and weather anomalies. Eutrophication leads to the rapid growth of red tide organisms. Some elements—iron and manganese, for example—may cause a change in salinity and may reduce the ability of a body of water to self-purify. When a region is affected by some or all of these factors, red tides may occur.

Concentration of Pollutants in Organisms

The concentration of metals, pesticides, and oil in organisms is gradually increasing. In some organisms, these pollutant levels are greater than allowable for consumption, resulting in decreased commercial value and even seafood poisoning and death.

Statistics from 1989 show that the concentrations of mercury, copper, and lead in commercial organisms exceeded acceptable levels at most sampling stations in the Bohai Sea, and the concentration of mercury in bivalves was 10 times the standard. Monitoring between 1981 and 1984 showed that the concentration of cadmium in crustaceans tripled, and lead and copper in fish and molluscs increased by 2 to 4 times.

Assessing Pollution in the Yellow Sea LME

When an ecosystem is severely affected by a phenomenon, it can respond in a variety of ways. In order to evaluate these responses properly, it is first important to assess the environmental conditions. The environmental conditions of a particular water body are determined by the pollutants introduced and also by its self-purification ability, which includes the processes of transportation, dilution, absorption, dissolution, and deposition. To properly assess these factors, mathematical models need to be developed and actual data need to be used to verify these models. Only after these models have been developed and tested can they be used to set water quality standards, develop regulations to reduce pollution discharges, and forecast the response of the ecosystem.

Pollution Inputs

Pollution inputs to the YSLME come mainly from wastewater discharges and atmospheric deposition, each affecting the ecosystem differently. Wastewater discharges lead to serious local pollution, whereas atmospheric deposition of pollutants affects the entire Yellow Sea ecosystem. For direct discharges, the pollution levels can be quantitatively assessed by monitoring the finite number of discharges. However, for atmospheric inputs it is necessary to establish a host of numerical models with varying time and space scales. For example, a sea–land wind model can be used to simulate the lateral transport of the short-distance inshore pollutant (within 100 km) whereas in the case of the open sea, a long-distance transport model is needed (Sang 1988). To assess air–sea exchange, solid deposition, and liquid deposition, an air–sea flux model must be used (Sang 1992; Pen and Qin 1992). This is exceedingly difficult to do since an enormous amount of actual data is needed to verify these models.

Study of the Self-Purification Ability of a Water Mass

To establish pollution discharge standards, we must assess the self-purification ability of the water mass. The study can be divided into two parts. The first part assesses the nearshore areas, such as estuaries or harbors, which have relatively serious pollution on a time scale of several days to years. The self-purification ability of these areas depends on local topography and such factors as meteorological, hydrographic, and biological conditions; thus, each water mass will distribute pollutants differently. The other part of the study is concerned with the water mass of the entire LME. The self-purification of an entire ecosystem has a time scale of several years to several decades.

To properly study the self-purification process of a water mass, a host of different models are needed, such as atmosphere advection–diffusion models, sea surface wind models, ocean advection–diffusion models, short-term wind-driven current models, long-term eddy-resolving circulation models, and so forth. In recent years, the Laboratory of Geophysical Fluid Dynamics (LAGFD) in the First Institute of Oceanography of SOA has developed some models for sea surface winds (She et al. 1994) and oil spill drift on the sea surface (Yuan et al. 1991), a model simulation of the cold-water mass in the Yellow Sea, and eddy-resolved ocean circulation and short-term wind-driven three-dimensional current models (Yuan et al. 1992). All these studies can be used in developing a model group for the simulation of water mass self-purification.

Conclusions and Suggestions

The interactions between pollution and the YSLME have multiple time–space scales. Many investigations have been carried out in recent decades. Some basic features of pollution in the YSLME have been identified and some numerical models have been developed to simulate pollution transport. However, many problems, such as the ecological effects of pollution, toxicology, air–sea fluxes of pollutants, secondary pollution, and the reasons for the elimination of benthic species, still need to be studied.

Figure 23-1 presents the pollution transport pathways in an ecosystem and shows areas where more information is needed in order to properly assess these processes in the YSLME.

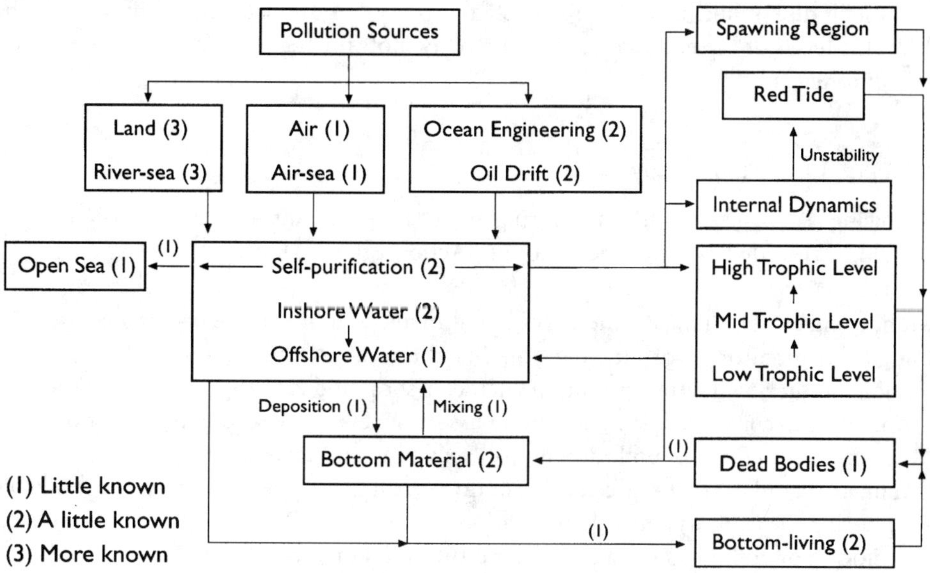

Figure 23-1. Pollution transport pathways in an ecosystem, showing the amount of information that is known concerning these processes in the Yellow Sea large marine ecosystem.

The main pollution sources in the YSLME are discharges from the mainland, atmospheric deposition, and pollution from coastal ocean engineering. Of these sources, very little research has been done on atmospheric deposition, which includes the deposition and horizontal transport of pollutants in the atmosphere and the air–sea interaction. The Environmental Protection Department has started monitoring pollution discharges on the mainland. Some monitoring has been done on the pollution resulting from ocean engineering (mainly petroleum exploration activities).

For some pollutants the Yellow Sea has the ability to purify itself through physical, biological, and chemical processes. The pollutant may be transferred out to sea, deposited onto the bottom, absorbed by organisms, and so forth. Very little is known about deposition or dispersal of pollutants. Also, models have been developed only for the physical processes that would affect purification. There has been no systematic monitoring of pollutants at the various trophic levels or at the boundary layers.

Besides the direct pollution sources described above, there are indirect pollution sources, mainly from bottom accumulation of pollutants and from decaying matter. Highly polluted sediments can release pollutants and cause increased concentrations in those areas. Vertical mixing in areas with highly polluted sediments can cause secondary pollution effects. Another cause of secondary pollution is the accumulation of dead organisms, as in a red tide event or in the death of many organisms over a large area. Secondary pollution processes need to be monitored and studied for the YSLME.

The ecological effects of pollution in an ecosystem are very complex because many pollutants can be bioaccumulated by organisms or can be passed onto higher trophic levels in the food chain. For example, in the YSLME some zooplankters, such as copepods and amphipods, bioaccumulate pollutants as a result of their feeding habits. These zooplankters are then the main food of fish higher in the food chain. This has serious consequences for both the biological community and the ecosystem as a whole.

Pollutants affect each organism differently, causing some populations to decline or disappear and others to flourish. This changes the ecological equilibrium and can alter the ecosystem structure. The existence of red tides and the reduction in the benthic species in Jiaozhou Bay are signals of such disequilibriums. It is obviously necessary to systematically monitor such changes in all trophic levels of the ecosystem. Unfortunately, not much attention has been paid to these changes until recently.

Suggestions for Further Study

The following are suggestions for monitoring and study programs for the YSLME. These suggestions could be regarded as supplements to the "Action Project Draft."

1. Monitoring of pollution accumulation in sediments and in benthic organisms.
2. Long-term monitoring of air pollution (its chemical constituents, aerosol spectrum, and vertical and horizontal transportation characteristics) and air–sea fluxes of pollutants.
3. Monitoring to collect red tide data and to assess the effects of secondary pollution.
4. Monitoring the effects of pollution on fish spawning and early life stages.
5. Conducting studies on the effects of pollution on all trophic levels and on the food chains.
6. Monitoring the capacity for self-purification.
7. Monitoring of pollutants from agricultural runoff and rivers.
8. Many types of models need to be developed to understand the fate of pollution in the ecosystem. Data are needed to develop and test these models.

References

Bronfman, A. M., Aldakimova, A. Y., Makarova, G. D., and Tolokoninikova, L. I. 1976. Characteristic features of the self-purification of waters and bottom of a shallow water sea. Okeanologiya 16(1): 91–97.

Chen, L., et al. 1992. Chemical Features of Aerosols over the Kuroshio Area (in Chinese). Papers in Kuroshio Investigation 4. Oceanology Publishing House, pp. 295–302.

Hoogweg, P., et al. 1991. North Sea strategies. Marine Pollution Bull. 23:57–61.

Liu, M., et al. 1988. Oceanology and Limnology. 19(6):778–786 (in Chinese).

Pen, H., and Qin, Y. 1992. Adv. Atmo. Sci. 16(5):622–630 (in Chinese).

Sun B., et al. 1984. Oceanology and Limnology. 15(5)439–448 (in Chinese).

Su W., et al. 1986. Adv. in Atmo. Sci. 10(2)138–144 (in Chinese).

Sang, J. 1988. Acta. Meteorologica Sinica. 45(3):358–362 (in Chinese).

Sang, J. 1992. Numerical Simulation in Atmosphere Diffusioning. Meteorology Publishing House,. pp. 161–202 (in Chinese).

Shao, M., and Wang, Z. 1992. Oceanology and Limnology 23(2):144–149 (in Chinese).

She, J., et al. 1994. A numerical typhoon surface wind model and its calibration. ACTA Oceanologica Sinica 13(4):47–55.

Su, W., et al. 1986. Adv. Atmo. Sci. 10(2):138–144 (in Chinese).

Sun, B., et al. 1984. Oceanology and Limnology 15(5):439-448 (in Chinese).

Wang, M. et al. 1984. Adv. Atmo. Sci. pp. 435–442 (in Chinese).

Wu, B. L., Sun, H., and Li, Y. 1992. A sharp decrease in the number of animal species in Jiaozhou Bay, Qingdao, China. Marine Pollution 24(4):216–217.

Yuan, Y., et al. 1991. Dynamic Numerical Model on Oil Spreading. Numerical Calculation in Physical Oceanography (in Chinese). Henan, China: Science and Technological Publishing House.

Yuan, Y., et al. 1992. The variation and movement of a kind of eddy in shelf slope water. Science in China (Series B) 35(8):1015–1024.

Zhang, S., and Liu, B. 1986. J. Oceano. of Huanhai and Bohai Sea. 4(1):78–82 (in Chinese).

Zheng, S., and Zhang, S. 1985. Oceanology and Limnology 16(2):121–126 (in Chinese).

V

Management Linkages

24

Management of the Great Barrier Reef as a Large Marine Ecosystem

Jon Brodie

The Great Barrier Reef

Statistics

The Great Barrier Reef (GBR) system covers an area of about 350,000 km^2 on the northeastern Australian continental shelf. It is a long, narrow system stretching 2,000 km along the coast, ranging from 50 km wide in the north to 200 km in the south and bounded by the coast to the west and the Coral Sea on the east (Figure 24-1). It encompasses the largest system of corals and related life forms anywhere in the world, with approximately 3,000 reefs. About 350 species of hard coral are found in the region, along with 1,500 species of fish, 240 species of seabirds, and at least 4,000 species of molluscs. Our knowledge of the taxonomy and distribution of the lesser-known groups of invertebrates such as echinoderms and worms is still very limited. In a recent study, 134 species of marine flatworms, over 90% of which were new species, were recorded from two adjacent reefs (Heron Island and One Tree Island Reefs), effectively increasing by a factor of 10 the total number of species of this order known from the whole GBR (Newman and Cannon 1994).

Habitats

Generally, the range of habitats found in the GBR is relatively uniform from north to south but varies across the continental shelf in a regular way (Hopley 1982). The principal habitats of the system have only existed in their present form since the sea level rose 10,000 years ago, flooding the shelf. Inshore the coastline is dominated by mangroves interspersed with areas of low-energy sandy beachlines and limited rocky shorelines. Immediately offshore, shallow seagrass beds are common (Lee Long et al. 1993), and in the north large areas of deep water (>10 m) seagrass are found farther offshore. The GBR lagoon floor is dominated by soft-bottomed communities of algae, sponges, and bryozoans (Birtles and Arnold 1988) interspersed with bare sand. In the north, extensive *Halimeda* sp. algal beds occupy the deeper offshore waters, their growth stimulated by nutrient-rich water upwelling from the Coral Sea (Drew and Abel 1988).

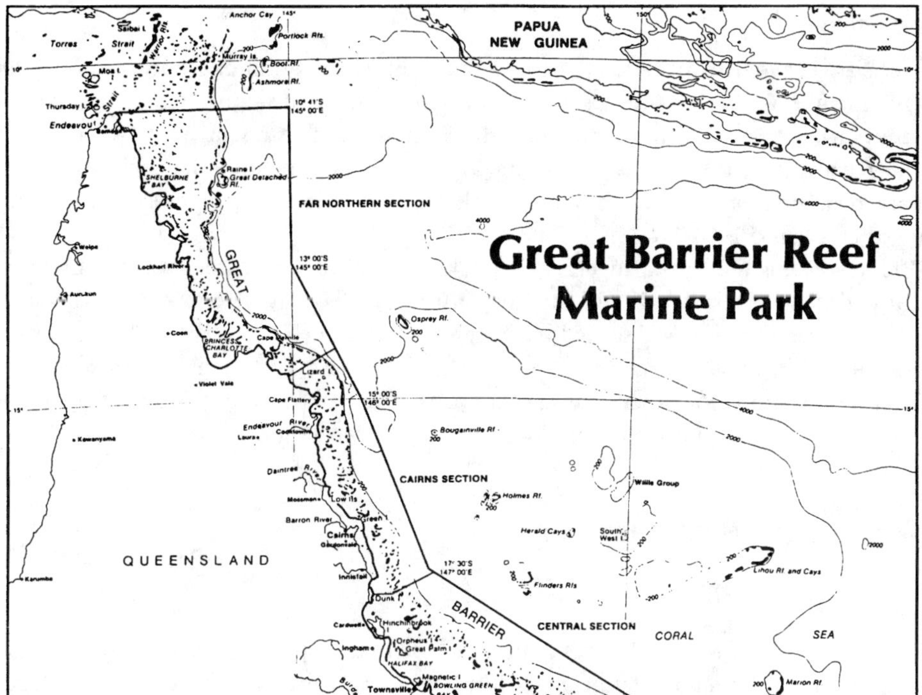

BRA Q136

Figure 24-1. Great Barrier Reef Marine Park.

The coral reefs of the GBR consist of two main types: the fringing reefs (approximately 760 reefs) that occur inshore around the continental islands, and those of the main reef (approximately 2,200 reefs) that occupy a band on the outer edge of the continental shelf. The latter group is often considered as either mid-shelf reefs on the inside of the band and adjacent to the GBR lagoon or outer-shelf reefs adjacent to the Coral Sea. The main reef does not form a continuous barrier but consists of individual reefs separated by interreefal waters. In some areas considerable passages exist, breaking the maze of reefs and joining the lagoon to the Coral Sea.

System Functions

Coral reefs are generally considered to do best in low nutrient conditions. Sources of nutrients to the GBR include Coral Sea surface water (nutrient poor), upwelling Coral Sea deep water (nutrient rich), terrestrial runoff, and atmospheric inputs, including nitrogen fixation by cyanobacteria (Furnas et al. 1994). Flushing of the GBR lagoon is limited by the enclosure formed by the main reef. Residence times of water in the lagoon, while not precisely known, may be prolonged (Wolanski 1994).

GBR ecosystems are linked together through larval transport. Almost all the biota present reproduce using a pelagic larval stage that may last from a few days to a few weeks. It is assumed from this that reefs are not self-seeded but receive most of their recruits from "upstream" reefs (James et al. 1990). Thus, the reefs exist in a connected system, and events on one reef may have significant implications for other "connected" reefs. The pattern of outbreaks of the crown-of-thorns starfish strikingly illustrates this connectivity (Moran 1986).

The Great Barrier Reef Marine Park

History

The Great Barrier Reef Marine Park is a multiple-use marine park established by the Australian federal government under legislation passed in 1975. The overriding objective of the legislation is the conservation of the Great Barrier Reef. The legislation established the Great Barrier Reef Marine Park Authority (the Authority) to manage the park, setting forth as the goal of the Authority "to provide for the wise use, protection and understanding of the Great Barrier Reef in perpetuity."

Multiple-Use Nature

The Great Barrier Reef Marine Park is not a park in the same sense as a terrestrial national park. It is a multiple-use protected area, and many uses, including extractive uses such as commercial fishing, are permitted in parts of the park. The one activity that is not permitted in any part of the park is mining. This proscription arose from the history of the park's formation: Potential mining was the trigger for the establishment of the park and was seen to be an unacceptable threat to the long-term existence of the GBR. Normal activities that occur in the park include commercial and recreational fishing, aquaculture, tourism, shipping traffic, research, and defense exercises.

Zoning

The principal policy tool used in the management of the park is zoning. The park is divided into four sections (Figure 24-1), each of which has a zoning plan that spatially regulates usage within the area. Zoning plans characteristically have five or six zones, each with defined objectives and uses; some uses require a permit. The zones and their uses are generally as follows:

General use: Trawling, other fishing, and most activities permitted.
Habitat protection: Trawling not permitted; other fishing and most activities permitted.
Conservation park: Trawling not permitted; limited line and spear fishing permitted.
Marine national park: Look but don't take.
Scientific research: Scientific research only.
Preservation: No entry.

More recent zoning provisions also place restrictions on the number of reefs within an area permitted to have structures, for example, tourist pontoons and fish-cage aquaculture facilities.

Issues and Their Management

Tourism

Tourism is the boom industry of the GBR region. A 40-fold increase in the number of tourists visiting the area in the period from 1946 to 1980 has been followed by continuous significant growth since then. In addition, with advances in the technology of high-speed ferries, large numbers of tourists are now able to visit even outer-shelf reefs; a number of reefs now have daily visitation rates of more than 500. The Authority has invested its major management presence in the control of tourism impacts. Monitoring programs have now shown that with good management, the environmental impacts of mass tourism are minimal. Management action has included the requirement for nutrient-removal tertiary treatment of sewage, supervised installation of well-engineered pontoons, diving and snorkeling behavioral codes, and fixed moorings to prevent anchor damage. Further development of codes of practice for tourism operators is now under way as a form of effective self-regulation.

Ballast Water

A number of substantial ports line the GBR coastline. These include large bulk shipment ports for the export of coal, alumina, and sugar. Ships entering these ports empty of cargo are ballasted with water collected in their last port of call. The ballast water is then discharged before the bulk cargo is loaded. This ballast water has been shown to contain organisms including bacteria, viruses, algal cells, plankton, and the larval forms of many invertebrates and fish. Invertebrates regularly detected in ballast water include echinoderms, polychaete worms, and molluscs (Jones 1991).

Globally, ballast water introductions have caused serious ecological and economic problems. Examples include Zebra mussels in the North American Great Lakes, the ctenophore *Mnemiopsis leidyi* in the Black Sea, toxic dinoflagellates (*Gymnodinium* sp. and *Alexandrium* sp.) in Australia and New Zealand, and the North Pacific starfish (*Asterias amurensis)* in Tasmania, Australia. It is possible that the relatively recent spread of red tide organisms throughout the Indo-Pacific region may be due to ballast water transport (Maclean 1989). In most cases the introduced species has few predators in the new environment and its population spirals out of control.

No undesirable introductions have been detected in the GBR region, but risk assessment and management options are now being studied as a response to this potential problem.

Crown-of-Thorns Starfish

The population of the crown-of-thorns starfish (*Acanthaster planci*) has exploded in waves of outbreaks on Indo-Pacific coral reefs since the mid-1960s (Birkeland and Lucas 1990). This coral-eating echinoderm has devastated reefs in many parts of the western Pacific region; many anthropogenic causes have been invoked to explain the outbreaks. The two most enduring have been overfishing of predators (fish or the triton shell) (Lassig and Engelhardt 1994) and enhanced survivorship of the larval stage of the animal (Brodie 1992) due to phytoplankton blooms associated with enhanced nutrient runoff from coastal development (agriculture, sewage, and soil erosion). Despite many years of substantial research effort, a conclusive answer as to whether the outbreaks are caused or enhanced by human activity and, if so, what the causes are has not emerged.

On the Great Barrier Reef the starfish has gone through two outbreak cycles in 1962–1974 and 1979–1992, with a third cycle apparently starting in 1994 (Lassig and Engelhardt 1994). In the second cycle about 500 reefs were affected, mostly in the central section of the GBR. Comprehensive monitoring of reefs for crown-of-thorns populations has been in place since 1985, which has allowed the detection of the apparent third cycle in its earliest phase. However, in view of the uncertainty as to whether the outbreaks are human induced or a natural part of the ecological variability of the GBR, the Great Barrier Reef Marine Park Authority has been reluctant to implement management action to control populations of the starfish. In any event, population control measures that have been attempted, most notably in Japan, have not been successful except in very small areas (Birkeland and Lucas 1990). At present, management on the GBR is restricted to small areas around tourism operations. Control is achieved using injection of a toxic chemical (copper sulfate or more recently sodium bisulfate) into the animals by divers (Lassig 1995).

Fishing

The principal commercial fishery of the GBR is otter trawling for prawns and scallops. Other commercial fisheries include line fishing for reef and pelagic stocks, crabbing, and inshore net fisheries. Recreational fisheries for bottom and pelagic stocks are also important, as are traditional fisheries of Aboriginal and Torres Strait Islanders. Fish stocks in the GBR are small, not at all comparable with temperate offshore fisheries.

Under the offshore constitutional settlement between the Australian states and the federal government, the management of most fisheries within the Great Barrier Reef Marine Park is the responsibility of the Queensland government. The Authority, however, with its aim of protecting the natural qualities and ecosystems of the GBR, has the role of ensuring that fishing does not have unacceptable ecological impacts on both the fished areas and the reef system as a whole.

The trawl fishery in the GBR targets tiger prawns (*Penaeus semisulcatus* and *P. esculentus*) and banana prawns (*P. merguiensis*) inshore, and king prawns (*P. longistylus* and *P. latisulcatus*) offshore. Scallops (*Amusium japonicum balloti*) are caught in the southern section of the park. The trawl fishery has about 900 commercial operators, and the annual catch is about 8,000 tonnes of product.

The Authority views trawling as potentially the most damaging form of commercial fishing. The technique is indiscriminate, with 70% of the total catch being nontarget species (bycatch). There are concerns that trawling may cause long-term modification to the seabed and unacceptable effects on the sessile bottom biota. Research in other areas of northern Australia shows that bottom communities can be substantially altered by trawling, and bottom-fish communities tend to be less diverse and smaller in trawled areas (Poiner and Harris 1985). In the park, 70% of the trawlable seabed is trawled at least once a year.

At present the Authority's management response to trawling is to exclude trawling from habitat protection zones. The areas closed to trawling represent about 20% of the park, but this is only about 5% of the trawlable area available in the park. In addition, a large-scale research program is now in progress to assess the environmental effects of trawling. The research focuses on comparisons between areas open to trawling versus closed areas, manipulative experiments to determine the real effects of trawling on benthic communities and assess rates of recovery, and determination of the fate of bycatch and its impact on local seabird populations.

The reef-fish line fishery in the GBR includes both commercial and recreational components; the main target species are the larger reef fish such as coral trout and snappers (*Plectropomus* sp., *Lutjanus* sp., and *Lethrinus* sp.). There are about 300 commercial operators, 24,000 recreational boats, and more than 150 charter boats. The commercial catch is about 4,000 tonnes, with a similar recreational and charter total catch.

The Authority permits line fishing in the vast majority of the park under zoning plans, with no distinction between commercial and recreational fishing. There is now some concern that fish stocks of the large predatory fish are being overexploited. Recent research results suggest there has been a 30% decline in mean fish size and 50% decline in catch per unit effort, coupled with a 25% increase in effort in the recreational fishery in recent years (Blamey and Hundloe 1991). However, other research does not support the case for declining stocks or sizes, and evidence of large variations in the resilience of different species is emerging.

Traditional fishing by indigenous people is confined to areas close to Aboriginal communities. Traditional hunting and fishing are permitted in all zones except Preservation Zones. Fishes, turtles, and dugong are hunted. Because turtles and dugong are recognized as threatened species in GBR waters, some concern regarding management of traditional hunting has been expressed. The Authority has developed a turtle and dugong strategy, and Aboriginal and Islander communities are now involved in the development of management plans for these animals.

Water Quality

There is growing realization that change in the water quality of terrestrial runoff is one of the most significant anthropogenic impacts on the Great Barrier Reef region (Bell 1991). Since European settlement, the watersheds of rivers in north and central Queensland have been extensively modified by forestry, urbanization, and agriculture, particularly sugarcane cultivation and grazing. Recognition of the potential problems of land degradation and subsequent downstream effects of agriculture occurred some time ago. Scientific evidence of the effects on the nutrient and sediment loads of the rivers is sparse, but research effort is increasing as the potential problems are recognized. Scientific debate continues as to the severity of the problem (Kinsey 1991a), but there is general agreement concerning the need for clarification of the principal sources and the initiation of management to minimize inputs.

Many coral reefs around the world are now in a state of alarming decline due to terrestrial runoff of sediments and nutrients and other factors (Wilkinson 1993). Deforestation, agriculture on steep slopes, and sewage discharges are the causative factors frequently implicated as problems for the Great Barrier Reef.

Excess nutrients can have a number of effects on coral and coral reef systems (Kinsey 1991b). Nitrogen and phosphorus are often limiting nutrients for the growth of phytoplankton, especially in warm, clear tropical waters where light is unlikely to be limiting. Thus, phytoplankton flourishes in nutrient-enhanced conditions, leading to decreased water clarity and reduced light for coral growth on the bottom. The increased phytoplankton crop also encourages the growth of filter-feeding organisms such as sponges, tube worms, and barnacles, which compete with coral for space as well as weaken the reef's physical matrix by boring. In addition, nutrients enhance the growth of turf and macroalgae that overgrow the coral, both competing for space and shading the colonies. Excessive phosphorus concentrations weaken the coral skeleton by making it grow with a less dense structure and making the colony more susceptible to damage from storm action (Rasmussen et al. 1993). A general reduction in calcification of the reef system also occurs (Kinsey and Davies 1979).

Recent studies using existing data in catchment models have quantified the principal sources of sediment and nutrients to the coastal catchments of Queensland (Moss et al. 1992). It is estimated that 15 million tonnes of sediment, 77 thousand tonnes of nitrogen, and 11 thousand tonnes of phosphorus are exported to the coastal zone via river discharge; most of this load enters the coastal waters of the GBR. Other significant findings are that grazing is a bigger contributor of nutrients than sugarcane cultivation and that sewage discharges are a minor component of the overall flux. Sewage discharges can be significant at local scales but contribute only a small percentage of the terrestrial nutrient flux to the coastal waters of the GBR. Sediment and nutrient delivery to the GBR from terrestrial discharge has increased by 4 times since European settlement of the adjacent coast, that is, in the last 130 years.

The principal nutrients lost from grazing lands are the nutrients naturally present in the soil as distinct from added fertilizer; the principal cause of this nutrient loss is forest removal and overgrazing. In the case of sugarcane cultivation, nutrients are lost as a combination of natural soil nutrients and added fertilizers, with fertilizer addition and loss far more important than in the grazing situation. Large increases in the use of fertilizers in the last 30 years have occurred in coastal catchments (Pulsford 1996), and significant loss of this material has occurred.

In recent years management changes in some industries have reduced nutrient runoff. The most notable example is green cane harvesting and trash blanketing in sugarcane cultivation, where major reductions in soil erosion and hence phosphorus loss can occur. In the case of sewage discharges, the present policy of the Authority is for tertiary nutrient-removal treatment or land irrigation of sewage effluents in the Great Barrier Reef Marine Park. The increasing use of land irrigation of effluents on the adjacent mainland has led to some reduction in direct fluxes to the coastal zone (Brodie 1995).

Opinions as to the spatial extent of terrestrial runoff across the continental shelf differ. Some studies suggest that terrigenous input reaches only halfway across the shelf, whereas others have found terrigenous marker chemicals extending right to the edge of the shelf break. In general, there appears to be an inner reefal area dominated by terrestrial sediment and an outer area dominated by carbonate sediment. Wolanski (1994) has shown, by modeling, how river plumes are normally constrained close to the coast by hydrodynamic conditions generated by the prevailing southeast wind regime. Under other wind conditions river plumes can reach the outer reef (Brodie and Mitchell 1992).

As research and monitoring programs have progressed in recent years, evidence of the scale and nature of eutrophication and sedimentation problems in the GBR has emerged. There is considerable anecdotal evidence that reefs, particularly inshore fringing reefs, are now muddier, and have less coral cover and more algal cover. With a lack of good long-term monitoring records from these reefs to support this evidence, the Authority has been collecting historical photographs of reefs (generally reef flats exposed at low tide) for comparison with current conditions. The comparisons show less coral cover now present on many fringing reef flats than in the period before 1950.

Evidence of eutrophication in the phytoplankton record is unclear. Studies that have repeated measurements of phytoplankton composition and abundance first done in 1928–1929 in a single area near Low Isles have found significant differences. The claim has been made that the differences show the system to be in a higher nutrient condition than at that time (Bell and Elmetri 1993). Results from broad-scale phytoplankton surveys in the GBR, on the other hand, show biomass and species composition consistent with an unaffected system (Furnas 1991; Liston et al. 1992; Brodie et al. 1997).

Corals grow in annual increments and hence leave a record in their skeleton of their growth characteristics in the year of deposition. Because coral growth is related to environmental conditions in the surrounding waters, coral skeletons retain a record of these conditions. Corals may attain ages of several hundred years, and thus interpretation of environmental conditions over this period is potentially available from cores taken from old coral colonies (Isdale 1984). Evidence from coral cores from near the Queensland coast suggests that coral growth conditions did change significantly about 50 years ago and that this can be correlated with land-use changes on the adjacent coast (Rasmussen et al. 1993). Some problems still remain in separating an anthropogenic source signal in the coral skeleton from natural variability in environmental conditions and the coral's response to those conditions (Lough et al. 1997).

Water-quality management in the Great Barrier Reef Marine Park involves policy decisions such as that requiring tertiary treatment of sewage effluents discharging into the Park (introduced in 1990), cooperative arrangements with Queensland government departments to reduce pollutant inputs, case-by-case management of activities such as dredging and spoil dumping, and enforcement of Australian government acts such as those regulating the dumping of substances from ships.

On a larger scale, it is hoped to reduce agricultural sources of sediment and nutrients through an Integrated Catchment Management (ICM) program. Coordinated through the ICM process, better land management methods such as stubble retention on crop lands, retention and rehabilitation of riparian zones and wetlands, vegetation management on grazing lands, and better fertilizer application technology will achieve the desired input reductions.

The reduction in nutrient fluxes from coastal catchments into the park is seen as the most important water-quality issue facing the Authority. The monitoring and modeling of such fluxes and the associated nutrient budgets are priority activities.

Monitoring

Monitoring in the GBR region has a long history, with significant data from as far back as the Great Barrier Reef Royal Society Expedition of 1929–1930 (Orr 1933). However, continuous, reliable data sets really only started to be collected in the last 15 years. Large data sets of oceanographic parameters, nutrients, and chlorophyll are available from the early 1980s to the present (e.g., Furnas 1991; Brodie et al. 1997). Monitoring of the condition of the benthic community started with broad-scale surveys of crown-of-thorns numbers and gross coral cover in the mid-1980s. It has progressed since 1991 to more detailed annual surveys of coral cover and composition on a large number of reefs (approximately 150). Monitoring also encompasses regular surveys of turtle, dugong, and seabird numbers, long-term sea temperature monitoring, and irregular surveys of pesticides and heavy metals in biota. Good long-term records of catches and catch per unit effort for the major fisheries do not exist, but more comprehensive monitoring was introduced in 1990.

The significance of the water-quality data and other data, particularly remotely sensed CZCS imagery for chlorophyll estimation, has been debated in recent years. Claims that the GBR lagoon is eutrophic (Bell 1991) are partially refuted by recently published chlorophyll data (as an indicator of phytoplankton abundance) for the period of 1975 to 1996 (Brodie et al. 1997). This data shows low mean chlorophyll concentrations and no upward trends during the period. However, data for the inshore GBR lagoon, the area likely to be most affected by terrestrial runoff, is sparse and the status of this area will have to be determined by further monitoring and research.

Conclusions

The Great Barrier Reef large marine ecosystem is fortunate in that it is subject to relatively low human impact at present, especially when compared with many other coral reef areas (Wilkinson 1993). The area is also able to be managed as a single unit by virtue of the Marine Park Act, although this unity is diluted because many of the human impacts affecting the system originate outside the park and are not under the direct control of the Authority. Multiple-use management of the park appears to have been successful so far and provides an example of both ecologically and economically sustainable use and development. In the future, successful mitigation of eutrophication associated with terrestrial runoff will test the indirect management power of the Authority in this field. Successful management of both the environmental impacts of fishing and conservation of fish stocks—something rarely achieved in any part of the world—will also be a severe test.

To further unify the management of the GBR region, a 25-year strategic plan for the Great Barrier Reef World Heritage Area was developed (Great Barrier Reef Marine Park Authority 1994). This binds Australian and Queensland local government agencies, industries, and community groups to a set of management objectives for the area. The plan was developed in a consultative fashion, with all these groups agreeing to support and implement its provisions.

References

Bell, P. R. F. 1991. Status of eutrophication in the Great Barrier Reef lagoon. Marine Pollution Bull. 23:89–93.

Bell, P. R. F., and Elmetri, I. 1993. Validation of 1928–1929 Phosphate and Microplankton Data for Low Isles as a Baseline for Eutrophication in the Great Barrier Reef Lagoon. Report to the Great Barrier Reef Marine Park Authority, Townsville.

Birkeland, C., and Lucas, J. S. 1990. *Acanthaster planci:* Major management problem of coral reefs. Boca Raton, FL: CRC Press.

Birtles, R. A., and Arnold, P. W. 1988. Distribution of Trophic Groups of Epifaunal Echinoderms and Molluscs in the Soft-Sediment Areas of the Central Great Barrier Reef Shelf. In Proceedings of the 6th International Coral Reef Symposium, Townsville, 3:325–332.

Blamey, R. K., and Hundloe, T. J. A. 1991. Characteristics of Recreational Boat Fishing in the Great Barrier Reef Region. Report to Great Barrier Reef Marine Park Authority, Townsville.

Brodie, J. E. 1992. Enhancement of larval and juvenile survival and recruitment in *Acanthaster planci:* A review. Austr. J. Marine Freshwater Res. 43(1):539–554.

Brodie, J. E. 1995. Management of Sewage Discharges in the Great Barrier Reef Marine Park. In O. Bellwood et al (eds.), Recent Advances in Marine Science and Technology '94. Townsville: James Cook University, pp 457–465.

Brodie, J., and Furnas, M. 1994. Long Term Monitoring Programs for Eutrophication and the Design of a Monitoring Program for the Great Barrier Reef. In Proceedings of the Seventh International Coral Reef Symposium, Guam, 1992, 1:77–84.

Brodie, J. E., and Mitchell, A. 1992. Nutrient Composition of the January 1991 Fitzroy River Plume. In G. T. Byron (ed.), Workshop on the Impacts of Flooding. Townsville: Great Barrier Reef Marine Park Authority Workshop Series No. 17, pp. 56–74.

Brodie, J. E., Furnas, M. J., Steven, A. D. L., Trott, L. A., Pantus, F. and Wright, M. 1997. Monitoring chlorophyll in the Great Barrier Reef lagoon: Trends and variability. In Proceedings of the Eighth International Coral Reef Symposium, Panama, 1996, 1:797–802.

Drew, E. A., and Abel, K. M. 1988. Studies on Halimeda. I. The distribution and species composition of Halimeda meadows throughout the Great Barrier Reef Province. Coral Reefs 6:195–205.

Furnas, M. 1991. Nutrient Status and Trends in Waters of the Great Barrier Reef. In D. Yellowlees (ed.), Land Use Patterns and Nutrient Loading of the Great Barrier Reef Region. Townsville: James Cook University, pp. 162–179.

Furnas, M. J., Mitchell, A. W., and Skuza, M. 1994. Nitrogen and Phosphorus Budgets for the Central Great Barrier Reef Shelf. Research Publication No. 36. Townsville: Great Barrier Reef Marine Park Authority.

Great Barrier Reef Marine Park Authority. 1994. A 25 Year Strategic Plan for the Great Barrier Reef World Heritage Area. Townsville: Great Barrier Reef Marine Park Authority.

Hopley, D. 1982. The Geomorphology of the Great Barrier Reef: Quaternary Development of Coral Reefs. New York: Wiley-Interscience.

Isdale, P. 1984. Fluorescent bands in massive corals record centuries of coastal rainfall. Nature 310:578–579.

James, M. K., Dight, I. I., and Day, J. C. 1990. Application of Larval Dispersal Models to Zoning in the Great Barrier Reef Marine Park. In Proceedings of the Fourth Pacific Congress on Marine Science and Technology, Tokyo.

Jones, M. M. 1991. Marine Organisms Transported in Ballast Water. Bulletin No. 11. Canberra: Bureau of Rural Resources.

Kinsey, D. W. 1991a. Can we resolve the nutrient issue for the reef? Search 22:119–121.

Kinsey, D. W. 1991b. Water Quality and its Effect on Reef Ecology. Land Use Patterns and Nutrient Loading of the Great Barrier Reef Region, Townsville: James Cook University, pp. 192–196.

Kinsey, D. W., and Davies, P. J. 1979. Effects of elevated nitrogen and phosphorus on coral reef growth. Limnol. Oceanogr. 24:935–940.

Lassig, B. 1995. Controlling Crown-of-Thorns Starfish. Townsville: Great Barrier Reef Marine Park Authority.

Lassig, B., and Englehardt, U. 1994. Crown-of-thorns starfish—Another outbreak looms. Search 25(3):66–68.

Lee Long, W. J., Mellors, J. E., and Coles, R. G. 1993. Seagrasses between Cape York and Hervey Bay, Queensland, Australia. Austr. J. Marine Freshwater Res. 44:19–31.

Liston, P., Furnas, M. J., Mitchell, A. W., and Drew, E. A. 1992. Local and mesoscale variability of surface water temperature and chlorophyll in the northern Great Barrier Reef, Australia. Continental Shelf Res. 10:907–922.

Lough, J. M., Barnes, D. J., and Taylor, R. B. 1997. Understanding growth mechanisms: The key to successful extraction of proxy climate records from corals. In Proceedings of the Eighth International Coral Reef Symposium, Panama, 1996, 2: 1697–1700.

Maclean, J. L. 1989. Indo-Pacific red tides, 1985–1988. Marine Pollution Bull. 20(7):304–310.

Moran, P. J. 1986. The Acanthaster phenomenon. Oceanogr. Marine Biol. Annu. Rev. 24:379–480.

Moss, A. J., Rayment, G. E., Reilly, N., and Best, E. K. 1992. A Preliminary Assessment of Sediment and Nutrient Exports from Queensland Coastal Catchments. Technical Report No. 4. Brisbane: Queensland Department of Environment and Heritage.

Newman, L. J., and Cannon, L. R. G. 1994. Biodiversity of tropical polyclad flatworms from the Great Barrier Reef, Australia. Memoirs of the Queensland Museum 36(1):159–163.

Orr, A. P. 1933. Physical and chemical conditions in the sea in the neighborhood of the Great Barrier Reef. In British Museum Great Barrier Reef Expedition 1928–1929, Science Report 2(3):37–86.

Poiner, I. R., and Harris, A. N. M. 1985. The Effect of Commercial Prawn Trawling on the Demersal Fish Communities of the South Eastern Gulf of Carpentaria. In Torres Strait Fisheries Seminar, pp. 239–261.

Pulsford, J. S. 1996. Historical nutrient usage in coastal Queensland river catchments adjacent to the Great Barrier Reef Marine Park. GBRMPA Research Publication No. 40. Townsville: Great Barrier Reef Marine Park Authority.

Rasmussen, C., Cuff, C., and Hopley, D. 1993. Evidence of Anthropogenic Disturbance Detailed in the Skeleton of Massive Corals of Australia's Great Barrier Reef. In Proceedings of the 7th International Coral Reef Symposium, Guam, 1992, 1:201–210.

Wachenfeld, D.R. 1997. Long-term trends in the status of coral reef-flat benthos—The use of historical photographs. In D. R. Wachenfeld et al. (eds.), State of the Great Barrier Reef World Heritage Area Workshop, Workshop Series No. 23. Townsville: Great Barrier Reef Marine Park Authority, pp. 134–148.

Wilkinson, C. R. 1993. Coral Reefs of the World Are Facing Widespread Devastation: Can We Prevent This Through Sustainable Management Practices? In Proceedings of the 7th International Coral Reef Symposium, Guam, 1992, 1:11–21.

Wolanski, E. 1994. Physical Oceanographic Processes of the Great Barrier Reef. Boca Raton, FL: CRC Press.

25

Borneo's Marine Ecosystem and the Greenhouse Risk Factor: A National Perspective

Arbain Hj. Kadri, Mazlin B. Mokhtar, Almah Bt. Awaluddin, and Saleem Mustafa

More than 70% of the Earth's surface is covered by saltwater. The water depth averages 3.8 km, which gives a volume of 1,370 × 10^6 km^3. This "inner space" harbors nearly 80% of the Earth's animal life and constitutes the single largest repository of organisms on the planet. The significance of ocean resources in terms of human requirements can hardly be overestimated. The enormity of the ocean and its capacity to support production of protein-rich food, pharmaceutically important natural products, and industrial raw materials have resulted in an enormous interest in exploitation of marine bounties. There can be no denying the fact that world economics, to a great extent, depend on the oceans. Human population growth and shortage of land resources are increasing human dependence on the ocean. The struggle for resources to support increasing populations contributed to the development of the exclusive economic zone (EEZ) concept to reduce the budding conflicts concerning oceanic bounty.

Another issue, this one of environmental concern, is complicating an already stressed situation. Ozone depletion and global warming are of international importance and require even greater international coordination and cooperation.

Borneo, the world's third largest island, lies in the Pacific at the equator. It is surrounded by a productive tropical sea. The island's terrain is mountainous, with one peak (Mount Kinabalu) attaining a height of 4,101 meters. The island is covered by tropical rain forests. The marine resources are traditionally exploited, with mariculture in nearshore areas gaining in importance. In some places along the coast the economic activity connected with exploitation of renewable saltwater resources is so high that the authors can define them as "intensive economic subzones" (IESs) within the internationally recognized EEZ. These IESs bear witness to the close interaction between the marine resources and society's sustainable use of them.

Sustainable Development and the Greenhouse Risk Factor

Resource sustainability is facing a challenge from human activities, both directly and indirectly. The modern concept of sustainable development is based on a holistic approach for which five essential operating elements have been identified by Haimes (1994):

1. Multiobjective analysis
2. Risk analysis, including risk of extreme events
3. Impact analysis
4. Consideration of multiple decision makers and constituencies
5. Accounting for interactions among a system's components and between the system and its environment

Assessment and management of risk, which is a measure of the probability and severity of adverse effects, are crucial to sustainable development. Often, assessment of a risk is easier than its management, but development without a program of risk management puts sustainability at stake.

The greenhouse effect is a natural phenomenon in which various gases (particularly water vapor, carbon dioxide, methane, and nitrous oxide) and cloud formations reflect the earth's long-wave, low-energy radiation back toward the earth rather than out to the atmosphere. Changes in the chemical composition of the atmosphere are enhancing the natural greenhouse effect and are being generated through human activities related to energy use, particularly fossil fuel burning. Changes in the atmospheric composition can have serious effects on a variety of earth's systems and can be expressed through increases in air temperature and thermal expansion of the oceans, among other things.

According to the Intergovernmental Panel on Climate Change (IPCC 1990), the global land and ocean warming in the past 100 years amounted to 0.5°C, and the anticipated warming rate is 0.3°C per decade. Rise in sea level, as a direct consequence of global warming, results from thermal expansion of water in the oceans and melting of ice on land (glaciers and poles). IPCC (1990) estimates put past century increases in sea level at 10 cm. Sea-level rise increases wave action and coastal erosion, threatening coastal vegetation and adding to the lashing power of waves.

Sea-level rise is of particular concern in Borneo for both maritime and coastal use of resources. Coastal erosion and inundation of land pose a serious challenge to the already established human interests. Inland freshwater aquaculture enterprises and land-based agriculture operations may face the problem of seawater invasion, with consequent increases in salinity to a level beyond the tolerance limit of the crops. The powerful winds and waves bashing the coast will enhance the erosion and continue changing the shoreline. Sediments eroded from one location will be deposited elsewhere, even from country to country.

With this variety of concerns relating to the effects of global change on resources, three solutions are suggested. First, on an international scale, a formula should be developed, based on scientific studies, that defines the origin of sediments in the event sediment deposits emerge as economically or strategically valuable assets. It should be the responsibility of the United Nations to settle claims of sovereignty as well as maritime limits. Second, there should be a concerted effort to study the quality and stability of shore deposits, coastal vegetation, area topography, pattern of rainfall, and inland water courses and to monitor currents, wave action, sea-level change, and related processes. This knowledge will help to determine the vulnerability of coasts to greenhouse effects and to guide suitable measures for strengthening coastal security. And finally, Mustafa et al. (1994) present the following action plan specific to Borneo:

1. Continued silt reinforcement of the coastline
2. Unhindered proliferation of mangrove forests and reforestation of degraded vegetation
3. Control over construction of dams and reservoirs that interfere with natural sediment transport
4. Preservation of mountain flora
5. Control of reclamation or any human activity that disrupts the coastline
6. Development of aquaculture enterprises in sheltered coastal bays

Ozone Depletion and International Actions

A second area of global environmental concern is ozone depletion. Ozone in the stratosphere (15–50 km) protects the surface of the earth from receiving high levels of ultraviolet (UV) radiation. UV radiation is of three types: UVA (320–400 nm), UVB (290–320 nm), and UVC (<290 nm). The UVC type is most lethal to cells, followed by UVA and UVB. Damage to this ozone umbrella by chlorofluorocarbons (CFCs) is exposing the life on this planet to increased UV radiation. DNA controls protein biosynthesis and determines the hereditary destiny of organisms and is very susceptible to UV radiation damage. The effect of increased UV levels on marine life is expected to be serious. Behrenfeld et al. (1993) observed inhibition of photosynthesis in the South Pacific Ocean by ambient radiation intensity. Setbacks to phytoplankton carbon fixation in the UV range of 290–320 nm were noted, and it was concluded that although an increase in UV radiation exposure will lead to more photodamage, photorepair and photoadaptation will occur but will require energy from the organism's allocation for growth and multiplication. These findings are consistent with the earlier work of Behrenfeld et al. (1992) showing a 20% decline in carbon fixation rates that resulted in a 9% decrease in biomass following UV exposure. Some 200 billion tons of carbon go through the photosynthetic process annually in the transformation of CO_2 and H_2O to organic compounds that support life on this planet. The consequences of photodamage on this process will be far reaching. Worrest (1982) and the U.S. Environmental Protection Agency (1987) also have documented UVB radiation damage to marine organisms, including fish, shrimp, crab, and plants, that affects their survival, growth, reproduction, and other functions.

Borneo's equatorial location exposes it to strong radiation and long light hours. Although there is some increased risk of UV damage, at the present time increased UV damage is more likely to occur nearer the polar regions, where ozone loss is greatest. Borneo's waters remain fairly well mixed, with no sign of a vertical salinity gradient or stratification. Also, the water is optically clear; hence, the phytoplankton can physically reorient their position and metabolically adjust, to some extent, to deal with changes in radiation. The sea surrounding Borneo shows no appreciable variation in thermal features and salinity. Limiting factors of the type that develop in stratified waters do not exist in Borneo. An organism's versatility depends greatly on its mobility. Those capable of migration and of overcoming ecological barriers can make their way to safer places. Others, particularly the sessile molluscs, which are lethargic and spend most of their life attached to substratum on the continental shelf or in the intertidal zone, will be at risk.

Tropical waters around Borneo are rich in plant and animal germplasm and sustain higher biodiversity than temperate seas. The rate and extent of global environmental change are endangering this important supply of natural variety. An ambitious program to preserve this biodiversity is suggested by the authors. This program includes the following items:

1. Monitoring of stratospheric ozone concentration and of greenhouse gases
2. Detecting possible changes in quantum or intensity of UV radiation at the earth's surface
3. Measuring penetration of UV radiation in the ocean
4. Determining possible changes in ozone and greenhouse gases as a sequel to control measures and determining the exact quantitative relationship between the two
5. Developing ozone-friendly substitutes for ozone-depleting chemicals, especially CFCs
6. Conducting laboratory studies on the possible actions of some chemicals for strengthening ozone integrity in the face of greenhouse gases and having a healing effect on the injured ozone layer (ozone chemotherapy)
7. Computer-based modeling and simulation work on the dynamics of change in ozone and major consequential processes on earth
8. Implementation of a carefully formulated ecodevelopment plan

To achieve success on a local scale in Borneo requires that the seriousness of the global environmental problems and their consequences be understood by all sectors of the community. General awareness and arousal of concern are necessary, along with a realignment of perspectives related to short-term personal and social gains at the expense of long-term environmental damage and loss. Changing this perspective and the actions that it supports is indeed difficult, but is required if we are to succeed in living sustainably with our marine resources.

References

Behrenfeld, M., Hardy, J. T., and Lee, H. 1992. Chronic effects of ultraviolet-B radiation on growth and cell volume of *Phaeodactylum tricornutum* (Bacillariophyceae). J. Phycol. 28:757–760.

Behrenfeld, M., Hardy, J., Gucinski, H., Hanneman, A., Lee, H., and Wones, A. 1993. Effects of ultraviolet-B radiation on primary production along latitudinal transacts in the South Pacific Ocean. Mar. Env. Res. 35:349–363.

Haimes, Y. Y. 1994. Risk Management: A Requisite for Sustainable Development. Proceedings of International Symposium on Water Resources Planning in a Changing World, Karlsruhe, Germany.

Intergovernmental Panel on Climate Change (IPCC). 1990. Strategies for Adaptation to Sea Level Rise. Coastal Zone Management Group Report. Rijkswaterstaat, The Hague: Intergovernmental Panel on Climate Change.

Mustafa, S., Awaluddin, A., and Mokhtar, M. 1994. Management of Coastal Water Resources of Borneo in the Face of Global Warming. Proceedings of International Symposium on Water Resources Planning in a Changing World, Karlsruhe, Germany.

Environmental Protection Agency (EPA). 1987. An Assessment of the Effects of Ultraviolet-B Radiation on Aquatic Organisms. In Assessing the Risks of Trace Gases That Can Modify the Stratosphere. Washington, DC: U.S. Environmental Protection Agency, pp. 1–13.

Worrest, R. C. 1982. Review of Literature Concerning the Impact of UV-B Radiation upon Marine Organisms. In J. Calkins (ed.), The Role of Solar Ultraviolet Radiation in Marine Ecosystems. New York: Plenum Press, pp. 429–458.

26

Large Marine Ecosystems: Assessment and Management

Kenneth Sherman

The long-term sustainability of coastal ecosystems as a resource for healthy economies in coastal nations appears to be diminishing. A growing awareness that the quality of the coastal ecosystems is being adversely affected by multiple driving forces has accelerated efforts by scientists and program managers to assess, monitor, and mitigate coastal stressors from an ecosystem perspective. The Intergovernmental Oceanographic Commission (IOC) of the United Nations Educational, Scientific, and Cultural Organization (UNESCO) is encouraging coastal nations to establish national programs for assessing and monitoring coastal ecosystems so as to enhance the ability of national and regional management organizations to develop and implement effective remedial programs for improving the quality of degraded ecosystems. This encouragement follows from the significant milestone achieved in June 1992 with the adoption by a majority of coastal countries of follow-on actions to the United Nations Conference on Environment and Development (UNCED) declaration on the ocean. This declaration called for the nations of the globe to (1) prevent, reduce, and control degradation of the marine environment so as to maintain and improve its life-support and productive capacities; (2) develop and increase the potential of marine living resources to meet human nutritional needs, as well as social, economic, and development goals; and (3) promote the integrated management and sustainable development of coastal areas and the marine environment.

Coastal Ecosystem Stress

Post-UNCED concern has been expressed over the deteriorating condition of the world's coastal ecosystems, which produce most of the world's living marine resources. Within the nearshore areas and extending seaward around the margins of the global land masses, coastal ecosystems are being subjected to increased stress from toxic effluents, habitat degradation, excessive nutrient loadings, harmful algal blooms, emergent diseases, fallout from aerosol contaminants, and episodic losses of living marine resources from pollution effects and overexploitation. Coastal pollution, changes in biodiversity, the degraded states of fish stocks, and the loss of coastal habitat generally are limiting achievement of the full economic potential of coastal ecosystems.

Present efforts to address these problems by local, regional, national, and international institutions responsible for resource stewardship have been less than successful. Informed decisions for ensuring the long-term development and sustainability of coastal marine resources can best be made when based on sound, scientifically derived options. For most coastal ecosystems, existing environmental data pertinent to studies of perturbations to habitats and populations at the species, population, community, and ecosystem levels are difficult to synthesize because of the spatially and temporally fragmented character of the variables in question and general lack of comparability among data sets. To overcome these shortcomings, there is a need for a more coherent and integrative assessment of the changing states of coastal ecosystems, from drainage basins to the adjacent marine ecosystems, that is directly linked to institutions responsible for the governance of the ecosystems. An essential component of an ecosystem management regime is the inclusion of a scientifically based strategy to monitor and assess the changing states and health of the ecosystem by tracking key biological and environmental parameters. From this perspective, marine ecosystem assessment and monitoring are defined as components of a management system that includes regulatory, institutional, and decision-making aspects relating to marine ecosystems. The system would include, therefore, a range of activities to provide management information about ecosystem conditions, contaminants, and resources at risk within the geographic extent of the ecosystem.

The Large Marine Ecosystem Assessment Strategy

An ecological framework that can serve as a basis for achieving the UNCED objectives is the large marine ecosystem (LME) concept. LMEs are increasingly being subjected to stress from growing exploitation of fish and other renewable resources, coastal zone damage, habitat losses, river basin runoff, dumping of urban wastes, and fallout from aerosol contaminants. These are regions of ocean space encompassing coastal areas from river drainage basins and estuaries on out to the seaward boundary of continental shelves and the seaward margins of coastal current systems. LMEs are relatively large regions on the order of 200,000 km^2 or larger, characterized by distinct bathymetry, hydrography, productivity, and trophically dependent populations. The theory, measurement, and modeling relevant to monitoring the changing states of LMEs are embedded in published reports on ecosystems with multiple steady states and on the pattern formation and spatial diffusion within ecosystems (AAAS 1986, 1989, 1990, 1991, 1993). Based on examination of the bathymetry, hydrography, productivity, and trophic linkages of marine populations, the spatial extents of 49 distinct LMEs from around the margins of the Atlantic, Pacific, and Indian Oceans have been described (Figure 26-1).

These LMEs produce 95% of the annual global fisheries biomass yields; reports on the changing states of biomass yields and health have been published for 29 of the LMEs (Sherman 1994). The assessments of the changing states of LMEs are based on information obtained from five operational modules that link science-based information to socioeconomic benefits for countries bordering on LMEs. The modules are focused on ecosystem productivity, fish and fisheries, pollution and health, socioeconomic conditions, and governance protocols.

Productivity Module

Productivity can be related to the carrying capacity of the ecosystem for supporting fish resources (Pauly and Christensen 1995). It has been reported that the maximum global level of primary productivity for supporting the average annual world catch of fisheries has been reached, and further large-scale unmanaged increases in fisheries yields from marine ecosystems are likely to be at trophic levels below fish in the marine food chain (Beddington 1995). Evidence of this effect appears to be corroborated by the recent changes in the species composition of the catches of fisheries from the East China Sea LME (Chen and Shen 1995). Measurement of ecosystem productivity can also serve as a useful indication of the growing problem of coastal eutrophication (NSTF 1993). In several LMEs, excessive nutrient loadings of coastal waters have been related to

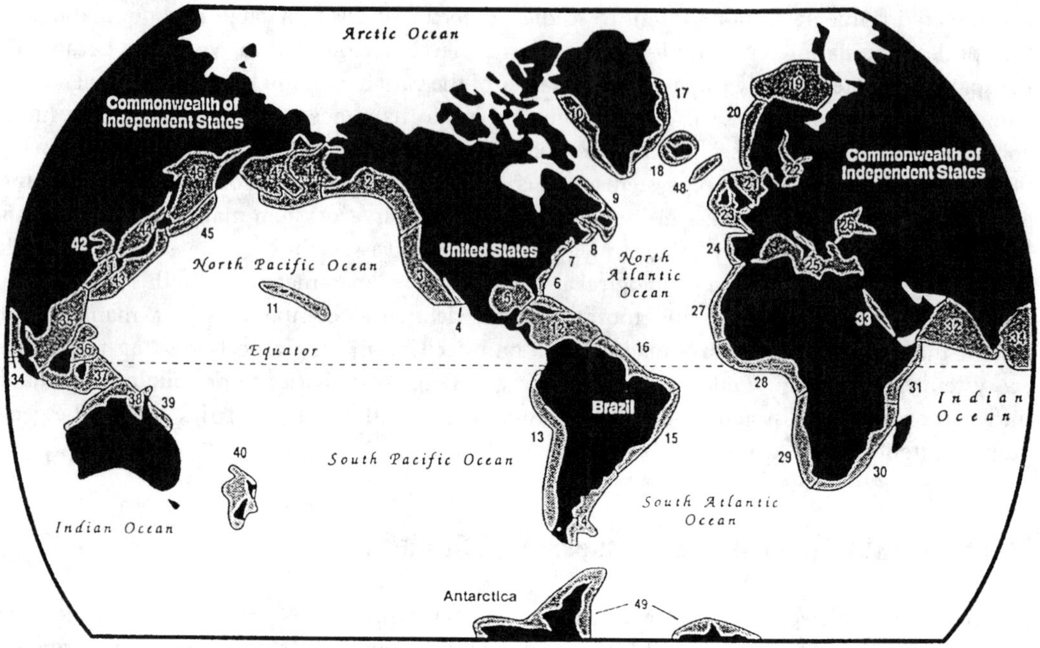

Figure 26-1. World map of large marine ecosystems.

1. Eastern Bering Sea
2. Gulf of Alaska
3. California Current
4. Gulf of California
5. Gulf of Mexico
6. Southeast U.S. Continental Shelf
7. Northeast U.S. Continental Shelf
8. Scotian Shelf
9. Newfoundland Shelf
10. West Greenland Shelf
11. Insular Pacific—Hawaiian
12. Caribbean Sea
13. Humboldt Current
14. Patagonian Shelf
15. Brazil Current
16. Northeast Brazil Shelf
17. East Greenland Shelf
18. Iceland Shelf
19. Barents Sea
20. Norwegian Shelf
21. North Sea
22. Baltic Sea
23. Celtic-Biscay Shelf
24. Iberian Coastal
25. Mediterranean Sea
26. Black Sea
27. Canary Current
28. Gulf of Guinea
29. Benguela Current
30. Agulhas Current
31. Somali Coastal Current
32. Arabian Sea
33. Red Sea
34. Bay of Bengal
35. South China Sa
36. Sulu-Celebes Seas
37. Indonesian Seas
38. Northern Australian Shelf
39. Great Barrier Reef
40. New Zealand Shelf
41. East China Sea
42. Yellow Sea
43. Kuroshio Current
44. Sea of Japan
45. Oyashio Current
46. Sea of Okhotsk
47. West Bering Sea
48. Faeroe Plateau
49. Antarctic

algal blooms that have been implicated in mass mortalities of living resources, emergence of pathogens (e.g., cholera, vibrios, red tides, paralytic shellfish toxins), and explosive growth of nonindigenous species (Epstein 1993).

The ecosystem parameters measured in the productivity module are zooplankton biodiversity and information on species composition, zooplankton biomass, water-column structure, photosynthetically active radiation (PAR), transparency, chlorophyll *a*, NO_2, NO_3, and primary production. The plankton of LMEs can be measured by deploying Continuous Plankton Recorder (CPR) systems from commercial vessels of opportunity (Glover 1967). The advanced plankton recorders can be fitted with sensors for temperature, salinity, chlorophyll, nitrate/nitrite, petroleum, hydrocarbons, light, bioluminescence, and primary productivity (Aiken 1981; Aiken and Bellan 1990; Williams and Aiken 1990; UNESCO 1992; Williams 1993), providing the means to monitor changes in phytoplankton, zooplankton, primary productivity, species composition and dominance, and long-term changes in the physical and nutrient characteristics of the LME, as well as longer-term changes relating to the biofeedback of the plankton to the stress of environmental change (Colebrook 1986; Dickson et al. 1988; Williams 1993).

The Fish and Fisheries Module

Changes in biodiversity among the dominant species within the fish communities of LMEs have resulted from excessive exploitation (Sissenwine and Cohen 1991), naturally occurring environmental shifts in climate regime (Bakun 1993), or coastal pollution (Mee 1992; Bombace 1993). Changes in the biodiversity of the fish community can generate cascading effects up the food chain to apex predators and down the food chain to plankton components of the ecosystem (Overholtz and Nicolas 1979; Payne et al. 1990). These three sources of variability in fisheries yield are operable in most LMEs. However, they can be described as primary, secondary, and tertiary driving forces in fisheries yields, contingent on the ecosystem under investigation. For example, in the Humboldt Current, Benguela Current, and California Current LMEs, the primary driving force influencing variability in fisheries yield is the influence of changes in upwelling strength (Bakun 1995, 1993; MacCall 1986; Crawford et al. 1989; Alheit and Bernal 1993); fishing and pollution effects are secondary and tertiary effects on fisheries yields in these ecosystems. In continental shelf LMEs, including the Yellow Sea and the Northeast U.S. Shelf, excessive fisheries effort has been the cause of large-scale declines in catch and changes in the biodiversity and dominance in the fish community (Tang 1993; Sissenwine 1986). In these ecosystems, pollution and environmental perturbation are of secondary and tertiary influence. In contrast, significant coastal pollution and eutrophication have been the principal factors driving the changes in fisheries yields of the Northwest Adriatic (Bombace 1993), the Black Sea (Mee 1992), and the near coastal areas of the Baltic Sea (Kullenberg 1986). Overexploitation and natural environmental changes are of secondary and tertiary importance. Consideration of the driving forces of change in biomass yield is important when developing options for management of living marine resources for long-term sustainability.

The fish and fisheries module includes fisheries-independent bottom trawl surveys and acoustic surveys for pelagic species to obtain time-series information on changes in biodiversity and abundance levels of the fish community. Standardized sampling procedures, when deployed from small calibrated trawlers, can provide important information on diverse changes in fish species. The fish catch provides biological samples for stock assessments, stomach analyses, and age, growth, fecundity, and size comparisons (ICES 1991); data for clarifying and quantifying multispecies trophic relationships; and samples for monitoring pathological conditions that may be associated with coastal pollution. The trawlers can also be used as platforms for obtaining water, sediment, and benthic samples for monitoring harmful algal blooms, virus vectors of disease, eutrophication, anoxia, and changes in benthic communities.

Pollution and Ecosystem Health Module

In several LMEs, pollution has been a principal driving force in changes of biomass yields. Assessing the changing status of pollution and the health of the entire LME is scientifically challenging. Ecosystem health is a concept of wide interest for which a single precise scientific definition is problematical. Methods to assess the health of LMEs are being developed from modifications to a series of indicators and indices described by several investigators (Costanza 1992; Rapport 1992; Norton and Ulanowicz 1992; Karr 1992). The overriding objective is to monitor changes in health from an ecosystem perspective as a measure of the overall performance of a complex system (Costanza 1992). The health paradigm is based on multiple-state comparisons of ecosystem resilience and stability (Holling 1986; Pimm 1984; Costanza 1992) and is an evolving concept.

Following the definition of Costanza (1992), to be healthy and sustainable an ecosystem must maintain its metabolic activity level, its internal structure and organization, and must be resistant to external stress over time and space scales relevant to the ecosystem. These concepts were discussed by panels of experts at two workshops convened in 1992 by the National Oceanic and Atmospheric Administration (NOAA 1993). Among the indices discussed by the participants were five that are being considered as experimental measures of changing ecosystem states and health: biodiversity, stability, yields, productivity, and resilience. The data from which to derive the experimental indices are obtained from time-series monitoring of key ecosystem parameters. An effort to validate the utility of the indices is under development by NOAA at the Northeast Fisheries Science Center. The ecosystem sampling strategy is focused on parameters relating to the resources at risk from overexploitation, species protected by legislative authority (marine mammals), and other key biological and physical components at the lower end of the food chain (plankton, nutrients, hydrography). The parameters of interest are described in Sherman (1994) and include zooplankton composition, zooplankton biomass, water-column structure, photosynthetically active radiation (PAR), transparency, chlorophyll *a*, NO_2, NO_3, primary production, pollution, marine mammal biomass, marine mammal composition, runoff, wind stress, seabird community structure, seabird counts, finfish composition, finfish biomass, domoic acid, saxitoxin, and paralytic shellfish poisoning (PSP). The experimental parameters selected incorporate the behavior of individuals and the resultant responses of populations and communities, as well as their interactions with the physical and chemical environment. The selected parameters provide a basis for comparing changing health status among ecosystems.

Fish, benthic invertebrates, and other biological indicator species are used in the pollution and ecosystem health module to measure pollution effects on the ecosystem, including the bivalve monitoring strategy of Mussel-Watch (Farrington et al. 1983), the pathobiological examination of fish (ICES 1988; O'Connor and Ehler 1991; White and Robertson 1996), and the estuarine and nearshore monitoring of contaminants in the water column, substrate, and selected groups of organisms.

Important components of the associated research to support the health assessment are the definition of routes of exposure of selected finfish and shellfish to toxic contaminants and the assessment of exposure to toxic chemicals by several life history stages. The routes of bioaccumulation and trophic transfer of contaminants are assessed, and critical life history stages and selected food-chain organisms are examined for a variety of parameters that indicate exposure to, and effects of, contaminants. Contaminant-related effects measured include diseases, impaired reproductive capacity, and impaired growth. Many of these effects can be caused by direct exposure to contaminants, or indirectly from alterations in prey organisms. The assessment of chemical contaminant exposure and effects on fisheries resources and food-chain organisms consists of a suite of parameters, including biochemical responses that are clearly linked to contaminant exposure coupled with measurements of organ disease and reproductive status that have been used in previous studies to establish links between exposure and effects. The specific suite of parameters measured will cover the same general responses and thus allow for comparable assessment of the physiological status of each species sampled as it relates to chemical contaminant exposure and effects at the individual species and population level. The implementation of protocols for assessing the frequency and effect of harmful algal blooms (Smayda 1991) and emergent diseases (Epstein 1993) are included in the pollution module.

The Socioeconomic Module

This module is characterized by its emphasis on practical applications of its scientific findings in managing the LME and on the explicit integration of economic analysis with the science-based assessments to ensure that prospective management measures are cost-effective. Economists and policy analysts will need to work closely with ecologists and other scientists to identify and evaluate management options that are both scientifically credible and economically practical.

The economic and management research will be closely integrated with the scientific research throughout, and is designed intentionally to respond adaptively to enhanced scientific information. This component of the LME approach to marine resources management was developed by the late James Broadus, former Director of the Marine Policy Center, Woods Hole Oceanographic Institution. It consists of six interrelated elements.

HUMAN FORCING FUNCTIONS. The natural starting point is a generalized characterization of the ways in which human activities affect the natural marine system and the expected sensitivity of these forcing functions to various types and levels of human activity. Population dynamics, coastal development, and land-use practices in the system's drainage basin are clear examples. Work integrating the efforts of natural and social scientists should concentrate further on resolving apparent effects (such as eutrophication-associated red tide events or changing fish population structures) that are confounded by cycles or complex dynamics in the natural system itself. Progress is possible, too, in achieving better characterizations of the way in which human forcing is mediated by alternate management options. Emphasis should be on isolating and quantifying those forcing activities (sewage discharge, agricultural runoff, fishing effort) likely to be expressed most prominently in effects on the natural system.

ASSESSING IMPACTS. Another natural element in the systemic approach is to estimate and even predict the economic impacts of unmanaged degradation in the natural system and, obversely, the expected benefits of management measures. Such assessment is a form of standard benefit–cost analysis, but it requires scientific information to describe the effects of human forcing so they may be quantified in economic terms. Initial analysis should focus on the social and economic sectors likely to experience the largest impacts: fishing, aquaculture, public health, recreation, and tourism.

FEEDBACKS. Collaborative effort should also be devoted to identifying and estimating the feedbacks of economic impacts into the human forcing function. Extensive coastal eutrophication, for example, associated with coastal development and runoff, might reduce the suitability of coastal areas for aquaculture production and increase its exposure to red tide damage, thereby putting a premium on capture fishery and increasing pressure on wild stocks. Similar feedbacks, both negative and positive, should be addressed and expressed in economic terms for all the major sectors.

ECOSYSTEM SERVICE AND THE VALUE OF BIODIVERSITY. Special consideration should be given to improved knowledge of how the natural system generates economic values. Many valuable services provided by natural systems are not traded in markets or included in planning evaluations, so extra care must be made to ensure that they are not sacrificed through ignorance. The services provided by coastal wetlands as nurseries for fisheries, natural pollution filters, and storm buffers are well-known examples that have particular relevance to coastal reclamation activities. Other examples are more subtle, including the importance of predator–prey relationships and the possibility of losing unrecognized "keystone" species in a valuable ecosystem. Experience indicates that aesthetic and recreational/tourist amenities are to be expected to have a growing economic value in the LMEs. A variety of sources of economic value arising from the natural diversity of the LME should be identified and assessed in regard to existing uses and potential management innovations.

ENVIRONMENTAL ECONOMICS. Many of the elements described in this section constitute topics in environmental economics. Specialists in that field attempt to estimate the economic values (both use and nonuse) associated with environmental resources and to identify the conditions associated with their optimal management (to derive the greatest net benefits for society). An important element is the collaboration between scholars from developing nations and those from the developed countries to transfer and adapt to the needs and techniques of environmental economics.

INTEGRATED ASSESSMENT. The ultimate objective is the integration of all the results achieved above, along with scientific characterizations of the LME, into a comprehensive analytic framework that will permit integrated assessment of human practices, effects, and management options in the region. Such work is at the forefront of recent research on the human dimensions of global environmental change as well as research on human interactions with natural marine systems.

Governance Module

The governance module is evolving based on case studies now under way among ecosystems that will be managed from a more holistic perspective in projects supported by the Global Environment Facility (GEF). These projects include the Yellow Sea ecosystem, where the principal effort is being undertaken by the People's Republic of China (Tang 1989) and South Korea (Zhang and Kim, this volume); the Gulf of Guinea LME, overseen by Ivory Coast, Ghana, Nigeria, Benin, and Cameroon (UNIDO 1997); and joint assessments by the governments of South Africa, Namibia, and Angola of the resources of the Benguela LME to guide the development of the shared resources of the ecosystem to ensure their long-term sustainability, particularly with regard to food security. The Great Barrier Reef Ecosystem is also being managed from a holistic ecosystem perspective (Bradbury and Mundy 1989; Kelleher 1993), along with the Northwest Australian Continental Shelf ecosystem (Sainsbury 1988), under management by the state and federal governments of Australia. Similarly, the Antarctic marine ecosystem is managed under the Commission for the Conservation of Antarctic Marine Living Resources (CCAMLR) and its 21-nation membership (Scully et al. 1986; Sherman and Ryan 1988). Movement toward ecosystem management is emerging for the North Sea (NSTF 1993), the Barents Sea (Eikeland 1992), and the Black Sea (Hey and Mee 1993).

Fast Track LME and Integrated Coastal Management Projects

A framework for linking science-based assessments of the changing states of coastal ecosystems to support the long-term sustainability of environmental quality and renewable resources while also guiding environmentally sound economic development of ecosystem resources and management practices is now emerging from a series of regional efforts aimed at cross-sectoral integration of assessments of coastal productivity, fish and fisheries, and pollution and ecosystem health with the socioeconomic and governance modules. The application of the modules is being supported, in part, by grants from the GEF in collaboration with national governments of countries bordering large marine ecosystems in Asia, Africa, Central and South America, and eastern Europe.

The GEF Operational Strategy calls for the development and implementation of projects in the International Waters Program that can achieve global benefits through the implementation by countries of more comprehensive approaches for restoring and protecting the international waters (IW) environment. The goal of the IW program "is to assist countries in making changes in the ways that human activities are conducted in different economic sectors so that the particular water body and its multi-country drainage basin can sustainably support the new patterns of human activities." The GEF has placed priority on changing sectoral policies and activities responsible for the most serious root causes of transboundary environmental problems and determining the expected baseline and additional actions needed to resolve each priority concern. Based on the countries' commitments to change sectoral policies or activities and to find baseline investments, the GEF may fund the agreed incremental cost of additional measures (GEF 1997). One of the focal areas for funding by the GEF is to mitigate stressors on large marine ecosystems and promote priority actions for improving environmental quality and the sustainable development of resources within LMEs important to the economic growth and food security of developing countries in Asia, Africa, Central and South America, and eastern Europe.

It is important to strengthen the linkages between science and management for the drainage basins and near coastal and geographic extent of the LMEs, and also to improve on the methodologies for ensuring that near coastal and drainage basin effects on the LME proper are included in the overarching strategy for a systems approach to a management strategy that includes (1) drainage basin, (2) near coastal, and (3) offshore coastal components of the LME.

Two model systems can be used to fill this pressing need for improved assessment strategies: the Batangas Bay model in the Philippines and the Xiamen Municipality model in China. Both were developed as demonstration projects in integrated coastal management (ICM). The projects are part of the Global Environment Facility/UNDP/International Maritime Organization's Regional Programs for the Prevention and Management of Marine Pollution in the East Asian Seas (Environment and Natural Resources Office of the Provincial Government of Batangas 1996). The approach provides a framework for management of the coast of the Batangas Bay region of the province of Batangas in the Philippines. It includes guidelines for the implementation of a core program of integrated waste management, water pollution abatement, conservation of stressed mangrove and coral reef areas, coastal tourism development, and improvements of the municipal fisheries. Special support programs addressing the interests of the stakeholders of the region have been developed and are being implemented for (1) development of legal and institutional mechanisms for ICM, (2) strengthening of provincial integrated planning and resource management, (3) improvement of policy support systems, (4) upgrading of monitoring and enforcement capabilities, (5) capacity building in technology transfer and coastal management, (6) community outreach, (7) establishment of a multisectoral information, education, and communication system, (8) expansion of research and extension activities, (9) establishment of a management information system, and (10) development of sustainable financing mechanisms. A list of the key environmental management issues that are the root causes of the declining environmental quality of the region is given in Table 26-1.

An action plan has been adopted by the provincial government to mitigate the coastal stressors. The plan is to be implemented by the Batangas Bay Council for Integrated Coastal Management, a special council created by legislation by the provincial government. This *governance body* will move forward the actions necessary to mitigate the stressors on the coastal region while also allowing for environmentally sustainable development of the natural resources of the region.

Table 26-1. Key environmental management issues in the declining environmental quality of the Bay of Batangas region. (Excerpted from Environment and Natural Resources Office of the Provincial Government of Batangas 1996.)

Area of concern	Issues	Contributory factors
IMPROPER SOLID WASTE COLLECTION AND DISPOSAL	1.1 Accumulation of uncollected solid wastes	• Low collection efficiency • Absence of incinerators • Lack of collection fleet/vehicles • Irregular street sweeping and open draining • Lack of incentives and penalties for waste minimization, segregation and recycling
	1.2 Inadequate solid waste disposal facilities	• Limited capacity and unhealthy condition of existing dump sites • Absence of dump sites for toxic and hazardous wastes • Absence of alternate landfill sites
WATER AND AIR POLLUTION	2.1 Disposal of industrial wastes in the bay	• Solid and liquid wastes from oil refineries, shipbuilding and fabrication, wood treatment, chemical companies, and fertilizer plants • Mine tailings and sediments from sand and gravel operations • Wastes from ships, especially domestic passenger vessels
	2.2 Disposal of domestic and agricultural wastes in the rivers and tributaries	• Organic wastes and chemicals from crop and livestock raising • Indiscriminate dumping of uncollected/untreated wastes by households and commercial establishments • Absence of sewage treatment plants
	2.3 Oil spills and oil discharges from industries	• Spills from ships, tanks, and pipes of oil refineries • Wastewater and chemicals from industries
	2.4 Gas and particulate emissions from industries and motor vehicles	• Absence of pollution-monitoring equipment/facilities • Absence of technical capability
DECLINING FISH HARVEST	3.1 Overfishing	• Encroachment of commercial fishing boats • Entry of small fishers from areas outside the coastal communities
	3.2 Decreasing effective fishing area in the bay	• Increasing number of sea vessels • Conversion of domestic port in Batangas City into international port • Restriction on fishing by some companies operating private wharves (oil companies)
	3.3 Habitat destruction	• Destructive fishing methods • Pollution from oil spill/discharges
IMPROPER MINING AND QUARRYING OPERATIONS	4.1 Application of environmentally destructive methods	• Open-pit mining
EXPANDING SHIPPING AND PORT DEVELOPMENT ACTIVITIES	5.1 Vessel traffic congestion	• Small water area of the bay • Absence of vessel traffic system • Inadequacy of vessel traffic-monitoring equipment/facilities
	5.2 Oil spill and ship collision	• Unregulated flow of cargo, passenger, and fishing vessels • Weak intersectoral cooperation in oil spill contingency plan implementation
	5.3 Marine pollution	• Oil spills and discharges from industries • Wastes disposed by domestic passenger ships in the bay
DETERIORATING SOCIOECONOMIC CONDITIONS OF PEOPLE IN COASTAL AREAS	6.1 Increasing population and settlements in coastal areas	• Immigration of industrial workers and their families • Displacement of some sectors, especially fisherfolk due to industrial development • Limited alternative livelihood opportunities
	6.2 Poor health and sanitation conditions	• Lack of environmental education and awareness • Improper household practices • Inadequate access to health services
LACK OF MULTI-SECTORAL PARTICIPATION IN ENVIRONMENT MANAGEMENT	7.1 Inadequate institutional and legal framework for participation	• Insufficient incentives and strategies for participation by private sector, NGOs, and community • Low understanding and appreciation of development–environment nexus • Lack of functional organization of coastal communities

This model system is expected to be introduced by the GEF/UNDP/IMO Regional Program to other candidate areas around the margins of the South China Sea LME. In addition, a parallel development for the coastal region of Xiamen, in the southwest of China, has been implemented and will serve as the model ICM demonstration system for the Yellow Sea LME. It is to be followed by a series of demonstration projects, to be known as the Fast Track ICM-LME (FTICM-LME) projects, around the margin of the Yellow Sea in China, South Korea, and North Korea, in support of the GEF-sponsored Yellow Sea LME project.

Efforts are under way to place greater focus on the linkage between scientific and societal needs and the utility of long-term coastal ocean assessments aimed at enhancing the long-term sustainability of marine resources and coastal areas and associated drainage basins. If the proposition for time-series assessments of changing ecosystem states is to be realized in this period of shrinking budgets, it would be in the best interests of science and economic development to be tightly linked in the endeavor. The basis for the linkage was emphasized not only in the UNCED declaration on the oceans, but also in a series of recent developments revolving around (1) global climate change, ozone depletion, and biodiversity issues; (2) the legal precedent for international cooperation implicit in the Law of the Sea; (3) a growing interest in marine ecosystems as regional units for marine research, monitoring, and management; and (4) renewed national interests in improving the health of degraded coastal ecosystems and depressed fish populations from over-exploitation.

Under the operational guidelines of the GEF International Waters Program, countries are encouraged to consider implementation of Fast Track projects that demonstrate new approaches, techniques, innovative technologies, and institutional arrangements for removing barriers to sustainable development, particularly with regard to projects for controlling land-based activities that degrade marine waters (GEF 1997). Efforts are presently under way by NOAA, in collaboration with the GEF and Dr. Chua Thia-Eng, Coordinator of the GEF/UNDP/IMO Regional Program for the East Asian Seas, to extend the model systems in place in Batangas Bay in the Philippines and Xiamen on the southeast coast of China to other GEF-supported planning and operational efforts under way in the Gulf of Guinea LME, Somali Current LME, Benguela Current LME, Bay of Bengal LME, and the Baltic Sea LME. The objective of these LME projects is in keeping with the GEF operational strategy for international waters, which is aimed at changing narrow national sectoral policies into broader ecosystem-wide Fast Track ICM-LME projects for improving and promoting the development and use of marine resources in a manner that promotes their long-term sustainability.

References

Aiken, J. 1981. The Undulating Oceanographic Recorder Mark 2. J. Plankton Res. 3:551–560.

Aiken, J., and Bellan, I. 1990. Optical Oceanography: An Assessment of a Towed Method. In P. J. Herring, A. K. Campbell, M. Whitfield, and L. Maddock (eds.), Light and Life in the Sea. London: Cambridge University Press, pp. 39–57.

Alheit, J., and Bernal, P. 1993. Effects of Physical and Biological Changes on the Biomass Yield of the Humboldt Current Ecosystem. In K. Sherman, L. M. Alexander, and B. D. Gold (eds.), Large Marine Ecosystems: Stress, Mitigation, and Sustainability. Washington, DC: AAAS Press, pp. 53–68.

American Association for the Advancement of Science (AAAS). 1986. Variability and Management of Large Marine Ecosystems. AAAS Selected Symposium 99. Boulder, CO: Westview Press.

American Association for the Advancement of Science (AAAS). 1989. Biomass Yields and Geography of Large Marine Ecosystems. AAAS Selected Symposium 111. Boulder, CO: Westview Press.

American Association for the Advancement of Science (AAAS). 1990. Large Marine Ecosystems: Patterns, Processes and Yields. Washington, DC: AAAS Press.

American Association for the Advancement of Science (AAAS). 1991. Food Chains, Yields, Models, and Management of Large Marine Ecosystems. Boulder, CO: Westview Press.

American Association for the Advancement of Science (AAAS). 1993. Large Marine Ecosystems: Stress, Mitigation, and Sustainability. Washington, DC: AAAS Press.

Bakun, A. 1993. The California Current, Benguela Current and Southwestern Atlantic Shelf Ecosystems: A Comparative Approach to Identifying Factors Regulating Biomass Yields. In K. Sherman, L. M. Alexander, and B. D. Gold (eds.), Large Marine Ecosystems: Stress, Mitigation, and Sustainability. Washington, DC: AAAS Press, pp. 199–221.

Bakun, A. 1995. A Dynamical Scenario for Simultaneous "Regime-Scale" Marine Population Shifts in Widely Separated LMEs of the Pacific. In Q. Tang and K. Sherman (eds.), The Large Marine Ecosystems of the Pacific Rim: A Report of a Symposium Held in Qingdao, People's Republic of China, 8–11 October 1994. Marine Conservation and Development Report. Gland, Switzerland: IUCN, pp. 47–72.

Beddington, J. R. 1986. Shifts in Resource Populations in Large Marine Ecosystems. In K. Sherman and L. M. Alexander (eds.), Variability and Management of Large Marine Ecosystems. AAAS Symposium 99. Boulder, CO: Westview Press, pp. 9–18.

Beddington, J. R. 1995. The primary requirements. Nature 374:213–214.

Bombace, G. 1993. Ecological and Fishing Features of the Adriatic Sea. In K. Sherman, L. M. Alexander, and B. D. Gold (eds.), Large Marine Ecosystems: Stress, Mitigation, and Sustainability. Washington, DC: AAAS Press, pp. 119–136.

Bottom, D. L., Jones, K. K., Rodgers, J. D., and Brown, R. F. 1989. Management of Living Resources: A Research Plan for the Washington and Oregon Continental Margin. NCRI-T-89-004. Newport, OR: National Coastal Resources Research and Development Institute.

Bradbury, R. H., and Mundy, C. N. 1989. Large-Scale Shifts in Biomass of the Great Barrier Reef Ecosystem. In K. Sherman and L. M. Alexander (eds.), Biomass Yields and Geography of Large Marine Ecosystems. AAAS Selected Symposium 111. Boulder, CO: Westview Press, pp. 143–167.

Chen, Y., and Shen, X. 1995. Changes in the Biomass of the East China Sea Ecosystem. In Q. Tang and K. Sherman (eds.), The Large Marine Ecosystems of the Pacific Rim: A Report of a Symposium Held in Qingdao, People's Republic of China, 8–11 October 1994. Marine Conservation and Development Report. Gland, Switzerland: IUCN, pp. 91–112.

Cicin-Sain, B. and Knecht, R. (eds.) 1995. Special Issue: Earth Summit Implementation: Progress Achieved on Oceans and Coasts, Ocean and Coastal Management 29(1–3).

Colebrook, J. M. 1986. Environmental influences on long-term variability in marine plankton. Hydrobiologia 142:309–325.

Colebrook, J. M., Warner, A. J., Proctor, C. A., Hunt, H. G., Pritchard, P., John, A. W. G., Joyce, D., and Barnard, R. 1991. 60 Years of the Continuous Plankton Recorder Survey: A Celebration. Plymouth, England: The Sir Alister Hardy Foundation for Ocean Science.

Costanza, R. 1992. Toward an Operational Definition of Ecosystem Health. In R. Costanza, B. G. Norton, and B. D. Haskell (eds.), Ecosystem Health: New Goals for Environmental Management. Washington, DC: Island Press, pp. 239–256.

Crawford, R. J. M., Shannon, L. V., and Shelton, P. A. 1989. Characteristics and Management of the Benguela as a Large Marine Ecosystem. In K. Sherman and L. M. Alexander (eds.), Biomass Yields and Geography of Large Marine Ecosystems. AAAS Selected Symposium 111. Boulder, CO: Westview Press, pp. 169–219.

Dickson, R. R., Kelly, P. M., Colebook, J. M., Wooster, W. S., and Cushing, D. H. 1988. North winds and production in the eastern North Atlantic. J. Plankton Res. 10:151–169.

Eikeland, P. O. 1992. Multispecies Management of the Barents Sea Large Marine Ecosystem: A Framework for Discussing Future Challenges. Lysaker, Norway: The Fridtjof Nansen Institute.

Environment and Natural Resources Office of the Provincial Government of Batangas. 1996. Strategic Environmental Plan for the Batangas Bay Region. MPP-EAS Technical Report No. 3. Quezon City, Philippines: GEF/UNDP/IMO Regional Programme for the Prevention and Management of Marine Pollution in the East Asian Seas.

Epstein, P. R. 1993. Algal blooms and public health. World Res. Rev. 5(2):190–206.

Farrington, J. W., Goldberg, E. D., Risebrough, R. W., Martin, J. H., and Bowen, V. T. 1983. U. S. "Musselwatch" 1976–1978: An overview of the trace metal, DDE, PCB, hydrocarbon and artificial radionuclide data. Environ. Sci. Technol. 17:490–496.

Global Environment Facility (GEF). 1997. GEF Operational Programs. Washington, DC: Global Environment Facility.

Glover, R. S. 1967. The Continuous Plankton Recorder survey of the North Atlantic. Symp. Zool. Soc. London 19:189–210.

Hayes, G. C., Carr, M. R., and Taylor, A. H. 1993. The relationship between gulf stream position and copepod abundance derived from the Continuous Plankton Recorder survey: Separating the biological signal from sampling noise. J. Plankton Res. 15:1359–1373.

Hey, E., and Mee, L. D. 1993. Black Sea. The ministerial declaration: An important step. Environ. Pollution Law 2315:215–217; 235–236.

Holling, C. S. 1986. The Resilience of Terrestrial Ecosystems: Local Surprise and Global Change. In W. C. Clark and R. E. Munn (eds.), Sustainable Development of the Biosphere. London: Cambridge University Press, pp. 292–317.

Holling, C. S. 1993. Investing in research for sustainability. Ecol. Appl. 3:552–555.

Intergovernmental Oceanographic Commission of UNESCO (IOC). 1992. The Use of the Large Marine Ecosystem Concept in the Global Ocean Observing System. Presented at the 25th Session of the IOC Executive Council, Paris, 10–18 March 1992. IOC/EC-XXV/Inf. 7.

International Council for the Exploration of the Sea (ICES). 1988. Results of the 1985 Baseline Study of Contaminants in Fish and Shellfish. Cooperative Research Report 131. Copenhagen, Denmark: ICES.

International Council for the Exploration of the Sea (ICES). 1991. Report of the Multispecies Assessment Working Group. ICES C.M. 1991/Assess 7.

Jossi, J. W., and Goulet, J. R. 1993. Zooplankton trends: U.S. northeast shelf ecosystem and adjacent regions differ from northeast Atlantic and North Sea. ICES J. Marine Sci. 50:303–313.

Jossi, J. W., and Smith, D. E. 1990. Continuous Plankton Records: Massachusetts to Cape Sable, NS and New York to the Gulf Stream, 1989. NAFO Ser. Doc. 90/66, pp. 1–11.

Karr, J. 1992. Ecological Integrity: Protecting Earth's Life Support Systems. In R. Costanza, B. G. Norton, and B. D. Haskell (eds.), Ecosystem Health: New Goals for Environmental Management. Washington, DC: Island Press, pp. 223–228.

Kelleher, G. 1993. Sustainable Development of the Great Barrier Reef as a Large Marine Ecosystem. In K. Sherman, L. M. Alexander, and B. D. Gold (eds.), Large Marine Ecosystems: Stress, Mitigation, and Sustainability. Washington, DC: AAAS Press, pp. 272–279.

Kullenberg, G. 1986. Long-Term Changes in the Baltic Ecosystem. In K. Sherman and L. M. Alexander (eds.), Variability and Management of Large Marine Ecosystems. AAAS Selected Symposium 99. Boulder, CO: Westview Press, pp. 19–32.

Levin, S. A. 1993. Approaches to Forecasting Biomass Yields in Large Marine Ecosystems. In K. Sherman, L. M. Alexander, and B. D. Gold (eds.), Large Marine Ecosystems: Stress, Mitigation, and Sustainability. Washington, DC: AAAS Press, pp. 36–39.

Likens, G. E. 1992. The Ecosystem Approach: Its Use and Abuse. In O. Kinne (ed.), Excellence in Ecology, Vol. 3. Oldendorf/Luhe, Germany: Ecology Institute.

MacCall, A. D. 1986. Changes in the Biomass of the California Current System. In K. Sherman and L. M. Alexander (eds.), Variability and Management of Large Marine Ecosystems. AAAS Selected Symposium 99. Boulder, CO: Westview Press, pp. 33–54.

Mangel, M. 1991. Empirical and Theoretical Aspects of Fisheries Yield Models for Large Marine Ecosystems. In K. Sherman, L. M. Alexander, and B. D. Gold (eds.), Food Chains, Yields, Models, and Management of Large Marine Ecosystems. Boulder, CO: Westview Press, pp. 243–261.

Mee, L. 1992. The Black Sea in crisis: A need for concerted international action. Ambio 21(4): 1278–1268.

National Coastal Monitoring Act (NCMA). 1992. 102nd Congress, 2nd Session, H.R. 2130 (29 October 1992). The National Oceanic and Atmospheric Administration Authorization Act of 1992.

National Oceanic and Atmospheric Administration (NOAA). 1993. Emerging Theoretical Basis for Monitoring the Changing States (Health) of Large Marine Ecosystems. Summary Report of Two Workshops: 23 April 1992, National Marine Fisheries Service, Narragansett, Rhode Island, and 11–12 July 1992, Cornell University, Ithaca, New York. NOAA Technical Memorandum NMFS-F/NEC-100.

North Sea Task Force (NSTF). 1991. Scientific Activities in the Framework of the North Sea Task Force. North Sea Environmental Report No. 4. London: North Sea Task Force, Oslo and Paris Commissions, International Council for the Exploration of the Sea.

North Sea Task Force (NSTF). 1993. North Sea Quality Status Report 1993. Oslo and Paris Commissions and ICES. Fredensborg, Denmark: Olsen and Olsen.

Norton, B. G., and Ulanowicz, B. G. 1992. Scale and biodiversity policy: A hierarchical approach. Ambio 21(3):244–249.

O'Connor, T. P., and Ehler, C. N. 1991. Results from the NOAA National Status and Trends Program on distribution and effects of chemical contamination in the coastal and estuarine United States. Environ. Monitoring Assess. 17:33–49.

Okemwa, E., Ntiba, M. J., and Sherman, K. (eds.). 1994. Status and Future of Large Marine Ecosystems of the Indian Ocean: A Report of the International Symposium and Workshop. A Marine Conservation and Development Report. Gland, Switzerland: IUCN.

Overholtz, W. J., and Nicolas, J. R. 1979. Apparent feeding by the fin whale *Balaenoptera physalus*, and humpback whale, *Megoptera novaeangliae*, on the American sand lance, *Ammodytes americanus*, in the Northwest Atlantic. Fish. Bull. U.S. 77:285–287.

Pauly, D., and Christensen, V. 1995. Primary production required to sustain global fisheries. Nature 374:255–257.

Payne, P. M., Wiley, D. N., Young, S. B., Pittmann, S., Clapham, P. J., and Jossi, J. W. 1990. Recent fluctuations in the abundance of baleen whales in the southern Gulf of Maine in relation to changes in selected prey. Fish. Bull. U.S. 88:687–696.

Pimm, S. L. 1984. The complexity and stability of ecosystems. Nature 307:321–326.

Rapport, D. J. 1992. What Is Clinical Ecology? In R. Costanza, B. G. Norton, and B. D. Haskell (eds.), Ecosystem Health: New Goals for Environmental Management. Washington, DC: Island Press, pp. 223–228.

Ricklefs, R. E. 1987. Community diversity: Relative roles of local and regional processes. Science 235:167–171.

Sainsbury, K. J. 1988. The Ecological Basis of Multispecies Fisheries, and Management of a Demersal Fishery in Tropical Australia. In J. A. Gulland (ed.), Fish Population Dynamics, 2nd ed. New York: John Wiley and Sons, pp. 349–382.

Scully, R. T. 1993. Convention on the Conservation of Antarctic Marine Living Resources. In K. Sherman, L. M. Alexander, and B. D. Gold (eds.), Large Marine Ecosystems: Stress, Mitigation, and Sustainability. Washington, DC: AAAS Press, pp. 242–251.

Scully, R. T., Brown, W. Y., and Manheim, B. S. 1986. The Convention for the Conservation of Antarctic Marine Living Resources: A Model for Large Marine Ecosystem Management. In K. Sherman and L. M. Alexander (eds.), Variability and Management of Large Marine Ecosystems. AAAS Symposium 99. Boulder, CO: Westview Press, pp. 281–286.

Sherman, K. (ed.). 1993. Emerging Theoretical Basis for Monitoring the Changing States (Health) of Large Marine Ecosystems. Summary Report of Two Workshops: 23 April 1992, National Marine Fisheries Service, Narragansett, Rhode Island, and 11–12 July 1992, Cornell University, Ithaca, New York. NOAA Technical Memorandum NMFS-F/NEC-100.

Sherman, K. 1994. Sustainability, biomass yields, and health of coastal ecosystems: An ecological perspective. Marine Ecol. Progr. Ser. 112:277–301.

Sherman, K., and Ryan, A. F. 1988. Antarctic marine living resources. Oceanus 31(2):59–63.

Sherman, K., and Solow, A. R. 1992. The Changing States and Health of a Large Marine Ecosystem. CM/L38 Session V.

Sherman, K., Grosslein, M., Mountain, D., Busch, D., O'Reilly, J., and Theroux, R. 1988. The Continental Shelf Ecosystem off the Northeast Coast of the United States. In H. Postma and J. J. Zilstra (eds.), Ecosystems of the World 27: Continental Shelves. Amsterdam: Elsevier, pp. 279–337.

Sissenwine, M. P. 1986. Perturbation of a Predator-Controlled Continental Shelf Ecosystem. In K. Sherman and L. M. Alexander (eds.), Variability and Management of Large Marine Ecosystems. AAAS Symposium 99. Boulder, CO: Westview Press, pp. 55–85.

Sissenwine, M. P., and Cohen, E. B. 1991. Resource Productivity and Fisheries Management of the Northeast Shelf Ecosystem. In K. Sherman, L. M. Alexander, and B. D. Gold (eds.), Food Chains, Yields, Models, and Management of Large Marine Ecosystems. Boulder, CO: Westview Press, pp. 107–123.

Smayda, T. 1991. Global Epidemic of Noxious Phytoplankton Blooms and Food Chain Consequences in Large Ecosystems. In K. Sherman, L. M. Alexander, and B. D. Gold (eds.), Food Chains, Yields, Models, and Management of Large Marine Ecosystems. Boulder, CO: Westview Press, pp. 275–308.

Tang, Q. 1989. Changes in the Biomass of the Yellow Sea Ecosystem. In K. Sherman and L. M. Alexander (eds.), Biomass Yields and Geography of Large Marine Ecosystems. AAAS Selected Symposium 111. Boulder, CO: Westview Press, pp. 7–35.

Tang, Q. 1993. Effects of Long-Term Physical and Biological Perturbations on the Contemporary Biomass Yields of the Yellow Sea Ecosystem. In K. Sherman, L. M. Alexander, and B. D. Gold (eds.), Large Marine Ecosystems: Stress, Mitigation, and Sustainability. Washington, DC: AAAS Press, pp. 79–93.

Tang, Q., and Sherman, K. (eds.). 1995. The Large Marine Ecosystems of the Pacific Rim: A Report of a Symposium Held in Qingdao, People's Republic of China, 8–11 October 1994. Marine Conservation and Development Report. Gland, Switzerland: IUCN.

UNIDO/UNDP/NOAA/UNEP. 1997. Gulf of Guinea Large Marine Ecosystem Project. Second Meeting of the Steering Committee, Cotonou, Bénin, 11–12 March 1997. Report GEF/GOG-LME/Sc-2.

United Nations Educational, Scientific, and Cultural Organization (UNESCO). 1992. Monitoring the Health of the Oceans: Defining the Role of the Continuous Plankton Recorder in Global Ecosystems Studies. The Intergovernmental Oceanographic Commission and the Sir Alister Hardy Foundation for Ocean Science. IOC/INF-869, SC-92/WS-8.

White, H. H., and Robertson, A. 1996. Biological Responses to Toxic Contaminants in the Northeast Shelf Large Marine Ecosystem. In K. Sherman, N. Jaworski, and T. Smayda (eds.), The Northeast Shelf Ecosystem: Assessment, Sustainability, and Management. Cambridge, MA: Blackwell Science, pp. 259–283.

Williams, R. 1993. Evaluation of New Techniques for Monitoring and Assessing the Health of Large Marine Ecosystems. In D. Rapport (ed.), NATO Advanced Research Workshop Evaluating and Monitoring the Health of Large-Scale Ecosystems. Berlin: Springer-Verlag, pp. 257–272.

Williams, R., and Aiken, J. 1990. Optical Measurements from Underwater Towed Vehicles Deployed from Ships-of-Opportunity in the North Sea. In H. O. Nielsen (ed.), Environment and Pollution Measurement, Sensor and Systems. Proc. SPIE 1269, pp. 186–194.

VI

Recommendations

27

Symposium Recommendations

Following the presentation and discussion of the scientific papers, a general planning session was convened to consider several recommendations developed during the symposium. Consensus on four recommendations was reached by the symposium participants:

1. The governments and institutions of the riparian countries of the Yellow Sea should support and become involved in the Sustainability and Protection of the Yellow Sea Large Marine Ecosystem project, which is being supported by the Global Environment Facility.
2. The participants at the meeting recognized that the South China Sea is one of the most biologically diverse areas anywhere and that recruitment of many important species on which coastal communities depend takes place in its offshore waters. The participants also took note that although coastal populations should be empowered to manage their marine living resources, it is recognized that processes governing the existence of marine living resources take place over a large range of spatial and temporal scales and that decision-making processes concerning management of resources therefore need to be developed in a way that recognizes these scales. The LME approach provides a framework for such a decision-making process and can be a vehicle for ensuring that there are linkages between biological, social, and economic sciences and management. In its acknowledgment of multiple scales of processes, the LME framework can contribute to harmonizing community-based decision making with ecosystem-scale dynamics.

   Therefore, the participants recommended that an LME activity be initiated for the South China Sea area. This should include the convening of a meeting on that system, with the focus on achieving a better understanding of the relationship between the scales of socioeconomic processes and the natural processes affecting marine living resources. Planning for such a meeting should take into account the work being carried out by LOICZ/IGBP Focus 4 for the South East Asian Region.
3. It is generally recognized that the apparent synchrony in marine ecosystem and population variations in widely separated LMEs distributed around the Pacific Basin and other areas of the world's oceans is most likely being generated by global climatic teleconnections. Achieving a better understanding of these linkages to resource variability represents one of the greatest current challenges to fishery science and to marine ecosystem management. Implementation of a number of LME-scale marine ecosystem monitoring programs distributed globally would be of great potential utility in diagnosing the nature of these linkages through comparative systems analyses and would thereby benefit global climatic research pertinent to marine ecosystem concerns.
4. The participants encouraged the nations of the region and subregions of the Pacific and their institutions to focus on LMEs with respect to compiling time-series data sets on ecosystem

components in order to carry out retrospective analyses of existing information and to provide a framework for future data gathering. In addition, the participants encouraged regional institutions (such as CPPS, SPREP, WESTPAC, ASEAMS, etc.) to foster LME-focused programs and projects, recognizing that this would promote better management as well as provide inputs to international activities, such as the establishment of the Global Ocean Observing System being fostered by IOC, WMO, UNEP, and other global and national institutions. In this regard, efforts should be made to ensure that all nations of the Pacific Rim adjacent to LMEs are brought into the process.

Index